D0536559

CLINICAL BIOCHEMISTRY AND THE SICK CHILD

Clinical Biochemistry and the Sick Child

EDITED BY

BARBARA E. CLAYTON

DBE, PhD, MD, Hon. DSc (Ed, Southampton),
FRCP, FRCPE, FRCPath

Honorary Research Professor in Metabolism,
University of Southampton

AND

JOAN M. ROUND

PhD, FIBMS, FRCPath

Consultant Clinical Scientist,
Department of Chemical Pathology,
UCL Hospitals, London

SECOND EDITION

OXFORD

Blackwell Scientific Publications

LONDON EDINBURGH BOSTON

MELBOURNE PARIS BERLIN VIENNA

Editorial Offices:
Osney Mead, Oxford OX2 0EL
25 John Street, London WC1N 2BL
23 Ainslie Place, Edinburgh EH3 6AJ
238 Main Street, Cambridge
Massachusetts 02142, USA
54 University Street, Carlton
Victoria 3053, Australia

Other Editorial Offices:
Librairie Arnette SA
1, rue de Lille
75007 Paris
France

Blackwell Wissenschafts-Verlag GmbH
Düsseldorfer Str. 38
D-10707 Berlin
Germany

Blackwell MZV
Feldgasse 13
A-1238 Wien
Austria

First published 1984
(*Chemical Pathology and the Sick Child*)
Second edition 1994

Set by Excel Typesetters Company, Hong Kong
Printed and bound in Great Britain
at the University Press, Cambridge

DISTRIBUTORS

Marston Book Services Ltd
PO Box 87
Oxford OX2 0DT
(*Orders*: Tel: 0865 791155
Fax: 0865 791927
Telex: 837515)

USA
Blackwell Scientific Publications, Inc.
238 Main Street
Cambridge, MA 02142
(*Orders*: Tel: 800 759-6102
617 876-7000)

Canada
Times Mirror Professional Publishing, Ltd
130 Flaska Drive
Markham, Ontario L6G 1B8
(*Orders*: Tel: 800 268-4178
416 470-6739)

Australia
Blackwell Scientific Publications Pty Ltd
54 University Street
Carlton, Victoria 3053
(*Orders*: Tel: 03 347-5552)

A catalogue record for this title
is available from the British Library

ISBN 0-632-03681-8

Library of Congress
Cataloging-in-Publication Data

Clinical biochemistry and the sick child/edited by Barbara E. Clayton and Joan M. Round. – 2nd ed.
p. cm.
Rev. ed. of: Chemical pathology and the sick child. 1984.
Includes bibliographical references and index.
ISBN 0-632-03681-8
1. Clinical biochemistry. 2. Pediatric pathology. I. Clayton, Barbara E. II. Round, Joan M. III. Chemical pathology and the sick child.
[DNLM: 1. Chemistry, Clinical – in infancy & childhood. QY 90 C6405 1994]
RJ49.4.C57 1994
618.92′007 – dc20

Contents

Contributors

PETER J. AGGETT MSc, FRCP, *Honorary Consultant Paediatrician, Norfolk and Norwich Hospital and Head of the Department of Nutrition, Diet and Health, Institute of Food Research, Norwich NR4 7UA, UK*

GUY T.N. BESLEY PhD, *Top Grade Biochemist, Willink Biochemical Genetics Unit, Royal Manchester Children's Hospital, Pendlebury, Manchester M27 4HA, UK*

D. JOHN BETTERIDGE BSc, PhD, MD, FRCP, *Consultant Physician and Diabetologist, Reader in Medicine, ULC Medical School, Mortimer Street, London W1A 6JJ, UK*

THIERRY BILLETTE DE VILLEMEUR MD, *Département de Pédiatrie, Hôpital Necker Enfants-Malades, 149 rue de Sèvres, 75743 Paris Cedex 15, France*

J. TREVOR BROCKLEBANK MD, BS, FRCP, *Senior Lecturer in Paediatrics and Paediatric Nephrologist, St James's University Hospital, Leeds LS9 7TF, UK*

CHARLES G.D. BROOK MA, MD, FRCP, DCH, *Professor of Paediatric Endocrinology, University College London, The Middlesex Hospital, Mortimer Street, London W1N 8AA, UK*

GARRY K. BROWN MB, BS, PhD, *Lecturer, Department of Biochemistry, University of Oxford, and Honorary Consultant, Department of Medical Genetics, The Churchill Hospital, Oxford, UK*

BARBARA E. CLAYTON DBE, PhD, MD, Hon. DSc (Ed, Southampton), FRCP, FRCPE, FRCPath, *Honorary Research Professor in Metabolism, University of Southampton, Southampton General Hospital, Tremona Road, Southampton SO9 4XY, UK*

MAUREEN CLEARY MBChB, MRCP, *Research Fellow Paediatrics, Willink Biochemical Genetics Unit, Royal Manchester Children's Hospital, Pendlebury, Manchester M27 4HA, UK*

ALAN CRAFT MD, FRCP, *Professor of Child Health and Paediatric Oncology, Consultant Paediatrician, Royal Victoria Infirmary, Queen Victoria Road, Newcastle upon Tyne NE1 4LP, UK*

SALLY C. DAVIES MB, MSc, FRCP, MRCPath, *Consultant Haematologist, Central Middlesex Hospital, Acton Lane, London NW10 7NS, UK*

HUGH TREVOR DELVES PhD, C. Chem, FRSC, *Honorary Reader in Analytical Science, University of Southampton, Clinical Biochemist Grade C, Director SAS Trace Element Unit, Southampton General Hospital, Southampton, SO9 4XY, UK*

JOSEPH M. GERTNER MB, MRCP, *Professor of Paediatrics, Cornell University Medical College, Program Director, Children's Clinical Research Center, 525 East 68th Street, New York NY 10021, USA*

ANNE GREEN MSc, MFSC, FRCPath, *Consultant Biochemist, Head of Department of Paediatric Clinical Chemistry, The Children's Hospital, Birmingham B16 8ET, UK*

JOSEPHINE HAMMOND MBBS, FRCP (Ed), FRCP (London), DCH, DObstRCOG, *Consultant Paediatrician, Queen Mary's Hospital for Children, Carshalton SM5 1AA, UK*

HUGO S.A. HEYMANS MD, PhD, *Professor of Paediatrics, Beatrix Children's Hospital, Oostersingel 59, 9713 EZ Groningen, The Netherlands*

JOHN W. HONOUR PhD, FRCPath, *Reader in Steroid Endocrinology, Department of Endocrinology, UCL Medical School, Mortimer Street, London W1A 6JJ, UK*

DAVID ISHERWOOD Dip CB, PhD, MA, *Department of Clinical Chemistry, Alder Hey Children's Hospital, Eaton Road, West Derby, Liverpool L12 2AP, UK*

MALCOLM J. JACKSON PhD, MRCPath, *Reader in Medicine, University of Liverpool, PO Box 147, Liverpool L69 3BX, UK*

JUNE K. LLOYD DBE, DSc (Hon), MD, FRCP, *Emeritus Professor of Child Health, University of London, 37 Allingham Street, London N1 8NX, UK*

ALEX P. MOWAT MB, ChB, FRCP, DCH, *Professor of Paediatric Hepatology and Consultant Paediatric Hepatologist, Department of Child Health, King's College Hospital, Denmark Hill, London SE5 8RS, UK*

ARNOLD MUNNICH MD, PhD, *Département de Pédiatrie, Hôpital Necker Enfants-Malades, 149 rue de Sèvres, 75743 Paris Cedex 15, France*

FLORENCE POGGI MD, *Département de Pédiatrie, Hôpital Necker Enfants-Malades, 149 rue de Sèvres, 75743 Paris Cedex 15, France*

JON PRITCHARD BA, MB, BChir, FRCP, *Consultant Paediatric Oncologist, The Hospitals for Sick Children, Great Ormond Street, London WC1N 3JH, UK*

A.T. PROUDFOOT BSc(Hons), FRCPE, *Director, Scottish Poisons Information Bureau, The Royal Infirmary, Edinburgh EH3 9YW, UK*

PAMELA G. RICHES PhD, FRCPath, *Professor, Department of Immunology, Charing Cross and Westminster Medical School, Westminster and Chelsea Hospital, 369 Fulham Road, London SW10 9NH, UK*

JOAN M. ROUND PhD, FIBMS, FRCPath, *Consultant Clinical Scientist, Department of Chemical Pathology, UCL Hospitals, Windeyer Building, Cleveland Street, London W1P 6DB, UK*

GILLIAN RUMSBY MSc, PhD, *Lecturer, Department of Chemical Pathology, UCL Hospitals, Windeyer Building, Cleveland Street, London W1P 6DB, UK*

IMDADALI B. SARDHARWALLA MBBS, FRCP (Ed), DCH, *Willink Biochemical Genetics Unit, Royal Manchester Children's Hospital, Pendlebury, Manchester M27 4HA, UK*

JEAN MARIE SAUDUBRAY MD, *Professor of Paediatrics, Department of Paediatrics, Hôpital Necker Enfants-Malades, 149 rue de Sèvres, 75743 Paris Cedex 15, France*

PAUL H. SCOTT BSc, PhD, *Principal Biochemist, Selly Oak Hospital, Birmingham B29 6JD, UK*

JOANNA SHELDON FIMLS, *Chief MLSO, Department of Immunology, Charing Cross and Westminster Medical School, Westminster and Chelsea Hospital, 369 Fulham Road, London SW10 9NH, UK*

JAN STERN PhD, FRCPath, *Consultant Clinical Scientist, Queen Mary's Hospital for Children, Wrythe Lane, Carshalton SM5 1AA, UK*

THOMAS L. TURNER MB, FRCP, *Consultant Paediatrician, Royal Hospital for Sick Children and The Queen Mother's Hospital, Yorkhill, Glasgow G3 8SJ, UK*

VALERIE WALKER MD, ChB, BSc (Hons), FRCPath, *Consultant in Chemical Pathology, Southampton General Hospital, and Honorary Clinical Senior Lecturer in Clinical Biochemistry, Southampton University Medical School, Southampton, UK*

J.A. WALKER-SMITH MD (Sydney), FRCP (Ed), FRCP (London), FRACP, *Professor of Paediatric Gastroenterology, Academic Department of Paediatric Gastroenterology, Queen Elizabeth Hospital for Children, Hackney Road, London E2 8PS, UK*

RONALD J.A. WANDERS PhD, *Associate Professor, Department of Clinical Biochemistry, University Hospital Amsterdam, Meibergdreef 9, 1105 AZ Amsterdam, The Netherlands*

BRIAN A. WHARTON MBA, MD, FRCP, FRCPE, FRCPG, DCH, *Old Rectory, Belbroughton, Worcestershire DY9 9TF, UK*

Preface to the Second Edition

It is now nearly 10 years since the first edition of *Chemical Pathology and the Sick Child* and we feel that in response to many requests a second edition is needed. Once again our aim will be to serve staff who are working outside major paediatric centres, in particular general paediatricians and all those who work in departments of chemical pathology/ clinical biochemistry.

In this edition we are incorporating an appendix which will give some guidance on reference ranges for common analyses; much of this section is an update of the small volume on reference ranges which we published with P. Jenkins in 1980. We particularly thank Anne Green, John Honour and David Isherwood for help to compile this section.

We have again been fortunate in our contributors, all of whom are experts in their field and we extend our grateful thanks to them all. As in the first edition, we have endeavoured through our contributors to discuss the more common disorders and clinical presentations which occur in infancy and childhood, but have been deliberately selective regarding conditions which are rarely seen.

We should also like to thank the many colleagues who have helped in numerous ways to make the publication of this volume possible, in particular Helen Harvey, Suzanne Heazlewood, Diana Burtinshaw, Anne Clayton and Caroline Sheard.

B.E. Clayton
J.M. Round

Note to the reader

SI units are used throughout this book. If the reader requires a conversion guide, the following is recommended: Laposata, M. (1992) *SI Unit Conversion Guide*. New England Journal of Medicine Books, Boston.

Preface to the First Edition

We have felt for some time that there was a need for a chemical pathology book which was concerned entirely with the problems of infants and children. Although we realize that there are specialist textbooks devoted to disorders of particular systems, they may not readily be available to staff who are working outside major centres. It is particularly for such staff that this book has been written. We hope that it will be of help to general paediatricians and all those who work in chemical pathology departments. We have endeavoured through our contributors to discuss the more common disorders which occur in infancy and childhood but have been deliberately selective regarding conditions which are rarely seen.

We have been very fortunate in our contributors, all of whom are experts in their own fields. To them we extend our grateful thanks. We should also like to thank the many colleagues who have helped us in numerous ways in completing this volume, and particularly Mrs B. Lloyd, Mrs V. Patel, Miss C. Sanders, Miss L. Snow and Miss J. Welch. We also thank Mrs J. Ross who so expertly compiled our index.

B.E. Clayton
J.M. Round

1: Introduction

A. GREEN & D. ISHERWOOD

Introduction

The statement often used by paediatricians and advocates of dedicated paediatric care is that 'the child is not a small adult'. This is amply illustrated by the differences in reference data for chemical analytes from those of adults, the wide variability of reference data within childhood, the availability and practicality of specimen collection (i.e. blood volume/timed urines) and the need for test protocols suitable for infants or small children.

This chapter addresses these issues and provides some practical guidance on specimen collection, reference data and undertaking of certain test protocols.

(a) Specimen collection

Blood

A clinical chemistry service based on micro methods is essential since a few '10-mL' blood specimens will remove a significant portion of the infant's total blood volume. This is particularly important for the premature neonate (Fig. 1.1). An added problem for the neonate is the high haematocrit, which means that compared with the older infant less plasma is available from the same blood volume (Table 1.1). The limited availability of the specimen means that the neonatologist or paediatrician needs to be cautious when making requests for investigations, and close liaison with the clinical chemist allows best use of a precious specimen.

Blood collection by capillary puncture

General

Capillary blood collection is not without risk, particularly in the neonate or young infant, and should therefore be minimized and planned in order to coordinate collection for both haematology and clinical chemistry. If specimens for both the clinical chemistry and haematology departments are to be collected, then the blood for the haematology department must be collected *first*. (*Beware* possible contamination from potassium ethylenediaminetetraacetate (EDTA).) This arrangement should minimize the extent to which the platelet count falls in blood shed after skin injury and should provide results similar to those from a venous specimen.

Every attempt should be made to collect clinical chemistry specimens without causing haemolysis. This *is* possible even when great pressure is used to obtain blood.

Capillary vs venous blood specimen

There are significant differences in the concentration of some analytes between capillary and venous blood. Of particular note are the higher potassium concentrations and potential for contamination from sweat in capillary specimens. The limitations of capillary blood should not be underestimated and whenever possible venous blood should be collected.

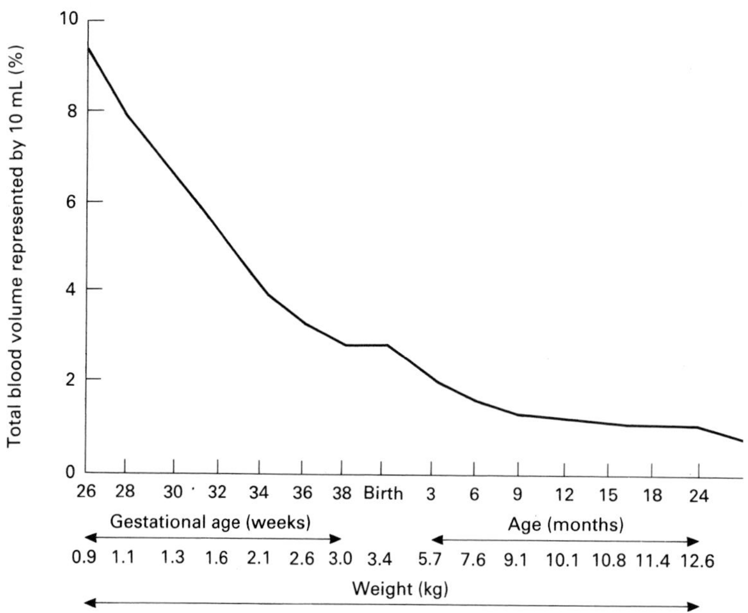

Fig. 1.1 Relationship of a 10-mL blood sample to total blood volume. The average blood volume per kilogram of body weight in premature babies is 115 mL/kg, in neonates is 80–110 mL/kg and in infants is 75–100 mL/kg. Redrawn from Werner, M. (ed) (1976) *Microtechniques for the Clinical Laboratory*. Wiley, New York.

Table 1.1 Haemoglobin, PCV and MCV in normal full-term infants

Days	No. cases	Hb (g/100 mL ± SD)	PCV (% ± SD)	MCV (μ^3 ± SD)
1	19	19.3 ± 2.2	61 ± 7.4	119 ± 9.4
2	19	19.0 ± 1.9	60 ± 6.4	115 ± 7.0
3	19	18.8 ± 2.0	62 ± 9.3	116 ± 5.3
4	10	18.6 ± 2.1	57 ± 8.1	114 ± 7.5
5	12	17.6 ± 1.1	57 ± 7.3	114 ± 8.9
6	15	17.4 ± 2.2	54 ± 7.2	113 ± 10.0
7	12	17.9 ± 2.5	56 ± 9.4	118 ± 11.2
Weeks				
1–2	32	17.3 ± 2.3	54 ± 8.3	112 ± 19.0
2–3	11	15.6 ± 2.6	46 ± 7.3	111 ± 8.2
3–4	17	14.2 ± 2.1	43 ± 5.7	105 ± 7.5
4–5	15	12.7 ± 1.6	36 ± 4.8	101 ± 8.1
5–6	10	11.9 ± 1.5	36 ± 6.2	102 ± 10.2
6–7	10	12.0 ± 1.5	36 ± 4.8	105 ± 12.0
7–8	17	11.1 ± 1.1	33 ± 3.7	100 ± 13.0
8–9	13	10.7 ± 0.9	31 ± 2.5	93 ± 12.0
9–10	12	11.2 ± 0.9	32 ± 2.7	91 ± 9.3
10–11	11	11.4 ± 0.9	34 ± 2.1	91 ± 7.7
11–12	13	11.3 ± 0.9	33 ± 3.3	88 ± 7.9

PCV, packed cell volume; MCV, mean cell volume.
Data reproduced by permission of Professor Matoth and the editor, from the paper by Matoth *et al.* (1971) *Acta Paediatr Scand* **60**, 317–323.

Selection and preparation of site for skin puncture

Capillary heel stab (neonate/infant)

Make sure that the baby is lying in a secure and comfortable position so that the heel is easily accessible. An area of heel that is free from previous puncture sites should be selected (Fig. 1.2). Check whether the ankle has been used for venous access, as there is a danger of reopening wounds. Punctures should not be performed on the posterior curvature of the heel where the bone is closest to the skin.

Capillary finger puncture (infant)

Above the age of 1 year the thumb or finger can be used (provided that there is sufficient flesh).

Before attempting to collect any blood, the site should be warm and well perfused – this may be achieved by rubbing, immersion in warm (<40°C) water or wrapping in a warm nappy or towel (also <40°C). A dirty heel or finger should be prewashed with soapy water, rinsed with clean water and thoroughly dried.

The site must be cleaned with an antiseptic solution (isopropyl alcohol – Sterets) and wiped completely dry with clean cotton wool to prevent haemolysis. Sterets can be used for neonates in an incubator; give the heel a quick wipe and then immediately remove the Steret from the incubator.

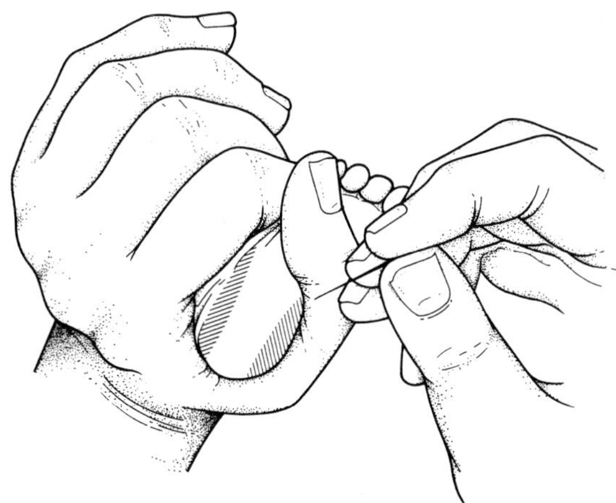

Fig. 1.2 Performing the heel prick. Recommended puncture sites are shown as hatched areas.

Skin puncture

Before performing the puncture, check that all the following necessary items are ready and in close proximity.

1 Steret.

2 Lancet.

3 Cotton wool.

4 Specimen tube, e.g. Sarstedt lithium heparin CB 300.

Check the name and registration number of the baby or child before starting the puncture.

Heel

Stand facing slightly away from the baby's head and hold the heel with the nearest hand. Grip the heel firmly by holding around the ankle with the index and middle fingers so that the sole of the foot is against the palm of the hand and the heel is exposed (see Fig. 1.2). The thumb is placed around the heel and pressure is controlled with the thumb and index finger. Whilst maintaining tension on the heel make the puncture as one continuous deliberate motion in a direction slightly off perpendicular to the puncture site. Release the pressure and wipe away the first drop of blood as it appears.

For a neonate the puncture depth *must not exceed* 2.4 mm, and in the newborn should preferably be no greater than 1.6 mm, to avoid penetration of the bone and hence the potential complication of osteomyelitis (Blumenfeld *et al.*, 1979; Meites, 1988). Several devices are available (e.g. Safety Flow Lancet or Microlance, Becton Dickinson (UK) Ltd).

Finger

Hold the infant's fingers flat, with the thumb or chosen finger slightly separated from the rest (Fig. 1.3). The phlebotomist's thumb and fingers should be placed so that pressure compresses the flesh of the infant's finger. Maintain the tension and prick the side of the finger just where it begins to curve

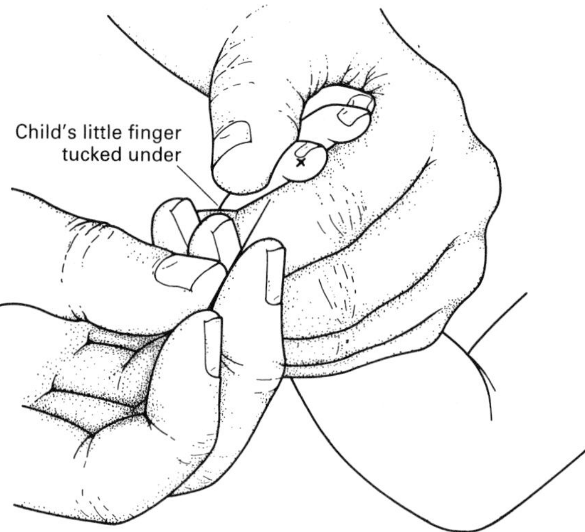

Fig. 1.3 Performing a skin puncture in an infant's finger. The 'x' shows the preferred puncture site position.

and about 3 mm from the nail bed; this is less painful for the patient and gives a better flow. Several devices are available (e.g. Microlance safety flow lancet, Beckton Dickinson; Glucolet 2, Bayer Diagnostics, UK). Release and wipe away the first drop of blood.

Blood collection

Hold the collection tube between thumb and forefinger. Obtain a drop of blood and touch the *side* of this drop which faces away from the patient against the inside of the tube which is furthest from the patient. 'Pluck' off the drop of blood and knock it down to the bottom of the tube by giving the tube a sharp tap on a hard surface. Using a minimum of pressure, obtain further drops of blood and touch them against the same site inside the neck of the tube. They will follow the track of the initial drop. Pressure around the heel/finger should be eased and then slowly and gently reapplied to allow more drops of blood to flow. Excessive 'milking' or 'massaging' is not recommended as this causes haemolysis. If blood should become smeared over the heel/finger *during* collection do not be tempted to scrape it off – wipe away with clean dry cotton wool (*not* a Steret!) and start again.

Using the little or another free finger of the hand holding the tube, periodically flick the blood gently in the tube. This is to ensure mixing of the blood with the anticoagulant. Blood will flow more easily if the heel is held low.

The blood volume needed is obviously dependent on which specific tests are required. This will vary for different laboratories and is dependent on the methods in use; it should be possible to obtain the most commonly requested tests (i.e. sodium, potassium, creatinine, bilirubin, calcium, aspartate aminotransferase (AST) and alanine aminotransferase (ALT)) from a total blood volume of 500 μL.

When the collection is finished, push the cap into the tube and invert gently several times to ensure mixing. Press a wad of dry clean cotton wool firmly against the puncture site and hold the baby's heel/finger above the body for a few minutes to stop the flow of blood. Plasters may be applied to the puncture site but are best avoided if possible; neonates have sensitive skin and a plaster that becomes detached may be ingested.

Make sure that the tube is labelled, either with a felt-tipped pen or with a narrow adhesive label as a 'flag' (i.e. wrap the label around the tube close to the top so that a portion overhangs at the side).

Notes

1 If there are skin problems seek advice from a member of the medical team before proceeding.
2 If a baby or child should appear to be in a worrying condition before or during a blood collection, inform the doctor or nurse in charge who will advise whether to proceed. Similarly, if the patient's condition appears to have changed after the procedure has been completed, inform the nurse or doctor in charge.

Emergency specimen collection for suspected inherited metabolic disorders

In a life-threatening situation where an inherited metabolic disorder is thought to be likely (either from the family history, results of preliminary investigations and/or clinical presentation), specimens should be collected for investigation. A laboratory with special expertise in metabolic disorders should be contacted to discuss appropriate investigations at the earliest opportunity and if possible premortem.

Whenever possible, specimens, particularly urine and blood, should be taken before death. Skin and tissue specimens are also best taken as biopsy specimens when the baby is still alive. If this is not possible, then they should be taken as soon as possible after death (see (3) and (4)). If any of the specimens are taken after death it is extremely important to record accurately both the time of death and when the specimens were taken. Appropriate storage, as detailed below, is essential.

1 *Urine.* However little is collected, urine is extremely useful. Ideally, 5–10 mL should be stored. It should be collected into a bottle with no preservative and stored deep frozen (−20°C or lower). If the sample is contaminated with blood, it should be centrifuged to remove cells before the supernatant is frozen.

2 *Blood.* A sample of blood (5–10 mL) should be collected in lithium heparin and another sample (0.5 mL) in fluoride; the plasma samples should be separated as soon as possible and stored deep frozen (−20°C). Store the packed red cells at +4°C (do not freeze). If DNA analysis is likely to be required, a further 5–10 mL whole blood (EDTA) in a plastic tube should be stored deep frozen (at least −20°C).

3 *Skin (for fibroblast culture).* Skin taken up to 24 h after death is likely to be viable *provided that it is not infected.* Take a skin sample and place it in a suitable transport medium (obtainable from most virology or cytogenetics departments). In an emergency, sterile isotonic saline can be used *but agar should not be used.* The specimen should be stored at +4°C before despatch, and *not frozen. Sterility is of paramount importance when taking skin biopsy specimens, especially at necropsy.*

If indicated:

4 *Tissue samples (liver, heart muscle, skeletal muscle).* These should be taken only if there is a strong clinical suspicion of a primary defect in one of these tissues. *It is very important that blood and urine specimens are also taken and not just tissue specimens.* Necropsy tissue samples are usually only suitable for biochemical analysis if taken *within 2 h of death.* Two or three needle biopsy specimens of tissue should be taken, placed in a plastic tube and snap frozen in liquid nitrogen (or solid carbon dioxide). The specimens should be stored deep frozen, as cold as possible.

Note that these samples are required for biochemical analysis only. Appropriate fixed samples may also be required for histological investigation.

5 *Cerebrospinal fluid.* Sometimes a cerebrospinal fluid sample may be useful. If the specimen is cloudy or blood stained, centrifuge and store the clear supernatant deep frozen.

Metabolic disorders are discussed in Chapter 4.

Urine

The young infant passes less than 500 mL urine per day (Table 1.2) and hence small losses from a timed collection are significant. Whenever possible, urinary analytes should be related to creatinine concentration to obviate the need for a timed collection. The use of liquid preservatives should be minimized, especially if an analyte concentration is not being related to creatinine concentration or measured as an excretion rate on a timed basis.

For neonates, the collection of urine into cotton wool balls is an effective way of accumulating several millilitres. For older infants, urine bags are necessary.

Table 1.2 Daily excretion of urine: reference (normal) values

Age	Volume (mL/day)
Full-term newborn, 1–2 days	15–60
2 months	250–450
6–8 months	400–500
1–2 years	500–600
2–4 years	600–750
5–7 years	650–1000
8–15 years	700–1500
Adult	1000–1600

After Ross Laboratories (1972) *Children Are Different*, 1st edn, p. 25. Columbus, Ohio.

Faeces

The collection of timed faecal specimens should not be necessary. Random collections, e.g. for reducing substances, are best collected on to a polythene sheet to ensure that the liquid portion of the specimen is not lost.

Test repertoire

The need to measure particular analytes and their turnaround time is determined by the clinical situation. The test repertoire for a laboratory serving a paediatric population will thus differ from that of the more general laboratory. Of particular note is the need for access to sweat testing and specialist metabolic, immunological and endocrinological investigations. Therapeutic drug monitoring and toxicology (see Chapter 7) have their own special needs and repertoire of analyses.

Reference data

Reference ranges are affected by many variables. These include the population from which they were obtained (i.e. well/sick, age, nutritional status, etc.), the condition of the subjects at the time of sampling (i.e. fasting, diet), the type and handling of the specimen collected and the analytical method used.

For ethical reasons and limitations of specimen volume, it is particularly difficult to establish ranges in completely normal children and babies. For this reason data have most often been collected from hospitalized children or those attending outpatient clinics or day wards. The available data ranges often relate to methods or technologies that have now been superseded and to small numbers of subjects for a particular age group. Not all laboratories are able to produce their own reference values and therefore they must rely on data produced by others. When doing this, one must be cautious to compare population, specimen type and collection, and analytical method. One cannot assume direct transferability of data from one laboratory to another, even when the same analysis is used, as there may be significant differences in the reference population.

The data provided in this section have been found by the authors to be of practical value in clinical situations. These data have been provided according to how they were expressed in the original article, e.g. mean ± 2 SD, 2.5–97.5 percentile range, and no attempt has been made to standardize them. A limited amount of information is provided about the subjects and methodology, where it is available and relevant; the reader is recommended to refer back to the reference for full details.

Subjects

The following definitions have been used:
neonate – first 4 weeks of age;
infant – 4 weeks to 2 years;
child – above 2 years.

Where it is known that a particular analyte concentration changes with age during childhood, and data are available, then age-specific data have been given.

Where an analyte, e.g. amylase, shows little difference across the whole of childhood, then a single 'childhood' range has been quoted.

Units

The international system of units (Système International, SI) has been used for conversion to conventional weight units; for a conversion guide see Laposata, M. (1992) *SI Unit Conversion Guide*. New England Journal of Medicine Books, Boston.

(b) Some protocols

Protocol for a prolonged fast—investigation of inherited metabolic disorders (see also Chapter 5)

Introduction

Some inherited metabolic disorders, and in particular fatty acid oxidation defects, can be difficult to diagnose in an 'unstressed' patient. If a patient is suspected to have this type of defect (i.e. fatty liver, history of hypoglycaemia or a hypoketotic or hyperketotic disorder) a prolonged fast with specimen collection can be helpful in leading to a diagnosis.

The procedure is potentially dangerous and must be carried out under strict medical supervision.

Patient selection

Patients must not be selected for this procedure until baseline metabolic investigations have been completed, i.e. plasma free fatty acids, 3-hydroxybutyrate, lactate, glucose, amino acids, and urinary organic acids and amino acids. The results of the above should be discussed with a specialist in metabolic disease before deciding to undertake a prolonged fast. It is important to ensure that basal investigations do not reveal a definitive abnormality before proceeding and that the patient is in a healthy and good nutritional state.

The procedure

Patient preparation

Insert a venous line prior to commencement of the fast. Patients under the age of 2 years should be catheterized.

The fast (water only should be given during this period)

The fast should commence after the 17.00-h evening meal and proceed for up to 24 h.

BM-Test Strip checking

The BM-Test Strip for glucose should be checked every 4 h until 06.00 h and then at least every hour thereafter. If the result is >4 mmol/L no further action is required, unless symptomatic. If the result is <4 mmol/L or the patient is symptomatic, request a quantitative glucose test, to be performed by the laboratory. If the result of the quantitative glucose test is <2.5 mmol/L, specimens should be collected for 'metabolites', cortisol and insulin, and the fast terminated. 'Metabolites' include lactate, free fatty acids and 3-hydroxybutyrate.

Urine collection

Collect *all urine* passed during the 24 h and store samples separately deep frozen, labelled with the time of collection (discuss storage arrangements with the laboratory). If the patient is catheterized collect urine in 2-hourly aliquots and freeze.

Blood specimen collection

Before commencing the test, contact the chemical pathology laboratory and discuss the correct anticoagulants.

The following blood specimens should be collected for laboratory analysis:

1 Pre-fast (i.e. between 14.00 and 15.00 h): free fatty acids, 3-hydroxybutyrate, glucose, lactate, liver enzymes and carnitine.

2 16-h fasting (i.e. 09.00 h) for metabolites.

3 20-h fasting (i.e. 13.00 h) for metabolites.

4 24-h fasting (i.e. 17.00 h) for metabolites.

5 At the end of the 24-h fasting period a normal meal or feed is given and a further blood specimen (1 mL in a fluoride tube for metabolites) is collected 1–1.5 h later.

If the patient becomes hypoglycaemic a blood specimen should be collected immediately, i.e. before dextrose is given. (Overdosage with dextrose is very dangerous.) The blood specimen is used for 'metabolites', cortisol and insulin. The blood should be taken to the laboratory immediately so that plasma for insulin analysis can be separated and stored deep frozen without delay.

The next urine passed must be collected. If the child is catheterized, continue to collect urine for the next 2 h after the fast has been discontinued.

It may be appropriate to fast some older patients for longer according to their particular clinical condition, and additional sampling should be considered.

Interpretation

Young children (<7 years) show an early decreased availability of glucose with an active ketogenesis after 15 h of fasting. In older children (>7 years) there is a later onset of fatty acid mobilization as a consequence of greater glycogen and gluconeogenic substrate storage.

When blood glucose falls below 3 mmol/L, ketone bodies concentrations exceed 1.8 mmol/L in the normal individual, whereas in those with fatty acid oxidation defects values are <0.8 mmol/L. The interpretation is less clear if glucose remains >3 mmol/L at the end of the fast (Bonnefont *et al.*, 1990).

Protein load

Introduction

Some inborn errors of metabolism present in an episodic manner and can be difficult to diagnose, particularly if blood and urine specimens are not obtained during an acute episode. Urea cycle disorders, in particular, can be exacerbated both clinically and biochemically by protein intake and a protein load may therefore assist in making a diagnossis.

Before carrying out the test:

1 Obtain a dietary assessment of the patient's daily protein intake and the weight of the patient.

2 Send a random urine specimen to the laboratory for metabolic investigations, i.e. amino acids and organic acids.

3 Obtain preload postprandial plasma ammonia* and plasma glutamine and glutamic acid concentrations. Many laboratories have special containers for ammonia. Contact the laboratory before proceeding.

Discuss these* results with the laboratory *before* proceeding with a protein load.

Protein load

It is wise to consult an experienced dietitian about this matter. A load of 1.5 g/kg body weight is given for children, and 1 g/kg for adults, provided that this does *not* exceed:

50 g protein;

the normal daily intake of the patient.

The load is usually given in the form of a milkshake, e.g. 'Build up', with additional protein supplements if necessary.

Patient preparation

The patient should be fasted for at least 4 h. This will depend on the age and the usual interval between meals/feeds. For older infants or children the load is usually undertaken after an overnight fast (i.e. 12 h). Close clinical supervision of the patient during the test is *essential*. The test lasts for 6 h after the protein load and no food or drinks, other than water, should be administered during this period.

Procedure

The exact investigations undertaken will depend on the clinical history and on the results of preload investigations (consult a specialist laboratory before proceeding).

Note: Blood *must* be transported to the laboratory immediately for ammonia estimation.

1 Basal samples, i.e. immediately preprotein load.
(a) Take blood for quantitative amino acids and ammonia.
(b) Collect a random urine specimen (5–10 mL) for amino acids, organic acids and orotic acid.

2 Protein load. Give the protein load and note whether any is refused or vomited. The patient must be observed during and following the load. Any change in the clinical condition, especially neurological symptoms, should be noted.

3 Collect all the urine passed for the next 6 h. Label each specimen carefully with the time of collection and store at −20°C (this may be difficult on a ward). Investigations will be undertaken for amino acids (qualitative), organic acids and orotic acid.

4 Postprandial blood specimen. At 1.5 h after the load, collect 5 mL blood into a lithium heparin tube for quantitative amino acids and ammonia.

Sweat test (sweat sodium and chloride)

Introduction

Cystic fibrosis is the most common serious genetic disease in Caucasians, with an incidence of approximately 1 in 2500 live births in the UK. It is a generalized disorder producing excessively viscous secretions owing to impaired chloride transport. The main symptoms at presentation are a failure to thrive, recurrent respiratory infections and pancreatic insufficiency leading to malabsorption. Sweat sodium and chloride concentrations are increased in cystic fibrosis, and their measurement provides the definitive test for diagnosis of the condition. DNA technology may in the future render this test obsolete (see also Chapter 2, p. 17).

Method

Sweating is induced locally by iontophoresis of a weak solution of pilocarpine on the flexor surface of the arm (Gibson & Cooke, 1959; Green *et al.*, 1985). Sweat is collected into a weighed filter paper, whilst evaporation is prevented by sealing with polythene. After reweighing the paper, the electrolytes sodium and chloride are extracted with a known amount of diluent. The concentrations of sodium and chloride are then determined. Reference data apply to sodium measurements by flame photometry and chloride measurements by colorimetric titration.

Reporting results

The following should be reported:

1 Time period (minutes) over which sweat was collected – this should be 20 min whenever possible.
2 Weight of sweat collected (must be more than 50 mg and preferably greater than 100 mg).
3 Sweat chloride (mmol/L).
4 Sweat sodium (mmol/L).

Notes

1 Sweat tests should only be undertaken by experienced personnel who perform tests on a regular basis.
2 Sweat sodium *and* chloride should be measured.
3 If a sweat weight of <50 mg or a sweat rate of <1 g/m^2/min is obtained, the sweat quantity is insufficient for accurate analysis and the results should not be reported.
4 All patients with borderline results should have the test repeated.
5 The diagnosis should always depend on the clinical findings as well as on the sweat test.
6 Sweat sodium and chloride increase with age in normal children (Kirk & Westwood, 1989; Kirk *et al.*, 1992).
7 Sweat chloride provides the better discrimination, particularly in older children.

Expected results in cystic fibrosis

The following criteria are considered to be consistent with cystic fibrosis:

1 Sweat sodium >60 mmol/L.
2 Sweat chloride >70 mmol/L.
3 Sweat chloride > sweat sodium.
4 Sum of Sweat sodium + chloride >140 mmol/L.

Reference data

Sweat sodium and chloride

Reference data:

Infants and children (1 month–5 years): sodium <50 mmol/L; chloride <50 mmol/L.

Above 5 years: older children may have sodium concentrations up to 60 mmol/L.

Results consistent with cystic fibrosis (1 month–16 years): **sodium >60 mmol/L; chloride >70 mmol/L.**

Results for sodium and chloride between **50 and 70 mmol/L** (or **sodium 60–70 mmol/L for older children**) should be regarded as *equivocal* and always repeated.

Method used:

Sweat collection after pilocarpine iontophoresis according to the Gibson–Cooke procedure.

Sodium – flame photometry.

Chloride – potentiometric.

Comments:

1 Sodium and chloride concentrations in normal children increase with age up to 12 years.
2 In babies and children with cystic fibrosis the chloride concentration is usually greater than the sodium concentration.
3 Sodium concentrations in patients with cystic fibrosis increase with age up to 12 years.

Acknowledgements

The authors would like to thank several members of their respective departments for providing some of the details in this chapter, in particular Kate Hall and Mary Anne Preece of Birmingham Children's Hospital.

References and further reading

Alstrom, T., Dahl, M., Grasbeck, R. *et al.* (1987) Recommendation for collection of skin puncture blood from children, with special reference to production of reference values. *Scand J Clin Lab Invest* **47**, 199–205.

Blumenfeld, T.A., Turi, G.K. & Blanc, W.A. (1979) Recommended site and depth of newborn heel skin punctures based on anatomical measurements and histopathology. *Lancet* **1**, 230–233.

Bonnefont, J.P., Specola, N.B., Vassault, A. *et al.* (1990) The fasting test in paediatrics: application to the diagnosis of pathological hypo- and hyperketotic states. *Eur J Pediatr* **150**, 80–85.

Gibson, L.E. & Cooke, R.E. (1959) A test for concentration of electrolytes in sweat in cystic fibrosis of the pancreas. *Paediatrics* **23**, 545–549.

Green, A., Dodds, P. & Pennock, C. (1985) A study of sweat sodium and chloride; criteria for the diagnosis of cystic fibrosis. *Ann Clin Biochem* **22**, 171–176.

International Federation of Clinical Chemistry. (1987) Approved recommendation (1987) on the theory of reference values. Part 5: Statistical standardisation of collected reference values. Determination of reference limits. *Clin Chim Acta* **170**, 813–832.

Kirk, J.M. & Westwood, A. (1989) Interpretation of sweat sodium results – the effect of patient age. *Ann Clin Biochem* **26**, 38–43.

Kirk, J.M., Keston, M., McIntosh, I. & Al Essa, S. (1992) Variation of sweat sodium and chloride with age in cystic fibrosis and normal populations: further investigations in equivocal cases. *Ann Clin Biochem* **29**, 145–152.

Meites, S. (1988) Skin puncture and blood collecting technique for infants: update and problems. *Clin Chem* **34**(9), 1890–1894.

Whitley, R.J. (1990) Editorial: Reference values in paediatric medicine. *Mayo Clinic Proc* **65**, 431–435.

2: The Impact of Molecular Biology and the New Genetics on Paediatric Chemical Pathology

G. RUMSBY

Introduction

The last 10 years have seen major advances in the field of molecular biology, both in the development of better diagnostic techniques and in our understanding of human disease. It has had a major impact on childhood diseases, providing earlier diagnosis, carrier detection and in some cases preclinical diagnosis in childhood of adult onset disorders. This chapter aims to give the reader an update on recent advances in molecular genetics, with particular reference to paediatric disorders, together with a brief description of techniques and their applications in paediatric clinical chemistry.

DNA structure

While the basic structure of deoxyribonucleic acid (DNA) will be familiar to many readers, it is perhaps worthwhile to give a brief résumé here. DNA consists of a backbone of sugar–phosphate repeating units attached, via the number 1 carbon of the deoxyribose sugar, to one of four bases: adenine (A), cytosine (C), guanine (G) or thymine (T). Two such strands are held together by hydrogen bonds between bases, adenine always pairing with thymine, and cytosine with guanine (base-pairing), to form the double helix (Fig. 2.1). The hydrogen bonds can be disrupted (denatured) by heat or a change in pH to produce single-stranded DNA, a process which can be reversed with restoration of the original double-stranded molecule by cooling or neutralization (reannealing).

The majority of DNA in the mammalian cell is in the nucleus, with some 0.5% in the mitochondria. The nuclear genome of a human haploid cell, i.e. an egg or sperm cell, contains about 3×10^9 base pairs (bp) of DNA arranged into 23 chromosomes – 22 autosomes and one sex chromosome. Fusion of the egg and sperm cell at fertilization yields a diploid cell with two copies of each chromosome and two sex chromosomes, either XX or XY. Each gene locus, with the exception of those on the non-homologous regions of the sex chromosomes in males, will consequently be represented twice. These alternative forms of the same gene are known as alleles and may be identical or different in sequence. Thus, an individual can be described as either homozygous (identical alleles) or heterozygous (different alleles).

Only about 2–3% of the genome codes for either a protein or a mature ribonucleic acid (RNA) product, the majority of DNA having no known function. When a gene is active and producing an RNA it is said to be expressed. Expression occurs in both a tissue-specific and temporal fashion. It is a complex process that is not yet understood fully and results in transcription, the process in which one DNA strand (the coding or antisense strand) is copied by an RNA polymerase. The coding regions or exons, which subsequently produce an RNA product, are not continuous but are interrupted by intervening sequences of DNA (introns) that are spliced out following transcription. RNA sequences that code for a protein product are subsequently translated into polypeptides, individual amino acids being specified by a three-nucleotide sequence (codon).

The mitochondrial genome is circular, approximately 16 kilobases (kb) in length and there are from two to 10 copies per mitochondrion. It is exclusively maternal in origin. The functions of the mitochondrial genome are discussed in Chapter 7(b).

There is a great deal of interindividual variation in

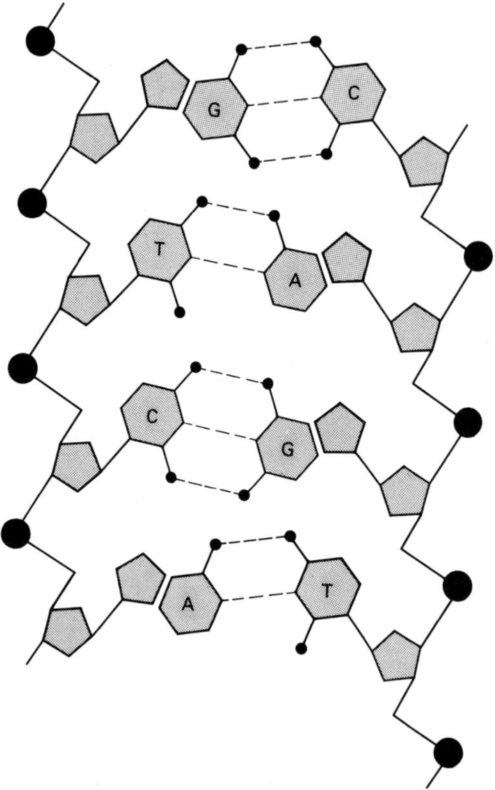

Fig. 2.1 Arrangement of bases in DNA. Adenine (A) pairs only with thymine (T), and cytosine (C) only with guanine (G). The broken lines represent hydrogen bonds. The backbone of each strand is formed of alternating deoxyribose sugar and phosphate groups.

the human genome and it has been estimated that on average 1 in every 250 nucleotides differs between allelic sequences. If two or more alleles occur at a single locus with a frequency of greater than 1%, this sequence variation is described as a polymorphism. Many of these differences will be in non-coding sequences and therefore will have no effect on the gene product, while some of those which occur in exons may be silent, i.e. although the nucleotide sequence may contain small differences, the resulting amino acid is not altered. Other changes are described as conservative, i.e. they code for an amino acid with very similar properties to the original and will therefore not affect the structure or function of the resultant protein. Changes that lead to marked alterations in the gene product are likely to cause disease.

Restriction endonucleases are enzymes that recognize specific sequences of nucleotides (typically 4–6 bp) and will cut the genome into millions of fragments, not in a random fashion but whenever the base sequence that defines their site occurs. The changes in DNA sequence which create polymorphisms may lead to the creation or loss of a restriction endonuclease recognition sequence (restriction fragment length polymorphisms, RFLPs). Other polymorphisms, which are useful in clinical diagnosis as they have many more allelic forms, arise from changes in the number of tandem repeats of a particular DNA sequence (variable numbers of tandem repeats, VNTRs). Polymorphisms are inherited in a Mendelian fashion and can therefore be used as genetic markers to map disease genes and for diagnostic purposes.

Genomic imprinting

The expression of some genes is dependent on their parental origin, i.e. whether they are inherited from the mother or from the father. This surprising finding begins to explain a number of diverse observations. For example, two quite different clinical syndromes, Prader–Willi and Angelman, are associated with deletions of the same region of chromosome 15 (Knoll *et al.*, 1989). Analysis subsequently revealed that Prader–Willi syndrome was the result of loss of part of a paternally derived chromosome 15 (Knoll *et al.*, 1989; Magenis *et al.*, 1990), whereas in Angelman syndrome the corresponding region of the maternal chromosome was deleted (Knoll *et al.*, 1989). Non-deletion cases have also been described in which both chromosomes were derived from a single parent (Nicholls *et al.*, 1989; Malcolm *et al.*, 1991). This situation, known as uniparental disomy, probably occurs when a pregnancy starts as a trisomy with two chromosomes from one parent and one from the other. Most trisomies are lethal, but occasionally one of the chromosomes is subsequently lost during early cell division and a viable disomic cell line results in which the remaining two chromosomes may be from a single parent.

The time at which chromosomes become imprinted with parental origin is not completely clear, although at least part of the modification appears to occur during meiosis and may be associated with changes in DNA methylation (Reik *et al.*, 1987). Sometimes, imprinting leads to the expression of an autosomal recessive disease in a child where only one parent is a carrier, as has been described in some cases of cystic fibrosis (Spence *et al.*, 1988; Voss *et al.*, 1989). There is some evidence in mice that imprinting may have a role in the regulation of fetal growth (Haig & Graham, 1991). Molecular markers are now available which can be used to distinguish the parental origin of chromosomes, so that many other disorders can be examined for possible parental origin effects which may in future help to explain anomalies in the transmission of certain diseases.

Techniques

All molecular biology techniques are dependent on the ability of single-stranded DNA molecules to pair with a second DNA molecule of complementary sequence in a process known as annealing or hybridization. Detailed discussion of these techniques is beyond the scope of this chapter and can be found in many modern textbooks (Sambrook *et al.*, 1989). This discussion will therefore be confined to those techniques used for the routine detection of genetic mutations.

Cloning

Cloning is the insertion of a DNA sequence into a plasmid or viral vector that can be subsequently introduced into a bacterial host. The vector with its DNA insert will replicate autonomously within the bacterium to produce multiple identical copies (clones) of the original DNA sequence (Fig. 2.2). The DNA can be recovered from the preparation and labelled, usually with radioactivity, to form a probe that can be used in hybridization assays to detect sequences with a high degree of sequence similarity to the probe.

Southern blot analysis

The Southern blot (Southern, 1975) provides a means of detecting and determining the size of a particular gene sequence from the millions of fragments generated by restriction enzyme digestion of genomic DNA. It can be used for the detection of RFLPs or to determine whether a major deletion or rearrangement has occurred within a gene (Fig. 2.3).

Polymerase chain reaction

The polymerase chain reaction (PCR) was developed in the mid 1980s (Saiki *et al.*, 1985) and has had an enormous impact on molecular biology. By means of enzymatic amplification a particular DNA sequence can be copied many thousands of times, producing large amounts of product without recourse to cloning (Fig. 2.4). The technique is ideally suited to paediatric investigations requiring only small quantities of DNA. PCR is particularly useful for the detection of single nucleotide changes (point mutations) in genomic DNA but also provides a rapid means of analysing RFLPs without the need for Southern blot analysis.

Dot blot analysis

This technique is frequently used for the detection of point mutations in PCR products. The amplified DNA is applied directly to a nylon membrane in the form of dots, without prior fractionation by electrophoresis. The probe, a string of nucleotides some 17–20 bases in length and usually labelled with ^{32}P, is complementary to the target DNA sequence and will reveal alleles that differ from the normal allele by a single nucleotide.

Clinical applications of DNA analysis

DNA analysis can be used to provide accurate prenatal and postnatal diagnosis and carrier detection for an ever-increasing number of genetic diseases (Table 2.1). Before genetic studies are to be considered, it is vital that an accurate patient and family history is taken and that an unambiguous biochemical diagnosis has been made. As we learn more about the molecular basis of genetic disease, it is becoming apparent that the clinical appearance (phenotype) that we see in a child can be caused by mutations at different gene loci. For example, virilizing congenital adrenal hyperplasia (CAH) presenting in childhood can be due to a deficiency

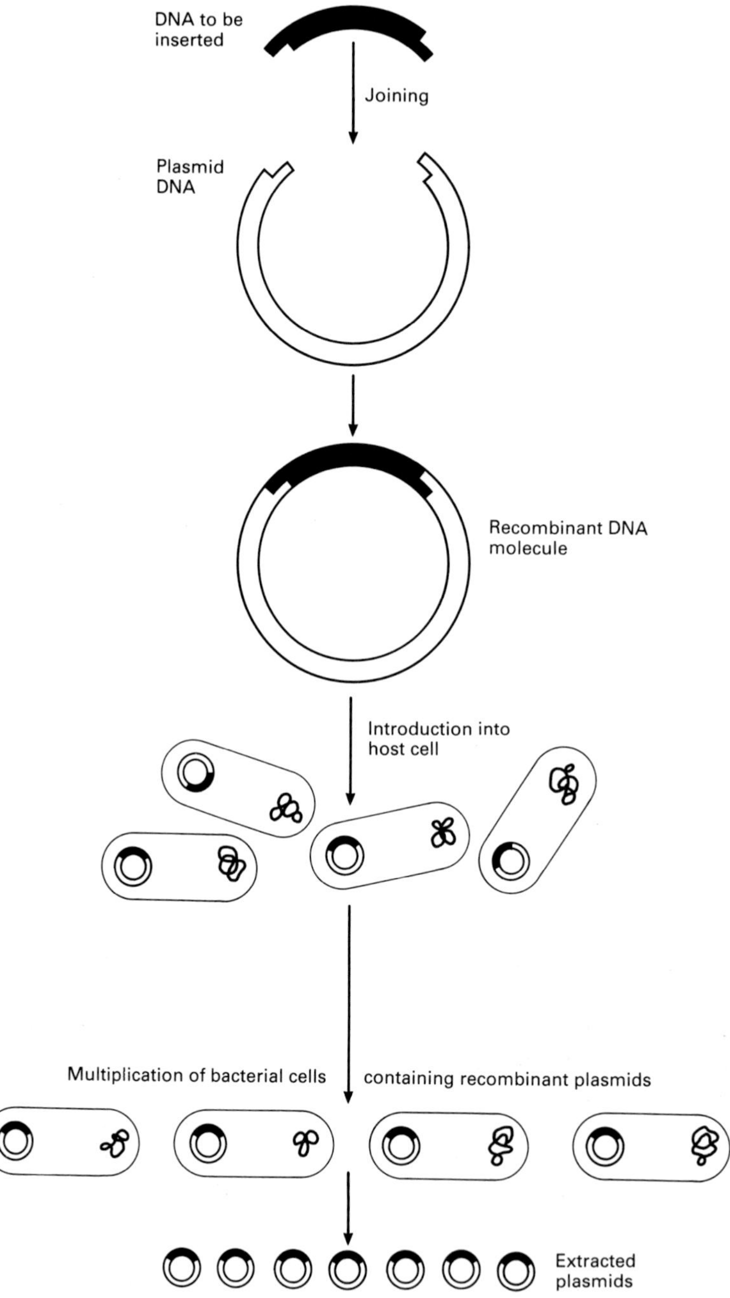

Fig. 2.2 Gene cloning into a plasmid vector.

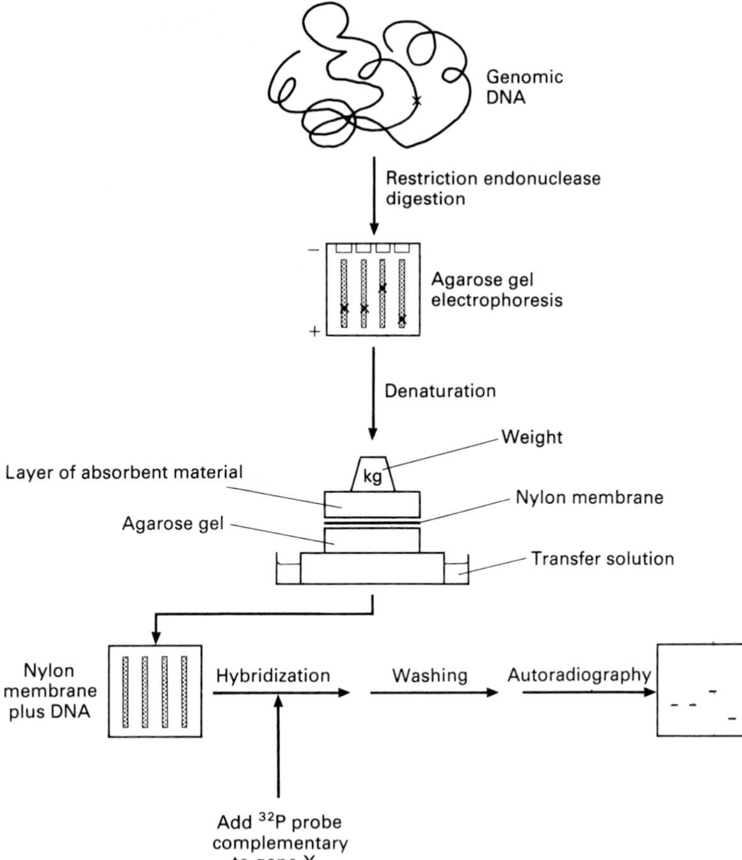

Fig. 2.3 Southern blot analysis.

of either the steroid 21-hydroxylase or 11-hydroxylase enzymes encoded by genes on chromosomes 6 and 8 respectively. The phenotype can also be a combination of a specific gene mutation modified by other genetic and environmental influences. For example, the most common mutation causing cystic fibrosis is the deletion of a 3-bp sequence encoding phenylalanine at codon 508 ($\triangle$F508) of the cystic fibrosis transmembrane regulator (CFTR) gene. Patients homozygous for this mutation typically have an early onset of the disease and pancreatic insufficiency (Johansen *et al.*, 1991) but with variable degrees of lung pathology (Santis *et al.*, 1990; Hamosh *et al.*, 1993).

It is often possible to trace a family history of dominantly inherited or X-linked disorders but a child presenting with an autosomal recessive disease is more usually born to a couple without any previously affected family members. Most DNA diagnostic tests require family studies and while maternity is seldom in doubt, except in the case of ovum donation, non-paternity may lead to misleading results. Human leucocyte antigen (HLA) typing or genetic fingerprinting can be used to clarify this situation, but this is not a routine investigation. Ideally, as soon as a child is identified as having an inherited disorder, a family history should be obtained and counselling about future pregnancies should be given. If prenatal diagnosis is possible, it is helpful to take blood samples from both the index case and the parents so that a genetic work-up can be done in advance of a possible second pregnancy. This action allows the family to be counselled in a calmer atmosphere than during early pregnancy

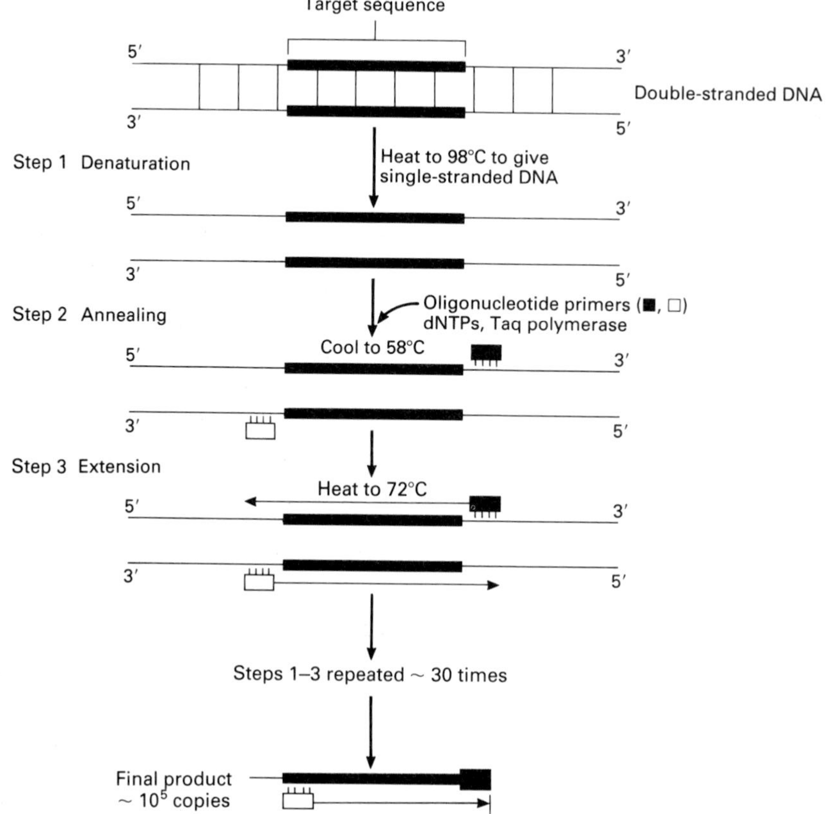

Fig. 2.4 Polymerase chain reaction.

and gives the laboratory more time to determine whether a family study is informative and which particular polymorphism(s) will be useful in future pregnancies. A number of diagnostic molecular genetics laboratories have now been established throughout the UK and form the National Consortium for Clinical Molecular Genetics Services (Clinical Molecular Genetics Society, 1993).

Postnatal diagnosis and carrier detection

For most genetic disorders a diagnosis can be reached using conventional biochemical tests. In some cases, however, DNA analysis may offer an improvement over current biochemical techniques. Cystic fibrosis, for example, is typically diagnosed by a sweat test and/or immunoreactive trypsinogen. Both these tests can give equivocal results and it has been found that looking for the common ΔF508 mutation helps to increase specificity and, in the case of screening programmes, to decrease the number of false positives (Ranieri *et al.*, 1991).

Carriers of genetic diseases frequently show no symptoms and their biochemistry is essentially normal or shows a large overlap with the normal population. There are exceptions: carriers of HbS (sickle trait) have the abnormal haemoglobin present in their blood but are asymptomatic. Carriers of X-linked disorders may, however, exhibit disordered metabolism; for example, carriers of ornithine transcarbamylase deficiency may have a degree of protein intolerance. DNA analysis can provide definitive carrier diagnosis for other siblings or first-degree relatives, and in the case of Duchenne muscular dystrophy this procedure has helped to assign a large number of females with affected siblings to

Table 2.1 Examples of diseases diagnosed prenatally using DNA methods

Disease
Adrenoleucodystrophy
α_1-Antitrypsin deficiency
Congenital adrenal hyperplasia (21-hydroxylase deficiency)
X-linked chronic granulomatous disease
Cystic fibrosis
Dihydropteridine reductase deficiency
Duchenne–Becker muscular dystrophy
Familial hypercholesterolaemia
Familial adenomatous polyposis coli
Fragile X mental retardation
Haemophilia A
Haemophilia B
Huntington's chorea
Isolated growth hormone deficiency
Lesch–Nyhan
Lowe's syndrome
Myotonic dystrophy
Norrie–Wiskott–Aldrich disease
Ornithine transcarbamylase deficiency
Osteogenesis imperfecta type IV
Phenylketonuria
Predisposition to hereditary retinoblastoma
Adult polycystic kidney disease
Sickle cell anaemia
α-Thalassaemia
β-Thalassaemia
Tuberose sclerosis
Wilson's disease
X-linked agammaglobulinaemia

either high- or low-risk groups; this is not possible using measurement of creatine kinase levels alone (Harris *et al.*, 1989).

Other diseases affecting selected populations, such as Tay–Sachs disease, which has a high incidence in the Ashkenazi Jewish population, may justify genetic screening programmes for carriers either instead of or in conjunction with enzyme analyses. A comparison of DNA-based and enzyme tests for Tay–Sachs disease revealed that DNA analysis offered an increased specificity (Triggs-Raine *et al.*, 1990). Population screening has also been proposed to detect carriers of diseases with a high frequency, such as cystic fibrosis, who represent 1 in 20 of all Caucasians. There are two major difficulties in screening for cystic fibrosis. Firstly, the most prevalent mutation ΔF508 occurs in only 70% of cystic fibrosis alleles from northern Europeans and significantly less in other populations (The Cystic Fibrosis Genetic Analysis Consortium, 1990); consequently, a significant number of carriers will not be detected. Secondly, as this approach involves a large number of people, screening and counselling would have to be incorporated into primary health-care schemes with consequent extra expense. The question then arises when such testing should be carried out. In one programme, women presenting in the antenatal clinic were targeted (Mennie *et al.*, 1992) and if they were found to be a carrier for one of six cystic fibrosis mutant gene alleles, testing was also offered to the partner. This prospective approach appears to be the most satisfactory provided that adequate counselling is readily available.

Preclinical diagnosis

By means of DNA analysis it is now possible to predict well in advance of the time of disease onset those members of a family who are likely to develop late-onset disorders such as Huntington's disease. Such presymptomatic testing has to be treated with caution as often there is no benefit to the family or individual in terms of treatment. There are sound arguments for using such molecular diagnoses for disorders that may present in childhood, for example adult polycystic kidney disease, and which would otherwise require all family members to have frequent assessments of renal function etc. Similarly, a clear molecular diagnosis that a child does not have Becker muscular dystrophy will relieve parental overanxiety. Otherwise, testing for late-onset disorders should be left until the child reaches the age of consent and can make his or her own decision (Harper & Clarke, 1990).

Sources of DNA for analysis

DNA is present in every nucleated cell, and therefore all genes, regardless of whether they are expressed in a tissue-specific manner, can be analysed in a blood sample. Carrier detection and postnatal genetic studies require whole blood collected into ethylenediaminetetraacetate (EDTA). It is important not to overfill the tubes and to mix them well to prevent clotting. Samples of buccal cells, obtained by rinsing the mouth, may be useful for analyses based on PCR techniques (Lench *et al.*, 1988). The stability of DNA greatly simplifies collection procedures, samples can be collected locally and posted at ambient temperatures.

For prenatal diagnosis of single gene defects, chorionic villus biopsy (CVS), taken at around 10–11 weeks of gestation, is the material of choice. In contrast to amniocentesis, CVS provides sufficient material for RFLP analysis without recourse to cell culture. The benefit of early diagnosis does, however, have to be balanced by the higher risk of miscarriage compared with amniocentesis (2% compared with 1% from amniocentesis) (Meade *et al.*, 1991) and an as yet unproven link between early placental biopsy and limb abnormalities (Firth *et al.*, 1991). Improvements in ultrasonography allow amniocentesis to be carried out at 12 weeks gestation or less and may prove to be the best option, but as yet there has been no randomized trial to compare the two procedures.

In the case of unexpected death in childhood, blood samples should be taken and stored at −70°C. Tissue samples taken at post mortem and stored as paraffin blocks have been used successfully for retrospective diagnosis of disease using PCR-based techniques (Miller *et al.*, 1992).

Diagnosis

There are two approaches to DNA diagnosis: either detection of the actual disease-causing mutation or the use of indirect methods in which a polymorphic site, or sites, close to or within the gene of interest acts as a marker for the disease-carrying chromosome.

Direct detection of mutations

Deletions

Deletion of a gene or part of a gene is a relatively rare cause of disease but occurs most often in the case of duplicated genes or gene families, e.g. the α-

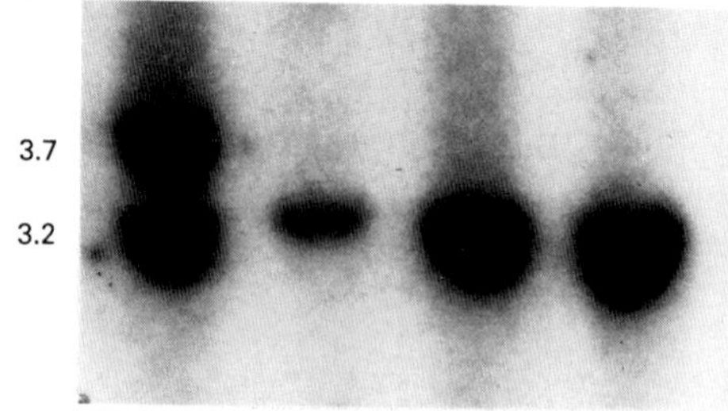

Fig. 2.5 The steroid 21-hydroxylase gene, *CYP21B*, is arranged in tandem with a pseudogene, *CYP21A*, on the short arm of chromosome 6. Reading from left to right, digestion of genomic DNA with the enzyme Taq I and hybridization with a probe for *CYP21* reveals two bands of 3.7 and 3.2 kb in a normal individual (lane 1), representing the *CYP21B* and *CYP21A* genes respectively. Deletion of *CYP21B* is detected by loss of the 3.7-kb fragment (lanes 2, 3 and 4).

globin (Orkin *et al.*, 1978), steroid 21-hydroxylase (White *et al.*, 1985; Rumsby *et al.*, 1986) and growth hormone gene loci (Phillips *et al.*, 1981), and in large genes such as dystrophin in which deletions account for approximately 40% or more of the mutations (Forrest *et al.*, 1987). Major deletions can be detected by Southern blot analysis as the absence or change in size of a restriction fragment, as illustrated in Fig. 2.5.

The dystrophin gene is the largest known gene in man, the coding sequence alone spreading over 14 kb. It is therefore difficult to detect deletions without carrying out multiple hybridizations using a number of probes spread across the gene. A multiplex PCR method has been devised in which nine deletion-prone exons are amplified simultaneously (Chamberlain *et al.*, 1988). Deletions in these exons can then be detected by visual inspection of the DNA product on an agarose gel (Fig. 2.6).

Southern blot analysis is unable to detect small deletions of less than about 50 bp, and PCR techniques are useful in this situation. For example, detection of the ΔF508 mutation in the CFTR gene utilizes amplification of exon 10 of this gene, followed by electrophoresis of the sample on a polyacrylamide gel that has a high resolving power. The ΔF508 genotype is visible after staining the gel with ethidium bromide and visualization on a UV light

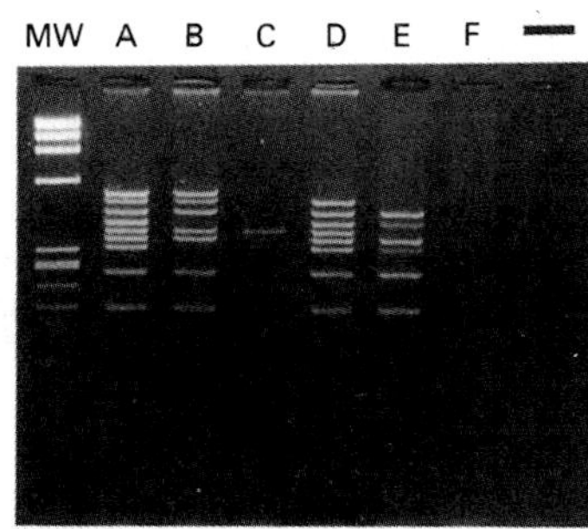

Fig. 2.6 Simultaneous amplification of several deletion-prone exons of the dystrophin gene in patients with Duchenne muscular dystrophy. Deletions were observed in samples A, B, C, E and F. (Adapted from Chamberlain *et al.*, 1988.)

box (Fig. 2.7), giving a product 3 bp smaller than exon 10 of the normal gene.

Detection of point mutations

Occasionally, a pathogenic point mutation coincides with a restriction enzyme recognition site and may therefore create or destroy the recognition sequence. For example, sickle cell anaemia is caused by a single A to T change (GAG → GTG) that causes the substitution of valine for glutamine in the protein. This mutation leads to the loss of a recognition site for the enzyme MstII and therefore alters the size of the DNA fragment that will hybridize to the β-globin probe (Fig. 2.8).

More commonly, the point mutation will not affect a restriction enzyme site and in this case it can be detected with an allele-specific oligonucleotide. A typical example of this type of approach is given by α_1-antitrypsin deficiency in which a G to A mutation at codon 342 leads to the replacement of glutamate by lysine and the formation of the Z variant of this protease inhibitor (Fig. 2.9). Where such mutations are prevalent (Table 2.2) they can be used for postnatal diagnosis or retrospective analysis.

Diseases caused by unstable DNA sequences

There are three gene loci currently known in which variation in tandem trinucleotide repeat sequences leads to apparent instability: the FMR-1 (fragile X mental retardation) gene, myotonic dystrophy and the androgen receptor (spinal and bulbar atrophy). Both fragile X syndrome and myotonic dystrophy have been shown to occur with increasing severity as they pass through subsequent generations of a family, a phenomenon known as genetic 'anticipation'. Elucidation of the molecular basis of both diseases has revealed the presence of repeated DNA

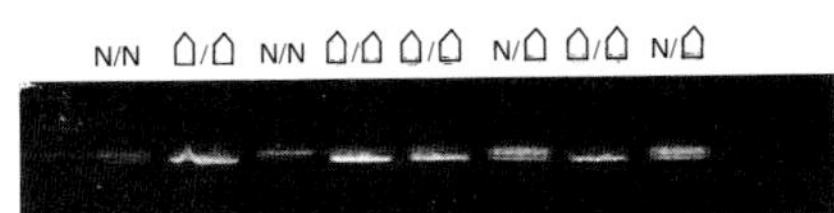

Fig. 2.7 Amplification of exon 10 of the cystic fibrosis transmembrane regulator (CFTR) gene using primers described by Mathew *et al.* (1989). Normal alleles (N) have a product of 50 bp. Alleles with a deletion of codon 508 (△) produce a 47-bp fragment.

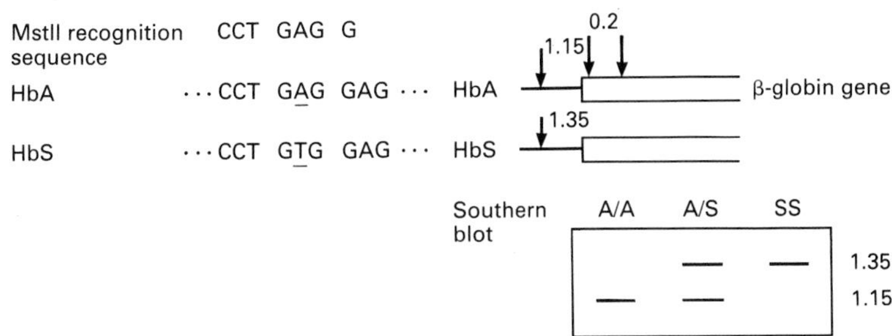

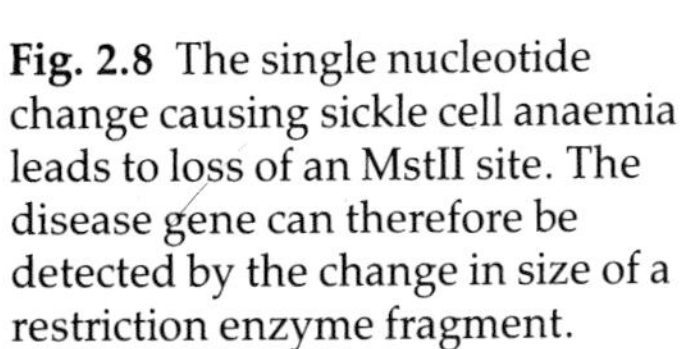
Fig. 2.8 The single nucleotide change causing sickle cell anaemia leads to loss of an MstII site. The disease gene can therefore be detected by the change in size of a restriction enzyme fragment.

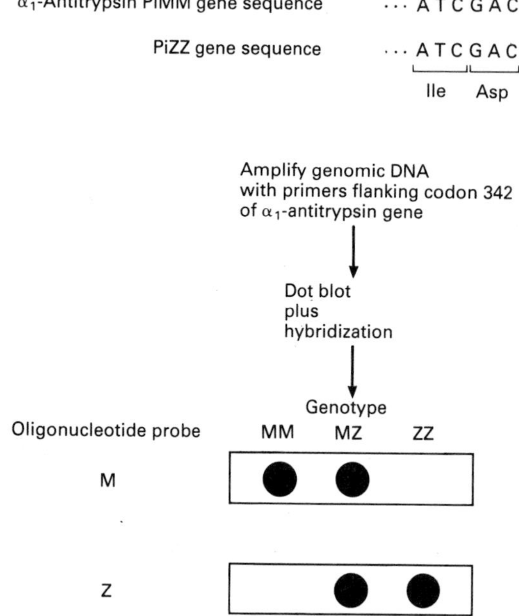

Fig. 2.9 Detection of single nucleotide changes in the α_1-antitrypsin gene by hybridization of amplified DNA with allele-specific oligonucleotides. The normal gene product is denoted PiMM and the Z variant PiZZ.

Table 2.2 Diseases in which a prevalent mutation has been found

Sickle cell anaemia
Steroid 21-hydroxylase deficiency
Medium-chain acyl-CoA dehydrogenase deficiency
α_1-Antitrypsin deficiency
Cystic fibrosis
Tay–Sachs
Lesch–Nyhan

sequences that increase in number particularly during female meiosis and, once they reach a particular size, can cause pathological effects.

Myotonic dystrophy is the most commonly encountered adult form of dystrophy (see Chapter 16) and has an autosomal dominant inheritance. The severity of this disease increases over multiple generations so that symptoms can range from mildly symptomatic adults to neonates with severe disease associated with hypotonia and retardation within a single family. The gene, which is on chromosome 19, encodes a putative protein kinase and has been found to contain a $(GCT)_n$ triplet in the 3′ untranslated region, where n varies from 5 to 30 in normal individuals and to values of 50 to 2000 or more in patients with myotonic dystrophy (Fu *et al.*, 1992; Harley *et al.*, 1992); the higher numbers correspond to an increasing severity of disease.

Fragile X syndrome is the most common cause of mental retardation in children and has a prevalence of 1 in 1000 schoolchildren (Webb *et al.*, 1986) (see also Chapter 20). It is associated with a fragile site on the long arm of the X chromosome at Xq27.3. The disorder is unusual in that 30% of carrier females show some degree of mental impairment, while 20% of males who carry a fragile X chromosome are phenotypically normal and are described as normal transmitting males (NTMs). These NTMs can pass the mutant allele to their daughters, who will also have no symptoms, but their grandsons are often affected. This finding, known as the Sherman paradox (Sherman *et al.*, 1985), can now be explained in terms of the number of $(CGG)_n$ triplets in the 5′ region of the FMR-1 gene, normal individuals having from 5 to 54 triplets, while NTMs have from 52 to more than 200 repeats, and affected males have more than 200 repeats (Fu *et al.*, 1991). The change in repeat numbers after transmission through a female meiosis may reflect unequal recombination at the fragile site during oogenesis.

Linkage analysis

The clinical expression of genetic disease can be the result of a variety of mutations in a gene and only rarely does a single mutation cause the disease or occur at a sufficiently high frequency to be diagnostically useful. It is therefore impractical to offer a clinical genetics service based on the detection of all possible mutations for each gene. In other diseases the identity of the gene is unknown, although it has been mapped to a particular chromosomal region. Linkage analysis can circumvent these problems by utilizing known polymorphisms close to or within the gene of interest. Provided that they are sufficiently close on the same chromosome (i.e. linked), these markers will be inherited along with the disease.

This approach, however, has two caveats:

1 It is not a direct measure of a mutation, merely a marker for the disease allele, and therefore linkage analysis can *only be used in family studies* where DNA is available from an affected child; this allows one to determine which particular form of the marker (polymorphism) is linked to the disease (phase of linkage).

2 There is a risk of a meiotic recombination event separating the marker polymorphism from the disease. This risk is very small in the case of an intragenic polymorphism, but can be significant when linked probes are used or when the gene is particularly large, for example the dystrophin gene.

The incidence of recombination at a particular locus can only be calculated by family studies and varies depending on the chromosome and the region of the chromosome. If a marker shows 5% recombination with the disease, the prediction that the fetus has the disease is only 95% certain. To reduce the occurrence of misdiagnosis due to recombination, flanking polymorphic markers can be used. In this situation, errors will only occur in the rare event of a double recombination. However, the further apart these bridging markers are, the greater the incidence of recombination between them and the less helpful they will be.

Figure 2.10 gives an example of the use of linkage analysis for the prenatal diagnosis of phenylketonuria. More than 10 mutations in the phenylalanine hydroxylase gene have been described in patients with phenylketonuria. Using an intragenic probe it is possible to offer prenatal diagnosis without prior knowledge of the pathological mutation (Woo *et al.*, 1983).

Paediatric malignancies

Cancers are generally the result of somatic gene mutations or a failure of gene regulation with environmental influences. However, some cancers, or rather the predisposition to develop cancer, are inherited in a Mendelian fashion.

Retinoblastoma is the most common ophthalmological malignancy in childhood and in about one-third of cases the predisposition to cancer is dominantly inherited from an affected parent (Vogel, 1979). Early detection of retinoblastoma ensures a better prognosis, but because the mutation is not fully penetrant, it requires a regular full ophthalmological assessment under anaesthetic. Using intragenic polymorphisms detected by hybridization with probes from within the genomic sequence of the retinoblastoma gene, it is now possible to determine those individuals at risk of retinoblastoma and allow clinical screening to focus on those patients at high risk of developing tumours (Onadim *et al.*, 1990).

Molecular genetic analysis has been particularly useful in the haematological malignancies for diagnosis, monitoring remission and detection of relapse (Rowley, 1990).

Prevention and cure of genetic disease

Prevention of a genetic disease can only be an option in families with a previously affected child or in those couples in whom population screening has revealed carrier status. Prevention can take the line of refusal to have any more children or prenatal

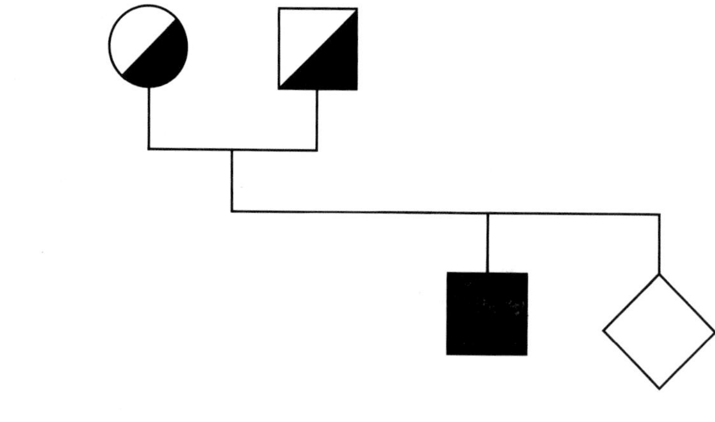

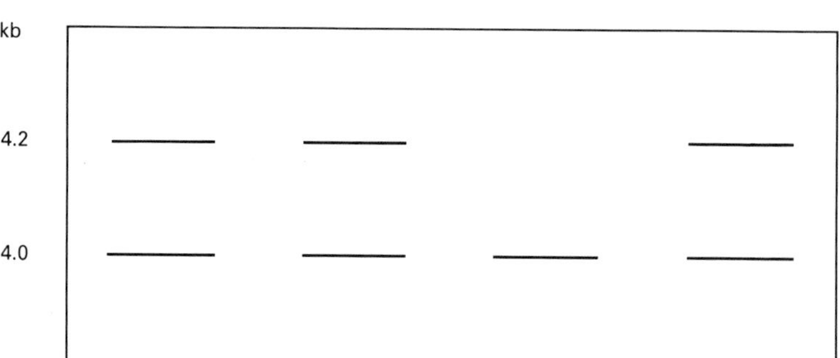

Fig. 2.10 Linkage analysis for prenatal diagnosis of phenylketonuria. In this particular family the disease is inherited with the 4-kb gene fragment detected by hybridization of Southern blots of Hind III digests of genomic DNA with a cDNA probe for phenylalanine hydroxylase. The fetus has inherited one mutant gene and one normal gene and is therefore predicted to be a carrier for the disease.

diagnosis with termination. Preimplantation diagnosis has recently been described for cystic fibrosis; a single cell from the eight-cell stage of the fertilized egg was removed, and diagnosis made by amplification of nuclear DNA prior to implantation of the embryo (Handyside *et al.*, 1992).

Advances in molecular genetics have identified a large number of single gene disorders and it is possible to express many of these genes in cultured cell lines. Thus, it is not unreasonable to wish to extend these findings to the treatment of disease in man. Much activity has been directed towards gene therapy in which either a normal gene is introduced to replace the function of the defective gene or an attempt is made to actually correct the original defect by site-directed recombination between the normal and defective gene. Genes can be inserted into either somatic cells, i.e. any body cell except a germ cell, or germ cells, where the gene is introduced into the fertilized egg. Somatic cell therapy raises no major ethical issues, being essentially the same as organ transplantation, and any genetic changes will be confined to the individual so treated. Germ-line therapy on the other hand will affect future generations and ethically this type of treatment is at present unacceptable in human subjects (Clothier, 1992).

The problems encountered in somatic cell therapy are, firstly, those of getting the gene into the appropriate tissue and into the appropriate subcellular localization where it will be effective and, secondly, to maintain adequate expression of the gene. Genes need to be cloned into a disabled viral vector, for example a retrovirus, to allow them to be taken up by the target tissue. The correction of haematological and pulmonary disorders, such as severe combined immunodeficiency and cystic fibrosis, have already been targeted. Those disorders currently treated by bone marrow transplantation are likely to be the first successes in this field.

Recombinant hormone production

The industrial production of a number of clinically useful proteins, such as growth hormone, factor VIII and erythropoietin, has been made possible by recombinant DNA technology. Despite the association of contaminated human-derived growth hormone and factor VIII with Creuzfeldt–Jacob disease and acquired immune deficiency syndrome (AIDS) respectively, these biosynthetic proteins have been welcomed by paediatricians. Biosynthetic proteins are available in unlimited supplies but are expensive and there is still much discussion with respect to who should have treatment and for how long (Brook, 1992) (see Chapter 9).

Conclusion

Perhaps the greatest achievement of the last 10 years has been the vast increase in our knowledge of the causes of human genetic disease. Technological improvements mean that screening for genetic disease is now possible, but along with these developments come major financial implications for primary health care and certain ethical questions not yet answered. For those individuals who are unfortunate enough to inherit a genetic disorder, the next 10 years may bring real benefits in terms of gene therapy.

References

Brook, C.G.D. (1992) Who's for growth hormone? *Lancet* **304**, 131–132.

Chamberlain, J.S., Gibbs, R.A., Ranier, J.E., Nguyen, P.N. & Caskey, C.T. (1988) Deletion screening of the Duchenne muscular dystrophy locus via multiplex DNA amplification. *Nucl Acids Res* **16**, 11141–11156.

Clinical Molecular Genetics Society (1993) *National Consortium for Clinical Molecular Genetics Services Handbook*. Clinical Molecular Genetics Society, Manchester.

Clothier, C. (Chairman) (1992) *Report of the Committee on the Ethics of Gene Therapy*. HMSO Books, London.

The Cystic Fibrosis Genetic Analysis Consortium (1990) Worldwide survey of the F508 mutation – report from the cystic fibrosis genetic analysis consortium. *Am J Hum Genet* **47**, 354–359.

Firth, H.V., Boyd, P.A., Chamberlain, P. *et al.* (1991) Severe limb abnormalities after chorion villus sampling at 56–66 days gestation. *Lancet* **337**, 762–763.

Forrest, S.M., Cross, G.S., Speer, A. *et al.* (1987) Preferential deletion of exons in Duchenne and Becker muscular dystrophies. *Nature* **329**, 638–640.

Fu, Y.-H., Kuhl, D.P.A., Pizzuti, A. *et al.* (1991) Variation of the CGG repeat at the fragile X site results in genetic instability: resolution of the Sherman paradox. *Cell* **67**, 1047–1058.

Fu, Y.-H., Pizzuti, A., Fenwick, R.G. *et al.* (1992) An unstable triplet repeat in a gene related to myotonic muscular dystrophy. *Science* **255**, 1256–1258.

Haig, D. & Graham, C. (1991) Genomic imprinting and the strange case of the insulin-like growth factor II receptor. *Cell* **64**, 1045–1046.

Hamosh, A., Corey, M. & the Cystic Fibrosis Genotype–Phenotype Consortium (1993) Correlation between genotype and phenotype in patients with cystic fibrosis. *N Engl J Med* **329**, 1308–1313.

Handyside, A.H., Lesko, J.G., Tarin, J.J., Winston, R.M.L. & Hughes, M.R. (1992) Birth of a normal girl after in vitro fertilization and preimplantation diagnostic testing for cystic fibrosis. *N Engl J Med* **327**, 905–909.

Harley, H.G., Rundle, S.A. & Reardon, W. (1992) Unstable DNA sequence in myotonic dystrophy. *Lancet* **339**, 1125–1128.

Harper, P.S. & Clarke, A. (1990) Should we test children for 'adult' genetic diseases? *Lancet* **335**, 1205–1206.

Harris, R., Elles, R., Craufurd, D. *et al.* (1989) Molecular genetics in the National Health Service in Britain. *J Med Genet* **26**, 219–225.

Johansen, H.K., Nir, M., Hoiby, N., Koch, C. & Schwartz, M. (1991) Severity of cystic fibrosis in patients homozygous and heterozygous for ΔF508 mutation. *Lancet* **337**, 631–634.

Knoll, J.H.M., Nicholls, R.D., Magenis, R.E. *et al.* (1989) Angelman and Prader–Willi syndromes share a common chromosome 15 deletion but differ in parental origin of detection. *Am J Med Genet* **32**, 285–290.

Lench, N., Stanier, P. & Williamson, R. (1988) Simple non-invasive method to obtain DNA for gene analysis. *Lancet* **i**, 1356–1358.

Magenis, R.E., Toth-Fejel, S., Allen, L.J. *et al.* (1990) Comparison of the 15q deletion in Prader–Willi and Angelman syndromes: specific regions, extent of deletions, parental origin and clinical consequences. *Am J Med Genet* **35**, 333–349.

Malcolm, S., Clayton-Smith, J., Nichols, M. *et al.* (1991) Uniparental paternal disomy in Angelman syndrome. *Lancet* **i**, 694–697.

Mathew, C., Roberts, R.G., Harris, A., Bentley, D.R. & Bobrow, M. (1989) Rapid screening for ΔF508 deletion in cystic fibrosis. *Lancet* **ii**, 1345–1346.

Meade, T.W., Ammala, P., Aynsley-Green, A. *et al.* (1991) Medical Research Council European trial of chorion villus sampling. *Lancet* **337**, 1491–1499.

Mennie, M.E., Gilfillan, A., Compton, M. *et al.* (1992) Prenatal screening for cystic fibrosis. *Lancet* **340**, 214–216.

Miller, M.E., Brooks, J.G., Forbes, N. & Insel, R. (1992) Frequency of medium chain acyl-CoA dehydrogenase deficiency G-985 mutation in sudden infant death syndrome. *Pediatr Res* **31**, 305–307.

Nicholls, R.D., Knoll, J.H.M., Butler, M.G. *et al.* (1989) Genetic imprinting suggested by maternal heterodisomy in non-deletion Prader–Willi syndrome. *Nature* **349**, 281–285.

Onadim, Z.O., Mitchell, C.D., Rutland, P.C. *et al.* (1990) Application of intragenic DNA probes in prenatal screening for retinoblastoma gene carriers in the United Kingdom. *Arch Dis Child* **65**, 651–656.

Orkin, S.H., Alter, B.P., Altay, C. *et al.* (1978) Applications of endonuclease mapping to the analysis and prenatal diagnosis of thalassemias caused by globin gene deletion. *N Engl J Med* **299**, 166–172.

Phillips, J.A., Hjelle, B.L., Seeburg, P.H. & Sachmann, M. (1981) Molecular basis for familial isolated growth hormone deficiency. *Proc Natl Acad Sci USA* **78**, 6372–6375.

Ranieri, E., Ryall, R.G., Morris, C.P. *et al.* (1991) Neonatal screening strategy for cystic fibrosis using immunoreactive trypsinogen and direct gene analysis. *Br Med J* **302**, 1237–1240.

Reik, W., Collick, A., Norris, M.L. *et al.* (1987) Genomic imprinting determines methylation of parental alleles in transgenic mice. *Nature* **328**, 248–251.

Rowley, J.D. (1990) The Philadelphia chromosome. A paradigm for understanding leukaemia. *Cancer* **65**, 2178–2184.

Rumsby, G., Carroll, M.C., Porter, R.R., Grant, D.B. & Hjelm, M. (1986) Deletion of the steroid 21-hydroxylase and complement C4 genes in congenital adrenal hyperplasia. *J Med Genet* **23**, 204–209.

Saiki, R.K., Scharf, S., Faloona, F. *et al.* (1985) Enzymatic amplification of β-globin genomic sequences and restriction site analysis for diagnosis of sickle cell anaemia. *Science* **230**, 1350–1354.

Sambrook, J., Fritsch, E.F. & Maniatis, T. (1989) *Molecular Cloning: A Laboratory Manual*, Vols 1, 2 & 3, 2nd edn. Cold Spring Harbor Laboratory, New York.

Santis, G., Osbourne, L., Knight, R.A. & Hodson, M.E. (1990) Linked marker haplotypes and the ΔF508 mutation in adults with mild pulmonary disease and cystic fibrosis. *Lancet* **335**, 1426–1429.

Sherman, S.L., Jacobs, P.A., Morton, N.E. *et al.* (1985) Further segregation analysis of the fragile X syndrome with special reference to transmitting males. *Hum Genet* **69**, 3289–3299.

Southern, E. (1975) Detection of specific sequences among DNA fragments separated by gel electrophoresis. *J Mol Biol* **98**, 503–517.

Spence, J.E., Perciaccante, R.G., Greig, G.M. *et al.* (1988) Uniparental disomy as a mechanism for human genetic disease. *Am J Hum Genet* **42**, 217–226.

Triggs-Raine, B.L., Feigenbaum, A.S.J., Natowicz, M. *et al.* (1990) Screening for carriers of Tay–Sachs disease amongst Ashkenazi jews. A comparison of DNA-based and enzyme-based tests. *N Engl J Med* **323**, 6–12.

Vogel, F. (1979) Genetics of retinoblastoma. *Hum Genet* **52**, 1–54.

Voss, R., Ben-Simon, E., Avital, A. *et al.* (1989) Isodisomy of chromosome 7 in patients with cystic fibrosis: could uniparental disomy be common in humans? *Am J Hum Genet* **45**, 373–380.

Webb, T.P., Bundey, S., Thake, A. & Todd, J. (1986) The frequency of fragile-X chromosome among school children in Coventry. *J Med Genet* **23**, 396–399.

White, P.C., Grossberger, D., Onufer, B.J. *et al.* (1985) Two genes encoding steroid 21-hydroxylase are located near the genes encoding the fourth component of complement in man. *Proc Natl Acad Sci USA* **82**, 1089–1093.

Woo, S.L.C., Lidsky, A.S., Guttler, F., Chandra, T. & Robson, K.J.H. (1983) Cloned human phenylalanine hydroxylase gene allows prenatal diagnosis and carrier detection of classical phenylketonuria. *Nature* **306**, 151–155.

3: The Newborn

B.A. WHARTON, P.H. SCOTT & T.L. TURNER

Introduction

This chapter describes the biochemical implications of routine neonatal care and presents a range of biochemical investigations in healthy newborn babies. It then approaches biochemical problems of the newborn in two ways: first, the investigation of those clinical presentations which may be a symptom of, or are accompanied by, biochemical upset, e.g. convulsions and diarrhoea; and second, the management and further investigation of abnormal biochemical signs, e.g. hyponatraemia and hypoglycaemia.

Clinical principles

Major principles of neonatal care

The main points of neonatal care are to keep the baby warm, provide adequate nutrition, prevent infection and ensure adequate oxygenation: these all have biochemical implications.

Warmth and prevention of hypothermia

Unfortunately, despite widespread knowledge of its existence, 'primary' hypothermia, i.e. due to an inadequate thermal environment, still occurs in the newborn. The most common time for it to occur is in the first few minutes of life in the labour room. It is almost always preventable. Occasionally, hypothermia results in hypoglycaemia (p. 34), which should be managed in the normal way. If the baby is warmed beneath an overhead radiant heater then attention to water balance is necessary to avoid hypernatraemia and dehydration.

Hypothermia often presents as a symptom of any illness, metabolic or otherwise, in the newborn, but has been a particular feature of babies who are later diagnosed as having 'kinky hair disease', an inborn error of copper transport. Rarely, hypothermia is a presenting feature of congenital hypothyroidism.

In very low birth-weight (VLBW) babies severe hypothermia is often associated with acute renal failure and disseminated intravascular coagulation (DIC) through mechanisms associated with hypoxia and/or hypotension.

Provision of adequate nutrition

Breast-fed babies should generally be fed immediately after birth, whilst still in the labour room, and frequently thereafter. If bottle fed, the normal healthy baby may be fed *ad libitum* and there is no need to prescribe specific amounts.

Low birth-weight babies are usually fed according to a schedule during the first 3–4 weeks of life. For babies of appropriate weight for the gestational age, 60 mL/kg body weight of either breast milk or an infant formula is given during the first 24 h; this increases by 15 mL/kg each day, to reach 200 mL/kg by day 10. Light-for-gestational-age babies receive 60 mL/kg during the first 24 h, increasing by 30 mL/kg each day, to reach 200 mL/kg by day 6. A full volume of 200 mL of a standard formula provides 3 g protein and 134 kcal/kg body weight (i.e. 2.25 g protein/100 kcal). The composition of expressed breast milk is variable, but on average 200 mL provides 2.4 g protein (nitrogen × 6.4) and 140 kcal. Table 3.1 shows the nutrients present in breast milk and typical modern formulas. Formulas specially designed for low birth-weight babies are increasingly used.

Generally they have a higher energy density and a higher nutrient:energy ratio than formulas for normal babies.

We do not use either nasojejunal or parenteral nutrition extensively but do so when respiratory support is necessary or gastric stasis occurs. Nasojejunal feeding is sometimes associated with a degree of malabsorption and continuous pump feeding may result in energy-rich fat being left behind in the syringe unless precautions are taken to prevent this. There are various well-described regimens for parenteral nutrition. We use a premixed solution of 10% dextrose/electrolytes and vitamins with Vamin N (Kabivitrum, Milton Keynes, UK) to provide 10 g glucose and 2.5 g protein/kg per day in the first week of life and higher thereafter. This is combined with a 10% fat solution (Intralipid) to provide up to 4 g/kg per day of fat from day 4 onwards, if the clinical condition allows.

Other regimens have been equally effective (McIntosh & Mitchell, 1990). Parenteral nutrition is not infrequently associated with glycosuria (insulin may be indicated), hypophosphataemia and moderate hyperammonaemia. The only permissable carbohydrate is glucose, there should not be a great excess of any one amino acid and the solution

Table 3.1 Composition of milks and formulas given to preterm babies. (a) Energy and major nutrients

	Human breast milk (expressed per 100 mL)	Infant formulas for normal babies: Example 1 (based on demineralized whey; per 100 mL)	Infant formulas for normal babies: Example 2 (based on skimmed milk; per 100 mL)	Preterm formulas (range observed in available products; per 100 mL)
Total solids (g/100 mL)		12.6	14	12.5–15.2
Energy (kcal)	70	67.6	70	74–81
(kJ)	293	283	293	310–339
Protein				
Total (g)	1.34	1.5	2.0	1.4–2.4
Casein (g)	0.4	0.6	1.2	0.6–1.0
Whey protein (g)	0.8	0.9	0.8	0.8–1.6
Taurine (mg)	4.8			0.0–5.1
Fat				
Total (g)	4.2	3.6	3.5	3.4–5.0
Butter fat (g)	–	–	2.1	0.0–2.8
Vegetable fat (g)	–	2.4	1.4	0.5–2.7
Medium chain triglyceride (g)		–	–	0.0–1.8
Other fat (g)		1.2	–	0.0–1.4
Polyunsaturated fatty acid (% total)	7.2	14.5	11.2	15–31
Fatty acids C8–C10 (% total)	1.5	3.3	2.6	2–50
Cholesterol (mg)	14	3.0		0.7–1.8
Choline (mg)	9			5.0–25.0
Carnitine (mg)	1			0.0–1.0
Carbohydrate				
Total (g)	7.0	7.2	7.7	6.3–9.7
Lactose (g)	7.0	7.2	5.7	2.3–8.7
Maltodextrin (g)	Trace	–	2.0	0.0–5.3
Glucose (g)	Trace	–	–	0.0–2.2
Amylose (g)	Trace	–	–	

Table 3.1 (b) Vitamins and minerals. After Wharton (1987)

	Human breast milk (expressed per 100 mL)	Infant formulas for normal babies: Example 1 (based on demineralized whey; per 100 mL)	Infant formulas for normal babies: Example 2 (based on skimmed milk; per 100 mL)	Preterm formulas (range observed in available products; per 100 mL)
Vitamins				
Retinol (μg)	60	79	50	0.55–150
Vitamin D (μg)	0.01	1.1	–	0.0–8.0
α-Tocopherol (μg)	0.35	638	1400	800–10 000
Vitamin K (μg)	1.5	5.8	–	0.0–7.0
Thiamin (μg)	16	71	40	37–100
Riboflavin (μg)	31	105	50	70–500
Nicotinic acid (μg)	230	528	170	210–2400
B_6 (μg)	6	42	17	23–200
B_{12} (μg)	0.01	0.11	0.04	0.1–0.5
Folic acid (μg)	5.2	5.3	4.2	5.0–50
Pantothenic acid (μg)	260	210	200	250–1500
Biotin (μg)	0.8	1.5	–	0.0–30
Inositol (mg)				3.0–130
Vitamin C (mg)	3.8	5.8	6.0	10–30
Minerals (atomic weight)				
Sodium (23) (mg)	15	15	24	17–60
(mmol)	0.6	0.7	1.0	0.7–2.6
Potassium (39) (mg)	60	56	70	60–100
(mmol)	1.5	1.4	1.8	1.5–2.4
Chlorine (35) (mg)	43	37	46	28–80
(mmol)	1.2	1.1	1.3	0.8–2.3
Calcium (40) (mg)	35	44	56	50–144
(mmol)	0.9	1.1	1.4	1.3–3.6
Magnesium (24) (mg)	2.8	5.3	6.3	5.0–15
(mmol)	0.12	0.22	0.26	0.2–0.6
Phosphorus (31) (mg)	15	33	46	31–72
(mmol)	0.03	1.1	1.5	1.0–2.3
Iron (56) (μg)	76	1270	500	40–1300
(μmol)	1.4	22.7	9.0	0.7–23.2
Copper (64) (μg)	39	50	42	10–200
(μmol)	0.61	0.78	0.65	0.2–3.1
Zinc (65) (μg)	295	370	600	100–1200
(μmol)	4.5	5.7	9.2	1.5–15.3
Manganese (55) (μg)	2.0	15.8	7.6	3.0–21
(μmol)	0.04	0.3	0.14	0.1–0.4
Selenium (79) (μg)	2–3		–	
(μmol)	0.03–0.04		–	
Iodine (127) (μg)	7	6.9	7.3	4.0–20
(μmol)	0.06	0.05	0.06	0.0–0.02
Fluorine (19) (μg)	5–30	10–20	–	
(μmol)	0.03–1.6	0.05–1.1	–	
Chromium (52) (ng)	340–430			
(nmol)	6.5–8.3			
Molybdenum (96) (μg)	0.5–25			
(μmol)	0.005–0.26			

should contain *all* amino acids. Old-style parenteral feed regimens (some of which are still used in adults), e.g. those including fructose, sorbitol, ethanol or very high glycine concentrations, are not appropriate in the newborn (or in childhood) as they may lead to a variety of biochemical disorders, e.g. hyperammonaemia, lactic acidosis and hyperuricacidaemia.

Prevention and early diagnosis of infection

An antigenic stimulus *in utero*, e.g. from an intrauterine infection, results in fetal production of IgM. In the cord blood of normal babies IgM is less than 200 mg/L, but frank intrauterine infection is usually associated with very high concentrations, e.g. 700–1500 mg/L. C-reactive protein estimation has been used increasingly as a marker of infection. Normal values may be less than 20 mg/mL. Values higher than 30 mg/mL are likely to be associated with established infection.

Findings in cerebrospinal fluid are more difficult to interpret in the newborn. Mild bleeding is common during birth and so xanthochromia with protein concentrations as high as 2.4 g/L may be found in normal babies (Naidoo, 1968).

A rare biochemical aid is the very high plasma activity of 'cardiac' enzymes, e.g. aspartate aminotransferase (AST) and lactate dehydrogenase, found in viral myocarditis, which is a rare cause of heart failure.

Apart from these specific considerations, however, severe infection may result in hypothermia, haemolysis, liver cell dysfunction, bacteraemic shock with hypotension, sick cell syndrome, peripheral circulatory failure, renal failure and thence a variety of biochemical abnormalities. Treatment with antibiotics may interfere with the Guthrie screen and may produce 'strange' unidentified compounds on urinary chromatograms. Neonatal infection is very common, primary metabolic disorders are not!

Management of hypoxia and cyanosis

Apart from changes in arterial oxygen tension, hypoxia results in a plethora of biochemical abnormalities, many of which persist for some time after successful treatment with oxygen and ventilatory support.

There are three major clinical presentations: ante and intrapartum hypoxia, postnatal hypoxia and suspected heart disease.

Ante and intrapartum hypoxia

An adequate airway with ventilation is the keystone of treatment. The administration of pure oxygen is not always essential and 40% oxygen or air alone may be adequate. There is rarely time for biochemical determinations at this acute stage.

If after 5 min of adequate therapy there is no attempt at spontaneous respiration, intravenous sodium bicarbonate 2 mmol/kg body weight with glucose 250 mg/kg (2.5 mL of 10% dextrose) may be given over 5–7 min whilst ventilation continues.

Evidence of mild-to-moderate renal failure is commonly seen following severe intrapartum hypoxia, and a number of babies develop hyperammonaemia (p. 41 and Chapter 5), probably as a result of hepatic ischaemia. Hypoxanthine metabolites have also been identified in the cerebrospinal fluid of infants experiencing intrapartum asphyxia.

Postnatal hypoxia

This is most commonly due to the respiratory distress syndrome (RDS) (hyaline membrane disease, pulmonary surfactant deficiency). In untreated babies there is hypoxaemia, hypercapnia and a mixed respiratory and metabolic acidosis.

The aim is to maintain a $P\text{O}_2$ of 8–10 kPa; if this cannot be achieved with an inspired oxygen concentration of less than 60% then mechanical support is indicated. Ideally, the $P\text{O}_2$ is monitored continuously by either an intravascular or a transcutaneous electrode so that blood analysis is performed only to calibrate the monitors and to determine $P\text{CO}_2$ and acid–base status. Because of the high concentration of HbF in VLBW babies, oxygen saturation monitoring may lack reliability, although its use is better established in more mature infants.

Other biochemical measurements, particularly for lactic acid, have been explored as a guide to therapy and prognosis, but they have little to offer over the routine measurement of arterial blood gases and acid–base. If there is marked hypercapnia as well, then mechanical treatment for the hypoxaemia often corrects the hypercapnia too, as alveolar ventilation improves. The determination of acid–base status is used more as an extra indicator of adequate respiratory support rather than as an indicator for alkali therapy. If severe metabolic acidosis is present together with hypoxaemia, the *main* treatment of the acidosis is adequate oxygenation. If the metabolic acidosis is not accompanied by hypoxaemia, it may be a 'leftover' metabolic complication of previously existing hypoxaemia; continued adequate oxygenation should dissipate it, but alkali may also be given. In these circumstances, however, the possibility of the acidosis being due to other causes, e.g. hypothermia or infection, must be considered.

There is currently extensive interest in the evaluation of both natural and artificial surfactant replacement and/or prophylactic treatment in the most severely affected babies. This has radically altered the risk of pneumothorax in such babies and has substantially reduced death rates. It has, however, had less effect in reducing longer term pulmonary complications and may have increased some (e.g. bronchopulmonary dysplasia).

There has also been interest in the role of thyroxine in the prevention of pulmonary surfactant deficiency disease. It may be that small doses of thyroxine given to mothers in preterm labour may encourage fetal lung maturation.

Babies who are severely ill with RDS develop hyperkalaemia and uraemia, probably reflecting tissue breakdown and renal failure. Initial management is aimed at the underlying hypoxaemia, not the secondary biochemical signs.

Suspected heart disease

Despite the marked hypoxaemia (e.g. below 5 kPa) seen in babies with *cyanotic congenital heart disease*, severe metabolic disturbance, such as acidosis, does not occur in all of them, presumably because oxygen saturation is still adequate for tissue respiration and the situation is stable (e.g. in Fallot's tetralogy). Metabolic acidosis in a baby with cyanotic congenital heart disease generally indicates the need for expert cardiological assessment rather than alkali therapy. In a few, however, metabolic acidosis requiring alkali therapy may develop while awaiting a palliative procedure.

The *hypoplastic left heart syndrome* may present as a shocked looking baby with acidosis and hypoglycaemia.

In persistent fetal circulation, either of the primary variety or provoked by meconium aspiration or diaphragmatic hernia, a severe metabolic acidosis frequently develops. Extracorporeal membrane oxidation (ECMO) currently offers a short-term treatment either to allow surgical intervention (in diaphragmatic hernia) or to reduce the need for ventilation.

A blue baby who does not have hypoxaemia may have *methaemoglobinaemia*. This is most commonly secondary to toxins, such as certain local anaesthetic agents given to the mother, but may be due to a rare inborn error of metabolism or haemoglobin M. A few drops of blood placed on a glass slide or filter paper look brown and do not become pink when oxygen is blown gently onto them. Precise concentrations of methaemoglobin are determined chemically and in blue newborns vary from 10 to 70%. Treatment involves the use of either methylene blue or ascorbic acid.

Clinical presentation of metabolic disorders
(see also Chapter 4)

Although some presentations will specifically suggest a metabolic disorder (e.g. coma occurring after a few days of life), unusual findings are more often non-specific, e.g. poor feeding or an abnormal weight chart. An unexplained previous neonatal death or consanguinity in the parents will raise clinical suspicion. Metabolic disorders must therefore be considered in any clinical presentation where the cause is not absolutely clear or when a symptom, e.g. hypothermia, is unusually prolonged. Often when non-specific symptoms have indicated the need for a 'septic screen' a simple metabolic screen should also be considered. In practice, if the 'simple' metabolic screen shown in Table 3.2, performed while the baby is on a *normal diet* is normal, it is unlikely that the symptoms are due to a primary

Table 3.2 Biochemical 'screening tests' for a suspected metabolic disorder

'Simple' metabolic screen often performed with a 'septic screen'
Plasma or blood
  Glucose, calcium, magnesium, sodium, potassium, urea, acid–base measurements, conjugated bilirubin, amino acid chromatography
Urine
  Reducing substances, glucose, ketones, bilirubin

Additional metabolic investigations if the cause of deterioration is not clear
Blood
  Ammonia
  Chloride (look for anion gap if acidotic)
  Lactic acid and pyruvic acid
  3-OH Butyrate, acetoacetate
  Uric acid
  Store heparinized plasma (2–5 mL at −20°C)
Urine
  Ketoacids (dinitrophenyl hydrazine test)
  Sulphite (Sulphitest, Merck, Alton, UK)
  Store urine (as much as possible)
Cerebrospinal fluid
  Store 0.5–1.0 mL at −20°C in case further investigations indicated, e.g. glycine

metabolic disorder. If other causes are unlikely, the more detailed extra metabolic investigations shown should be performed.

Screening by chromatography for hyperamino-acidaemia will detect non-specific biochemical signs of many inborn errors (e.g. hyperglycinaemia in the organic acidaemias). Although plasma glutamine may be raised, the hyperammonaemias are often not accompanied by simply determined biochemical signs, so that if they are suspected (e.g. in a comatose baby) plasma ammonia must be determined. If symptoms of unknown cause persist, it is worthwhile repeating the simple biochemical screen, even if it was originally normal, since some primary metabolic disorders cause clinical symptoms before the secondary biochemical signs become apparent. The investigations are considered in more detail in the sections below.

Clinical interpretation of biochemical signs

Clearly, interpretation is impossible unless the specimen has been taken adequately, stored in a suitable container and transported to the laboratory properly, e.g. specimens of blood taken via intravascular catheters may easily be contaminated (usually diluted) by the solution infused to keep the catheter open. Not all biochemical disturbances indicate a primary metabolic disorder; metabolic acidosis is more commonly due to hypoxia or infection than to (say) a renal tubular abnormality. Hypoglycaemia may merely reflect inadequate feeding in a susceptible baby.

When considering the aetiology of any abnormal biochemical sign, consider whether it may be explained by either an unsatisfactory specimen or one of the common causes of secondary biochemical disorders, such as hypoxia, infection, hypothermia, unsuitable or inadequate enteral or parenteral feeding, or some other iatrogenic cause, e.g. maternal drugs. If not, then a primary metabolic disturbance can be considered.

Reference data

A number of reference data for proteins important in immunology are available from specialist laboratories. Some reference data for low birth-weight and full-term infants are given in Appendix A1, Table A1.3; see also Chapter 22.

Jaundice

Jaundice is the most common cause of biochemical investigation in the newborn, although in the majority it amounts to no more than a determination of plasma unconjugated and total bilirubin.

Pathophysiology

So-called physiological jaundice has usually been attributed to normal postnatal haemolysis and immaturity of the glucuronyl transferase enzyme system. Other factors must play a role, since in preterm babies the enzyme does not approach full functional capacity until the end of the first month, by which time the jaundice has long since gone. One such factor is the enterohepatic circulation of bilirubin. Any condition causing this to be exaggerated, e.g. inadequate feeding, impaired movement along the gut with delayed passage of meconium, intestinal atresia or Hirschsprung's disease, is often accompanied by jaundice. Another important factor is the amount of other non-haem pigments, such as myoglobin, which require breakdown.

Neonatal jaundice is common but always requires careful, albeit usually rapid, consideration because of the small but definite risk of kernicterus, i.e. encephalopathy due to unconjugated bilirubin. It is too simplistic to assess the risk of kernicterus merely in terms of the plasma concentration of unconjugated bilirubin. Other neurological insults, such as hypoxia, anaemia and acidosis, are common in the neonatal period and may render the blood–brain barrier more permeable to bilirubin whilst, in addition, clearance mechanisms are impaired. Conditions which lead to an increased proportion of free bilirubin, i.e. not bound to albumin, are also important. These include a low plasma albumin concentration (e.g. in the preterm neonate), dissociation of bilirubin from its high-affinity binding site on albumin (e.g. acidosis or high concentrations of free fatty acids) and competition from other substances (e.g. sulphonamides) for the binding sites.

Indications for an immediate reduction in plasma bilirubin by exchange transfusion therefore vary with the clinical circumstances: low gestational age, a short time since birth or the presence of anaemia or acidosis will all suggest the need for exchange transfusion at plasma bilirubin concentrations which would be tolerated in the term baby. In practice, a concentration of less than 250 μmol/L, even in a very ill immature baby, is rarely cause for concern, whereas most paediatricians would proceed to exchange transfusion if the unconjugated bilirubin concentration was greater than 400 μmol/L, even in a fit 1-week-old normal-sized baby. Between these limits individual consideration is necessary. Various objective measurements are available, e.g. free bilirubin, reserve albumin-binding capacity and albumin–bilirubin affinity. They are not used widely in Britain but a detailed review of their clinical use is given by Lee and Gartner (1978).

'Routine' management (see Fig. 3.1)

Investigation of a raised bilirubin concentration which is unusually early or high is essential and it is initiated once the bilirubin concentration has reached the phototherapy action line. Prolonged jaundice is usually considered to have occurred when the bilirubin concentration is either continuing to rise or failing to subside after the first week.

Conjugated hyperbilirubinaemia, whenever it occurs, requires early and, if necessary, extensive investigation since if surgical intervention is necessary it should be within the first 6 weeks of life.

Jaundiced babies who have a bilirubin concentration less than that requiring phototherapy may benefit from enhanced enteral feeds (+20% requirements) because this interrupts the enterohepatic circulation of bilirubin. Once phototherapy is commenced, 'insensible' fluid losses lead to a mandatory increase in fluid intake.

Investigations are designed to detect in *unusually early jaundice* abnormal haemolysis (Rhesus, ABO, glucose-6-phosphate dehydrogenase (G6PD) deficiency), in *unusually severe jaundice* abnormal haemolysis, infection or any cause of conjugated hyperbilirubinaemia (infection or metabolic), and in *unusually prolonged jaundice* all the preceding conditions and hypothyroidism.

Prolonged jaundice

In most instances the bilirubin is all unconjugated; there is no suggestion of haemolysis, anaemia or infection; and after demonstrating that thyroid-stimulating hormone (TSH) and T_4 are normal it is reasonable not to investigate further for some weeks, so long as the child continues to thrive. Breast-feeding jaundice due to steroid substances in the breast milk, which interfere with conjugation of bilirubin, seldom merits interruption of breast feeding. This course of events is commonly seen in mildly preterm babies who are breast fed.

Familial non-haemolytic jaundice will also present in this way and is currently classified into two types: the type I disorder (Crigler–Najjar) is very rare, is an autosomal recessive condition and kernicterus is

Investigation of jaundice

Early jaundice (phototherapy line reached before 48 h)

Maternal	Group and antibodies
Infant	Group, Coombs, bilirubin, full blood count
Other	Glucose-6-phosphate dehydrogenase Mediterranean and non-Caucasian males glucose-1-phosphate uridyl transferase if Clinitest positive
Urine	Clinitest, Clinistix, urine microscopy

Prolonged unconjugated jaundice (greater than 175 μmol/L)

Infant	Thyroid function tests, aspartate aminotransferase (AST), alanine aminotransferase (ALT) Throat swab and urine for cytomegalovirus (CMV)

Conjugated hyperbilirubinaemia (greater than 50 μmol/L)

As in prolonged unconjugated jaundice plus:

Exclusion of infection (especially urinary tract infection)
α_1-antitrypsin phenotyping
Toxoplasma IgM, rubella IgM, syphilitic serology, hepatitis A and B, human immunodeficiency virus screen
Immunoreactive trypsin/sweat test
α-glutamyl transferase, alkaline phosphate

Ultrasound of portal tracts and gall bladder
Bone-marrow aspirate (for Gaucher's disease)
IDA scan (to demonstrate hepatic uptake and bile flow)

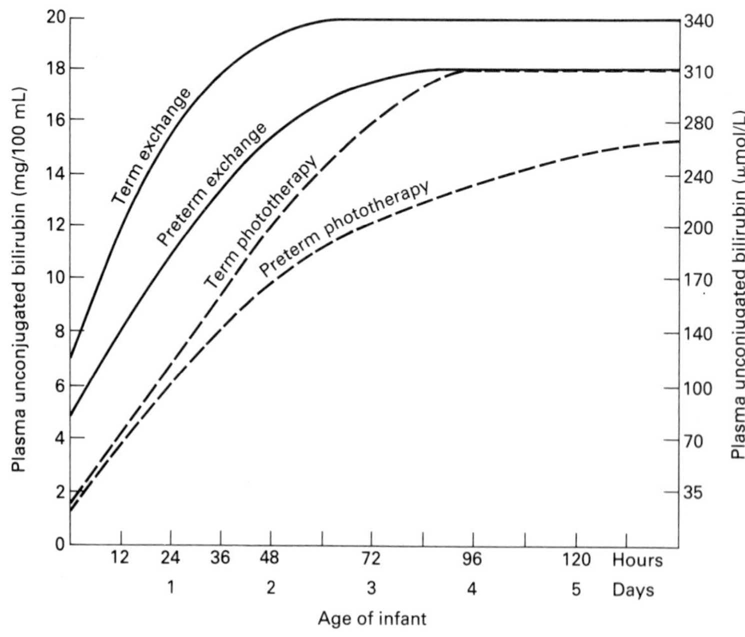

Fig. 3.1 Neonatal jaundice: investigation of jaundice, and criteria for exchange transfusion and phototherapy. Note that other criteria, especially clinical state, may invalidate this guide to therapy.

almost inevitable; type II is an autosomal dominant condition and kernicterus does not occur, presumably because there is some residual glucuronyl transferase activity. Some instances of type II may be, in effect, double heterozygote forms of Gilbert's disease. Gilbert's disease is similar to type II but enzyme activity is about 50% of normal and *when bilirubin production rates are normal* the plasma bilirubin remains below 100 μmol/L; it is, therefore, rarely diagnosed in the newborn.

Conjugated hyperbilirubinaemia

A marked increase in conjugated bilirubin is most commonly a complication of either Rhesus haemolytic disease (the inspissated bile syndrome) or parenteral nutrition. If these have been excluded, infective and metabolic causes of hepatocellular dysfunction should be sought.

Metabolic disorders presenting with cholestasis can be divided into four groups: (a) amino acid disorders, e.g. tyrosinaemia; (b) lipid disorders, e.g. Wolman's, Niemann–Pick, Gaucher's disease; (c) carbohydrate disorders, e.g. galactosaemia, fructosaemia, glycogenosis IV; and (d) uncharacterized, e.g. α_1-antitrypsin deficiency, cystic fibrosis, iron or copper load (see Fitzgerald (1988) for review). Of these the most important emergency diagnosis is galactosaemia (see section on Hypoglycaemia), whilst the most common disorder presenting in this way is α_1-antitrypsin deficiency. Interpretation of plasma concentrations of this protein in the neonatal period is not straightforward. In normal babies (phenotype PiM or PiMZ) the protein concentration increases with gestational age, reaches a peak at 1 week postnatally, and then declines to reach adult levels by 4 weeks (see Appendix A1, Table A1.3). Plasma levels in the neonatal period do not provide a definitive method of identifying deficient babies with the potentially pathological PiZZ or Pinull phenotype. A concentration of 1.0 g/L or less during the first month of life indicates possible deficiency of the pathological type and phenotyping should be performed (Morse, 1978a,b). The Dubin–Johnson and Rotor syndromes may present in this way and in these conditions plasma concentrations of other substances excreted by the liver, e.g. alkaline phosphatase and 5-nucleotidase, are often not increased.

The presence of a heart murmur or other abnormalities raises the possibility of trisomy 13, 18, the Alagille syndrome (odd face, systolic murmur) or arteriohepatic dysplasia (hypoplasia or stenosis of pulmonary arteries – main or peripheral, some with odd facies).

Commonly, all aetiological and clinical clues and all investigations are negative and the diagnosis is either 'cryptogenic hepatitis' or biliary atresia. Whilst the baby is being investigated, moderately raised concentrations of conjugated bilirubin often fall to normal and so it is reasonable to observe the baby but for no longer than a month. At the end of this time, if there is no evidence of biochemical improvement more definitive investigations should be considered, e.g. hepatic ultrasound, technetium-99m iminodiacetic acid scan (IDA) cholescintography, bone marrow aspiration (for Gaucher's disease) and liver biopsy with or without mini-laparotomy. A fuller description is provided by Mowat (1987), and see Chapter 13.

The severely ill child with jaundice

The jaundice may be a 'red herring' and represent no more than the 'physiological' causes of jaundice in a baby who is ill from some other disorder.

If the bilirubin is almost wholly unconjugated, Rhesus incompatibility may cause severe illness, mainly because of the anaemia. If conjugated bilirubin is present as well, Rhesus incompatibility is still a possible cause, but in addition, infection (particularly with Gram-negative bacteria) and metabolic abnormalities such as galactosaemia should be considered. The occurrence of a biochemical abnormality does not necessarily indicate a metabolic aetiology; septicaemia commonly leads to acidosis; a severe hepatitis of infective aetiology may lead to relative galactose intolerance with galactosuria (although not sufficient to give 2% reducing substances) and tyrosinaemia, usually with accompanying raised plasma concentrations of phenylalanine and methionine.

When the diagnosis is in doubt, specimens should be obtained for microbiological and biochemical studies. The baby should be treated with antibiotics, in case there is bacterial sepsis, and with 'clear fluids', as a means of limiting substrate lest there is a metabolic disease.

Treatment of unconjugated hyperbilirubinaemia

Treatment of unconjugated hyperbilirubinaemia with phototherapy and exchange transfusion has a number of biochemical implications.

Phototherapy

Possible mechanisms whereby phototherapy reduces plasma concentrations of unconjugated hyperbilirubin include direct photo-oxygenation of the bilirubin and subsequent hydrolysis into more water-soluble shorter length pyrroles, and the formation in the skin of 'photobilirubin', which more easily dissociates from skin-binding sites and can be excreted *directly* into bile in its unconjugated form. Extra-sensory water losses from the skin and in the stool are increased and we provide additional water.

Exchange transfusion

Exchange transfusion may be indicated immediately after birth in Rhesus (Rh) incompatibility, particularly if the cord haemoglobin concentration is below 12 g/100 mL and/or the cord bilirubin concentration is above 85 μmol/L. In these circumstances, to treat the anaemia and wash out bilirubin and antibody, we usually perform a double-volume (160 mL/kg body weight) exchange with concentrated Rh −ve red cells containing as the anticoagulant citrate phosphate dextrose. The haematocrit of the transfused cells is lowered to 65% with fresh frozen plasma if necessary. The cells are ABO compatible with the mother's blood. In rare instances of cross-matching difficulty, cells frozen in liquid nitrogen may be resuspended in fresh frozen plasma; although resuspension in normal saline is acceptable in an adult, this is not suitable for neonatal exchange transfusion. We do not find it necessary to give calcium routinely during the exchange, and if severe acidosis is present, attention to its cause, e.g. hypoxia or hypothermia, is preferable to boluses of sodium bicarbonate, since they add to the hyperosmolality in the blood. In hydropic conditions a single-volume exchange may be all that can be tolerated initially.

With the anaemia corrected, the indication for subsequent exchanges is usually the plasma bilirubin concentration, which is interpreted in the light of the general condition of the baby, e.g. maturity, other biochemical abnormalities, etc. (see section on Pathophysiology, p. 30).

Other treatments

Phenobarbitone (8 mg/kg body weight per day), which induces glucuronyl transferase, and oral agar (1 g/kg per day), which blocks the enterohepatic recirculation of bilirubin, may be used to reduce the plasma bilirubin concentration. Their action is generally too slow to be of much value in acute neonatal management, but they have been used in the treatment of the inborn errors of unconjugated bilirubin metabolism (Poland *et al.*, 1972; Arrowsmith *et al.*, 1975).

Hypoglycaemia

Hypoglycaemia was recognized as a common problem in the newborn about 20 years ago, but changes in feeding regimens since then have reduced its prevalence considerably. A major review is given by Cornblath and Schwartz (1991) and a brief one by Rayner (1982); reference should also be made to Chapter 5.

Definition and recognition

Using whole blood specimens, Cornblath and Schwartz (1991) have redefined hypoglycaemia as <2.2 mmol/L in the first 24 h and <2.2–2.8 mmol/L thereafter in neonates of any gestation or birth weight. A recent survey of UK neonatologists (Koh & Vong, 1992) indicates that the majority accept these updated values for their management protocols. These limits are clinically useful in that symptoms truly due to the hypoglycaemia *per se* are most unlikely to occur at blood glucose concentrations above this level, but in practice an adequate feeding regimen almost always ensures a blood glucose concentration greater than 2.5 mmol/L and we view with suspicion even a small baby whose blood glucose remains below 2.5 mmol/L for more than 6 h.

Plasma or blood glucose should be determined every 4 h during the first 2 days of life in all neonates weighing less than 2.25 kg, babies of known diabetic mothers, any baby sufficiently ill, whatever the cause, to warrant admission to a special or intensive care nursery and babies who, although not of low birth weight, are, nevertheless, light for the gestational age. Babies in the last category, e.g. a 2.3-kg baby born at 41 weeks who is breast fed, may easily slip through the net. A stick test is usually performed at the cot-side and if the result is low a chemical determination of glucose should be made.

The rate of glycolysis exhibited by erythrocytes from preterm and full-term babies is higher than that of the adult (Meites & Saniel-Banrey, 1979). This affects the stability of blood glucose in samples of blood collected from venous or capillary sites. The use of fluoride as a preservative may be inadequate and recent work has shown that even in combination with sodium iodoacetate (fluoride 17.7 g/L, iodoacetate 5.0 g/L) there was a 4% loss of glucose in 1 h, rising to 10% in 5 h. It is therefore important to minimize the time taken for such specimens to reach the laboratory and to be analysed.

Most laboratories now use glucose oxidase-based methods. However, it has been suggested that only the hexokinase/G6PD method is sufficiently accurate for the satisfactory diagnosis of hypoglycaemia in newborn babies. Glucose distributes itself throughout the aqueous phase of whole blood, so plasma glucose values may be up to 20% higher than the equivalent whole blood values when the haematocrit is high. Erring on the side of safety we accept an estimation of *blood* glucose <2.2 mmol/L as hypoglycaemia, even though determination of the plasma glucose would probably be higher.

Hypoglycaemia in the small baby

The most probable diagnosis is that the light-for-gestational-age baby has poor glycogen stores, but if an adequate feeding regimen has been used the diagnosis should not be accepted too readily and the possibility of infection, hypothermia, perinatal hypoxia or a high insulin state, e.g. the newborn of a missed diabetic mother, should be considered. The combination of hypoglycaemia with metabolic acidosis should raise the suspicion of infection, perinatal hypoxia or occasionally an inborn error of carbohydrate or amino acid metabolism.

Management of symptomatic hypoglycaemia

If at any time symptoms such as collapse, irritability, apnoea or convulsions occur with a blood or plasma glucose <2.2 mmol/L, dextrose should be given intravenously. A suitable procedure would be as follows:

1 If possible, at the time of venous cannulation, collect blood for later assay (if necessary) of insulin and other metabolites, and blood for determination of acid–base status; arrange to collect urine to test for reducing substances and glucose.

2 Administer 25% dextrose in a dosage of 4 mL/kg (i.e. 1 g dextrose) over 1–2 min into a peripheral vein, followed by a 10% dextrose drip at about 5 mL/kg per h to maintain the blood glucose at 2.5–5.5 mmol/L.

3 Note whether symptoms promptly abate as the blood glucose rises; the occurrence of symptoms due to hypoglycaemia *per se* is unusual on modern feeding regimens. More commonly, hypoglycaemia and symptoms occur as associated effects of some other cause, e.g. intracranial problems, in which case the symptoms are not alleviated as the blood glucose rises. Nevertheless, treatment to maintain a normal blood glucose level should be continued.

4 Once the immediate emergency is over, determine acid–base status, if this has not already been done, and consider whether the baby could have an infection; if there are congenital abnormalities, particularly those affecting the mid-line or genitalia, obtain blood for later assay of growth hormone, cortisol, T_4 and TSH, since panhypopituitarism and cortisol deficiency are rare causes of neonatal hypoglycaemia.

5 If large volumes of >10% dextrose are necessary to maintain the blood glucose above 2.5 mmol/L or if after 12 h it is proving difficult to wean the baby from intravenous dextrose to enteral feeds, add oral prednisone and consider the possibility of a high insulin state.

6 Wean slowly, replacing intravenous dextrose by equal volumes of enteral feed over 2–3 days.

7 Recently, it has been suggested that oral glucose gel rubbed into the buccal mucosa may be as effective as intravenous dextrose in the emergency treatment of symptomatic hypoglycaemia. The ease and speed of application are obvious advantages but we have no long-term experience of its use.

Management of asymptomatic hypoglycaemia

A suitable regime would be:

During the first 6 h of life. Blood glucose <2.2 mmol/L, which is quite common; feed hourly or continuously.

After the first 6 h. Blood glucose persistently <2.2 mmol/L, consider alternative diagnoses, begin intravenous treatment and proceed as for management of symptomatic hypoglycaemia.

Other causes of a reduced hepatic glucose release

The most common causes are perinatal hypoxia and infection, and these diagnoses should be carefully considered. A hypoplastic left heart may present as hypoglycaemia with acidosis.

Less common, but important, causes are inborn errors of metabolism, e.g. *glycogen storage disease type 1* may present at this age with hypoglycaemia and metabolic acidosis; after a few days, however, these biochemical symptoms may disappear as the baby takes frequent feeds and so the diagnosis is delayed (Fernandes *et al*., 1969; Hufton & Wharton, 1982).

If the baby is receiving a formula containing fructose or sucrose (a few are available in continental Europe) then *fructose intolerance* or fructose-1,6-bisphosphatase deficiency may present similarly. *Galactosaemia* also presents in this way but in our limited experience of this condition in the neonatal period, evidence that the baby is ill (e.g. vomiting, jaundice) precedes the hypoglycaemia. Urine should be tested for reducing substances and glucose in any episode of hypoglycaemia which is not easily explained, and when interpreting the result care must be taken that the baby has received reasonably normal amounts of lactose in the previous 24 h. If there is doubt about the diagnosis and the baby is ill, a 'clear fluid' dextrose and electrolyte regimen should be given and blood taken for determination of galactose-1-phosphate uridyl transferase activity in the red cells. It may be dangerous to reintroduce lactose in an attempt to obtain meaningful results on urine.

Inborn errors of metabolism affecting the branched-chain amino acids, such as maple syrup urine disease (MSUD) and propionic acidaemia, often cause hypoglycaemia, but never as an isolated finding. Nevertheless, if hypoglycaemia is not readily explained the plasma amino acid pattern should be screened.

Endocrine deficiencies, such as panhypopituitarism, septo-optic dysplasia or adrenocortical insufficiency, may occur. They are sometimes associated with 'mid-line' or genital abnormalities, e.g. cleft palate and lip or small penis, and are sometimes familial (Moncrieff *et al*., 1972). An antenatal clue of adrenal hypoplasia is the *very* low urinary oestriol excretion by the mother. Plasma growth hormone concentrations are normally much higher throughout the neonatal period than in adults.

High insulin states

Most commonly this occurs in the baby of the inadequately controlled diabetic mother. As antenatal control has improved, these babies present fewer problems concerned with blood glucose and management is often directed more towards respiratory difficulties, jaundice occasionally hypocalcaemia and the increased incidence of congenital malformations. Babies should, if possible, be fed as though they were light for gestational age, even though they may be heavy. Blood glucose must be monitored and hypoglycaemia, whether asymptomatic or symptomatic, should be managed as for the 'small baby' above. Occasionally, a diabetic mother produces a poorly grown baby and such babies often have persisting hypoglycaemia requiring intravenous therapy and corticosteroids. High insulin output also occurs in severe Rhesus incompatibility and may lead to marked rebound hypoglycaemia about an hour after an exchange transfusion with acid citrate–dextrose blood.

Less common, but very serious, causes of hyperinsulinism presenting in the newborn and persisting thereafter are β-cell tumours, i.e. nesidioblastosis and islet cell adenoma. An early clue is the amount of intravenous glucose needed to maintain a normal blood glucose, e.g. >8 mg/kg per min. A better indication (so long as the precaution of taking blood for the determination of plasma insulin before emergency intravenous therapy has been observed) is a plasma insulin of 10 U/mL in the presence of a blood glucose below 2.3 mmol/L. This may not always be achieved and it has been suggested that blood 3-hydroxybutyrate and serum free fatty acid are more satisfactory indicators. If the glucose level is below 2.2 mmol/L then a 3-hydroxybutyrate level below 1.1 mmol/L or a free fatty acid level below 0.46 mmol/L suggests hyperinsulinaemia.

Hyperinsulinism may also occur in Beckwith's syndrome, i.e. large tongue, visceromegaly and omphalocele, and in the infant giant syndrome.

Reducing substances present in urine

Glucose is the most common reducing substance in urine. It is most commonly iatrogenic and due to excessive amounts of intravenous dextrose. It occurs in ill small babies with excessive plasma glucose concentrations, presumably because their insulin output is limited. Renal glycosuria may be an indication of generalized renal tubular disease.

A non-glucose reducing substance in the urine will usually be either an antibiotic metabolite (e.g. cephalosporin) or galactose since newborn babies are no longer fed sucrose and they should not receive fructose or sorbitol intravenously. Galactosuria occurs in galactosaemia. If at the time the urine is tested the baby has received a reasonable amount of galactose in the previous 24 h (e.g. 5 g/kg body weight, i.e. 10 g of lactose, or 150 mL/kg of breast milk or most formula milks) 2% galactosuria would be expected. However, the baby has commonly been ill and has been taken off milk feeds by the time the urine is tested, so interpretation is difficult. Galactosuria may also occur whenever there is severe liver cell dysfunction, from whatever cause, but usually it will not be more than 1%. These babies are also very ill and so galactosaemia is a possibility. A 'false positive' urinary test for galactosaemia occurs when profuse diarrhoea fluid contaminates a urine sample and the lactose present in the stool is detected as a urinary non-glucose reducing substance.

Hypocalcaemia and hypercalcaemia

Symptoms due to hypocalcaemia, such as irritability, twitching and convulsions, rarely occur if the plasma total calcium is more than 1.8 mmol/L (7.2 mg/100 mL). When ionized calcium measurements are available it seems that symptoms rarely occur if the concentration is above 0.8 mmol/L (3.2 mg/100 mL). This estimation is available in only a few centres, however, and specimens which must be taken anaerobically cannot be sent away, as they deteriorate rapidly. Corrections for plasma albumin are not usually made in the neonatal period but may require consideration in some babies with severe hypoproteinaemia. Substantial reviews of calcium metabolism in childhood are given by Tsang *et al.* (1981) and Gertner (1990); reference should also be made to Chapter 12.

Early-onset hypocalcaemia (birth–72 h)

This usually occurs in very small or ill babies, e.g. following hypoxia or in the newborn of a diabetic mother. The aetiology is uncertain but is thought to represent an imbalance of calcitonin and parathormone. Preterm infants have a delayed postnatal rise in parathormone and those with hypocalcaemia have very low or undetected levels of parathormone. In addition, preterm and asphyxiated infants have been found to have elevated levels of calcitonin. When there is asphyxia or when catabolism is increased, hyperphosphataemia can occur from tissue protein breakdown, and hence cause hypocalcaemia. Prevention is important. Very small babies weighing <1500 g rapidly develop hypocalcaemia if they are not given calcium. If they are not receiving normal volumes of milk or formula, administered clear fluids, whether intravenous or enteral, should provide 1 mmol (40 mg) calcium and 0.5 mmol

(12 mg) magnesium per kg body weight daily.

It is difficult to know whether symptoms are due to the hypocalcaemia *per se* or to other associated problems. If, in the presence of continued convulsions, the plasma calcium is below 1.8 mmol/L, it is useful firstly to take blood for magnesium and phosphate determinations, and secondly to set up a *slow* intravenous drip of calcium, preferably with heart-rate monitoring so that if bradycardia occurs the infusion can be slowed; 1 mL/kg per h of 10% calcium gluconate (0.22 mmol (9 mg) calcium) is a reasonable starting rate. Rapid intravenous injections of calcium salts are dangerous and the aim is to reduce the amount of calcium given intravenously as soon as possible. The plasma calcium should then be maintained at 2–2.5 mmol/L (8–10 mg/100 mL) by adding calcium and magnesium to clear fluid given intravenously or to an enteral feed, starting with the amounts described above for prevention. Generally, a convulsion at this age is more likely to be due to another cause. In view of the potential danger of the procedure, therefore, rapid intravenous calcium is rarely, if ever, indicated without first establishing that there is indeed hypocalcaemia.

The plasma phosphate concentration would usually be normal or low. It should be noted that normal concentrations are higher than in adults (see Appendix A1, Table A1.3) and haemolysis may lead to a spuriously raised value. A raised level suggests an unusual cause, e.g. accidental overload with phosphate during intravenous nutrition or hypoparathyroidism, although this rarely causes *early* hypocalcaemia.

Late-onset hypocalcaemia
(5–10 days or sometimes later)

Some years ago, convulsions around the seventh day of life were most commonly due to hypocalcaemia precipitated by the high phosphate intake of a neonate receiving unmodified cows' milk. Now that more babies are breast fed and infant formulas contain quite low concentrations of phosphate, this cause has almost disappeared. Today, rather than indicating an unsuitable postnatal diet, late-onset hypocalcaemia more probably indicates a poor antenatal vitamin D status, particularly if the mother is Asian. Less commonly it is due to one of the varieties of congenital hypoparathyroidism.

The following scheme of management is suggested. In most instances the hypocalcaemia is only a transient problem and there is no need to progress beyond step (4).

1 Confirm that the baby's plasma phosphate concentration is raised. If it is not, the hypocalcaemia is not that of the typical late variety, but more probably a non-specific sign of infective or metabolic disease. Also determine plasma magnesium, since this is frequently below normal as well. Exclude uraemia as a cause.

2 If the baby is having fits, treat with:

(a) Intramuscular magnesium sulphate (0.2 mL/kg 50% solution per dose). This may be repeated at 12-h intervals for a further two doses. This is more effective treatment than either oral calcium gluconate or phenobarbitone. Alternatively, control seizures with intravenous 10% calcium gluconate (1 mL/kg), very slowly if absolutely necessary; it rarely is.

(b) Give a low-phosphate diet, i.e. breast milk or a demineralized whey formula. This is an essential step, and attempts to raise the plasma calcium with calcium supplements, whilst the plasma phosphate remains raised, are often unsuccessful.

(c) Give oral supplements of calcium (3 mmol/kg body weight in 24 h (calcium gluconate 1.4 g)). The main effect of the calcium is to precipitate the dietary phosphate as an insoluble salt in the intestine.

(d) If the baby is Asian we usually give vitamin D immediately (calciferol 25 μg (1000 iu) daily) on the assumption that in the majority of instances the hypocalcaemia reflects a poor maternal and fetal vitamin D status. It is wise to save some of the baby's serum so that if the hypocalcaemia is prolonged the serum level of 25-hydroxycholecalciferol (25-hydroxyvitamin D) can be checked prior to treatment. Serum must be stored at −20°C. The plasma calcium level usually rises within 72 h, but on occasion may take up to 10 days. Treatment is slowly withdrawn when the plasma calcium is normal, and the usual prophylactic dose of vitamin D (7 μg) is continued.

3 Seek evidence of frank rickets in the baby;

X-ray the skull and a knee. If rickets is present, 25-hydroxycholecalciferol should be continued for about a month and then reduced to 7 µg daily. Lesions of the ends of long bones (admittedly somewhat different from rickets) also occur in syphilis which, like any severe neonatal illness, may be associated with the early onset of hypocalcaemia.

4 Seek circumstantial evidence of *maternal osteomalacia*, i.e. plasma calcium level low or normal, phosphate low or normal, alkaline phosphatase raised. If these investigations are delayed until 1 week postpartum, the placental alkaline phosphatase has largely disappeared from the maternal circulation and the activity should be within the normal adult range. If this preliminary screening suggests osteomalacia, decisions must then be made whether to confirm the diagnosis in the mother, e.g. undertake X-rays of bones, bone biopsy and determination of alkaline phosphatase isoenzymes, to accept a simple deficiency of diet and sunlight as the cause, or whether to investigate further for evidence of malabsorption and renal tubular disease. In practice, it is probably reasonable to accept the 'simple' diagnosis for an Asian mother but to investigate a Caucasian mother further. It is wise to save some of the mother's plasma whatever the result, so that concentrations of 25-hydroxycholecalciferol and parathyroid hormone (PTH) may be determined if the hypocalcaemia does not respond to the simple measures described in (2). If the neonatal hypocalcaemia is a reflection of maternal vitamin D deficiency, the serum 25-hydroxycholecalciferol level will be below 10 ng/mL in both mother and baby, but normal concentrations have been found in adults with frank osteomalacia. This investigation may very occasionally suggest maternal hyperparathyroidism.

5 If the baby's plasma calcium level is not corrected within 10 days following the treatment described in (2), the possibility of *congenital hypoparathyroidism* requires careful consideration. During further investigations it may be possible to achieve a normal plasma calcium by increasing the dosage of 25-hydroxycholecalciferol. Occasionally, doses of up to 50 µg/kg per day are necessary. These higher doses induce the formation of the active 1,25-dihydroxycholecalciferol, even in the absence of PTH, but care is necessary to avoid hypercalcaemia and thence nephrocalcinosis. Experience with more active metabolites is limited but 1,25-dihydroxycholecalciferol in doses of 30–80 ng/kg/day has been used in older children.

6 Seek evidence of the *Di George syndrome* (i.e. absent thymus, absent parathyroids). If there is no thymic shadow on the chest X-ray, tests of cellular immunity should be pursued and a fresh blood transfusion avoided if possible in view of the small, but definite, risk of a graft-versus-host reaction. Abnormalities of the heart, particularly of the great vessels, are often associated with abnormal development of the parathyroid glands. Plasma PTH levels in these patients are low.

7 Obtain blood for a determination of PTH. Most of the available assays are used mainly for the diagnosis of hyperparathyroidism and are much less reliable in the lower range. Hypoplasia of the parathyroids, whether isolated or part of the Di George syndrome, is associated with a low concentration of the hormone, despite the hypocalcaemia (estimated on the same sample as that used for the hormone assay), but other causes of hypocalcaemia lead to hormone concentrations in the upper normal or high range. Ideally, the baby should show a normal response to exogenous PTH.

8 *Pseudohypoparathyroidism* should be considered, i.e. biochemical signs of hypoparathyroidism but with normal parathyroid glands producing large amounts of hormone. Children with the type 1 disorder are short and stocky with a round face and short fourth and fifth metacarpals, so that they cannot make a complete set of knuckles. There is often a family history of the disorder indicating 'autosomal dominant with variable penetrance'. PTH levels are high normal or high but there is no metabolic response to administered PTH. The hormone normally increases renal tubular cyclic adenosine monophosphate (AMP) and inhibits tubular phosphate reabsorption so that urinary phosphate and cyclic AMP increase. It also promotes a natriuresis and decreases reabsorption of bicarbonate, which can create a mild hyperchloraemic acidosis. The kidney of the newborn is initially unresponsive to PTH, but towards the end of the first week of life urinary cyclic AMP and phosphate excretion increase in response to PTH. In the type 2 disorder there are no anatomical abnormalities; there is a normal increase in urinary cyclic AMP following exogenous PTH but *no* increase in urinary phosphate. Investigations using exogenous PTH require detailed consultation with a specialized laboratory. Hypoparathyroidism and pseudohypoparathyroidism are discussed further in Chapter 12.

Osteopetrosis can rarely present as either early or late hypocal caemia.

Other causes of hypocalcaemia

Hypocalcaemia may occur during and following exchange transfusion with acid citrate–dextrose blood and may also be a non-specific sign of metabolic disturbance, e.g. in hypernatraemia and the organic acidaemias.

Hypercalcaemia

Hypercalcaemia is unusual in the newborn. In retrospect, the infant with the severe variety of infant idiopathic hypercalcaemia (Williams syndrome) or familial hypocalciuric hypercalcaemia or primary hereditary (dominant) hyperparathyroidism (Spiegel *et al.*, 1977) may have had symptoms which go back to the neonatal period, but the diagnosis is rarely made then unless a plasma calcium is determined for some other reason.

Three other diagnoses should be considered: (a) the phosphate depletion syndrome, particularly in VLBW babies receiving a parenteral nutrition regimen that does not contain sufficient phosphorus; (b) disseminated or occasionally lone malignancy, e.g. neuroblastoma or hepatoblastoma due either to the osteolytic effect of secondaries or to tumour production of parathyroid hormone related protein (PTHrP), prostaglandin (PG) E_2 or osteoclast-stimulating factor (Tsang *et al.*, 1981); and (c) transient hyperparathyroidism secondary to maternal hypoparathyroidism (Glass & Barr, 1981).

Neurological symptoms

Convulsions and other abnormal movements

The more common metabolic causes of convulsions include hypoglycaemia, hypocalcaemia, hypomagnesaemia and occasionally hypernatraemia or hyponatraemia (see other sections of this chapter).

Inborn errors of metabolism affecting mainly organic substances usually cause other prominent symptoms as well, such as vomiting or coma. There are exceptions, however, e.g. B_6-dependent convulsions (it is worth investigating for this condition only if hitherto intractable convulsions are controlled by pyridoxal (pyridoxine) up to 300 mg/24 h), and

Table 3.3 Neonatal neurological distress due to metabolic disease

Types	Clinical symptoms	Acidosis	Ketosis	Hyperlactacidaemia	Hyperammonaemia	Most frequent diagnoses
I	Neurological distress	0	+	0	0	Maple syrup urine disease
II	Neurological distress	+	+	0	+	Organic acidurias
III	Neurological distress	+	+	+	0	Congenital lactic acidaemias
IVA	Neurological distress	0	0	0	+	Urea cycle defects
IVB	Neurological distress	0	0	0	0	Non-ketotic hyperglycinaemia, sulphite oxidase deficiency Peroxisomal disorders Respiratory chain defects

sulphite oxidase and/or xanthine oxidase deficiency (clues are a very low plasma concentration of uric acid and the sulphur amino acids).

Abnormal 'extra-pyramidal' movements may occur following neonatal kernicterus, but in the neonatal period irritability and floppiness, rather than abnormal movements, are prominent. The Lesch–Nyhan syndrome has a raised plasma uric acid from birth but abnormal movements are not apparent till later. Plasma uric acid is well above adult levels in normal babies in the first week, falling to the adult concentration within 2 weeks (Wharton *et al.*, 1971). Many metabolic errors affecting the central nervous system lead to dystonic movements but are often associated with an abnormal conscious state (see Coma, below).

Coma with and without seizures and/or vomiting

Intracranial haemorrhage and/or hypoxic brain damage are much the commonest causes, and usually cause symptoms from birth. Although creatine kinase is often raised following head injuries in adults it does not help in the diagnosis of intracranial injury in the perinatal period (Wharton *et al.*, 1971). Metabolic causes should be suspected in babies who seem well initially and then deteriorate after a few days as their protein intake increases, but the occasional metabolic disorder may cause coma within the first 24 h of life, e.g. non-ketotic hyperglycinaemia, sulphite oxidase deficiency (Table 3.3). Suspicion of a metabolic cause is strengthened if: (a) there is nothing in the perinatal history to suggest hypoxia/haemorrhage (e.g. Apgar score is normal, endotracheal intubation is unnecessary); (b) lumbar puncture and ultrasound scan of the head are normal; (c) there has been a previous unexplained neonatal death; (d) there is parental consanguinity; or (e) one of the simple metabolic screens discussed previously is positive. Surprisingly, some metabolic errors may present with neurological symptoms before the peripheral evidence of a profound biochemical disorder is apparent; for example, we have seen one baby with propionic acidaemia who was already comatose and on a ventilator for 24 h before metabolic acidosis occurred. When symptoms persist it is worth repeating all the simple screening tests.

'Blind' treatment may have to be instituted before a definite diagnosis is made (see below). The differential diagnosis of neonatal neurological distress is considered in more detail in Chapter 4.

Coma with acidosis

An organic acidaemia or one of the pyruvate/lactate acidoses (see Metabolic acidosis) is possible, but intraventricular haemorrhage and meningitis are more common. Ketoacidosis suggests MSUD or methylmalonic, propionic or isovaleric acidurias. Other rare organic acidurias are less likely to have ketosis and often present much later (see Chapter 6).

Coma with a haematological disorder

Coma associated with blood abnormalities such as thrombocytopenia or neutropenia is most probably due to overwhelming infection, but organic acidaemias with secondary hyperglycinaemia may present in this way. The combination of anaemia, acidosis and coma is most commonly due to haemorrhage with hypoxia but may be a presentation of 5-oxoprolinuria.

Hyperammonaemia

Hyperammonaemia occurs most commonly following perinatal hypoxia probably due to hepatic ischaemia, when there is an excessive protein intake, particularly during intravenous feeding, or as a transient phenomenon in preterm babies. It may occur variably as a non-specific biochemical symptom in various metabolic disorders, e.g. organic acidaemias, periodic hyperlysinaemia. Urinary tract infection with stasis is an occasional cause. Reye's syndrome rarely occurs in the newborn. If these causes are excluded and the ammonia concentration is very high (more than three times normal) and ketosis is absent, then one of the inborn errors of the urea cycle is likely, especially if the blood urea is also low. These are discussed further in Chapter 5.

Coma without any other metabolic symptom, such

as hypoglycaemia, acidosis or hyperammonaemia, may occur in hypervalinaemia.

Management of a suspected metabolic disorder

If a metabolic error is suspected, the screening tests described already should be performed, and since a lumbar puncture will usually be performed an aliquot of cerebrospinal fluid (CSF) should be deep frozen for further examination if indicated, e.g. if non-ketotic hyperglycinaemia is a possibility following screening for plasma amino acids. In this condition the plasma glycine concentration is between two and 10 times normal, CSF glycine is more than 10 times normal and the plasma:CSF glycine ratio, which is usually at least 30, is much reduced (blood and CSF should be collected at the same time).

Whilst waiting for the results, most babies should receive antibiotics, since severe infection cannot usually be excluded. If the baby is very ill, dextrose–electrolytes only should be given, but before doing so plasma and urine should be collected for more detailed biochemical analysis, e.g. quantitative estimation of amino acids, organic acids and short-chain fatty acids. Although a high carbohydrate intake is the ultimate aim, it should be introduced gradually over 48–72 h as large amounts introduced suddenly will often result in glycosuria and lead to secondary metabolic problems. Babies should be ventilated as necessary.

If the baby is *in extremis* the aim of treatment is to: (a) remove the potentially toxic metabolite; (b) prevent further production of the metabolite by excluding possible substrates from the diet and reducing the breakdown of body proteins; and (c) induce imperfect enzyme activity by supplying large amounts of coenzyme. The toxic metabolite may be removed by peritoneal dialysis, haemofiltration, haemodialysis or exchange transfusion. The first sample of blood removed should be saved so that analyses of enzymes in red cells or white cells may be performed later, if indicated.

Dietary limitation of a suspected substrate is achieved by giving a reasonable energy, protein-free (initially) diet by intravenous or enteral route. If the latter is possible, glucose polymers may be helpful as a source of carbohydrate. An empirical vitamin cocktail may be given in an attempt to provide a coenzyme: thiamine 50 mg, riboflavin 50 mg, biotin 10 mg, nicotinamide 100 mg, pyridoxine 50 mg, folic acid 5 mg, ascorbic acid 300 mg – all 8-hourly, plus B_{12} 1 mg daily. L-Carnitine (100 mg/kg per day) should also be added. If death is inevitable and no diagnosis has been reached, it is reasonable to introduce normal feeds, stop the vitamin therapy for 24 h and then collect plasma, urine and possibly CSF for freezing. It may be helpful to make a preparation of leucocytes.

Ideally, a skin biopsy for fibroblast culture should be taken and sometimes a liver biopsy is helpful. Liver biopsies taken sometime after death are generally not helpful. Perry (1981) has described the method of autopsy in suspected disorders of amino acid metabolism (see Chapters 1 and 6).

Hypotonia

Many metabolic disturbances result in marked hypotonia (usually without paralysis) and a similar biochemical approach to that described in the section on Coma (p. 40) might be used, even when the baby is fully conscious.

Conditions not mentioned in that section and which include floppiness as a symptom are hypercalcaemia, renal tubular acidosis and hyperthyroidism. The neurolipidoses (e.g. Tay–Sachs), mucopolysaccharidoses (e.g. Hunter–Hurler) and mucolipidoses, although causing hypotonia, usually present later in infancy. Mucolipidosis type II (I-cell disease) may cause a floppy baby with decreased intrauterine movements, but the abnormal appearance and bones are more obvious features. The hypotonia of Pompe's disease (glycogenosis type II) may be apparent before heart failure occurs; the enzyme defect can be demonstrated in leucocytes. Peroxisomal metabolic disorders, such as Zellweger syndrome and neonatal adrenoleucodystrophy, present with very early and marked hypotonia (Stephenson, 1988). Prader–Willi syndrome, as characterized by abnormalities of free fatty acids and chromosome 15, may also present as floppiness.

Plasma creatine kinase activity might be determined, but in practice its value at this age in the differential diagnosis of hypotonia is limited. In normal babies the activity is 'high' with a peak at about 24 h, thereafter falling rapidly to reach 'adult' levels within 10 days; levels in low birth-weight babies are slightly lower (Wharton *et al.*, 1971). Boys with Duchenne muscular dystrophy have markedly raised plasma concentrations of the enzyme in the neonatal period but the disease does not cause hypotonia at this time. Rarer autosomal recessive varieties of muscular dystrophy may present with floppiness in the early months of life (Wharton, 1965), and have a moderately raised enzyme activity. A slightly raised plasma enzyme activity, even with profound hypotonia, does not necessarily indicate a myopathy since this may occur with neural causes, e.g. Werdnig–Hoffman disease (see Chapter 16). Hypothyroidism should be excluded.

Diarrhoea and dehydration

The main biochemical implication of diarrhoea is the assessment and management of dehydration and the associated electrolyte disorders, but there are a few aetiological points.

Aetiology

Infection is the most common aetiology and the episode is usually short lived.

If diarrhoea persists for more than a week the most likely cause is *secondary* food intolerance, e.g. secondary to gastroenteritis and/or cows' milk protein intolerance. Simple exclusion of lactose is rarely sufficient and modification of dietary protein and sometimes fat is often necessary, e.g. using a hydrolysed protein or individually prescribed modular formula. Chronic intractable diarrhoea of undetermined cause presenting in the newborn, particularly in Asian babies or where it is familial, is often very difficult to manage (Candy *et al.*, 1981).

Surgical conditions may present with, or be complicated by, diarrhoea. Malabsorption of monosaccharides and disaccharides, secondary to malrotation, Hirschsprung's disease and following surgery of the gut in the newborn, for whatever cause, is well described, and a secondary chloride-losing diarrhoea following surgery may occur (Aaronson, 1971).

Continuing diarrhoea may rarely be due to an inborn error of intestinal function. Chloride-losing diarrhoea is relatively common in Finland but does occur elsewhere (Launiala *et al.*, 1968). It should be considered as a diagnosis if there has been maternal polyhydramnios, which may be due to intrauterine diarrhoea, and there is alkalosis in the baby, since most diarrhoea states leading to dehydration are accompanied by acidosis. Investigation shows that the concentration of chloride in the stool fluid is greater than the sum of the concentrations of sodium and potassium in the fluid. A similar defect but involving Na^+/H^+ exchange and resulting in metabolic acidosis and hypokalaemia has been described (Booth *et al.*, 1985).

Sucrase–isomaltase deficiency is unlikely to cause problems or be diagnosed in the newborn since sucrose is not usually included in the diet. Truly congenital lactase deficiency and glucose–galactose malabsorption are very rare but do cause severe neonatal diarrhoea.

Certain inborn errors of intermediary metabolism have diarrhoea as a feature, e.g. lysinuric protein intolerance, 3-hydroxy-3-methylglutaryl aciduria and dicarboxylic aciduria. We have no experience of these; they seem to present later in infancy and diarrhoea is often accompanied by other serious signs such as hypoglycaemia, acidosis and coma.

Management of dehydration

Mild dehydration may be managed with 'oral rehydration therapy'. In the newborn it is wise to prescribe minimum volumes to be taken (e.g. 150–200 mL/kg per day) rather than rely on appetite. If this cannot be achieved or dehydration is more severe, intravenous therapy is indicated. There are numerous intravenous regimens for managing dehydration. A didactic regimen of four steps has been described (Tripp *et al.*, 1977). This regimen is adequate for most causes of dehydration, e.g. due to gastroenteritis or the adrenogenital syndrome. Usually it can be used irrespective of the biochemical findings, but modifications may be based on a knowledge of electrolytes and acid–base results.

1 Initially infuse 50 mL/kg body weight of N/2 saline (75 mmol/L) in dextrose over 4 h. If the baby has collapsed, about 10 mL/kg of this initial infusion may be given rapidly (over 30 min) as plasma.
2 After 4 h: begin to infuse 200 mL/kg body weight of N/5 saline (30 mmol/L) in dextrose over 24 h. If by 4 h the baby has already passed at least 4 mL/kg body weight of urine, potassium may be added to the infusion (20 mmol/L), step (3) can be omitted and step (4) commenced. If metabolic acidosis was present initially it should now be improving. If it is not, reconsider the diagnosis and the cause of acidosis.
3 After 6 h: if by this time the baby has passed a urine volume of at least 6 mL/kg (since 0 h), continue as in step (2), adding potassium up to 20 mmol/L. If this urine volume has not been achieved:
(a) continue infusion as in step (2) and give frusemide (5 mg/kg) intramuscularly;
(b) if during the next 4 h (6–10 h) urine excretion exceeds 4 mL/kg body weight, continue as in step (2), but be prepared to repeat steps (3) a–c if the urine flow of 6 mL/kg every 6 h is not maintained;
(c) if a urine volume of 1 mL/kg per h is still not achieved following frusemide administration, consider administering intravenous mannitol (0.75 g/kg body weight) and/or the regimen for acute renal failure. If at any time frusemide or mannitol is given, it is helpful to measure both urine and plasma osmolality and sodium, urea and creatinine concentrations after the commencement of treatment. A urine: plasma ratio for osmolality of less than 1 and urine: plasma ratios for urea and creatinine of less than 5 suggest established renal failure, as does a fractional sodium excretion greater than 2.5% (see section on Acute renal failure). When calculating a urine: plasma ratio for creatinine care should be taken to express both quantities in millimoles.
4 From 28–30 h onwards: gradually change from intravenous N/5 saline (30 mmol/L) in dextrose with potassium, to oral glucose electrolyte solution and gradually reintroduce milk or formula so that in 3–5 days a full-strength full-volume diet is achieved. Reducing substances in a watery stool are easily detected with a 'Clinitest' tablet (Ames Division, Miles Laboratories, UK); care must be taken when collecting the stool to obtain the stool fluid, for example by using a polythene sheet. The test should be performed immediately the stool is obtained since faecal bacteria will continue to ferment the sugars. Breast-fed term babies receiving 11 g lactose per kg body weight daily (i.e. 150 mL/kg per day of milk containing 7.2 g lactose per 100 mL) commonly excrete 0.25% reducing substances (i.e. a green 'Clinitest') and concentrations below 0.5% may be ignored. If the diarrhoea is thought to be due to gastroenteritis it is also possible to ignore stool reducing substances, regardless of their concentration, during the first 48 h of therapy and to press on with the management regimen described above.

If stool reducing substances do persist in significant amounts (i.e. >0.5% for longer than 48 h), then secondary food intolerance or one of the rarer causes of continuing diarrhoea should be considered. There is little point in attempting carbohydrate tolerance tests (rise in plasma glucose following oral loads of particular carbohydrates) at this stage. They may precipitate profuse diarrhoea and even if abnormal do not determine a primary cause or a secondary effect.

Table 3.4 Causes of acute renal failure in the newborn

Prerenal
Hypoxia ± shock
Dehydration of whatever cause
Renal
Renal vein thrombosis (often secondary to dehydration, shock, hypoxia, complication of uncontrolled maternal diabetes)
Renal arterial occlusion (complication of umbilical artery catheterization)
Acute tubular necrosis (often secondary to a prerenal cause)
Renal dysplasia and hypoplasia
Varieties of polycystic disease
Postrenal
Urethral valves or stricture
Prune belly syndrome (absent abdominal musculature)

Although renal agenesis is an obvious cause of renal failure, these babies usually die from pulmonary hypoplasia well before significant azotaemia has occurred.

Acute renal failure

Acute renal failure results in azotaemia and oliguria (arbitrarily taken as less than 24 mL/kg in 24 h), but non-oliguric renal failure may also occur. Now that babies receive much lower protein intakes than previously, a raised plasma urea concentration (>5 μmol/L) indicates some degree of renal failure of whatever variety. The plasma creatinine concentration rises immediately after birth, reaching a peak as high as 115 μmol/L in preterm babies during the first 48 h of life, but thereafter falls to less than 60 μmol/L by the end of the neonatal period. A creatinine concentration greater than 100 μmol/L after the first 48 h of life is abnormally high.

The following formula is sometimes used to estimate the glomerular filtration rate (GFR) from a plasma creatinine measurement:

$$\text{GFR} = \frac{k \times \text{length (cm)}}{\text{plasma creatinine } (\mu\text{mol/L})}$$

The value of k for low birth-weight infants is 30 and for term normal-weight babies is 41, calculated from Schwartz *et al.* (1987).

Aetiology

Table 3.4 shows some causes of acute renal failure. Most causes of neonatal renal failure are prenatal and, even in established intrinsic renal failure, hypoxia, shock and hypotension are the major causes (Anand *et al.*, 1978; Norman & Asadi, 1979).

Perinatal hypoxia with or without hypotension is the most common cause of acute renal failure in the newborn (Dauber *et al.*, 1976), but since most babies are only moderately affected and recover, this cause rarely appears on the list prepared by a nephrology service, where developmental problems such as renal dysplasia, varieties of polycystic disease and urinary obstruction predominate. A period of azotaemia (and sometimes hyponatraemia) together with proteinuria and microscopic haematuria is common in babies following a perinatal insult. Dehydration for whatever reason is an important treatable cause.

Management

Whenever oliguria and azotaemia are noted, the possibility of dehydration should be considered carefully. Signs of dehydration in the newborn are difficult to assess, particularly in the light-for-gestational-age marasmic baby, but serial weights are often available; only 3% of normal healthy term babies lose more than 8% of their body weight. If urine and blood are available for analysis then the concentrations and the urinary (U): plasma (P) ratio may be determined. Thus, a urinary sodium below 20, urinary osmolality >350 and ratios for osmolality >1.5, urea or creatinine >10 and fractional sodium excretion (U:P ratio for sodium ÷ U:P ratio for creatinine expressed as a percentage) <2.5% favour a 'prerenal' cause for the renal failure and indicate the need for restoration of the circulation.

When urine is not available, a molar plasma urea: creatinine ratio greater than 80 suggests prerenal failure, so long as the urea concentration is greater than 10 mmol/L and creatinine is greater than 0.2 mmol/L (Mathew *et al.*, 1980; Gaudio & Siegel, 1987). Urethral obstruction in the newborn usually results in a large bladder.

It is stressed that prerenal failure is the most common variety in the newborn. Therefore, if this is a possibility, even though doubt may exist, it is reasonable to follow the regimen given for dehydration, with the use of frusemide as indicated. This will help to differentiate between a prerenal and intrinsic renal cause (see section on Diarrhoea and dehydration).

If true acute renal failure is established, fluid intake is reduced to 2 mL/kg body weight per h. This may be given as a standard formula to which extra fat and carbohydrate have been added to boost the energy intake, but in practice this stage is rarely reached in the newborn, and if it is progression is so rapid that dialysis is usually indicated.

Hyponatraemia

We regard significant hyponatraemia as a plasma sodium concentration below 130 mmol/L. It presents in four major ways, although there is considerable overlap.

The ill baby requiring intensive care

These babies often have indwelling cannulae or catheters and may be receiving nutrition parenterally. Contamination of blood samples taken via indwelling cannulae or catheters may cause spuriously low concentrations owing to dilution, or occasionally spuriously high ones owing to contamination with sodium heparin. The regimen for parenteral nutrition should incorporate a fat-free period of 4 h so that blood for electrolytes may be taken at the end of this period; otherwise, hyperlipidaemia may lead to spuriously low sodium concentrations. Plasma osmolality should be normal, however, as should serum sodium measured with an ion-selective electrode.

If these causes of 'pseudohyponatraemia' are excluded, then inappropriate antidiuretic hormone (ADH) production, which would in infants be primarily hypothalamic in origin, should be considered. Another cause is the 'sick cell' concept, which involves increased cell permeability allowing efflux of usually non-diffusible solute.

The possibility of sodium deficiency and the consequent need for replacement requires urgent consideration, but the ill baby requiring intensive care is a prime candidate for hyponatraemia resulting from causes other than sodium deficiency:

1 Excessive amounts of water (as 5% dextrose) may have been given to the mother during labour.

2 Diminished venous return may arise either from a raised intrathoracic pressure due to inappropriately high continuous positive airways pressure (CPAP), or from positive end-expired pressure (PEEP) during ventilation leading to inappropriate ADH secretion.

3 Intracranial injury (haemorrhage or oedema following hypoxia) involving the hypothalamus may lead to truly inappropriate secretion of ADH or sick cells.

4 Intravenous nutrition may result in hyperglycaemia with a secondary fall in plasma sodium to maintain normal osmolality (it may also cause glycosuria resulting in polyuria, dehydration and sodium deficiency – see below).

5 Hypoxia, intracranial injury and severe infection (often secondary to the various foreign bodies in the baby, such as an endotracheal tube, arterial monitoring cannulae or catheters – particularly 'main line' central venous catheters) may lead to the sick cell syndrome. In this condition the sick cells leak organic solutes and the extracellular sodium is diluted in an attempt to maintain osmolality.

It is, nevertheless, often impossible to confidently exclude true sodium depletion as the cause of the hyponatraemia. A moderately raised plasma urea level (5–9 mmol/L) favours sodium *deficiency*. In practice, the measurement of urinary sodium concentration is not always useful. It is very low in healthy newborn babies and so cannot be taken as certain evidence of deficiency. Although a high urinary concentration in the presence of hyponatraemia (e.g. urine sodium >10 mmol/L when plasma sodium <130 mmol/L) occurs in renal or endocrine causes of sodium deficiency, it may also occur in inappropriate ADH secretion. Accurate daily weighing is invaluable. Generally, hyponatraemia with loss of weight indicates sodium and water depletion, whereas hyponatraemia with weight gain indicates inappropriate ADH/sick cell syndrome; however, the extraneous paraphernalia of intensive care often make accurate weighing impossible. When in doubt it is reasonable to proceed as follows:

1 If the blood urea is very raised (>10 mmol/L) proceed as for acute renal failure with a view to differentiating between prerenal or renal causes. If the blood urea is only moderately raised (5–9 mmol/L) and the baby seems to be passing urine reasonably, i.e. at least four wet nappies a day, regard the sodium deficiency as mild and give additional sodium and water, e.g. 150–200 mL/kg daily of N/5 saline (30 mmol/L) in 10% dextrose (i.e. 4.5–6 mmol Na/kg).

2 If the blood urea and plasma creatinine levels are normal, review all the possible factors which may result in hyponatraemia without sodium deficiency and proceed as necessary; has there been excessive water administration, are CPAP or PEEP values unnecessarily high, are hyperglycaemia and glycosuria occurring, is there undiagnosed hypoxia or untreated infection? This review is the most important aspect of management. If one of these causes is very likely, first deal with the underlying cause and then restrict fluid, e.g. 80 mL/kg daily, which if given as N/5 saline (30 mmol/L) in 10%

dextrose will provide 2.4 mmol Na/kg body weight.

3 If there is uncertainty it may be necessary to 'sit on the fence'. Restrict the fluid intake to 120 mL/kg body weight of N/5 saline (30 mmol/L) in 10% dextrose; this regimen provides moderate fluid restriction in case of inappropriate ADH secretion, while providing a sodium intake (3.6 mmol/kg per day) in excess of normal requirements. Since this regime does 'sit on the fence' it should not be suitable for any cause, but in practice it does seem to be useful! If this regimen is followed by a rise in blood urea, then the probable diagnosis is sodium deficiency which has not been corrected, and more fluid and sodium are required. If the blood urea remains the same or falls and the plasma sodium falls, then inappropriate ADH/sick cell syndrome is more likely. Further restriction of both fluid and sodium is necessary and treatment should be aimed at the underlying cause.

4 If hyperglycaemia or glycosuria is present, the amount of glucose in the intravenous regimen should be reduced or insulin should be considered (0.5–1.0 unit/kg 12-hourly subcutaneously). Although a glucose–insulin regimen has been used in the management of sick cell syndrome in older children and adults, we counsel caution of its use in the newborn, except *en passant* during the management of hyperglycaemia secondary to parenteral nutrition.

Sodium deficiency due to excessive losses

This may occur at any time. It is usually associated with dehydration and is due to excessive loss from either the gastrointestinal or the renal tract.

Gastrointestinal losses

These are most commonly due to infective gastroenteritis but are occasionally due to inborn errors of absorption of chloride or sugar (see section on Diarrhoea and dehydration). Dangerous losses due to vomiting are uncommon in the newborn but pyloric stenosis can occur in the first week of life and intermittent obstruction of the upper gastrointestinal tract presents as vomiting, e.g. malrotation of the intestine with Ladd's bands causing intermittent duodenal obstruction or an ectopic pancreas in the wall of the duodenum. Generally, the gastrointestinal disorders which lead to hyponatraemia will have clear additional presenting symptoms, e.g. vomiting, diarrhoea and abdominal distensions, but very severe watery diarrhoea can be overlooked because it is mistaken for urine.

Gastrointestinal losses of sodium are often associated with hypokalaemia since there is a concomitant loss of potassium in the stool fluid, a response by the renal tubules to the alkalosis of vomiting or chloride-losing diarrhoea, and an effect of hyperaldosteronism secondary to the sodium loss. Renin and aldosterone concentrations are therefore both raised and the urinary sodium: potassium ratio is below 1. Management is described in the section on Diarrhoea and dehydration.

Adrenocortical insufficiency (see also Chapter 10)

Symptoms due to severe sodium deficiency occur at 7–14 days of age, perhaps because the effect of maternal salt-retaining hormones is disappearing. An early clue before this is the finding of ambiguous genitalia. This may indicate masculinization of a genotypic female owing to 21-hydroxylase deficiency (the most common). Alternatively, it may indicate the rarer feminization of a genotypic male owing to 20,22-desmolase or 3β-hydroxysteroid dehydrogenase deficiency. About 50% of babies with the 21-hydroxylase defect and all babies with the other defects have life-threatening salt loss. Investigation of a baby with ambiguous genitalia and before salt loss has presented is described later.

Deficiencies of 21-hydroxylase in a male, of 20,22-desmolase and 3β-hydroxysteroid dehydrogenase in the female, and specific defects of aldosterone biosynthesis, i.e. 18-hydroxylase or 18-hydroxysteroid dehydrogenase deficiency, do not result in ambiguous genitalia however, and so the diagnosis may not be considered until the baby becomes very ill with hyponatraemia dehydration.

A clue at the time of the acute illness is the presence of hyperkalaemia, reflecting an aldosterone deficiency. More sophisticated investigation shows a raised renin level (normal mean 1.18 pmol A1/mL per h at 6 days) and a *low* aldosterone (normal mean 1248 pmol/L on day 6).

If the baby is ill and the diagnosis of adrenocortical

insufficiency is suspected, routine treatment for severe dehydration should be instituted. Biochemical measurements required for diagnostic purposes are described in Chapter 10.

Except in the classical 21-hydroxylase deficiency, it may be impossible to reach an exact biochemical diagnosis during the clinical emergency. If in doubt, it is better to risk treating the baby with steroids (which may prove to have been unnecessary) and possibly confuse the diagnosis than for the baby to die; the diagnosis can always be sorted out later.

Adrenocortical insufficiency may rarely be due to adrenal hypoplasia. This may be isolated, associated with pituitary hypoplasia or associated with xanthomatosis (Wolman's disease). Oestriol excretion by the mother during pregnancy is very low and may provide an antenatal clue.

Renal losses

Pseudohypoaldosteronism, probably due to autosomal recessive inheritance of tissue insensitivity (Cheek & Perry, 1958), presents like the salt-losing crisis of the adrenogenital syndrome. Hyperkalaemia is present but the plasma and urinary glucocorticoid and androgenic steroids are normal, whilst aldosterone secretion is excessive. Treatment is with large amounts of sodium, e.g. 10–20 mmol/kg per day; salt-retaining corticosteroids have no effect.

In some cases of glomerular failure, due to whatever cause, salt loss is sufficiently excessive to cause hyponatraemia. In older children this is often referred to as salt-losing nephropathy and is associated, for example, with polycystic disease and dysplasia, or obstructive uropathy. Some babies who have had a hypoxic insult pass through a brief azotaemic stage with mild proteinuria and microscopic haematuria, and then develop hyponatraemia: we suspect this is the same phenomenon.

Excessive sodium loss can occur as one facet of a multiple disorder of tubular function (Fanconi syndrome) and in any renal tubular acidosis. Bartter's syndrome may be associated with hyponatraemia together with the more constant biochemical features of the syndrome, i.e. alkalosis and hypokalaemia. This triad of biochemical signs may raise the question of whether an upper gastrointestinal abnormality has caused vomiting, but vomiting is not a particular feature of Bartter's syndrome. When the cause of sodium deficiency is a renal disorder, sodium loss may be relatively slow and so it is possible to demonstrate unequivocal evidence of an inappropriate loss of sodium, i.e. a urinary excretion greater than the dietary intake in the presence of hyponatraemia. This is often impossible in the renal losses that are secondary to adrenocortical insufficiency because the baby is seriously ill. Intravenous therapy with sodium is imperative, and thereafter urinary sodium concentrations are difficult to interpret.

Late hyponatraemia in low birth-weight babies

This occurs usually from the second week of extrauterine life in otherwise well preterm babies born at 34 weeks gestation or earlier.

Such babies are often growing quite rapidly and the plasma sodium concentration may be very low (e.g. <120 mmol/L) in the face of apparently good health. There are probably two mechanisms involved. First, the rapid growth velocity outstrips the dietary supply of sodium. Calculations (using the factorial method) of sodium requirement by the rapidly growing babies are higher than can be obtained from either breast milk or most modern infant formulas (see section on Very low birth-weight babies). Second, renal tubular conservation of sodium in the presence of the hyponatraemia is relatively immature. This is probably due to an immaturity of mechanisms within the renal tubule itself. Preterm babies have a higher fractional excretion of sodium at any given filtered load than do term babies, even though aldosterone secretion is high (see Appendix A1, Table A1.6; and Spitzer & Aperia, 1992).

Heart failure and hyponatraemia

Hyponatraemia in babies with heart failure is most commonly the result of overvigorous diuretic therapy, and is usually associated with a raised blood urea. If potassium supplements are not given there may also be hypokalaemia and alkalosis.

Hyponatraemia may, however, be the result of the heart failure itself leading to a sick cell syndrome: this is a biochemical sign of severe failure and in these circumstances the blood urea is often moderately raised. Finally, many babies in heart failure will be ill ones requiring intensive care so that many of the factors discussed above will be acting. It may therefore be difficult to decide whether the hyponatraemia is due to the heart failure, the treatment, or some associated problem, and whether to relax diuretic therapy. Biochemical investigation rarely helps to solve this dilemma.

If hyponatraemia and uraemia occur when the baby is clearly not in heart failure it is reasonable to assume tentatively that there is sodium depletion. Restrict the fluid intake (120 mL/kg), stop the diuretics and give additional sodium, e.g. 4 mmol/kg body weight. The sodium level should rise, and when (if) diuretics again become necessary a milder regimen should be introduced. If the hyponatraemic baby is in definite heart failure and this is due to a patent ductus arteriosus, surgical or pharmacological closure by indomethacin may be considered. If it is inappropriate to give indomethacin, or if it fails, then empirically we have used the following regimen: (a) reduce fluid intake to 120 mL/kg; (b) increase the strength of a standard demineralized whey formula by one-third (104 kcal and 1 mmol Na/kg body weight daily); and (c) substitute a 'milder' diuretic, e.g. chlorthiazide 50 mg or spironolactone instead of frusemide.

Ambiguous genitalia

The causes of ambiguous genitalia and their diagnosis are discussed in Chapter 10. The essential feature of early management is to determine whether the ambiguous genitalia are due to congenital adrenal hyperplasia, in which case some babies are in grave danger from a salt-losing crisis (see section on Adrenocortical insufficiency).

Management in the first week

Evidence of other abnormalities is sought; for example, trisomy 13, campomelic dysplasia and Smith–Lemli–Opitz syndrome can include genital ambiguities. Determination of genotypic sex (by nuclear sexing, fluorescent Y chromosome staining, or full karyotype) should be arranged urgently. A full karyotype is necessary to detect such abnormalities as 46 XX/46 XY, or 45 X/46 XY chimerism, or 48 XXY. It is often possible to make a reasonable clinical 'estimate' of genotypic sex. Most babies born with ambiguous genitalia, but without a palpable gonad, are virilized XX babies. A palpable gonad is almost always a testicle and so the baby with a gonad and ambiguous genitalia is an incompletely masculinized XY baby. The possibility of virilizing drugs having been given to the mother is explored. An early genitogram and ultrasound search for Müllerian structures (Fallopian tubes, uterus and upper vagina) are valuable but an experienced ultrasonographer is essential.

Meanwhile, preparation for, and early detection of, potential salt loss proceeds. The baby is weighed daily. There is little point in performing serial plasma sodium determinations since the concentration remains normal until immediately prior to the crisis. Hyperkalaemia may be an earlier clue but the possibility of haemolysis in a heel-prick sample complicates the interpretation. Biochemical determinations are discussed in Chapter 10. Pagon (1987) describes a diagnostic approach to early and later management of ambiguous genitalia.

Later management

If salt loss does not occur, or has been adequately treated, then a more leisurely anatomical and biochemical approach proceeds (see Chapter 10).

Although assignment of gender may be psychosocially urgent, life-threatening events are not imminent.

Other plasma electrolyte abnormalities

Hypokalaemia

As an acute neonatal problem, the most common causes are excessive losses from either the intestine in gastroenteritis or from the kidney as a complication of diuretic therapy; the cause is usually clear.

The mild renal failure commonly seen with perinatal hypoxia may be followed by hypokalaemia, presumably because there is a diuresis. Potassium loss from the tubules may occur in any state leading to metabolic alkalosis, e.g. vomiting, or hyperaldosteronism secondary to a deficiency of total body sodium. Finally, if potassium is inadvertently omitted from a total intravenous regimen hypokalaemia develops within 24–48 h in VLBW babies (less than 1.5 kg). Treatment with oral potassium (e.g. 5 mmol/kg body weight daily) for all these causes is simple. Whether extra water and sodium are required must also be considered carefully.

The differential diagnosis of chronic hypokalaemia is little different from that in older children, i.e. mainly losses from the gut or kidney secondary to alkalosis. Disorders that affect tubular function and lead to a potassium leak together with acidosis include the Fanconi syndrome, e.g. due to cystinosis (presentation in the neonatal period is unusual) and any distal renal tubular acidosis. Potassium loss and alkalosis occur in Liddle's syndrome together with hypertension (not reported before 10 months of age), and in Bartter's syndrome in which there is no hypertension (Gill, 1992).

Generally, hypokalaemia with metabolic alkalosis is unusual in newborn babies. If chronic hypercapnia, diuretic therapy without potassium supplements, marked vomiting and chloride-losing diarrhoea can be excluded, then a renal tubular disorder is likely (see Chapter 11; Linshaw, 1987). Primary hyperaldosteronism is rare in childhood.

Hyperkalaemia

Hyperkalaemia is commonly spurious as a result of haemolysis of a heel-prick sample of capillary blood.

True hyperkalaemia is frequently associated with uraemia. This may be a feature of renal failure or occur as a result of adrenal insufficiency, which in itself causes sodium loss, dehydration and uraemia. Adrenal insufficiency should be considered particularly when the baby has become ill during the second week of life or when there are ambiguous genitalia. If there is difficulty in deciding the aetiology of hyperkalaemia and uraemia in an ill baby it is reasonable to treat with a regimen for dehydration while investigating further. This is suitable initial treatment for adrenal insufficiency, whereas if the cause is renal the treatment will help differentiate the pre- and intrinsic renal varieties.

A baby who is severely ill with hypoxia or infection may have hyperkalaemia, with only mild elevations of blood urea, probably reflecting tissue injury and thence protein catabolism. A blood transfusion given too rapidly through very small needles may be followed by haemolysis and hyperkalaemia.

If the *true* plasma potassium concentration (a venous sample will probably be necessary to be certain) is more than 8 mmol/L, then lowering of the concentration as an emergency is achieved by giving 1 U soluble insulin and 2 g dextrose intravenously.

Hypomagnesaemia

Although the mean minus two standard deviations for plasma magnesium concentration in the newborn is 0.5 mmol/L, concentrations below this may not be associated with any symptoms (Harvey *et al.*, 1970).

'Isolated' hypomagnesaemia, i.e. without any other biochemical abnormality, may present with convulsions but is rare (Dooling & Stern, 1967).

It frequently accompanies late-onset hypocalcaemia and treatment with magnesium as well as calcium may be necessary to achieve a normal calcium level (see section on Hypocalcaemia).

It may also accompany hypokalaemia due to excessive losses of both metals from the gut during gastroenteritis, particularly in a malnourished marasmic baby, or from the renal tubules in a rare hereditary tubular disorder (Gitelman *et al.*, 1969).

Hypernatraemia

Hypernatraemia almost always occurs with fluid depletion. It is therefore associated with excessive weight loss or inadequate weight gain and is most commonly due either to use of an overhead radiant heater or to gastroenteritis. Urinary osmolality will initially be high but as acute renal failure develops the urine may become more dilute. Some instances of isonatraemic dehydration with acidosis may be converted into hypernatraemic dehydration if attempts are made to correct the acidosis with high concentrations of sodium bicarbonate without correcting the dehydration (see section on Management of dehydration).

Hypernatraemia also occurs in fluid depletion due to diabetes insipidus, but urinary osmolality and specific gravity remain low. Pituitary diabetes insipidus may be associated with other signs of severe intracranial pathology, such as microcephaly and convulsions. It also can occur as a complication of maternal lithium treatment and as a primary hereditary defect. Nevertheless, it is only occasionally diagnosed in the newborn and hereditary (sex-linked) nephrogenic diabetes insipidus presents later in infancy, although theoretically there seems to be no reason why it should not be present in the newborn.

Hypernatraemia without primary fluid depletion can occur in feeding accidents, when excessive amounts of sodium are added to feeds but even then urinary losses of water, in an attempt to excrete the unnecessary sodium, will be excessive and so lead to secondary dehydration.

Acid–base abnormalities

The technology of blood acid–base determination has advanced greatly, and today results can be obtained with relative ease and speed. As a result, the importance of a good-quality specimen may be overlooked. If the service is provided by the laboratory, then the specimens collected will be capillary and care must be taken in the training of staff. Unless a good flow of blood is achieved from a well-perfused limb the results will be worthless. If the blood-gas machine is on a special-care unit and used by medical staff, many specimens will be arterial. Here again, training is essential, although the margin for error is much reduced with machines which are self-calibrating and have a diagnostic fault-finding capability. It is important, however, to stress the need for an ongoing quality assessment programme.

It is very easy to introduce air bubbles into the blood sample on collection or on introduction to the machine. This often gives a spurious result and the various components of the acid–base profile will lack internal consistency.

Metabolic acidosis

A base excess of below 8 mmol/L is abnormal. Moderate degrees of metabolic acidosis, 'late metabolic acidosis', were very common in preterm babies fed on formulas containing 3 g or more of unmodified cows' milk protein per 100 kcal, largely because of the endogenous acid load (i.e. acid formed from metabolism of protein, particularly the sulphur amino acids), but also reflecting some renal limitation in acid excretion (see Kildeburg and Winters (1972) and Berger *et al.* (1978) for reviews). The use of demineralized whey formulas has reduced the prevalence of this phenomenon considerably, but it should still be considered amongst the diagnoses. Unsuitable intravenous feeding re-

Table 3.5 Metabolic acidosis – causes and other features

Preliminary assessment
Hypoxia
  Arterial and tissue
  Often hypercapnia too
Dehydration
Azotaemia
Infection

Over-production
Inborn errors of branched-chain amino acid metabolism
  Ketosis, abnormal amino acid pattern
Inborn errors of carbohydrate metabolism
  Lactate raised, hypoglycaemia
  Urinary reducing substances may be present
Lactate/pyruvate acidoses
  Normal amino acid pattern or hyperalaninaemia
  Lactate raised

Under-excretion
Glomerular disease
  Azotaemic renal failure
Tubular disease (hyperchloraemia present)
  Proximal
    Variable urine pH
    May be part of general tubular diseases with leakage of other substances
  Distal
    Urine pH always >6
    Transient and permanent varieties

gimens, particularly ones containing fructose or sorbitol, will cause acidosis, but these carbohydrates should not be used in the newborn. The diagnosis may be approached in three ways (see Table 3.5).

Preliminary assessment

By far the most common cause of metabolic acidosis in the newborn is hypoxia, and after this infection and dehydration should be considered carefully before going on to consider intrinsically metabolic causes. Intrapartum asphyxia or postnatal hypoxia, e.g. due to RDS, will usually be accompanied by a respiratory acidosis too, but this and the hypoxaemia may quickly disappear when adequate ventilation is achieved, leaving only the metabolic acidosis. Therefore, in assessing the possibility of hypoxia as a cause of metabolic acidosis the history may be more important than the contemporary blood gases. Infection may not be very severe and the baby not yet ill, but mild acidosis develops; certainly we have diagnosed urinary tract infections following further investigations of a mild persistent acidosis. Tissue hypoxia, despite normal arterial blood gases, may occur in hypothermia and other causes of a poor peripheral circulation, e.g. following an intraventricular haemorrhage or with the hypoplastic left heart syndrome. These diagnoses will usually be suspected on other grounds too, but may cause some confusion.

If the common neonatal causes (hypoxia, dehydration and infection) are excluded, then intrinsic metabolic causes may be suspected. The simple metabolic screen shown in Table 3.2 may be performed together with a plasma chloride determination. Generally, if the plasma chloride is raised, the amino acid pattern is normal and there is no ketonuria, renal tubular disease should be considered. If, however, the plasma chloride is low or normal, over-production of an 'undetermined anion' is probably responsible for the acidosis and this should be sought.

Over-production of acid

Most of the inborn errors of amino acid metabolism that result in over-production of acid have other serious symptoms, including coma, apnoea and convulsions, and these may occur before marked acidosis is apparent, e.g. MSUD and organic acidaemias. There is usually ketonuria and the plasma amino acid screen is abnormal. Alternatively, some other biochemical sign will give a diagnostic clue, e.g. a reducing substance in the urine in galactosaemia. The combination of metabolic acidosis with anaemia occurs in 5-oxoprolinuria and this may at first be ascribed to perinatal shock and haemorrhage.

If none of these symptoms are present and the preliminary metabolic screen is negative, lactic acidosis should be considered. Pyroglutamic aciduria is a rare possibility causing metabolic acidosis and little else. A raised fasting plasma lactate (1.4 mmol/L) occurs in hypoxia, tissue hypoxia, dehydration and infection, and so these diagnoses should again be considered.

Glycogen storage disease type 1 results in lactic acidosis and the diagnosis may be easily missed (see Chapter 6). The congenital lactate/pyruvate acidoses are usually accompanied by hyperalaninaemia and ketosis. In this group of disorders neurological symptoms may be intermittent (e.g. pyruvate carboxylase deficiency and pyruvate dehydrogenase deficiency) but some have a dysmorphic face; others are associated with severe progressive neurological deterioration, e.g. Leigh's encephalopathy (see Chapter 4).

Reduced excretion of acid

The acidosis may be due to renal failure, leading to azotaemia and often mild hyponatraemia, secondary to a reduction in the number of nephrons.

If the plasma urea and creatinine concentrations are normal and hyperchloraemia is present, renal tubular acidosis is probable. At this stage it is useful to collect three random urine samples for pH, protein, sugar, and sodium and potassium concentrations, and as each urine sample is passed, to take blood for the determination of acid–base status and sodium, potassium and chloride levels. In practice, we sometimes arrange for the blood sample to also be used for a lactate determination.

Distal renal tubular acidosis (classical or type I) is

associated with hyperchloraemia; an alkaline urine (pH > 6), irrespective of the blood standard bicarbonate concentration (but only when the standard bicarbonate is below 18 mmol/L should any significance be attached to a urine pH > 6); a low urinary sodium: potassium ratio (probably reflecting mild secondary hyperaldosteronism); and often mild hypokalaemia. Correction of the acidosis is easily achieved, usually requiring much less than 5 mmol/kg daily of bicarbonate. Permanent primary distal renal tubular acidosis (Butler Allbright syndrome) may present in the neonatal period and one variant is associated with nerve deafness. Commonly, however, the problem is a transient one in preterm babies and the disorder often seems to have disappeared by the time investigations are complete. It was more common some years ago, and was probably due to intoxication with vitamin D or mercury.

Proximal renal tubular acidosis (type II) is also associated with hyperchloraemia but the urine pH may fall to 4 or 5 as the blood standard bicarbonate falls below 18 mmol/L; usually, the urinary sodium: potassium ratio is normal. Correction requires more than 5 mmol/kg daily of bicarbonate. The lesion may be a primary single hereditary one, but more usually is part of a multiple tubular dysfunction owing to, for example, cystinosis, galactosaemia, fructosaemia, cyanotic congenital heart disease or renal infarcts following the withdrawal of umbilical artery catheters. In practice, proximal renal tubular acidosis, whether primary or secondary, is rarely diagnosed in the newborn. The diagnosis and management of the renal tubular acidoses are discussed in detail by Rodriguez-Soriano (1990).

Alkalosis

An increased standard bicarbonate or base excess occurs most commonly as metabolic compensation for a raised $P\text{CO}_2$. Apart from this, alkalosis is rare in the newborn; if found, the accuracy of the result should be questioned and an iatrogenic cause considered, e.g. administration of alkali.

If such causes are excluded, conditions leading to excessive loss of anions should be considered; they are often accompanied by hyponatraemia, hypochloraemia and hypokalaemia, e.g. excessive vomiting, chloride-losing diarrhoea (see section on Hyponatraemia) and Bartter's syndrome.

The association of alkalosis with certain clinical presentations may point to the diagnosis, e.g. with vomiting consider upper gastrointestinal obstruction or occasionally Bartter's syndrome; with diarrhoea consider chloride-losing diarrhoea; with drowsiness consider hyperammonaemia of whatever cause. Very prolonged gentamicin therapy has resulted in alkalosis in older children but is not described in the newborn.

Oedema and hydrops

Hydrops fetalis

Table 3.6 shows the causes of hydrops fetalis. In reaching a diagnosis the following questions should be asked: (a) is the baby anaemic? (b) could the baby be suffering from an intrauterine infection? and (c) is there evidence of disease of the heart, kidney or liver? In most instances the cause is severe Rhesus haemolytic disease but if this is excluded the most important chemical investigations are the plasma albumin and urinary protein.

Heavy proteinuria and hypoalbuminaemia indicate the nephrotic syndrome. This may be of the idiopathic variety (rare outside Finland) owing to a specific nephropathy (diffuse mesangial sclerosis), or secondary to intrauterine infection, including treatable syphilis (of which the only other manifestation might be osteitis), renal vein thrombosis and, very rarely, a nephroblastoma.

In the Finnish variety of congenital nephrotic syndrome the placenta is more than 25% of the fetal weight and the proteinuria is initially highly selective. Hyperlipidaemia is present, so spuriously low plasma concentrations of sodium may be obtained. If untreated, death almost always occurs within a year, mainly due to infection in the absence of uraemia. However, the results of aggressive nutritional support, salt, poor albumin infusion, early nephrectomy and transplantation are encouraging (Hallman & Rapola, 1978; Hoyer & Anderson, 1981).

Table 3.6 Causes of hydrops fetalis. After Etches and Lemons (1979)

Severe chronic anaemia in utero
Rhesus incompatibility
Homozygous α-thalassaemia (usually preterm stillbirths)
ABO incompatibility (unusual)
Chronic bleeding into mother or other twin
G6PD deficiency

Intrauterine infections
Syphilis, cytomegalovirus, toxoplasmosis, Chagas disease, leptospirosis

Cardiovascular disorders
Any variety of congenital heart disease
Angiomatous malformations in placenta or fetus
Arrhythmias: congenital heart block, SVT, flutter
Myocarditis, e.g. Coxsackie b

Renal tract disorders
Congenital nephrotic syndrome
Urinary ascites due to lower urinary tract obstruction
Polycystic disease
Renal vein thrombosis

Liver disease
Congenital cirrhosis, hepatitis, vascular tumours
Storage disorders including Gaucher, Sialidosis, GM_1-gangliosidosis, Wolman's disease, Niemann–Pick disease

Others
Pulmonary lymphangiectasia (early death due to respiratory failure)
Cystic adenomatoid malformation of lung, pulmonary sequestration
Achondroplasia
Neuroblastoma, sacrococcygeal teratoma
Turner's syndrome, Down's syndrome, Edwards' syndrome
Chylous ascites (no peripheral oedema)

Idiopathic
Up to 50% of non-immunological hydrops

G6PD, glucose-6-phosphate dehydrogenase;
SVT, supraventricular tachycardia.

Postnatal oedema

Mild peripheral oedema is common in preterm babies. 'Biochemical' causes of pitting oedema should be considered only after common perinatal complications have been excluded as a cause, e.g. hypothermia, intrapartum hypoxia (possibly due to a shift of fluid from the mother to the baby, and also increased capillary permeability), cardiac failure, severe chronic anaemia *in utero* and iatrogenic overload.

Iatrogenic overload with sodium readily causes oedema in the newborn. It occurs in various circumstances: as an accident; because hyponatraemia due to inappropriate ADH secretion/sick cell syndrome has incorrectly been ascribed to sodium deficiency and excessive sodium supplements have been given; when a diagnosis of oliguric renal failure has been delayed; when signs of marasmus are interpreted as signs of profound dehydration (the most experienced of us may do this) and inappropriate rehydration is given; or when large amounts of sodium bicarbonate are given to correct a metabolic acidosis without correcting the underlying cause, such as dehydration.

All the causes of hydrops fetalis shown in Table 3.5 should be considered. Evidence of hypoproteinaemia and proteinuria is again sought. Hypoproteinaemia occurs in the nephrotic syndrome and also occurs in the rare forms of congenital cirrhosis and the rare neonatal, possibly familial, variety of protein-losing enteropathy (Cottom *et al.*, 1961). Neonatal ascites may be the presenting sign in liver storage disorders (Table 3.5), usually with a large liver. We have seen plasma albumin concentrations as low as 25 g/L in oedematous preterm babies in whom the most probable cause of the oedema was iatrogenic sodium overload. It seems, therefore, that little diagnostic significance should be attached to 'low' plasma albumin unless it is below 20 g/L.

Hypocalcaemia, which may result from various causes, may be associated with neonatal oedema. The pathogenesis is unclear (Benson & Parsons, 1964; Chiswick, 1971). Late oedema, sometimes in association with a haemolytic anaemia and thrombocytosis, is described in vitamin E deficiency (Hassan *et al.*, 1966). The ratio in the diet of vitamin E to polyunsaturated fatty acids is adequate in modern infant-feeding formulas, so that this syndrome should now be rare. Hypocupraemia is also associated with moderate peripheral oedema in VLBW babies (Sutton *et al.*, 1985).

Hypertension

Hypertension diagnosed in the newborn has usually presented as heart failure and is most commonly secondary to coarctation of the aorta. Renal vascular and cystic disease are the next most common, and iatrogenic hypertension occurs in fluid and sodium overload during acute renal failure.

Metabolic causes are rare. They include over-production of catecholamines by neuroadrenal tissue in a phaeochromocytoma or neuroblastoma, or over-production of aldosterone or other substances with salt-retaining properties, e.g. 11β-hydroxylase or 17α-hydroxylase deficiency of the adrenal cortex. In practice, 17α-hydroxylase deficiency does not present until there is delayed puberty because of deficient oestrogen production. The 11β-hydroxylase deficiency results in early hypertension but is more usually diagnosed in the newborn only during the investigation of ambiguous genitalia of an XX baby. Liddle's syndrome (pseudohyperaldosteronism) results in hypertension and hypokalaemic alkalosis but has not been diagnosed before the age of 10 months (see section on Hypokalaemia).

Bone lesions

Frank *rickets* may occasionally be present *at birth* as a result of severe maternal osteomalacia (see section on Hypocalcaemia). It also occurs some weeks after birth in VLBW babies who are beginning to show rapid catch-up growth, and in a few the ribs may be so soft that there is respiratory difficulty. Various factors may play a role, e.g. inadequate intake of vitamin D (particularly by babies receiving breast milk), an inadequate conversion in the liver and/or kidney to form the active 1,25-dihydroxycholecalciferol and, most importantly, a dietary deficiency of phosphate and/or calcium. A raised alkaline phosphatase without bone lesions in a rapidly growing low birth-weight baby may be an indication for giving extra calciferol (e.g. 25 μg daily) on the basis that the raised activity represents 'subclinical' rickets, but such a policy has not been assessed adequately. Some paediatricians prescribe this amount routinely to *all* VLBW babies. The situation is complex since, under normal circumstances, rapid growth is physiologically associated with higher concentrations of alkaline phosphatase. Similarly, the place of 1α- or 1,25-hydroxycalciferol is not yet clear. The most important aspect of treatment is to increase the calcium and phosphorus intake. A phosphorus depletion syndrome has been described – plasma phosphate below 1.5 mmol/L, urinary phosphorus below 1.0 mmol/L, hypercalciuria (urinary calcium more than 0.15 mmol/kg per day) and osteopenia. The biochemical abnormalities, including the hypercalciuria, disappear on phosphate supplements of 1 mmol/kg per day (Senterre & Salle, 1982). See reviews by Tsang *et al.* (1981) and Gertner (1990).

Bone X-rays with some features resembling florid rickets occur in *congenital hypophosphatasia*. Symptoms more commonly occur in mid-infancy but the full severe syndrome may be apparent in the neonatal period: hypotonia, very soft skull, demineralization of bones with flaring and irregularities of the long ends of the bones, and shortening and bowing of the bones. Alkaline phosphatase activity is extremely low, calcium and phosphorus levels are normal or high and there is an excess of phosphoethanolamine in the urine. Few babies survive (Goyer, 1963).

Primary hyperparathyroidism in the newborn is rare; *transient neonatal hyperparathyroidism* secondary to maternal hypoparathyroidism is a little more common. Both present as bone deformities, demineralization and a periosteal reaction. The plasma calcium is not necessarily raised in the secondary variety and serum alkaline phosphatase activity may initially be normal. Recorded plasma PTH concentrations in both varieties were increased by four times (Spiegel *et al.*, 1977; Glass & Barr, 1981).

Periosteal lesions, together with anaemia, neutropenia and occasionally oedema, owing to *copper deficiency* have been described in a few VLBW babies, and in one baby a fractured femur occurred (Blumenthal *et al.*, 1980). Reported cases have had a very low plasma copper concentration, e.g. 0.7–2.5 μmol/L (stated normal 11–25), and sometimes a raised alkaline phosphatase and mild hypocalcaemia (Sutton *et al.*, 1985).

Neuroblastoma and other malignancies may present

in the newborn, sometimes resulting in hypercalcaemia proceeding to uraemia owing to nephrocalcinosis. *Mucolipidosis type II* (I cell disease) may be apparent in the newborn with hepatosplenomegaly, a coarse 'funny' face and 'dysostosis multiplex'. There is no excess mucopolysacchariduria, but arylsulphatase A, hexoseaminidase, α- and β-galactosidase, α-mannosidase and α-fucosidase activities are markedly raised. Plasma acid phosphatase is normal.

In other bone lesions apparent in the newborn, e.g. osteogenesis imperfecta, thanatophoric dwarfism and occasionally achondroplasia, the biochemical defect is unclear and none of the presently available biochemical determinations are abnormal.

Very low birth-weight babies

Babies weighing less than 1500 g at birth are immature biochemically, often ill from such complications as hypoxia, hypothermia and infection, and even when 'healthy' their rapid catch-up growth imposes various metabolic strains.

Early critical days

Problems at this time have already been dealt with in other parts of this chapter; see sections on Early-onset hypocalcaemia, Hypoglycaemia, Hyperammonaemia, Metabolic acidosis, Bone lesions, etc.

Thyroid function

Screening tests for hypothyroidism may be more difficult to interpret than in the normal-sized baby. Serum concentrations of thyroxine and triiodothyronine are lower, and are even more reduced in the very ill baby (Uhrmann *et al.*, 1981).

Fisher (1990) describes three 'thyroid syndromes' in preterm or ill babies.

1 Transient hypothyroxinaemia: low T_4 and free T_4, basal thyroid stimulating hormone (TSH) and response of TSH to thyrotrophin-releasing hormone (TRH) are normal; the condition corrects over 4–8 weeks without treatment.

2 Transient primary hypothyroidism: low serum T_4 and high TSH. This is probably due to relative iodine deficiency; recovery over 2–3 months occurs but treatment with thyroxine and iodine is recommended. Conversely, a similar syndrome may occur due to 'blocking' of the thyroid gland when babies are exposed to excess iodine from skin antiseptics.

3 Non-thyroidal illness: any neonatal illness inhibits conversion of T_4 to T_3 and limits T_4 binding by thyroglobulin (TBG); TSH and TBG are often low; treatment is aimed at the underlying illness.

Hyperaminoacidaemia and hyperaminoaciduria

(see Chapters 4 and 6)

Eight factors should be considered when interpreting plasma and urinary amino acid concentrations. The first five are particularly important in very small babies but apply to all newborns.

1 *Adequacy of specimen*. Difficulty in collecting blood may distort the result because of haemolysis. Plasma protein should be precipitated within an hour or even earlier if concentrations of the sulphur amino acids are required. Loss of amino acids can ensue from storage of samples from which all protein has not been removed.

2 *Reference range of normal*. This should be adequately defined, e.g. age, maturity, diet, etc., since these factors affect the result (see below).

3 *Maturity*. The more immature have higher plasma, and hence higher urinary, concentrations of many amino acids, particularly phenylalanine, tyrosine and methionine. In addition, the short proximal tubule reduces reabsorption so that physiological aminoaciduria occurs in all newborn babies.

4 *Diet*. Higher protein intakes, particularly of unmodified cows' milk protein (mainly casein), cause higher plasma concentrations of phenylalanine, tyrosine and methionine (Valman *et al.*, 1971) and an increased urinary excretion of cystathionine. Intravenous amino acid solutions obviously directly affect the result.

5 *Growth velocity*. Rapid growth rates (e.g. 30 g/day in babies below 2 kg birth weight) result in lower plasma concentrations of many amino acids, particularly the branched-chain ones.

6 *The possibility of an inborn error of amino acid*

metabolism. A specific pattern involving one amino acid or a characteristic group of amino acids is seen. This usually stands out on semi-quantitative screening without the need for exact quantitative determination.

7 *The possibility of other metabolic errors causing a secondary change in amino acid concentrations*. This includes hyperalaninaemia in pyruvate/lactate acidoses, hyperglycinaemia in the organic acidaemias, hyperglutamicacidaemia in the hyperammonaemias, and a mixed aminoaciduria due to generalized tubular dysfunction, e.g. from cystinosis, galactosaemia or vitamin D deficiency.

8 *The possibility of liver disease*. Plasma phenylalanine, methionine and particularly tyrosine are raised in hepatocellular disease, irrespective of its aetiology.

Population screening tests for hyperaminoacidaemia were commonly positive in preterm babies receiving a very high protein intake from unmodified cows' milk (e.g. 7 g protein/kg; 5.3 g/100 kcal). This is seen much less commonly; indeed, it occurs much less commonly in term babies also, following the substantial changes in infant-feeding practice which occurred in the 1970s (Walker *et al.*, 1981).

Catch-up growth

The factorial method of estimating nutritional requirements suggests that breast milk contains inadequate amounts of protein, calcium and sodium to sustain a reasonably normal body composition. Infant formulas specially designed for the low birthweight baby are available.

In any case, the factorial calculations depend on our limited knowledge of the body composition of the fetus, with little allowance for individual variation. In practice, therefore, management is aimed at the recognition of: (a) deficiencies because the diet is inadequate for rapid catch-up growth; and (b) excesses secondary to the baby's immature biochemical systems. The following points might be useful. If the baby is receiving breast milk only for more than about 2 weeks, it is useful to check regularly the levels of plasma sodium, phosphate, alkaline phosphatase and, if possible, copper. In babies receiving a formula these investigations may also be useful, but, in addition, if the protein intake is greater than that given to term babies (e.g. more than 3 g/kg body weight or greater than 2.25 g/100 kcal), screening for metabolic acidosis, hyperaminoacidaemia and hyperammonaemia may be indicated.

These suggestions are made more with the aim of increasing our biochemical awareness of the problems in VLBW babies, rather than as essential investigations.

References

Aaronson, I. (1971) Secondary chloride-losing diarrhoea. Observations on stool electrolytes in infants after bowel surgery. *Arch Dis Child* **46**, 479–482.

Anand, S.K., Northway, J.D. & Crussi, G.F. (1978) Acute renal failure in newborn infants. *J Pediatr* **92**, 985–988.

Arrowsmith, W.A., Payne, R.B. & Littlewood, J.M. (1975) Comparison of treatments for congenital nonobstructive non-haemolytic hyperbilirubinaemia. *Arch Dis Child* **50**, 197–199.

Benson, P.F. & Parsons, V. (1964) Hereditary hyperparathyroidism presenting with oedema in the neonatal period. *Quart J Med* **33**, 197–208.

Berger, H.M., Scott, P.H., Kenward, C., Scott, P. & Wharton, B.A. (1978) Milk pH, acid base status and growth in babies. *Arch Dis Child* **53**, 926–930.

Blumenthal, I., Lealman, G.T. & Franklyn, P.P. (1980) Fracture of the femur, fish odour, and copper deficiency in a preterm infant. *Arch Dis Child* **55**, 229–231.

Booth, I.W., Strange, G., Murev, H., Fenton, T.R. & Milla, P.J. (1985) Defective jejunal brush Na^+H^+ exchange: a cause of congenital secretory diarrhoea. *Lancet* **1**, 1066–1069.

Candy, D.C.A., Larcher, V.F., Cameron, D.J.S. *et al.* (1981) Lethal familial protracted diarrhoea. *Arch Dis Child* **56**, 15–23.

Cheek, D.B. & Perry, J.W. (1958) A salt-wasting syndrome in infancy. *Arch Dis Child* **33**, 252–255.

Chiswick, M.L. (1971) Association of oedema and hypomagnesaemia with hypocalcaemic tetany in the newborn. *Br Med J* **3**, 15–18.

Cornblath, M. & Schwartz, R. (1991) Hypoglycaemia in the neonate. In: *Disorders of Carbohydrate Metabolism in Infancy* (eds M. Cornblath & R. Schwartz), 3rd edn, pp. 87–123. Blackwell Scientific Publications, Boston.

Cottom, D.G., London, D.R. & Wilson, B.D. (1961) Neonatal oedema due to exudative enteropathy. *Lancet* **ii**, 1009–1012.

Dauber, I.M., Krauss, A.M., Symchych, P.S. & Auld, P.A.M. (1976) Renal failure following perinatal anoxia. *J Pediatr* **88**, 851–855.

Dooling, E.C. & Stern, I. (1967) Hypomagnesaemia with convulsions in a newborn infant. Report of a case associated with maternal hypophosphatemia. *Can Med Assoc J* **97**, 827–828.

Etches, P.C. & Lemons, J.A. (1979) Non-immune hydrops fetalis. *Paediatrics* **64**, 326–330.

Fernandes, J., Huijing, F. & Van de Kamer, J.H. (1969) A screening method for liver glycogen diseases. *Arch Dis Child* **44**, 311–317.

Fisher, D.A. (1990) Euthyroid low thyroxine (T_4) and triiodothyronine (T_3) states in prematures and sick neonates. *Pediatr Clin N Am* **37**, 1297–1312.

Fitzgerald, J.F. (1988) Cholestatic disorders of infancy. *Pediatr Clin N Am* **35**, 357–374.

Gaudio, K.M. & Siegel, N.J. (1987) Pathogenesis and treatment of acute renal failure. *Pediatr Clin N Am* **34**, 771–787.

Gertner J. (1990) Disorders of calcium and phosphorus homeostasis. *Pediatr Clin N Am* **37**, 1441–1465.

Gill, J. (1992) Disorders of renal transport of sodium potassium, magnesium and calcium. In: *Pediatric Kidney Disease* (ed. C.M. Edelmann), 2nd edn, pp. 1973–1988. Little Brown, Boston.

Gitelman, H.J., Graham, J.B. & Welt, L.G. (1969) A new familial disorder characterised by hypokalaemia and hypomagnesaemia. *Ann NY Acad Sci* **162**, 856.

Glass, E.J. & Barr, D.G.D. (1981) Transient hyperparathyroidism secondary to maternal pseudohypoparathyroidism. *Arch Dis Child* **56**, 565–567.

Goyer, R.A. (1963) Ethanolamine phosphate excretion in a family with hypophosphatasia. *Arch Dis Child* **38**, 205–208.

Hallman, N. & Rapola, J. (1978) Congenital nephrotic syndrome. In: *Pediatric Kidney Disease* (ed. C.M. Edelmann), pp. 711–717. Little Brown, Boston.

Harvey, D.R., Cooper, L.V. & Stevens, J.F. (1970) Plasma calcium and magnesium in newborn babies. *Arch Dis Child* **45**, 506–509.

Hassan, H., Hashim, S.A., Van Itallie, T.B. & Sebrell, W.H. (1966) Syndrome in premature infants associated with low plasma vitamin E levels and high polyunsaturated fatty acid diet. *Am J Clin Nutr* **19**, 147–157.

Hoyer, J.R. & Anderson, C.E. (1981) Congenital nephrotic syndrome. *Clin Perinatol* **8**, 333–346.

Hufton, B. & Wharton, B.A. (1982) Type 1 glycogen storage disease presenting in the neonatal period. *Arch Dis Child* **4**, 309–311.

Kildeburg, P. & Winters, R.W. (1972) Infant feeding and blood acid base status. *Pediatrics* **49**, 801–802.

Koh, T.H.H.G. & Vong, S.K. (1992) Definition of neonatal hypoglycaemia, 1986 and 1992. In: *Proceedings of Neonatal Society, June 1992* (abstract).

Launiala, K., Perheentupa, J., Pasternack, A. & Hallman, N. (1968) Familial chloride diarrhoea–chloride malabsorption. *Bibl Paediatr* **87**, 137–149.

Lee, K.S. & Gartner, L.M. (1978) Bilirubin binding by plasma proteins: a critical evaluation of methods and clinical implications. *Rev Perinat Med* **2**, 319–343.

Linshaw, M. (1987) Potassium homeostasis and hypokalaemia. *Pediatr Clin N Am* **34**, 649–682.

McIntosh, N. & Mitchell, V. (1990) A clinical trial of 2 parenteral nutrition solutions in neonates. *Arch Dis Child* **65**, 692–699.

Mathew, O.P., Jones, A.S., James, E. *et al.* (1980) Neonatal renal failure: usefulness of diagnostic indices. *Pediatrics* **65**, 57–60.

Meites, S. & Saniel-Banrey, K. (1979) Preservation, distribution, and assay of glucose in blood with special reference to the newborn. *Clin Chem* **25**, 531–534.

Moncrieff, M.W., Hill, D.S., Archer, J. & Arthur, L.J.H. (1972) Congenital absence of pituitary gland and adrenal hypoplasia. *Arch Dis Child* **47**, 136–137.

Morse, J.O. (1978a) Alpha-1-antitrypsin deficiency. *N Engl J Med* **299**, 1045–1048.

Morse, J.O. (1978b) Alpha-1-antitrypsin deficiency. *N Engl J Med* **299**, 1099–1105.

Mowat, A.P. (1987) *Liver Disorders in Childhood*, 2nd edn, pp. 37–88. Butterworth, Sevenoaks.

Naidoo, B.T. (1968) The cerebrospinal fluid in the healthy newborn infant. *SA Med J* **42**, 933–935.

Norman, M.E. & Asadi, F.K. (1979) A prospective study of acute renal failure in the newborn infant. *Pediatrics* **63**, 475–479.

Pagon, R.A. (1987) Diagnostic approach to the newborn with ambiguous genitalia. *Pediatr Clin N Am* **34**, 1017–1030.

Perry, T.L. (1981) Autopsy investigation of disorders of amino acid metabolism. In: *Laboratory Investigation of Fetal Disease* (ed. A.J. Barson), pp. 429–451. J. Wright, Bristol.

Poland, R.C., Avery, G.B., Goetcherian, E. & Odell, G.B. (1972) Treatment of Crigler – Najjar syndrome with agar. *Pediatr Res* **6**, 377A.

Rayner, P.H. (1982) Endocrine emergencies in the early weeks. In: *Topics in Perinatal Medicine II* (ed. B.A. Wharton), pp. 114–125. Pitman Medical, London.

Rodriguez-Soriano, J. (1990) Renal tubular acidosis. *Ped Nephrol* **4**, 268–275.

Rokicki, W., Forest, M.G., Loras, B., Bonnet, H. & Bertrand, J. (1990) Free cortisol of human plasma in the first three months of life. *Biol Neonat* **57**, 21–29.

Schwartz, G.J., Brion, L.P. & Spitzer, A. (1987) The use of plasma creatinine concentration for estimating glomerular filtration rate in infants, children and adolescents. *Pediatr Clin N Am* **34**, 571–590.

Senterre, J. & Salle, B. (1982) Calcium and phosphorus economy of the preterm infant. *Acta Paed Scand Suppl* **296**, 85–92.

Spiegel, A.M., Harrison, H.E. & Marx, S.J. (1977) Neonatal primary hyperparathyroidism with autosomal dominant inheritance. *Pediatrics* **90**, 269–272.

Spitzer, A. & Aperia, A. (1992) Sodium and water homeostasis. In: *Pediatric Kidney Disease* (ed. C.M. Edelmann), 2nd edn, pp. 93–126. Little Brown, Boston.

Stephenson, J.B.P. (1988) Inherited peroxisomal disorders involving the nervous system. *Arch Dis Child* **63**, 767–770.

Sutton, A., Harvie, A., Cockburn, F., Farquharson, J. & Logan, R.W. (1985) Copper deficiency in the preterm infant of very low birthweight. *Arch Dis Child* **60**, 644–651.

Tripp, J.H., Wilmers, M.J. & Wharton, B.A. (1977) Gastroenteritis: a continuing problem of child health in Britain. *Lancet* **ii**, 233–236.

Tsang, R.C., Greer, F. & Steichen, J.J. (1981) Perinatal metabolism of vitamin D. *Clin Perinatol* **8**, 287–306.

Uhrmann, S., Marks, K.H., Maisels, M.J. *et al.* (1981) Frequency of transient hypothyroxinaemia in low birth weight infants. *Arch Dis Child* **56**, 214–217.

Valman, H.B., Brown, R.J.K., Palmer, T., Oberholzer, V.G. & Levin, B. (1971) Protein intake and plasma amino acids of infants of low birth weight. *Br Med J* **4**, 789–791.

Walker, V., Clayton, B.E., Ersser, R.S. *et al.* (1981) Hyperphenylalaninaemia of various types among three-quarters of a million neonates tested in a screening programme. *Arch Dis Child* **56**, 759–764.

Wharton, B.A. (1965) An unusual variety of muscular dystrophy. *Lancet* **i**, 248–249.

Wharton, B.A. (1987) *Nutrition and Feeding of Preterm Infants*, pp. 6–9. Blackwell Scientific Publications, Oxford.

Wharton, B.A., Bassi, U., Gough, G. & Williams, A. (1971) Clinical value of plasma creatine kinase and uric acid levels during first week of life. *Arch Dis Child* **46**, 356–362.

4: The Newborn with Suspected Metabolic Disease: an Overview

F. POGGI, T. BILLETTE DE VILLEMEUR,
A. MUNNICH & J.M. SAUDUBRAY

Introduction

At the present time there are over 300 known human diseases which are due to inborn errors of metabolism and this number is constantly increasing as new concepts and new techniques become available for identifying biochemical phenotypes. Among them, about 100 present in the neonatal period or very early infancy. However, the incidence of inborn errors may well be underestimated as diagnostic errors are frequent. Despite the relative abundance of new case reports, there is considerable evidence that many of these disorders remain misdiagnosed. The clinical diagnosis of inborn errors of metabolism in neonates may at times be difficult for a number of reasons:

1 In acute situations, many physicians think that because individual inborn errors are rare, they should be considered only after more common conditions (like sepsis) have been excluded.

2 The neonate has an apparently limited repertoire of responses to severe overwhelming illness and the predominant clinical signs and symptoms are non-specific: poor feeding, lethargy, failure to thrive, etc. It is certain that many patients with such defects succumb in the newborn period without having received a specific diagnosis, death often having been attributed to sepsis or other common causes.

3 Classical autopsy findings in such cases are often non-specific and unrevealing. Infection is often suspected as the cause of death, but sepsis is the common accompaniment of metabolic disorders.

4 Many general practitioners and paediatricians only think of inborn errors of metabolism in inadequate and very non-specific clinical circumstances, like psychomotor retardation, hypotonia or seizures. Conversely, they ignore many highly specific symptoms that are excellent keys to the diagnosis.

5 Although most genetic metabolic errors are hereditary and transmitted as recessive disorders, the majority of cases appear sporadically, because of the small size of sibships in developed countries.

6 Finally, good detection of inborn errors of metabolism relies in part on screening programmes but depends primarily on a high index of clinical suspicion and coordinated (integrated) access to expert laboratory services. This makes it an absolute necessity to teach primary-care physicians a simple method of clinical screening and protocols for collecting samples before deciding to initiate sophisticated metabolic investigations.

7 The growing complexity of such investigations needs a tight collaboration between physicians, geneticists and biochemists, and to gather together clinical data and biological findings in metabolic networks. Each time it is possible, patients should be referred to specialized reference centres.

This chapter deals with the overview of the clinical keys to the diagnosis of inherited metabolic disorders in neonates. It is largely based upon our personal experience over 20 years of more than 600 identified patients evaluated at the metabolic and genetic services clinic at Hôpital des Enfants-Malades in Paris. Some parts of this chapter have already been published (Saudubray *et al.*, 1989).

General pathophysiological considerations

In most of the inborn errors considered in this chapter, the basic biochemical lesion either affects one metabolic pathway, which is common to a large number of cells or organs (e.g. energy deficiency in mitochondrial disorders, or storage diseases due to lysosomal disorders), or, when the metabolic block is restricted to one specific organ, gives rise to humoral and systemic consequences (e.g. hyperammonaemia in urea cycle defects or hypoglycaemia in hepatic glycogenosis). For this reason presenting symptoms can be very diverse. The central nervous system is frequently involved. At an advanced stage, many non-specific secondary abnormalities are observed and these can render correct diagnosis difficult.

As far as physiopathology is concerned, metabolic disorders can be divided into three groups which can be helpful for diagnostic purposes.

1 Diseases that disturb the synthesis or catabolism of complex molecules. In this first group of disorders, symptoms are permanent, progressive, independent of intercurrent events and without relationship to food intake.

All lysosomal disorders belong to this category in which deficiencies lead to the progressive accumulation of undigested substrates, usually complex polymers that cannot be hydrolysed normally. The polymers accumulate within lysosomes where they can be seen by use of optic and electronic microscopy (storage disorders). The tissues that are affected are those in which the substance in question is normally catabolized in the largest amounts (circulating lymphocytes, fibroblasts, liver, spleen, conjunctiva, bone marrow, intestinal mucosa, etc.). Conversely, the demonstration of accumulated circulating substrates is rarely available. Only a few lysosomal disorders are clinically expressed in the neonatal period.

In disorders of peroxisomal biogenesis, many anabolic functions are disturbed, including plasmalogen (a major myelin constituent), cholesterol and bile acid biosynthesis. Generalized peroxisomal β-oxidation deficiency results in a variety of disturbances, which are still poorly understood, because of an overlapping in function between the peroxisomes and other organelles, including the mitochondria and endoplasmic reticulum. Whatever the exact physiopathology may be, these multiple and complex biochemical abnormalities result in a striking disorder of neuronal migration with malformations and severe neurological dysfunction. In peroxysomal disorders, there is no polymer accumulated in the blood cells and main viscera. A useful frequent marker to the diagnosis is the accumulation of very long-chain fatty acids in the plasma. Most peroxisomal disorders present in the neonatal period.

Another group of disorders is formed by mutations involving intracellular trafficking and processing of secretory proteins such as α_1-antitrypsin deficiency or the more recently described carbohydrate-deficient glycoprotein syndrome (Jaeken *et al.*, 1991). Such mutations are difficult to demonstrate because they do not manifest simply as enzyme deficiency states. Their diagnosis lies in the measurement in the plasma of specific protein(s) like α_1-antitrypsin, glycosylated transferrin, thyroid-binding globulin or total serum glycoproteins.

2 Inborn errors of intermediary metabolism that lead to an acute or progressive endogenous intoxication secondary to an accumulation of toxic compounds proximal to the metabolic block. Symptoms correlate directly to food intake and to nutritional status.

Amino acidopathies (such as maple syrup urine disease (MSUD) or tyrosinaemia type I), most of the organic acidurias (methylmalonic, propionic, isovaleric, etc.), congenital urea cycle defects and sugar intolerances (galactosaemia, fructosaemia) belong to this group (see also Chapters 5 and 6). All the conditions in this group present clinical similarities, including a symptom-free interval, clinical signs of 'intoxication', acute (such as vomiting, lethargy, coma, liver failure) or chronic (such as progressive developmental delay or a failure to thrive), and frequent humoral disturbances (acidosis, ketosis, hyperammonaemia, etc.). The biochemical diagnosis is easy and mostly relies on chromatography for plasma and urine amino acids or organic acids. Treatment of these disorders requires toxin removal (blood exchange transfusion or haemodialysis, peritoneal dialysis, special diets).

3 Inborn errors of intermediary metabolism in which symptoms are at least partly due to the deficiency in energy production or utilization processes, ensuing distally from a defect in the liver, myocardium, muscle or brain.

Congenital lactic acidaemias (pyruvate carboxylase (PC) or pyruvate dehydrogenase (PDH) deficiency), fatty acid oxidation defects and mitochondrial respiratory chain disorders belong to this group (see also Chapters 5 and 6). These diseases present an overlapping clinical spectrum which sometimes results also in part from the accumulation of toxic compounds in addition to the deficiency in energy production. Frequent symptoms common to this group include hypoglycaemia, hyperlactacidaemia, severe generalized hypotonia, myopathy, cardiomyopathy, failure to thrive, cardiac failure, circulatory collapse, sudden infant death syndrome and malformations; the latter suggest that the abnormal processes affected the fetal energy pathways (Clayton & Thompson, 1988). The recent abundance of reports on inborn errors of the respiratory chain emphasizes the amazing clinical diversity of these disorders and illustrates the ubiquitous role of energy processes at every age and in every organ (Munnich *et al.*, 1992). Treatment of these disorders (if there is one) would require adequate energy replacement.

According to this very general pathophysiological classification, and despite the large number of inborn errors of intermediary metabolism which can be observed in the neonatal period, it is possible to assign most patients to one of five clinico-biochemical syndromes. The following clinical method of screening is easy to implement and prospectively we have found it very reliable over the past 20 years. It provides considerable cost–benefit advantages over techniques such as mass screening for acute metabolic errors, and is much more rapid, allowing specific treatment to be undertaken within hours in these critically ill neonates.

It also highlights that despite the growing sophistication of biochemical investigations, the crucial step of early clinical suspicion and diagnosis lies in the hands of the primary-care physician (Saudubray *et al.*, 1989).

Clinical approach to neonates with suspected metabolic defects

The clinical diagnosis can be schematized into the following three steps.

Presenting signs: apparently non-specific symptoms

The neonate has a limited repertoire of responses to severe illness and at first glance presents non-specific symptoms such as respiratory disorders, hypotonia, poor sucking reflex, vomiting, diarrhoea, dehydration, lethargy and seizures, all symptoms which could be attributed easily to infection or some other common cause. If there are siblings, the death of affected siblings may have been falsely attributed to sepsis, heart failure or intraventricular haemorrhage, and it is important to review critically clinical records and autopsy reports when they are available.

Apparently non-specific symptoms are included in a very evocative clinical context

In the 'intoxication' type of metabolic distresses, an extremely evocative clinical setting is the course of a full-term baby born after a normal pregnancy and delivery who, after an initial symptom-free period during which the baby appears completely normal, deteriorates for no apparent reason and does not respond to symptomatic therapy. The interval between birth and clinical symptoms may range from hours to weeks, depending on the nature of the metabolic block and the environment. In organic acidaemias and urea cycle defects, the duration of the interval does not necessarily correlate with the protein content of the feeding.

Investigations routinely performed in all sick neonates, including chest X-ray, cerebrospinal fluid (CSF) examination, bacteriological studies and cerebral ultrasound, all yield normal results. This unexpected and 'mysterious' deterioration of a child after a normal initial period is the most important signal of the presence of an inherited disease of the 'intoxication' type. If such a deterioration is present, a careful re-evaluation of the child's condition is warranted. Signs previously interpreted as non-specific manifestations of neonatal hypoxia, infection or other common diagnoses take on a new significance in this context.

In 'energy deficiencies', however, the clinical presentation is less evocative and displays variable severity. A careful reappraisal of the child is always warranted.

Reappraisal of the neonate: clinical approach to inborn errors according to the main presenting sign

Neurological deterioration

Indeed, most inborn errors of intermediary metabolism of the 'intoxication' type or 'energy deficiency' type are brought to a doctor's attention because of neurological deterioration. In the 'intoxication' type of metabolic distresses, the initial symptom-free interval varies in duration between the conditions. Typically, the first reported sign is poor sucking and feeding, after which the baby sinks into an unexplained coma, despite supportive measures. At a more advanced stage, neurovegetative problems with respiratory disorders, hiccups, apnoea, bradycardia and hypothermia can appear. In the comatose state, many of these conditions have characteristic changes in muscle tone and involuntary movements. Generalized hypertonic episodes with opisthotonos are frequent, and boxing or pedalling movements as well as slow limb elevations, spontaneously or upon stimulation, are observed. Most non-metabolic causes of coma are associated with hypotonia, so that the presence of 'normal' peripheral muscle tone in a comatose baby reflects a relative hypertonia. Another neurological pattern suggesting metabolic disease is axial hypotonia and limb hypertonia with large-amplitude tremors and myoclonic jerks, which are often mistaken for convulsions. An abnormal urine and body odour are present in some diseases in which volatile metabolites accumulate (the most important examples are the maple syrup odour of MSUD and the sweaty feet odour of isovaleric acidaemia and type II glutaric acidaemia) (see also Chapter 6).

If one of the preceding risk factors is present, metabolic disorders should be given a high diagnostic priority and should be investigated simultaneously with other diagnostic considerations.

In the 'energy deficiencies', clinical presentation is less evocative and displays a more variable severity. In many conditions, there is no free interval. The most frequent symptoms are a severe generalized hypotonia, hypertrophic cardiomyopathy, rapidly progressive neurological deterioration, with possible dysmorphia or malformations. However, in contrast to the 'intoxication' group, lethargy and coma are rarely inaugural signs. Hyperlactacidaemia with or without metabolic acidosis is a very frequent symptom.

Only a few lysosomal disorders with 'storage' symptoms are expressed in the neonatal period. By contrast, most of the peroxisomal disorders present immediately after birth with dysmorphia and severe neurological dysfunction.

Seizures

True convulsions occur late and inconsistently in inborn errors of intermediary metabolism, with the exception of pyridoxine-dependent seizures (Bankier *et al.*, 1983) and some cases of non-ketotic hyperglycinaemia, sulphite oxidase (SO) deficiency (Mises *et al.*, 1982; Wadman *et al.*, 1983), and peroxisomal disorders where there may be important inaugural elements in the clinical presentation. In contrast, newborns with MSUD, organic acidurias and urea cycle defects rarely experience seizures in the absence of pre-existing stupor or coma, or hypoglycaemia. The electroencephalogram (EEG) often shows a periodic pattern in which bursts of intense activity alternate with nearly flat segments (Mises *et al.*, 1982).

Convulsions are the unique symptom in pyridoxine-dependent seizures. This rare disorder should be considered with all refractory seizures in infants under 1 year of age. The clinical response to vitamin B_6 administration is the best way of making the diagnosis. If it is available without delay, an EEG should be carried out during the injection of pyridoxine; an immediate improvement in the tracing may be seen (Bankier *et al.*, 1983).

Hypotonia

Hypotonia (Table 4.1) is a very common symptom in sick neonates. Whereas many non-metabolic inherited diseases can give rise to severe generalized

Table 4.1 Hypotonia in the neonatal period and early infancy

Leading symptoms	Other signs	Age of onset	Diagnosis
Evocative clinical context (dysmorphia, bone changes, visceral symptoms, malformations)	Main bone changes, hypomineralization, rachitic deformations	Congenital to early infancy	Hypophosphatasia (large fontanelle, wide cranial suture) Calciferol metabolism defects Osteogenesis imperfecta (large skull, hyperlaxity, blue sclerae)
	Predominant dysmorphia, psychomotor retardation, malformations	Congenital	Peroxisomal disorders (Zellweger and variants, chondrodysplasia punctata) Lowe syndrome (cataract) Chromosomal abnormalities Other polymalformative syndromes (mainly those with muscular dystrophy: Walker Warburg, Fukuyama, muscular dystrophy, etc.)
	Storage disorders, hepatosplenomegaly, ascites, oedema, coarse facies, vacuolated lymphocytes	Congenital	Lysosomal disorders: Niemann–Pick type A GM_1-gangliosidosis (Landing) Galactosialidosis Sialidosis type II I cell disease
	Hepatomegaly, cholestatic jaundice, failure to thrive	Congenital	Peroxisomal disorders
	Cataract, tubulopathy	Congenital	Lowe syndrome (X-linked) Respiratory chain disorders
	Cardiomyopathy, macroglossia, vacuolated lymphocytes	Neonatal to early infancy	Pompe's disease Respiratory chain disorders
Neurological neonatal distresses (see also Table 4.13)	With ketosis and ketoacidosis (types I and II)	Neonatal to first month	MSUD Organic acidurias
	With hyperlactacidaemia (type III)	Neonatal to first month	Congenital hyperlactacidaemias (PC, PDH, Krebs, respiratory chain)
	With hyperammonaemia (type IVa)	Neonatal to first days	Urea cycle defects Triple H Lysinuric protein intolerance Transient hyperammonaemia
	Without ketoacidosis or hyperammonaemia (type IVb), seizures, myoclonias	Congenital to first days	Non-ketotic hyperglycinaemia Sulphite oxidase deficiency Peroxisomal disorders (acyl-CoA oxidase, Zellweger variants)
Apparently isolated at birth	Massive generalized hypotonia, amimia, fetal distress, hydramnios, arthrogryposis, respiratory failure	Congenital	Sever fetal neuromuscular disease Steinert (myotonia in mother) Myasthenia (prostigmin test) Congenital myopathy Hereditary sensitivo-motor neuropathy (abnormal conduction nerve velocity) Familial dysautonomy Congenital dystrophy (CK elevated)

continued

Table 4.1 *Continued*

Leading symptoms	Other signs	Age of onset	Diagnosis
	Predominant proximal amyotrophy, paralysis, areflexia	Congenital to first weeks	Werdnig Hoffmann (SMA type I)
	Mental retardation, secondary obesity	Congenital	Prader–Willi syndrome (frequent)
	Stridor, dystonia, nystagmus	First months	Pelizaeus Merzbacher (X-linked)

MSUD, maple syrup urine disease; PC, pyruvate carboxylase; PDH, pyruvate dehydrogenase; CK, creatine kinase; SMA, spinal muscular atrophy.

neonatal hypotonia (mainly all severe fetal neuromuscular disorders), indeed only a few inborn errors of metabolism present in neonates with isolated or predominant hypotonia. Discounting disorders in which hypotonia is included in a very evocative clinical context of major bone changes, dysmorphia, malformations or visceral symptoms (see Table 4.1), the most severe metabolic hypotonias are observed in hereditary hyperlactacidaemias, respiratory chain disorders, urea cycle defects, non-ketotic hyperglycinaemia (NKH), SO deficiency and peroxisomal disorders (PZO). In all these circumstances, the diagnosis is mostly based upon the association with the central hypotonia of lethargy, coma, seizures and neurological symptoms in NKH, SO deficiency and PZO, and of characteristic metabolic changes in congenital lactic acidosis and urea cycle disorders (hyperammonaemia) (see also Chapters 3 and 5). Severe forms of Pompe's disease (α-glucosidase deficiency) can at first mimic respiratory chain disorders when generalized hypotonia is associated with cardiomyopathy. However, Pompe's disease does not start strictly in the neonatal period. In reality, these two disorders are very different since Pompe's disease is associated with highly suggestive electrocardiogram (ECG) changes (short PR interval, large amplitude of QRS) and vacuolated lymphocytes, whereas respiratory chain disorders present with hypertrophic dilated cardiomyopathy and lactic acidosis (see section on Cardiomyopathy, below). Evocative cranio-facial dysmorphism is present only in typical Zellweger syndrome, but can be very moderate or even absent in variant forms of PZO such as neonatal adrenoleucodystrophy, infantile Refsum disease or acyl-CoA oxidase deficiency. A classical error made by inexperienced physicians is to misdiagnose Zellweger syndrome with Down's syndrome or other chromosomal aberrations. Lowe syndrome should be systematically considered in boys who present with congenital cataracts, tubulopathy and minor facial dysmorphism. Finally, one of the most frequent diagnoses is Prader–Willi syndrome where hypotonia is central and apparently isolated at birth. This diagnosis can only be confirmed through the demonstration of chromosome-15 deletion or maternal heterodisomy of this chromosome, or in the second year of life when polyphagia and considerable obesity become manifest. As shown in Table 4.1, no one demonstrated inborn error of metabolism occurs in neonates as strictly isolated hypotonia.

Liver failure, ascites and oedema

Four clinical groups of metabolic disorders can present with liver failure, ascites and oedema in the neonatal period or early in infancy (Table 4.2).

1 Lysosomal disorders that present either as non-immune hydrops fetalis syndrome, or early in the neonatal period with much clinical evidence of storage disorders (hepatosplenomegaly, dysostosis multiplex, vacuolated circulating lymphocytes, cherry red spot and coarse facies) (see also Chapter 7(c)).

2 Severe congenital haemolytic anaemias can give rise to hydrops fetalis syndrome. The best known is

Table 4.2 Liver failure, ascites and oedema

Leading symptoms	Other signs	Age of onset	Diagnosis
Hydrops fetalis	Storage disorders (hepatosplenomegaly, coarse facies, dysostosis multiplex)	Congenital	Landing, Niemann–Pick A, C Sialidosis type II Galactosialidosis Mucopolysaccharidosis type VII (Sly)
	Haemolytic syndromes		Barth haemoglobin Glycolytic enzymopathies (rare)
With isolated liver failure Jaundice, haemorrhagic syndrome, hepatosplenomegaly, hypoglycaemia, cytolysis	*E. coli* sepsis, cataract, tubulopathy	First days	Galactosaemias (transferase and epimerase deficiencies)
	Vomiting, dizziness, tubulopathy	First days	Fructosaemias (in case of fructose-containing diet)
	Haemolytic anaemia	Congenital to first days	Neonatal haemochromatosis (with low transaminases)
	Tubulopathy, haemolytic anaemia	First weeks to first months	Tyrosinaemia type I
	Hypoglycaemia	After weaning	Fructose intolerance Fructose bisphosphatase deficiency
Liver failure with neurological or muscular symptoms	Hypotonia, failure to thrive, cardiomyopathy, developmental delay, lactic acidosis	Neonatal to infancy	Respiratory chain disorders Glycogenosis type IV (hepatic fibrosis, cirrhosis)
Storage disorders Hepatosplenomegaly, cherry red spot, dysostosis multiplex, vacuolated lymphocytes	Coarse facies, failure to thrive, 'hepatitis'	Congenital to first weeks	Mucopolysaccharidosis type VII Niemann–Pick A and C (acute forms)
	Coarse facies, decerebration, spasticity, seizures	Congenital to first weeks	Landing, Sialidosis II Galactosialidosis
Hepatosplenomegaly, vacuolated lymphocytes	Failure to thrive, vomiting, diarrhoea, adrenal calcifications	First weeks (rare in neonatal period)	Wolman's disease

Barth haemoglobin hydrops fetalis syndrome owing to the absence of a functional haemoglobin α chain.

3 The most frequent disorders observed in the neonatal period and early infancy and presenting as isolated liver failure syndrome (jaundice, haemorrhagic syndrome, hypoglycaemia, hepatosplenomegaly, elevated transaminases) are fructosaemia and galactosaemia. Fructose bisphosphatase (FBP) deficiency can present, although rarely, with hepatocellular insufficiency. In tyrosinaemia type I, liver failure becomes obvious only at 3 or 4 weeks of age.

It is also recommended that one should search for α_1-antitrypsin deficiency and cystic fibrosis, although these disorders present in neonates as hepatic cholestasis rather than as hepatocellular insufficiency. Neonatal haemochromatosis seems to be not so rare and should be systematically considered, but there is no easy method to prove this diagnosis before death (Barnard & Manci, 1991; Knisely, 1992), although, as a rule, transaminase activity levels are low in this disorder (Knisely, 1992).

4 A last group of disorders which presents in early infancy with liver failure is associated with severe neurological dysfunction. It encompasses glycogenosis type IV, owing to a branching enzyme deficiency, and respiratory chain disorders with predominant hepatocellular dysfunction. In both conditions, liver dysfunction is included in a neurological picture with severe hypotonia, developmental delay, cardiomyopathy and hyperlactacidaemia. In respiratory chain defects, diagnosis can be suspected on hyperlactacidaemia with an elevated lactate:pyruvate (L:P) ratio and hyperketonaemia; a definite conclusion relies on the demonstration of a specific defect in the respiratory chain which can be tissue specific and restricted to the liver.

In our experience, hepatic presentations of inherited fatty acid oxidation disorders and urea cycle defects comprise acute steatosis or Reye's syndrome with normal bilirubin, only a slightly prolonged prothrombin time, and moderate elevation of transaminases, rather than true liver failure with ascites and oedema. We have recently observed a patient with severe generalized peroxisomal dysfunction who presented in the first month of life with a failure to thrive, hepatomegaly and liver failure associated with retinitis pigmentosa, and without obvious dysmorphia or neurological signs.

One must emphasize that there are frequent difficulties in investigating patients with severe hepatic failure. At an advanced stage, many non-specific symptoms, which are secondary consequences of the disturbance of liver intermediary metabolism, can be present. Melituria (galactosuria, glycosuria, fructosuria), hyperammonaemia, hyperlactacidaemia, short fast hypoglycaemia, hypertyrosinaemia (>200 μmol/L) and hypermethioninaemia (sometimes higher than 500 μmol/L) are the signs most commonly encountered in advanced hepatocellular insufficiencies. Succinyl acetone excretion seems to be highly specific of tyrosinaemia type I, owing to fumaryl acetoacetase deficiency, but its absence does not rule out this diagnosis. Investigations of parents can be very helpful for the diagnosis of galactosaemias, mainly when the patient has been transfused (measurement of galactose uridyl transferase and epimerase in parents' erythrocytes).

Hepatomegaly and hepatosplenomegaly

Hepatomegalies (Table 4.3) as the result of inborn errors of metabolism can be considered within three clinical circumstances: (a) with manifestations of hepatocellular necrosis; (b) with cholestatic jaundice; and (c) with no pre-eminent hepatic dysfunction (Odièvre, 1991).

In the first circumstance, the clinical picture is characterized by some degree of jaundice, oedema, ascites, tendency to bleeding and hepatic encephalopathy. In neonates and early in infancy, four disorders should first be considered: not only the classical diagnosis of galactosaemias, fructosaemia and tyrosinaemia type I, but also respiratory chain disorders (Cormier *et al.*, 1991; Odièvre, 1991; Parrot-Roulard *et al.*, 1991; Vilaseca *et al.*, 1991).

When cholestatic jaundice is present, an α_1-antitrypsin deficiency is the most frequent cause in the neonatal period. Neonatal haemochromatosis (Barnard & Manci, 1991; Knisely, 1992), inborn errors of bile acid synthesis (Clayton, 1991), Niemann–Pick type C (presenting as giant cell hepatitis) and inborn errors of peroxisome metabolism (as Zellweger syndrome, infantile Refsum disease or trihydroxycoprostanoyl-CoA oxidase deficiency) should also be considered.

When hepatomegaly is the only or major symptom of liver disease, a number of metabolic conditions can be involved. In such circumstances, a major clinical key to the diagnosis is the consistency of the liver and the characteristics of its surface. When the liver displays a firm or rock-hard consistency, whether associated with liver dysfunction or not, different causes of metabolic cirrhosis should first be considered. According to the age of onset, they include tyrosinosis type I, galactosaemia, glycogenosis type IV, severe neonatal haemochromatosis, α_1-antitrypsin deficiency, Wilson disease and cystic fibrosis.

When the liver consistency is normal or soft, a large list of diagnoses has to be considered. The presence of a splenomegaly associated with hepatomegaly suggests lysosomal disorders in which massive enlargement of the spleen is more often seen than in portal hypertension. Coarse facies, bone changes, joint stiffness, ocular symptoms, vacuolated lymphocytes and neurological deterioration are strongly suggestive of mucolipidosis and mucopolysaccharidosis. A failure to thrive, anorexia, poor feeding, severe diarrhoea, hypotonia,

Table 4.3 Permanent hepatomegaly or hepatosplenomegaly

Symptoms		Age of onset	Diagnosis
With manifestations of hepatocellular necrosis			
Jaundice (mild to severe), oedema, ascites, bleeding tendency, possible neurological signs, intercurrent sepsis, elevated transaminases, disturbed hepatocellular functions (liver failure), hypoglycaemia, hyperammonaemia, possible haemolytic anaemia		Neonatal to early infancy	Galactosaemia (uridyl transferase and epimerase) Fructose intolerance Tyrosinaemia type I (after at least 3 weeks) Respiratory chain disorders Neonatal haemochromatosis (congenital, low transaminases) (see also Table 4.2)
With cholestatic jaundice			
Yellow-brown urine, jaundice, light or acholic stools		Neonatal	α_1-Antitrypsin deficiency (frequent) Byler disease (severe intrahepatic familial cholestasis) Inborn error of bile acid metabolism Neonatal haemochromatosis (congenital) Galactosaemias, fructosaemias (rare) Niemann–Pick type C Classical Zellweger and variants
		Early infancy	Infantile Refsum disease Trihydroxycoprostanoyl-CoA oxidase deficiency
With hepatomegaly as only or major revealing symptom of liver disease			
With firm rock-hard liver consistency with or without splenomegaly and portal hypertension (hepatic fibrosis, cirrhosis)		Early infancy	Tyrosinosis type I Galactosaemia, fructosaemia Glycogenosis type IV Neonatal haemochromatosis
With normal or soft liver consistency With splenomegaly	'Storage' disorders: coarse facies, bone changes, joint stiffness, vacuolated lymphocytes, ocular symptoms, progressive neurological dysfunction	Neonatal to early infancy	Landing Galactosialidosis (early infantile) Sialidosis type II I cell disease (mucolipidosis type II)
	Severe failure to thrive: anorexia, poor feeding, neurological deterioration	Early infancy	Niemann–Pick type A (interstitial pneumonia) Farber disease (interstitial pneumonia, hoarseness, skin nodules, hyperthermia) Gaucher's type II (acute neuronopathy, spasticity, opisthotonos, vegetative state)

continued

Table 4.3 *Continued*

Symptoms		Age of onset	Diagnosis
	Chronic diarrhoea: failure to thrive, anorexia, poor feeding, hypotonia	Early infancy	Wolman's disease (adrenal gland calcifications, vacuolated lymphocytes) Lysinuric protein intolerance (leuconeutropenia, recurrent metabolic attacks, hyperammonaemia) Chronic granulomatous disease (repeated infections, dermatitis, stomatitis, hyperthermia, inflammatory bowel disease) Glycogenosis type Ib (repeated infections, dermatitis, stomatitis, hyperthermia, inflammatory bowel disease, severe hypoglycaemia)
With normal or soft liver consistency Without splenomegaly	Fasting hypoglycaemia ± metabolic acidosis, hyperlactacidaemia, failure to thrive, hypotonia	Neonatal to late infancy	Glycogenosis types Ia and b Glycogenosis type III Fructose bisphosphatase deficiency Glycogen synthetase deficiency
	Severe hypotonia, cardiomyopathy, failure to thrive	Early infancy	Pompe's disease Respiratory chain disorders Phosphorylase-b kinase deficiency Carbohydrate-deficient glycoprotein syndrome

hyperthermia and frequent infections are frequent presenting signs in Niemann–Pick type A, Farber, Gaucher's type II and Wolman's disease; they also occur in chronic granulomatous disease, glycogenosis type Ib and lysinuric protein intolerance, which present mostly in early infancy.

When hepatomegaly is not associated with splenomegaly, two main clinical circumstances can be considered: those with fasting hypoglycaemia usually suggest glycogenosis types I and III, and FBP deficiency.

Hepatomegaly in a context of severe hypotonia, cardiomyopathy or a failure to thrive is frequently seen in Pompe's disease, respiratory chain defects or other rare disorders listed in Table 4.3.

Severe diarrhoea

Severe diarrhoea (Table 4.4) owing to inborn errors of metabolism can be divided into four clinical categories.

The first presents with severe watery diarrhoea, resulting in dehydration, and occurs immediately after birth, after weaning or after starch dextrins have been added to the diet. It encompasses the very rare disorders of congenital chloride diarrhoea, glucose–galactose malabsorption, lactase and sucrase–isomaltase deficiencies and the less severe diarrhoeic syndrome observed in acrodermatitis enteropathica.

In a second category, clinical features are dominated by chronic diarrhoea associated with fat-soluble vitamin malabsorption syndrome, steatorrhoea, striking hypocholesterolaemia and osteopenia; all symptoms present early in infancy. Inborn errors of bile acid metabolism presenting with unexplained neonatal cholestatic jaundice (Clayton, 1991) and infantile Refsum disease (Poll-The *et al.*, 1987a,b; Mandel *et al.*, 1992) are the most frequent disorders in this category. Respiratory chain disorders and, among them, Pearson syndrome owing to mitochondrial DNA deletion can

Table 4.4 Severe diarrhoea and failure to thrive

Leading symptoms	Other signs	Age of onset	Diagnosis
Severe watery diarrhoea, attacks of dehydration	No meconium, non-acidic diarrhoea, metabolic alkalosis, hypochloraemia and hypochloruria	Congenital to infancy	Congenital chloride diarrhoea
	Acidic diarrhoea, reducing substances in stools, glucosuria	Neonatal	Glucose–galactose malabsorption Lactase deficiency
	Acidic diarrhoea, reducing substances in stools, after weaning or after starch dextrins are added to the diet	Neonatal to infancy	Sucrase–isomaltase deficiency
	Skin lesions, alopecia (late onset), failure to thrive, less severe diarrhoea	Neonatal or after weaning	Acrodermatitis enteropathica
Fat-soluble vitamins malabsorption, severe hypocholesterolaemia, osteopenia, steatorrhoea	Unexplained cholestatic jaundice	Neonatal to infancy	Bile acid synthesis defects Infantile Refsum disease
	Hepatomegaly, hypotonia, slight mental retardation, retinitis pigmentosa	First month to early infancy	Infantile Refsum disease
	Pancreatic insufficiency, neutropenia, pancytopenia, lactic acidosis	First month to early infancy	Pearson syndrome Respiratory chain defects (mitochondrial DNA deletion)
Severe failure to thrive, anorexia, poor feeding With *predominant hepatosplenomegaly*	Severe hypoglycaemia, inflammatory bowel disease, neutropenia, recurrent infections, hepatomegaly without splenomegaly	Neonatal to early infancy	Glycogenosis type Ib (glucose-6-phosphate carrier deficiency)
	Hypotonia, vacuolated lymphocytes, adrenal gland calcifications	Neonatal	Wolman's disease
	Neutropenia, thrombopenia, osteopenia, recurrent attacks of hyperammonaemia, interstitial pneumonia, orotic aciduria	Neonatal to early infancy	Lysinuric protein intolerance
Severe failure to thrive, anorexia, poor feeding With *megaloblastic anaemia*	Stomatitis, infections, peripheral neuropathy, intracranial calcifications	Infancy	Congenital folate malabsorption
	Severe pancytopenia, vacuolization of marrow precursors, exocrine pancreas insufficiency, lactic acidosis	Neonatal	Pearson syndrome Respiratory chain defects (mitochondrial DNA deletion)
Severe failure to thrive, anorexia, poor feeding, vomiting, frequent infections	Chronic ketoacidosis, leucopenia	Early infancy	Organic acidurias
	Chronic hyperammonaemia	Early infancy	Urea cycle defects

also be revealed early in infancy by severe intestinal malabsorption and exocrine pancreatic insufficiency, although these symptoms are in general included in a severe picture of pancytopenia, non-regenerative macrocytic anaemia or multivisceral failure (Rotig *et al.*, 1990).

A third group of patients presents with severe failure to thrive, poor feeding, hypotonia and frequent infections associated with obvious hepatosplenomegaly. Although their pathogeny is very diverse, three inherited disorders can share these clinical features: (a) glycogenosis type Ib, due to a defective microsomal glucose-6-phosphate carrier, with severe hypoglycaemia, neutropenia and inflammatory bowel disease; (b) Wolman's disease, due to lipase deficiency, with vacuolated lymphocytes and adrenal gland calcifications; and (c) lysinuric protein intolerance with leuconeutropenia, thrombopenia and recurrent attacks of hyperammonaemia.

The clinical presentation of the fourth group is dominated by the finding of non-regenerative megaloblastic anaemia, which may be associated with pancytopenia. Congenital folate malabsorption, Pearson syndrome and, later in infancy, transcobalamin II (TC II) and intrinsic factor deficiency are the best known disorders of this category.

Finally, chronic diarrhoea with failure to thrive, anorexia, hypotonia and recurrent infections can be the first clinical expression, early in infancy, of subacute forms of organic acidurias, urea cycle defects and adenosine deaminase deficiency.

Dehydration

Besides dehydration due to digestive causes, some inborn errors of metabolism can present with recurrent attacks of dehydration secondary to polyuria, hyperventilation or hypersudation. In some disorders, recurrent attacks of dehydration can be the presenting sign. According to the main accompanying symptoms (severe diarrhoea, salt wasting, ketoacidosis or failure to thrive), dehydration due to inborn errors of metabolism can be classified as shown in Table 4.5.

Cardiomyopathy, cardiac failure, heartbeat disorders and cot death (Table 4.6)

Discounting mucolipidosis type II, which can give rise to a cardiomyopathy integrated in an already known and evident clinical picture, there are three main groups of inborn errors that can present with cardiomyopathy as the revealing or predominant sign: (a) Pompe's disease; (b) respiratory chain disorders; and (c) fatty acid oxidation defects. In most cases, cardiomyopathy is of a hypertrophic and hypokinetic type.

1 Pompe's disease is an already well-known disorder in which cardiomyopathy appears early in infancy and is associated with severe muscle hypotonia, macroglossia, evocative ECG changes and circulating vacuolized lymphocytes. The severity of cardiac damage and of gross motor dysfunction contrasts sharply with the normal mental development.

2 In respiratory chain disorders and Krebs cycle deficiency (α-ketoglutarate dehydrogenase), cardiomyopathy is most often associated with hypotonia, myopathy, diverse neurological signs and developmental delay. Hyperlactacidaemia after meals with a high L:P ratio and ketosis are important keys in screening for these disorders, although hyperlactacidaemia can at times be moderate or even absent. 3-Methylglutaconic aciduria of unknown origin has been frequently found in patients exhibiting cardiomyopathy and lactic acidaemia. Some of these patients are probably affected with misdiagnosed respiratory chain disorders (see Table 4.11).

Recent observations suggest that some respiratory chain disorders are tissue specific and are expressed only in the myocardium, as already found in phosphorylase-b kinase deficiency (Mitzuta *et al.*, 1984; Servidei *et al.*, 1988). This highlights the interest in measuring respiratory chain enzyme activities directly on an endomyocardium biopsy each time it is possible. The new multisystemic disorder carbohydrate-deficient glycoprotein syndrome can sometimes present in infancy with cardiac failure owing to pericardial effusions and cardiac tamponade (Jaeken *et al.*, 1991).

3 Among the hitherto known hereditary defects of

Table 4.5 Acute neonatal dehydration

Leading symptoms	Other signs	Age of onset	Diagnosis
With severe diarrhoea: 'digestive causes' (see also Table 4.4)	Severe watery acidic diarrhoea, glucosuria	Neonatal	Glucose–galactose malabsorption Congenital lactase deficiency
	Hydramnios, no meconium, severe watery non-acidic diarrhoea, metabolic alkalosis, hypokalaemia, hypochloraemia	Congenital	Congenital chloride diarrhoea
	Severe watery diarrhoea	After weaning or when sucrose or starch dextrins are added to the diet	Sucrase–isomaltase deficiency
With ketoacidosis: 'organic acidurias'	Polyuria, polypnoea, hyperglycaemia, glycosuria	Neonatal to first month	Diabetic coma (rare) Methylmalonic aciduria Propionic acidaemia Isovaleric aciduria Hydroxyisobutyric aciduria
With failure to thrive, anorexia, poor feeding, polydipsia, polyuria: 'renal tubular dysfunction'	Hypernatraemia, vomiting, psychomotor retardation, spasticity, X-linked	Neonatal to first month	Nephrogenesis diabetes insipidus
	Hyperchloraemia, metabolic acidosis, alkaline urine pH	Early in infancy	RTA type I (distal): nephrocalcinosis, hypokalaemia RTA type II (proximal): Fanconi syndrome, acidic urine pH when profoundly acidotic Respiratory chain defects RTA type IV: hyperkalaemia (associated with adrenogenital syndrome or pseudo-hypoaldosteronism) and obstructive uropathy
With salt-losing syndrome: 'adrenal dysfunctions'	Severe hyponatraemia, ambiguous genitalia in girls	End of first week of life	Congenital adrenal hyperplasias: 21-hydroxylase deficiency (frequent) β-hydroxysteroid dehydrogenase, very severe (rare) Desmolase deficiency, very severe (rare)
	Without ambiguous genitalia		Hypoaldosteronism

RTA, renal tubular acidosis.

Table 4.6 Cardiomyopathy (primary, non-obstructive), cardiac failure and heartbeat disorders

Leading symptoms	Other signs	Age of onset	Diagnosis
Integrated in an evident clinical picture			
With 'storage' signs, coarse facies, dysostosis multiplex, osteoporosis, hepatomegaly, corneal opacities, inguinal hernias, developmental delay, vacuolated lymphocytes		Neonatal to infancy	Mucolipidosis type II (I cell disease)
Predominant or revealing symptom			
With hypotonia, muscle weakness, failure to thrive	Ketoacidosis, hyperlactacidaemia, developmental delay	First month to early infancy	Respiratory chain disorders (complex I, II, IV) 3-Methylglutaconic aciduria (unknown origin) Ketoglutarate dehydrogenase deficiency
	Macroglossia, hepatomegaly, vacuolated lymphocytes, ECG changes (short PR interval, large amplitude of QRS)	First month to early infancy	Pompe's disease (glycogenosis type II due to α-glucosidase deficiency)
	Myopathy, hypoglycaemia	Neonatal to infancy	Fatty acid oxidation disorders (MAD, LCAD, LCHAD, TL, LCTP) Phosphorylase-b kinase
With hypoketotic hypoglycaemia, 'hepatic' signs, Reye's syndrome		Neonatal to childhood	Fatty acid oxidation disorders (MAD, TL, CPT II, LCAD, LCHAD)
With heartbeat disorders, cardiac arrest, collapsus, sudden death		Neonatal to early infancy	Fatty acid oxidation disorders (MAD, CPT II, LCAD, LCHAD, LCTP, VLCAD) Respiratory chain disorders
With macrocytic megaloblastic anaemia	Haemolytic uraemic syndrome, ketoacidosis, retinopathy, multivisceral failure	First month to early infancy	Cobalamin metabolism defects (CbLc)
With tamponade, pericardial effusions, multiorgan failure	Muscular weakness, stroke-like attacks, peculiar fat pads, thick and sticky skin	Neonatal to infancy	Carbohydrate-deficient glycoprotein syndrome

continued on p. 74

Table 4.6 *Continued*

Leading symptoms	Other signs	Age of onset	Diagnosis
Apparently primitive heartbeat disorders (without evidence of cardiomyopathy)	Hyperkalaemia, hyponatraemia	Neonatal to infancy	Adrenal dysfunctions (see also Table 4.5)
	Hypoparathyroidism, hypocalcaemia, prolonged QT interval	Infancy	Congenital hypo- and pseudo-hypoparathyroidism
	Collapsus, heart-rate trouble, hypoketotic hypoglycaemia	Neonatal to infancy	Fatty acid oxidation defects (CPT II, LCAD, LCHAD, LCTP, VLCAD)

ECG, electrocardiogram; MAD, multiple acyl-CoA dehydrogenase; LCAD, long-chain acyl-CoA dehydrogenase; LCHAD, long-chain 3-hydroxy-acyl-CoA dehydrogenase; TL, translocase; LCTP, trifunctional enzyme; CPT II, carnitine palmitoyl transferase II; VLCAD, very long-chain acyl-CoA dehydrogenase; CbLc, cobalamin metabolism defect group c.

fatty acid oxidation, seven can be revealed by cardiomyopathy and/or heartbeat disorders (auriculoventricular block, bundle-branch blocks, ventricular tachycardia). In general, severe forms of multiple acyl-CoA dehydrogenase (glutaric aciduria type II owing to electron transfer flavoprotein (ETF) or electron transfer flavoprotein dehydrogenase (ETFDH) deficiency) and long-chain fatty acid disorders involving carnitine palmitoyl transferase II (CPT II), translocase (TL), long-chain acyl-CoA dehydrogenase (LCAD), long-chain 3-hydroxy-acyl-CoA dehydrogenase (LCHAD) and trifunctional enzyme (LCTP) deficiencies are severe, begin early in infancy or even in the neonatal period, and can be revealed by neonatal death, cardiac arrest or collapsus which can be easily misdiagnosed as toxic shock or idiopathic sudden infant death syndrome. All patients with fatty acid oxidation disorders and who display cardiomyopathy in the neonatal period are permanently exposed to the risk of unexpected death, despite therapy. Cardiomyopathy due to idiopathic systemic carnitine deficiency secondary to a carnitine transport defect does not strike in the neonatal period.

The diagnosis work-up of fatty acid oxidation defects relies on the urinary organic acid profile; plasma and urine carnitine and acylcarnitine determination; loading tests; a fasting test; and whole fatty acid oxidation studies on fresh lymphocytes or intact fibroblasts.

Initial biochemical approach

Once clinical suspicion of an inborn metabolic error is aroused, general supportive measures and laboratory investigations must be undertaken immediately (Table 4.7). Abnormal urine odours can best be detected on a drying filter paper or by opening a container of urine which has been left closed at room temperature for a few minutes. Although serum ketone bodies reach 0.5–1 mmol/L in early neonatal life, acetonuria is an important sign of a metabolic disease and is rarely, if ever, observed in a normal newborn (Settergren *et al.*, 1976). Its presence is always abnormal in neonates (see Table 4.8). The dinitrophenylhydrazine (DNPH) test screens for the presence of α-keto acids, such as are found in MSUD. The DNPH test can be considered significant only in the absence of glucosuria and acetonuria, which also react with DNPH. Hypocalcaemia and elevated or reduced blood glucose levels are frequently present in metabolic diseases. The physician should be wary of attributing marked neurological dysfunction merely to these findings. The general approach to persistent hypoglycaemia is given in Table 4.9.

The metabolic acidosis of organic acidurias is usually accompanied by an elevated anion gap. The urine pH should be below 5; otherwise, renal acidosis is a consideration (see Metabolic acidosis; Table 4.8). Ammonia and lactic acid should be

Table 4.7 Initial investigations

	Basic investigations	Specific investigations
Urine	Smell (special odour) Look (special colour) Acetone (Acetest, Ames) Reducing substances (Clinitest, Ames) Keto acids (DNPH) pH (pHstix, Merck) Sulfitest (Merck) Brand reaction Electrolytes (Na^+, K^+) Uric acid (search for *hypo*uricuria)	Urine collection: collect separately each fresh micturition and place in fridge Freezing: freeze at −20°C samples collected before treatment and an aliquot collected 24 h after treatment Do not use samples without having taken expert metabolic advice
Blood	Blood cell count Electrolytes (search for anion gap) Glucose, calcium Blood gases (pH, P_{CO_2}, HCO_3H, P_{O_2}) Uric acid (search for *hypo*uricaemia) Prothrombin time Transaminases (and other 'liver tests') Ammonaemia Lactic, pyruvic acids 3OHbutyrate, acetoacetate Free fatty acids	Plasma heparinized 5 mL at −20°C Blood on filter paper (as 'Guthrie' test) Whole blood: 10–15 mL collected in EDTA and frozen (for molecular biology studies)
Miscellaneous	Lumbar puncture Chest X-ray Cardiac echography, ECG Cerebral ultrasound, EEG	Skin biopsy (fibroblasts culture) CSF (1 mL frozen) Postmortem: liver, muscle biopsies (macroscopic fragment frozen at −70°C) Autopsy

DNPH, dinitrophenylhydrazine; EDTA, ethylenediaminetetraacetate; ECG, electrocardiogram; EEG, electroencephalogram.

determined systematically in newborns at risk. An elevated ammonia level can in itself induce respiratory alkalosis; hyperammonaemia with keto-acidosis suggests an underlying organic acidaemia (Table 4.10). Elevated lactic acid levels in the absence of infection or tissue hypoxia are a significant finding. Moderate elevations (3–6 mmol/L) are often observed in organic acidaemias and in the hyperam-monaemias; levels greater than 10 mmol/L are frequent in hypoxia. A normal serum pH does not exclude hyperlactacidaemia, as neutrality is usually maintained until levels of 5 mmol/L are present. It is important to measure lactate (L), pyruvate (P), 3-hydroxybutyrate (3OHB) and acetoacetate (AA) levels as often as possible by collecting into a special tube, provided by the laboratory, a plasma sample which has been immediately deproteinized at the bedside. This is necessary in order to preserve cyto-plasmic and mitochondrial redox states for the measurement of L:P and 3OHB:AA ratios, respectively (see Hyperlactacidaemias; Table 4.11). Some organic acidurias induce granulocytopenia and thrombocytopenia, which may be mistaken for sepsis (see Pancytopenia, thrombopenia and leuconeutropenia; Table 4.12).

The storage of adequate amounts of plasma urine and CSF is an important element in diagnosis. The utilization of these precious samples should be care fully planned after taking advice from specialists in inborn errors of metabolism. Although not available in most hospital laboratories, some sophisticated investigations (such as amino acid or organic acid

Table 4.8 Neonatal metabolic acidosis (pH < 7.3; P_{CO_2} < 30 mmHg; HCO_3 < 15 mmols/L)

With ketosis	Hyperglycaemia (glucose >7 mmol/L)	Hyperammonaemia (>100 μmol/L)	Branched-chain organic acidurias (MMA, PA, IVA)	Neurological signs, dehydration, thrombopenia, leucopenia
		Ammonia normal or low (<30 μmol/L)	Diabetes, ketolysis defects (OATD)	Dehydration, hyperventilation
	Normoglycaemia	Hyperlactacidaemia	Congenital lactic acidosis (PC, MCD)	Hyperventilation, only a few neurological signs
			Respiratory chain disorders	Hypotonia, myocardiopathy
			3-Hydroxyisobutyric or other organic acidurias	Neurological signs
		Normal lactate	Ketolytic defects (OATD)	See above
			Organic acidurias	
	Hypoglycaemia (glucose <2 mmol/L)	Hyperlactacidaemia	Gluconeogenesis defects (FBP, G6P, GS)	Hepatomegaly
			Respiratory chain defects	Liver failure
		Normal lactate	MMA, PA, IVA	See above
			Adrenal insufficiency	Dehydration, collapsus, salt losing, hyponatraemia, ambiguous genitalia
Without ketosis	Hyperlactacidaemia (lactate >5 mmol/L)	Normal glucose	Pyruvate dehydrogenase deficiency	Minor facial dysmorphia, normal L : P ratio, corpus callosum agenesia
		Hypoglycaemia	Fatty acid oxidation disorders, HMGCoA lyase	Hypotonia, cardiomyopathy, acute cardiac symptoms, sudden death, hyperammonaemia
			FBP, G6P	Hepatomegaly
	Normal lactate	Normal glucose	Renal tubular acidosis I and II	Hyperchloraemia, alkaline urine pH
			Pyroglutamic aciduria	Anion gap, acid urine pH

MMA, methylmalonic aciduria; PA, propionic aciduria; IVA, isovaleric aciduria; OATD, succinyl-CoA transferase deficiency; FBP, fructose bisphosphatase; G6P, glucose-6-phosphatase; GS, glycogen synthetase; L : P, lactate : pyruvate; HMGCoA, hydroxymethylglutaryl-CoA.

chromatography) are available in many places. It is important to insist, however, that any reference laboratory used for this purpose provides not only prompt test results and reference ranges, but also an interpretation of abnormal results (Burton, 1987).

If the baby dies, an adequate diagnosis is still important in order to make adequate genetic counselling possible. A postmortem protocol for the diagnosis of genetic disease has been proposed and includes the taking of urine and serum samples, fibroblasts culture (premortem if possible) and muscle and liver biopsies (three or more of 1 cm^3 each, stored frozen on dry ice or in liquid nitrogen) (Kronick *et al.*, 1983).

Once the above clinical and laboratory data have been assembled, specific therapeutic recommendations can be made. This process is completed within 2 or 4 h and often precludes long waiting periods for sophisticated diagnostic results. On the basis of this evaluation, most patients can be classified into one of five groups (Table 4.13).

Clinical approach to aetiologies in metabolic diseases

According to the major clinical presentations (neurological deterioration 'intoxication' type, neurological deterioration 'energy deficiency' type,

Table 4.9 Persistent neonatal hypoglycaemia

Main orientating symptom	Other clinical signs	Diagnosis
Hepatomegaly with severe liver failure, hepatic necrosis (see this symptom)	Permanent short fast hypoglycaemia, moderate hyperlactacidaemia, hypoglycaemia improves easily with glucose infusion	Galactosaemia Fructosaemia Neonatal haemochromatosis Respiratory chain defects Tyrosinaemia type I (after 1 month)
With isolated hepatomegaly, lactic acidosis, ketosis	Short fast hypoglycaemia improves easily with glucose infusion. No response to glucagon	Glycogenosis type Ia,b Fructose bisphosphatase deficiency
Without hepatomegaly	Severe hypoglycaemia despite glucose infusion, no ketoacidosis	Hyperinsulinism (focal, diffuse) (dramatic response to glucagon) Growth hormone deficiency
	With ketoacidosis, dehydration, hypoglycaemia improves easily with glucose infusion	Organic acidurias Adrenal insufficiency (hyponatraemia, salt losing)
	Without ketosis, variable acidosis (lactic), muscle, cardiac symptoms, moderate cytolysis and slight hepatic insufficiency, hypoglycaemia recurs when glucose intake is decreased	Fatty acid oxidation disorders HMGCoA lyase deficiency ACTH unresponsive Adrenal insufficiency

HMGCoA, hydroxymethylglutaryl-CoA; ACTH, adrenocorticotrophic hormone.

storage disorders, cardiac injury and liver dysfunction) and to the proper use of the laboratory data described above, most patients can be assigned to one of five syndromes (Table 4.13). In our experience, type I (MSUD), type II (organic acidurias), type IVa (urea cycle defects) and non-ketotic hyperglycinaemia (the most common disease in type IVb) encompass more than 65% of the newborn with inborn errors of intermediary metabolism. The experienced clinician will, of course, have to interpret carefully the metabolic data, especially in relation to the time when they were collected and the treatments which were used. It is important to insist on the need to collect at the same time all the biological data listed in Table 4.7. Some very significant symptoms (such as metabolic acidosis, and especially ketosis) can be moderate and transient, largely depending on the symptomatic therapy. Conversely, at an advanced stage, many non-specific abnormalities (such as respiratory acidosis, severe hyperlactacidaemia, secondary hyperammonaemia) can disturb the primitive truth of the biological pattern. This is particularly true in disorders with a rapid fatal course, like urea cycle disorders in which the initial near-constant presentation of hyperammonaemia with respiratory alkalosis and without ketosis shifts rapidly to a rather non-specific picture associated with acidosis and hyperlactacidaemia (Saudubray *et al.*, 1989).

Type I: neurological distress 'intoxication type' with ketosis

Type I is represented by MSUD. It is one of the commonest amino acidopathies. After a symptom-free interval of 4 to 5 days, feeding difficulties develop and the neonate gradually becomes comatose. Generalized hypertonic episodes with opisthotonos and boxing and pedalling movements as well as slow limb elevations are constant. A maple syrup odour is present and the urine DNPH test is strongly positive, whereas urine tests for acetone may be negative. None of our patients with MSUD have had an initial blood pH less than 7.3. The diagnosis is confirmed by serum amino acid chromatography, which displays an elevation of the

Table 4.10 Neonatal hyperammonaemia

With metabolic acidosis (pH < 7.3; $P\text{CO}_2$ < 20 mmHg; BD < −10)	With ketosis (Acetest +/+++)	Organic acidurias (MMA, PA, IVA): neurological signs, leucopenia, thrombopenia Pyruvate carboxylase deficiency: lactic acidosis, high L:P ratio, hyperventilation, few neurological signs Respiratory chain defects (?): hyperammonaemia is possible but is rarely predominant
	Without ketosis (Acetest −)	HGMCoA lyase deficiency: severe fasting hypoglycaemia, characteristic organic aciduria Fatty acid oxidation disorders (GA II, CPT II, TL, LCAD): hypotonia, liver symptoms, cardiomyopathy, hypoglycaemia, hyperlactacidaemia Organic acidurias after glucose infusion Urea cycle defects at an advanced stage (with non-specific lactic acidosis)
Without metabolic acidosis	With hypoglycaemia (glucose <2 mmol/L) without ketosis	Transient neonatal hyperammonaemia: premature/dysmature baby, respiratory distress syndrome Fatty acid oxidation defects: see above
	Without hypoglycaemia	Urea cycle defects (OTC, CPS, ASS, citrullinaemia, triple H syndrome, acetyl glutamate synthetase): deep coma, severe hyper- and hypotonia, seizures; no ketosis Fatty acid oxidation defects (see above); no ketosis Organic acidurias after glucose infusion; ketosis variable MSUD (with ketosis, DNPH +)

BD, base deficit; MMA, methylmalonic aciduria; PA, propionic aciduria; IVA, isovaleric aciduria; L:P, lactate:pyruvate; HMGCoA, hydroxymethylglutaryl-CoA; GA II, glutaric aciduria type II; CPT II, carnitine palmitoyl transferase II; TL, translocase; LCAD, long-chain acyl-CoA dehydrogenase; OTC, ornithine transcarbamylase; CPS, carbamyl phosphate synthetase; ASS, arginosuccinic synthetase; MSUD, maple syrup urine disease; DNPH, dinitrophenylhydrazine.

branched-chain amino acids leucine (usually higher than 2 mmol/L), valine and isoleucine, and the presence of alloisoleucine.

Type II: neurological distress 'intoxication type' with ketoacidosis

Type II, neurological distress 'intoxication type' with ketoacidosis and hyperammonaemia, encompasses many of the organic acidurias. Between the ages of 1 and 4 days, these neonates develop feeding difficulties and deteriorate into a coma over hours to days. In contrast to patients with MSUD, these patients are acutely ill, dehydrated, acidotic with an increased anion gap and they are often hypothermic. The usual approach to such patients is first to exclude adrenogenital syndrome by carefully checking serum electrolytes, searching for high potassium and low sodium concentrations. Truncal hypotonia and peripheral hypertonia are seen but large-amplitude tremors of the limbs are the dominant abnormal movements. In isovaleric acidaemia, a potent 'sweaty feet' odour is present. The urine in the acute phase is positive for ketones, but this highly significant sign can be transient. Neutropenia and thrombocytopenia are also commonly observed and can contribute to the confusion with sepsis. Hyperammonaemia, sometimes as high as that observed in urea cycle defects, is a constant finding (Cathelineau *et al.*, 1981). Moderate hypocalcaemia is also frequent. In some patients, we have observed severe hyperglycaemia (greater than 15 mmol/L) with glycosuria before treatment with glucose. These data associated with dehydration and keto-

Table 4.11 Severe hyperlactacidaemia (>7 mmol/L) with metabolic acidosis

Without ketosis	L : P ratio high (>30)	'Acquired' hyperlactacidaemias due to tissular anoxia: collapsus, cardiac arrest, septicaemia, meningitis, toxic shock, heart hypoplasia, hepatic failure, ventricular haemorrhage Advanced stage of congenital hyperammonaemia: profound coma, hypo- and hypertonia, seizures, hyperammonaemia Severe forms of fatty acid oxidation defects (glutaric aciduria type II, CPT II, LCAD): hypotonia, cardiac failure, cardiomyopathy, severe hypoglycaemia Respiratory chain disorders: neurological signs, hypotonia, cardiomyopathy, hepatic failure, tubulopathy, macroglossia
	L : P ratio normal or low (<12)	Pyruvate dehydrogenase deficiency: minor facial dysmorphia, corpus callosum agenesia, no severe neurological sign at the onset, hyperlactacidaemia majorated by high glucose intake
With ketosis	Ketosis and hyperlactacidaemia persist or even increase, despite glucose infusion	Pyruvate carboxylase deficiency and possible Krebs cycle defects: hyperventilation, few neurological signs at the beginning, high L : P ratio contrasts with low 3OHB : AA ratio, hyperammonaemia, hypercitrullinaemia Respiratory chain defects: severe hypotonia, cardiomyopathy, neurological signs, hepatic failure, tubulopathy, macroglossia, high L : P and 3OHB : AA ratios, 3-methylglutaconic aciduria and excretion of Krebs cycle intermediates
	Ketosis and hyperlactacidaemia decreases after glucose infusion	Organic acidurias (MMA, PA, IVA, 3-hydroxyisobutyric): neurological distress, leucopenia, thrombopenia, hyperammonaemia Gluconeogenesis defects (glycogenosis type I, fructose bisphosphatase deficiency): hepatomegaly, severe fasting, hypoglycaemia

L : P, lactate : pyruvate; CPT II, carnitine palmitoyl transferase II; LCAD, long-chain acyl-CoA dehydrogenase; 3OHB : AA, 3-hydroxybutyrate : acetoacetate; MMA, methylmalonic aciduria; PA, propionic aciduria; IVA, isovaleric aciduria.

acidosis can suggest neonatal diabetes. The metabolic acidosis and other humoral disturbances observed in organic acidaemias may have adverse consequences on many different organ systems and may lead to a variety of erroneous diagnoses (Burton, 1987).

In addition to methylmalonic (MMA), propionic (PA) and isovaleric (IVA) acidurias, a large number of rare organic acidurias, presenting usually with neurological distress and metabolic acidosis, have been described in recent years as organic acid analysis techniques have become more widely available and reliable (Chalmers & Lawson, 1982). Among them, glutaric aciduria type II (GA II) or multiple acyl-CoA dehydrogenase (MAD) deficiency (Goodman *et al.*, 1987) (a variety of long-chain fatty acid oxidation defects), and hydroxymethylglutaryl-CoA lyase (HMGCoA lyase) deficiency (Wysocki & Hahnel, 1986) have many similarities with MMA, PA and IVA, except that ketosis is absent and hypoglycaemia is frequent. Patients with MAD deficiency have a 'sweaty feet' odour similar to that seen in IVA, severe lactic acidosis and can have congenital defects (Goodman *et al.*, 1987). Very rare conditions in this group are succinyl-CoA transferase deficiency (OATD) (Tildon & Cornblath, 1972; Saudubray *et al.*, 1987), biotin-dependent multiple carboxylase deficiency (MCD) due to holoenzyme synthetase deficiency (Burri *et al.*, 1981), short-chain fatty acyl-CoA dehydrogenase deficiency (Amendt *et al.*, 1987), 3-methylglutaconicuria (Divry *et al.*, 1987) and glycerol kinase deficiency (Francke *et al.*, 1987), which all display ketoacidosis. Pyroglutamic aciduria is a rare condition that can start in the first days of life with a severe metabolic acidosis but without ketosis or abnormalities of the blood glucose, lactate and ammonia. There are no severe neurological signs and the clinical picture mimics renal tubular acidosis (RTA).

The final diagnosis of all these organic acidurias is made by identifying specific abnormal metabolites by gas chromatography and mass spectroscopy

Table 4.12 Pancytopenia, thrombopenia and leuconeutropenia

	Major clinical findings	Age of onset	Disorders
Pancytopenia	With preponderant macrocytic or megaloblastic non-regenerative anaemia	First month	Pearson syndrome due to mitochondrial DNA deletion (failure to thrive, chronic diarrhoea, multivisceral failure, lactic acidosis, vacuolization of marrow precursors)
		Early infancy	Transcobalamin II deficiency (TC II) (diarrhoea, malabsorption, infection, stomatitis)
		First weeks	Methylcobalamin metabolism defects: CbL E–CbL G (failure to thrive, developmental delay)
	With ketoacidosis, hypotonia, failure to thrive, attacks of coma	Neonatal to infancy	MMA, PA, IVA, MCD (frequent infections, hyperammonaemia) Mevalonic aciduria
	With chronic diarrhoea, failure to thrive	Neonatal to infancy	Folate malabsorption TC II deficiency Pearson syndrome Mevalonic aciduria
Preponderant or isolated leuconeutropenia	Recurrent attacks of hypoglycaemia, hepatomegaly, failure to thrive, frequent infections	First month to early infancy	Glycogenosis type Ib
	Failure to thrive, osteoporosis, recurrent attacks of hyperammonaemia, interstitial pneumonia, hepatosplenomegaly, chronic diarrhoea, orotic acid excretion (moderate), thrombopenia	After weaning, early in infancy	Lysinuric protein intolerance
	Thrombopenia, hyperammonaemia, ketoacidosis, hypotonia, attacks of coma	Infancy	Organic acidurias (MMA, PA, IVA) Mevalonic aciduria
Preponderant thrombopenia	Haemolytic uraemic syndrome, ketoacidosis, cardiomyopathy, megaloblastic anaemia	First weeks to first months	Cobalamin metabolism defect (CbL C)
	Leuconeutropenia, hyperammonaemia, ketoacidosis	Infancy	Organic acidurias (MMA, PA, IVA)

CbL, cobalamin metabolism defect; MMA, methylmalonic aciduria; PA, propionic aciduria; IVA, isovaleric aciduria; MCD, multiple carboxylase deficiency.

of blood and urine. Free carnitine plasma concentrations are always decreased, with abnormal excretion of specific acylcarnitines. By contrast, plasma and urine amino acid chromatography are often normal or non-specific with, for example, a profile showing a slight increase in glycine.

Type III: lactic acidosis with neurological distress 'energy deficiency' type

The clinical presentation of these babies is very diverse. Unlike the previous disease category in which moderate acidosis is noted during the evalu-

Table 4.13 Five neonatal types of inherited 'metabolic distress'

Type	Clinical type	Acidosis/ketosis	Other signs	Most usual diagnosis	Elective methods of investigation
I	Neurological distress 'intoxication' type, abnormal movements, hypertonia	Acidosis 0 DNPH +++ Acetest 0/±	NH_3 N or $\nearrow$ ± Lactate N Blood count N Glucose N Calcium N	MSUD (special odour)	Amino acid chromatography (plasma, urine)
II	Neurological distress 'intoxication' type, dehydration	Acidosis ++ Acetest ++ DNPH 0/±	NH_3 $\nearrow$ +/++ Lactate N or $\nearrow$ ± Blood count: leucopenia thrombopenia Glucose N or $\nearrow$ Calcium N or $\searrow$	Organic acidurias (MMA, PA, IVA, MCD) Ketolytic defects (Tables 4.5, 4.8, 4.10, 4.12)	Organic acid chromatography by GLCMS (urine, plasma) Carnitine (plasma) Carnitine esters (urine, plasma)
	Neurological distress 'energy deficiency' type with hepatic disturbances	Acidosis ++/± Acetest 0 DNPH 0	NH_3 $\nearrow$ ±/++ Lactate $\nearrow$ /++± Blood count N Glucose $\searrow$ +/++ Calcium N or $\searrow$ +	Fatty acid oxidation and ketogenesis defects (Tables 4.6, 4.9)	See above Loading test Fasting test Fatty acid oxidation studies on lymphocytes or fibroblasts
III	Neurological distress 'energy deficiency' type, polypnoea, hypotonia	Acidosis +++/+ Acetest ++/0 Lactate +++/+	NH_3 NO_2 $\nearrow$ ± Blood count: anaemia or N Glucose N or $\searrow$ ± Calcium N	'Congenital lactic acidosis' (PC, PDH, Krebs cycle, respiratory chain) MCD (Tables 4.1, 4.6, 4.11)	Plasma redox potential states (L:P, OHB:AA ratios) Organic acid chromatography (urines) Polarographic studies Enzyme assays (muscle, lymphocytes or fibroblasts)
IVa	Neurological distress 'intoxication' type, moderate hepatic disturbances, hypotonia, seizures, coma	Acidosis 0 (alkalosis) Acetest 0 DNPH 0	NH_3 $\nearrow$ +/+++ Lactate NO_2 $\nearrow$ + Blood count N Glucose N Calcium N	Urea cycle Triple H Fatty acid oxidation defects (GA II, CPT II, LCAD, LCHAD) (Table 4.10)	AAC (plasma, urine) Orotic acid (urine) Liver or intestine enzyme studies (CPS, OTC)
IVb	Neurological distress, seizures, myoclonic jerks, severe hypotonia	Acidosis 0 Acetest 0 DNPH 0	NH_3 N Lactate N Blood count N Glucose N	NKH SO ± XO Pyridoxine dependency Peroxisomal disorders (Table 4.1)	AAC (NKH, SO) VLCFA, phytanic acid in plasma (PZO)
IVc	Storage disorders, coarse facies, hepatosplenomegaly, ascites, hydrops fetalis, macroglossia, bone changes, cherry red spot, vacuolated lymphocytes	Acidosis 0 Acetest 0 DNPH 0	NH_3 N Lactate N Blood count N Glucose N Hepatic signs	GM_1-gangliosidosis ISSD (sialidosis type II) I cell disease Niemann–Pick type IA MPS VII Galactosialidosis (Tables 4.1–4.3)	Enzyme studies (lymphocytes, fibroblasts)
V	Hepatomegaly, hypoglycaemia	Acidosis ++/+ Acetest +	NH_3 N Lactate $\nearrow$ +/++ Blood count N Glucose $\searrow$ ++	Glycogenosis types I and III Fructose bisphosphatase (Tables 4.3, 4.9)	Fasting test Loading test Enzyme studies (lymphocytes, fibroblasts)
	Hepatomegaly, jaundice, liver failure, hepatocellular necrosis	Acidosis +/0 Acetest +/0	NH_3 NO_2 $\nearrow$ + Lactate $\nearrow$ +/++ Glucose NO_2 $\searrow$ ++	Fructosaemia, galactosaemias Tyrosinosis type I Neonatal haemochromatosis Respiratory chain disorders (Tables 4.2, 4.3, 4.9)	Enzyme studies (fructosaemia, galactosaemia) Organic acids and enzyme studies (tyrosinaemia type I)

continued on p. 82

Table 4.13 *Continued*

Type	Clinical type	Acidosis/ketosis	Other signs	Most usual diagnosis	Elective methods of investigation
	Hepatomegaly, cholestatic jaundice, ± failure to thrive, ± chronic diarrhoea	Acidosis 0 Ketosis 0	NH_3 N Lactate N Glucose N	α_1-Antitrypsin Inborn errors of bile acid metabolism Peroxisomal disorders (Tables 4.3, 4.4)	Protein electrophoresis Organic acid chromatography (plasma, urine, duodenal juice) VLCFA, phytanic acid (plasma), pipecolic acid (plasma, urine)
	Hepatosplenomegaly, 'storage' signs, ± failure to thrive, ± chronic diarrhoea	Acidosis 0 Ketosis 0	NH_3 N Lactate N or ↗ ± Glucose N	Storage disorders (Tables 4.3, 4.4)	Oligosaccharides, sialic acid Mucopolysaccharides (urine) Enzyme studies (leucocytes, fibroblasts)

N, normal; 0, absent (acidosis) or negative (acetest, DNPH); ↗, elevated; ↘ decreased; DNPH, dinitrophenylhydrazine; MSUD, maple syrup urine disease; MMA, methylmalonic aciduria; PA, propionic aciduria; IVA, isovaleric aciduria; MCD, multiple carboxylase deficiency; GLCMS, gas–liquid chromatography–mass spectroscopy; PC, pyruvate carboxylase; PDH, pyruvate dehydrogenase; L:P, lactate:pyruvate; OHB:AA, 3-hydroxybutyrate:acetoacetate; GA II, glutaric aciduria type II; CPT II, carnitine palmitoyl transferase II; LCAD, long-chain acyl-CoA dehydrogenase; LCHAD, long-chain 3-hydroxyacyl-CoA dehydrogenase; AAC, amino acid chromatography; CPS, carbamyl phosphate synthetase; OTC, ornithine transcarbamylase; NKH, non-ketotic hyperglycinaemia; SO, sulphite oxidase; XO, xanthine oxidase; VLCFA, very long-chain fatty acid; PZO, peroxisomal disorders; GM_1, GM_1-gangliosidosis; ISSD, infantile sialic acid storage disease; MPS, mucopolysaccharidosis. For an explanation of ±, +, etc., see Table 4.14.

Table 4.14 An explanation of symbols used in Table 4.13

		Blood		
Symbol	Explanation	NH_3 (μmol/L)	Lactate (mmol/L)	Glucose (mmol/L)
N	Normal	<80	<1.5	3.5–5.5
±	Slight	80–100	1.5–3	3.5–2
+	Moderate	100–150	3–7	(3.5–2)
++	Marked	150–300	7–10	2–1
+++	Massive	>300	>10	<1

ation of an acutely ill comatose neonate, the main medical preoccupation in group III patients is the acidosis itself, which clinically may be surprisingly well tolerated. However, the acidosis can at times be mild. The blood pH is usually normal until lactate concentrations of 5 mmol/L are reached. An elevated anion gap exists, and can be explained in part by the presence of equimolar amounts of lactic acid in the blood. Often, in the absence of adequate treatment, the acidosis recurs soon after bicarbonate therapy.

If a high lactic acid concentration is found, it is urgent to rule out readily treatable causes, especially hypoxia. Ketosis is present in most of the primary lactic acidaemias but is absent in acidosis secondary to tissue hypoxia. MCD may present as lactic acidosis, and biotin therapy is indicated in all patients with lactic acidosis of unknown cause after baseline blood and urine samples have been taken. Primary lactic acidoses form a complex group (Robinson & Sherwood, 1984). A definite diagnosis is often elusive and is attempted with specific enzyme assays and by considering metabolite levels, redox potential states and fluxes under fasting and fed conditions (see Hyperlactacidaemias; Table 4.11). Defects most frequently demonstrated are PC deficiency, PDH deficiency, respiratory chain disorders (complex I and IV) and MCD, which has already been discussed with group II diseases. Many cases remain unexplained. The PC deficiencies that we have investigated have had a stereotypical biochemical pattern with an elevated L:P ratio contrasting with a reduced 3OHB:AA ratio, moderate citrullinaemia and hyperammonaemia and hepatic dysfunction. This pattern has been called 'French phenotype' by Robinson *et al.* (1987). Respiratory chain disorders are frequently observed in the neonatal period. Since the early descriptions (Van Biervliet *et al.*, 1977), a number of cases have been described (Di Mauro *et al.*, 1987). The most frequent symptoms are severe generalized hypotonia (Table 4.1), dilated cardiomyopathy of hypokinetic type (Table 4.6), rapid neurological deterioration, respiratory failure and severe lactic acidaemia. Some patients displayed facial dysmorphia and malformations, as has also been observed in infants with a deficiency of the PDH complex (Aleck *et al.*, 1988; Clayton & Thompson, 1988).

Type IV

Type IVa: neurological distress 'intoxication' type with hyperammonaemia and without ketoacidosis: urea cycle defects

As mentioned above, this group of patients is one of the most important among neonatal inborn errors of metabolism. Primary hyperammonaemias due to urea cycle defects have a variable symptom-free interval, which lasts sometimes only a matter of hours. A brief hypertonic period and hiccups may occur, after which a profound hypotonic coma rapidly develops and cardiocirculatory function may be compromised. Coagulation factor depletion, elevated serum aminotransferases and hepatomegaly are frequent. The blood ammonia rises precipitously to levels of 400–2000 mol/L or more. Respiratory alkalosis (pH $>$ 7.4) and moderate hyperlactacidaemia are frequently observed. An important diagnostic clue to separate urea cycle defects from organic acidurias with hyperammonaemia is the universal absence of ketonuria. As already stated, at an advanced stage, neurovegetative disorders rapidly give rise to non-specific findings including acidosis and hyperlactacidaemia (Table 4.10).

The two principal urea cycle disorders, which have non-diagnostic amino acid chromatograms, are ornithine transcarbamylase (OTC) deficiency and carbamyl phosphate synthetase (CPS) deficiency. The former is the only sex-linked congenital hyperammonaemia, the others being autosomal recessive. Massive orotic acid excretion is present and the blood citrulline level is very low. CPS deficiency is initially detected by the negative findings of a non-specific amino acid chromatogram and the absence of orotic acid excretion. Enzyme diagnosis by liver biopsy is the only definitive diagnostic technique. Citrullinaemia, arginosuccinic aciduria and argininaemia are diagnosed by amino acid chromatography which demonstrates the accumulation of citrulline, arginosuccinate or arginine respectively. In the 'triple H' syndrome, hyperammonaemia, hyperornithinaemia and homocitrullinaemia are present. An especially important diagnostic consideration is transient hyperammonaemia of the neonate, in which the patient, often premature and having mild respiratory distress syndrome, develops a deep coma and severe hyperammonaemia, which disappears permanently if initial treatment is successful (Ballard *et al.*, 1978).

Fatty acid oxidation disorders can also, though rarely, present in the neonatal period with hyperammonaemia and can mimic urea cycle disorders. They are mostly associated with hypoglycaemia, hepatic dysfunction, muscular and cardiac symptoms or sudden infant death.

Type IVb: neurological deterioration 'energy deficiency' type without ketoacidosis and without hyperammonaemia

Non-ketotic hyperglycinaemia displays a very typical, although non-pathognomonic, pattern. It is characterized by coma, hypotonia and myoclonic jerks at birth or after a few hours in a neonate who did not experience perinatal hypoxia. A burst suppression EEG pattern (Mises *et al.*, 1982) is always present. The diagnosis rests upon the demonstration of elevated serum glycine levels and especially an elevated CSF: serum glycine ratio. The very rare D-glyceric acidaemia shares most of the signs of non-ketotic hyperglycinaemia, including hyperglycinaemia in some cases, whereas others present different patterns with metabolic acidosis.

The clinical spectrum of SO and combined SO and xanthine oxidase (XO) deficiencies (Wadman *et al.*, 1983) includes hypotonia, seizures, myoclonic jerks, microcephaly and dysmorphic features. Lens dislocation may occasionally be noted as early as during the first month of life. This disorder is probably under-diagnosed as its clinical pattern shares many similarities with common acute fetal distress. In combined SO and XO deficiencies owing to an abnormal molybdenum cofactor, the uric acid concentration is very low in plasma and urine, which is a very useful tool for diagnosis. In both SO deficiencies, sulphites are found in fresh urine in high concentrations (test with Sulfitest, Merck). Amino acid chromatography, performed on immediately deproteinized plasma and on fresh urine, shows a specific profile with high sulphite concentrations in the form of sulphocysteine, whereas the cystine concentration is close to zero.

The common symptoms of PZO presenting in the neonatal period are an absence of a symptom-free interval, severe generalized hypotonia, early-onset epileptic seizures and cranio-facial dysmorphism (Poll-The *et al.*, 1987a). Some patients can display as inaugural signs hepatomegaly, jaundice, liver failure and a failure to thrive without marked neurological dysfunction. Retinitis pigmentosa should be systematically searched for. Diagnosis requires special investigations including very long-chain fatty acid, plasmalogen, phytanic, pipecolic and bile acids acid determinations in plasma, urine and fibroblasts. The most frequent conditions are Zellweger syndrome and neonatal adrenoleucodystrophy. Many other rare variants have been recently described.

In most diseases of intermediary metabolism, convulsions are a late manifestation. However, they may be important elements in the clinical presentation of most patients with non-ketotic hyperglycinaemia, SO deficiency and PZO described above.

Type IVc: storage disorders without metabolic disturbances

Only a few lysosomal disorders are expressed clinically in the neonatal period. They can be associated with hydrops fetalis, neonatal ascites and oedema (GM_1-gangliosidosis, Gaucher's disease, mucopolysaccharidosis type VII (MPS VII), sialidosis, galactosialidosis, sialuria, Niemann–Pick type C) (see Tables 4.1–4.3).

Type V: hepatomegaly and liver dysfunction

The clinical presentation of type V diseases is different from the preceding ones.

In a first group of disorders, hypoglycaemic seizures are often the presenting sign, and hepatomegaly, ketosis and lactic acidosis are present. The baby improves dramatically with intravenous glucose administration. The main diseases of this group are glucose-6-phosphatase (G6P) deficiency (type I glycogen storage disease), glycogenosis type III and fructose-1,6-bisphosphatase deficiency. Until now, the clinical presentation of phosphoenolpyruvate carboxykinase deficiency has not been clearly defined in the neonatal period. Marked hepatocellular dysfunction is uncommon, but may occur, especially in fructose-1,6-bisphosphatase deficiency.

In a second group of disorders, like tyrosinaemia type I, galactosaemia, fructosaemia, (if the diet contains fructose) or neonatal haemochromatosis, hypoglycaemia is usually an incidental finding in a clinical setting dominated by jaundice and other evidence of liver failure with hepatocellular necrosis. Severe neonatal hepatocellular insufficiency associated with hyperlactacidaemia has been recently described as the presenting sign in respiratory chain disorders (Cormier *et al.*, 1991). Gluconeogenesis is active only in the fasting state; thus, the patient becomes hypoglycaemic only during fasting, and is safe from harm if a continuous supply of oral or intravenous glucose is provided. Often, these infants become symptomatic only when feeding intervals are increased at several months of age. Exploration of these conditions involves enzyme assays in fibroblasts, blood cells or liver tissue, and must often be performed under expert supervision, as precipitous drops in the blood glucose level are frequent.

In a third group of disorders, the clinical presentation is dominated by an isolated cholestatic jaundice with hepatomegaly. α_1-Antitrypsin deficiency, inborn errors of bile acid metabolism or variant forms of PZO should be first considered.

References

Aleck, K.A., Kaplan, A.M., Sherwood, W.G. *et al.* (1988) In utero central nervous system damage in pyruvate dehydrogenase deficiency. *Arch Neurol* **45**, 987–989.

Amendt, B.A., Greene, C., Sweetman, L. *et al.* (1987) Short chain acyl-CoA dehydrogenase deficiency. Clinical and biochemical studies in two patients. *J Clin Invest* **79**, 1303–1309.

Ballard, R.A., Vinocour, B., Reynolds, J.W. *et al.* (1978) Transient hyperammonemia of the preterm infant. *N Engl J Med* **299**, 920–925.

Bankier, A., Turner, M. & Hopkins, I.J. (1983) Pyridoxine dependent seizures. A wider clinical spectrum. *Arch Dis Child* **58**, 415–418.

Barnard, J.A. III & Manci, E. (1991) Idiopathic neonatal iron-storage disease. *Gastroenterology* **101**, 1420–1427.

Burri, B.J., Sweetman, L. & Nyhan, W.L. (1981) Mutant holocarboxylase synthetase: evidence for the enzyme defect in early biotin-responsive multiple carboxylase deficiency. *J Clin Invest* **68**, 1491–1495.

Burton, B.K. (1987) Inborn errors of metabolism: the clinical diagnosis in early infancy. *Pediatrics* **79**, 359–369.

Cathelineau, L., Briand, P., Ogier, H. *et al.* (1981) Occurrence of hyperammonemia in the course of 17 cases of methylmalonic acidemia. *J Pediatr* **99**, 279–280.

Chalmers, R.A. & Lawson, A.H. (1982) *Organic Acids in Man*, pp. 1–523. Chapman and Hall, London.

Clayton, P.T. (1991) Inborn errors of bile acid metabolism. *J Inher Metab Dis* **14**, 478–496.

Clayton, P.T. & Thompson, E. (1988) Dysmorphic syndromes with demonstrable biochemical abnormalities. *J Med Genet* **25**, 463–472.

Cormier, V., Rustin, P., Bonnefont, J.P. *et al.* (1991) Hepatic failure in disorders of oxidative phosphorylation with neonatal onset. *J Pediatr* **119**, 951–954.

Di Mauro, S., Bonilla, E., Zeviani, M. *et al.* (1987) Mitochondrial myopathies. *J Inher Metab Dis* **10** (Suppl. 1), 113–128.

Divry, P., Vianey-Liaud, C., Mory, O. *et al.* (1987) A methylglutaconic aciduria familial neonatal form with fatal onset. *J Inher Metab Dis* **10** (Suppl. 2), 286–289.

Francke, U., Harper, J.F., Darras, B.T. *et al.* (1987) Congenital adrenal hypoplasia, myopathy, and glycerol kinase deficiency: molecular genetic evidence for deletions. *Am J Hum Genet* **40**, 212–227.

Goodman, S.I., Frerman, F.E. & Loehr, J.P. (1987) Recent progress in understanding glutaric acidemias. *Enzyme* **38**, 76–79.

Jaeken, J., Stibler, H. & Hagberg, B. (1991) The carbohydrate-deficient glycoprotein syndrome: a new inherited multisystemic disease with severe nervous system involvement. *Acta Paed Scand* **375** (Suppl.), 5–71.

Knisely, A.S. (1992) Neonatal hemochromatosis. *Adv Pediatr* **39**, 383–402.

Kronick, J.B., Scriver, C.R., Goodyer, P.R. *et al.* (1983) A perimortem protocol for suspected genetic disease. *Pediatrics* **71**, 960–963.

Mandel, H., Meiron, D., Schutgens, R.B.H. *et al.* (1992) Infantile Refsum disease: gastrointestinal presentation of a peroxisomal disorder. *J Ped Gastr Nutr* **14**, 83–85.

Mises, J., Moussalli-Salefranque, F., Laroque, M.L. *et al.* (1982) EEG findings as an aid to the diagnosis of neonatal non-ketotic hyperglycinemia. *J Inher Metab Dis* **5** (Suppl. 2), 117–120.

Mitzuta, K., Kashimoto, E., Tsutou, A. *et al.* (1984) A new type of glycogen storage disease caused by deficiency of cardiac phosphorylase b kinase. *Biochem Biophys Res Commun* **119**, 582–587.

Munnich, A., Rustin, P., Rötig, A. *et al.* (1992) Clinical aspects of mitochondrial disorders. *J Inher Metab Dis* **15**, 448–455.

Odièvre, M. (1991) Clinical presentation of metabolic liver disease. *J Inher Metab Dis* **14**, 526–530.

Parrot-Roulaud, F., Carré, M., Lamirau, T. *et al.* (1991) Fatal neonatal hepatocellular deficiency with lactic acidosis: a defect of the respiratory chain. *J Inher Metab Dis* **14**, 289–292.

Poll-The, B.T., Saudubray, J.M., Ogier, H. *et al.* (1987a) Clinical approach to inherited peroxisomal disorders. In: *Human Genetics* (eds F. Vogel & K. Sperling), pp. 345–351. Springer-Verlag, Berlin.

Poll-The, B.T., Saudubray, J.M., Ogier, H. *et al.* (1987b) Infantile Refsum disease: an inherited peroxisomal disorder. Comparison with Zelwegger syndrome and neonatal adrenoleukodystrophy. *Eur J Pediatr* **146**, 477–483.

Robinson, B.H. & Sherwood, W.G. (1984) Lactic acidemia. *J Inher Metab Dis* **7** (Suppl. 1), 69–73.

Robinson, B.H., Oei, J., Saudubray, J.M. *et al.* (1987) The French and North American phenotypes of pyruvate carboxylase deficiency. Correlation with biotin-containing protein by 3H-biotin incorporation, 35Sstreptavidin labeling, and Northern blotting with a cloned cDNA probe. *Am J Hum Genet* **40**, 50–59.

Rotig, A., Cormier, V., Blanche, S. *et al.* (1990) Pearson's marrow-pancreas syndrome. *J Clin Invest* **86**, 1601–1608.

Saudubray, J.M., Ogier, H., Bonnefont, J.P. *et al.* (1989) Clinical approach to inherited metabolic diseases in the neonatal period: a 20-year survey. *J Inher Metab Dis* **12** (Suppl. 1), 25–41.

Saudubray, J.M., Specola, N., Middleton, B. *et al.* (1987) Hyperketotic states due to inherited defects of ketolysis. *Enzyme* **38**, 80–90.

Servidei, S., Metlay, L.A., Chodosh, J. *et al.* (1988) Fatal infantile cardiopathy caused by phosphorylase b kinase deficiency. *J Pediatr* **113**, 82–85.

Settergren, G., Lindblad, B.S. & Persson, B. (1976) Cerebral blood flow and exchange of oxygen, glucose, ketone bodies, lactate, pyruvate and amino acids in infants. *Acta Paed Scand* **65**, 343–353.

Tildon, J.T. & Cornblath, M. (1972) Succinyl-CoA:3-ketoacidosis in infancy. *J Clin Invest* **51**, 493–498.

Van Biervliet, J.P.G.M., Bruinvis, L., Ketting, D. *et al.* (1977) Hereditary mitochondrial myopathy with lactic acidemia, a DeToni–Fanconi–Debré syndrome and a defective respiratory chain in voluntary muscle. *Pediatr Res* **1**, 1088–1093.

Vilaseca, M.A., Briones, P., Ribes, A. *et al.* (1991) Fatal hepatic failure with lactic acidemia, Fanconi syndrome and defective activity of succinate:cytochrome C reductase. *J Inher Metab Dis* **14**, 285–288.

Wadman, S.K., Duran, M., Breemer, F.A. *et al.* (1983) Absence of hepatic molybdenum cofactor: an inborn error of metabolism leading to a combined deficiency of sulphite oxidase and xanthine dehydrogenase. *J Inher Metab Dis* **6**, 78–83.

Wysocki, S.J. & Hahnel, R. (1986) 3-Hydroxy-3-methylglutaryl-coenzyme A lyase deficiency. *J Inher Metab Dis* **9**, 225–233.

5: Hypoglycaemia and Hyperammonaemia

V. WALKER

(a) Hypoglycaemia

Introduction

Hypoglycaemia is a common paediatric problem, particularly among newborns. There are many more causes than in adults. Most lead to inadequate glucose production from glycogen and gluconeogenesis when fasting. Hyperinsulinism is relatively uncommon, except among insulin-treated diabetics. A prompt diagnosis is essential and must be achieved safely, with economic use of available tests. This requires knowledge of the homeostatic mechanisms which maintain a normal blood glucose in early life.

Definition

After the first 72 h of life (term babies) and first week (preterm babies), hypoglycaemia is generally defined as a plasma or capillary whole blood glucose concentration of less than 2.5 mmol/L. Lower concentrations are found frequently among the newborn, and published definitions of neonatal hypoglycaemia vary widely (Koh *et al.*, 1988b). This is currently a highly controversial issue (Cornblath *et al.*, 1990). There is a growing consensus, however, that levels above 2.5 mmol/L are desirable neonatally: neurophysiological disturbances (Koh *et al.*, 1988a) and increased plasma adrenaline (Pryds *et al.*, 1990) have been observed at lower concentrations, and preterm babies with plasma levels below 2.5 mmol/L for several days had lower developmental scores when tested at 18 months (Lucas *et al.*, 1988) (see also Chapter 3).

Symptoms and pathology

The brain is an obligate glucose user, with the ketone bodies, acetoacetate and 3-hydroxybutyrate acting as a major alternative fuel only in prolonged fasting. The brain of a term newborn may need 5.5 mg/kg per min of glucose, and that of a 6-year-old at least 3 mg/kg per min. This represents 60–80% of the daily output of glucose from the liver (Bier *et al.*, 1977). Moderate hypoglycaemia may evoke a significant stress response and behavioural changes, but if prolonged and severe, fits, coma and permanent neurological damage result. Clinical signs in the newborn are non-specific and include jitteriness, hypotonia, irregular respirations, apnoea and, if severe, convulsions. This risk is enhanced by other factors found frequently in hypoglycaemic disorders: lack of ketones, hypoxia, changes in cerebral blood flow and the presence of toxic metabolites. Excessive release of excitatory neurotransmitters has been postulated to cause hypoglycaemic symptoms (Sieber & Traystman, 1992). They cannot be explained by energy failure, which is a late event (Cornblath *et al.*, 1990).

Normal control of glucose homeostasis

Plasma glucose is maintained from glucose absorbed from the gut and, in the postabsorptive (fasting) state, by hydrolysis of glycogen and synthesis from 3-carbon substrates, pyruvate, lactate, alanine and glycerol (gluconeogenesis) (Figs 5.1 & 5.2). Gluconeogenesis becomes increasingly important as the glycogen reserves are depleted. Using tracer studies, glucose production rates, 3 to 6 h after feeding, were estimated to be (mg/kg per min, mean

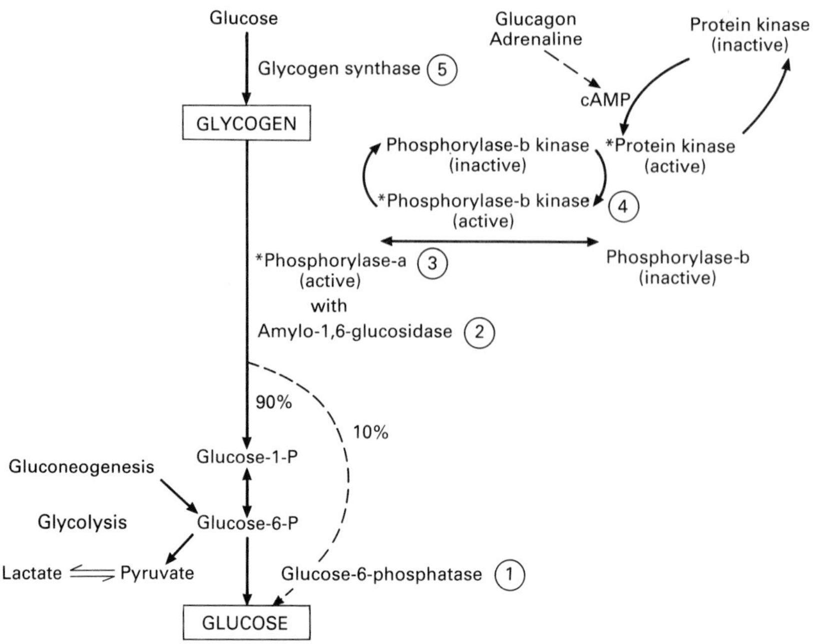

Fig. 5.1 Hepatic glycogen metabolism. The circles 1 to 5 indicate enzymes which may be deficient in hepatic glycogenoses associated with hypoglycaemia. The *active* forms of enzymes marked with * are phosphorylated; the *inactive* forms are dephosphorylated. After Aynsley-Green *et al.* (1977), Dunger and Leonard (1982), Fernandes (1990), Shin (1990), Smit *et al.* (1990) and Maire *et al.* (1991).

± SE): neonates: preterm, 5.46 ± 0.31; term, 6.07 ± 0.46; children under 6 years: 7.1 ± 0.27; late childhood: 5.4 ± 0.28; adults: 2.28 ± 0.23. It was proposed that the large brain to body mass of young children explained their high glucose requirement (Bier *et al.*, 1977).

Glycogenolysis and gluconeogenesis are activated by changes in substrate concentrations and by the hormones glucagon, adrenaline, cortisol and growth hormone (indirectly) in the presence of low insulin levels. An increased glucagon:insulin ratio is particularly important. Activity of the key gluconeogenic enzyme, glucose-1,6-bisphosphatase, is tightly controlled by a recently discovered compound, glucose-2,6-bisphosphate (Pilkis *et al.*, 1986; Hers, 1990; Van den Berghe, 1991).

Mobilization and β-oxidation of endogenous fatty acids are critical for gluconeogenesis (Girard, 1990), through enhanced production of acetyl-CoA and reduced nicotinamide adenine dinucleotide ($NADH_2$) in the liver: acetyl-CoA is a cofactor for pyruvate carboxylase; acetyl-CoA and $NADH_2$ inhibit pyruvate dehydrogenase, and thereby direct pyruvate to gluconeogenesis; $NADH_2$ is available for gluconeogenesis. In addition, glycerol, released during lipolysis, is a gluconeogenic substrate.

Unlike other organs (except erythrocytes), the brain does not need insulin to take up glucose. Low levels of insulin during fasting, together with insulin resistance, attributed to actions of adrenaline and growth hormone, divert glucose from peripheral tissues. This helps to conserve glucose for the brain. As fasting continues, the brain requirement for energy is met increasingly by ketones produced from fat oxidation (Robinson & Williamson, 1980).

Developmental changes

In utero, glucose, amino acids, glycerol, ketones and free fatty acids (FFAs) cross the placenta. Growth hormone and insulin do not. Almost all fetal glu-

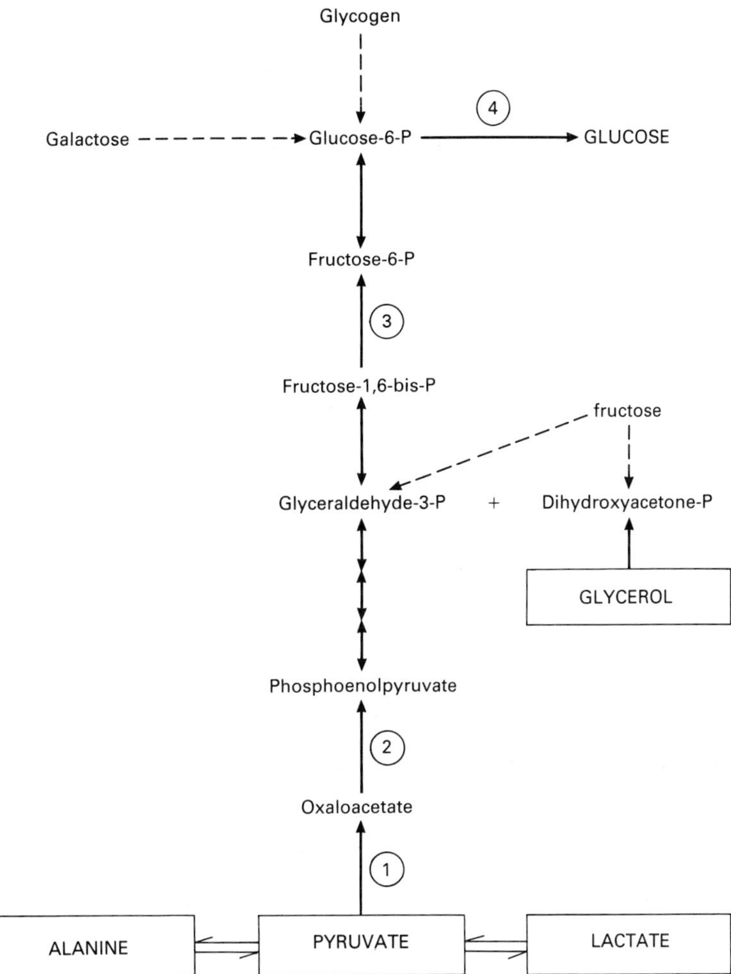

Fig. 5.2 The gluconeogenic pathway showing the four enzymes unique to this pathway. 1, Pyruvate carboxylase; 2, Phosphoenolpyruvate carboxykinase; 3, fructose-1,6-bisphosphatase; 4, glucose-6-phosphatase. The gluconeogenic substrates are shown in boxes.

cose is maternal in origin. Metabolic processes are dominated by anabolism: glycogen synthesis is activated by insulin and increases significantly in the third trimester. Gluconeogenesis and ketogenesis are normally very low in fetal liver, reflecting low levels of gluconeogenic enzymes, particularly phosphoenolpyruvate carboxykinase (PEPCK), and decreased hepatic receptors for glucagon (Blackburn & Loper, 1992; Girard *et al.*, 1992).

With birth, the maternal glucose supply is lost, secretion of glucagon and catecholamines increases and of insulin decreases, probably reflecting increased sympathetic nervous system stimulation by stresses associated with birth. Receptors for insulin decrease and for glucagon increase, and the high glucagon : insulin ratio induces transcription of the PEPCK gene (Blackburn & Loper, 1992; Girard *et al.*, 1992). Gluconeogenesis, glycogenolysis, lipolysis and ketogenesis increase dramatically. Fat oxidation contributes significantly to the term baby's energy in the first 72 h of life, reflected in a fall in respiratory quotient from 1.0 to 0.7. It then declines rapidly. Ketogenesis may be limited in preterm babies (Hawdon *et al.*, 1992). Blood glucose levels of well, milk-fed newborns fall to a nadir at around 4–6 h of age (mean (SEM): term babies 3.2 (0.3) mmol/L; preterm 2.5 (0.3) mmol/L) and then stabilize (Hawdon *et al.*, 1992). With regular feeding, cyclical

Table 5.1 Plasma concentrations of intermediary metabolites in normal fasting children

Duration of fast (h) (Source)	Glucose (mmol/L)	Lactate (mmol/L)	Free fatty acids (mmol/L)	3-Hydroxybutyrate (mmol/L)	Acetoacetate (µmol/L)
1–12 months					
15 (B)	4.7 (3.9–5.3)	1.8 (1.1–2.3)	1.0 (0.5–1.6)	0.4 (0.1–1.0)	
20 (B)	3.9 (3.5–4.6)	1.3 (0.9–1.8)	0.9 (0.6–1.3)	1.1 (0.5–2.3)	
24 (B)	3.6 (2.7–4.5)	1.4 (0.8–2.4)	1.3 (1.1–1.6)	1.8 (1.1–2.8)	
24 (S)	2.9 (0.3)	2.0 (0.3)	1.5 (0.2)	2.5 (0.5)	240 (110)
1–8 years					
14 (L)	4.3	0.86	1.0	0.3	115
15 (B)	4.4 (3.5–4.8)	1.0 (0.8–1.5)	1.1 (0.6–1.5)	0.6 (<0.1–0.9)	
20 (B)	3.5 (2.8–4.3)	1.1 (0.5–1.7)	1.7 (0.9–2.6)	1.8 (0.8–2.6)	
24 (B)	3.3 (2.8–3.8)	1.2 (0.7–1.6)	2.1 (1.1–2.8)	2.5 (1.7–3.2)	
24 (L)	3.5	1.75	2.0	2.1	551
30 (H)	2.9 (0.2)			3.7 (0.4)	
36 (K)	2.7 (0.4)		2.3 (0.7)	3.8 (0.4)	
6–18 years					
14 (L)	4.4	0.8	0.1	0.1	93
15 (B)	4.7 (4.4–4.9)	0.9 (0.6–0.9)	0.7 (0.2–1.1)	0.1 (<0.1–0.3)	
20 (B)	4.3 (3.8–4.9)	0.7 (0.6–0.9)	1.0 (0.6–1.3)	0.4 (<0.1–0.8)	
24 (B)	3.8 (3.0–4.3)	0.7 (0.4–0.9)	1.4 (1.0–1.8)	0.9 (0.5–1.3)	
36 (K)	3.8 (0.7)		1.5 (0.4)	2.1 (0.7)	
40 (L)	3.4	1.6	1.8	3.7	723

Sources and ranges shown: B, Bonnefont *et al.* (1990), mean and 10th to 90th percentiles; H, Haymond *et al.* (1982), mean (SEM); K, Kerr *et al.* (1983), mean (SD); L, Lamers *et al.* (1985), median; S, Stanley and Baker (1976), mean (SEM).

changes in insulin and glucose are established by around 5 days of age in term babies, but later in preterm infants. Secretion of digestive hormones and peptides during feeding promotes development of the enteroinsular axis.

Normal young children tolerate fasting less well than those over 6 or 7 years of age, probably because their glycogen reserves are smaller, and their glucose requirement greater. There is an active ketogenesis from 15 h of fasting (Bonnefont *et al.*, 1990), and hormonal and metabolic responses are greater in younger than older children (Kerr *et al.*, 1983). This is apparent in Table 5.1.

Causes of hypoglycaemia

The major causes are summarized in Table 5.2.

Transient neonatal hypoglycaemia

Transient hypoglycaemia is common during the first 72 h of life, particularly among babies of low birth weight. Plasma glucose concentrations below 1.6 mmol/L were recorded in as many as 28% of babies with birth weight below 1850 g, and levels less than 2.6 mmol/L in 67% (Lucas *et al.*, 1988). The risk is increased if babies are born small for gestational age because of intrauterine growth retardation. Hypoglycaemia occurs because the amount of glucose supplied by feeding, glycogenolysis and gluconeogenesis does not meet the requirements. Many factors may contribute to a reduced supply: delay in instituting oral or intravenous feeding, because of clinical problems after delivery; low glycogen stores at birth; the time

Table 5.2 Causes of hypoglycaemia

Inadequate hepatic glycogenolysis and/or gluconeogenesis
Transient neonatal hypoglycaemia
Idiopathic ketotic hypoglycaemia
Inherited metabolic defects:
Glycogen storage diseases
Gluconeogenic defects
Defects of fat oxidation and ketogenesis
Other organic acid disorders
Hereditary fructose intolerance
Galactosaemia
Tyrosinaemia type I
Severe liver damage
Acute liver failure
Reye's syndrome
Hormone deficiencies
Glucocorticoids
Growth hormone
Poisoning with ethanol, methanol or salicylates
Miscellaneous
Falciparum malaria
Hyperinsulinism
Transient neonatal
Infants of diabetic mothers
Intrapartum maternal glucose infusion
Maternal β-sympathomimetics for preterm labour
Erythroblastosis fetalis
Birth-asphyxiated preterm babies
Beckwith–Wiedemann syndrome
Nesidioblastosis
Administration of insulin; hypoglycaemic drugs; quinine
Insulin autoantibodies

lag before gluconeogenesis is fully activated by induction of PEPCK; low sensitivity of the liver to glucagon stimulation soon after birth (Girard, 1990; Blackburn & Loper, 1992; Hawdon *et al.*, 1992). In addition, glucose requirements may be increased. Babies who become hypoxic during labour, and are asphyxiated at birth, often have unrecordably low blood glucose levels within minutes of delivery. In hypoxic conditions, glucose is metabolized anaerobically by glycolysis, yielding only two molecules of adenosine triphosphate (ATP) per molecule of glucose compared with 36 molecules from aerobic oxidation. The glycogen reserves are rapidly consumed. In addition, fatty acid catabolism, ketogenesis and ketone utilization, all oxygen-dependent processes, are depressed. Glucose demand is higher when the respiratory rate is rapid because of lung problems, in septic babies, if the red cell mass is increased (polycythaemia) and in hypothermia. Several of these factors may contribute to the hypoglycaemia found commonly in babies with cyanotic heart disease.

It is essential to monitor frequently the blood glucose of babies at risk of hypoglycaemia – at least 3- to 4-hourly initially. Glucose oxidase stick tests are convenient and acceptable for this, providing they are not out of date and are used correctly, and low values are confirmed by a laboratory test on liquid blood. For a detailed account of the management of neonatal hypoglycaemia, see Chapter 3.

Idiopathic ketotic hypoglycaemia

This has become a less frequent diagnosis in recent years. It affects underweight children aged 1.5 to 7 years. A high proportion have been born small for gestational age (Dahlquist *et al.*, 1979). Typically, they present after a prolonged fast, perhaps because they have missed breakfast. They are irritable, drowsy and may have fits. They smell ketotic, have ketonuria and are often acidotic (plasma bicarbonate around 15 mmol/L). Plasma FFAs and ketones are high; lactate, pyruvate and alanine are low; and insulin is appropriately suppressed. Similar falls in glucose and increases in FFAs and ketones have been observed in normal fasted young children, and it seems likely that ketotic hypoglycaemia is merely an exaggeration of the normal fasting response.

There is no evidence for an endocrine insufficiency or a gluconeogenic defect. The probable explanation is depletion of 3-C substrate molecules for gluconeogenesis (Senior & Wolfsdorf, 1979). Some glucose that leaves the liver is oxidized completely by the tissues (especially brain) to carbon dioxide, and carbon is thus lost from the body. A considerable proportion, however, is oxidized incompletely to 3-C intermediates by glycolysis, particularly in muscle. These return to the liver as pyruvate, lactate and alanine, and, with energy provided by fat metabolism, are resynthesized to glucose. Recycled carbohydrate metabolites and not amino acid skeletons are the main gluconeogenic substrates. When glycogen reserves are poor, as in underweight children, less glucose is released from the liver during fasting and the return of 3-C intermediates for resynthesis to glucose will be very reduced.

If samples were not collected during the hypoglycaemic episode, the ketogenic response may be monitored during a supervised fast (see p. 99). An alternative, now used infrequently, has been to monitor the response to a ketogenic diet (low calorie/high fat) given for 24 h. Plasma FFAs, ketones and cortisol are raised and insulin is suppressed (Dahlquist *et al.*, 1979).

The glycogen storage disorders

These are inherited disorders in which there is excessive accumulation of normal or abnormal glycogen in the liver and/or muscles. At least 10 different glycogenoses have been identified to date (Shin, 1990). Deficiency of α-1,4-glucosidase (Pompe's disease) is different from the other conditions because it is a lysosomal disorder and does not cause metabolic derangements.

Four glycogen storage disorders (GSDs), which all involve the liver, may be associated with hypoglycaemia, as may glycogen synthase deficiency (Hers *et al.*, 1989; Fernandes, 1990). Their key clinical and biochemical features are summarized in Table 5.3.

The most serious are defects of the glucose-6-phosphatase system, since these prevent production of glucose by gluconeogenesis as well as glycogenolysis (see Fig. 5.1). The enzyme defect may involve the active site for hydrolysis, which is in the endoplasmic reticulum (type Ia), or one of the translocases needed to transport substrates to it from the cytoplasm (types Ib and Ic). All have a similar clinical presentation, dominated by a very poor tolerance to fasting, with recurrent episodes of severe hypoglycaemia with lactic acidaemia, reflecting enhanced pyruvate production. Plasma ketones are inexplicably low. Triglycerides are high, reflecting increased synthesis, and cholesterol is often raised, but it is uncertain whether the risk of atheroma is increased. Hyperuricaemia is due mainly to increased purine degradation and partly to competition with lactic acid for renal excretion. Gout may occur in older children. The kidneys are enlarged, and the glomerular filtration rate is increased in early life. Later, focal glomerulosclerosis develops, with proteinuria, and may progress to renal failure. The kidney histology and clinical course closely resemble those of insulin-dependent diabetes mellitus. Adenomas develop in the liver during the second and third decades and, rarely, may become malignant. Plasma α-fetoprotein is monitored regularly to detect this transformation. Cerebral function is normal, unless prolonged hypoglycaemic episodes have occurred.

Treatment demands aggressive therapy to ensure a continuous supply of dietary glucose. The aim is to maintain the plasma glucose level above 2.5 mmol/L during the day, and around 3.5–4.0 mmol/L at night, with plasma lactate before meals at 3.0–6.0 mmol/L. Normal biochemistry is seldom achieved. Until the end of the adolescent growth spurt, this necessitates overnight feeding with glucose polymers via a nasogastric tube with an infusion pump. Glucose polymers and uncooked corn starch are useful sources of slow-release glucose during the day.

A deficiency of amylo-1,6-glucosidase (debrancher enzyme) leads to storage of an abnormal glycogen with short outer chains (limit dextrin), mainly in the liver but also in skeletal and heart muscle. Because only glycogenolysis, but not gluconeogenesis, is impaired, the disorder is less severe than GSD type I. Nevertheless, recurrent hypoglycaemia may be a problem, particularly neonatally. The risk decreases in later childhood. In contrast to GSD type I, plasma lactate is normal during hypoglycaemia and ketones are raised appropriately. Paradoxically, plasma insulin may also be elevated. The abnormal glycogen may cause liver damage, in a minority of cases progressing to cirrhosis. High alanine aminotransferase (ALT) levels favour this form of GSD. This is the only hepatic GSD associated with raised creatine kinase. The mechanism underlying hypercholesterolaemia is unknown (Maire *et al.*, 1991).

Deficiencies of the hepatic phosphorylase system

Table 5.3 Hepatic glycogen storage disorders that cause hypoglycaemia

Disorder/enzyme deficiency	Clinical features	Abnormal biochemistry results	Response to glucagon	Tissue for diagnosis
Ia: glucose-6-phosphatase Ib: glucose-6-phosphate translocase Ic: phosphotranslocase	Severe recurrent hypoglycaemia Hepatomegaly Poor growth Bleeding tendency Later: liver adenomas, rarely hepatoma, gout, osteoporosis, focal glomerulosclerosis with proteinuria and renal failure Ib (in addition): recurrent infections, some have inflammatory bowel disease, acute leukaemia	Severe fasting hypoglycaemia with high lactate and low ketones Triglycerides very high Cholesterol often raised Uric acid often raised ALT: small increase (some) GFR: increased Later: proteinuria, possibly renal failure	*Fasting* No increase in glucose; may fall Lactate high throughout and may rise	Liver, preferably fresh to distinguish Ib
III: amylo-1,6-glucosidase (debrancher)	Recurrent hypoglycaemia, tends to abate in older children Enormous hepatomegaly; may progress to cirrhosis Early mild muscle weakness; may progress Cardiomyopathy (some) Poor growth (some)	Fasting hypoglycaemia (variable severity), with appropriate ketosis and normal lactate Cholesterol high (striking) Triglycerides raised ALT raised CK often raised	*Fasting* No change in glucose or lactate *2 h after carbohydrate meal* Glucose increases (but often by <2.0 mmol/L); lactate rises, then falls	Red blood cells: glycogen increased Red cells, white cells or fibroblasts for enzyme
Hepatic phosphorylase system				
VI: hepatic phosphorylase and phosphorylase-b kinase*	Fasting hypoglycaemia unusual, and abates in childhood Marked hepatomegaly; disappears by puberty Phosphorylase kinase deficiency generally X-linked and occurs in boys	Mild fasting hypoglycaemia, with appropriate ketosis and normal lactate Cholesterol often raised Triglycerides often raised	*Fasting* Variable increase in glucose; lactate not increased *2 h after carbohydrate meal* Sometimes increased glucose response	Red blood cells: phosphorylase kinase Liver: phosphorylase
O: glycogen synthase	Only two cases to 1989 Recurrent hypoglycaemia No or minimal hepatomegaly Poor growth	Severe fasting hypoglycaemia with ketosis and normal lactate Postprandial hyperglycaemia with lactic acidosis	*Fasting* No increase in glucose or lactate *2 h after carbohydrate meal* Increased glucose; fall in lactate	Liver

ALT, alanine aminotransferase; GFR, glomerular filtration rate; CK, creatine kinase. * Different classification numbers are used for phosphorylase-b-kinase deficiency.

may be due to defects either of phosphorylase itself (autosomal recessive) or of its activator, phosphorylase-b kinase (most often X-linked, affecting boys). In childhood there is marked liver enlargement and mild fasting hypoglycaemia may occur, but is unusual. This and the hepatomegaly often abate around puberty (Fernandes, 1990; Smit *et al.*, 1990; Maire *et al.*, 1991). Glycogen synthase deficiency seems to be extremely rare. Here, there is a lack, not an excess, of stored glycogen. Children present with severe fasting hypoglycaemia with ketosis. Plasma lactate is raised postprandially (Aynsley-Green *et al.*, 1977).

If GSD is suspected, fasting and postprandial (1–2 h) blood samples should be taken for glucose, lactate and 3-hydroxybutyrate; fasting blood samples for lipids; and blood samples for uric acid, liver function tests and full blood count. The diagnosis may then be evident. If not, a glucagon test may help. Glucagon stimulates glycogenolysis by activating phosphorylase-b kinase, by an indirect mechanism, causing an increase in plasma glucose in normal fasting children. In some forms of hepatic GSD there is little or no response, as shown in Table 5.3. Children are fasted for a variable period, according to their fasting tolerance, which may be very brief, and are monitored closely. An intravenous line is sited to provide emergency access. Intramuscular glucagon (20 μg/kg body weight) is given and blood taken at 0, 15, 30, 45, 60 and 120 min for glucose and at 0, 30 and 60 min for lactate. Proposed criteria for an abnormal response are an increase of plasma glucose of less than 2.0 mmol/L, or a plasma lactate concentration exceeding 2.4 mmol/L, fasting or after glucagon (Dunger & Leonard, 1982). Others have used a higher dose (30 μg/kg) and different criteria for an abnormal response. Repeating the test 2 h after a carbohydrate meal helps to differentiate debrancher deficiency. Affected children have a poor fasting response, because most of their glycogen is in the limit dextrin form, which they cannot hydrolyse. With feeding, new straight chains of glucose are added to this. These are hydrolysed by phosphorylase after injection of glucagon. Unfortunately, the glucagon test may give both false positive and false negative results (Dunger & Leonard, 1982).

Inherited disorders of gluconeogenesis

Defects of gluconeogenesis are very rare and usually present in early infancy. The problem here is that the Cori cycle is defective: 3-C intermediates returning to the liver cannot be resynthesized to glucose. Fasting hypoglycaemia is associated with increased blood lactate, pyruvate, alanine, ketones and FFAs and often a metabolic (lactic) acidosis.

Most of the enzyme steps in gluconeogenesis are reversible and used in glycolysis, except for the four indicated in Fig. 5.2, catalysed by pyruvate carboxylase, PEPCK, fructose-1,6-bisphosphatase and glucose-6-phosphatase. Deficiencies of the glucose-6-phosphatase complex cause GSD type I. Surprisingly, hypoglycaemia is *not* a consistent finding in pyruvate carboxylase deficiency, in which the clinical presentation is dominated by neurological damage attributed to a lack of oxaloacetate for the tricarboxylic acid cycle (de Vivo & di Mauro, 1990).

Deficiencies of fructose-1,6-bisphosphatase and PEPCK both cause life-threatening hypoglycaemic episodes in newborns, who have low glycogen reserves, and in older infants when reserves are depleted by fasting. Their defective gluconeogenic pathway cannot then maintain the blood glucose. The liver is enlarged in both disorders. Management is with regular carbohydrate feeding (Baerlocher, 1990).

Defects of fat oxidation and ketogenesis

During fasting, long-chain fatty acids (C14 to C22) are mobilized from the fat depots by lipolysis. Any defect in the process by which these acids are transported into the mitochondria of liver cells and degraded by β-oxidation will lead to deficiencies of the products acetyl-CoA, acetoacetate, 3-hydroxybutyrate and $NADH_2$. This will cause: (a) a local shortage in the liver of acetyl-CoA and $NADH_2$, essential to drive gluconeogenesis and to generate ATP for hepatic biosynthetic and metabolic processes; and (b) a deficiency of ketones for export to the brain, muscles, myocardium and kidney, as an alternative energy source to glucose. Deficiency of hydroxymethylglutaryl-CoA (HMGCoA) lyase, needed for the last stage of ketogenesis, similarly leads to deficiency of ketones.

The first inherited defect of fat oxidation was reported only in 1976 (Gregersen *et al.*, 1976). Defects have since been recognized at all steps of the pathway (Vianey-Liaud *et al.*, 1987; Pollitt, 1989; Roe & Coates, 1989; Bennett, 1990; Bartlett *et al.*, 1991). Their biochemistry is discussed in Chapter 6. Nine of them may be associated with fasting intolerance. Affected children are in danger when they become calorie depleted, generally during an intercurrent infection when they refuse feeds. The blood glucose is maintained initially from liver glycogen; however, this is rapidly depleted. As the fast continues, fatty acids are mobilized and transported to the liver. Here, the normal processes fail. Depending upon the site of the block, long-chain fatty acids may undergo little, or only partial, mitochondrial β-oxidation. An acute disturbance of liver function results: the liver becomes infiltrated with fat and enlarges rapidly; blood clotting is disturbed; plasma transaminases and ammonia increase. However, the most important consequences are grossly inadequate gluconeogenesis with a falling blood glucose, and poor ketone production. The brain is therefore deprived of its two fuels, and a severe brain disturbance (encephalopathy) develops within a few hours, because membrane pumps fail and brain oedema develops. Other organs which depend heavily on fatty acids and ketones are often damaged as well: the heart (cardiomyopathy), skeletal muscles and kidneys. Partially degraded fatty acids accumulate rapidly in the liver. Some are shortened (down to 6 C) by the fat oxidation system of peroxisomes, but this does not generate ATP. Large amounts are converted by microsomal enzymes to hydroxy and dicarboxylic acids. Conjugates are formed with glycine, glucuronic acid and, importantly, carnitine (Roe *et al.*, 1985; Roe & Coates, 1989). Products of chain length C12 and less are excreted by the kidney. Longer derivatives are not.

Medium-chain acyl-CoA dehydrogenase (MCAD) deficiency is the commonest disorder (Touma & Charpentier, 1992). Affected children appear normal and do not have liver enlargement or any overt metabolic disturbance when well and feeding regularly. They may never have symptoms. However, if they are calorie deprived, they develop an illness very much like Reye's syndrome (p. 111), generally over only a few hours, with irritability, vomiting and increasing drowsiness which progresses to coma. Typically, children are admitted moribund, often unrousable, and have severe hypoglycaemia. The peak age of the first acute episode is 14 months and 40% have died as a result, usually because of the brain disturbance. Emergency treatment is with intravenous glucose, which provides energy, shuts off fat metabolism and rapidly reverses the metabolic disturbance. Careful dietary management should prevent further episodes. The risk of acute illness decreases, but persists, after around 6 years of age, when children are less dependent on fat metabolism.

Other disorders in this group more often present with hypoglycaemic episodes in early infancy; for example, long-chain acyl-CoA dehydrogenase (LCAD) and 3-hydroxyacyl-CoA dehydrogenase deficiencies (Treem *et al.*, 1991; Wanders *et al.*, 1991).

At the time of hypoglycaemia, plasma ketones are inappropriately low and FFAs elevated. A ratio of FFA to 3-hydroxybutyrate exceeding 2.0 is highly indicative of a fat defect, values being less than 1.0 normally and in most other hypoglycaemic disorders. Plasma concentrations of total carnitine are often reduced, but of greater significance is a fall in the percentage of free carnitine (for example to less than 10%), reflecting a large increase in the esterified fraction (Bennett, 1990). Plasma aspartate aminotransferase (AST), ALT and, often, ammonia are raised, and there may be a metabolic acidosis, but this is often mild. There is a gross urinary organic acid disturbance and, frequently, ketonuria. This is probably because damage to ketone-using organs in the advanced stage of illness decreases their capacity to use the small amounts of ketones which are produced. It is important not to exclude the diagnosis on the basis of ketonuria. All these abnormalities disappear rapidly after glucose administration. It is therefore crucial to collect appropriate samples on presentation. If this opportunity is lost, a closely supervised fast may be necessary to establish the diagnosis (Bonnefont *et al.*, 1990; Bartlett *et al.*, 1991) (see p. 99). However, this is a hazardous procedure, which has caused deaths in two children, and should not be undertaken lightly.

Other organic acidurias

During acute illness, propionic, methylmalonic and isovaleric acidurias may be associated with either hyper- or hypoglycaemia (Ogier *et al.*, 1990). In methylmalonic aciduria, hypoglycaemia has been attributed to inhibition of pyruvate carboxylase (hence, gluconeogenesis) by methylmalonyl-CoA (Rosenberg, 1983). Severe to moderate fasting hypoglycaemia is often observed in both classic and variant types of maple syrup urine disease, associated with undetectable insulin. The postulated mechanism is reduced availability of gluconeogenic substrates (Haymond *et al.*, 1973).

Hereditary fructose intolerance and galactosaemia

Hereditary fructose intolerance is due to a deficiency of fructose-1-phosphate aldolase B, an enzyme expressed exclusively in the liver, kidney cortex and intestine. The problem is that fructose-1-phosphate cannot be cleaved to enter the glycolytic and gluconeogenic pathways (Fig. 5.2). It accumulates and sequesters phosphate and, perhaps by direct inhibition and ATP depletion, inhibits phosphorylase activity (and hence glycogenolysis), and gluconeogenesis (Cox, 1988; Odièvre, 1990). Symptoms, which only occur when fructose and sucrose are introduced into the diet, include vomiting, diarrhoea and, infrequently, severe postprandial hypoglycaemia, with pallor, fits and shock. Affected children have hepatomegaly and thrive poorly. Later, they develop an aversion to sweet foods and the symptoms improve. Treatment is strict dietary exclusion.

Acute hypoglycaemic episodes are associated with severe hypophosphataemia and lactic acidosis. Plasma urate and magnesium are raised because ATP, adenosine diphosphate (ADP) and adenosine monophosphate (AMP) are degraded and associated magnesium is released. When the diet contains fructose, there is fructosuria. Liver function tests and clotting are abnormal, particularly in infancy, and there is evidence of proximal renal tubular dysfunction – glucosuria, proteinuria, phosphaturia, generalized amino aciduria and renal tubular acidosis.

A closely supervised intravenous fructose load test may establish the diagnosis, but must only be carried out when children are well, on a fructose-free diet, and with intravenous glucose at hand. A dose of 0.2 g/kg for infants, and 0.25 g/kg for older children, is given over 2 to 4 min. The response is positive if hypoglycaemia develops 30 to 60 min after the dose and plasma phosphate falls. Oral fructose load tests may cause severe pain and shock and must be avoided. Enzyme confirmation requires a liver biopsy.

Hypoglycaemia may rarely occur in galactosaemia. The mechanism is unknown, but probably reflects disturbed liver function. The glycaemic response to glucagon is impaired in patients receiving galactose (Segal, 1983) (see Chapter 13).

Hereditary tyrosinaemia type I

In this autosomal recessive disorder there is a deficiency of fumarylacetoacetate hydrolase, an enzyme of the tyrosine degradation pathway. The disorder is characterized by progressive liver disease and renal tubular defects, with hypophosphataemic rickets. A diffuse hyperplasia of the pancreatic islets has been reported in some cases, and hyperinsulinism may account for hypoglycaemia, which occurs occasionally. In the acute, neonatally presenting form, the plasma tyrosine and methionine levels are raised, phosphate and potassium are low, and there is tyrosyluria, amino aciduria, glucosuria and phosphaturia. In the chronic form, presenting after 1 year, the plasma tyrosine level may be only marginally raised and the methionine level normal, and the renal tubular dysfunction less severe. The diagnosis is established by demonstrating increased succinylacetone on urinary organic acid analysis, or in blood with an enzyme test, and is confirmed by showing the enzyme defect in lymphocytes or fibroblasts (Kvittingen, 1986; Halvorsen, 1990).

Hyperinsulinism

Insulin lowers blood glucose by promoting peripheral uptake and stimulating glycolysis and glycogen synthesis. It also inhibits lipolysis and ketogenesis. Under fasting conditions, excessive inappropriate insulin secretion may be extremely damaging to the brain since it is deprived of both its fuels – glucose and ketones. Prolonged hypoglycaemia due to hyperinsulinism causes permanent brain damage (Aynsley-Green, 1981; Spitz *et al.*, 1992). When hypoglycaemic, the plasma insulin is inappropriately high. Many define this as >10 mU/L, but others have used 7 or 12 mU/L as a cut-off level. FFAs are low (<0.46 mmol/L), as is 3-hydroxybutyrate (<1.1 mmol/L) (Stanley & Baker, 1976; Spitz *et al.*, 1992). Plasma glucose increases in response to glucagon injection (Aynsley-Green *et al.*, 1981).

Transient hyperinsulinism in the newborn

In many cases of neonatal hyperinsulinism the disturbance is transient and one of the predisposing factors listed in Table 5.2 is evident. It is suspected if a glucose infusion rate of more than 10 mg/kg per min is required to maintain normoglycaemia, the normal requirement being 4–6 mg/kg per min (Aynsley-Green *et al.*, 1981). Babies whose insulin secretion was excessive *in utero* are generally large, since insulin is the major fetal growth hormone. Macrosomia correlates with cord C-peptide concentrations (Stenninger *et al.*, 1991). The response to glucagon is blunted in hyperinsulinaemic newborns, perhaps because of the delayed development of glucagon receptors (Blackburn & Loper, 1992). Insulin injection at delivery markedly impaired the postnatal increase in liver PEPCK, and delayed the onset of gluconeogenesis (Girard, 1990). A delay in normal perinatal adaptation may account for transient hyperinsulinism observed in some birth-asphyxiated preterm babies (Collins *et al.*, 1990).

Babies born to mothers with insulin-dependent or gestational diabetes, whose glycaemic control was suboptimal, may become hypoglycaemic, often between 1 and 4 h of life. A normal plasma glucose is generally attained within a few days, although hypoglycaemia may be more prolonged. Hyperinsulinism occurs because an increased placental transfer of glucose leads to fetal hyperglycaemia and chronic stimulation of fetal islet tissue. A further adverse factor is impaired glucagon secretion after delivery (Bloom & Johnston, 1972). These babies do badly if they suffer birth asphyxia, since they cannot meet the greatly increased glucose demands and do not produce ketones.

Islet cell hyperplasia also accounts for hyperinsulinism in babies with severe Rhesus incompatibility, but the mechanism is unknown. Babies with Beckwith–Wiedemann syndrome are large, have a large tongue and, sometimes, exomphalos. Around 50% develop symptomatic hypoglycaemia, which is usually transient. Islet cell tissue is increased. The number of somatostatin-producing cells is increased, as is extractable somatostatin (Gerver *et al.*, 1991). Raised growth hormone and somatomedin levels have been detected at birth, suggesting a possible disorder of growth-regulating peptides (Aynsley-Green, 1989). Transient hyperinsulinism may also be iatrogenic. Maternal glucose infusions during labour may cause fetal hyperglycaemia that stimulates insulin, and inhibits glucagon, secretion. β-Sympathomimetics, given to the mother for preterm labour, cross the placenta rapidly and may stimulate fetal insulin secretion (Blackburn & Loper, 1992). Finally, abrupt termination of intravenous glucose infusion of a baby may cause rebound hypoglycaemia, the plasma glucose falling more rapidly than insulin when glucose is withdrawn.

Persistent hyperinsulinism

After 1 month of age, hyperinsulinism is the commonest cause of severe recurrent hypoglycaemia in the first year of life (Stanley & Baker, 1976). Brain damage is a frequent consequence (Spitz *et al.*, 1992). The cause is a maldevelopment of the pancreatic islet tissue, called nesidioblastosis. The islet cells are increased in number and bud off from pancreatic ducts, so that the islets are anatomically disorganized. All four types of islet endocrine cells are involved – insulin, glucagon, somatostatin and pancreatic polypeptide. There is an uncontrolled release of insulin, which has been attributed to loss of cell contact, a defect in glucose recognition by β-cells and somatostatin deficiency (Aynsley-Green, 1981; Upp *et al.*, 1987). The normal release of insulin in response to leucine (abundant in the milk protein casein) is exaggerated, and children formerly categorized as having 'leucine-sensitive hypoglycaemia' (which is *not* a pathological entity) almost certainly had this disorder. Although most cases have appeared sporadically, there have been several familial cases, and it is probably an autosomal recessive disorder (Woolf *et al.*, 1991). It has, rarely, caused sudden infant death (Polak & Wigglesworth, 1976).

Around 80% of babies present with symptomatic hypoglycaemia in the first 3 days of life, and the others before 6 months. The diagnosis is made from analyses of blood glucose, insulin and ketones when the baby is hypoglycaemic, and from the demonstration of a glycaemic response to glucagon (Antunes *et al.*, 1990). Glucose tolerance tests cause rapid clearance of glucose and profound hypoglycaemia and are dangerous, as are leucine and arginine provocation tests (Aynsley-Green *et al.*, 1981). They have no place in diagnosis.

Medical treatment is with diazoxide and chlorothiazide, which inhibit glucose-stimulated insulin release. However, if this is ineffective, early surgery to remove 95% of the pancreas is required (Spitz *et al.*, 1992). Preoperative treatment with a somatostatin analogue may help to stabilize blood glucose, but tolerance has developed to long-term therapy (Hawdon *et al.*, 1990).

Insulin autoantibodies

An unusual cause of hypoglycaemia is associated with insulin autoantibodies. It seems to be relatively common in Japan, but extremely rare elsewhere. It is believed that hypoglycaemia develops when insulin is released from binding antibody (Meschi *et al.*, 1992).

Hormone deficiencies

Glucocorticoids promote gluconeogenesis by increasing the supply of gluconeogenic amino acids from peripheral tissues, increasing the synthesis of PEPCK, activating glucose-6-phosphatase and having permissive effects on glucagon actions and on lipolysis (Orth *et al.*, 1992). Growth hormone stimulates lipolysis (Goodman, 1968).

Both growth hormone and cortisol deficiency may cause hypoglycaemia in children, and particularly in the newborn. Neonatal hypoglycaemia may be an important early sign of endocrine disorders and may contribute to neurological damage if diagnosis is delayed (Stanhope & Brook, 1985). Causes include panhypopituitarism, isolated growth hormone deficiency, adrenal hypoplasia and congenital adrenal hyperplasia. Important clues in neonates are micropenis and mid-line developmental abnormalities, indicating intrauterine hypopituitarism, and pigmented external genitalia, seen in congenital adrenal hyperplasia. Familial selective glucocorticoid insufficiency is an inherited, probably autosomal recessive, disorder which presents with severe recurrent hypoglycaemia at the age of 1–2 years. The pathogenesis of this condition is still unknown (Forest, 1989).

Acute liver failure and Reye's syndrome

In the absence of accelerated glucose use, there has to be extensive damage to the liver to cause hypoglycaemia. It is most likely when massive liver destruction occurs rapidly, as in fulminant viral hepatitis or toxic hepatitis, and probably results from hepatocyte damage (Cryer, 1992). Although plasma insulin levels have been raised in liver failure, there was also insulin resistance (Vilstrup *et al.*, 1986). Reye's syndrome is discussed on p. 111. The presentation is very like that of fat oxidation defects, with acute liver and brain disturbances. Transient hypoglycaemia is common and results from defective gluconeogenesis.

Drug- and alcohol-induced hypoglycaemia

Insulin-induced hypoglycaemia is a significant risk in children with insulin-dependent diabetes mellitus. It is commonly precipitated by exercise, decreased food intake and errors relating to insulin dosage. HbA_{1c} tends to be lower among children who have recurrent severe hypoglycaemic episodes than among those who do not (Daneman *et al.*, 1989). Severe hypoglycaemia may also occur in children having diagnostic insulin stress tests for growth hormone deficiency. Treatment of such crises with excessive amounts of 50% dextrose may cause acute fluid shifts and brain oedema, with irreversible brain damage. Guidelines for safe management have been proposed (Shah *et al.*, 1992). On rare occasions, insulin is administered to children with malign intent. Estimation of the plasma C-peptide is useful if there is uncertainty whether hypoglycaemia is due to administered insulin or to excessive endogenous secretion. The antimalarial drug, quinine (but not chloroquine), increases pancreatic insulin secretion. This exacerbates the tendency to hypoglycaemia in falciparum malaria, because of the enormous requirement of the parasite in red blood cells for glucose for glycolysis (White *et al.*, 1987).

Salicylate may produce hypoglycaemia in children. The cause is probably multifactorial: salicylates uncouple oxidative phosphorylation and thereby ATP generation – this would stimulate increased glucose use for glycolysis; there may be increased demand for glucose because of restlessness and hyperventilation; the central emetic effect may cause nausea and vomiting, and decrease food intake (Meredith & Vale, 1981).

Profound hypoglycaemia may occur when children accidentally ingest alcohol some hours after a meal. Ethanol inhibits gluconeogenesis, as well as the cortisol and growth hormone responses to hypoglycaemia. It delays the adrenaline response, but does not affect glucagon secretion. Plasma ethanol measurements at presentation may not be markedly raised and correlate poorly with blood glucose (Cryer, 1992). Patients who are particularly dependent upon gluconeogenesis to maintain the fasting blood glucose, for example with GSD type III, should be advised not to drink.

Investigation of hypoglycaemia: acute samples and fasting tests

For the neonatal period, see also Chapter 3.

Children presenting acutely with hypoglycaemia are often dangerously ill and possibly moribund. In such an emergency, a detailed history and examination may have to be deferred. Too often, blood is taken only for glucose, and possibly electrolytes, before glucose is given. When the child has recovered, it may be very difficult to establish a diagnosis without recourse to a fasting test, with the associated risks.

It is extremely important to collect appropriate samples at the time of the acute event. These will provide a diagnosis, or at least indicate which group of disorders is likely (Surtees & Leonard, 1989). Tests not routinely available may be sent away for analysis, if indicated, when the preliminary results are available. Blood *should always* be taken for: glucose, insulin, growth hormone, cortisol, lactate, amino acids (with alanine quantification, if possible), 3-hydroxybutyrate, FFAs and liver function tests. If a fat oxidation defect is likely, blood should also be taken in case carnitine analyses are required. The first urine passed *must always* be tested for ketones and saved for organic acid analysis. Urea and electrolytes, and perhaps blood gas and acid–base status, will be needed for management. Plasma ammonium should be measured in cases with a Reye's syndrome-like illness. Table 5.4 summarizes the key diagnostic biochemical features during hypoglycaemia in different disorders.

If the results are not helpful, or the chance was missed, or there is uncertainty about whether reported episodes at home were hypoglycaemic or not, there may be no alternative but to investigate the metabolic response to fasting. The differential diagnosis is most often hyperinsulinism, a possible fat oxidation defect or idiopathic ketotic hypoglycaemia. Fasting tests are potentially dangerous. They should be avoided in the first month of life, if possible. They must only be carried out after the child has recovered from the acute illness and has been feeding for several days at least. The test must be closely monitored, including stick test measurements of blood glucose, and should be arranged so that the later stages of the fast occur during the day. An intravenous cannula should be sited for access in emergency, and the test must be terminated promptly if the child becomes hypoglycaemic or abnormally drowsy.

There can be no definitive protocol: the duration of fasting must be decided on an individual basis, knowing the fasting tolerance of the child. Guide-

Table 5.4 Key biochemical abnormalities in blood at the time of hypoglycaemia

Hyperinsulinism	Insulin inappropriately high; ketones, FFAs, lactate all low
Idiopathic ketotic hypoglycaemia	FFAs and ketones high; insulin, lactate, alanine all low
Hepatic glycogen storage disorders	
Defects of glucose-6-phosphatase (Ia, Ib, Ic)	Lactate high; ketones and insulin low
Debrancher deficiency (III)	Lactate normal; appropriate ketosis, insulin sometimes inappropriately high
Phosphorylase deficiency and phosphorylase-b kinase deficiency	Normal lactate; appropriate ketosis; insulin low
Glycogen synthase deficiency	Normal lactate; ketones high; insulin low
Defects of gluconeogenesis	Lactate, alanine, ketones, FFAs all raised; insulin low; metabolic acidosis
Defects of fat oxidation and ketogenesis	FFAs raised; ketones low; FFA:3-hydroxybutyrate ratio raised; total and free carnitine low; free:esterified carnitine ratio low; insulin low (*Note: may be ketonuria*)
Acute liver failure and Reye's syndrome	ALT and AST raised; raised ammonia; clotting problems; ketones and insulin variable

FFA, free fatty acid; ALT, alanine aminotransferase; AST, aspartate aminotransferase.

lines differ a little (Bonnefont *et al.*, 1990; Bartlett *et al.*, 1991). In general, fasting should not exceed 8 h in infants under 1 year, 16 h in those from 1 to 5 years old, 20 h in older children and 24 h in adolescents. At the end of the fast, or when hypoglycaemia supervenes, blood and urine samples are collected, as outlined above for acute hypoglycaemic episodes. It is helpful too to take additional samples for glucose, 3-hydroxybutyrate and FFAs 4 and 2 h before the end of the test to monitor the course of ketogenesis.

Published reference data for fasting tests are scattered. Responses for normal children have been accumulated in Table 5.1. The following cut-off values may help to define an abnormal response when hypoglycaemic (see this chapter): plasma insulin >10 mU/L (alternatively >7 or >12 mU/L), lactate >2.0 mmol/L, 3-hydroxybutyrate <1.1 mmol/L, FFAs <0.4 mmol/L, FFA:3-hydroxybutyrate ratio >2.0, alanine >480 µmol/L. Reference ranges for carnitine vary among laboratories. A plot of $\log_{10}$ of total ketones (acetoacetate plus 3-hydroxybutyrate) against FFAs has been found to be discriminatory for hypoketonaemic disorders (Bartlett *et al.*, 1991).

(b) Hyperammonaemia

Introduction

Ammonia is toxic. Prolonged or recurrent hyperammonaemia causes irreversible brain damage and may be lethal if severe. In childhood, it has a variety of causes, but the most serious are inherited disorders involving the urea cycle. It is essential to diagnose hyperammonaemia and its probable cause quickly, in order to institute appropriate treatment.

Biochemistry of ammonia, glutamine and the urea cycle

Ammonia (NH_3) is a weak base, in equilibrium with the ammonium (NH_4^+) ion. At physiological pH, ammonia accounts for less than 5% of the total NH_3/NH_4^+. This is fortunate, since cell membranes are five times more permeable to ammonia than NH_4^+ (Souba, 1987). Ammonia is generated during metabolism in all organs. Large amounts are produced in skeletal muscle, particularly during

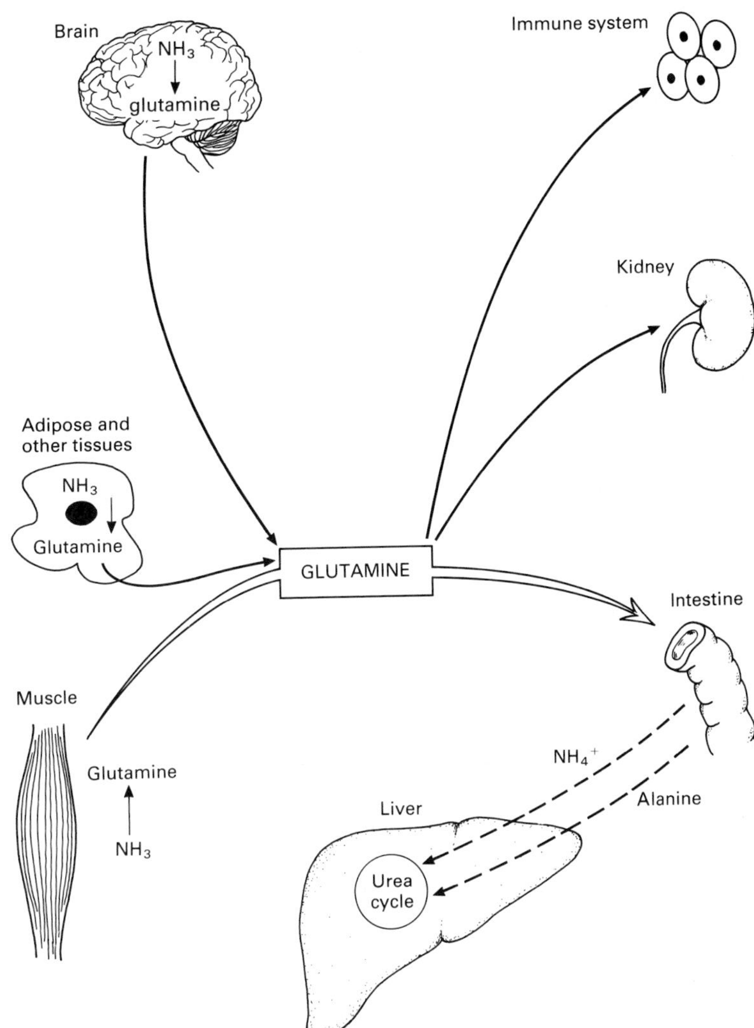

Fig. 5.3 The role of glutamine in the short-term buffering of ammonia produced in tissues.

exercise, mainly from deamidation of AMP in the purine nucleotide cycle, and some from amino acid catabolism. These processes, and glutamate dehydrogenase activity, contribute to ammonia production in the brain (Cooper & Plum, 1987). The kidney releases large amounts of ammonia from glutamine into the renal tubules to promote hydrogen ion excretion. Some diffuses back into the bloodstream, increasing renal vein concentrations (Warter *et al.*, 1983). In addition, ammonia is absorbed from the gut lumen, where it is produced by bacterial metabolism of amino acids from foods and desquamated cells and, to a small extent, by hydrolysis of urea in gut secretions. However, plasma concentrations of ammonium in the systemic circulation are normally very low (<40 μmol/L; Green, 1988; Brusilow & Horwich, 1989). This is achieved by a highly coordinated process (Fig. 5.3) (Harris & Crabb, 1986; Cooper & Plum, 1987; Meijer *et al.*, 1990).

Most ammonia generated in the tissues is initially detoxified by amidation of glutamate to form glutamine. More than 80% of the body's glutamine is in muscle. Glutamine is then released into the bloodstream. Some is extracted by the kidneys, and some is used as a respiratory substrate by cells of the

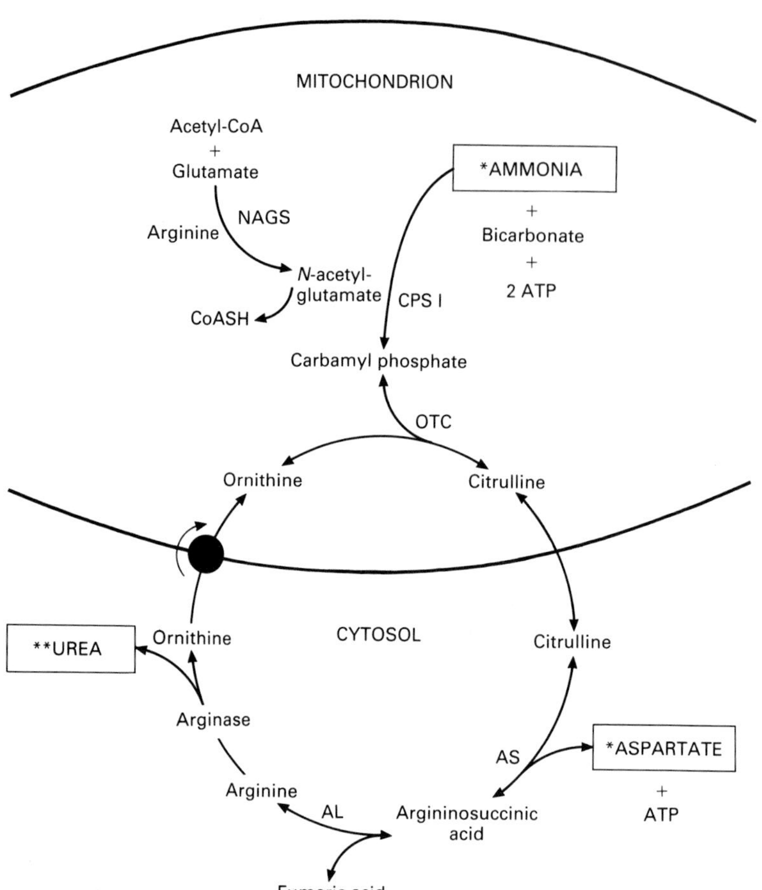

Fig. 5.4 The urea cycle. NAGS, *N*-acetylglutamate synthetase; CPS I, carbamylphosphate synthetase I; OTC, ornithine transcarbamylase; AS, argininosuccinate synthetase; AL, argininosuccinate lyase; black circle and arrow represent the ornithine transport system.

immune system. Most, however, is taken up by intestinal epithelial cells which oxidize it for energy and incorporate the waste nitrogen into alanine or citrulline, or release it as ammonia. These are carried by the portal bloodstream to the liver, along with ammonium absorbed from the gut lumen. The concentration of ammonium in the portal blood is five to 20 times that in the systemic circulation (Mounger & Branson, 1972). Ammonium and alanine are extracted by the periportal liver cells and their nitrogen is incorporated into urea. Ammonium which escapes this process is detoxified by cells in the centre of the liver lobules by incorporation into glutamine. Citrulline passes through the liver and is taken up by the kidneys.

The urea cycle is the major route through which the body detoxifies ammonia and removes surplus amino-group nitrogen from the body. Around 80% of excreted nitrogen is in the form of urea. The cycle comprises five enzymes (Fig. 5.4): two, carbamylphosphate synthetase I (CPS I) and ornithine transcarbamylase (OTC), are intramitochondrial; the other three, argininosuccinic acid synthetase (AS), argininosuccinic acid lyase (AL) and arginase, are cytosolic. CPS I has an absolute requirement for an allosteric activator, *N*-acetylglutamate, produced by *N*-acetylglutamate synthetase (NAGS). The full complement of enzymes is expressed only in the liver, which is the only organ that can produce urea. The direct substrates of the cycle are ammonia (not NH_4^+), bicarbonate and aspartate. Ornithine acts as a carrier and is not degraded. From animal studies, around 33% of ammonia is from NH_4^+ in portal blood, 6–13% from deamidation of glutamine by periportal liver cells, 20% from amino-acid amino groups transferred to glutamate and then released by glutamate dehydrogenase, and the remainder from hepatic artery ammonium and ammonia released directly from amino acids such as asparagine, glycine, serine and threonine (Meijer *et al.*, 1990).

Aspartate is the main route by which nitrogen from alanine enters the urea cycle. Alanine carries large amounts of 3-C units from muscle to liver for gluconeogenesis (p. 89). The surplus amino groups are transferred first to 2-oxoglutarate by ALT, forming glutamate which is transported into the mitochondria. AST next transfers the amino groups to oxaloacetate, and the aspartate formed then passes into the cytosol to enter the urea cycle. Nitrogen from other amino acids is also introduced via aspartate. The rate of urea synthesis is regulated precisely, and normally flux through the cycle is around 20–50% of maximal capacity. Short-term regulation is at the CPS I step. It is mediated by *N*-acetylglutamate, the concentration of which reflects the availability of acetyl-CoA, glutamate and arginine, the activator of NAGS. Long-term control is achieved through altering enzyme production (Mehler, 1986; Brusilow & Horwich, 1989; Meijer *et al.*, 1990). The enzymes of the urea cycle develop early in the fetus. Human fetal liver can synthesize urea by 16 weeks of gestation (Räihä & Suihkonen, 1968).

Toxicity of ammonia

Severe hyperammonaemia in the newborn causes a gross neurological disturbance ('encephalopathy'). This commences with poor feeding and irritability, but progresses rapidly with increasing lethargy, vomiting, hypothermia, fits, abnormal movements and muscle tone, hyperventilation (often with respiratory alkalosis), coma and death. Intracranial pressure increases and at postmortem examination the brain is oedematous and the astrocytes are swollen. In older infants and children, the features vary with severity: lethargy, irritability and confusion, somnolence, vomiting, unsteady gait (ataxia), slurred speech, fits, coma and death. Recurrent or chronic hyperammonaemia results in mental retardation, fits and often protein aversion and poor growth. There is scarring (gliosis) of the brain, with loss of neurones (Brusilow & Horwich, 1989).

Normally, brain ammonium concentrations are 1.5–3 times higher than blood concentrations. Some is produced by brain metabolism and some enters by diffusion from the blood and cerebrospinal fluid (CSF). This contribution increases as blood concentrations rise. There is only one detoxification mechanism: amidation of glutamate to glutamine by glutamine synthetase, located largely in astrocytes. In hyperammonaemic states, the capacity of this 'enzymatic barrier' becomes saturated and brain ammonium rises progressively. This will occur rapidly if acute ammonia loading is superimposed on a hyperammonaemic state, and will be exacerbated by alkalosis. A rapid increase in intracellular glutamine might create an osmotic gradient and water shifts, accounting for brain oedema in acute hyperammonaemia. Hyperammonaemia interferes with many electrophysiological properties of the brain, affecting both excitatory and inhibitory postsynaptic potentials. It also decreases brain concentrations of the excitatory neurotransmitters, aspartate and glutamate (majority of studies), and may alter receptor binding of γ-aminobutyric acid (GABA). Both electrophysiological and neurotransmitter disturbances probably contribute to the symptoms. So, too, may disturbances of the mitochondrial and cytoplasmic redox states. The popular idea that hyperammonaemia would deplete 2-oxoglutarate and brain energy has not been substantiated by *in vivo* studies using ^{31}P nuclear magnetic resonance (NMR), which did not detect changes in ATP and creatine phosphate (Cooper & Plum, 1987; Meijer *et al.*, 1990). Ammonium is not known to be toxic to other tissues. Skeletal muscle takes up ammonium when plasma levels are increasing, and incorporates it into glutamine, providing an important defence mechanism (Souba, 1987).

Definition and causes of hyperammonaemia

With carefully collected samples from resting subjects, and good analytical methods, the venous plasma ammonium concentration after 1 month of age is 15–40 μmol/L, as in adults, and there is little difference between arterial and venous concentrations (Colombo *et al.*, 1984; Green, 1988; Brusilow & Horwich, 1989). Venous levels are increased by muscle activity, as for example in struggling babies. They are lower than arterial levels in hyperammonaemic states because of extraction by muscles. Higher values are common during the first few days of life, reflecting physiological disturbances: up to around 100 μmol/L in apparently healthy term babies. Higher values have been reported in preterm babies compared with term babies (Batshaw & Brusilow, 1978). Values up to 90 μmol/L may persist for some weeks if low birth-weight babies are fed milks with high protein or a high casein:whey ratio (Räihä *et al.*, 1976). As a rough guide, concentrations exceeding 150 μmol/L in newborns should be investigated if there is not an obvious explanation, such as sampling after birth asphyxia. Small increases occur non-specifically in sick older children, where a concentration above 80 μmol/L, without obvious cause, requires full investigation (Bachmann, 1990).

Three factors cause hyperammonaemia, alone or in combination: defective function of the urea cycle; shunting of portal blood away from the liver; and an excessive load of ammonia. Table 5.5 lists the most important causes in childhood.

Table 5.5 Causes of hyperammonaemia

Inherited metabolic defects
Urea cycle defects
Triple 'H' syndrome
Lysinuric protein intolerance
Organic acid defects:
Fatty acid β-oxidation defects
Hydroxymethylglutaryl-CoA (HMGCoA) lyase deficiency
Propionic, methylmalonic and isovaleric acidaemias
2-Methylacetoacetyl-CoA thiolase (β-ketothiolase) deficiency
Biotinidase deficiency
Glutaric acidaemia type I
Pyruvate dehydrogenase and pyruvate carboxylase deficiencies
Reye's syndrome
Liver failure
Circulatory insufficiency:
Birth asphyxia, shock, septicaemia, near-miss cot death, right-sided heart failure
Transient hyperammonaemia of the newborn
Proteus urinary infections
Acute leukaemia
Drugs:
Sodium valproate
Salicylate poisoning
Asparaginase
Toxins:
Aflatoxin
Hypoglycin-A
Pesticides
Margosa oil
Poor samples or sample processing
Parenteral nutrition

Inherited defects of the urea cycle

Inherited defects of all five urea cycle enzymes occur. OTC deficiency is X-linked. The others are autosomal recessive. Two cases of NAGS deficiency have also been reported. (For reviews see: Brusilow (1985a,b); Brusilow and Horwich (1989); Bachmann (1990).) Except for arginase deficiency, hyperammonaemia is the commonest presentation. The severity depends upon the residual activity of the deficient enzyme, the load of ammonia for excretion and the extent of depletion of urea cycle substrates. The most severe defects present with overwhelming neonatal hyperammonaemia. Typically, the baby is born at term after an uneventful pregnancy and delivery, and appears well for the first 24 to 48 h of life. Then, there is increasing lethargy with poor feeding, and a rapidly developing encephalopathy which progresses to coma (p. 103). Intracranial or pulmonary haemorrhage may occur. The disorder may be misdiagnosed as septicaemia, respiratory distress or cerebral haemorrhage. Some inherited organic acid disorders have a similar presentation and may be associated with hyperammonaemia. Parental consanguinity, previous unexplained sibling deaths or male deaths in the pedigree (OTC deficiency) are pointers to an inherited defect. Untreated, these babies die or survive with mental retardation and cerebral palsy. Less severe deficiencies present later, characteristically with episodic hyperammonaemia (p. 103). Recurrent vomiting may be attributed to a gastrointestinal disturbance or to psychosomatic 'cyclical vomiting', behaviour disturbances to drug abuse, and encephalopathic episodes to viral encephalitis (Drogari & Leonard, 1988). Avoidance of dietary protein is a useful clue. Hyperammonaemia is often precipitated by intercurrent childhood illnesses, when endogenous protein catabolism is increased, or by a higher dietary protein load, for example with weaning. Children with very mild defects may be asymptomatic but they are still at risk of hyperammonaemic episodes. These have rarely been precipitated by treatment with sodium valproate (p. 113).

There is little to distinguish CPS I, OTC, AS and AL deficiencies clinically, although the family history may point to OTC deficiency. Abnormal fragile hair (trichorrhexis nodosa) occurs in 50% of children, but not in neonates, with argininosuccinic aciduria, and may be due to prolonged arginine deficiency (Bachmann, 1990). It has rarely been found in AS deficiency. Around 35% of children have hepatomegaly at presentation – this may be gross in neonates with AL deficiency. Arginase deficiency is distinct from the other defects: hyperammonaemic episodes may occur but are rare, and the plasma ammonium may be normal otherwise. Instead, there is a chronic progressive neurological disorder with mental retardation and spasticity. This may be due to mild episodic hyperammonaemia, hyperargininaemia or neurotransmitter disturbances (Brusilow & Horwich, 1989).

Routine biochemistry at presentation is not diagnostic. Newborns frequently, but not invariably, have a respiratory alkalosis and a very low plasma urea, with normal or raised creatinine. Plasma urea is often normal in late-presenting disorders. Liver transaminases (ALT and AST) are commonly raised in the plasma, and may remain so after the acute illness. Plasma ammonium is generally >300 μmol/L in hyperammonaemic coma, and often much higher in neonates. The defect is localized by quantitative analysis of amino acids in plasma and urine and of urinary orotic acid (Table 5.6). Non-specific changes in plasma amino acids in the hyperammonaemic defects are: raised glutamine and alanine, and reduced arginine. Plasma and urinary concentrations of citrulline are low in CPS I and OTC deficiencies, extremely high in AS deficiency and high in AL deficiency. Argininosuccinic acid and its anhydride (which forms spontaneously *in vitro*) in plasma and urine are diagnostic of AL deficiency. In arginase deficiency, plasma arginine is grossly elevated and there is an amino aciduria with increased excretion of arginine, ornithine, lysine, citrulline and glutamine. Substituted guanidines (for example, guanidinoacetic acid) have also been reported in this disorder (Brusilow, 1985a).

In those defects in which mitochondrial synthesis of carbamyl phosphate from ammonia is normal, but further metabolism is blocked, carbamyl phosphate accumulates and leaks into the liver cell cytoplasm. Here, it joins the pool of carbamyl phosphate normally synthesized from glutamine by the cytosolic enzyme CPS II as a precursor for pyrimidine synthesis (Fig. 5.5). Production and urinary excretion of orotic acid and other pyrimidines, uracil, uridine, and pseudouridine is considerably increased (Bachmann & Colombo, 1980; Brusilow & Horwich, 1989; Bachmann, 1990). Orotic acid excretion may be increased markedly in OTC, AS and arginase deficiencies, barely raised in AL deficiency, and is normal in NAGS and CPS I deficiencies. Rarely, orotic acid crystalluria has occurred in OTC and AS deficiencies.

Enzyme confirmation of NAGS and CPS I deficiencies and (if indicated) OTC deficiency requires liver biopsy tissue. Deficiencies of AS, AL and arginase can be confirmed on red blood cells or fibroblasts (Table 5.6). Prenatal diagnosis of AS and AL deficiencies is available with enzyme analyses of chorionic villus samples or cultured amniocytes, and sometimes has been corroborated by metabolite analysis of amniotic fluid. Fetal red cell analysis is potentially useful for arginase deficiency. In some families with CPS I and OTC deficiencies, DNA analysis of chorionic villi is informative (Brusilow & Horwich, 1989).

OTC deficiency

Most OTC mutations lead to a severe deficiency of the enzyme in hemizygous boys. They develop fulminating hyperammonaemia (500–2000 μmol/L) in the first days of life and this is usually lethal (Hauser

Table 5.6 Inherited defects of the urea cycle, triple 'H' syndrome and lysinuric protein intolerance

Defect	Clinical	Abnormal biochemistry	Therapeutic agents	Diagnostic tissue
N-Acetylglutamate synthetase deficiency	Two cases: one died of hyperammonaemia at 8 days; one, episodic hyperammonaemia, died at 9 years	*Plasma*: raised glutamine and alanine, low ornithine and arginine,* citrulline low *Urine*: orotic acid normal at presentation; variable increase with therapy	Citrulline or arginine; *N*-carbamylglutamate	Liver
Carbamylphosphate synthetase I deficiency	Most, severe neonatal illness; a few, late onset	*Plasma*: amino acids as;* citrulline low *Urine*: orotic acid low	Citrulline or arginine; sodium benzoate ± phenylacetate or phenylbutyrate	Liver Prenatal: CVS for DNA
Ornithine transcarbamylase (OTC) deficiency	X-linked: Hemizygous males: most, lethal neonatal illness; some, mild variants, episodic illness Heterozygous females: wide variation, including neonatal illness, episodic hyperammonaemia, no symptoms; protein aversion common	When hyperammonaemic: *Plasma*: amino acids as;* citrulline low *Urine*: increased orotic acid and other pyrimidines *Asymptomatic carriers*: biochemistry often normal; stress tests: allopurinol (urinary orotidine); protein load (orotic acid)	Citrulline or arginine; benzoate ± phenylacetate or phenylbutyrate	Liver Prenatal: CVS for DNA (80% proven carrier mothers); fetal sexing
Argininosuccinate synthetase deficiency (citrullinaemia)	Acute neonatal or episodic illness; a few, abnormal brittle hair (trichorrhexis nodosa)	*Plasma*: amino acids as;* raised citrulline (1000–5000 μmol/L) *Urine*: gross citrullinuria; orotic acid raised	Arginine	Fibroblasts Prenatal: CVS or amniocytes for enzyme; amniotic fluid citrulline
Argininosuccinate lyase deficiency (argininosuccinic aciduria)	Neonatal: acute; often gross hepatomegaly; abnormal facies; hair normal Late presenting: varies; episodic illness to minimal symptoms; 50% trichorrhexis nodosa	*Plasma*: amino acids as;* raised citrulline (100–300 μmol/L) and argininosuccinic acid *Urine*: argininosuccinic acid with anhydrides, and citrullinuria; orotic acid raised (may be argininosuccinic aciduria only)	Arginine	Red cells Fibroblasts Prenatal: CVS or amniocytes for enzyme; amniotic fluid arginino-succinate

continued

Table 5.6 *Continued*

Defect	Clinical	Abnormal biochemistry	Therapeutic agents	Diagnostic tissue
Arginase deficiency	Developmental delay; progressive spasticity; mental retardation; symptomatic hyperammonaemia infrequent	*Plasma*: ammonium and glutamine normal, or small increase; arginine raised (up to 1500 μmol/L) *Urine*: raised cystine, arginine, ornithine, lysine, citrulline and glutamine; orotic acid high; substituted guanidines present	None	Red blood cells Liver Prenatal: fetal red blood cells, potentially
Triple 'H' syndrome: ornithine transport into mitochondria	Varies: neonatal; developmental delay with fits; no symptoms; protein aversion common; may be clotting disorder (factors VII and X deficient)	*Plasma*: raised ornithine, glutamine, alanine; low lysine *Urine*: homocitrulline, raised ornithine (if severe), often increased orotic acid	Citrulline	Fibroblasts No prenatal
Lysinuric protein intolerance: renal and intestinal transport of ornithine, arginine and lysine	50% of patients from Finland; periodic hyperammonaemia, failure to thrive, hepatosplenomegaly, sparse hair, lung disease, protein aversion	*Plasma*: ammonium: fasting normal, postprandial raised; raised: glutamine, alanine, citrulline; low: ornithine, arginine, lysine; raised: LDH, ferritin, thyroid-binding globulin *Urine*: raised lysine (massive), ornithine, arginine; orotic acid (postprandial)	Citrulline	None No prenatal

* Raised glutamine and alanine, low ornithine and arginine.
CVS, chorionic villus sample; LDH, lactate dehydrogenaze.

et al., 1990). Less often, the mutation causes production of a kinetically abnormal enzyme and partial OTC deficiency. Presentation is later with intermittent hyperammonaemia (Batshaw *et al.*, 1986; Drogari & Leonard, 1988; Wendel *et al.*, 1989). The OTC gene has been mapped to the short arm of the X chromosome. Around 30% of cases arise as new mutations. Gene deletions have been identified in 10% of affected boys, and in a small number of others point mutations at a restriction enzyme cleavage site. In the majority, the gene abnormality is not demonstrable. Restriction fragment length polymorphisms (RFLPs) have been identified at the OTC locus and are used to track the abnormal gene in family studies by linkage analysis (Brusilow & Horwich, 1989).

In girls, one of each pair of X chromosomes is inactivated randomly in all cells, so that only one X chromosome is expressed (Lyon, 1962). Girls who are heterozygous for a mutant OTC gene express the abnormal gene in some liver cells, and the normal one in the others. The severity of the defect depends on how many liver cells express the normal gene (Ricciuti *et al.*, 1976), and whether the mutation causes a severe or a milder variant abnormality. As many as 18% of carriers have presented

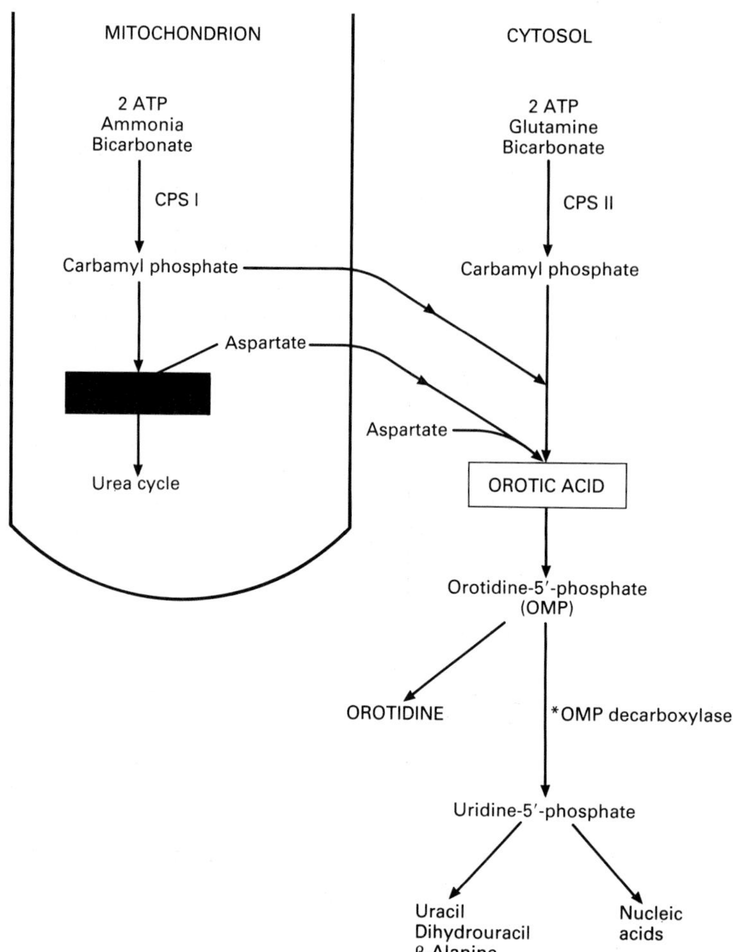

Fig. 5.5 Enhanced production of orotic acid and other pyrimidines in urea cycle defects indicated by the dark band.* Oritidine-5′-phosphate (OMP) decarboxylase is the enzyme inhibited by the allopurinol metabolite, oxypurinol ribonucleotide (p. 109). CPS, Carbamylphosphate synthetase.

with hyperammonaemia. Of these, around 20% died and half the survivors had serious neurological sequelae (Batshaw *et al.*, 1986; Rowe *et al.*, 1986). Two women presented after childbirth (Arn *et al.*, 1990). Even asymptomatic girls may have evidence of mild cerebral dysfunction with careful testing (Batshaw *et al.*, 1980), and are at risk of hyperammonaemic episodes.

Carrier detection for OTC deficiency

It is important to identify carriers in OTC-deficient kindred, firstly because they are at risk of hyperammonaemia, and secondly because they may transmit the abnormal gene to half their own children. A mother who has had one affected child may not be a carrier – there may have been a new mutation. For more than 80% of confirmed carriers, prenatal diagnosis is now possible by DNA analysis: in a minority of cases by identification of a gene deletion or point mutation; in the majority by linkage analysis using RFLPs to track the abnormal gene.

Pedigree analysis alone may indicate carrier status. More often, it is necessary to look for biochemical evidence. Many carriers have a normal plasma ammonium level and orotic acid excretion. However, with a stress test, excessive pyrimidine production may be demonstrable. In protein load tests, 1 g/kg of oral protein is given after an overnight fast, as a high-protein meal or commercial formula feed. Plasma ammonium is measured basally and at 1, 2 and 4 h after loading, and urine is collected for 12 h as individual samples or 4-hourly pools. A positive response is an increased orotate excretion and rise in plasma ammonium; for example: orotate >4.4 mmol/mol creatinine (>6.0 μg/mg creatinine) and ammonium >46 μmol/L (>3 SDs above normal mean basal levels) (Batshaw *et al.*, 1989). These tests are useful and specific, but produce around 10% false negative results. Lethargy and vomiting may occur, and it is advisable to have intravenous sodium benzoate ready, in case severe hyperammonaemic symptoms develop. Oral alanine load tests may produce false positive results and are not recommended for carrier testing (Batshaw *et al.*, 1989; MacKenzie *et al.*, 1989). Another approach is to use allopurinol to demonstrate increased pyrimidine production (Hauser *et al.*, 1990). A metabolite of this drug, oxopurinol monophosphate, inhibits orotidine monophosphate decarboxylase (Fig. 5.5). Accumulating orotidine monophosphate is dephosphorylated, and urinary orotidine excretion is increased. Urine is collected for 24 h after a dose of 300 mg allopurinol has been administered. The test is safe and specific, but produces 10% false negative results. Few laboratories currently measure orotidine.

Management of urea cycle defects: principles

Acute hyperammonaemic coma is a medical emergency and aggressive treatment must be instituted quickly. The aims are to: limit the nitrogen load on the urea cycle; maximize any residual activity of the cycle; remove ammonium; and provide alternatives to urea for waste nitrogen excretion. Until an organic acidaemia has been excluded, carnitine is often supplied as well. For detailed regimens, refer to Brusilow and Horwich (1989) and Bachmann (1990). Oral protein and amino acid infusions are stopped and glucose is infused to supply calories and limit endogenous protein catabolism. Oral lactulose is given to acidify the gut contents, lower the NH_3:NH_4^+ ratio and thereby reduce ammonia absorption, and oral antibiotics are given to decrease the gut bacterial flora. Arginine is infused; firstly, to replenish ornithine in the urea cycle and, secondly, to drive the cycle by activating NAGS and indirectly, therefore, CPS I. Ammonium is removed by haemodialysis or arteriovenous haemofiltration (Sperl *et al.*, 1992), which work more quickly than peritoneal dialysis. When the amino acid and orotic acid results are available, and the site of the defect has been localized, nitrogen excretion is promoted in alternative forms to urea. In AS and AL deficiencies, the urea cycle intermediates which accumulate, citrulline and arginine succinate, are rapidly cleared by the kidney. By increasing the arginine supplement, to provide a continuous source of the carrier ornithine, large amounts of waste nitrogen can be excreted as these intermediates. In NAGS, CPS I and OTC deficiencies a different approach is required. After excluding an organic acid defect, sodium benzoate is infused intravenously. This is esterified to benzoyl-CoA, which conjugates with glycine to form hippuric acid, a harmless compound cleared rapidly by the kidneys. Glycine is constantly replenished using nitrogen derived from alanine and glutamate (Fig. 5.6). Used in the recommended doses, sodium benzoate is non-toxic. Accidental overdose has caused cardiovascular collapse, pyrexia and severe metabolic acidosis. Plasma concentrations should not exceed 2000 μmol/L (Bachmann, 1990). Benzoate assays are not widely available for monitoring, however. Phenylacetate is also infused in emergencies, but smells unpleasant. Its precursor, phenylbutyrate, is less offensive and can be given orally. By conjugating with glutamine, phenylacetate removes two nitrogen atoms per molecule as phenylacetylglutamine.

Similar principles are adopted for long-term treatment. Dietary protein is restricted to reduce plasma ammonium to normal levels (if possible) but still permit growth. Essential amino acids, adequate calories, vitamins and trace elements must be provided. Arginine is given in AS and AL deficiencies. Citrulline is often a preferred alternative in CPS I and OTC deficiencies, since it not only is a source of arginine and ornithine, but also condenses with aspartate. Oral sodium benzoate and phenylbutyrate improve protein tolerance. Plasma amino acids are monitored closely: preprandial arginine

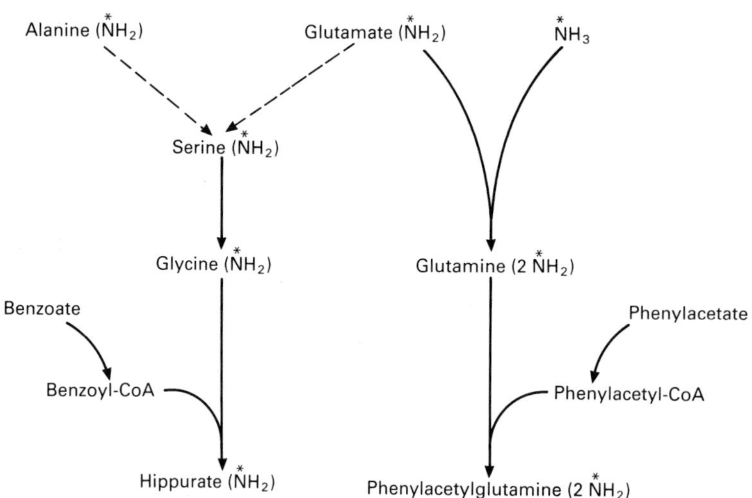

Fig. 5.6 Diversion of waste amino acid nitrogen from the urea cycle by benzoate and phenylacetate. (*NH$_2$) represents the amino groups conjugated with benzoate or phenylacetate for excretion.

should be 100–150 μmol/L; threonine around the upper normal limit; glycine above 100 μmol/L on benzoate therapy; glutamine may increase before hyperammonaemia develops. Urinary orotic acid is helpful in monitoring treatment for OTC deficiency.

The outcome for newborns rescued from severe hyperammonaemic coma is poor, particularly if it is prolonged. Survivors are then severely brain damaged (Msall *et al.*, 1984). When it has been possible to prevent hyperammonaemia from birth, normal development has been achieved. In the late-onset disorders, careful treatment has improved growth, behaviour and nutrition, and reduced the frequency of hyperammonaemic episodes.

Triple 'H' syndrome and lysinuric protein intolerance

In these two inherited autosomal recessive disorders, the intramitochondrial concentration of ornithine is very low. This limits urea cycle function. Both may present with episodic hyperammonaemia of variable severity (Table 5.6). In triple 'H' syndrome (hyperammonaemia, hyperornithinaemia, homocitrullinuria), there is probably a defect of the carrier protein which transports ornithine into mitochondria. Carbamyl phosphate accumulates: some combines with lysine, in lieu of ornithine, producing homocitrulline which is excreted in the urine; and some stimulates pyrimidine synthesis so that orotic acid excretion increases. Treatment follows the principles for urea cycle enzyme defects. Citrulline supplements have been given to increase intramitochondrial ornithine, but their value is disputed (Shih, 1990; Tuchman *et al.*, 1990). Patients with lysinuric protein intolerance have a defect in renal and intestinal transport of the basic amino acids, ornithine, arginine and lysine. Their plasma concentrations are low, but urinary excretion is high with, in particular, massive excretion of lysine. Plasma ammonium and urinary orotic acid may be normal when fasting, but increase for several hours after a protein meal. Treatment is with oral citrulline and protein restriction, if necessary. The prognosis is good (Rajantie, 1981; Simell, 1989).

Organic acid disorders

Severe hyperammonaemia complicates a range of inherited organic acid defects during episodes of acute decompensation (Table 5.5). Urinary orotic acid is not increased (Bachmann & Colombo, 1980). In most cases, the probable explanation is a decrease in *N*-acetylglutamate, the allosteric activator of CPS I. Abnormal organic acid intermediates accumulating in the mitochondria trap CoASH, forming acyl-CoA derivatives. This leads to a deficiency of acetyl-CoA, needed for *N*-acetylglutamate synthesis. In addition, some acyl-CoA molecules (for example, propionyl-CoA) may inhibit NAGS directly (Rosenberg & Fenton, 1989; Meijer *et al.*, 1990). Hyperammonaemia is uncommon in primary lactic acidoses. It has been reported in very severe deficiencies of pyruvate dehydrogenase, probably because of a depletion of intramitochondrial acetyl-CoA (Matsuo *et al.*, 1985; Brown *et al.*, 1987). It occurs also in patients with severe pyruvate carboxylase deficiency (cross-reacting material (CRM) negative), in association with raised plasma alanine, citrulline and lysine. The explanation here is an inability to carboxylate pyruvate to oxaloacetate, which normally carries nitrogen into the urea cycle as aspartate for disposal (Fig. 5.4). Citrulline and ammonia accumulate (Robinson, 1985).

Reye's syndrome

This is a rare disorder of young children in which there is a severe non-infective brain disturbance (encephalopathy) associated with a fatty liver and deranged liver function (Hall, 1986). When first reporting the syndrome, Reye *et al.* (1963) commented that it was unlikely that the aetiology would be the same in all cases. (The urea cycle defects were only just being recognized at this time, and the first inherited organic aciduria was reported 4 years later.)

Typically, a previously healthy child has a common viral infection, which apparently takes a normal course. At 3 to 5 days from the onset, however, there is profuse effortless vomiting and lethargy, progressing rapidly to delirium and coma, sometimes with fits. Mortality in the British Isles has been around 40% and in the USA 20%, perhaps reflecting a difference in the age of affected children: between 1981 and 1986, the median was 14 months, British Isles; 8–9 years, USA (Hall, 1986, 1990). In the acute illness, the liver is enlarged and laden with fat. Glycogen is depleted. On electron microscopy, the liver mitochondria are swollen, with expansion of the matrix space and granularity of the matrix. Peroxisomes and endoplasmic reticulum are increased (Partin *et al.*, 1971). Activities of mitochondrial enzymes, including CPS I and OTC (urea cycle), glutamate dehydrogenase, pyruvate carboxylase (gluconeogenesis), citrate synthetase, succinic dehydrogenase and pyruvate dehydrogenase, are low. Measured cytosolic enzymes are normal (de Vivo, 1978; Robinson *et al.*, 1978). The brain is oedematous, and there may be fat infiltration of the heart, skeletal muscles and kidney. Plasma transaminases are raised (three to 30 times normal), bilirubin seldom exceeds 17 μmol/L, there may be hypoglycaemia, and ammonium is transiently (24–48 h) raised to between two and 20 times normal. Plasma concentrations of short- and medium-chain fatty acids are raised and urinary excretion of medium-chain dicarboxylic acids, from incomplete fat oxidation, may be increased (Volk, 1981; Heubi *et al.*, 1987; Brown & Imam, 1991). Creatine kinase is often raised. Plasma amino acids are deranged and there are reported increases, particularly, in glutamine, alanine, lysine, ornithine and α-aminobutyric acid (Romshe *et al.*, 1981). However, the liver disturbance is evanescent and within 3 or 4 days the histology returns to normal, even if coma persists. Criteria used for defining Reye's syndrome in the UK National Surveillance Scheme (Hall, 1990) are: a child under 16 years with an unexplained non-inflammatory encephalopathy and one or more of: serum hepatic transaminases > three times the upper limit of normal; plasma ammonium > three times the normal upper limit; characteristic fatty infiltration of the liver. This rather broad definition would include a range of disorders.

The cause of the syndrome has attracted much speculation. It has become clear that a significant proportion of children presenting with a Reye's syndrome-like illness have an inherited metabolic defect (Table 5.7) (Robinson, 1987; Greene *et al.*, 1988; Brown & Imam, 1991; Green & Hall, 1992), with metabolic decompensation triggered by the viral infection. These conditions are not identified without appropriate investigation and have probably been missed in the past. Presentation under 3 years (Forsyth *et al.*, 1991), recurrent episodes and a family history of unexplained sibling death should alert suspicion. In a minority of other children the cause has been exposure to a known hepatic toxin (Starko & Mullick, 1983): aflatoxin, a fungal food contaminant (Thailand); margosa oil (India and Malaysia); hypoglycin in unripe ackee apples (West Indies); and pesticides (Canada). A few have had salicylate poisoning. In others, a non-metabolic

Table 5.7 Metabolic disorders that may present as a Reye's syndrome-like illness

Inherited metabolic defects
Fatty acid β-oxidation defects
3-Hydroxy-3-methylglutaryl-CoA (HMGCoA) lyase deficiency
Propionic, methylmalonic and isovaleric acidaemias
2-Methylacetoacetyl-CoA thiolase (β-ketothiolase) deficiency
Glutaric aciduria type I
Urea cycle defects
Hereditary fructose intolerance
Drugs and toxins
Sodium valproate (idiosyncratic reaction)
Poisoning with: salicylate; hypoglycin A from unripe ackee apples; aflatoxin; margosa oil; pesticides

cause, such as sepsis, was found in retrospect. Where no cause can be identified, it is proposed that a viral infection triggers an abnormal reaction in an individual whose susceptibility is increased by genetic factors and/or an exogenous toxin. Salicylates, used frequently to treat feverish children, were implicated because they disturb mitochondrial functions *in vitro*. Six case-control studies in the USA and UK supported a possible association (Hall, 1990). Although some studies had design faults, recommendations were issued in the USA (1982) and UK (1986) to avoid giving aspirin to feverish children under 12 years.

The number of reported cases of Reye's syndrome with no identifiable cause has decreased from a peak of 79 in 1983 to 14 in 1990/91 (British Paediatric Surveillance Unit, 1991). This may be due to more careful investigation at presentation and, perhaps, to discontinuation of aspirin use in young children. It is essential to investigate thoroughly for an inherited metabolic disorder all children with a Reye's syndrome-like illness (see p. 114). Plasma salicylate analysis will pick up the rare cases of aspirin poisoning.

Acute and chronic liver failure

Hyperammonaemia complicates acute and chronic liver failure. In both, cell destruction leads to decreased urea cycle capacity and loss of glutamine synthetase activity. In cirrhosis, a more important factor usually is development of a collateral circulation of blood from the portal system to the inferior vena cava, bypassing the liver. Uptake of ammonia by skeletal muscle and detoxification to glutamine affords some protection (Souba, 1987). In acute hepatic failure, plasma methionine, tyrosine, phenylalanine and lysine are often raised, as are glutamine and alanine (Romshe *et al.*, 1981).

Hypoxia and ischaemia

Plasma ammonium is raised transiently in newborns who have suffered severe fetal distress and birth asphyxia (178–565 μmol/L in one series; Goldberg *et al.*, 1979). Plasma transaminases are usually raised too. An increased nitrogen load and hepatic ATP depletion may be contributory (see below). Children who are shocked, have had prolonged hypotension, or who have been resuscitated from a near-miss cot death may similarly have high levels (Briddon & Oberholzer, 1986). Right-sided heart failure causing necrosis of hepatocytes also increases ammonium levels. This has been attributed to loss of protective glutamine synthetase activity (Meijer *et al.*, 1990).

Transient hyperammonaemia of the newborn (THAN)

This is a rare overwhelming illness that develops in the first 24 h of life. It is associated with respiratory distress, most often owing to hyaline membrane disease, and severe hyperammonaemia (possibly >1000 μmol/L) causing coma (Ballard *et al.*, 1978; Hudak *et al.*, 1985; Tokatli *et al.*, 1991). Affected babies are almost always preterm, although it does occur rarely in term infants. They have not suffered birth asphyxia. Around 30% have died as a result of hyperammonaemia and complications of prematurity. With prompt aggressive treatment, plasma ammonium falls rapidly and remains normal, and the outlook for survivors has generally been good. A few have had serious neurological sequelae. The diagnosis can be made only after excluding inherited urea cycle defects and organic acid disorders.

The cause remains speculative. Plasma citrulline is moderately raised, but urea cycle enzymes have been normal when measured. Immaturity is an unlikely explanation, since the urea cycle is active at 16 weeks of gestation (Räihä & Suihkonen, 1968). Tissue hypoxia/ischaemia might be responsible: firstly, by increasing the nitrogen load for excretion because of enhanced degradation of purine nucleotides; and, secondly, through depletion of liver ATP which is a cofactor for CPS I and AS in the urea cycle. A fall in hepatic blood flow, because of shunting through the ductus venosus, would exacerbate liver ischaemia (Hudak *et al.*, 1985).

Urinary tract infections

Rarely, hyperammonaemic coma develops in children who have an anatomical abnormality of the lower urinary tract which causes urinary stasis and predisposes to infection. If the bacterium is a urease-producing proteus, ammonia is produced from urea and the urine becomes alkaline (pH 8–9). Because the p*K*a of NH_3 is 9.15, the ratio of NH_3 to NH_4^+ increases to around 1:1. Free ammonia readily diffuses into the bloodstream. Most children have had the prune belly syndrome, a congenital absence or deficiency of the abdominal musculature, accompanied by a large hypotonic bladder and dilated and tortuous ureters. A neurogenic bladder has been another cause (Samtoy & DeBeukelaer, 1980; Kuntze *et al.*, 1985; Diamond *et al.*, 1989).

Leukaemia

Treatment with asparaginase for acute leukaemia has caused hyperammonaemic coma. This enzyme hydrolyses the amide group of asparagine (Leonard & Kay, 1986). Hyperammonaemia has also occurred in leukaemic patients not receiving this drug. Accelerated catabolism may have been contributory (Watson *et al.*, 1985).

Treatment with valproic acid

Valproic acid, 2-n-propylpentanoic acid, is an 8-C branched-chain fatty acid which is used widely as a safe antiepileptic drug. It is metabolized by the liver, after conjugation to valproyl-CoA, mainly by fatty acid β-oxidation (Cotariu & Zaidman, 1988). Some is excreted in urine as valproyl carnitine. A small increase in plasma ammonium occurs commonly during treatment, in the absence of clinical liver dysfunction. It is commoner in young children, especially if they are on multiple drugs. It is reversible and usually asymptomatic, although drowsiness has been reported (Cotariu & Zaidman, 1988; Thom *et al.*, 1991). In normal adults stressed by an alanine load, valproate decreased urea synthesis, increased plasma ammonium and reduced the urinary orotic acid response. The findings indicate that valproate inhibits CPS I (Hjelm *et al.*, 1986), probably because of decreased production of its activator, *N*-acetylglutamate (Coude *et al.*, 1981). Valproyl-CoA may sequester CoASH and thereby reduce the availability of acetyl-CoA. Valproate also increases ammoniagenesis from glutamine in the kidney, and renal vein ammonium (Warter *et al.*, 1983).

Two serious complications of valproate therapy, associated with hyperammonaemia, have been reported. A small number of patients with unsuspected inherited urea cycle disorders have developed acute hyperammonaemia during treatment. These have included boys hemizygous for a variant OTC defect (Tripp *et al.*, 1981; Kennedy & Cogswell, 1989), girls heterozygous for OTC deficiency (Batshaw & Brusilow, 1982; Hjelm *et al.*, 1986) and an adult with argininosuccinic aciduria (Morgan *et al.*, 1987). In these cases, the further 'brake' by valproate on a defective urea cycle precipitated acute hyperammonaemia. Clearly, a careful medical and family history is mandatory before starting valproate treatment.

Fatal hepatotoxicity with a fatty liver is the second, very rare complication (Gerber *et al.*, 1979; Ware & Millward-Sadler, 1980; Cotariu & Zaidman, 1988). The estimated occurrence is 1 in 37 000 of patients of all ages, treated only with sodium valproate. However, the frequency increased to 1 in 500 in children under 2 years receiving polytherapy, and with neurological defects in addition to epilepsy (Dreifuss *et al.*, 1987). Presentation is with a Reye's syndrome-like illness, often following a viral infection. Starvation may be a precipitating factor. It appears to be idiosyncratic, is not dose dependent and occurs early in treatment. It cannot be predicted from liver function tests. It may be caused by inhibition of fatty acid oxidation by the metabolite 4-en-valproic acid. This is structurally similar to the toxin hypoglycin (Cotariu & Zaidman, 1988). Reduced availability of carnitine may be a further factor (Thom *et al.*, 1991).

Parenteral nutrition with crystalline L-amino acids

Unlike casein or fibrin hydrolysates, the crystalline L-amino acid preparations available now for intravenous nutrition have a low ammonium content. Nevertheless, their use still causes a moderate asymptomatic increase in plasma ammonium, up to around twice the upper normal limit, in both preterm and term babies (Shohat *et al.*, 1984; Coran & Drongowski, 1987; Puntis *et al.*, 1988). Higher levels may occur with sepsis (Thomas *et al.*, 1982). Routine ammonium measurement is unnecessary during parenteral nutrition, but should be undertaken if unexplained lethargy develops.

Investigation of hyperammonaemia

Because of the serious damaging effects of hyperammonaemia if it is not treated quickly (Msall *et al.*, 1984), a hospital serving newborn and paediatric patients must be able to offer analysis of plasma ammonium at any time. Reliable methods are now available, which require only small volumes of plasma. Cation exchange resin procedures are too slow for emergency use, but enzymatic methods, a dry film method (Kodak Ektachem) and ammonium-specific electrodes are generally satisfactory alternatives. Older analyses based on alkalinization and microdiffusion may be unreliable (Green, 1988; Brusilow & Horwich, 1989; da Fonseca-Wollheim, 1990a). With care to avoid contamination by ammonium in water, laboratory reagents, sample collection tubes and reaction vials, accurate analysis should not be a problem. Spurious results are much more likely to arise from unsuitable blood samples or inappropriate sample preparation. Ammonium is increased in blood from a struggling or fitting child because of release from muscles; it is higher in capillary than arterial or venous blood, because of tissue damage, platelet aggregation and contamination during collection, and it is raised in haemolysed samples and in serum compared with plasma (Beddis *et al.*, 1980; Colombo *et al.*, 1984). The main practical problem, however, is that it is released in substantial amounts *in vitro* by enzymatic hydrolysis of glutamine, adenosine and other amido compounds in blood. This starts within minutes of sample collection, and continues even after centrifugation of plasma (da Fonseca-Wollheim, 1990a). Deamidation of glutamine may be accelerated in samples with increased γ-glutamyltransferase (da Fonseca-Wollheim, 1990b). For ammonium estimation, venous blood should be collected into ethylenediaminetetraacetate (EDTA) or sodium or lithium heparin, preferably transported on water ice and centrifuged within 15 min, and no longer than 30 min. Analyses should be carried out immediately. If they are not, plasma should be stable at −70°C for up to 3 weeks (Green, 1988; da Fonseca-Wollheim, 1990a).

Ammonium should be measured in any *newborn* with unexplained lethargy, neurological disturbance, fits, vomiting or hyperventilation, especially if the baby was well initially, in the first 24 to 72 h of life. Later, measurement is indicated in any child presenting with a Reye's syndrome-like illness or unexplained coma, with episodes of unsteadiness (ataxia), vomiting, drowsiness, disorientation or bizarre behaviour, or with severe epilepsy. Because of the range of inherited metabolic disorders that may precipitate hyperammonaemic coma and Reye's syndrome, it is essential that appropriate samples are collected at presentation for biochemical analyses. Guidelines are presented in Table 5.8. Results to provide a working diagnosis should be available within 24 h, and a more definitive answer within 2–3 days. A laboratory with a limited repertoire should seek outside specialist help at an early stage. Emergency management must be instituted as soon as blood samples have been taken. The diagnosis of a probable urea cycle defect should be confirmed by enzyme analysis, and blood collected for future DNA studies, even if probes for the defect are not available now. If a child

Table 5.8 Samples from an encephalopathic child with hyperammonaemia or a Reye's syndrome-like illness

Blood for:	(Ammonium)
	Glucose
	Acid–base status
	*3-Hydroxybutyrate and free fatty acids
	Salicylate
	*Carnitine
	Lactate
	Clotting studies
Urine for:	Ketones
	Organic acids
	Amino acids
	Orotic acid
If moribund:	Heparinized blood:
	Red cells washed in physiological or phosphate-buffered saline and frozen;
	plasma frozen
	EDTA blood (not centrifuged) for DNA
	Skin biopsy for fibroblast culture
	Needle liver biopsies for:
	Urea cycle enzymes (liquid nitrogen)
	routine histology
	electron microscopy

* Analyses may be deferred, pending results of other metabolite analyses.
EDTA, ethylenediaminetetraacetate.

is moribund, appropriate tissues must be collected before death. The findings will be needed for prenatal diagnosis of subsequent pregnancies. In less acute situations, preliminary investigation should include liver function tests and analyses of plasma amino acids and urinary amino, organic and orotic acids.

References

Antunes, J.D., Geffner, M.E., Lippe, B.M. & Landaw, E.M. (1990) Childhood hypoglycemia: differentiating hyperinsulinemic from nonhyperinsulinemic causes. *J Pediatr* **116**, 105–108.

Arn, P.H., Hauser, E.R., Thomas, G.H. *et al.* (1990) Hyperammonemia in women with a mutation at the ornithine carbamoyltransferase locus. *N Engl J Med* **322**, 1652–1655.

Aynsley-Green, A. (1981) Nesidioblastosis of the pancreas in infancy. *Develop Med Child Neurol* **23**, 372–379.

Aynsley-Green, A. (1989) Hypoglycaemia. In: *Clinical Paediatric Endocrinology* (ed. C.G.D. Brook), 2nd edn, pp. 618–637. Blackwell Scientific Publications, Oxford.

Aynsley-Green, A., Polak, J.M., Bloom, S.R. *et al.* (1981) Nesidioblastosis of the pancreas: definition of the syndrome and the management of the severe neonatal hyperinsulinaemic hypoglycaemia. *Arch Dis Child* **56**, 496–508.

Aynsley-Green, A., Williamson, D.H. & Gitzelmann, R. (1977) Hepatic glycogen synthetase deficiency. *Arch Dis Child* **52**, 573–579.

Bachmann, C. (1990) Urea cycle disorders. In: *Inborn Metabolic Diseases* (eds J. Fernandes, J.-M. Saudubray & K. Tada), pp. 211–228. Springer-Verlag, Berlin.

Bachmann, C. & Colombo, J.P. (1980) Diagnostic value of orotic acid excretion in heritable disorders of the urea cycle and in hyperammonemia due to organic acidurias. *Eur J Pediatr* **134**, 109–113.

Baerlocher, K. (1990) Disorders of gluconeogenesis. In: *Inborn Metabolic Diseases* (eds J. Fernandes, J.-M. Saudubray & K. Tada), pp. 113–123. Springer-Verlag, Berlin.

Ballard, R.A., Vinocur, B., Reynolds, J.W. *et al.* (1978) Transient hyperammonemia of the preterm infant. *N Engl J Med* **299**, 920–925.

Bartlett, K., Aynsley-Green, A., Leonard, J.V. & Turnbull, D.M. (1991) Inherited disorders of mitochondrial β-oxidation. In: *Inborn Errors of Metabolism* (eds J. Schaub, F. Van Hoof & H.L. Vis), pp. 19–41. Nestlé Nutrition Workshop Series, 24. Vevey/Raven Press Ltd., New York.

Batshaw, M.L. & Brusilow, S.W. (1978) Asymptomatic hyperammonemia in low birthweight infants. *Pediatr Res* **12**, 221–224.

Batshaw, M. & Brusilow, S. (1982) Valproate-induced hyperammonemia. *Ann Neurol* **11**, 319–321.

Batshaw, M.L., Msall, M., Beaudet, A.L. & Trojak, J. (1986) Risk of serious illness in heterozygotes for ornithine transcarbamylase deficiency. *J Pediatr* **108**, 236–241.

Batshaw, M.L., Naylor, E.W. & Thomas, G.H. (1989) False positive alanine tolerance test results in heterozygote detection of urea cycle disorders. *J Pediatr* **115**, 595–598.

Batshaw, M.L., Roan, Y., Jung, A.L., Rosenberg, L.A. & Brusilow, S.W. (1980) Cerebral dysfunction in asymptomatic carriers of ornithine transcarbamylase deficiency. *N Engl J Med* **302**, 482–485.

Beddis, J.R., Hughes, E.A., Rosser, E. & Fenton, J.C.B. (1980) Plasma ammonia levels in newborn infants admitted to an intensive care baby unit. *Arch Dis Child* **55**, 516–520.

Bennett, M.J. (1990) The laboratory diagnosis of inborn errors of mitochondrial fatty acid oxidation. *Ann Clin Biochem* **27**, 519–531.

Bier, D.M., Leake, R.D., Haymond, M.W. *et al.* (1977)

Measurement of 'true' glucose production rates in infancy and childhood with 6,6-dideuteroglucose. *Diabetes* **26**, 1016–1023.

Blackburn, S.T. & Loper, D.L. (1992) Carbohydrate, fat, and protein metabolism. In: *Maternal, Fetal, and Neonatal Physiology: A Clinical Perspective*, pp. 583–613. WB Saunders, Philadelphia.

Bloom, S.R. & Johnston, D.I. (1972) Failure of glucagon release in infants of diabetic mothers. *Br Med J* **IV**, 453–454.

Bonnefont, J.P., Specola, N.B., Vassault, A. *et al.* (1990) The fasting test in paediatrics: application to the diagnosis of pathological hypo- and hyperketotic states. *Eur J Pediatr* **150**, 80–85.

Briddon, A. & Oberholzer, V.G. (1986) Plasma amino acid patterns in critically ill children. *J Inher Metab Dis* **9** (Suppl. 2), 254–256.

British Paediatric Surveillance Unit (1991) *Annual Report*.

Brown, G.K., Scholem, R.D., Hunt, S.M., Harrison, J.R. & Pollard, A.C. (1987) Hyperammonaemia and lactic acidosis in a patient with pyruvate dehydrogenase deficiency. *J Inher Metab Dis* **10**, 359–366.

Brown, J.K. & Imam, H. (1991) Interrelationships of liver and brain with special reference to Reye syndrome. *J Inher Metab Dis* **14**, 436–458.

Brusilow, S.W. (1985a) Inborn errors of urea synthesis. In: *Genetics and Metabolic Disease in Pediatrics* (eds J.K. Lloyd & C.R. Scriver), pp. 140–165. Butterworth, London.

Brusilow, S.W. (1985b) Disorders of the urea cycle. *Hosp Prac* **20**, 65–72.

Brusilow, S.W. & Horwich, A.L. (1989) Urea cycle enzymes. In: *The Metabolic Basis of Inherited Disease* (eds C.R. Scriver, A.L. Beaudet, W.S. Sly & D. Valle), 6th edn, pp. 629–663. McGraw-Hill, New York.

Collins, J.E., Leonard, J.V., Teale, D. *et al.* (1990) Hyperinsulinaemic hypoglycaemia in small for date babies. *Arch Dis Child* **65**, 1118–1120.

Colombo, J.P., Peheim, E., Kretschmer, R., Dauwalder, H. & Sidiropoulos, D. (1984) Plasma ammonia concentrations in newborns and children. *Clin Chim Acta* **138**, 283–291.

Cooper, A.J.L. & Plum, F. (1987) Biochemistry and physiology of brain ammonia. *Physiol Rev* **67**, 440–519.

Coran, A.G. & Drongowski, R.A. (1987) Studies on the toxicity and efficacy of a new amino acid solution in pediatric parenteral nutrition. *J Parent Ent Nutr* **11**, 368–377.

Cornblath, M., Schwartz, R., Aynsley-Green, A. & Lloyd, J.K. (1990) Hypoglycemia in infancy: the need for a rational definition. *Pediatrics* **85**, 834–837.

Cotariu, D. & Zaidman, J.L. (1988) Valproic acid and the liver. *Clin Chem* **34**, 890–897.

Coude, F.X., Rabier, D., Cathelineau, L. *et al.* (1981) A mechanism for valproate induced hyperammonemia. *Pediatr Res* **15**, 974–975.

Cox, T.M. (1988) Hereditary fructose intolerance. *Quart J Med* **68**, 585–594.

Cryer, P.E. (1992) Glucose homeostasis and hypoglycemia. In: *Williams Textbook of Endocrinology* (eds J.D. Wilson & D.W. Foster), 8th edn, pp. 1223–1253. WB Saunders, Philadelphia.

Dahlquist, G., Gentz, J., Hagenfeldt, L. *et al.* (1979) Ketotic hypoglycemia of childhood – a clinical trial of several unifying etiological hypotheses. *Acta Paed Scand* **68**, 649–656.

Daneman, D., Frank, M., Perlman, K., Tamm, J. & Ehrlich, R. (1989) Severe hypoglycemia in children with insulin-dependent diabetes mellitus: frequency and predisposing factors. *J Pediatr* **115**, 681–685.

Diamond, D.A., Blight, A & Ransley, P.G. (1989) Hyperammonemic encephalopathy: a complication associated with the prune belly syndrome. *J Urol* **142**, 361–362.

Dreifuss, F.E., Santilli, N., Langer, D.H. *et al.* (1987) Valproic acid hepatic fatalities: a retrospective review. *Neurology* **37**, 379–385.

Drogari, E. & Leonard, J.V. (1988) Late onset ornithine carbamoyl transferase deficiency. *Arch Dis Child* **63**, 1363–1367.

Dunger, D.B. & Leonard, J.V. (1982) Value of the glucagon test in screening for hepatic glycogen storage disease. *Arch Dis Child* **57**, 384–389.

Fernandes, J. (1990) The glycogen storage diseases. In: *Inborn Metabolic Diseases* (eds J. Fernandes, J.-M. Saudubray & K. Tada), pp. 69–85. Springer-Verlag, Berlin.

da Fonseca-Wollheim, F. (1990a) Preanalytical increase of ammonia in blood specimens from healthy subjects. *Clin Chem* **36**, 1483–1487.

da Fonseca-Wollheim, F. (1990b) Deamidation of glutamine by increased plasma γ-glutamyltransferase is a source of rapid ammonia formation in blood and plasma specimens. *Clin Chem* **36**, 1479–1482.

Forest, M.G. (1989) Adrenal steroid deficiency states. In: *Clinical Paediatric Endocrinology* (ed. C.G.D. Brook), 2nd edn, pp. 368–406. Blackwell Scientific Publications, Oxford.

Forsyth, B.W., Shapiro, E.D., Horwitz, R.I., Viscoli, C.M. & Acampora, D. (1991) Misdiagnosis of Reye's-like illness. *Am J Dis Child* **145**, 964–966.

Gerber, N., Dickinson, R.G., Harland, R.C. *et al.* (1979) Reye-like syndrome associated with valproic acid therapy. *J Pediatr* **95**, 142–144.

Gerver, W.J.M., Menheere, P.P.C.A., Schaap, C. & Degraeuwe, P. (1991) The effects of a somatostatin analogue on the metabolism of an infant with Beckwith–Wiedemann syndrome and hyperinsulinaemic hypoglycaemia. *Eur J Pediatr* **150**, 634–637.

Girard, J. (1990) Metabolic adaptations to change of nutrition at birth. *Biol Neonat* **58**, 3–15.

Girard, J., Ferré, P., Pégorier, J.-P. & Duée, P.-H. (1992) Adaptations of glucose and fatty acid metabolism during perinatal period and suckling–weaning transition. *Physiol Rev* **72**, 507–562.

Goldberg, R.N., Cabal, L.A., Sinatra, F.R., Plajstek, C.E. & Hodgman, J.E. (1979) Hyperammonemia associated with perinatal asphyxia. *Pediatrics* **64**, 336–341.

Goodman, H.M. (1968) Multiple effects of growth hormone on lipolysis. *Endocrinology* **83**, 300–308.

Green, A. (1988) When and how should we measure plasma ammonia? *Ann Clin Biochem* **25**, 199–209.

Green, A. & Hall, S.M. (1992) Investigation of metabolic disorders resembling Reye's syndrome. *Arch Dis Child* **67**, 1313–1317.

Greene, C.L., Blitzer, M.G. & Shapira, E. (1988) Inborn errors of metabolism and Reye syndrome: differential diagnosis. *J Pediatr* **113**, 156–159.

Gregersen, N., Lauritzen, R. & Rasmussen, K. (1976) Suberylglycine excretion in the urine from a patient with dicarboxylic aciduria. *Clin Chim Acta* **70**, 417–425.

Hall, S.M. (1986) Reye's syndrome and aspirin: a review. *J R Soc Med* **79**, 596–598.

Hall, S.M. (1990) Reye's syndrome and aspirin: a review. *Br J Clin Pract* **44** (Suppl. 70), 4–11.

Halvorsen, S. (1990) Tyrosinemia. In: *Inborn Metabolic Diseases* (eds J. Fernandes, J.-M. Saudubray & K. Tada), pp. 199–209. Springer-Verlag, Berlin.

Harris, R.A. & Crabb, D.W. (1986) Metabolic interrelationships. In: *Textbook of Biochemistry* (ed. T.M. Devlin), pp. 539–559. John Wiley & Sons, New York.

Hauser, E.R., Finkelstein, J.E., Valle, D. & Brusilow, S.W. (1990) Allopurinol-induced orotidinuria. *N Engl J Med* **322**, 1641–1645.

Hawdon, J.M., Ward Platt, M.P. & Aynsley-Green, A. (1992) Patterns of metabolic adaptation for preterm and term infants in the first neonatal week. *Arch Dis Child* **67**, 357–365.

Hawdon, J.M., Ward Platt, M.P., Lamb, W.H. & Aynsley-Green, A. (1990) Tolerance to somatostatin analogue in a preterm infant with islet cell dysregulation syndrome. *Arch Dis Child* **65**, 341–343.

Haymond, M.W., Karl, I.E., Clarke, W.L., Pagliara, A.S. & Santiago, J.V. (1982) Differences in circulating gluconeogenic substrates during short-term fasting in men, women, and children. *Metabolism* **31**, 33–42.

Haymond, M.W., Karl, I.E., Feigin, R.D., de Vivo, D. & Pagliara, A.S. (1973) Hypoglycemia and maple syrup urine disease: defective gluconeogenesis. *Pediatr Res* **7**, 500–508.

Hers, H.-G. (1990) Mechanisms of blood glucose homeostasis. *J Inher Metab Dis* **13**, 395–410.

Hers, H.-G., Van Hoof, F. & De Barsy, T. (1989) Glycogen storage diseases. In: *The Metabolic Basis of Inherited Disease* (eds C.R. Scriver, A.L. Beaudet, W.S. Sly & D. Valle), 6th edn, pp. 425–452. McGraw-Hill, New York.

Heubi, J.E., Partin, J.C., Partin J.S. & Schubert, W.K. (1987) Reye's syndrome: current concepts. *Hepatology* **7**, 155–164.

Hjelm, M., De Silva, L.V.K., Seakins, J.W.T., Oberholzer, V.G. & Rolles, C.J. (1986) Evidence of inherited urea cycle defect in a case of fatal valproate toxicity. *Br Med J* **292**, 23–24.

Hjelm, M., Oberholzer, V., Seakins, J., Thomas, S. & Kay, J.D.S. (1986) Valproate-induced inhibition of urea synthesis and hyperammonaemia in healthy subjects. *Lancet* **II**, 859.

Hudak, M.L., Jones, M.D. & Brusilow, S.W. (1985) Differentiation of transient hyperammonemia of the newborn and urea cycle enzyme defects by clinical presentation. *J Pediatr* **107**, 712–719.

Kennedy, C.R. & Cogswell, J.J. (1989) Late onset ornithine carbamoyl transferase deficiency in males. *Arch Dis Child* **64**, 638.

Kerr, D.S., Hansen, I.L. & Levy, M.M. (1983) Metabolic and hormonal responses of children and adolescents to fasting and 2-deoxyglucose. *Metabolism* **32**, 951–959.

Koh, T.H.H.G., Aynsley-Green, A., Tarbit, M. & Eyre, J.A. (1988a) Neural dysfunction during hypoglycaemia. *Arch Dis Child* **63**, 1353–1358.

Koh, T.H.H.G., Eyre, J.A. & Aynsley-Green, A. (1988b) Neonatal hypoglycaemia – the controversy regarding definition. *Arch Dis Child* **63**, 1386–1388.

Kuntze, J.R., Weinberg, A.C. & Ahlering, T.E. (1985) Hyperammonemic coma due to proteus infection. *J Urol* **134**, 972–973.

Kvittingen, E.A. (1986) Hereditary tyrosinemia type I – an overview. *Scand J Clin Lab Invest* **46** (Suppl. 184), 27–34.

Lamers, K.J.B., Doesburg, W.H., Gabreëls, F.J.M. *et al.* (1985) The concentration of blood components related to fuel metabolism during prolonged fasting in children. *Clin Chim Acta* **152**, 155–163.

Leonard, J.V & Kay, J.D.S. (1986) Acute encephalopathy and hyperammonaemia complicating treatment of acute lymphoblastic leukaemia with asparaginase. *Lancet* **i**, 162–163.

Lucas, A., Morley, R. & Cole, T.J. (1988) Adverse neurodevelopmental outcome of moderate neonatal hypoglycaemia. *Br Med J* **297**, 1304–1308.

Lyon, M.F. (1962) Sex chromatin and gene action in the mammalian X-chromosome. *Am J Hum Genet* **14**, 135–148.

MacKenzie, A.E., MacLeod, H.L., Heick, H.M.C. & Korneluk, R.G. (1989) False positive results from the alanine loading test for ornithine carbamoyltransferase deficiency heterozygosity. *J Pediatr* **115**, 605–608.

Maire, I., Baussan, C., Moatti, N., Mathieu, M. & Lemonnier, A. (1991) Biochemical diagnosis of hepatic glycogen storage diseases: 20 years French experience. *Clin Biochem* **24**, 169–178.

Matsuo, M., Ookita, K., Takemine, H., Koike, K. & Koike, M. (1985) Fatal case of pyruvate dehydrogenase deficiency. *Acta Paed Scand* **76**, 140–142.

Mehler, A.H. (1986) Amino acid metabolism, 1: general pathways. In: *Textbook of Biochemistry with Clinical Correlations* (ed. T.M. Devlin), 2nd edn, pp. 437–452. John Wiley & Sons, New York.

Meijer, A.J., Lamers, W.H. & Chamuleau, R.A.F.M. (1990) Nitrogen metabolism and ornithine cycle function. *Physiol Rev* **70**, 701–748.

Meredith, T.J. & Vale, J.A. (1981) Salicylate poisoning. In: *Poisoning: Diagnosis and Treatment* (eds J.A. Vale & T.J. Meredith), pp. 97–103. Update Books, London.

Meschi, F., Dozio, N., Bognetti, E. *et al.* (1992) An unusual case of recurrent hypoglycaemia: 10-year follow up of a child with insulin autoimmunity. *Eur J Pediatr* **151**, 32–34.

Morgan, H.B., Swaiman, K.F. & Johnson, B.D. (1987) Diagnosis of argininosuccinic aciduria after valproic acid-induced hyperammonemia. *Neurology* **37**, 886–887.

Mounger, E.J. & Branson, A.D. (1972) Ammonia encephalopathy secondary to ureterosigmoidostomy: a case report. *J Urol* **108**, 411–412.

Msall, M., Batshaw, M.L., Suss, R., Brusilow, S.W. & Mellits, E.D. (1984) Neurologic outcome in children with inborn errors of urea synthesis. *N Engl J Med* **310**, 1500–1505.

Odièvre, M. (1990) Disorders of fructose metabolism. In: *Inborn Metabolic Diseases* (eds J. Fernandes, J.-M. Saudubray & K. Tada), pp. 107–112. Springer-Verlag, Berlin.

Ogier, H., Charpentier, C. & Saudubray, J.-M. (1990) Organic acidemias. In: *Inborn Metabolic Diseases* (eds J. Fernandes, J.-M. Saudubray & K. Tada), pp. 271–299. Springer-Verlag, Berlin.

Orth, D.N., Kovacs, W.J. & De Bold, C.R. (1992) The adrenal cortex. In: *Williams Textbook of Endocrinology* (eds J.D. Wilson & D.W. Foster), 8th edn, pp. 489–619. WB Saunders, Philadelphia.

Partin, J.C., Schubert, W.K. & Partin, J.S. (1971) Mitochondrial ultrastructure in Reye's syndrome (encephalopathy and fatty degeneration of the viscera). *N Engl J Med* **285**, 1339–1343.

Pilkis, S.J., Fox, E., Wolfe, L. *et al.* (1986) Hormonal modulation of key hepatic regulatory enzymes in the gluconeogenic/glycolytic pathway. *Ann NY Acad Sci* **478**, 1–19.

Polak, J.M. & Wigglesworth, J.S. (1976) Islet cell hyperplasia and SIDS. *Lancet* **ii**, 570–571.

Pollitt, R.J. (1989) Disorders of mitochondrial β-oxidation: prenatal and early postnatal diagnosis and their relevance to Reye's syndrome and sudden infant death. *J Inher Metab Dis* **12** (Suppl. 1), 215–230.

Pryds, O., Christensen, N.J. & Friis-Hansen, B. (1990) Increased cerebral blood flow and plasma epinephrine in hypoglycemic, preterm neonates. *Pediatrics* **85**, 172–176.

Puntis, J.W.L., Green, A., Preece, M.A., Ball, P.A. & Booth, I.W. (1988) Hyperammonaemia and parenteral nutrition in infancy. *Lancet* **ii**, 1374–1375.

Räihä, N.C.R. & Suihkonen, J. (1968) Development of urea-synthesizing enzymes in human liver. *Acta Paed Scand* **57**, 121–124.

Räihä, N.C.R., Heinonen, K., Rassin, D.K. & Gaull, G.E. (1976) Milk protein quantity and quality in low-birthweight infants: 1. Metabolic responses and effects on growth. *Pediatrics* **57**, 659–674.

Rajantie, J. (1981) Orotic aciduria in lysinuric protein intolerance: dependence on the urea cycle intermediates. *Pediatr Res* **15**, 115–119.

Reye, R.D.K., Morgan, G. & Baral, J. (1963) Encephalopathy and fatty degeneration of the viscera; a disease entity in childhood. *Lancet* **ii**, 749–752.

Ricciuti, F.C., Gelehrter, T.D. & Rosenberg, L.E. (1976) X-chromosome inactivation in human liver: confirmation of X-linkage of ornithine transcarbamylase. *Am J Hum Genet* **28**, 332–338.

Robinson, A.M. & Williamson, D.H. (1980) Physiological roles of ketone bodies as substrates and signals in mammalian tissues. *Physiol Rev* **60**, 143–187.

Robinson, B.H. (1985) The lacticacidemias. In: *Genetic and Metabolic Disease in Pediatrics* (eds J.K. Lloyd & C.R. Scriver), pp. 111–139. Butterworth, London.

Robinson, B.H., Taylor, J., Cutz, E. & Gall, D.G. (1978) Reye's syndrome: preservation of mitochondrial enzymes in brain and muscle compared with liver. *Pediatr Res* **12**, 1045–1047.

Robinson, R.O. (1987) Differential diagnosis of Reye's syndrome. *Develop Med Child Neurol* **29**, 110–120.

Roe, C.R. & Coates, P.M. (1989) Acyl-CoA dehydrogenase deficiencies. In: *The Metabolic Basis of Inherited Disease* (eds C.R. Scriver, A.L. Beaudet, W.S. Sly & D. Valle), 6th edn, pp. 889–914. McGraw-Hill, New York.

Roe, C.R., Millington, D.S., Maltby, D.A. *et al.* (1985) Diagnostic and therapeutic implications of medium-chain acylcarnitines in the medium-chain acyl-CoA dehydrogenase deficiency. *Pediatr Res* **19**, 459–466.

Romshe, C.A., Hilty, M.D., McClung, H.J., Kerzner, B. & Reiner, C.B. (1981) Amino acid pattern in Reye syndrome: comparison with clinically similar entities. *J*

Pediatr **98**, 788–790.

Rosenberg, L.E. (1983) Disorders of propionate and methylmalonate metabolism. In: *The Metabolic Basis of Inherited Disease* (eds J.B. Stanbury, J.B. Wyngaarden, D.S. Fredrickson, J.L. Goldstein & M.S. Brown), 5th edn, pp. 474–497. McGraw-Hill, New York.

Rosenberg, L.E. & Fenton, W.A. (1989) Disorders of propionate and methylmalonate metabolism. In: *The Metabolic Basis of Inherited Disease* (eds C.R. Scriver, A.L. Beaudet, W.S. Sly & D. Valle), 6th edn, pp. 821–844. McGraw-Hill, New York.

Rowe, P.C., Newman, S.L. & Brusilow, S.W. (1986) Natural history of symptomatic partial ornithine transcarbamylase deficiency. *N Engl J Med* **314**, 541–547.

Samtoy, B. & DeBeukelaer, M.M. (1980) Ammonia encephalopathy secondary to urinary tract infection with *Proteus mirabilis*. *Pediatrics* **65**, 294–297.

Segal, S. (1983) Disorders of galactose metabolism. In: *The Metabolic Basis of Inherited Disease* (eds J.B. Stanbury, J.B. Wyngaarden, D.S. Fredrickson, J.L. Goldstein & M.S. Brown), 5th edn, pp. 167–191. McGraw-Hill, New York.

Senior, B. & Wolfsdorf, J.I. (1979) Hypoglycemia in children. *Pediatr Clin N Am* **26**, 171–180.

Shah, A., Stanhope, R. & Matthew, D. (1992) Hazards of pharmacological tests of growth hormone secretion in childhood. *Br Med J* **304**, 173–174.

Shih, V.E. (1990) Hyperornithinemias. In: *Inborn Metabolic Diseases* (eds J. Fernandes, J.-M. Saudubray & K. Tada), pp. 229–239. Springer-Verlag, Berlin.

Shin, Y.S. (1990) Diagnosis of glycogen storage disease. *J Inher Metab Dis* **13**, 419–434.

Shohat, M., Wielunsky, E. & Reisner, S.H. (1984) Plasma ammonia levels in preterm infants receiving parenteral nutrition with crystalline L-amino acids. *J Parent Ent Nutr* **8**, 178–180.

Sieber, F.E. & Traystman, R.J. (1992) Special issues: glucose and the brain. *Crit Care Med* **20**, 104–114.

Simell, O. (1989) Lysinuric protein intolerance and other cationic aminoacidurias. In: *The Metabolic Basis of Inherited Disease* (eds C.R. Scriver, A.L. Beaudet, W.S. Sly & D. Valle), 6th edn, pp. 2497–2513. McGraw-Hill, New York.

Smit, G.P.A., Fernandes, J., Leonard, J.V. *et al.* (1990) The long-term outcome of patients with glycogen storage diseases. *J Inher Metab Dis* **13**, 411–418.

Souba, W.W. (1987) Interorgan ammonia metabolism in health and disease: a surgeon's view. *J Parent Ent Nutr* **11**, 569–579.

Sperl, W., Geiger, R., Maurer, H. *et al.* (1992) Continuous arteriovenous haemofiltration in a neonate with hyperammonaemic coma due to citrullinaemia. *J Inher Metab Dis* **15**, 158–159.

Spitz, L., Bhargava, R.K., Grant, D.B. & Leonard, J.V. (1992) Surgical treatment of hyperinsulinaemic hypoglycaemia in infancy and childhood. *Arch Dis Child* **67**, 201–205.

Stanhope, R. & Brook, C.G.D. (1985) Neonatal hypoglycaemia: an important early sign of endocrine disorders. *Br Med J* **291**, 728–729.

Stanley, C.A. & Baker, L. (1976) Hyperinsulinism in infants and children: diagnosis and therapy. In: *Advances in Pediatrics* (ed. L.A. Barness), Vol. 23, pp. 315–355. Year Book Medical Publishers, Chicago.

Starko, K.M. & Mullick, F.G. (1983) Hepatic and cerebral pathology findings in children with fatal salicylate intoxication: further evidence for a casual relation between salicylate and Reye's syndrome. *Lancet* **i**, 326–331.

Stenninger, E., Schollin, J. & Aman, J. (1991) Neonatal macrosomia and hypoglycaemia in children of mothers with insulin-treated gestational diabetes mellitus. *Acta Paed Scand* **80**, 1014–1018.

Surtees, R. & Leonard, J.V. (1989) Acute metabolic encephalopathy: a review of causes, mechanisms and treatment. *J Inher Metab Dis* **12**, 42–54.

Thom, H., Carter, P.E., Cole, G.F. & Stevenson, K.L. (1991) Ammonia and carnitine concentrations in children treated with sodium valproate compared with other anticonvulsant drugs. *Develop Med Child Neurol* **33**, 795–802.

Thomas, D.W., Sinatra, F.R., Hack, S.L. *et al.* (1982) Hyperammonemia in neonates receiving intravenous nutrition. *J Parent Ent Nutr* **6**, 503–506.

Tokatli, A., Coskun, T. & Özalp, I. (1991) Fifteen years experience with 212 hyperammonaemic cases at a metabolic unit. *J Inher Metab Dis* **14**, 698–706.

Touma, E.H. & Charpentier, C. (1992) Medium chain acyl-CoA dehydrogenase deficiency. *Arch Dis Child* **67**, 142–145.

Treem, W.R., Stanley, C.A., Hale, D.E., Leopold, H.B. & Hyams, J.S. (1991) Hypoglycemia, hypotonia, and cardiomyopathy: the evolving clinical picture of long-chain acyl-CoA dehydrogenase deficiency. *Pediatrics* **87**, 328–333.

Tripp, J.H., Hargreaves, T., Anthony, P.P. *et al.* (1981) Sodium valproate and ornithine carbamyl transferase deficiency. *Lancet* **i**, 1165–1166.

Tuchman, M., Knopman, D.S. & Shih, V.E. (1990) Episodic hyperammonemia in adult siblings with hyperornithinemia, hyperammonemia, and homocitrullinuria syndrome. *Arch Neurol* **47**, 1134–1137.

Upp, J.R., Ishizuka, J., Lobe, T.E. *et al.* (1987) Somatostatin secretion in cultured human islet cells from patients with nesidioblastosis: a compensatory mechanism? *J Pediatr Surg* **22**, 1185–1186.

Van den Berghe, G. (1991) The role of the liver in metabolic homeostasis: implications for inborn errors of metabolism. *J Inher Metab Dis* **14**, 407–420.

Vianey-Liaud, C., Divry, P., Gregersen, N. & Mathieu, M. (1987) The inborn errors of mitochondrial fatty acid oxidation. *J Inher Metab Dis* **10**, 159–198.

Vilstrup, H., Iversen, J. & Tygstrup, N. (1986) Glucoregulation in acute liver failure. *Eur J Clin Invest* **16**, 193–197.

de Vivo, D.C. (1978) Reye syndrome: a metabolic response to an acute mitochondrial insult. *Neurology* **28**, 105–108.

de Vivo, D.C. & di Mauro, S. (1990) Disorders of pyruvate metabolism, the citric acid cycle, and the respiratory chain. In: *Inborn Metabolic Diseases* (eds J. Fernandes, J.-M. Saudubray & K. Tada), pp. 130–133. Springer-Verlag, Berlin.

Volk, D.M. (1981) Reye's syndrome: an update for the practicing physician. *Clin Pediatr* **20**, 505–511.

Wanders, R.J.A., Ijlst, L., Duran, M. *et al.* (1991) Long-chain 3-hydroxyacyl-CoA dehydrogenase deficiency: different clinical expression in three unrelated patients. *J Inher Metab Dis* **14**, 325–328.

Ware, S. & Millward-Sadler, G.H. (1980) Acute liver disease associated with sodium valproate. *Lancet* **ii**, 1110–1113.

Warter, J.M., Brandt, C., Marescaux, C. *et al.* (1983) The renal origin of sodium valproate-induced hyperammonemia in fasting humans. *Neurology* **33**, 1136–1140.

Watson, A.J., Karp, J.E., Walker, J.G. *et al.* (1985) Transient idiopathic hyperammonaemia in adults. *Lancet* **ii**, 1271–1274.

Wendel, U., Wieland, J., Bremer, H.J. & Bachmann, C. (1989) Ornithine transcarbamylase deficiency in a male: strict correlation between metabolic control and plasma arginine concentration. *Eur J Pediatr* **148**, 349–352.

White, N.J., Miller, K.D., Marsh, K. *et al.* (1987) Hypoglycaemia in African children with severe malaria. *Lancet* **i**, 708–711.

Woolf, D.A., Leonard, J.V., Trembath, R.C., Pembrey, M.E. & Grant, D.B. (1991) Nesidioblastosis: evidence for autosomal recessive inheritance. *Arch Dis Child* **66**, 529–530.

6: Inherited Organic Acid Disorders

V. WALKER

Introduction

Organic acids are carbon-containing acidic compounds. Most organic acids in physiological fluids are carboxylic acids or their glycine or glucuronide conjugates. Also included are hydroxyl compounds such as phenols or cresols. By definition, compounds with a primary amino group, detectable by conventional amino acid analysers, are excluded. Organic acids are produced continuously in the body as intermediates in the metabolism of amino acids, carbohydrates, lipids, biogenic amines and some drugs. They also derive from dietary constituents and from the actions of microflora on gastrointestinal contents. Normally, they do not accumulate in the body, since they are rapidly converted to non-acidic end products, or are excreted as water-soluble metabolites in urine. However, if they are produced in excess, or if their metabolism is prevented by an inherited enzyme defect, concentrations increase in tissues, blood and urine.

The first inherited defect was recognized in two children with an illness characterized by excretion of large amounts of isovaleric acid in their urine, identified by gas chromatography–mass spectrometry (Tanaka *et al.*, 1966). Tanaka *et al.* postulated, and later proved (Tanaka *et al.*, 1976), that this was due to a deficiency of the enzyme isovaleryl-CoA dehydrogenase. Since then, a large number of other inherited defects have been identified, many with variant forms (Chalmers & Lawson, 1982; Goodman, 1986; Chalmers, 1987; Scriver *et al.*, 1989). These are clinically important conditions. The most severe cause a fulminating and fatal illness in the first days of life. Others, such as the congenital lactic acidoses (Chapter 7(b)), fumarase deficiency (Walker *et al.*, 1989b) and glutaric aciduria type I (Goodman & Frerman, 1989; Przyrembel, 1990), produce irreversible and/or progressive neurological damage. Defects of fatty acid metabolism may present with either a life-threatening Reye's syndrome-like illness in apparently normal children, or cardiomyopathy, and have caused sudden death in infancy. Prompt diagnosis of organic acid defects is important. The condition may be one that is treatable: delay often leads to irreversible brain damage, or death. If the disorder is not treatable, the parents will need genetic counselling about the risks for future pregnancies. Prenatal diagnosis is available for many of the defects, but can be undertaken only when the precise diagnosis is known.

Collectively, organic acid defects are certainly more common than phenylketonuria (around 1 in 10 000 live births in the UK), although the overall incidence is unknown. One of the fat oxidation defects alone, medium-chain acyl-CoA dehydrogenase (MCAD) deficiency, may occur as often as 1 in 6400 (Matsubara *et al.*, 1991) or 1 in 13 400 live births (Blakemore *et al.*, 1991). Congenital lactic acidoses seem also to be relatively common. Neonatal screening programmes have found an incidence of around 1 in 50 000 live births for biotinidase deficiency (Dunkel *et al.*, 1989; Wolf & Heard, 1989), 1 in 50 000 each for the mild and severe forms of methylmalonic aciduria (Mahoney & Bick, 1987), 1 in 100 000 for tyrosinaemia type I (Chalmers, 1987) and 1 in 69 000 to 1 in 290 000 for maple syrup urine disease in different communities in the USA (Danner & Elsas, 1989).

Alerting clues to an inherited organic acid defect

Inherited organic acid defects present in many different ways. Often, metabolic acidosis is mild or absent. It is important to be alert to these conditions so that the diagnosis may be made quickly. Table 6.1 lists some important clues and Table 6.2 major presenting features and their causative defects. Neurological disturbances are extremely common features. They may be due directly to the defect, or secondary to brain damage during acute illness. They are seldom characteristic and occur in many more non-inherited conditions.

Diagnosis

Inherited organic acid defects are diagnosed by biochemical analyses.

Preliminary biochemical tests and samples to collect in suspected inherited organic disorders

Essential preliminary tests include: * arterial blood gases and acid–base status; biochemical profile, including electrolytes and liver function tests; plasma glucose and plasma amino acids. Urine must be saved for organic acids and amino acids. The best sample is the *first* urine passed at the time of acute admission. Diagnostic abnormalities often disappear quickly following intravenous glucose or other resuscitation. A urine bag should be applied at once and, if possible, at least 5 mL urine collected. This should be placed in a sterile urine bottle and frozen, without preservative. The next sample of urine passed should also be saved. Other tests which may be indicated are plasma ammonium (for example in cases of a Reye's syndrome-like illness), and plasma and CSF lactate if primary lactic acidosis is suspected (see Chapter 7(b)). It is useful to store plasma in case carnitine analysis is required, especially if there is a cardiomyopathy. * Confirmation of a diagnosis generally requires tissue enzyme analyses (by a referral laboratory), usually of blood cells or cultured skin fibroblasts. Increasingly now, DNA analysis provides information which may be useful in subsequent pregnancies, and blood should be stored for this purpose.

Table 6.1 Alerting clues to an inherited organic acid defect

Clinical
Siblings with known inborn errors, similar illness or unexplained deaths
Parental consanguinity
Rapid unexplained deterioration of a newborn who was well at birth
Clinical improvement when protein feeds withdrawn and glucose substituted
Unexplained metabolic acidosis
Ketonuria in a sick neonate
Abnormal odour
Unexplained encephalopathy (acute brain disturbance)
Hypoglycaemia without obvious cause
Reye's syndrome-like illness
Abnormal muscle tone
Severe unexplained fits
Persistent vomiting with no anatomical cause
Cardiomyopathy
Unexplained liver disease
Congenital malformations
Unexplained failure to thrive
Intermittent coma or lethargy
Unsteady gait – ataxia
Additional laboratory clues
Hyperammonaemia
Hyperglycinaemia
Positive ferric chloride test (Phenistix, Ames Laboratories)
Haematological disturbances
Postmortem findings
Fat infiltration of liver and, possibly, heart or kidneys

If death is imminent and an organic acid (or other metabolic) defect is a strong possibility, but there has been no time for diagnosis, samples should be collected as indicated above (*–*). In addition, 2–3 mL 'spare' plasma should be frozen in case of need, and 5 mL blood collected into an ethylenediaminetetraacetate (EDTA) tube, to be saved for future DNA analysis. The whole blood may be stored for several weeks at 4°C pending DNA extraction, or long term at −70°C. A skin biopsy should be taken to establish a fibroblast culture. During working hours this is placed immediately in tissue culture medium. In an emergency outside of working hours it may be stored overnight in a dry sterile container at 4°C (*not* frozen).

Table 6.2 Clinical features of organic acid defects and alerting clues from laboratory tests

Clinical

Severe ketoacidosis
Propionic acidaemia, methylmalonic aciduria, MSUD, isovaleric acidaemia, 3-methylcrotonyl carboxylase and multiple carboxylase deficiencies, 2-methylacetoacetyl-CoA thiolase deficiency, (pyruvate carboxylase deficiency)

Severe acidosis with mild or absent ketosis
MADD (severe form), HMGCoA lyase deficiency, glutathione synthetase deficiency, (congenital lactic acidoses)

Reye's syndrome-like illness
Fatty acid β-oxidation defects, HMGCoA lyase deficiency, propionic and isovaleric acidaemias, methylmalonic aciduria, 2-methylacetoacetyl-CoA thiolase deficiency, glutaric aciduria type I

Cot death
Fatty acid β-oxidation defects, HMGCoA lyase deficiency, MSUD

Abnormal odour
MSUD (burnt sugar); tyrosinaemia type I (cabbage-like); isovaleric acidaemia and MADD (sweaty feet, cheesy); multiple carboxylase, 3-methylcrotonyl-CoA carboxylase, HMGCoA lyase (tom-cats' urine)

Odd facial appearance
Neonatal MADD, fumarase and glycerol kinase deficiencies (some congenital lactic acidoses)

Macrocephaly (large head)
Neonatal MADD, glutaric aciduria type I, aspartoacylase and HMGCoA lyase deficiencies, L-2-OHglutaric aciduria

Abnormal skin and/or hair
Rash and alopecia: multiple carboxylase deficiencies (both forms); alopecia: isovaleric acidaemia; epidermolysis: propionic acidaemia and methylmalonic aciduria; hyperkeratosis, palm and sole erosions: tyrosinaemia type II

Neurological problems
Very common: include developmental delay, fits, problems with balance, gait, coordination, muscle tone, speech; very few diagnostic: one exception, glutaric aciduria type I: progressive movement disorder with dysarthria, dystonia, choreoathetosis

Eye abnormalities
Optic atrophy: aspartoacylase and biotinidase deficiencies, 3-methylglutaconic aciduria, (respiratory chain); squint: MSUD, hereditary orotic aciduria, (respiratory chain); corneal erosions and plaques: tyrosinaemia type II; keratoconjunctivitis: biotinidase deficiency

Deafness
Biotinidase deficiency, 3-methylglutaconic aciduria

Hepatomegaly
Fatty acid β-oxidation defects, HMGCoA lyase deficiency, sometimes propionic acidaemia and methylmalonic aciduria, fumarase deficiency, mevalonic aciduria, (fructose 1,6-bisphosphatase deficiency; glycogen storage disease type I)

Cardiac problems
Cardiomyopathy: MADD, LCAD, LCHAD, CPT II (one case), carnitine and carnitine translocase deficiencies, tyrosinaemia type I, chronic forms of methylmalonic aciduria and propionic acidaemia, (respiratory chain); valvular damage and aneurysms: alkaptonuria; malformations: hereditary orotic aciduria

Renal problems
Cysts: MADD; chronic renal failure: methylmalonic aciduria; Fanconi syndrome: tyrosinaemia type I, (respiratory chain); crystalluria/stones: primary hyperoxaluria types I and II; hereditary orotic aciduria

Severe hypotonia/muscle weakness
LCAD, LCHAD, SCAD, MADD, carnitine and carnitine translocase deficiencies; succinic semialdehyde dehydrogenase and multiple carboxylase deficiencies, methylmalonic aciduria (*cbl*C group), D-glyceric aciduria (respiratory chain)

continued on p. 124

Table 6.2 *Continued*

Laboratory tests
Hypoglycaemia
Fatty acid β-oxidation defects, HMGCoA lyase and multiple carboxylase deficiencies, MSUD, propionic and isovaleric acidaemias, methylmalonic aciduria, tyrosinaemia type I (gluconeogenic defects, glycogen storage disease type I)
Hyperglycinaemia
Propionic acidaemia, methylmalonic aciduria (*mut*^{0}, *mut*$^-$, *cbl*A and *cbl*B forms)
Hyperammonaemia
Fatty acid β-oxidation defects, HMGCoA lyase, 2-methylacetoacetyl-CoA thiolase and biotinidase deficiencies, propionic and isovaleric acidaemias, methylmalonic aciduria, glutaric aciduria type I (pyruvate dehydrogenase and pyruvate carboxylase deficiencies)
Positive ferric chloride test (Phenistix, Ames, Miles Ltd)
MSUD (grey–green), tyrosinaemia types I and II (green), alkaptonuria (dark brown)
Haematological abnormalities
Anaemia: mevalonic aciduria, propionic and isovaleric acidaemias and methylmalonic aciduria (*mut*^{0}, *mut*$^-$, *cbl*A and *cbl*B) (acute); megaloblastic anaemia: methylmalonic aciduria (*cbl*C, *cbl*D, defects of vitamin B_{12} absorption and transport), hereditary orotic aciduria; haemolytic: glutathione synthetase deficiency
Immunological defects
Biotinidase deficiency, hereditary orotic aciduria, isovaleric and propionic acidaemias and methylmalonic aciduria (acute)

Included (in parentheses) are primary lactic acid disorders when they might cause diagnostic confusion.
MSUD, maple syrup urine disease; LCAD, MCAD, SCAD, MADD and LCHAD, long-, medium- and short-chain, multiple, and long-chain 3-hydroxyacyl-CoA dehydrogenase deficiencies, respectively; CPT, carnitine palmityl transferase.

Appropriate samples must also be collected from children who have died from a cot death, if there is any reason to suspect an inherited defect: for example, if the child was vomiting or drowsy during the hours before death, or had had encephalopathic illnesses previously; if there had been unexplained sibling deaths; or if postmortem findings, such as a fatty liver, suggested a possible fat oxidation defect. If possible, urine should be expressed or aspirated from the bladder, or squeezed from a wet nappy. If urine is not available, eye fluid collected during the postmortem examination may contain diagnostic metabolites. A skin biopsy should be taken after the skin has been cleaned thoroughly with alcohol. Fibroblasts will often grow, even from skin collected as long as 24 h after death.

Analysis of organic acids and acylcarnitines

Appendix 6.1 summarizes the most important diagnostic urinary organic acids and their carnitine conjugates. Many other abnormalities are also found (Chalmers & Lawson, 1982; Scriver *et al.*, 1989).

Organic acids

Detection of the full range of diagnostic organic acids, which have diverse chemical structures, is analytically demanding. Unquestionably, the best available profiling methods use capillary gas chromatography–mass spectrometry (GC–MS) (Goodman & Markey, 1981; Chalmers & Lawson, 1982). Organic acids must first be isolated from biological fluids. Solvent extraction procedures are used widely. For quantitative studies, particularly of the more polar compounds, an anion exchange procedure may be preferable. The acids are then derivatized for gas chromatography. Because organic acids are secreted into the glomerular filtrate by the kidneys, urine is usually the best fluid for diagnosis. However, the procedures are applicable to plasma and other fluids. Analysis of eye fluid, collected at postmortem examination, may sometimes provide a diagnosis in a dead baby, when urine is not available (Bennett *et al.*, 1987b; Mills *et al.*, 1990). Generally, diagnoses are made by observing *qualitative* disturbances in the organic acid profile. Quantitative age-related reference data are available for comparison when quantification is necessary, although much of this was obtained using packed columns (Chalmers & Lawson, 1982). For data on the urine of newborns see Tracey *et al.* (1981), Sann *et al.* (1987) and Walker and Mills (1989). A wide range of acids are detectable in normal urine (Tuchman & Ulstrom, 1985). Quantitative organic acid analysis of amniotic fluid, collected between 12 and 18 weeks of gestation, has been used for prenatal diagnosis of several organic acid defects, including methylmalonic aciduria, propionic acidaemia, tyrosinaemia type I and MADD (Jakobs, 1989). These analyses complement enzyme studies of fetal cells.

High performance liquid chromatography (HPLC) has been investigated as an alternative analytical approach, in view of its lower cost and ease of use (Bulusu *et al.*, 1991). Abnormal metabolites, present in organic acid defects, have been detected in plasma (Daish & Leonard, 1985) and urine (Chong *et al.*, 1989), enabling diagnosis. However, the full range of diagnostic compounds found with use of GC–MS was often not observed. Most fatty acid defects would not be detectable by HPLC methods involving ultraviolet detection, and these analyses are therefore not suitable as primary diagnostic procedures. Some organic acid disorders can be detected by high-resolution proton nuclear magnetic resonance (NMR) spectroscopy (Iles *et al.*, 1985; Iles & Chalmers, 1988; Brown *et al.*, 1989), but the method is insensitive.

It is important to be aware of the many acquired disturbances of urinary organic acids which occur, since these are encountered far more frequently than those due to inherited defects, and they may cause confusion (Appendix 6.2, p. 145). For example, changes in severe fasting ketosis may lead to misdiagnosis of a fat oxidation defect (Greter *et al.*, 1980; Niwa, 1986), or severe hypoxia/ischaemia to maple syrup urine disease (Landaas & Jakobs, 1977; Walker & Mills, 1992).

Carnitine and acylcarnitines

Plasma free and total carnitine is usually measured with a radioenzymatic procedure applied to plasma before and after alkaline hydrolysis. Total acylcarnitines are calculated from the difference (McGarry & Foster, 1976). By including a precipitation step, long-chain and short- plus medium-chain acylcarnitines may be measured, but separation is often incomplete (Fishlock *et al.*, 1984). Sophisticated analyses and equipment are needed to identify individual acylcarnitines. HPLC analyses have proved unsatisfactory, unless coupled with MS (Millington, 1986).

Diagnosis of fat oxidation defects

The large number of tests for these disorders reflects the diagnostic difficulties (p. 139).

The defects: biochemistry, management and outcome

With a few exceptions, inherited organic acid disorders fall into three groups: defects of amino acid metabolism, defects of fatty acid oxidation, and the congenital lactic acidoses, in which raised lactic acid is the major biochemical abnormality (Chapter 7(b)). Organic acid abnormalities also occur in some peroxisomal disorders (Chapter 7(a)). Neither the lactic acidoses nor peroxisomal defects will be considered here, except when they enter the differential diagnosis of an organic acid defect. Disorders have been selected to present an overview of the clinical features and metabolic derangements encountered, and the principles of management. Appendix 6.1, p. 143 summarizes the most important diagnostic urinary organic acid abnormalities. Many other metabolites are also found. For detailed reviews see: Chalmers and Lawson (1982), Chalmers (1987), Scriver *et al.* (1989) and Fernandes *et al.* (1990).

Defects of branched-chain amino acid metabolism, propionic acidaemia and methylmalonic aciduria

A significant proportion of organic acid defects involve the pathways for catabolism of the branched-chain amino acids leucine, valine and isoleucine (Fig. 6.1). These three amino acids are first transaminated to oxo-acids, which are then decarboxylated by branched-chain 2-oxo-acid dehydrogenase. From then on, catabolism proceeds by separate pathways in which most reactions are reversible. Propionyl-CoA is a key metabolite in the degradation of valine and isoleucine, as well as in that of threonine, methionine, odd-chain fatty acids and cholesterol. It is also formed from propionic acid absorbed from the gut, where it is produced by the microflora (Bain *et al.*, 1988; Walter *et al.*, 1988).

Propionic acidaemia

In its classical form, propionic acidaemia is one of the most devastating organic acid defects (Wolf *et al.*, 1981; Rosenberg & Fenton, 1989; Ogier *et al.*, 1990). It results from a deficiency of propionyl-CoA carboxylase. This enzyme consists of four sets of two subunits ($\alpha_4\beta_4$). Biotin is an essential cofactor and is attached to the α subunits by an enzyme, holocarboxylase synthetase. The vast majority of defects affect the α or β subunits. Rarely, there is a problem with the biotin cofactor and then, generally, other biotin-dependent carboxylases are defective as well. Because of the defect, propionyl-CoA accumulates rapidly. This is toxic, partly because it competes with acetyl-CoA, which is structurally similar. It combines with oxaloacetate in the tricarboxylic acid cycle (forming methylcitrate); inhibits the pyruvate dehydrogenase complex; probably inhibits *N*-acetylglutamate synthetase, needed to produce the activator of the first step of the urea cycle (Chapter 5(b)); and replaces acetyl-CoA as the primer for long-chain fatty acid synthesis, leading to the formation of odd-chain fatty acids. In addition, it causes a *deficiency* of acetyl-CoA, because it traps coenzyme A in the mitochondria. Lack of acetyl-CoA contributes to decreased *N*-acetylglutamate synthesis and urea cycle activation, and decreases activity of pyruvate carboxylase and therefore production of oxaloacetate and gluconeogenesis (Chapter 5(a)). Initially, *carnitine* is able to protect against coenzyme A trapping. In addition to its well-recognized role in long-chain fatty acid oxidation, this compound takes up acyl groups from accumulating intramitochondrial short-chain acyl-CoA molecules:

$$\text{carnitine + acyl-CoA} \underset{\text{carnitine acetyltransferase}}{\rightleftharpoons} \text{acylcarnitine + coenzyme A}$$

Liberated coenzyme A is recycled and acylcarnitine diffuses into the plasma and is excreted in the urine, if small enough for glomerular filtration. This is an important detoxification process in many organic acid defects. However, it becomes overwhelmed in propionic acidaemia: large amounts of carnitine are lost in urine as propionylcarnitine, and a state of carnitine depletion may develop (Roe & Bohun, 1982; Chalmers *et al.*, 1984).

As well as these *direct* effects of propionyl-CoA, abnormal acids and ketones, derived from propionyl-CoA or from isoleucine and valine catabolism, accumulate. Severe ketoacidosis supervenes rapidly. One or more metabolites interferes with glycine cleavage or glycine–serine interconversion, causing increased plasma and urinary glycine.

Most patients present with fulminating ketoacidosis in the newborn period. Typically, they are normal at birth and appear well for the first 24–48 h of life. Then they suck poorly, refuse feeds, may vomit, become increasingly lethargic and hypotonic, and may have fits, hypothermia, respiratory distress, bradycardia and hepatomegaly, and coma supervenes. Many die. Those who survive are severely brain damaged and have frequent ketoacidotic episodes, precipitated by the catabolism associated with normally trivial intercurrent illnesses, or by increased dietary protein. Some children have a later presentation with recurrent attacks of ketoacidosis with lethargy, neurological disturbances, coma and brain damage. Others have developmental retardation without apparent ketoacidotic episodes. A minority, identified during family studies, have been asymptomatic. At acute presentation, there is severe ketoacidosis, hyperammonaemia, hyperglycinaemia and hyperglycinuria, and often either hyper- or hypoglycaemia. Plasma total and free carnitine levels are low, with an increased proportion of acylcarnitines. Blood counts often show neutropenia and thrombocytopenia, because of marrow suppression by toxic metabolites.

Methylmalonic aciduria

Methylmalonic aciduria results from a deficiency of methylmalonyl-CoA mutase, a vitamin B_{12}-dependent enzyme (Mahoney & Bick, 1987; Fenton & Rosenberg, 1989). This arises either because of a

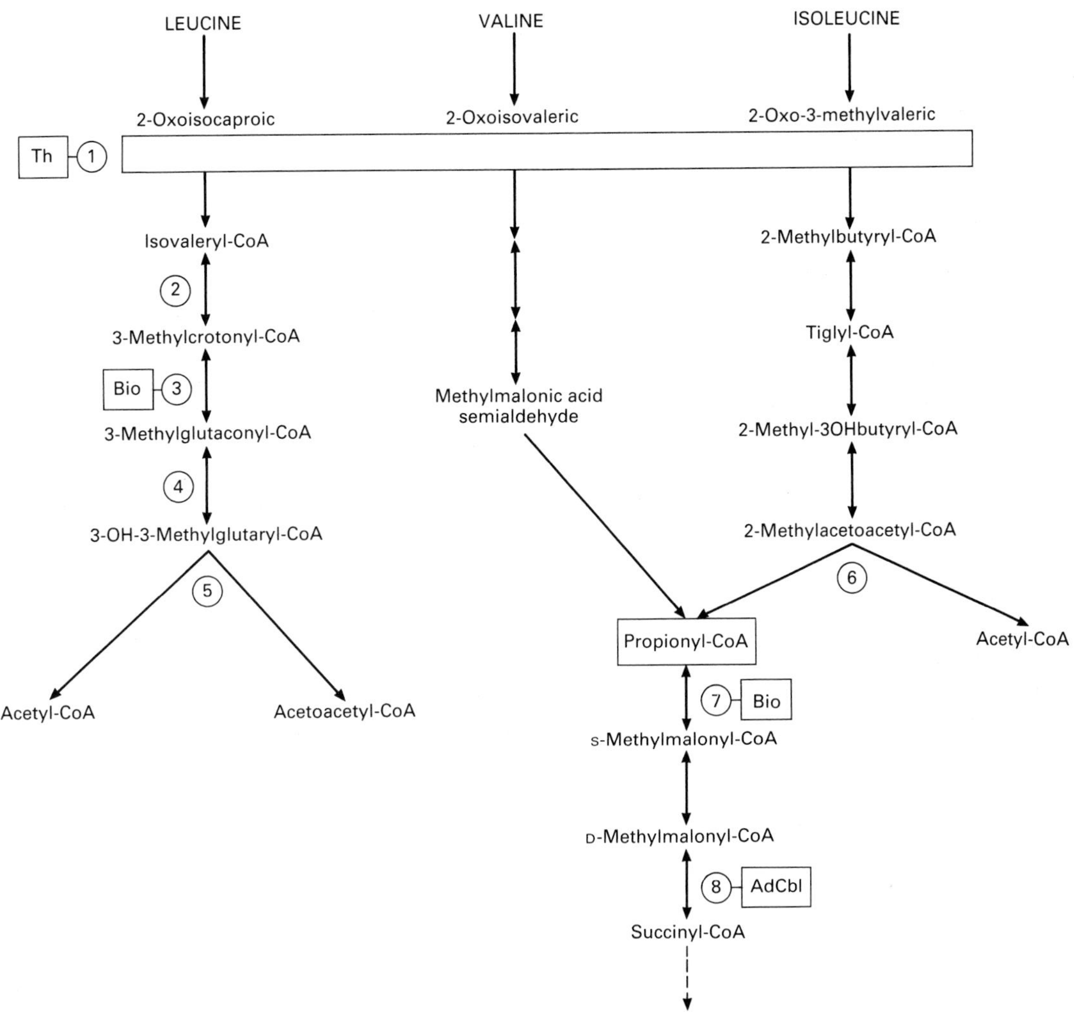

Fig. 6.1 Catabolism of the branched-chain amino acids showing the inherited enzyme defects which cause organic aciduria. Enzymes which have a vitamin cofactor are indicated in boxes: Th, thiamine pyrophosphate; Bio, biotin; AdCbl, adenosylcobalamin. Enzymes are circled: 1, branched-chain 2-oxo-acid dehydrogenase; 2, isovaleryl-CoA dehydrogenase; 3, 3-methylcrotonyl-CoA carboxylase; 4, 3-methylglutaconyl-CoA hydratase; 5, 3-hydroxy-3-methylglutaryl-CoA lyase; 6, 2-methylacetoacetyl-CoA thiolase (β-ketothiolase); 7, propionyl-CoA carboxylase; and 8, methylmalonyl-CoA mutase. OH, Hydroxyl.

defect of the apoenzyme itself (subgroups mut^0, no activity, and mut^-, partial activity) or because of a deficiency of the enzyme's essential cofactor, adenosylcobalamin. Production of this cofactor is a multistep process: the dietary precursor, vitamin B_{12} (cobalamin, Cbl) is first linked to intrinsic factor in the gut lumen, then taken up by ileal mucosal cells. Here, Cbl is bound to transcobalamin II and released into the portal circulation. The complex attaches to specific receptors present on the surface of many cell types, and is taken into the cells by endocytosis. Cbl is released into the cytoplasm. Some enters the mitochondria, where its cobalt atom undergoes two reductions (CoIII to CoI). It is then adenosylated to adenosylCbl (AdoCbl). Some Cbl is methylated in the cytosol to form a different vitamin B_{12} cofactor, *methyl*cobalamin (MeCbl). This is the coenzyme of N^5-methyltetrahydrofolate: homocysteine methyltransferase (methionine synthetase), which methylates homocysteine to methionine. Ten different inherited defects are known which affect Cbl metabolism. Two affect Cbl absorption, one Cbl transport (transcobalamin II deficiency) and three (classified as *cbl*C, *cbl*D and *cbl*F mutations) the early stages of intracellular Cbl processing. In all of these six mutations, production of *both* AdoCbl and MeCbl is decreased, causing combined deficiencies of methylmalonyl-CoA mutase and methionine synthetase. Methylmalonic aciduria and homocystinuria result. Two mutations affect only AdoCbl production (*cbl*A and *cbl*B). These cause methylmalonic aciduria *without* homocystinuria. (The two remaining mutations decrease MeCbl only and do not cause methylmalonic aciduria.)

The clinical course is variable. Apoenzyme defects (*mut* groups) often present with overwhelming neonatal illness, as described for propionic acidaemia. The babies are normal at birth but, within 3 or 4 days of protein feeding, deteriorate rapidly. There is severe ketoacidosis, increased plasma ammonium and glycine, leucopenia and thrombocytopenia. Plasma glucose is often low, probably because accumulating methylmalonyl-CoA inhibits pyruvate carboxylase (Rosenberg, 1983). Large amounts of methylmalonic acid, together with many metabolites, including propionylcarnitine, characteristic of propionic acidaemia, are excreted in the urine. Without aggressive treatment, survivors have recurrent ketoacidosis, growth delay, mental retardation and a high mortality rate (Matsui *et al.*, 1983). *cbl*A and *cbl*B group patients present with similar ketoacidotic illnesses and laboratory abnormalities, but seldom neonatally. Of *cbl*A patients 90% (but <40% of *cbl*B) respond clinically and biochemically to vitamin B_{12}, usually given as hydroxocobalamin. Acute metabolic illness is unusual in patients with combined deficiencies of AdoCbl and MeCbl (*cbl*C, *cbl*D, *cbl*F groups): less methylmalonic acid is excreted and plasma ammonium and glycine are not increased. The largest group of these patients (*cbl*C) present with failure to thrive, feeding problems, developmental delay and, sometimes, fits. Most have megaloblastic anaemia and some develop haemolytic anaemia. Vitamin B_{12} therapy decreases methylmalonic acid excretion but does not improve the neurological outcome, suggesting that the methylation defect is the major factor in brain damage. Supplements of methionine, betaine, tetrahydrofolate and MeCbl have not prevented neurological deterioration (Mahoney & Bick, 1987). The outlook is poor, with early death. Patients with absorptive defects of Cbl present with megaloblastic anaemia, as do babies with transcobalamin II deficiency who, in addition, fail to thrive and develop neurological symptoms. Methylmalonic aciduria and homocystinuria are usual. There is a good response to vitamin B_{12} treatment.

Methylmalonic aciduria (generally modest) occurs in patients with a *dietary deficiency* of vitamin B_{12}, and this should always be considered. It occurs in breast-fed babies of mothers with low vitamin B_{12} reserves, for example vegans (Higginbottom *et al.*, 1978; personal observations).

Maple syrup urine disease

Maple syrup urine disease is a heterogeneous disorder due to a defect in the multienzyme complex, branched-chain 2-oxo-acid dehydrogenase (Danner & Elsas, 1989; Wendel, 1990). This has three catalytic components, E_1, E_2 and E_3, and a kinase and phosphatase which inactivate and activate the complex, respectively. The E_3 component (lipoamide dehydrogenase) is identical to that in pyruvate dehydrogenase and 2-oxoglutarate dehydrogenase. Thiamine pyrophosphate is an essential cofactor which is believed to stabilize the complex, particularly the E_1 component. There is a spectrum of clinical severity. The worst affected present neonatally. Babies usually appear well for 4–7 days, but then deteriorate rapidly, as in propionic acidaemia. The urine classically smells sweet, but this may be a later feature. There is severe ketoacidosis and sometimes hypoglycaemia. Plasma and urinary branched-chain amino acids are grossly elevated, plasma leucine to 2–5 mmol/L, and alloisoleucine is present. There is a gross organic aciduria, dominated by 2-oxo and 2-hydroxy branched-chain acids. Without aggressive treatment, survivors are brain damaged and have repeated ketoacidosis, generally precipitated by infection. Others present later with episodic ketoacidosis, drowsiness, slurred speech and ataxia and, sometimes, coma and death, or have developmental delay. A few have been asymptomatic. A very small number of patients with a less severe presentation have had a thiamine-responsive defect (Duran & Wadman, 1985). They responded clinically and biochemically to 10–20 mg thiamine daily, in addition to dietary treatment. Deficiency of the E_3 component of branched-chain 2-oxo-acid, pyruvate and 2-oxoglutarate dehydrogenases occurs rarely, and leads to both branched-chain ketonuria and lactic acidosis (Robinson, 1985).

Other defects of branched-chain amino acid metabolism

Isovaleric acidaemia, 3-methylcrotonyl-CoA carboxylase deficiency (isolated) and 2-methylacetoacetyl-CoA thiolase (β-ketothiolase) deficiency all typically present with severe ketoacidosis (Sweetman, 1989; Ogier *et al.*, 1990); isovaleric acidaemia often has a neonatal presentation similar to propionic acidaemia. Intermittent episodes may lead to coma. Vomiting is a feature of isovaleric acidaemia and (sometimes with haematemesis and diarrhoea) of 2-methylacetoacetyl-CoA thiolase deficiency. Acute illnesses are associated with a strong smell of sweaty feet in isovaleric acidaemia and of tom-cats' urine in 3-methylcrotonyl-CoA carboxylase deficiency. In contrast, a deficiency of 3-methylglutaconic hydratase causes only mild problems (speech retardation). 3-Methylglutaconic aciduria also occurs with normal hydratase activity (the defect is still unknown) and is associated with serious brain damage. Deficiency of the last enzyme of leucine catabolism, 3-hydroxy-3-methylglutaryl-CoA (HMGCoA) lyase, is a very serious disorder, since this enzyme is also responsible for releasing ketones from fat oxidation in the liver. It presents like the fat oxidation defects and is described with them (p. 131).

Multiple carboxylase deficiency

Two inherited defects cause combined deficiencies of the body's four carboxylases: acetyl-CoA, pyruvate, propionyl-CoA and 3-methylcrotonyl-CoA carboxylases (Bartlett *et al.*, 1985; Wolf *et al.*, 1986; Dunkel *et al.*, 1989; Wolf & Heard, 1989; Baumgartner, 1990). Both involve the biotin cofactor essential for enzyme activity. Both are treatable. The B vitamin, biotin, is released from dietary protein in the gut lumen as lysyl-biotin (biocytin). This is hydrolysed by biotinidase in the pancreatic secretions, and free biotin is absorbed from the gut. It is carried in the blood bound to albumin (and possibly to *circulating* biotinidase, synthesized by the liver). Biotin is taken up by tissue cells, and bound covalently to the carboxylase apoenzymes by holocarboxylase synthetase. In time, the carboxylases are degraded by lysosomal enzymes and lysyl-biotin is released into the blood. Circulating biotinidase removes the lysyl groups and most of the free biotin is reused, the rest being excreted in the urine.

Deficiency of *holocarboxylase synthetase* generally presents before 3 months of age, with a severe acidotic illness, poor feeding, hypotonia and fits, and may progress rapidly to coma and death. There may be alopecia (lack of hair) and skin rash. Immunodeficiency leads to fungal infections, especially thrush. Plasma lactate is raised and there is an organic aciduria (Appendix 6.1). The urine may smell of tom cats. These biochemical disturbances resolve with biotin treatment (10 mg/day) and clinical symptoms improve.

In *biotinidase* deficiency, severe depletion of biotin develops because biotin recycling is impaired. Decreased availability of biotin from food and excessive renal losses may contribute. Symptoms appear after 3 months of age with fits, ataxia, developmental delay, hearing loss, blindness due to optic atrophy, conjunctivitis, fungal infections, skin rash and often striking alopecia – children may be almost bald. Most have acidosis with a diagnostic organic aciduria. Serum biotinidase is extremely low, as is biotin. The biochemical and clinical disturbances respond to 10 mg biotin per day, and if started early the outcome is good. Hearing loss and blindness are not reversed. Biotin depletion recurs within days if treatment is stopped. Screening programmes of neonates for biotinidase deficiency have measured the enzyme in blood spots (Dunkel *et al.*, 1989; Wolf & Heard, 1989).

Principles of management of defects of branched-chain amino acid metabolism, propionic acidaemia and methylmalonic aciduria; outcome

Episodes of severe ketoacidosis are a medical emergency which must be treated promptly and aggressively (Table 6.3). This is particularly true in the newborn period. For detailed guidelines see Danner and Elsas (1989), Ogier *et al.* (1990) and Wendel (1990). It is essential to stop all protein immediately and to minimize endogenous protein catabolism by providing a good supply of calories and treating infection. Some have given insulin, cautiously, with glucose. Oral metronidazole reduces the numbers of propionate-producing bacteria in the gut, and hence the amount of absorbed propionate. It is helpful in propionic acidaemia and methylmalonic aciduria. Exchange transfusions or peritoneal dialysis may be necessary to remove toxic metabolites in maple syrup urine disease and propionic acidaemia. Methylmalonic acid and isovalerylglycine are readily cleared by the kidneys, providing the patients are well hydrated. Glycine supplements are given in isovaleric acidaemia to accelerate removal of isovaleric acid as isovalerylglycine. L-Carnitine promotes excretion of toxic acids as acylcarnitines and replenishes carnitine reserves. A trial of pharmacological doses of vitamins is worthwhile in those conditions with vitamin-responsive variants, although neonatally presenting illnesses seldom respond. Hydroxocobalamin may need to be given for 3 weeks for an effect in methylmalonic aciduria. Some have used a multivitamin cocktail in acute situations, generally comprising biotin, thiamine, riboflavin, ascorbic acid, vitamin B_{12}, nicotinamide and pyridoxine. This is not harmful, but a more focused therapeutic approach should be possible with speedy return of biochemical results. The response to vitamins should be carefully reappraised later (Leonard & Daish, 1985). As soon as metabolic control is being regained (within 2–3 days), small amounts of protein are reintroduced, to prevent the catabolism associated with protein malnutrition. The aim is to restrict the intake of toxic amino acids and, at the same time, to increase the protein intake gradually to a level to permit normal growth and development. For example, in maple syrup urine disease, a milk (babies) or amino acid mixture (children) is given to supply the recommended dietary allowance of all the essential and non-essential amino acids except

Table 6.3 Principles of management of defects of branched-chain amino acid metabolism, propionic acidaemia and methylmalonic aciduria

Objective	Methods
Provide supportive care	Correct acidosis Rehydrate Ventilate if necessary
Remove the source of amino acid intoxication	Stop all protein/amino acids Promote anabolism by supplying calories: infuse glucose ± insulin Limit catabolism – treat infections Give oral metronidazole – propionic acidaemia, methylmalonic aciduria
Remove toxins	Hydrate Exchange transfusions or peritoneal dialysis Supply carnitine, intravenously then orally Supply glycine (isovaleric acidaemia)
Activate defective enzymes	Try: Hydroxocobalamin (methylmalonic aciduria) once daily Biotin (multiple carboxylase deficiency) 10 mg/day

valine, leucine and isoleucine. A small calculated amount of natural protein is added to supply enough branched-chain amino acids for needs, without reaching toxic concentrations. The natural protein intake is constantly readjusted to keep the plasma leucine level at 100–500 μmol/L, and valine and isoleucine slightly higher than normal. An increased allowance is needed during growth spurts. With such a restricted diet, supplements of vitamins, minerals and trace elements are essential. Calories must be adequate. In propionic acidaemia and methylmalonic aciduria, a low-protein, high-calorie diet is used, or a diet which aims to remove specific toxic amino acids. In these conditions, oral L-carnitine is given as a long-term supplement. In isovaleric acidaemia both L-carnitine and glycine supplements are used. Vitamin treatment is lifelong for those patients with responsive disorders. In multiple carboxylase deficiencies, this is usually the only treatment needed. Dietary management of the most severe defects is extremely difficult, and the children are intolerant of even very low protein intakes.

In general, patients with a fulminating neonatal illness have had a poor outcome, with early death or severe brain damage with neurological disabilities, including cerebral palsy and mental retardation. Survivors often have recurrent ketoacidosis, serious feeding problems and poor growth (Naughten *et al.*, 1982; Leonard *et al.*, 1984; Rousson & Guibaud, 1984; Mahoney & Bick, 1987; Wendel, 1990). However, prompt diagnosis and immediate institution of aggressive treatment has improved the prognosis considerably in maple syrup urine disease and methylmalonic aciduria. Some children have developed normally and attend normal schools. Nevertheless, the IQ of recently treated patients with maple syrup urine disease was lower than that of unaffected family members, and neurological deficits were common. There was a clear association with severity and duration of the neonatal illness (Wendel, 1990; Nord *et al.*, 1991). Clearly, prompt treatment is essential to obtain the best possible results. More children are being identified now with mild variant forms of these disorders – their outcome is much better. In general, early-treated patients with vitamin-responsive disorders do well. This is not true, however, for patients with methylmalonic aciduria with the *cbl*C mutation.

Inherited defects of mitochondrial fatty acid β-oxidation

Long-chain fatty acid oxidation is essential to provide energy for contraction of heart muscle and skeletal muscle, particularly during prolonged exercise. It is also essential during fasting, since it produces ketones in the liver and promotes gluconeogenesis. The ketones are released and used as an energy source by the brain, kidneys and skeletal muscle, and blood glucose is maintained (Chapter 5(a)). Ketogenesis only occurs in the liver.

Mitochondrial β-oxidation

Long-chain fatty acids (C14–C22) are carried by albumin in plasma. They enter liver and muscle cells, probably by both a carrier mechanism and diffusion. In the cytoplasm, the acyl group is first combined with coenzyme A and then transferred to a carrier molecule, carnitine, by the enzyme carnitine palmityl transferase I (CPT I) for transport across the mitochondrial wall by carnitine translocase. It is released into the mitochondrial matrix as acyl-CoA by carnitine palmityl transferase II (CPT II) and carnitine is recycled (Fig. 6.2a). Short- (<C6) and medium- (C6–C12) chain fatty acids enter the mitochondria directly, without the carnitine carrier, and combine with coenzyme A in the matrix. The acyl-CoA then enters the spiral of β-oxidation, in which four groups of enzymes work sequentially to release pairs of carbon units as acetyl-CoA (Fig. 6.2b). In the liver, some acetyl-CoA is incorporated into HMGCoA, from which acetoacetate is released by HMGCoA lyase. This enzyme also hydrolyses HMGCoA from leucine catabolism (Fig. 6.1). With each turn of the spiral, the fatty acyl group is shortened by two carbon atoms. The first step of β-oxidation removes hydrogen from the C2–C3 bond. This is catalysed by long-chain acyl-CoA dehydrogenase (LCAD) until the chain length reaches C12 to C14; MCAD takes over to C4 to C6 and, finally, short-chain acyl-CoA dehydrogenase (SCAD) completes the degradation. All three dehydrogenases

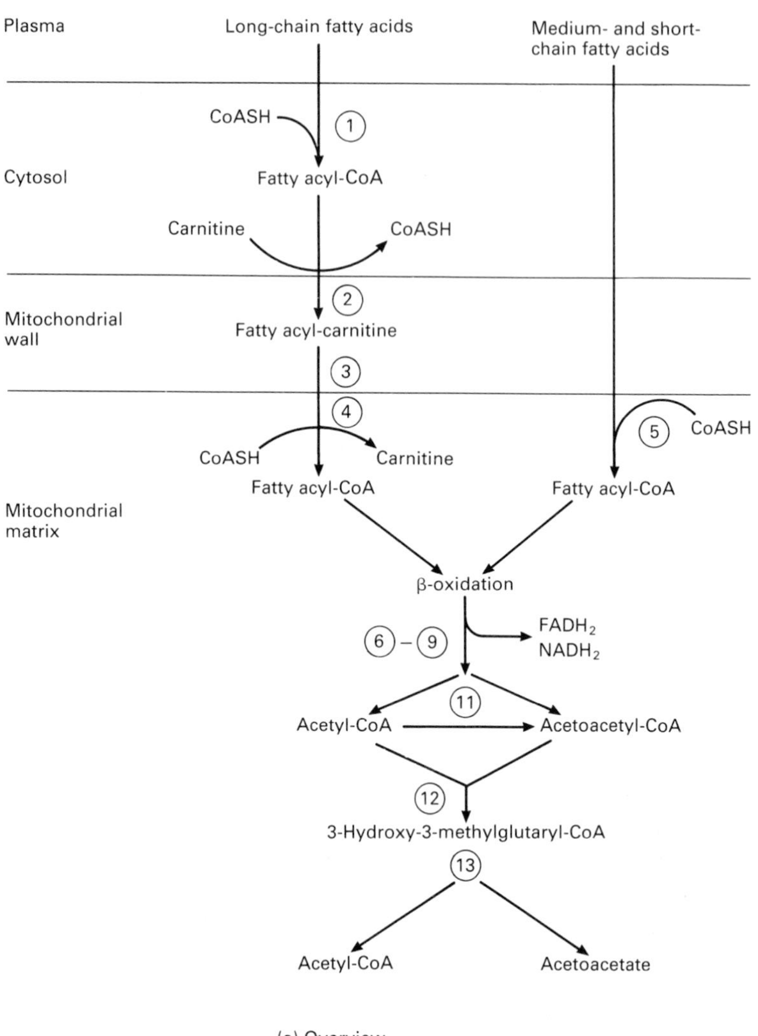

Fig. 6.2 (a) and (b) Transport of fatty acids into mitochondria, β-oxidation and ketogenesis in liver cells. Enzymes are circled: 1, cytoplasmic acyl-CoA synthase; 2, carnitine palmityl transferase I; 3, carnitine translocase; 4, carnitine palmityl transferase II; 5, acyl-CoA synthase; 6, fatty acyl-CoA dehydrogenases (long-, medium- and short-chain); 7, enoyl-CoA hydratase; 8, 3-hydroxyacyl-CoA dehydrogenase; 9, 3-oxoacyl-CoA thiolase; 10, ETF: ubiquinone oxidoreductase; 11, acetyl-CoA acetyltransferase; 12, 3-hydroxy-3-methylglutaryl-CoA (HMGCoA) synthase; 13, HMGCoA lyase. ETF, Electron transfer flavoprotein; CoASH, coenzyme A; CoQ, ubiquinone; *n*, number of carbon atoms in fatty acyl chains.

have flavin adenine dinucleotide (FAD) as the coenzyme and, like two similar mitochondrial enzymes (isovaleryl-CoA and glutaryl-CoA dehydrogenases), transfer electrons via electron transfer flavoprotein (ETF) to ETF: ubiquinone oxidoreductase (ETF-QO) and thence to ubiquinone and the electron transport chain. There are long- and short-chain-length-specific enoyl-CoA hydratases and 3-hydroxyacyl-CoA dehydrogenases (Gregersen, 1985a; Vianey-Liaud *et al.*, 1987; Roe & Coates, 1989; Stanley, 1990; Bartlett *et al.*, 1991).

Carnitine has a central role in mitochondrial β-

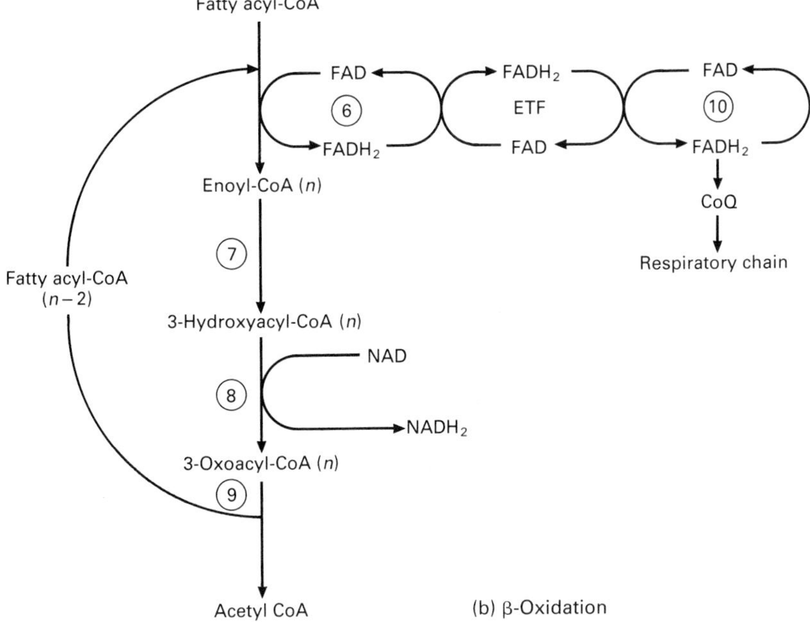

Fig. 6.2 *Continued*

oxidation. It is obtained from the diet and synthesized in the liver. It is taken up by skeletal and cardiac muscle by a transport system which is also present in the small intestine and renal tubules. Of the body carnitine 98% is in muscle and myocardium (Editorial, 1990). In addition to its role in fatty acid oxidation, carnitine is now known to have an extremely important function of recycling coenzyme A in the mitochondria.

Extramitochondrial fatty acid catabolism

In liver (but not muscle), considerable fatty acid catabolism also occurs outside the mitochondria when the fat influx is high during fasting. It is particularly active when mitochondrial fatty acid oxidation is defective. It results from the interaction of two cytoplasmic organelles, peroxisomes and microsomes (Gregersen *et al.*, 1983; Gregersen, 1985a; Vianey-Liaud *et al.*, 1987). Peroxisomes have an enzyme system for β-oxidation of saturated and unsaturated very long-chain (C24 and above) and long-chain (C14–C22) fatty acids. Different enzymes are involved from those of mitochondria, carnitine transport is not required, and no ATP is produced. Long-chain fatty acids are oxidized to medium-chain acids, mainly C10–C12, but with small amounts of C8 and C6 fatty acids. These products have two fates (Fig. 6.3a). Some enter the mitochondria and are degraded to acetyl-CoA. Some C10 to C12 acids transfer to microsomes, where the ω or ω-1 carbon atom is hydroxylated by cytochrome P_{450} mono-oxygenases. ω-Hydroxy acids are then oxidized to C10–C12 dicarboxylic acids (Fig. 6.4).

These in turn have two fates (Fig. 6.3b): some are transported by carnitine into the mitochondria, enter the β-oxidation spiral and are shortened to succinyl-CoA. Some transfer back to peroxisomes and are oxidized to C6, C8 and C10 dicarboxylic acids: adipic, suberic and sebacic acids, respectively. These are excreted in urine. C10–C12 ω-1-hydroxy acids are chain-shortened by peroxisomes to 5-hydroxyhexanoic and 7-hydroxyoctanoic acids and are excreted. In simple fasting ketosis, large amounts of medium-chain dicarboxylic acids are excreted in urine, *together* with large amounts of ketones. In

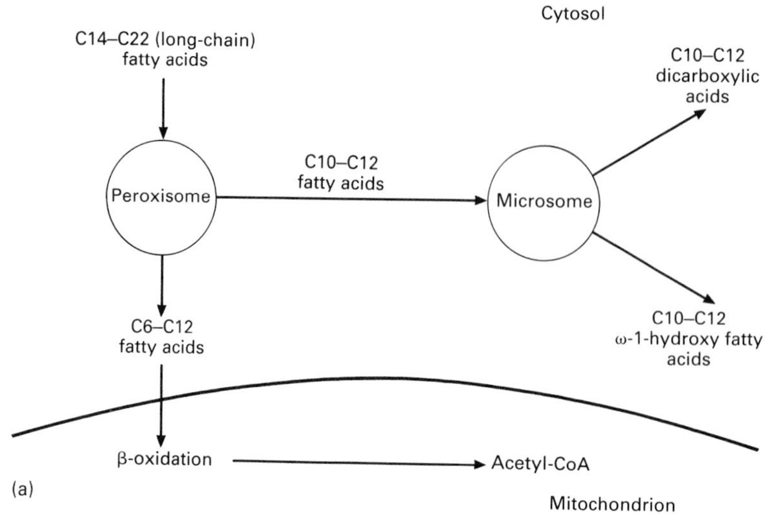

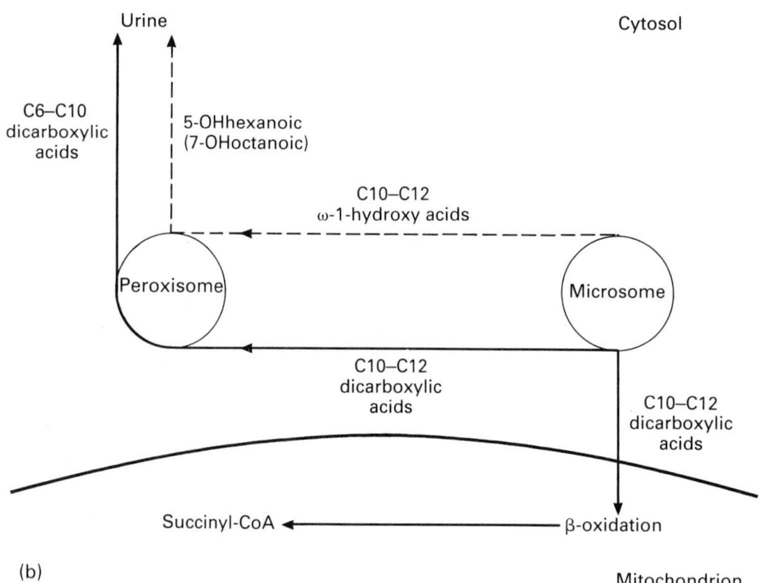

Fig. 6.3 (a) Peroxisomal oxidation of long-chain fatty acids and production of C10–C12 dicarboxylic and ω-1-hydroxy fatty acids by microsomes. (b) Fate of C10–C12 dicarboxylic and ω-1-hydroxy fatty acids.

inherited mitochondrial fat oxidation defects, there is dicarboxylic aciduria with minimal, or inappropriately low, ketonuria.

Inherited defects of mitochondrial β-oxidation

The first defects to be characterized were muscle carnitine deficiency (Engel & Angelini, 1973), muscle CPT deficiency (DiMauro & DiMauro, 1973) and MCAD deficiency (Gregersen *et al.*, 1976). Others have been recognized since (reviewed by: Vianey-Liaud *et al.* (1987), Pollitt (1989), Roe and Coates (1989), Angelini (1990), Bennett (1990) and Stanley (1990)). Some seem to be rare, but MCAD deficiency

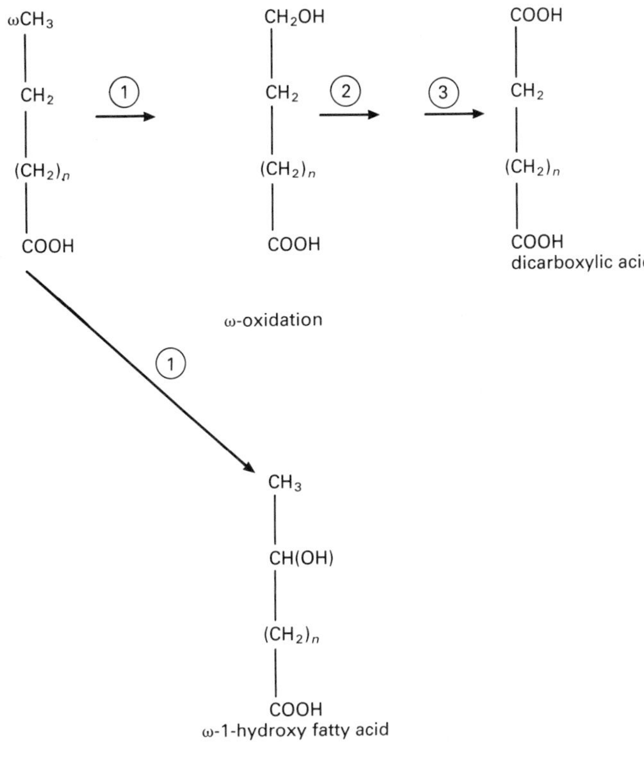

Fig. 6.4 Microsomal ω and ω-1 oxidation of fatty acids. 1, Cytochrome P_{450} mono-oxygenase; 2, alcohol dehydrogenase; 3, aldehyde dehydrogenase.

is emerging as one of the commonest inborn errors of metabolism among populations of northern European descent. The main clinical and biochemical features are summarized in Table 6.4. In multiple acyl-CoA dehydrogenase deficiency (MADD), leucine and lysine catabolism are defective, as well as fat oxidation, because activities of isovaleryl-CoA and glutaryl-CoA dehydrogenase *and* LCAD, MCAD and SCAD are all low. This is due to a defect of either ETF or ETF-QO, which are essential for all five flavoprotein dehydrogenases (Gregersen, 1985b; Frerman & Goodman, 1989). The condition presents as a severe neonatal form, or as a milder disorder (ethylmalonic aciduria) with a later onset. HMGCoA lyase deficiency impairs both leucine catabolism and hepatic ketogenesis (Gibson *et al.*, 1988; Sweetman, 1989). Carnitine deficiency is most often an acquired disturbance, secondary to an organic acid disorder in which abnormal acyl-CoA intermediates accumulate in mitochondria. However, *primary* carnitine deficiency is now a proven entity, which is due to a defect of the carnitine transport system (Stanley, 1990). In some, this affects muscle carnitine uptake predominantly. Muscle carnitine levels are low, but plasma levels are normal (Colin *et al.*, 1987). In others (systemic carnitine deficiency), muscle, renal and, probably, intestinal transport are all defective, so that gut and renal losses of carnitine are excessive. Plasma and muscle carnitine levels are both very low (Chapoy *et al.*, 1980; Tripp *et al.*, 1981; Treem *et al.*, 1988; Editorial, 1990).

Clinical features

Defective fat oxidation may be a very dangerous handicap, particularly for young children who depend heavily on fat metabolism during fasting. These disorders commonly present with life-threatening or fatal illness. One serious problem is

Table 6.4 Inherited defects of mitochondrial β-oxidation of fatty acids

Defect	Clinical presentation	Diagnosis	Treatment
LCAD	Early, some neonatal *Episodic encephalopathy with hypoglycaemia and low ketones, cardiorespiratory arrest; hypertrophic cardiomyopathy, muscle weakness, (cot death)* Hepatomegaly	*Acute*: plasma: †low glucose, mild acidosis, ketones low, high FFA: ketones, low total and free carnitine, raised acylcarnitine, high ammonium, raised transaminases†; C6–C12 dicarboxylic aciduria *Non-acute*: fasting test, lipid load, tissue enzyme	‡Avoid fasting Diet: low fat, high carbohydrate, MCT supplements; L-carnitine‡
LCHAD	Early, some neonatal As above, * to * Chronic progressive liver disease, some cirrhosis, failure to thrive, late pigmentary retinopathy	*Acute*: plasma: as above, † to †; C6–C12 dicarboxylic aciduria with dominance of OH-dicarboxylic acids (C6–C14) *Non-acute*: fasting test, lipid load, 3-phenylpropionic acid load, tissue enzyme	As above, ‡ to ‡
MCAD	First episode, 2 months to 4 years; neonatal rare; episodic hypoglycaemia with encephalopathy; normal between. 12% asymptomatic Risk less after 6 years	*Acute*: plasma: as above, † to †; raised octanoic and *cis*-4-decenoic acids and medium-chain acylcarnitines; urine: C6–C12 dicarboxylic acids, acylglycines, medium-chain acylcarnitines *Non-acute*: plasma *cis*-4-decenoic acid, plasma and urine acylcarnitines, DNA, 3-phenylpropionate load, fasting test, carnitine load	Avoid fasting; glucose if feeding interrupted (no MCTs)
SCAD (infantile) (three cases)	Neonatal: poor feeding, vomiting, normal development (1); death at 3 days (1); muscle weakness, failure to thrive, developmental delay (1)	Plasma: glucose normal, acidosis (2), low free carnitine, raised acylcarnitines; urine: organic aciduria, tissue enzyme	Symptomatic: avoid fasting; L-carnitine ? Low-fat diet (no MCTs)
MADD:			
Severe – no congenital anomalies	Present, 24–48 h: hypotonia, respiratory distress, hypoglycaemia, acidosis, hepatomegaly, vomiting, cardiomyopathy, sweaty feet odour. Die at few months, (cot death)	Plasma: as above, † to †, but severe acidosis; generalized amino acidaemia; organic aciduria; abnormal acylcarnitines, plasma and urine; tissue enzymes	Symptomatic; avoid fasting Diet: high carbohydrate, low fat, low protein; L-carnitine Trial of riboflavin (no MCTs)
Severe with congenital anomalies	As above, plus cystic kidneys, facial anomalies, hypospadias		

continued

Table 6.4 *Continued*

Defect	Clinical presentation	Diagnosis	Treatment
Later onset: ethylmalonic adipic aciduria	Onset: few weeks to adult life; episodic vomiting, hypoglycaemia, acidosis (some), coma, (cot death) Some: myopathy	*Acute*: plasma: as above, † to † and often raised sarcosine; organic aciduria; abnormal acylcarnitines, plasma and urine *Non-acute*: often no abnormalities; some: raised urine 2-OHglutaric acid, fasting test, tissue enzymes	
HMGCoA lyase	Onset: neonatal or first 12 months; episodic hypoglycaemia with encephalopathy, acidosis, vomiting, hypotonia, hepatomegaly, (cot death)	*Acute*: plasma: as above, † to †, but severe acidosis and often very high ammonium Organic aciduria, tissue enzyme	Avoid fasting Diet: restrict fat and leucine; L-carnitine
Hepatic CPT I	Onset: neonatal or first months; recurrent fasting hypoglycaemia	*Acute*: plasma: as above, † to †, free carnitine low, acylcarnitines raised; urine: organic acids normal or mild C6–C12 dicarboxylic aciduria *Non-acute*: fasting test, lipid load, tissue enzyme	Avoid fasting Diet: low fat with MCTs
CPT II	Onset: after 5 years; recurrent rhabdomyolysis after exercise, fasting, cold (One case (infant): cardiomyopathy, hypoglycaemia)	*Acute*: plasma: raised CK; urine: myoglobinuria, organic acids normal, muscle histology and enzyme (*Acute*: as for CPT I plus raised CK)	Avoid fasting, cold, heavy exercise
Carnitine translocase (one case)	Neonatal; cardiomyopathy, hypotonia, episodic hypoglycaemia; died at 3 years	*Acute*: plasma: as above, † to †, and raised CK; mild C6–C12 dicarboxylic aciduria with fasting	Avoid fasting Diet: low fat, high carbohydrate; L-carnitine
Carnitine deficiency			
Systemic	Onset: first months; hypotonia, cardiomyopathy, cardiorespiratory arrest, episodic hypoglycaemia with hepatomegaly	*Acute*: plasma: as above, † to †; carnitine very low; mild C6–C12 dicarboxylic aciduria; low muscle and liver carnitine; fibroblasts: carnitine transport	L-carnitine
Myopathic	Early childhood: muscle weakness, cardiomyopathy	*Plasma*: carnitine normal; no organic aciduria; low muscle carnitine; fibroblasts: carnitine transport	L-carnitine

LCAD, LCHAD, MCAD, SCAD and MADD, long-chain, long-chain 3-hydroxy-, medium-chain and short-chain acyl-CoA dehydrogenase and multiple acyl-CoA dehydrogenase deficiency, respectively; HMGCoA lyase, 3-hydroxy-3-methylglutaryl-CoA lyase; CPT, carnitine palmityl transferase; FFA, free fatty acid; MCT, medium-chain triglyceride; CK, creatine kinase; DNA, deoxyribonucleic acid.

See Appendix 6.1, p. 143 for details of organic acidurias.

hypertrophic cardiomyopathy, which causes heart failure or cardiorespiratory arrest. It occurs in MADD, LCAD (Hale *et al.*, 1985; Treem *et al.*, 1991), long-chain 3-hydroxyacyl-CoA dehydrogenase (LCHAD) (Duran *et al.*, 1991; Wanders *et al.*, 1991), CPT II (Demaugré *et al.*, 1991 – one infant), systemic and muscle carnitine, and carnitine translocase (Stanley *et al.*, 1992) deficiencies, and also (rarely) neonatally in MCAD deficiency (Walker *et al.*, 1990b).

Other children have died because provision of glucose and ketones by the liver during fasting has been hopelessly inadequate to meet the needs of the brain. It occurs when affected children have fed poorly or missed feeds, often during some normally trivial infective illness, such as gastroenteritis. At first, glucose is released from liver glycogen. However, as this supply wanes with continued fasting, long-chain fatty acids are mobilized from the fat depots and transported to the liver as an alternative source of fuel. Because β-oxidation is defective, little of the incoming fat can be degraded to completion. The amounts of acetyl-CoA, reduced nicotinamide adenine dinucleotide ($NADH_2$) and ketones produced are grossly reduced, compared with normal fasting children, and so is glucose synthesis by gluconeogenesis. The small amount of glucose produced is used up quickly. As the supply of glucose and ketones falls, brain function is disturbed, membrane pumps fail and the brain becomes oedematous. An early symptom of this is effortless vomiting. Drowsiness, restlessness and confusion develop and progress to coma over a few hours. There may be fits. At the same time, liver function is disturbed because of diminished hepatic energy production. Some of the incoming fatty acids are incorporated into triglycerides, and this fat infiltration frequently leads to acute liver enlargement. Fat may also accumulate in the heart, kidneys and skeletal muscles. At presentation, there is usually severe hypoglycaemia, but the fall in blood glucose is a relatively late event. Without intervention, children die. The illness generally develops over 1 to 3 days. It has all the clinical features of Reye's syndrome (p. 111), and has almost certainly been mislabelled as this in the past.

Episodic hypoketonaemic hypoglycaemia often starts early in infancy in MADD (Goodman *et al.*, 1982; Frerman & Goodman, 1989), and in deficiencies of LCAD (Hale *et al.*, 1985; Treem *et al.*, 1991), LCHAD (Duran *et al.*, 1991), HMGCoA lyase (Gibson *et al.*, 1988; Sweetman, 1989), hepatic CPT I (Bougnères *et al.*, 1981) and carnitine translocase (Stanley *et al.*, 1992). In MADD and HMGCoA lyase deficiency there is also severe metabolic acidosis, and in MADD often an unpleasant odour of sweaty feet because of accumulating short-chain fatty acids. MCAD deficiency seldom presents neonatally: the first episode generally occurs between 2 months and 4 years (mean 13.5 months) (Touma & Charpentier, 1992). Children with this defect are developmentally normal and well, when not hypocaloric. Indeed, around 15% never have overt symptoms. However, about 25% die during an acute episode and a few have sustained permanent brain damage when treatment was delayed. Cardiomyopathy has not been a feature, except neonatally. The risk of hypoglycaemic episodes decreases after around 6 years in this disorder, as children become less dependent on ketogenesis.

Rarely, inherited fat oxidation defects have led to unexpected death during the night (cot death) – in some cases because of hypoketotic hypoglycaemia, in others because of cardiorespiratory arrest due to cardiomyopathy. Deaths have been reported for MCAD, LCAD, LCHAD, CPT II and HMGCoA lyase deficiencies, MADD and ethylmalonic aciduria (Editorial, 1986; Emery *et al.*, 1988; Leonard *et al.*, 1988; Pollitt 1989, 1990; Ribes *et al.*, 1990; Demaugré *et al.*, 1991). It was suggested that fat oxidation defects might account for as many as 7% of cot deaths (Howat *et al.*, 1985), but subsequent studies have shown this to be an overestimate, and 1–2% is probably more likely (Touma & Charpentier, 1992). In retrospect, the babies have usually been unwell and fed poorly for a day or so before dying. At postmortem examination, fat infiltration of the liver and/or heart is usually evident, although this is not invariable in MCAD deficiency (personal observation).

Muscle weakness is a chronic feature of several defects. Muscle CPT (possibly CPT II deficiency (Bougnères *et al.*, 1981)) is unique in causing recurrent rhabdomyolysis. Symptoms usually develop in adult life but have occurred in childhood after 5 years of age (Angelini, 1990).

Management

For those disorders associated with recurrent hypoketotic hypoglycaemia, it is essential to avoid fasting by ensuring regular feeding. During intercurrent illnesses with loss of appetite, parents are instructed to give glucose drinks. If these are refused, there is vomiting, the child becomes floppy or less responsive, prompt hospital admission is mandatory so that glucose may be given via vein or nasogastric tube (Stanley, 1990). Dietary restrictions are usually unnecessary for MCAD and primary carnitine deficiency, but some fat restriction is usual for the other defects, together with protein restriction in MADD and HMGCoA lyase deficiency. Medium-chain triglycerides are useful supplements in LCAD and LCHAD deficiencies, but are contraindicated in MADD, MCAD and SCAD deficiencies. Oral L-carnitine is the treatment for primary carnitine deficiency and is given long term in most other defects (except MCAD deficiency) in which secondary carnitine depletion occurs. Riboflavin (100–300 mg/day), as a precursor of the FAD coenzyme, should be tried in MADD. It is seldom successful in severe forms of the disorder, but may be effective in ethylmalonic aciduria.

Biochemical diagnosis

At presentation with *acute encephalopathy*, the following findings for *plasma* are common: very low glucose, inappropriately low ketones (3-hydroxybutyrate < 1.1 mmol/L), increased free fatty acids (FFAs), high FFA:3-hydroxybutyrate ratio (generally > 2.0, normal fasting < 1.0), low free carnitine and often low total carnitine with proportionately increased acylcarnitines, raised ammonium and raised liver transaminase activities (aspartate aminotransferase (AST) and alanine aminotransferase (ALT)), generally three to 10 times the upper limit of normal (Bennett, 1990). Special tests help in the differential diagnosis: in MCAD deficiency, plasma concentrations of octanoic, decanoic and *cis*-4-decenoic acids are increased. *cis*-4-Decenoic acid is a metabolite of linoleic acid, and an increase is pathognomonic of MCAD deficiency, either as an isolated defect or part of MADD (Duran *et al.*, 1988; Heales *et al.*, 1991). There are diagnostic abnormalities of plasma acylcarnitines, detectable with sophisticated mass spectrometry; for example, increased medium-chain acylcarnitines (particularly octanoylcarnitine) in MCAD deficiency and long-chain acylcarnitines in LCAD deficiency (Roe & Coates, 1989). Currently, few have access to these analyses.

In *urine*, ketones may be increased, but at inappropriately low concentrations for the severity of hypoglycaemia. Organic acid analysis shows a medium-chain (C6–C12) dicarboxylic aciduria, which is usually gross, although relatively mild changes (if any) were found for defects of the carnitine cycle: systemic carnitine, hepatic CPT I and carnitine translocase deficiencies (Chapoy *et al.*, 1980; Bougnères *et al.*, 1981; Demaugré *et al.*, 1988; Treem *et al.*, 1988). In some defects there are diagnostic changes in the organic acid profile, for example MADD and HMGCoA lyase deficiency (Appendix 6.1, p. 143). In MCAD deficiency, three substrates of MCAD which accumulate in mitochondria (hexanoyl-, suberyl- and 3-phenylpropionyl-CoA) are detoxified by mitochondrial glycine-*N*-acylase to the corresponding glycine conjugates. These are excreted in urine and are pathognomonic of MCAD deficiency, isolated or as part of MADD (Gregersen *et al.*, 1976; Rinaldo *et al.*, 1989). Phenylpropionic acid is produced in the gut by bacteria. Young milk-fed babies with MCAD deficiency may not excrete 3-phenylpropionylglycine (Bennett, 1990). In LCHAD deficiency there are large amounts of 3-hydroxydicarboxylic acids (C6–C14), which collectively exceed C6–C12 non-hydroxylated dicarboxylic acids (Pollitt, 1990; Przyrembel *et al.*, 1991). However, interpretation may be difficult since lesser amounts of 3-hydroxy acids are also excreted in severe simple fasting ketosis (Appendix 6.2, p. 145). There are also characteristic disturbances of urinary acylcarnitines in some defects. Long-chain and 3-hydroxy long-chain acylcarnitines are not excreted in the urine. The dicarboxylic aciduria disappears very quickly with glucose administration, and acylcarnitine excretion falls sharply.

Diagnosis when children are not encephalopathic can be extremely difficult, except in neonatal MADD and HMGCoA lyase deficiency, which have a persistent organic aciduria. Diagnostic acylglycines and acylcarnitines are often detectable in urine from children with MCAD deficiency when they are well, but the concentrations are generally very low and sophisticated analyses are required. In the other defects there is no persisting organic aciduria, even when there is symptomatic cardiomyopathy. Tests are selected from a range of investigations to uncover and characterize a defect (Table 6.4).

1 Plasma total, free and esterified acylcarnitine levels. Levels are extremely low in primary systemic carnitine deficiency. In other defects free carnitine is reduced and acylcarnitines are increased proportionately. Total carnitine levels are low when the acylcarnitines are small enough for renal clearance.

2 Analysis of individual plasma or urinary acylcarnitines. Abnormalities are often detectable, but few centres have the necessary diagnostic facilities. An oral carnitine load (100 mg/kg) promotes urinary excretion of medium-chain acylcarnitines during the 6 h post-load, and has been used as a safe test for MCAD deficiency in asymptomatic children (Roe *et al.*, 1985, 1986).

3 Quantitative analysis of plasma *cis*-4-decenoic acid. High values occur in children with MCAD deficiency, even when they are well (Heales *et al.*, 1991).

4 DNA analysis. So far, tests are available only for MCAD deficiency. Up to 85% of affected individuals of northern European descent are homozygous for a single point mutation of an A to G nucleotide at position 985 of the MCAD gene – 'G985 mutation'. Other mutations have been identified, but these are individually uncommon (Blakemore *et al.*, 1991; Matsubara *et al.*, 1991; Yokota *et al.*, 1991).

5 A 3-phenylpropionate loading test for MCAD (and MADD) and LCHAD deficiencies. Fasting is unnecessary. Urine is collected for 6 h after a load of 25 mg 3-phenylpropionic acid per kg body weight has been given orally (in babies via a nasogastric tube). Normal subjects oxidize this compound to benzoic acid, which is conjugated with glycine and excreted as hippuric acid. In MCAD and LCHAD deficiencies, oxidation is incomplete and 3-phenylpropionylglycine and 3-hydroxy-3-phenylpropionylglycine, respectively, are excreted, as well as hippuric acid (Seakins & Rumsby, 1988; Duran *et al.*, 1991).

6 Fasting tests. When there are no other diagnostic clues, it may be necessary to carry out a carefully supervised fast to demonstrate a defect (see p. 99). The aims are to examine the ketogenic response to fasting and induce dicarboxylic aciduria, which may have diagnostic features. Plasma glucose is monitored throughout and 3-hydroxybutyrate, FFAs, total and free carnitine and urinary organic acids are monitored at the onset of hypoglycaemia or the termination of fasting. It is useful to take additional samples for FFAs and 3-hydroxybutyrate at 4 and 2 h before the planned end of the fast. In fat oxidation defects, the FFA:3-hydroxybutyrate ratio is high (>2.0 and often >4.0) (Bennett, 1990; Bonnefont *et al.*, 1990), and there is only a small rise in 3-hydroxy-butyrate (to <1.1 mmol/L). Coma and two deaths have resulted from fasting tests. They must only be undertaken when children are well, and should be avoided in neonates.

7 Oral loading tests with long-chain polyunsaturated fatty acids. This test is similarly reserved for cases where no diagnostic metabolites are demonstrable by other means. After an overnight fast of up to 10–12 h, according to the child's known fasting tolerance, 1.5 g polyunsaturated sunflower oil per kg body weight is given. Plasma glucose and ketones are measured basally and 1, 2 and 3 h after loading, and urine is analysed for organic acids (Saudubray, 1991). Experience with this test is still limited. It seems to be safe, but should probably be avoided in MCAD deficiency, for which other tests are available, because of the potential risk of toxicity from accumulating medium-chain fatty acids.

8 Enzyme analyses are used to establish a diagnosis and are applied to cultured skin fibroblasts or tissue biopsies. Decreased rates of catabolism of radiolabelled fatty acid substrates demonstrate a defect somewhere in the β-oxidation pathway. ^{3}H release assays for long-, medium- and short-chain fatty acid oxidation are used widely. Specific assays for individual enzymes are available, but are more demanding. Those for individual dehydrogenases require ETF, which is in short supply (Pollitt, 1989; Bennett, 1990).

Presymptomatic diagnosis in the newborn

With early diagnosis and careful feeding it should be possible to prevent hypoglycaemic encephalopathy. All newborns have an accelerated turnover of fatty acids during the first 72 h of life; this is the result of hormonal changes at birth and is not a response to hypoglycaemia. It was predicted that babies with a fat oxidation defect might have an abnormal dicarboxylic aciduria at this time, which would enable them to be identified (Bennett *et al.*, 1986). Using gas chromatography–mass spectrometry, this has been found in MCAD deficiency (Bennett *et al.*, 1987a; Walker *et al.*, 1990a,b). However, the diagnostic acylglycines may not be detected using gas chromatography alone (Bennett *et al.*, 1990), or if sample collection is delayed after the first 72 h. Normal babies have a dicarboxylic aciduria perinatally, which may cause confusion (Downing *et al.*, 1989). HMGCoA lyase deficiency and MADD may be detected by their characteristic organic aciduria. Alternative approaches for MCAD deficiency are analysis of DNA or *cis*-4-decenoic acid in blood spots. Both analyses might be applicable to neonatal screening for this defect (Editorial, 1991; Matsubara *et al.*, 1991). More fatty acid oxidation defects, and some other organic acid disorders, can be detected by analysis of acylcarnitines in blood spots. This can be done using two mass spectrometers in series (tandem mass spectrometry), and amino acids may be measured simultaneously. With this method at least 15 disorders of amino acid and fatty acid catabolism are detectable in blood spots, and its potential for large-scale neonatal screening is being evaluated (Millington *et al.*, 1990, 1991). The analytical instruments are extremely expensive.

Other organic acid defects

There are many other organic acid defects (see Appendix 6.1, p. 143). Three are mentioned here to demonstrate the range of pathological disturbances encountered.

Glutathione synthetase deficiency

This has two sequelae: it causes a massive accumulation of 5-oxoproline (pyroglutamic acid) and it leads to tissue depletion of glutathione. Glutathione is normally a source of sulphydryl groups; is an important reducing agent, for example in deoxyribonucleotide synthesis; detoxifies peroxides and protects against free radical damage; and detoxifies many foreign compounds. Patients present with severe persistent metabolic acidosis from birth. Red cell haemolysis occurs as a result of oxidative membrane damage, and neonatal jaundice and anaemia result. The haemolytic tendency persists but seldom necessitates transfusion in older children. Impaired granulocyte function predisposes to recurrent infections. Progressive brain damage occurs, but severity varies widely. It has been attributed to oxygen free radical damage. Treatment is with alkali to correct the acidosis and with α-tocopherol (vitamin E), which improves granulocyte function. Other therapies aimed at restoring glutathione levels, or providing alternative sources of sulphydryl groups, have been unsuccessful. Treatment with oxidant drugs must be avoided (Jellum *et al.*, 1970; Larsson *et al.*, 1985; Meister & Larsson, 1989; Larsson, 1990).

Succinic semialdehyde dehydrogenase deficiency

This deficiency is a disorder of metabolism of the neurotransmitter γ-aminobutyric acid (GABA). It leads to the accumulation of succinic semialdehyde which, in turn, is reduced to 4-hydroxybutyric acid and excreted in the urine. This compound is normally present at very low concentrations in the brain, but is not normally detectable in body fluids. In some cases, free GABA is increased in the cerebrospinal fluid (CSF). Clinical features are non-progressive ataxia, hypotonia, psychomotor delay, autistic features and convulsions. There is no metabolic acidosis. It seems likely that a disturbance of neurotransmission underlies the symptoms, but it is uncertain whether this is due to 4-hydroxybutyric acid, GABA, or both. Inhibition of succinic semialdehyde synthesis by the antiepileptic drug, γ-vinyl GABA (vigabatrin) has improved symptoms

(Rating *et al.*, 1984; Jaeken *et al.*, 1990; Jakobs *et al.*, 1992).

Aspartoacylase deficiency

This defect has been recognized only recently (Kvittingen *et al.*, 1986; Hagenfeldt *et al.*, 1987). Normally, the enzyme hydrolyses *N*-acetylaspartate, a compound found in large amounts in the brain. Its function there is unknown. It may be a source of acetate for synthesis of structural brain lipids, or serve as a depot for the neurotransmitter aspartate. It is reported that *N*-acetylaspartate is present only in the brain (Matalon *et al.*, 1988), but there is limited evidence for this. Large amounts of *N*-acetylaspartate are excreted in the urine of affected individuals, and CSF concentrations are high, but there is not a metabolic acidosis. This defect is now known to be the cause of Canavan disease, a devastating form of leucodystrophy, recognized clinically for many years. Symptoms develop within the first months of life: head enlargement (macrocephaly), severe mental retardation, spasticity, blindness, and sometimes deafness. The white matter of the brain appears spongy because of extensive demyelination. There is no treatment (Divry *et al.*, 1988; Matalon *et al.*, 1989; Jakobs *et al.*, 1991).

Appendix 6.1 Inherited organic acid defects: major urinary organic acid abnormalities†

Defect	Major urinary organic acid abnormalities
Branched-chain amino acid, propionic and methylmalonic acid metabolism	
Branched-chain 2-oxo-acid dehydrogenase (maple syrup urine disease)	2-Oxoisocaproic, 2-oxo-3-mevaleric, 2-oxoisovaleric and corresponding hydroxy acids
Lipoamide dehydrogenase (E_3)	2-OHIsovaleric (*2-oxoisocaproic), 2-OHbutyric; raised lactic, pyruvic and 2-oxoglutaric acids
Isovaleryl-CoA dehydrogenase (isovaleric acidaemia)	Isovalerylglycine, 3-OHisovaleric, isovalerylcarnitine
3-Methylcrotonyl-CoA carboxylase	3-OHIsovaleric, 3-mecrotonylglycine
3-Meglutaconic aciduria	
Hydratase deficient	3-Meglutaconic, 3-OHisovaleric
Hydratase normal	3-Meglutaconic, 3-meglutaric
3-OH-3-Meglutaryl-CoA lyase	3-OH-3-Meglutaric, 3-meglutaconic, 3-meglutaric, 3-OHisovaleric
2-Meacetoacetyl-CoA thiolase (β-ketothiolase)	2-me-3-OHButyric, 2-meacetoacetic (*tiglylglycine)
Propionyl-CoA carboxylase (propionic acidaemia)	3-OHPropionic, methylcitric, 3-OHvaleric, 3-oxovaleric, propionylcarnitine (*tiglylglycine, propionylglycine, 2-me-3-OHbutyric)
Multiple carboxylase: holocarboxylase synthetase or biotinidase	3-OHIsovaleric, 3-mecrotonylglycine, 3-OHpropionic, methylcitric (*tiglylglycine), raised lactic
Methylmalonyl-CoA mutase (methylmalonic aciduria)	
mut^0, *mut*$^-$, *cbl*A, *cbl*B	Methylmalonic, 3-OHpropionic, methylcitric, tiglylglycine, propionylcarnitine
*cbl*C, *cbl*D, *cbl*F	Methylmalonic
B_{12} transport and absorption defects	Methylmalonic (*methylcitric)
Fatty acid β-oxidation defects	
LCAD	Saturated and monounsaturated C6–C12 dicarboxylic acids, 5-OHhexanoic
MCAD	As above plus hexanoyl-, suberyl- and phenylpropionylglycines, octanoylcarnitine
SCAD	Ethylmalonic, methylsuccinic
MADD	
Severe: ETF, or ETF: ubiquinone oxidoreductase (glutaric aciduria type II)	As for MCAD and SCAD plus glutaric, 2-OHglutaric, 3-OHisovaleric, isobutyryl-, isovaleryl- and 2-mebutyryl-glycines and carnitines, glutarylcarnitine
Mild (ethylmalonic aciduria)	Ethylmalonic, methylsuccinic, adipic (*hexanoylglycine)
LCHAD	As for LCAD plus saturated and unsaturated C6–C14 3OH-dicarboxylic acids
Hepatic CPT	Mild C6–C12 dicarboxylic aciduria or normal
Muscle CPT II	None reported
Carnitine translocase	Small increases in adipic, suberic, sebacic
Systemic carnitine deficiency	Mild C6–C12 dicarboxylic aciduria or normal

continued on p. 144

Appendix 6.1 *Continued*

Defect	Major urinary organic acid abnormalities
Muscle carnitine deficiency	None reported
Miscellaneous	
Alanine : glyoxylate aminotransferase (primary hyperoxaluria type I)	Oxalic, glycolic, glyoxylic
Aspartoacylase (Canavan disease)	*N*-Acetylaspartic
Cytosolic acetoacetyl-CoA thiolase	3-OHButyric, acetoacetic
Fumarase	Fumaric (massive) (*increased succinic, 2-oxoglutaric)
Fumarylacetoacetate hydrolase (tyrosinaemia type I)	4-OHPhenyllactic, 4-OHphenylpyruvic, succinylacetone, *N*-acetyltyrosine
Glutaryl-CoA dehydrogenase (glutaric aciduria type I)	Glutaric, 3-OHglutaric, glutaconic (variable)
Glutathione synthetase	5-Oxoproline (= pyroglutamic acid)
D-Glyceric dehydrogenase (primary hyperoxaluria type II)	Oxalic, L-glyceric
D-Glyceric kinase (?) – precise defect unknown	D-Glyceric
Glycerol kinase	Increased glycerol
Homogentisic acid oxidase (alkaptonuria)	Homogentisic acid
D-2-OHglutaric aciduria	D-2-OHGlutaric
L-2-OHglutaric aciduria	L-2-OHGlutaric
Mevalonate kinase	Mevalonic (detected as mevalonolactone)
Mitochondrial acetoacetyl-CoA thiolase (one case)	3-OHButyric, lactic, citric
3-Oxo-acid-CoA transferase	3-OHButyric, acetoacetic, 3-OHisovaleric
2-Oxoadipic dehydrogenase	2-Oxoadipic, 2-OHadipic
5-Oxoprolinase	5-Oxoproline (pyroglutamic)
Succinic semialdehyde dehydrogenase	4-OHButyric
Tyrosine aminotransferase (tyrosinaemia type II)	4-OHPhenyllactic, 4-OHphenylpyruvic, *N*-acetyltyrosine
Uridine-5′-monophosphate synthetase (hereditary orotic aciduria)	Orotic acid

† Excludes primary lactic acidoses and peroxisomal defects, except primary hyperoxaluria type I.
Only the major abnormalities are shown.
OH, hydroxyl; me, methyl; LCAD, MCAD, SCAD, MADD, LCHAD, long-chain, medium-chain, short-chain, multiple, and long-chain 3-hydroxyacyl-CoA dehydrogenase, respectively; CPT, carnitine palmityl transferase; ETF, electron transfer factor.
* Metabolites found often, but not invariably.

Appendix 6.2 Acquired organic acidurias

Causes	Increased organic acids	References
Clinical		
Ketosis: fasting and diabetic	Acetoacetic, 3-OHbutyric, saturated and unsaturated C6–C12 dicarboxylic and 3-OHdicarboxylic acids, 5-OHhexanoic, 3-OHisovaleric, 3-OHisobutyric, 2-me-3-OHbutyric	Greter *et al.* (1980); Niwa (1986)
Hypoxia–ischaemia	Lactic	
Hypoxia–ischaemia with increased catabolism	Lactic, 2-OHbutyric, 2-OHisovaleric, 3-OHisovaleric, 2-me-3-OHbutyric, branched-chain oxo-acids	Landaas and Pettersen (1975); Landaas and Jakobs (1977); Walker and Mills (1992)
Newborns (well)	Saturated and unsaturated C6–C12 dicarboxylic and 3-OHdicarboxylic acids	Downing *et al.* (1989)
Abnormal gut microflora: small bowel resection and/or abnormal colonization	D-Lactic acid	Garcia *et al.* (1984); Haan *et al.* (1985)
	Glutaric, 3-hydroxypropionic	McCabe *et al.* (1982)
	propan 1,3-diol	Pollitt *et al.* (1987)
	2,3-Butanediol and acetoin	Walker *et al.* (1989a)
Hirschsprung's disease	Methylmalonic	Flannery *et al.* (1988)
Dietary		
Medium-chain triglycerides and pregestimil	Saturated C6–C12 dicarboxylic acids	Mortensen and Gregersen (1980)
Nutramigen	5-Oxoproline	Jellum (1977)
Vitamin B_{12} deficiency	Methylmalonic (methylcitric, some)	Higginbottom *et al.* (1978)
Intravenous amino acids with high phenylalanine (preterms)	4-OHPhenyllactic, 4-OHphenylpyruvic, phenyllactic, *N*-acetyltyrosine	Walker and Mills (1990)
Drugs (many) e.g.		
sodium valproate	Valproate metabolites	Cotariu and Zaidman (1988)
chloral hydrate	Glucuronide	
salicylate	2-Hydroxyhippurate	
propylene glycol (drug solvent)	Propylene glycol (1,2-propanediol)	
Artefacts		
Bacterial contamination	Succinic, benzoic	Chalmers and Lawson (1982)
Terminal/postmortem changes	Lactic, succinic, *N*-acetylaspartic	

OH, hydroxyl; me, methyl.

References

Angelini, C. (1990) Defects of fatty-acid oxidation in muscle. In: *Clinical Endocrinology and Metabolism*, Vol. 4. *Muscle Metabolism* (eds J.B. Harris & D.M. Turnbull), pp. 561–582. Baillière Tindall, London.

Bain, M.D., Borriello, S.P., Tracey, B.M. *et al.* (1988) Contribution of gut bacterial metabolism to human metabolic disease. *Lancet* **I**, 1078–1079.

Bartlett, K., Aynsley-Green, A., Leonard, J.V. & Turnbull, D.M. (1991) Inherited disorders of mitochondrial β-oxidation. In: *Inborn Errors of Metabolism* (eds J. Schaub, F. Van Hoof & H.L. Vis), pp. 19–41. Nestlé Nutrition Workshop Series, 24. Vevey/Raven Press Ltd., New York.

Bartlett, K., Ghneim, H.K., Stirk, H.-J. & Wastell, H. (1985) Enzyme studies in biotin-responsive disorders. *J Inher Metab Dis* **8** (Suppl. 1), 46–52.

Baumgartner, R. (1990) Biotin-responsive multiple carboxylase deficiency. In: *Inborn Metabolic Diseases* (eds J. Fernandes, J.-M. Saudubray & K. Tada), pp. 311–320. Springer-Verlag, Berlin.

Bennett, M.J. (1990) The laboratory diagnosis of inborn errors of mitochondrial fatty acid oxidation. *Ann Clin Biochem* **27**, 519–531.

Bennett, M.J., Allison, F., Pollitt, R.J. *et al.* (1987a) Prenatal diagnosis of medium-chain acyl-CoA dehydrogenase deficiency in family with sudden infant death. *Lancet* **I**, 440–441.

Bennett, M.J., Coates, P.M., Hale, D.E. *et al.* (1990) Analysis of abnormal urinary metabolites in the newborn period in medium-chain acyl-CoA dehydrogenase deficiency. *J Inher Metab Dis* **13**, 707–715.

Bennett, M.J., Pollitt, R.J., Land, J.M., Turner, M.J. & Cheetham, C.H. (1987b) Lethal multiple acyl-CoA dehydrogenation deficiency with dysmorphic features. *J Inher Metab Dis* **10**, 95–96.

Bennett, M.J., Variend, S. & Pollitt, R.J. (1986) Screening siblings for inborn errors of fatty acid metabolism in families with a history of sudden infant death. *Lancet* **II**, 1470.

Blakemore, A.I.F., Singleton, H., Pollitt, R.J. *et al.* (1991) Frequency of the G985 MCAD mutation in the general population. *Lancet* **337**, 298–299.

Bonnefont, J.P., Specola, N.B., Vassault, A. *et al.* (1990) The fasting test in paediatrics: application to the diagnosis of pathological hypo- and hyperketotic states. *Eur J Pediatr* **150**, 80–85.

Bougnères, P.-F., Saudubray, J.-M., Marsac, C. *et al.* (1981) Fasting hypoglycemia resulting from hepatic carnitine palmitoyl transferase deficiency. *J Pediatr* **98**, 742–746.

Brown, J.C.C., Mills, G.A., Sadler, P.J. & Walker, V. (1989) ^{1}H NMR studies of urine from premature and sick babies. *Magnetic Res Med* **11**, 193–201.

Bulusu, S., Mills, G.A. & Walker, V. (1991) Analysis of organic acids in physiological fluids by high performance liquid chromatography. *J Liquid Chromat* **14**, 1757–1777.

Chalmers, R.A. (1987) Disorders of organic acid metabolism. In: *The Inherited Metabolic Diseases* (ed. J.B. Holton), pp. 141–214. Churchill Livingstone, Edinburgh.

Chalmers, R.A. & Lawson, A.M. (1982) *Organic Acids in Man*. Chapman and Hall, London.

Chalmers, R.A., Roe, C.R., Stacey, T.E. & Hoppel, C.L. (1984) Urinary excretion of L-carnitine and acylcarnitines by patients with disorders of organic acid metabolism: evidence for secondary insufficiency of L-carnitine. *Pediatr Res* **18**, 1325–1328.

Chapoy, P.R., Angelini, C., Jann Brown, W. *et al.* (1980) Systemic carnitine deficiency – a treatable inherited lipid storage disease presenting as Reye's syndrome. *N Engl J Med* **303**, 1389–1394.

Chong, W.K., Mills, G.A., Weavind, G.P. & Walker, V. (1989) High performance liquid chromatographic method for the rapid profiling of plasma and urinary organic acids. *J Chromat* **487**, 147–153.

Colin, A.A., Jaffe, M., Shapira, Y. *et al.* (1987) Muscle carnitine deficiency presenting as familial fatal cardiomyopathy. *Arch Dis Child* **62**, 1170–1172.

Cotariu, D. & Zaidman, J.L. (1988) Valproic acid and the liver. *Clin Chem* **34**, 890–897.

Daish, P. & Leonard, J.V. (1985) Rapid profiling of plasma organic acids by high performance liquid chromatography. *Clin Chim Acta* **146**, 87–91.

Danner, D.J. & Elsas, L.J. (1989) Disorders of branched chain amino acid and keto acid metabolism. In: *The Metabolic Basis of Inherited Disease*, Vol. 1 (eds C.R. Scriver, A.L. Beaudet, W.S. Sly & D. Valle), 6th edn, pp. 671–692. McGraw-Hill, New York.

Demaugré, F., Bonnefont, J.-P., Colonna, M. *et al.* (1991) Infantile form of carnitine palmitoyltransferase II deficiency with hepatomuscular symptoms and sudden death. *J Clin Invest* **87**, 859–864.

Demaugré, F., Bonnefont, J.-P., Mitchell, G. *et al.* (1988) Hepatic and muscular presentations of carnitine palmitoyl transferase deficiency: two distinct entities. *Pediatr Res* **24**, 308–311.

DiMauro, S. & DiMauro, P.M. (1973) Muscle carnitine palmityltransferase deficiency and myoglobinuria. *Science* **182**, 929–931.

Divry, P., Vianey-Liaud, C., Gay, C. *et al.* (1988) *N*-Acetylaspartic aciduria; report of three new cases in children with a neurological syndrome associating macrocephaly and leukodystrophy. *J Inher Metab Dis* **11**, 307–308.

Downing, M., Rose, P., Bennett, M.J., Manning, N.J. & Pollitt, R.J. (1989) Generalised dicarboxylic aciduria:

a common finding in neonates. *J Inher Metab Dis* **12** (Suppl. 2), 321–324.

Dunkel, G., Scriver, C.R., Clow, C.L. *et al.* (1989) Prospective ascertainment of complete and partial serum biotinidase deficiency in the newborn. *J Inher Metab Dis* **12**, 131–138.

Duran, M. & Wadman, S.K. (1985) Thiamine-responsive inborn errors of metabolism. *J Inher Metab Dis* **8** (Suppl. 1), 70–75.

Duran, M., Bruinvis, L., Ketting, D., de Klerk, J.B.C. & Wadman, S.K. (1988) *cis*-4-Decenoic acid in plasma: a characteristic metabolite in medium-chain acyl-CoA dehydrogenase deficiency. *Clin Chem* **34**, 548–551.

Duran, M., Wanders, R.J.A., de Jager, J.P. *et al.* (1991) 3-Hydroxydicarboxylic aciduria due to long-chain 3-hydroxyacyl-coenzyme A dehydrogenase deficiency associated with sudden neonatal death: protective effect of medium-chain triglyceride treatment. *Eur J Pediatr* **150**, 190–195.

Editorial (1986) Sudden infant death and inherited disorders of fat oxidation. *Lancet* **II**, 1073–1075.

Editorial (1990) Carnitine deficiency. *Lancet* **I**, 631–633.

Editorial (1991) Medium chain acyl CoA dehydrogenase deficiency. *Lancet* **338**, 544–545.

Emery, J.L., Howat, A.J., Variend, S. & Vawter, G.F. (1988) Investigation of inborn errors of metabolism in unexpected infant deaths. *Lancet* **II**, 29–31.

Engel, A.G. & Angelini, C. (1973) Carnitine deficiency of human skeletal muscle with associated lipid storage myopathy: a new syndrome. *Science* **173**, 899–902.

Fenton, W.A. & Rosenberg, L.E. (1989) Inherited disorders of cobalamin transport and metabolism. In: *The Metabolic Basis of Inherited Disease*, Vol. 2 (eds C.R. Scriver, A.L. Beaudet, W.S. Sly & D.Valle), 6th edn, pp. 2065–2082. McGraw-Hill, New York.

Fernandes, J., Saudubray, J.-M. & Tada, K. (eds) (1990) *Inborn Metabolic Diseases*. Springer-Verlag, Berlin.

Fishlock, R.C., Bieber, L.L. & Snoswell, A.M. (1984) Sources of error in determinations of carnitine and acylcarnitine in plasma. *Clin Chem* **30**, 316–318.

Flannery, D., Lafer, C.Z. & Roesel, R.A. (1988) Gut bacterial metabolism. *Lancet* **2**, 225–226.

Frerman, F.E. & Goodman, S.I. (1989) Glutaric acidemia type II and defects of the mitochondrial respiratory chain. In: *The Metabolic Basis of Inherited Disease*, Vol. 1 (eds C.R. Scriver, A.L. Beaudet, W.S. Sly & D. Valle), 6th edn, pp. 915–931. McGraw-Hill, New York.

Garcia, J., Smith, F.R. & Cucinell, S.A. (1984) Urinary D-lactate excretion in infants with necrotizing enterocolitis. *J Pediatr* **104**, 268–270.

Gibson, K.M., Breuer, J. & Nyhan, W.L. (1988) 3-Hydroxy-3-methylglutaryl-coenzyme A lyase deficiency: review of 18 reported patients. *Eur J Pediatr* **148**, 180–186.

Goodman, S.I. (1986) Inherited metabolic disease in the newborn: approach to diagnosis and treatment. In: *Advances in Pediatrics, 33* (eds L.A. Barness, A.M.F. Bongiovanni, G. Morrow, F. Oski & A.M. Rudolph), pp. 197–224. Year Book Medical Publishers, Chicago.

Goodman, S.I. & Frerman, F.E. (1989) Organic acidemias due to defects in lysine oxidation: 2-ketoadipic acidemia and glutaric acidemia. In: *The Metabolic Basis of Inherited Disease*, Vol. 1 (eds C.R. Scriver, A.L. Beaudet, W.S. Sly & D. Valle), 6th edn, pp. 845–853. McGraw-Hill, New York.

Goodman, S.I. & Markey, S.P. (eds) (1981) *Diagnosis of Organic Acidemias by Gas Chromatography–Mass Spectrometry*. Alan R. Liss, New York.

Goodman, S.I., Stene, D.O., McCabe, E.R.B. *et al.* (1982) Glutaric acidemia type II: clinical, biochemical, and morphologic considerations. *J Pediatr* **100**, 946–950.

Gregersen, N. (1985a) The acyl-CoA dehydrogenation deficiencies. *Scand J Clin Lab Invest* **45** (Suppl. 174), 11–60.

Gregersen, N. (1985b) Riboflavin-responsive defects of β-oxidation. *J Inher Metab Dis* **8** (Suppl. 1), 65–69.

Gregersen, N., Lauritzen, R. & Rasmussen, K. (1976) Suberylglycine excretion in the urine from a patient with dicarboxylic aciduria. *Clin Chim Acta* **70**, 417–425.

Gregersen, N., Mortensen, P.B. & Kølvraa, S. (1983) On the biologic origin of C_6–C_{10}-dicarboxylic and C_6–C_{10}-ω-1-hydroxy monocarboxylic acids in human and rat with acyl-CoA dehydrogenation deficiencies: *in vitro* studies on the ω- and ω-1-oxidation of medium-chain (C_6–C_{12}) fatty acids in human and rat liver. *Pediatr Res* **17**, 828–834.

Greter, J., Lindstedt, S., Seeman, H. & Steen, G. (1980) 3-Hydroxydecanedioic acid and related homologues: urinary metabolites in ketoacidosis. *Clin Chem* **26**, 261–265.

Haan, E., Brown, G., Bankier, A. *et al.* (1985) Severe illness caused by the products of bacterial metabolism in a child with a short gut. *Eur J Pediatr* **144**, 63–65.

Hagenfeldt, L., Bollgren, I. & Venizelos, N. (1987) *N*-Acetylaspartic aciduria due to aspartoacylase deficiency – a new aetiology of childhood leukodystrophy. *J Inher Metab Dis* **10**, 135–141.

Hale, D.E., Batshaw, M.L., Coates, P.M. *et al.* (1985) Long-chain acyl coenzyme A dehydrogenase deficiency: an inherited cause of nonketotic hypoglycemia. *Pediatr Res* **19**, 666–671.

Heales, S.J.R., Woolf, D.A., Robinson, P. & Leonard, J.V. (1991) Rapid diagnosis of medium-chain acyl CoA dehydrogenase deficiency by measurement of *cis*-4-decenoic acid in plasma. *J Inher Metab Dis* **14**, 661–667.

Higginbottom, M.C., Sweetman, L. & Nyhan, W.L. (1978) A syndrome of methylmalonic aciduria, homocystinuria, megaloblastic anaemia and neurological abnormalities in a vitamin B_{12}-deficient breast-fed infant of a strict vegetarian. *N Engl J Med* **299**, 317–323.

Howat, A.J., Bennett, M.J., Variend, S., Shaw, L. & Engel, P.C. (1985) Defects of metabolism of fatty acids in the sudden infant death syndrome. *Br Med J* **290**, 1771–1773.

Iles, R.A. & Chalmers, R.A. (1988) Nuclear magnetic resonance spectroscopy in the study of inborn errors of metabolism. *Clin Sci* **74**, 1–10.

Iles, R.A., Hind, A.J. & Chalmers, R.A. (1985) Use of proton nuclear magnetic resonance spectroscopy in detection and study of organic acidurias. *Clin Chem* **31**, 1795–1801.

Jaeken, J., Casaer, P., Haegele, K.D. & Schechter, P.J. (1990) Review: normal and abnormal central nervous system GABA metabolism in childhood. *J Inher Metab Dis* **13**, 793–801.

Jakobs, C. (1989) Prenatal diagnosis of inherited metabolic disorders by stable isotope dilution GC–MS analysis of metabolites in amniotic fluid: review of four years experience. *J Inher Metab Dis* **12** (Suppl. 2), 267–270.

Jakobs, C., Michael, T., Jaeger, E., Jacken, J. & Gibson, K.M. (1992) Further evaluation of Vigabatrin therapy in 4-hydroxybutyric aciduria. *Eur J Pediatr* **151**, 466–468.

Jakobs, C., Ten Brink, H.J., Langelaar, S.A. *et al.* (1991) Stable isotope dilution analysis of *N*-acetylaspartic acid in CSF, blood, urine and amniotic fluid: accurate postnatal diagnosis and the potential for prenatal diagnosis of Canavan disease. *J Inher Metab Dis* **14**, 653–660.

Jellum, E. (1977) Profiling of human body fluids in healthy and disease states using gas chromatography and mass spectrometry, with special reference to organic acids. *J Chromat* **143**, 427–462.

Jellum, E., Kluge, T., Börresen, H.C., Stokke, O. & Eldjarn, L. (1970) Pyroglutamic aciduria – a new inborn error of metabolism. *Scand J Clin Lab Invest* **26**, 327–335.

Kvittingen, E.A., Guldal, G., Borsing, S. *et al.* (1986) *N*-Acetylaspartic aciduria in a child with progressive cerebral atrophy. *Clin Chim Acta* **158**, 217–227.

Landaas, S. & Jakobs, C. (1977) The occurrence of 2-hydroxyisovaleric acid in patients with lactic acidosis and ketoacidosis. *Clin Chim Acta* **78**, 489–493.

Landaas, S. & Pettersen, J.E. (1975) Clinical conditions associated with urinary excretion of 2-hydroxybutyric acid. *Scand J Clin Lab Invest* **35**, 259–266.

Larsson, A. (1990) Disorders of the gamma glutamyl cycle. In: *Inborn Metabolic Diseases* (eds J. Fernandes, J.-M. Saudubray & K. Tada), pp. 331–336. Springer-Verlag, Berlin.

Larsson, L., Wachtmeister, L., von Wendt, L. *et al.* (1985) Ophthalmological, psychometric and therapeutic investigation in two sisters with hereditary glutathione synthetase deficiency (5-oxoprolinuria). *Neuropediatrics* **16**, 131–136.

Leonard, J.V. & Daish, P. (1985) Evaluation of cofactor responsiveness. *J Inher Metab Dis* **8** (Suppl. 1), 17–19.

Leonard, J.V., Daish, P., Naughten, E.R. & Bartlett, K. (1984) The management and long term outcome of organic acidaemias. *J Inher Metab Dis* **7** (Suppl. 1), 13–17.

Leonard, J.V., Green, A., Holton, J.B. & Bartlett, K. (1988) Inborn errors of metabolism and unexpected infant deaths. *Lancet* **II**, 854.

McCabe, E.R.B., Goodman, S.I., Fennessey, P.V. *et al.* (1982) Glutaric, 3-hydroxypropionic, and lactic aciduria with metabolic acidemia, following extensive small bowel resection. *Biochem Med* **28**, 229–236.

McGarry, J.D. & Foster, D.W. (1976) An improved and simplified radioisotopic assay for the determination of free and esterified carnitine. *J Lipid Res* **17**, 277–281.

Mahoney, M.J. & Bick, D. (1987) Recent advances in the inherited methylmalonic acidemias. *Acta Paed Scand* **76**, 689–696.

Matalon, R., Kaul, R., Casanova, J. *et al.* (1989) Aspartoacylase deficiency: the enzyme defect in Canavan disease. *J Inher Metab Dis* **12** (Suppl. 2), 329–331.

Matalon, R., Michals, K., Sebesta, D. *et al.* (1988) Aspartoacylase deficiency and *N*-acetylaspartic aciduria in patients with Canavan disease. *Am J Med Genet* **29**, 463–471.

Matsubara, Y., Narisawa, K., Tada, K. *et al.* (1991) Prevalence of K329E mutation in medium-chain acyl-CoA dehydrogenase gene determined from Guthrie cards. *Lancet* **338**, 552–553.

Matsui, S.M., Mahoney, M.J. & Rosenberg, L.E. (1983) The natural history of the inherited methylmalonic acidemias. *N Engl J Med* **308**, 857–861.

Meister, A. & Larsson, A. (1989) Glutathione synthetase defects and other disorders of the γ-glutamyl cycle. In: *The Metabolic Basis of Inherited Disease*, Vol. 1 (eds C.R. Scriver, A.L. Beaudet, W.S. Sly & D. Valle), 6th edn, pp. 855–868. McGraw-Hill, New York.

Millington, D.S. (1986) New methods for the analysis of acylcarnitines and acylcoenzyme A compounds. In: *Mass Spectrometry in Biomedical Research* (ed. S.J. Gaskell), pp. 97–114. John Wiley & Sons, Chichester.

Millington, D.S., Kodo, N., Norwood, D.L. & Roe, C.R. (1990) Tandem mass spectrometry: a new method for acylcarnitine profiling with potential for neonatal screening for inborn errors of metabolism. *J Inher Metab Dis* **13**, 321–324.

Millington, D.S., Kodo, N., Terada, N., Roe, D. & Chace, D.H. (1991) The analysis of diagnostic markers of genetic disorders in human blood and urine using tandem mass spectrometry with liquid secondary ion mass spectrometry. *Int J Mass Spect Ion Process* **111**, 211–228.

Mills, G.A., Walker, V., Ashton, M.R., Manning, N.J. &

Pollitt, R.J. (1990) Vitreous humour organic acids in medium chain acyl-CoA dehydrogenase deficiency. *J Inher Metab Dis* **13**, 239–240.

Mortensen, P.B. & Gregersen, N. (1980) Medium-chain triglyceride medication as a pitfall in the diagnosis of non-ketotic C_6–C_{10}-dicarboxylic acidurias. *Clin Chim Acta* **103**, 33–37.

Naughten, E.R., Jenkins, J., Francis, D.E.M. & Leonard J.V. (1982) Outcome of maple syrup urine disease. *Arch Dis Child* **57**, 918–921.

Niwa, T. (1986) Metabolic profiling with gas chromatography–mass spectrometry and its application to clinical medicine. *J Chromat* **379**, 313–345.

Nord, A., Van Doorninck, W.J. & Greene, C. (1991) Developmental profile of patients with maple syrup urine disease. *J Inher Metab Dis* **14**, 881–889.

Ogier, H., Charpentier, C. & Saudubray, J.-M. (1990) Organic acidemias. In: *Inborn Metabolic Diseases* (eds J. Fernandes, J.-M. Saudubray & K. Tada), pp. 271–299. Springer-Verlag, Berlin.

Pollitt, R.J. (1989) Disorders of mitochondrial β-oxidation: prenatal and early postnatal diagnosis and their relevance to Reye's syndrome and sudden infant death. *J Inher Metab Dis* **12** (Suppl. 1), 215–230.

Pollitt, R.J. (1990) Clinical and biochemical presentation in 20 cases of hydroxydicarboxylic aciduria. In: *Fatty Acid Oxidation: Clinical, Biochemical, and Molecular Aspects* (eds K. Tanaka & P.M. Coates), pp. 495–502. Alan R. Liss, New York.

Pollitt, R.J., Fowler, B., Sardharwalla, I.B., Edwards, M.A. & Gray, R.G.F. (1987) Increased excretion of propan-1,3-diol and 3-hydroxypropionic acid apparently caused by abnormal bacterial metabolism in the gut. *Clin Chim Acta* **169**, 151–158.

Przyrembel, H. (1990) Defects of lysine degradation. In: *Inborn Metabolic Diseases* (eds J. Fernandes, J.-M. Saudubray & K. Tada), pp. 300–310. Springer-Verlag, Berlin.

Przyrembel, H., Jakobs, C., Ijlst, L., de Klerk, J.B.C. & Wanders, R.J.A. (1991) Long-chain 3-hydroxyacyl-CoA dehydrogenase deficiency. *J Inher Metab Dis* **14**, 674–680.

Rating, D., Hanefeld, F., Siemes, H. *et al.* (1984) 4-Hydroxybutyric aciduria: a new inborn error of metabolism. I. Clinical review. *J Inher Metab Dis* **7** (Suppl. 1), 90–92.

Ribes, A., Briones, P., Vilaseca, M.A., Baraibar, R. & Gairi, J.M. (1990) Sudden death in an infant with 3-hydroxy-3-methylglutaryl-CoA lyase deficiency. *J. Inher Metab Dis* **13**, 752–753.

Rinaldo, P., O'Shea, J.J., Welch, R.D. & Tanaka, K. (1989) Stable isotope dilution analysis of n-hexanoylglycine, 3-phenylpropionylglycine and suberylglycine in human urine using chemical ionization gas chromatography/mass spectrometry selected ion monitoring. *Biomed Environ Mass Spect* **18**, 471–477.

Robinson, B.H. (1985) The lacticacidemias. In: *Genetic and Metabolic Disease in Pediatrics* (eds J.K. Lloyd & C.R. Scriver), pp. 111–139. Butterworth, London.

Roe, C.R. & Bohun, T.P. (1982) L-Carnitine therapy in propionic acidaemia. *Lancet* **I**, 1411–1412.

Roe, C.R. & Coates, P.M. (1989) Acyl-CoA dehydrogenase deficiencies. In: *The Metabolic Basis of Inherited Disease*, Vol. 1 (eds C.L. Scriver, A.L. Beaudet, W.S. Sly & D. Valle), 6th edn, pp. 889–914. McGraw-Hill, New York.

Roe, C.R., Millington, D.S., Maltby, D.A. *et al.* (1985) Diagnostic and therapeutic implications of medium-chain acylcarnitines in the medium-chain acyl-CoA dehydrogenase deficiency. *Pediatr Res* **19**, 459–466.

Roe, C.R., Millington, D.S., Maltby, D.A. & Kinnebrew, P. (1986) Recognition of medium-chain acyl-CoA dehydrogenase deficiency in asymptomatic siblings of children dying of sudden infant death or Reye-like syndromes. *J Pediatr* **108**, 13–18.

Rosenberg, L.E. (1983) Disorders of propionate and methylmalonate metabolism. In: *The Metabolic Basis of Inherited Disease* (eds J.B. Stanbury, J.B. Wyngarden, D.S. Fredrickson, J.L. Goldstein & M.S. Brown), 5th edn, pp. 474–497. McGraw-Hill, New York.

Rosenberg, L.E. & Fenton, W.A. (1989) Disorders of propionate and methylmalonate metabolism. In: *The Metabolic Basis of Inherited Disease*, Vol. 1 (eds C.L. Scriver, A.L. Beaudet, W.S. Sly & D. Valle), 6th edn, pp. 821–844. McGraw-Hill, New York.

Rousson, R. & Guibaud, P. (1984) Long term outcome of organic acidurias: survey of 105 French cases (1967–1983). *J Inher Metab Dis* **7** (Suppl. 1), 10–12.

Sann, L., Divry, P., Cartier, B., Vianey-Laud, C. & Maire, I. (1987) Ketogenesis in hypoglycemic neonates. *Biol Neonat* **52**, 80–85.

Saudubray, J.-M. (1991) Verbal communication. Reported in: *Inborn Errors of Metabolism* (eds J. Schaub, F. Van Hoof & L. Vis), p. 37. Nestlé Nutrition Workshop Series. Vevey/Raven Press Ltd., New York.

Scriver, C.R. Beaudet, A.L., Sly, W.S. & Valle, D. (eds) (1989) *The Metabolic Basis of Inherited Disease*, Vols 1 & 2, 6th edn. McGraw-Hill, New York.

Seakins, J.W.T. & Rumsby, G. (1988) The use of phenylpropionic acid as a loading test for medium-chain acyl-CoA dehydrogenase deficiency. *J Inher Metab Dis* **11** (Suppl. 2), 221–224.

Stanley, C.A. (1990) Disorders of fatty acid oxidation. In: *Inborn Metabolic Diseases* (eds J. Fernandes, J.-M. Saudubray & K. Tada), pp. 395–410. Springer-Verlag, Berlin.

Stanley, C.A., Hale, D.E., Berry, G.T. *et al.* (1992) A

deficiency of carnitine–acylcarnitine translocase in the inner mitochondrial membrane. *N Engl J Med* **327**, 19–23.

Sweetman, L. (1989) Branched chain organic acidurias. In: *Metabolic Basis of Inherited Disease*, Vol. 1 (eds C.R. Scriver, A.L. Beaudet, W.S. Sly & D. Valle), 6th edn, pp. 791–819. McGraw-Hill, New York.

Tanaka, K., Budd, M.A., Efron, M.L. & Isselbacher, K.J. (1966) Isovaleric acidaemia: a new genetic defect of leucine metabolism. *Proc Natl Acad Sci USA* **56**, 236–242.

Tanaka, K., Mandell, R. & Shih, V.E. (1976) Metabolism of [1-^{14}C] and [2-^{14}C] leucine in cultured skin fibroblasts from patients with isovaleric acidemia. *J Clin Invest* **58**, 164–172.

Touma, E.H. & Charpentier, C. (1992) Medium chain acyl-CoA dehydrogenase deficiency. *Arch Dis Child* **67**, 142–145.

Tracey, B.M., Dore, C.J., Lawson, A.M., Watts, R.W.E. & Chalmers, R.A. (1981) Urinary organic acids in normal full-term newborns aged 1–7 days. *J Inher Metab Dis* **4**, 63–64.

Treem, W.R., Stanley, C.A., Finegold, D.A., Hale, D.E. & Coates, P.M. (1988) Primary carnitine deficiency due to a failure of carnitine transport in kidney, muscle, and fibroblasts. *N Engl J Med* **319**, 1331–1336.

Treem, W.R., Stanley, C.A., Hale, D.E., Leopold, H.B. & Hyams, J.S. (1991) Hypoglycemia, hypotonia, and cardiomyopathy: the evolving clinical picture of long-chain acyl-CoA dehydrogenase deficiency. *Pediatrics* **87**, 328–333.

Tripp, M.E., Katcher, M.L., Peters, H.A. *et al.* (1981) Systemic carnitine deficiency presenting as familial endocardial fibroelastosis. *N Engl J Med* **305**, 385–390.

Tuchman, M. & Ulstrom, R.A. (1985) Urinary organic acids in health and disease. In: *Advances in Pediatrics, 32* (ed. L.A. Barness), pp. 469–506. Year Book Medical Publishers, Chicago.

Vianey-Liaud, C., Divry, P., Gregersen, N. & Mathieu, M. (1987) The inborn errors of mitochondrial fatty acid oxidation. *J Inher Metab Dis* **10** (Suppl. 1), 159–198.

Walker, V. & Mills, G.A. (1989) Urinary organic acid excretion by babies born before 33 weeks of gestation. *Clin Chem* **35**, 1460–1466.

Walker, V. & Mills, G.A. (1990) Metabolism of intravenous phenylalanine by babies born before 33 weeks of gestation. *Biol Neonat* **57**, 155–166.

Walker, V. & Mills, G.A. (1992) Effects of birth asphyxia on urinary organic acid excretion. *Biol Neonat* **61**, 162–172.

Walker, V., Mills, G.A., Hall, M.A. & Lowes, J.A. (1989a) Carbohydrate fermentation by gut microflora in preterm neonates. *Arch Dis Child* **64**, 1367–1373.

Walker, V., Mills, G.A., Hall, M.A. *et al.* (1989b) A fourth case of fumarase deficiency. *J Inher Metab Dis* **12**, 331–332.

Walker, V., Mills, G.A. & Radford, M. (1990a) Diagnosis of medium chain acyl-CoA dehydrogenase (MCAD) deficiency in neonates. *Lancet* **I**, 1288–1289.

Walker, V., Mills, G.A., Weavind, G.P., Hall, M.A. & Johnston, P.G.B. (1990b) Diagnosis of medium chain acyl-CoA dehydrogenase (MCAD) deficiency in an asymptomatic neonate. *Ann Clin Biochem* **27**, 267–269.

Walter, J.H., Leonard, J.V., Thompson, G.N., Halliday, D. & Bartlett, K. (1988) Gut bacterial metabolism. *Lancet* **ii**, 226.

Wanders, R.J.A., Ijlst, L., Duran, M. *et al.* (1991) Long-chain 3-hydroxyacyl-CoA dehydrogenase deficiency: different clinical expression in three unrelated patients. *J Inher Metab Dis* **14**, 325–328.

Wendel, U. (1990) Disorders of branched-chain amino acid metabolism. In: *Inborn Metabolic Diseases* (eds J. Fernandes, J.-M. Saudubray & K. Tada), pp. 263–270. Springer-Verlag, Berlin.

Wolf, B. & Heard, G.S. (1989) Disorders of biotin metabolism. In: *The Metabolic Basis of Inherited Diseases*, Vol. 2 (eds C.R. Scriver, A.L. Beaudet, W.S. Sly & D. Valle), 6th edn, pp. 2083–2103. McGraw-Hill, New York.

Wolf, B., Heard, G.S., Jefferson, L.G. *et al.* (1986) Neonatal screening for biotinidase deficiency: an update. *J Inher Metab Dis* **9** (Suppl. 2), 303–306.

Wolf, B., Hsia, Y.E., Sweetman, L. *et al.* (1981) Propionic acidemia: a clinical update. *J Pediatr* **99**, 835–846.

Yokota, I., Coates, P.M., Hale, D.E., Rinaldo, P. & Tanaka, K. (1991) Molecular survey of a prevalent mutation ^{985}A- to -G transition, and identification of five infrequent mutations in the medium-chain acyl-CoA dehydrogenase (MCAD) gene in 55 patients with MCAD deficiency. *Am J Hum Genet* **49**, 1280–1291.

7: Inborn Errors of Cellular Organelles

H.S.A. HEYMANS, R.J.A. WANDERS, G.K. BROWN & G.T.N. BESLEY

(a) Peroxisomal disorders

H.S.A. HEYMANS & R.J.A. WANDERS

Introduction

In humans 15 different inherited peroxisomal disorders, caused by an impairment of one or more peroxisomal functions, are known. They are usually subdivided into three groups deppending upon the extent of peroxisomal dysfunction (see Table 7.1). This classification is of little clinical significance but it provides a systematic pathophysiological insight. In fact, a Zellweger phenotype may be found in patients with a generalized (classic Zellweger syndrome), multiple (Zellweger-like syndrome) or single (pseudo-Zellweger syndrome, i.e. peroxisomal thiolase deficiency) loss of peroxisomal functions.

The Zellweger syndrome had been described by Bowen *et al.* (1964) and Passarge and McAdams (1967) but it was not until 1973 that Goldfischer *et al.* demonstrated the characteristic absence of recognizable peroxisomes in the liver and kidneys. Meanwhile, De Duve and his collaborators (for a review see De Duve and Baudhuin (1966)) had demonstrated the presence of oxidases and catalase in peroxisomes.

The ultimate realization that Zellweger syndrome is a disorder of the peroxisome came in the early 1980s when it was found that Zellweger patients show elevated levels of very long-chain fatty acids (VLCFA) (notably C26:0) in plasma and strongly decreased plasmalogen levels in erythrocytes and tissues (Brown *et al.*, 1982; Heymans *et al.*, 1983, 1984).

Peroxisomes: metabolic functions

β-Oxidation of fatty acids and fatty acid derivatives

One of the most important functions of peroxisomes in humans is the β-oxidation of fatty acids and fatty acid derivatives. As in mitochondria, peroxisomal β-oxidation proceeds via successive steps of dehydrogenation, hydration, dehydrogenation and thiolytic cleavage. Before fatty acids and fatty acid derivatives can be subjected to β-oxidation, activation to a coenzyme-A (CoA) ester must occur. This is brought about by a variety of fatty acid-activating enzymes present in each cell. As a result of one cycle of β-oxidation, a fatty acyl-CoA ester is shortened by two carbon atoms. Accordingly, many cycles of β-oxidation are required to break up completely a fatty acyl-CoA ester into its acetyl-CoA units.

The first step in peroxisomal β-oxidation, i.e. the dehydrogenation of acyl-CoA esters to their *trans*-2-enoyl-CoA esters, is brought about by at least three different enzymes referred to as palmitoyl-CoA oxidase, pristanoyl-CoA oxidase and trihydroxycholestanoyl-CoA oxidase. From the limited information available most fatty acyl-CoA esters are handled by palmitoyl-CoA oxidase. Since this enzyme is not reactive with pristanoyl-CoA and trihydroxycholestanoyl-CoA, these substrates require the active involvement of pristanoyl-CoA oxidase and trihydroxycholestanoyl-CoA oxidase (see Mannaerts and van Veldhoven (1992) for review).

The subsequent steps in peroxisomal β-oxidation are believed to be carried out by a multifunctional protein incorporating enoyl-CoA hydrolase, 1,3-hydroxyacyl-CoA dehydrogenase, enoyl-CoA isom-

Table 7.1 Classification of peroxisomal disorders

Group A: generalized loss of peroxisomal functions
Cerebrohepatorenal (Zellweger) syndrome
Neonatal adrenoleucodystrophy (NALD)
Infantile Refsum disease (IRD)
Hyperpipecolic acidaemia (HPA)
Group B: multiple loss of peroxisomal functions
Rhizomelic chondrodysplasia punctata (RCDP)
Zellweger-like syndrome
Group C: single loss of peroxisomal functions
X-linked adrenoleucodystrophy or one of its phenotypic variants (X-ALD)
Acyl-CoA oxidase deficiency (pseudo-NALD)
Bifunctional protein deficiency
Peroxisomal thiolase deficiency (pseudo-Zellweger)
Dihydroxyacetonephosphate acyltransferase deficiency (DHAPAT deficiency) (pseudo-RCDP)
Glutaryl-CoA oxidase deficiency
Phytanic acid storage disease (Refsum disease)
Hyperoxaluria type I
Acatalasaemia

erase activity and a specific peroxisomal thiolase. The presence of enoyl-CoA isomerase and 2,4-dienoyl-CoA reductase activity in peroxisomes enables them to remove the double bond in unsaturated fatty acids, thus allowing their oxidation.

Although peroxisomes and mitochondria are both capable of fatty acid β-oxidation, they serve different purposes in the cell. Mitochondrial fatty acid β-oxidation is responsible for the oxidation of most (>95%) of the fatty acids absorbed from food. Although the peroxisomal system is not so important for energy production from fatty acids, it catalyses the chain shortening of a distinct set of substrates which cannot be handled by mitochondria. Peroxisomes are obligatory for the β-oxidation of the following compounds:

1 VLCFAs (C24:0 and C26:0).
2 Di- and trihydroxycholestanoic acids.
3 Pristanic acid.
4 Long-chain dicarboxylic acids.
5 Certain prostaglandins.
6 Certain leucotrienes.
7 12- and 15-hydroxyeicosatetraenoic acids.
8 Certain mono- and polyunsaturated acids.

Known peroxisomal disorders are shown in Table 7.1. β-Oxidation of VLCFAs in peroxisomes first requires activation via a specific activating enzyme present in the peroxisomes. This enzyme activity is deficient in X-linked adrenoleucodystrophy (X-ALD) patients. The VLCFA-CoA ester is subsequently oxidized by palmitoyl-CoA oxidase, the multifunctional protein and peroxisomal thiolase. Di- and trihydroxycholestanoic acids are formed from cholesterol via a complex set of reactions (Hanson *et al.*, 1979). Subsequently, di- and trihydroxycholestanoic acids (DHCA and THCA) are activated to their CoA esters via a specific cholestanoyl-CoA synthetase. DHCA-CoA and THCA-CoA are then degraded by peroxisomal β-oxidation involving the specific enzyme cholestanoyl-CoA oxidase, the multifunctional protein and thiolase. Accordingly, the pathways of VLCFA β-oxidation and DHCA/THCA β-oxidation differ with regard to the first two steps, i.e. the activation and oxidase reaction, and converge at the level of the multifunctional protein and thiolase (see Wanders *et al.* (1990a) for review).

Ether-phospholipid biosynthesis

Peroxisomes play an essential role in ether-phospholipid biosynthesis since the two enzyme activities responsible for the introduction of the ether bond in ether-phospholipids, i.e. dihydroxyacetonephosphate acyltransferase (DHAPAT) and alkyldihydroxyacetonephosphate synthase (alkyl DHAP synthase), are localized in peroxisomes. The main end products of this biosynthetic pathway in humans are plasmalogens which are widely distributed in mammalian cell membranes and make up as much as 80–90% of ethanolamine phospholipids in the white matter of the brain. The physiological role of ether lipids has not yet been clarified, except for platelet-activating factor (PAF), which is thought to play a part in asthma, anaphylaxis, inflammation and other reactions (see Van den Bosch *et al.* (1992) for review).

Glyoxylate catabolism

Deficiency of the peroxisomal enzyme alanine: glyoxylate aminotransferase in hyperoxaluria type I is associated with severe clinical consequences resulting from the accumulation of glyoxylate and, especially, oxalate, which is formed from glyoxylate in a reaction catalysed by lactate dehydrogenase (see Danpure (1989) for review). Peroxisomes, at least in humans, are the main site of glyoxylate degradation.

L-Pipecolic acid degradation

Peroxisomes also play an indispensable role in the degradation of L-pipecolic acid, a metabolite of L-lysine, since L-pipecolic acid oxidase is a peroxisomal enzyme in humans. Although isolated L-pipecolic acid oxidase deficiency has not been described yet, several instances of an accumulation of L-pipecolic acid in the absence of other abnormalities have been reported (see Danks *et al.*, 1975; Roesel *et al.*, 1991).

Biosynthesis of certain polyunsaturated fatty acids such as docosahexaenoic acid

Based on the finding that docosahexaenoic acid levels are deficient in erythrocytes and tissues from Zellweger patients, it has been suggested that peroxisomes are involved in the biosynthesis of docosahexaenoic acid catalysing one of the intermediary reactions, i.e. the Δ4-desaturation step (Martinez, 1989, 1990). In addition, peroxisomes are also involved in hydrogen peroxide metabolism, cholesterol and dolichol synthesis, polyamine catabolism and other metabolic processes. Whether phytanic acid oxidase activity is confined to peroxisomes is still a matter for discussion.

Biogenesis of peroxisomes

Peroxisomes, like mitochondria and chloroplasts, arise by growth and division of pre-existing peroxisomes, rather than by budding from the endoplasmic reticulum. They may transiently or permanently be interconnected to form a peroxisomal reticulum containing elements resembling smooth endoplasmic reticulum.

Unlike mitochondria and chloroplasts, peroxisomes contain no DNA, so that all peroxisomal proteins must be coded for by nuclear genes. All peroxisomal proteins investigated so far, including soluble matrix proteins, core proteins and integral membrane proteins, are synthesized on free ribosomes and are imported post-translationally into peroxisomes. Almost all are synthesized in their final size, except for oxoacyl-CoA thiolase, which is synthesized as a 44-kDa precursor protein. After transfer into the peroxisomes it is converted to its 41-kDa mature form. As with other proteins destined for particular subcellular sites, peroxisomal proteins must possess topogenic signals to direct them to peroxisomes. One such topogenic signal is the serine–lysine–leucine (SKL) motif (or one of its variants) present at the C-terminus of a number of peroxisomal proteins (see Subramani (1992) for review). The fact that the SKL motif, or one of its variants, is not present in all peroxisomal proteins indicates that such proteins must use some other targeting signal. Recent results from different laboratories (Osumi *et al.*, 1991; Swinkels *et al.*, 1991) have shown that a non-SKL type of topogenic signal is located at the amino-terminal end of peroxisomal β-oxoacyl-CoA thiolase. Whether there are any other topogenic signals is unknown.

The peroxisomal disorders: clinical and biochemical characteristics

Deficiency of peroxisomes is associated with a wide spectrum of clinical abnormalities, ranging from severe in Zellweger syndrome to much milder in infantile Refsum disease (IRD) (Poll-The *et al.*, 1987). The severity of clinical abnormalities may reflect the extent of peroxisomal malfunction, as shown by Schrakamp *et al.* (1988), who studied *de novo* plasmalogen biosynthesis in patients' fibroblasts, and Lazarow and Moser (1989), who studied VLCFA levels. The latter reported that, on average, plasma C26:0 levels are lower in patients affected by IRD compared with those with Zellweger syndrome and neonatal adrenoleucodystrophy (NALD) (Kelley *et al.*, 1986). In both studies the differences did not reach statistical significance and it is doubtful whether the diagnosis of Zellweger syndrome, NALD, IRD or hyperpipecolic acidaemia (HPA) can be made unequivocally on the basis of plasma VLCFA levels and/or plasmalogen biosynthesis in fibroblasts.

Biochemically, patients affected by a group A disorder, i.e. a disorder of peroxisome biogenesis, also called peroxisomal deficiency disorder, show a great number of abnormalities which all follow logically from the functions of peroxisomes in humans, as described earlier (see Table 7.2).

Group B comprises rhizomelic chondrodysplasia punctata (RCDP) and Zellweger-like syndrome. RCDP is clinically characterized by a disproportionately short stature, primarily affecting the proximal parts of the extremities, typical facial appearance, congenital contractures, characteristic ocular involvement, severe growth deficiency and mental retardation. Although death in early infancy has been described, most patients survive beyond the first year of life, sometimes well into their second decade. Four distinct abnormalities have been found in RCDP, including a deficient activity of DHAPAT, alkyl DHAP synthase and phytanic acid oxidase. Furthermore, peroxisomal thiolase occurs in an abnormal molecular form (Lazarow & Moser, 1989). Recently, a patient with all the clinical signs and symptoms of RCDP but with a single deficiency of DHAPAT only was described (Wanders *et al.*, 1992c). The identification of this new peroxisomal disorder (pseudo-RCDP) exemplifies the functional importance of ether-phospholipids, although no unique function has yet been identified for these lipids. In Zellweger-like syndrome, an entity clinically indistinguishable from Zellweger syndrome, abnormalities in ether-phospholipid synthesis and peroxisomal β-oxidation are found, despite the fact that peroxisomes are abundantly present in the liver.

Group C includes X-ALD as the most frequent disorder. The clinical presentation of X-ALD is highly variable (Moser *et al.*, 1987). In some patients the clinical presentation may only begin in adolescence or adult life as in adrenomyeloneuropathy (AMN). Childhood X-ALD patients typically display behavioural, visual and/or auditory disturbances as well as an abnormal gait. The disease usually culminates within a few years in dementia, blindness, quadriplegia and death. Darkening of the skin, secondary to adrenal dysfunction, may develop. Adrenal insufficiency may precede the development of neurological symptoms and remain the only clinical abnormality.

Biochemically, patients with X-ALD show elevated VLCFA levels (notably C24:0 and C26:0) due to an impaired peroxisomal β-oxidation of these fatty acids.

Apart from X-ALD, there are other peroxisomal disorders, clinically resembling peroxisomal deficiency disorders but with a biochemical defect restricted to the peroxisomal β-oxidation pathway only. These include acyl-CoA oxidase deficiency (Poll-The *et al.*, 1988; Wanders *et al.*, 1990b), bifunctional protein deficiency (Watkins *et al.*, 1989) and peroxisomal thiolase deficiency (Schram *et al.*, 1987), all reported in single cases only.

Apart from these disorders with a clearly defined defect in peroxisomal β-oxidation, a number of patients with a similar clinical presentation have been described with a defect in peroxisomal β-oxidation of unknown aetiology (Clayton *et al.*, 1988; Naidu *et al.*, 1988; Barth *et al.*, 1990). Using complementation analysis, the precise enzymatic defect can be identified in these patients, as recently shown in the case of the patient described by Clayton *et al.* (1988) (see Wanders *et al.*, 1992a).

Other diseases belonging to group C are hyperoxaluria type I (alanine:glyoxylate aminotransferase deficiency (see Danpure (1989) for review), glutaryl-CoA oxidase deficiency in a variant case of glutaric aciduria type I (Bennett *et al.*, 1991), DHAPAT deficiency in a variant case of RCDP (Wanders *et al.*, 1992c) and acatalasaemia, a relatively innocuous disease.

It is still disputed whether or not phytanic acid storage disease (Refsum disease) is a peroxisomal disorder. Nevertheless, Refsum disease in its classic form displays a characteristic set of four abnormalities, i.e. retinitis pigmentosa, peripheral neuropathy, cerebellar ataxia and elevated cerebrospinal (CSF) protein levels, although it must be stressed that this classic tetrad of abnormalities is not observed in every patient.

Finally, some patients have been described with clinical characteristics reminiscent of a peroxisomal disorder and with abnormal bile acids, i.e. di- and trihydroxycholestanoic acids only (Christensen *et*

Table 7.2 Biochemical characteristics of the peroxisomal disorders

Parameter measured	Type of peroxisomal disorder								
	Peroxisome deficiency disorders	Zellweger-like syndrome	X-linked ALD	Acyl-CoA oxidase deficiency	Bifunctional protein deficiency	Thiolase deficiency	Di/trihydroxy-cholestanoic acidaemia	RCDP	Pseudo-RCDP
Metabolites in body fluids									
Very long-chain fatty acids	Elevated	Elevated	Elevated‡	Elevated	Elevated	Elevated	Normal	Normal	Normal
Bile acid intermediates	Elevated	Elevated	Normal	Normal	Elevated	Elevated	Elevated	Normal	Normal
Pipecolic acid	Elevated*	NA	Normal	Normal	Normal	Normal	Normal	Normal	Normal
Phytanic acid	Elevated†	NA	Normal	Normal	Elevated§	Elevated§	Elevated§	Elevated†	Normal
Pristanic acid	Elevated†	NA	Normal	Normal	Elevated§	Elevated§	Elevated§	Normal	Normal
Plasmalogen synthesis									
DHAPAT	Deficient	Deficient	Normal	Normal	Normal	Normal	Normal	Deficient	Deficient
Alkyl DHAP synthase	Deficient	NA	Normal	Normal	Normal	Normal	Normal	Deficient	Normal
De novo synthesis	Impaired	Deficient	Normal	Normal	Normal	Normal	Normal	Deficient	Deficient
Peroxisomes									
Hepatic peroxisomes	Deficient	Present	Present	Present	Present	Present	Present	Present	Present
Particle-bound catalase	Deficient	NA	Present	Present	Present	Present	Present	Present	Present
Peroxisomal β-oxidation									
Activity with C26:0	Deficient	Deficient	Deficient	Deficient	Deficient	Deficient	Deficient	Normal	Normal
Enzyme proteins									
Acyl-CoA oxidase	Deficient	Deficient	Normal	Deficient	Normal	Normal	Normal	Normal	Normal
Bifunctional protein	Deficient	Deficient	Normal	Normal	Deficient	Normal	Normal	Normal	Normal
Peroxisomal thiolase	Deficient	Deficient	Normal	Normal	Normal	Deficient	Normal	Abnormal¶	Normal

* Age dependent; † age and diet dependent; ‡ elevated except in some cases; § may be normal depending on age and diet; ¶ abnormal molecular form, 44 kDa rather than 41 kDa.

ALD, adrenoleucodystrophy; RCDP, rhizomelic chrondrodysplasia punctata; NA, not analysed; DHAPAT, dihydroxyacetonephosphate acyltransferase; DHAP, dihydroxyacetonephosphate.

al., 1990; Przyrembel *et al.*, 1990; Wanders *et al.*, 1991). Further studies will be required to reveal whether the underlying biochemical abnormalities in these patients are indeed a consequence of peroxisomal dysfunction.

Clinical recognition of peroxisomal disorders

The clinical presentation of patients affected by a particular peroxisomal disorder is highly variable and may depend strongly on the patient's age. This is true not only for X-ALD but also for the disorders of peroxisome biogenesis with such diverse disease entities as Zellweger syndrome and IRD. Recent results have shown that even as well defined an entity as RCDP shows strong heterogeneity. Indeed, Smeitink *et al.* (1992) identified a patient with bone dysplasia with all the biochemical abnormalities of RCDP but lacking characteristic features such as calcific stippling, rhizomelic shortening of the upper extremities, etc. According to Monnens and Heymans (1987), biochemical investigations of peroxisomal functions should be undertaken in all patients showing two or more of the following abnormalities.

Craniofacial abnormalities.

Neurological abnormalities (hypotonia, seizures, nystagmus, hearing deficiencies, white matter degeneration).

Ocular abnormalities (cataract, chorioretinopathy, extinguished electroretinography (ERG), optic nerve dysplasia/atrophy).

Hepatological abnormalities (hepatomegaly, liver function disturbances, fibrosis/cirrhosis).

Skeletal abnormalities (calcific stippling, rhizomelic shortening of the limbs).

Gastrointestinal abnormalities.

In a recent retrospective study, Theill *et al.* (1992) concluded that the combined presence of at least three major symptoms (present in >75% of the affected patients) and one or more minor symptoms in a particular patient warrants biochemical investigation of peroxisomal functions. Major symptoms include psychomotor retardation, hypotonia, impaired hearing, low/broad nasal bridge, abnormal ERG and hepatomegaly. Minor symptoms include large fontanelle, shallow orbital ridges, epicanthus, anteverted nostrils and retinitis pigmentosa.

Biochemical identification of peroxisomal disorders: postnatal diagnosis

If discussion is restricted to those peroxisomal disorders in which there is neurological involvement, which includes all the disorders listed in Table 7.1, except hyperoxaluria type I and acatalasaemia, it is clear that there is an accumulation of VLCFAs in all disorders except RCDP, pseudo-RCDP (DHAPAT deficiency), di/trihydroxycholestanoic acidaemia, glutaryl-CoA oxidase deficiency and Refsum disease (see Table 7.1) (Schutgens *et al.*, 1989; Wanders *et al.*, 1988). Accordingly, VLCFA analysis of plasma or serum has generally been regarded as a good screening method for peroxisomal disorders. VLCFA analysis is usually carried out by means of gas chromatography with or without mass spectrometry. Most laboratories use the procedure developed by Moser and coworkers (see Moser & Moser, 1991). This procedure is rather laborious.

Recently, Onkenhout *et al.* (1989) described a simple one-step procedure for the determination of plasma VLCFAs; this seems to be superior to the other methods used (for discussion see Moser & Moser, 1991). Experience in our own laboratory over the last few years has shown that VLCFA analysis is a reliable method for screening for those peroxisomal disorders in which VLCFA β-oxidation is impaired.

We have recently found that plasma VLCFA levels may be completely normal in some cases of X-ALD. In two patients showing all the clinical signs and symptoms of X-ALD, we found normal plasma VLCFA levels but additional studies in fibroblasts revealed strongly abnormal values, showing that VLCFA β-oxidation was deficient in these patients; this was confirmed by direct C26:0 β-oxidation activity measurements (Wanders *et al.*, 1992b). This suggests that great care is required in the interpretation of VLCFA levels and that even in the absence of elevated VLCFA levels additional studies in fibroblasts are necessary for patients showing clinical signs and symptoms suggestive of X-ALD or one of its phenotypic variants.

If VLCFA levels have been found to be abnormal in plasma from a particular patient, additional studies will have to be undertaken to establish the underlying basis for the VLCFA accumulation. This may include the following types of analysis.

Whole blood (EDTA) (5 mL)

(A) Plasma

1 VLCFAs.*

2 Di- and trihydroxycholestanoic acids.*

3 Pipecolic acid.*

4 Phytanic acid.*

5 Pristanic acid.*

(B) Platelets, leucocytes

6 DHAPAT activity.*

(C) Erythrocytes

7 Plasmalogen levels.*

Skin fibroblasts

8 VLCFAs.*

9 VLCFA β-oxidation (C26:0)

10 Plasmalogen biosynthesis.*

11 Plasmalogen levels.

12 Catalase latency.

13 DHAPAT activity.*

14 Alkyl DHAP synthase activity.

15 Phytanic acid β-oxidation activity.*

16 Pristanic acid β-oxidation activity.

17 Catalase immunofluorescence.

18 Immunoblotting.*

Liver biopsy specimen

(i) IN FIXATIVE

19 Peroxisome analysis (DAB staining).

20 Immunogold-labelling studies.

(ii) IN LIQUID N_2

21 Pipecolic acid oxidation.*

22 THCA-CoA β-oxidation.

23 Immunoblotting.*

If plasma VLCFA levels are normal, additional studies are necessary. In practice, only a limited number of assays (marked with *) are performed. It is only necessary to study cultured skin fibroblasts in patients in whom there is a suspicion of X-ALD or one of its phenotypic variants.

Treatment of peroxisomal disorders

Recently, promising results have been described in the treatment of some peroxisomal disorders with a single loss of peroxisomal functions (group C). In the majority of peroxisomal disorders, however, major abnormalities develop *in utero* and include irreversible changes of the central nervous system. The implications of treating such patients should be considered carefully.

Classical Refsum disease

Since the accumulation of phytanic acid is considered to be the basic abnormality in Refsum disease and since phytanic acid is exclusively of exogenous origin, the elimination of phytanic acid and its precursors from the diet should prevent further accumulation. Reduction of phytanic acid levels by dietary treatment has now been firmly established. In a few patients normal levels have been achieved but in most instances the plasma level plateaus at a moderately elevated level. The response in plasma phytanic acid levels may be delayed for months after initiating the diet owing to mobilization from tissue stores. In several clinics periodic plasmapheresis or plasma exchange has been used to reduce body stores and helps to keep plasma levels low. In patients with a good fall in plasma phytanic acid levels, the peripheral neuropathy has been arrested and other symptoms have regressed. Improvement in nerve conduction velocity has been demonstrated in a number of cases with return to normal in some. Lenz *et al.* (1979), who studied biopsies before and after 2 years of dietary treatment, reported arrest of demyelination and considerable remyelination and regeneration. Muscle strength and gait have improved and sensory deficits have receded. Furthermore, ichthyosis may regress after the start of treatment. Vision, hearing and central nervous system functions usually do not improve, although further deterioration is prevented. Prior to the use of dietary therapy, one-half of untreated patients died before 30 years of age, whereas now almost all patients on dietary therapy survive. In summary, every effort should be made to identify patients as early as possible in order to institute treatment immediately and avoid irreversible damage (see Steinberg (1989) for review).

Hyperoxaluria type I

Treatment of primary hyperoxaluria type I is directed towards decreasing oxalate production by inhibiting its synthesis and increasing oxalate solubility at a given urinary concentration of oxalate. Most of the efforts have concentrated on ways to increase the solubility of calcium oxalate. High fluid intake and alkalinization of the urine remain the mainstays of this approach. Excessive fluid intake is necessary to assist excretion of the enormous amounts of endogenously produced oxalate. Other helpful therapies may include the use of magnesium oxide. Furthermore, haemodialysis can remove large quantities of oxalate and its precursors. Attempts to try and reduce the production of oxalate with succinimide, allopurinol, calcium carbimide and isocarbazide have been unsuccessful.

It is important that pyridoxine should be tried in every patient. Pyridoxine at a usual daily dose of 1000 mg/m^2 body surface area can substantially reduce the production and excretion of oxalate, although most patients have pyridoxine-resistant forms of the disease. The efficacy of pyridoxine is probably directly related to the extent to which alanine glyoxylate aminotransferase is deficient. If there is some residual enzyme activity, high levels of pyridoxal-phosphate, which is obligatory in the enzyme reaction as a coenzyme, may allow residual enzyme activities to operate optimally. In this way, flux through alanine glyoxylate aminotransferase may be stimulated considerably, leading to a reduced production of oxalate; this has been observed in a minority of patients.

In those cases in which alanine glyoxylate aminotransferase is fully deficient, pharmacological doses of pyridoxine show no effect. These patients will usually develop renal failure and require renal transplantation. The overall success rate of this treatment is low owing to the fact that the biochemical defect is situated in the liver and not in the kidneys. As a consequence, renal transplantation gives only temporary relief, the new organ inevitably becoming obstructed by further depositions of calcium oxalate (see Danpure (1989), Hillman (1989) and Latta and Brodehl (1990) for reviews). Definite correction of the metabolic lesion requires liver transplantation. Preliminary results suggest that this is indeed the treatment of choice in pyridoxine-resistant forms of hyperoxaluria type I (Watts *et al.*, 1991).

X-linked adrenoleucodystrophy

Adrenal replacement therapy

Almost all affected boys and 60% of men with AMN have an impaired adrenal reserve. It is essential to test adrenocortical function and provide replacement therapy as necessary, since untreated patients with X-ALD may succumb to an adrenal crisis. Replacement therapy does not appear to alter the course of neurological deterioration (Moser *et al.*, 1987).

Dietary therapy

Based upon the success of dietary restriction of phytanic acid in classical Refsum patients, a diet low in C26 : 0 was tried in X-ALD patients. This approach was not successful, however, presumably owing to the fact that in contrast to phytanic acid, VLCFAs are not only available from exogenous sources but also arise *de novo* from the elongation of long-chain fatty acids.

There is some evidence that treatment of presymptomatic, but not symptomatic, patients with oleic acid and erucic acid (*cis*-13-docosaenoic acid) may prevent deterioration (Rizzo *et al.*, 1986, 1987, 1989; Moser *et al.*, 1987; Uziel *et al.*, 1991).

The somewhat disappointing results of dietary treatment of X-ALD patients with symptoms has led to a search for other potential options, such as bone marrow transplantation. An encouraging result has recently been reported in an 8-year-old ALD patient who showed mild neurological disability at the time of transplant (Aubourg *et al.*, 1990). Plasma C26:0 levels were subsequently found to normalize completely. More importantly, 2 years after the transplant the neurological deficit had cleared, magnetic resonance imaging (MRI) studies are normal and intellectual function is equal to that of an unaffected twin brother. Only the future will reveal whether bone marrow transplantation will be equally favourable for other ALD patients with mild neurological involvement.

Zellweger syndrome and other disorders of peroxisome biogenesis

So far, only a few patients suffering from a disorder of peroxisome biogenesis have been treated, since most of them are already profoundly damaged at birth. As described above, Martinez (1989, 1990) has found that docosahexaenoic acid levels are profoundly deficient in patients suffering from a disorder of peroxisome biogenesis, such as Zellweger syndrome, neonatal ALD or IRD. This, together with the fact that docosahexaenoic acid has been implicated to play a major role in brain and eye function, led to the suggestion that administration of docosahexaenoic acid might be beneficial for affected patients. This is now being evaluated.

Nature of the primary defect in Zellweger syndrome: identification of the first Zellweger gene

In 1986, Schram *et al.* made the important observation that, at least in the case of the peroxisomal β-oxidation enzymes, synthesis of the proteins occurs normally in Zellweger fibroblasts and is followed by rapid degradation in the cytosol. This could be due either to the absence of a peroxisomal membrane or to a defect in the transfer of peroxisomal proteins. Several lines of evidence, including the identification of empty membrane structures called 'peroxisomal ghosts' in Zellweger fibroblasts (Santos *et al.*, 1988), suggest that the primary defect in the Zellweger syndrome and the other peroxisome deficiency disorders is at the level of the machinery required to translocate newly synthesized peroxisomal proteins across the peroxisomal membrane.

Fujiki and coworkers (Tsukamoto *et al.*, 1991) have recently succeeded in restoring peroxisome biogenesis in a Chinese-hamster-ovary (CHO) mutant with a deficiency of peroxisomes by transfecting these cells with rat-liver cDNA; they have identified the factor responsible for correction as a 35-kDa peroxisomal membrane protein. The function of the protein, which the authors refer to as peroxisome-assembly factor 1 (PAF-1), is as yet unknown. By undertaking complementation studies between this CHO mutant and fibroblasts from patients with a deficiency of peroxisomes, which were found to belong to nine different complementation groups, it was found that one of the cell lines is also deficient in PAF-1 (Shimozawa *et al.*, 1992).

(b) Mitochondrial diseases

G.K. BROWN

Introduction

Disorders of mitochondrial function are increasingly recognized as significant causes of metabolic and neurological disease in infancy and childhood. Diagnosis of these conditions can be difficult as a result of great variation in clinical presentation, tissue involvement, pathological change and the degree of biochemical abnormality. There is no single test which will establish the diagnosis of mitochondrial dysfunction unequivocally in all patients and results of a number of different types of investigations may be needed to build up a convincing case. In this section, the better defined patterns of mitochondrial disease in infancy and childhood will be described, together with investigations which can be used to help make the diagnosis. The clinical aspects will follow a brief summary of the relevant biochemistry and genetics of the mitochondrion.

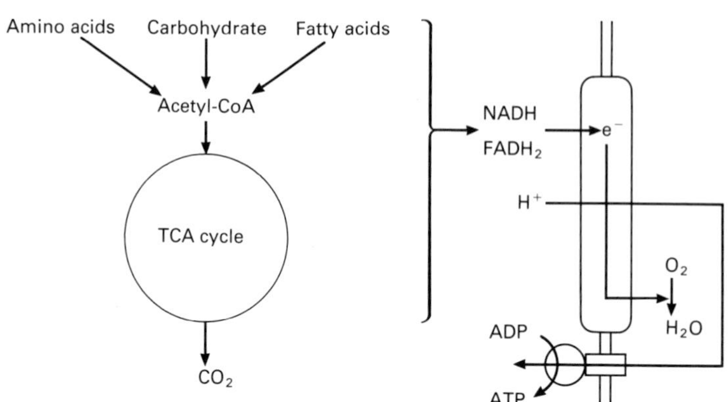

Fig. 7.1 The overall scheme of mitochondrial energy production. The catabolic pathways for carbohydrate, fatty acids and amino acids converge with the formation of acetyl-CoA, and the complete oxidation of this compound to carbon dioxide and water generates the bulk of the energy available from these nutrients. In the individual pathways of substrate oxidation, and in the tricarboxylic acid (TCA) cycle, reduced nicotinamide adenine dinucleotide (NADH) and flavin adenine dinucleotide ($FADH_2$) are generated and these are reoxidized in the electron transport chain of the inner mitochondrial membrane. When electrons from the reduced cofactors pass through the electron transport chain complexes, protons are pumped from the mitochondrial matrix into the intermembrane space, creating an electrochemical gradient. Discharge of this gradient through the adenosine triphosphate (ATP) synthase complex leads to the production of ATP.

The central pathways of energy metabolism

The major energy-generating reactions in the cell are concentrated in the mitochondrion, where the pathways of carbohydrate, fatty acid and amino acid oxidation converge with generation of the common metabolic intermediate, acetyl-CoA. This compound is then the substrate of the tricarboxylic acid (TCA) cycle, whose products are CO_2 and the reduced cofactors nicotinamide adenine dinucleotide (NADH) and flavin adenine dinucleotide ($FADH_2$). Reduced cofactors produced in the TCA cycle and the separate pathways of substrate oxidation are subsequently reoxidized in the electron transport chain (ETC) of the inner mitochondrial membrane. As the electrons generated by reoxidation of cofactors pass through the protein complexes of the ETC, a proton gradient is established across the inner membrane which is used to drive adenosine triphosphate (ATP) synthesis. The electrons are ultimately transferred to oxygen, which is reduced to water. This water and the CO_2 generated by the TCA cycle from acetyl-CoA represent the end products of substrate oxidation.

The relationship between these processes is shown in Fig. 7.1. Impaired mitochondrial ATP generation can occur as a consequence of defects in substrate oxidation (Robinson *et al.*, 1980; Stanley *et al.*, 1983), the TCA cycle (Zinn *et al.*, 1986), the ETC (Zeviani *et al.*, 1986; Wallace *et al.*, 1988a), maintenance of the proton gradient (Luft *et al.*, 1962) or ATP synthesis (Holt *et al.*, 1990). Disorders of the TCA cycle and ATP synthase appear to be rare and the great majority of defects in mitochondrial energy production are due to mutations affecting the pyruvate dehydrogenase (PDH) complex, the fatty acid uptake and β-oxidation pathways or the ETC complexes. Defects of fatty acid oxidation, which commonly present as fasting hypoglycaemia, are considered in Chapter 5. This section will be restricted to mitochondrial diseases involving the ETC complexes, the ATP synthase and the PDH complex. After a brief consideration of the structure

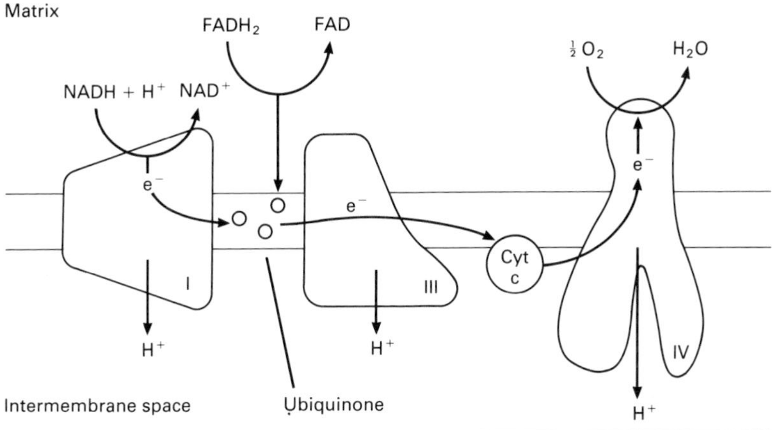

Fig. 7.2 The mitochondrial electron transport chain. The electron transport chain is composed of three large protein complexes embedded in the inner membrane. These complexes, numbered I, III and IV, are connected by smaller, more mobile components which are present at considerably higher molar concentration. Complexes I and III are joined by the small hydrophobic compound, ubiquinone, while complexes III and IV are joined by the small protein, cytochrome c. Reduced nicotinamide adenine dinucleotide (NADH), which is generated from a large number of different dehydrogenase enzymes, is reoxidized by complex I. Electrons from the reoxidation of flavin adenine dinucleotide ($FADH_2$) enter the electron transport chain at the level of ubiquinone. $FADH_2$ is derived mainly from succinate dehydrogenase of the tricarboxylic acid (TCA) cycle and the acyl-CoA dehydrogenases of the fatty acid β-oxidation pathway. The electrons pass via complex III and cytochrome c to the terminal component of the chain, complex IV or cytochrome oxidase, and thence to molecular oxygen with the formation of water. Each of the electron transport chain complexes contributes to the proton electrochemical gradient and each can result in the production of one molecule of adenosine triphosphate (ATP) per pair of electrons transferred.

and function of these mitochondrial components, clinical, biochemical and pathological consequences of their dysfunction will be described.

The protein complexes of the mitochondrial inner membrane

The process of oxidative phosphorylation, the central energy-generating mechanism of the cell, is performed by a group of integral membrane protein complexes in the mitochondrial inner membrane. This inner membrane is a highly specialized structure which also acts as a selectively permeable barrier to control the internal environment of the mitochondrion and allow energy transduction from the proton gradient generated by the ETC. The structural organization of the inner membrane is of fundamental importance for energy generation and its disruption is a major factor in many mitochondrial diseases. The ETC is composed of three large multi-subunit protein complexes linked by two smaller mobile components, ubiquinone (coenzyme Q) and cytochrome c (Fig. 7.2). In addition, the inner membrane contains the succinate dehydrogenase complex of the TCA cycle and the ATP synthase complex.

The structure and function of the protein complexes of the ETC have been extensively reviewed (Hatefi, 1985). Complex I, or NADH dehydrogenase, is composed of at least 25 protein subunits and has a molecular weight of approximately 800 000. It contains multiple electron carriers, including flavin mononucleotide (FMN) and a number of iron–sulphur centres, and passes electrons from NADH to ubiquinone. Complex I is inhibited by rotenone and this characteristic is used to distinguish its activity from non-specific NADH reoxidation in mitochondrial preparations.

Ubiquinol, the reduced form of ubiquinone, passes electrons to complex III, the cytochrome bc_1 complex. Ubiquinone is a low-molecular-weight hydrophobic compound which is present in the membrane at a much higher molar concentration than the protein complexes and is able to act as a highly mobile electron carrier. Complex III contains nine or 10 subunits with a molecular weight of 250 000. The electron-transferring components are two b-type haem groups, b_{562} and b_{566}, cytochrome c_1 and a non-haem iron–sulphur centre. Electrons from complex III are passed to a second low-molecular-weight component, cytochrome c, and thence to complex IV, also known as cytochrome oxidase.

The terminal complex of the ETC, cytochrome oxidase, reduces molecular oxygen to water. The complex has about 13 subunits and a molecular weight of approximately 500 000. The oxidation/reduction reactions involve copper atoms associated with two haem groups, a and a_3. The activity of this complex is strongly inhibited by cyanide.

Complex II of the ETC is in fact the succinate dehydrogenase of the TCA cycle. This is a relatively small flavoprotein with four protein subunits and a molecular weight of 140 000. In addition to FAD, the complex also contains iron–sulphur centres and a b-type cytochrome. Electrons from complex II are transferred directly to ubiquinone and so only pass through two of the ETC complexes.

The final major component of the mitochondrial inner membrane is complex V, the ATP synthase. This multienzyme complex is composed of at least 12 different subunits arranged in three domains: F_1, F_0 and the connecting stalk. The F_1 domain forms a globular extension which projects into the mitochondrial matrix and is the catalytic component for ATP synthesis and release. The F_0 segment is a transmembrane domain with a proton channel for coupling the proton gradient, generated by the ETC complexes, to ATP production. The stalk contains several different protein subunits, one of which confers oligomycin sensitivity on the complex. Discharge of the proton electrochemical gradient through the ATP synthase leads to the release of ATP synthesized from adenosine diphosphate (ADP) and inorganic phosphate at the active site of the F_1 domain.

The three major protein complexes of the ETC all contribute to the proton electrochemical gradient across the inner mitochondrial membrane with a potential yield of one molecule of ATP synthesized for each pair of electrons transferred by each complex. The yield of ATP is therefore three molecules per molecule of NADH reoxidized, and two molecules of ATP from the reoxidation of the $FADH_2$ generated from the fatty acid β-oxidation and succinate dehydrogenase (complex II).

The pyruvate dehydrogenase complex

Aerobic oxidation of carbohydrates and some glucogenic amino acids begins with the transport of pyruvate into the mitochondrion and its conversion to acetyl-CoA. This reaction, catalysed by the PDH complex, is a major metabolic control point and the activity of the complex is regulated by a cycle of phosphorylation and dephosphorylation. These reactions are catalysed by specific kinase and phosphatase enzymes whose activity is regulated in response to metabolite concentrations within the mitochondrion.

The PDH complex itself is comprised of multiple copies of five different protein subunits and these form the four functional entities which carry out the complex reaction sequence (Patel & Roche, 1990). There are three well-defined enzyme components, E1 or pyruvate dehydrogenase, E2 or dihydrolipoyl transacetylase, and E3 or dihydrolipoyl dehydrogenase (Fig. 7.3). In addition, there is a protein component X which is also involved in acetylation reactions and may be involved in the interaction between the E1 and E2 enzymes. The E2 enzyme forms the core of the complex with the E1 and E3 components attached to the surface. The E1 enzyme is itself a complex structure, a heterotetramer of 2α and 2β subunits. The E1α subunit contains the pyruvate and thiamine pyrophosphate binding sites as well as the phosphorylation sites at which the activity of the whole complex is controlled. Acetyl-CoA produced by the activity of the PDH complex is further oxidized in the TCA cycle, and the NADH

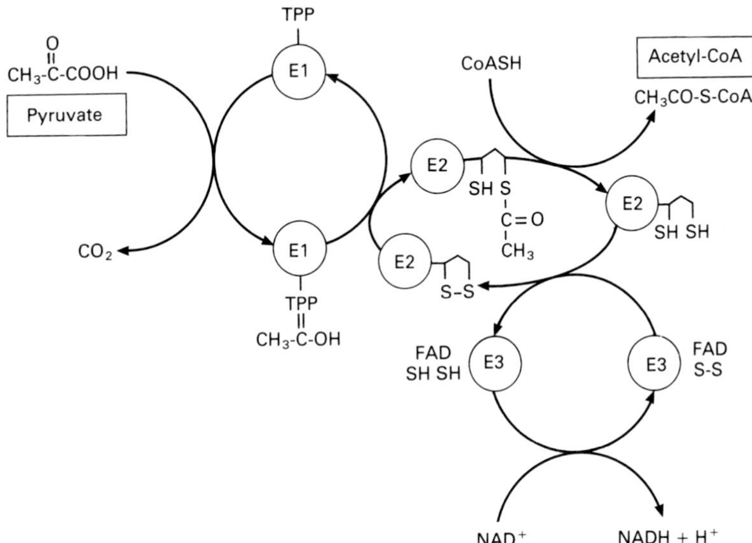

Fig. 7.3 The pyruvate dehydrogenase complex. The conversion of pyruvate to acetyl-CoA requires three enzyme components: E1 (pyruvate dehydrogenase), E2 (dihydrolipoyl transacetylase) and E3 (dihydrolipoyl dehydrogenase). The E1 enzyme is a heterotetramer of 2α and 2β subunits. Pyruvate binds to the E1α subunit through the thiamine pyrophosphate (TPP) cofactor and the acetyl group generated from its oxidative decarboxylation is passed via the lipoic acid side-chain of E2 to CoA. The reduced lipoic acid is reoxidized by E3 for further rounds of acetyl transfer.

which is also generated is reoxidized in the electron transport chain.

Phosphorylation of the PDH complex by the specific PDH kinase leads to inactivation. The activity of the kinase is stimulated by high ATP:ADP, acetyl-CoA:CoA and NADH:NAD^+ ratios (Linn *et al.*, 1969). Although reactivation of the complex by the specific phosphatase is stimulated by both Mg^{2+} and Ca^{2+}ions, fluctuations in the intramitochondrial Ca^{2+} concentration are more important *in vivo* as they can occur in response to a variety of extracellular hormone signals acting through changes in cytoplasmic calcium (McCormack *et al.*, 1990).

The genetic basis of mitochondrial disease

Much of the complexity of the clinical and biochemical features of mitochondrial diseases is a direct consequence of the genetic organization of this organelle. The mitochondrion is unique among mammalian cytoplasmic organelles in containing its own genetic material. However, the mitochondrial genome is extremely restricted and encodes only a small number of polypeptides of the ETC and some components of the mitochondrial transcription and translation apparatus. The majority of mitochondrial components are derived from nuclear genes and these products have to be transported into the organelle for assembly into functional complexes, some of which also contain mitochondrially encoded subunits.

Some disorders of mitochondrial function clearly result from nuclear gene mutations. Defects of the PDH complex (Hansen *et al.*, 1991) and the enzymes of fatty acid β-oxidation (Kelly *et al.*, 1990) fall into this class. Other cases also appear to be a consequence of a primary nuclear gene abnormality, although they are associated with mutations of mitochondrial DNA (mtDNA) in the form of deletions and duplications (Holt *et al.*, 1988; Poulton *et al.*, 1989). There is a third group of mitochondrial diseases, however, in which the primary mutation is in the mtDNA itself (Wallace *et al.*, 1988a; Goto *et al.*, 1990; Holt *et al.*, 1990; Shoffner *et al.*, 1990). In these cases, the special features of mitochondrial genetics play a major role in determining the nature of the illness and are also an important consideration in diagnosis.

The human mitochondrial genome is a circular DNA molecule of 16 569 base pairs which encodes 13 polypeptides, 22 transfer ribonucleic acids (tRNAs) and two ribosomal RNAs (rRNAs) (Fig. 7.4) (Anderson *et al.*, 1981). The polypeptides specified are all subunits of the ETC and ATP synthase:

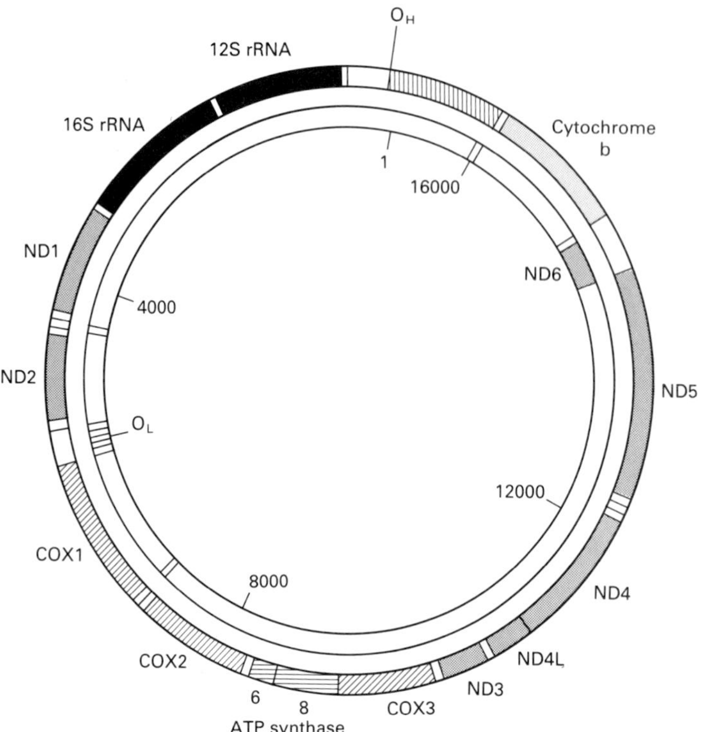

Fig. 7.4 The human mitochondrial genome. Human mitochondrial DNA is a circular molecule of 16 569 base pairs. Both strands contain coding information and they are distinguished as the heavy (H) and light (L) strands on the basis of their nucleotide content. The mitochondrial DNA encodes two ribosomal ribonucleic acids (rRNAs), 12S and 16S, 22 transfer RNAs (shown as small white segments) and 13 protein-coding regions. These comprise seven subunits of complex I (ND1–6 and ND4L), three subunits of complex IV (COX1–3), cytochrome b of complex III and subunits 6 and 8 of the adenosine triphosphate (ATP) synthase. Positions in the mitochondrial genome are defined by base-pair number, starting near the origin of heavy-strand replication (O_H) and proceeding anticlockwise.

seven subunits of complex I, apocytochrome b of complex III, three subunits of complex IV and two subunits of the ATP synthase. The mitochondrial genome is highly compact with only a small region of non-coding sequence around the origin of replication, and the coding regions are not interrupted by introns. Transcription of the genome by a nuclear-encoded RNA polymerase generates large polycystronic RNA molecules from both DNA strands, and these are subsequently processed by cleavage to yield the final mRNA, tRNA and rRNA products. Independent mitochondrial tRNAs are necessary as the genetic code in the mitochondrion differs slightly from that used in the nucleus. By convention, the sequence of mtDNA is numbered in an anticlockwise direction from a point close to the origin of heavy strand replication.

There are multiple copies of mtDNA (probably 2–10) in each mitochondrion and many mitochondria in each cell, so each cell contains several thousand mtDNA molecules. At present, no mechanisms for DNA repair have been defined in the mitochondrion and this may be responsible for the observed high mutation rate of mtDNA compared with that of nuclear DNA. In most individuals, all mtDNA molecules are the same (defined as homoplasmy), whereas patients with mitochondrial disease associated with mtDNA mutations usually have a mixture of mutant and wild-type molecules (heteroplasmy). In this situation, the multiplicity of mtDNA molecules can lead to complex patterns of segregation with different proportions of normal and mutant mtDNA found in different tissues (Wallace, 1986). The phenotype and rate of progression of mitochondrial disease appear to be determined to a considerable extent by the segregation of mutant mtDNA molecules, combined with varying thresholds for dysfunction in different tissues depending on oxidative energy requirements and functional reserves (Howell, 1983), and on selective proliferation of mutant mtDNA in some tissues (Hayashi *et al.*, 1991).

The other major characteristic of the mitochondrial genome is that it is transmitted between generations only by the mother, probably because at fertilization only the head of the sperm enters the egg and the mitochondria in the mid-piece are lost. The characteristic pattern of maternal inheritance is a feature of those mitochondrial diseases in which there is a primary mtDNA mutation (Giles *et al.*, 1980).

The enzymes and enzyme complexes of the mitochondrial matrix are composed exclusively of nuclear-encoded subunits. In the case of the PDH complex, there are nine different gene products involved in the three enzyme activities, protein X and the specific kinase and phosphatase. Of the genes for PDH subunits which have been mapped in the human genome, all are autosomal except for the gene for the E1α subunit, which is located on the short arm of the X chromosome (Brown, R.M. *et al.*, 1989; Patel & Roche, 1990).

Clinical features of mitochondrial disease

In principle, any metabolically active tissue can be affected in mitochondrial disease. Indeed, the diagnosis may be suspected when combinations of symptoms and signs suggest multiple organ involvement for which there is no other recognized anatomico-pathological connection. Nevertheless, by far the most common organs to be affected are muscle and the central nervous system. Clinical features which are commonly found in patients with mitochondrial disease are summarized as follows.

System	*Clinical abnormalities*
Skeletal muscle	Proximal weakness, progressive external ophthalmoplegia, ptosis.
Central nervous system	Myoclonus, ataxia, seizures, dementia, sensorineural deafness, optic atrophy, cortical blindness, hemiparesis, hemianopia, central respiratory failure, headache.
Heart	Cardiomyopathy, arrhythmia, heart block.
Eyes	Pigmentary retinopathy, cataracts.
Kidney	Renal tubular dysfunction, Fanconi syndrome.
Gastrointestinal tract	Jaundice, hepatocellular failure, feeding difficulties, episodic vomiting, malabsorption, exocrine pancreatic dysfunction.
Metabolic/ endocrine	Lactic acidosis, short stature, diabetes mellitus, hypoparathyroidism, hypogonadism, diabetes insipidus.
Bone marrow	Sideroblastic anaemia, bone marrow aplasia.

In patients with clinical findings suggestive of a mitochondrial disease, the family history may be very important in supporting the diagnosis. However, a clear family history of maternal inheritance is only found in disorders where there is a primary defect in the mitochondrial genome. Many cases of mitochondrial disease are sporadic and familial disorders may be the result of nuclear gene mutations, in which case one of the typical patterns of Mendelian inheritance may be apparent.

The manifestations of mitochondrial disease are highly variable, but some combinations of clinical, biochemical and genetic abnormalities occur sufficiently commonly to be defined as relatively distinct entities. It should be remembered, however, that the distinction between the different conditions is often blurred and not all the characteristics of a particular disease may be present. In addition, different components of individual disorders may evolve and progress in quite different ways in different patients. Defined disorders of the ETC will be considered first, in approximate order of age at presentation, followed by PDH deficiency.

Fatal infantile myopathy

This is characterized by early onset of a profound myopathy with respiratory insufficiency. It is usually associated with severe lactic acidosis and often renal tubular dysfunction, cardiomyopathy or liver disease (DiMauro *et al.*, 1980; Boustany *et al.*, 1983; Bresolin *et al.*, 1985; Zeviani *et al.*, 1986). Many patients have marked cytochrome oxidase deficiency, which may be generalized but is often restricted to the affected tissues. An underlying genetic defect has not yet been defined.

In addition to the fatal myopathy, there is a transient cytochrome oxidase deficiency, which is a benign condition restricted to muscle. This presents at birth with severe myopathy and lactic acidosis, but there is spontaneous clinical improvement and the biochemical abnormalities return to normal (DiMauro *et al.*, 1983; Zeviani *et al.*, 1987). It has been suggested that the benign and fatal myopathies may be differentiated by differences in immunoreactive protein (Tritschler *et al.*, 1991).

Severe infantile multisystem mitochondrial disease

Severe infantile multisystem mitochondrial disease is a fatal condition which overlaps to a considerable extent with fatal infantile myopathy but is characterized by more variable tissue expression. Onset is in infancy and although myopathy is the most common feature, liver disease and nephropathy are often present, together with lactic acidosis of variable degree. Muscle biopsy reveals ragged red fibres and absent cytochrome oxidase histochemical staining. Some cases appear to have an isolated deficiency of cytochrome oxidase, others have a general deficiency of all ETC complexes. In this latter group, a number of patients have been shown to have a reduction in the mtDNA content of their muscle fibres, usually to less than 20% of normal (Moraes *et al.*, 1991; Tritschler *et al.*, 1992). On the basis of the limited number of families studied, autosomal recessive inheritance has been proposed for this form.

Pearson's syndrome

Onset of this disorder is usually during infancy, with development of a refractory sideroblastic anaemia with vacuolation of bone marrow precursor cells. There is also exocrine pancreatic insufficiency and some patients develop insulin-dependent diabetes or hepatocellular disease. Lactic acidosis may be present but is not a prominent feature. Patients who survive may go on to develop features of the Kearns–Sayre syndrome (see below). Heteroplasmy for large deletions of mtDNA is demonstrable in muscle and various other tissues (McShane *et al.*, 1990; Rotig *et al.*, 1990).

Leigh's syndrome

A subacute neurodegenerative disorder, Leigh's syndrome most commonly develops during infancy with psychomotor regression, seizures, brainstem abnormalities, weakness and hypotonia. Characteristic pathological findings include focal necrotic lesions in the brainstem associated with local capillary proliferation (Montpettit *et al.*, 1971). These pathological changes appear to represent a common response to a defect in energy metabolism which can result from several different underlying causes. Leigh's syndrome has been described in patients with systemic cytochrome oxidase deficiency (Van Costa *et al.*, 1991), PDH deficiency (Kretzschmar *et al.*, 1987) (see below) and a point mutation in the mitochondrial gene for subunit 6 of the ATP synthase (Tatuch *et al.*, 1992).

The genetic basis of Leigh's syndrome is complex, reflecting the multiple possible underlying defects. Many cases of Leigh's syndrome resulting from isolated cytochrome oxidase deficiency are sporadic but there have been reports of multiple affected siblings and some cases may be inherited as an autosomal recessive trait. Most cases of Leigh's syndrome owing to PDH deficiency are due to defects in the E1α subunit and the gene for this is X-linked. The ATP synthase mutation is maternally transmitted. The defined causes of Leigh's syndrome account for only a small proportion of recognized cases, and in the remaining patients inheritance is probably autosomal recessive in the majority, with some evidence for additional X-linked forms (Benke *et al.*, 1982).

Chronic progressive external ophthalmoplegia and Kearns–Sayre syndrome

These disorders are closely related aetiologically (Berenberg *et al.*, 1977). They may present at any age from early childhood and cases of Kearns–Sayre syndrome are characteristically sporadic. Symptoms are generally well established by the second decade with progressive external ophthalmoplegia, ptosis and proximal myopathy as common features. In addition, patients with Kearns–Sayre syndrome often have a pigmentary retinopathy and may develop heart block or cerebellar ataxia (Kearns & Sayre, 1958). Ragged red fibres are almost always found on muscle biopsy and the disease is associated with heteroplasmy for mtDNA deletion. While deletions of widely differing size and position have been described in different patients, a common 4.9-kb deletion between base pairs 8469 and 13 447 is found in up to 30% (Moraes *et al.*, 1989). In addition, a few families have been described with multiple different deletions in individual family members. This predisposition to mtDNA deletion is inherited as an autosomal dominant trait (Zeviani *et al.*, 1989).

Mitochondrial encephalomyopathy, lactic acidosis and stroke-like episodes

Mitochondrial encephalomyopathy, lactic acidosis and stroke-like episodes (MELAS) results from primary mtDNA mutations affecting the gene for the mitochondrial tRNA for leucine, with the most common mutation being an A to G transition at base pair 3243 (Goto *et al.*, 1990). Early growth and development are usually normal until the onset during childhood of recurring episodes of stroke-like attacks, leading to hemiparesis, hemianopia or cortical blindness. There may also be recurrent episodes of vomiting and seizures. Blood and/or CSF concentrations of pyruvate and lactate may be elevated (Pavlakis *et al.*, 1984). Cerebral computerized tomography (CT) scans may show focal hypodensities, calcification of the basal ganglia and generalized atrophy (Allard *et al.*, 1988). MRI is particularly sensitive for the diagnosis of this condition and may demonstrate multifocal hyperintense signals in the cortical grey matter, with limited involvement of the adjacent underlying white matter (Matthews *et al.*, 1991). These lesions may persist, but some resolve and new areas may be affected during subsequent episodes. Cerebral manifestations are most prominent in this disease and muscle involvement may be minimal. In this regard, many patients do not have ragged red fibres on muscle biopsy, although a strongly positive histochemical reaction for succinate dehydrogenase in vascular endothelial cells is said to be characteristic (Sakuta & Nonaka, 1989). Patients with MELAS are heteroplasmic for the mtDNA mutation with varying proportions of mutant DNA molecules in different tissues, and the disease is maternally transmitted.

Myoclonic epilepsy and ragged red fibres

Myoclonic epilepsy and ragged red fibres (MERRF) can develop at any time from childhood onwards and is characterized by myoclonic epilepsy and action myoclonus, with associated cerebellar ataxia, deafness, seizures and dementia in many patients. Muscle biopsy reveals ragged red fibres and the blood lactate concentration may be elevated (Rosing *et al.*, 1985). This disease is also maternally inherited and results from a primary mtDNA mutation, in this case a single base substitution in the gene for the tRNA for lysine at base pair 8344 (Shoffner *et al.*, 1990).

Leber's hereditary optic atrophy

Although this condition does not often present during childhood, early-onset cases are being described with increasing frequency, and with the elucidation of the underlying genetic defects there is increasing presymptomatic investigation of relatives of affected individuals. The most common presentation is of rapid and painless loss of vision in one eye during the late teens, with similar changes developing in the other eye after a short interval (Nikoskelainen *et al.*, 1987). After the loss of vision, the major clinical sign is optic atrophy. A characteristic peripapillary microangiopathy is found presymptomatically and in many asymptomatic relatives. Leber's optic atrophy is often associated with point mutations in the mitochondrial gene for subunit 4 of complex I, most commonly a G to A substitution at position 11 778 (Wallace *et al.*, 1988a), although a number of other mutations have been described. These mutations are all maternally transmitted but expression of the disease is much more common in males than in females.

Pyruvate dehydrogenase deficiency

This is the only mitochondrial matrix enzyme defect which will be considered in this chapter. PDH deficiency is the most commonly defined cause of primary lactic acidosis in infancy and childhood (Robinson & Sherwood, 1984) but many patients do not have significant systemic metabolic abnormalities (Brown *et al.*, 1988). The spectrum of clinical presentation ranges from severe lactic acidosis in the neonatal period to a chronic neurodegenerative disease with gross cerebral atrophy, ventricular dilatation and a characteristic pattern of developmental anomalies in the central nervous system (Brown, G.K. *et al.*, 1989; Ho *et al.*, 1989; Robinson *et al.*, 1989). These include agenesis of the corpus callosum, absence of the medullary pyramids and ectopia of the inferior olivary nuclei (Chow *et al.*, 1987). Intermediate presentations towards the 'metabolic' end of the spectrum are characterized by episodic lactic acidosis, often associated with cerebellar ataxia. On the neurological side are patients with subacute neurodegeneration, often with brainstem involvement characteristic of Leigh's syndrome (Kretzschmar *et al.*, 1987).

Most cases of PDH deficiency are due to a primary defect in the gene for the E1α subunit and although this gene is located on the X chromosome, males and females are affected at approximately equal frequency. The disease is almost always sporadic and due to a new germ-line mutation arising in one of the parents. The underlying mutation has been identified in a number of cases and is different in almost all patients (Dahl *et al.*, 1992).

Investigation of mitochondrial disease

In spite of considerable progress in recent years, the diagnosis of mitochondrial disease remains unsatisfactory in many cases. This is due to a combination of variability in clinical presentation, the multiplicity of biochemical and genetic abnormalities, and difficulties in assessing mitochondrial function using *in vitro* assays. Although there are some cases with well-defined clinical features and readily detectable genetic abnormalities, many patients cannot be easily categorized and evidence for a mitochondrial defect must be assembled from a range of clinical, biochemical and pathological investigations. These may include *in vivo* spectroscopy and imaging, *in vitro* biochemical analyses and detailed morphological studies. Some of these techniques are not widely available and necessitate the referral of patients to specialized centres. Nevertheless, there are many investigations which can be performed by clinical biochemistry laboratories and these can either help to establish the diagnosis or provide sufficient evidence to suggest that additional specialized investigation is not warranted. Although major metabolic pathways are involved in mitochondrial disease, a number of patients do not have significant systemic metabolic disturbance. In these cases, there is a risk that results of biochemical investigations may not be considered sufficiently abnormal for the possibility of mitochondrial disease to be entertained.

Routine biochemical studies

The metabolic hallmark of mitochondrial disease is the accumulation of lactic acid. This may be a primary consequence of the enzyme defect, as in PDH deficiency, or a reflection of impaired aerobic metabolism, as in ETC deficiencies, in which case greatly enhanced anaerobic glucose oxidation and lactate production are required to sustain cellular ATP levels. Although many patients with mitochondrial disease do have significant lactic acidosis, at least at some times during the course of their illness, it is rarely severe, except in some cases of PDH deficiency. Lactate estimation is therefore neither a sensitive nor a specific diagnostic test.

Technical difficulties with the measurement of lactate in body fluids are a potential source of problems in interpretation. The major problem is artificially high blood lactate concentration as a result of difficulties with sample collection (Braybrooke *et al.*, 1975). However, this is now such a well-recognized complication that it should rarely prove misleading. Other body fluids, such as urine and CSF, are not subject to the same acute fluctuations in response to patient agitation and vascular stasis and should be collected routinely in cases of suspected mitochondrial disease.

Most methods for lactate estimation are based on enzymatic analysis using lactate dehydrogenase. This is available as a standard method for automated clinical chemistry analysers and as a Kodak Ektachem dry-slide method. Ion exchange chromatographic methods have been described and are particularly suitable for CSF analysis. Finally, recent developments of lactate-specific electrodes offer the possibility of rapid blood lactate estimations in conjunction with blood gas analysis (Toffaletti, 1991).

The presence of an elevated blood lactate concentration certainly supports the diagnosis of all forms of mitochondrial disease in the appropriate clinical setting, and other causes of lactic acidosis (e.g. gluconeogenic defects) will usually be excluded by the history and clinical examination. Often the lactic acid accumulation associated with mitochondrial disease is subacute or chronic and is therefore well compensated. Blood gas and bicarbonate analysis rarely reveal a profound metabolic acidosis.

A more difficult diagnostic problem arises in patients with a convincing clinical presentation, but normal blood lactate. In such cases, the possibility of episodic metabolic disturbance or restricted tissue distribution of the biochemical abnormality must be considered. Although many patients with PDH deficiency or ETC defects have persistent lactic acidosis, in some cases this may be quite mild and may only become clinically significant when associated with intercurrent illness (Driscoll *et al.*, 1987; Brown, G.K. *et al.*, 1989). A number of patients have episodes of severe lactic acidosis punctuated by periods in which the lactate is well within the normal range. When metabolic acidosis is not significant, a clue to the metabolic basis of the disorder may be obtained by estimation of the urine lactate: creatinine ratio. This is often raised, even when a spot blood lactate is within the normal range, owing to the integrating effect of excretion of lactate over a period of hours (Dunger & Leonard, 1984).

In a significant proportion of patients with predominantly neurological dysfunction, systemic lactic acidosis is never a feature and repeated blood lactate estimations fall within, or just above, the normal range. In these cases, estimation of lactate in CSF is extremely helpful in directing attention towards a possible metabolic basis for the neurological abnormalities, as there is often a significant discrepancy between the blood and CSF levels (Brown *et al.*, 1988). This analysis is also relatively specific as few other conditions cause a raised CSF lactate (status epilepticus and meningitis are the main ones which might be relevant in this context) (Eross *et al.*, 1981).

In many laboratories, samples are analysed for lactate and pyruvate together and much has been made of the lactate : pyruvate ratio as a specific indicator of particular disorders. High lactate : pyruvate ratios may indeed be found in patients with ETC defects, especially those with significant impairment of complex I activity. Likewise, a disproportionate increase in pyruvate may suggest a primary defect in pyruvate metabolism. However, the lactate : pyruvate ratio is so variable in all mitochondrial diseases (quite apart from technical difficulties with pyruvate estimation owing to its instability in biological samples) that this index should never be considered as definitive evidence for any specific form of mitochondrial disease, nor should it be used as a reason to dismiss a particular diagnosis.

Routine biochemical investigations may also demonstrate abnormal function of various organs and the pattern of involvement may provide important clues: an unusual combination of affected tissues is a characteristic of mitochondrial disease, especially in infants and young children. Because of the generally progressive nature of mitochondrial disease, biochemical dysfunction may be detectable some time before significant symptoms develop. Organs which are commonly involved can be assessed by routine biochemical tests and certain patterns of biochemical abnormalities are highly suggestive of mitochondrial disease.

Liver involvement is usually identified by elevated transaminases, alkaline phosphatase and bilirubin. The elevations may be quite modest but some patients develop severe hepatocellular disease and may die in liver failure. Various forms of renal tubular dysfunction have been defined in patients with mitochondrial disease; however, the most characteristic is the Fanconi syndrome with glycosuria, amino aciduria, hypophosphataemia and renal tubular acidosis. Secondary disturbances of plasma electrolytes are common in this situation.

Muscle disease is associated with elevated plasma creatine kinase, although if elevated the levels are much lower than those observed in muscular dystrophy or inflammatory muscle disease.

Endocrine disorders may be prominent in some patients, with insulin-dependent diabetes being the most common manifestation (Poulton *et al.*, 1989). Hypoparathyroidism has been described, particularly in patients with Kearns–Sayre syndrome (Pellock *et al.*, 1978). Malabsorption, due to either exocrine pancreatic dysfunction or direct involvement of the intestine, may be revealed by increased faecal fat.

Specific biochemical investigations

Clinical and biochemical evidence of a mitochondrial disorder usually leads to more specific investigation aimed at identifying the defective component(s). In general, these studies are too specialized for the routine biochemical laboratory and are performed only in a few reference and research centres. Tissue samples from the patient are required and the most suitable is skeletal muscle, commonly obtained from the quadriceps femoris or deltoid by open or needle biopsy. With an open muscle biopsy, sufficient tissue can be obtained for isolation of mitochondria for polarographic analysis and analysis of cytochrome content. Measurement of oxygen uptake in the presence of different substrates and inhibitors of the ETC may allow the definition of sites of functional impairment (Byrne & Trounce, 1985) and the characteristic oxidized and reduced absorption spectra of the cytochromes of complexes III and IV and cytochrome c can be used to identify any deficiency of these components (Hatefi, 1985).

Assays of individual ETC complexes have been developed and can be used to complement polarographic studies (Wallace *et al.*, 1988b). In addition, these are the only methods suitable for small needle biopsy samples from which insufficient material is obtained for mitochondrial isolation. Although the principles behind the assay methods are well defined, variations in sample preparation and assay conditions make comparison of results from different laboratories difficult, if not impossible. However, even with standardized methods, there is a remarkable lack of correlation between clinical presentation, underlying genetic abnormality and the pattern and degree of functional impairment of the complexes. On the basis of assays of ETC components, isolated deficiency of each ETC component has been described as well as all possible combinations. Abnormal results in these assays certainly contribute to the diagnosis of mitochondrial disease but do not distinguish between different forms (Holt *et al.*, 1989).

The definition of defects in mitochondrial components can be extended in some cases by immunochemical analysis. Antibodies to the various complexes and their individual subunits have been generated and these can be used in immunoblotting or *in situ* binding studies to demonstrate deficiencies of particular structural components (Miyabayashi *et al.*, 1987; Schapira *et al.*, 1988). While these studies may provide useful supportive evidence of a mitochondrial defect, their value is limited by the poor correlation between patterns of immunological reactivity, functional defects and clinical presentations.

Cultured fibroblasts and transformed lymphoid cells have generally proved unsuitable for studies of mitochondrial diseases involving mtDNA mutations. It appears that selective pressures acting on proliferating cells lead to preferential loss of mutant mtDNA molecules. The main disorders which can be identified using cultured fibroblasts are nuclear gene defects such as PDH deficiency (Wicking *et al.*, 1986) and systemic cytochrome oxidase deficiency (Miyabayashi *et al.*, 1987). Attempts have been made to devise screening tests for mitochondrial disease using cultured fibroblasts from a skin biopsy in which rates of $^{14}CO_2$ production from [^{14}C]-pyruvate and the ratio of lactate to pyruvate secreted into the culture medium are measured. These studies are less invasive than muscle biopsy and may provide useful information in some cases (Robinson *et al.*, 1990). In particular, they are quite effective in demonstrating defects in PDH. However, the limited and variable expression of mitochondrial dysfunction in these cells makes them unsuitable for general screening purposes.

Morphological studies

Whenever a muscle biopsy is obtained for investigation of mitochondrial disease, preparations should always be made for both biochemical and morphological studies. Ideally, the latter will include histochemistry and electron microscopy. There may be non-specific histological changes such as accumulation of neutral lipid and glycogen; however, the characteristic pathological finding in skeletal muscle in patients with many forms of mitochondrial disease is the 'ragged red fibre'. In transverse sections of muscle stained with modified Gomori trichrome stain, aggregates of abnormal mitochondria at the periphery of the fibres stain intensely red to give the ragged appearance. An intense subsarcolemmal succinate dehydrogenase reaction may be considered as a 'ragged red fibre equivalent' (Karpati *et al.*, 1991). Ragged red fibres have been described in all mitochondrial diseases involving the ETC but are not found with defects of matrix enzymes such as PDH and the fatty acid β-oxidation enzymes.

In conditions where they are found, the frequency and distribution of ragged red fibres vary widely, in keeping with the variable and progressive nature of the clinical and biochemical abnormalities in mitochondrial disease. When neurological abnormalities predominate, as in MELAS, they may be absent or present in only small numbers (Shapiro *et al.*, 1975). Their frequency increases with time and they may not be apparent in early stages of the disease. The interpretation of muscle pathology may prove difficult as occasional ragged red fibres may be found in normal individuals with increasing age, and in some patients with mitochondrial disease, ragged red fibres cannot be demonstrated, even with repeated biopsies. The evolution and limited distribution of ragged red fibres highlight the importance of repeat muscle biopsy in patients with a convincing clinical presentation and negative pathological findings.

The most useful histochemical reactions for the investigation of mitochondrial disease are NADH dehydrogenase, succinate dehydrogenase and cytochrome oxidase staining. Ragged red fibres are usually cytochrome oxidase negative (Byrne *et al.*, 1985). Cytochrome oxidase negative fibres may also be seen without aggregation of abnormal mitochondria and a small number of such fibres will be readily detectable by histochemical analysis, even though they would not be sufficient to reduce significantly cytochrome oxidase activity if the same muscle sample was homogenized for an enzyme assay. Cytochrome oxidase is not deficient in some patients with mitochondrial disease and is not a sensitive diagnostic test.

Abnormal mitochondrial morphology is best defined by electron microscopy. A variety of changes has been described, including large variations in size, shape and position of the mitochondria, together with abnormal organization of the inner membrane with loss of the normal pattern of cristae and replacement with concentric lamellae. In addition, there may be large crystalline inclusions within the mitochondrial matrix (DiMauro *et al.*, 1985). While these changes are characteristic, there is again considerable variation, and absence of these features does not exclude a diagnosis of mitochondrial disease.

Genetic analysis

In a number of mitochondrial diseases, underlying or associated genetic changes have been defined and these provide a highly sensitive and specific addition to the investigations. Two types of mutation in mtDNA have been recognized: large structural rearrangements (deletions and duplications) (Holt *et al.*, 1988; Poulton *et al.*, 1989) and point mutations (Wallace *et al.*, 1988a; Goto *et al.*, 1990; Holt *et al.*, 1990; Shoffner *et al.*, 1990). The former can be identified by Southern blotting using appropriate regional probes which detect different segments of the mitochondrial genome (Moraes *et al.*, 1989). Single base changes are most conveniently detected following amplification of the appropriate region by the polymerase chain reaction (PCR). In some cases, the mutation can be detected directly as it leads to the formation or loss of a restriction enzyme site (Wallace *et al.*, 1988a); in other cases, a restriction site polymorphism can be generated at the mutation site using mismatched PCR (Zeviani *et al.*, 1991). Genetic analysis may be complicated by heteroplasmy for the mutant mtDNA. As with clinical and biochemical changes, there is wide variation in the level of mutant mtDNA in different individuals and in different tissues from the same patient. In some samples, the level of the mutant molecules may be too low to detect unless sensitive amplification methods are used.

In vivo investigations

Because much of the current investigation of mitochondrial disease involves the invasive procedure of muscle biopsy, attempts have been made to develop techniques which can delineate the extent of tissue involvement and provide an assessment of the level of mitochondrial function *in vivo*. Many of these methods require expensive equipment, which again is available in only a few centres, and although they may be helpful in diagnosis, they are at present most useful for following the course of established disease and monitoring the effects of potential therapies.

Gross pathological changes in the central nervous system are a characteristic feature of a number of mitochondrial diseases and these are best defined by CT and MRI (Allard *et al.*, 1988; Matthews *et al.*, 1991). The abnormalities are highly variable; for example, changes ranging from gross cerebral atrophy, ventricular dilatation and agenesis of the corpus callosum (Chow *et al.*, 1987) to localized lesions in the brainstem and basal ganglia (Kretzschmar *et al.*, 1987) may be found in different patients with PDH deficiency. Characteristic focal cortical lesions associated with the stroke-like episodes of the MELAS syndrome are detectable by MRI (Matthews *et al.*, 1991).

Nuclear magnetic resonance (NMR) spectroscopy can be used as a non-invasive technique to define the degree of functional mitochondrial impairment in both skeletal muscle and brain. In skeletal muscle, ^{31}P NMR spectroscopy reveals an abnormal resting state with a relatively high inorganic phosphate content and a reduced concentration of the high-energy intermediate, creatine phosphate (Matthews *et al.*, 1990). After exercise, there is delayed recovery of the resting creatine phosphate level (Arnold *et al.*, 1985). Proton NMR spectroscopy has been used to demonstrate localized accumulation of lactate in the brain in patients with mitochondrial disease (Detre *et al.*, 1991).

Management of mitochondrial disease

The treatment of most forms of mitochondrial disease remains unsatisfactory. In most cases there is no specific treatment and only supportive measures can be provided. In some patients with defects of the ETC, particularly those involving complex III, attempts have been made to improve energy production by bypassing the site of the block using artificial electron acceptors, such as vitamin K, ascorbic acid (Elef *et al.*, 1984; Argov *et al.*, 1986) and coenzyme Q (Bendahan *et al.*, 1992). These have been claimed to produce some clinical and biochemical improvement (Bresolin *et al.*, 1990) but recent, more extensive reports have cast doubts on their general efficacy (Matthews *et al.*, 1992).

Treatment of PDH deficiency has been attempted in some cases using low-carbohydrate, high-fat ketogenic diets, sometimes with the administration of dichloroacetate (Aynsley-Green *et al.*, 1984). This drug inhibits the specific PDH kinase which inactivates the PDH complex by phosphorylation (Whitehouse *et al.*, 1974). While these measures can reverse the biochemical abnormalities to a certain extent, they do not affect the course of the disease in most cases because the central nervous system function is irreversibly impaired as a result of maldevelopment *in utero*.

In the absence of effective therapy, there is a demand for antenatal diagnosis in subsequent pregnancies in families with affected individuals. The extreme variability of biochemical indices of mitochondrial disease greatly limits the usefulness of approaches based on an assessment of mitochondrial function; however, definition of the underlying genetic defects in a number of different mitochondrial diseases opens up possibilities of direct genetic analysis for antenatal diagnosis. While this will allow precise identification of the presence or absence of a disease-associated mutation in the fetus, it will not enable accurate prediction of the course and outcome of the disease in affected individuals, particularly those with primary mtDNA mutations (Harding *et al.*, 1992). The complex segregation of the multiple mitochondrial genomes in different tissues and the varying thresholds for organ dysfunction make this impossible to determine with our current level of understanding. The concepts relating to the transmission and expression of mitochondrial disease can prove quite difficult to convey, and specialized genetic counselling of families at risk is essential.

Some patients with mitochondrial disease present with a well-recognized pattern of clinical features and have clearly demonstrable biochemical and pathological changes and a specific mutation in the mtDNA. The diagnostic process in such patients is relatively straightforward and the presence of a primary mitochondrial defect easily accepted. Unfortunately, a significant number of patients do not fit into this category and the variability in their clinical, biochemical and genetic abnormalities makes for a considerable diagnostic challenge. Although much is now known about mitochondrial disease, the full spectrum of this group of disorders is far from defined, and this, coupled with difficulties in establishing a definitive diagnosis in many cases, makes it impossible to determine the true incidence. Faced with all this uncertainty, coordinated investigation by clinicians, radiologists, clinical biochemists and muscle pathologists will often be required to provide a compelling case for a diagnosis of primary mitochondrial disease.

(c) Lysosomal disorders

G.T.N. BESLEY

Introduction

Lysosomal disorders are those conditions in which a defect in lysosomal function leads to the progressive accumulation of specific substrate(s), which are often complex and partially degraded macromolecules. The disorders are inherited in a recessive manner, usually autosomal, and most are caused by deficiency of a specific lysosomal enzyme activity. Others are due to defects in lysosomal enzyme processing, deficiency of a specific enzyme/substrate activator or stabilizing protein, or to a defective lysosomal transport protein.

At present, some 40 different lysosomal disorders are known in humans. The reader is recommended to refer to recent reviews (Benson & Fensom, 1985; Scriver *et al.*, 1989; Neufeld, 1991). As with most metabolic defects, lysosomal disorders are rare in the general population, with estimated incidences in the region of 1 in 25 000 to 1 in 100 000 for the more common conditions. However, some disorders occur more frequently, particularly among certain populations. For example, Tay–Sachs disease and Gaucher's disease type I would have frequencies of approximately 1 in 3000 live births amongst Ashkenazi Jews. Within some inbred communities the incidence of certain conditions has been found to be considerably higher. Unfortunately, there are at present no simple screening tests for these disorders; nevertheless, a number of characteristic clinical features should alert the physician to investigate for a lysosomal disorder (see Table 7.3). Before discussing these conditions in some detail, it is worth briefly considering lysosomal function and the mechanism by which lysosomal enzymes are normally processed and transferred to this organelle.

The lysosome

Lysosomes form part of a complex intracellular digestive system responsible (through the action of specific hydrolytic enzymes) for the stepwise degradation of macromolecules.

The organelle is present in all mammalian cells except the mature erythrocyte and shows considerable heterogeneity in terms of size and number per cell. Bounded by a single limiting membrane, the lysosome contains a number of hydrolytic enzymes, including glycosidases, lipases, proteases and nucleases, which are maintained in an acid environment by an energy-dependent proton pump. Macromolecules are taken up into the lysosome by one of four mechanisms: (a) pinocytosis, where extracellular fluid is pinched off into a vesicle; (b) phagocytosis, the ingestion of larger particles into vesicles, especially into macrophages; (c) receptor-mediated endocytosis, where molecules (ligands)

Table 7.3 Clinical features commonly associated with lysosomal disorders

Phenotype	Disorder
Mental and motor retardation, loss of learned skills	Most lysosomal storage disorders apart from non-neuropathic variants, Fabry, MPS I (Scheie), MPS IV and MPS VI
Loss of white matter on brain scan, slow nerve conduction velocity	Krabbe, metachromatic leucodystrophy
Hepato(spleno)megaly	GM_1-gangliosidosis, Wolman's, Gaucher's, Niemann–Pick, most glycoproteinoses, mucolipidoses, mucopolysaccharidoses and Pompe's
Cardiomegaly	Pompe's, GM_1-gangliosidosis, MPS I
Coarse facial features, bone changes on X-ray, dysostosis multiplex	GM_1-gangliosidosis, most glycoproteinoses, mucolipidoses and mucopolysaccharidoses
Eye involvement	
Cherry red spot	GM_1-gangliosidosis, GM_2-gangliosidosis (Sandhoff and Tay–Sachs), Niemann–Pick type A, sialidosis, galactosialidosis
Blindness, corneal clouding	Krabbe, metachromatic leucodystrophy, Fabry, aspartylglucosaminuria, mannosidosis, mucolipidoses II, III and IV, MPS I, IV and VI
Angiokeratoma	Fabry, fucosidosis, sialidosis, galactosialidosis, mannosidosis
Vacuolated lymphocytes	GM_1-gangliosidosis, Wolman's, Pompe's, aspartylglucosaminuria, fucosidosis, mannosidosis, sialic acid storage disorders, sialidosis, galactosialidosis
Foam/storage cells in marrow	Gaucher's, Niemann–Pick, Wolman's, GM_1-gangliosidosis, mucolipidosis IV, galactosialidosis

MPS, mucopolysaccharide disorder.

bind to a cell-surface receptor which invaginates to form a special vesicle to become incorporated into the lysosome; and (d) the cell's own constituents may be taken up by a process of autophagy.

Lysosomal enzymes

Some 70 different lysosomal enzymes are known; all appear to be glycoproteins. Most act at the terminal end of a complex substrate, functioning in a sequential manner to release monomeric units which may diffuse out of the lysosome. Some monomers may, however, require a specific transport protein for export. The enzymes are optimally active at an acid pH, usually around pH 5, and show a high degree of substrate specificity in terms of the terminal group and its linkage; the remaining substrate structure may not be important. Thus, many lysosomal enzyme activities may be conveniently measured using simple and synthetic spectrophotometric or fluorimetric substrates. Many laboratories now use fluorimetric substrates where the terminal group is linked through a phenolic hydroxyl group of 4-methylumbelliferone to provide a highly sensitive, water-soluble and synthetic substrate. Although assays with these 4-methylumbelliferyl substrates are very simple and convenient, there is usually some loss in specificity compared with activities measured with the natural substrate. Residual activities in enzyme-deficient patients may therefore be somewhat higher with these substrates. However, it is usually possible to manipulate assay conditions to improve enzyme specificity.

Lysosomal enzymes undergo a series of modifications during their maturation to a fully functional enzyme within the lysosome (Kornfeld & Mellman,

1989). During the last 10 years extensive work has been carried out to define these steps, and this has contributed much to our understanding of the defects in lysosomal disorders. Most lysosomal enzymes are synthesized as high-molecular-weight precursors that pass to the endoplasmic reticulum, where the signal peptide is removed. The newly formed polypeptide undergoes a sequence of glycosylation changes as it passes from the endoplasmic reticulum to the Golgi apparatus. At least one oligosaccharide chain is transferred from a lipid carrier to the amino group on an asparagine residue of the nascent polypeptide, i.e. it is *N*-glycosylated. The oligosaccharide may be highly branched and contain a number of mannose residues. An important step is the transfer of a phosphate group to position 6 of a terminal mannose, which allows the enzyme to bind to the mannose-6-phosphate receptor. This step allows lysosomal enzymes to be selectively taken up into the lysosome, where further modification takes place to form the mature and functionally active enzyme.

Lysosomal disorders

The clinical spectrum of these conditions is remarkably varied. Some patients present in the neonatal period with hydrops fetalis (Machin, 1989). Others, even with the same enzyme deficiency, may present in late adulthood (Bauman *et al.*, 1991). However, for most conditions there are some fairly characteristic clinical features, the recognition of which will help direct investigations at least towards a group of diseases. Most patients show normal development during the first weeks or months of life, before progressive deterioration, resulting from the underlying storage process within the lysosomes, manifests itself. Unfortunately, there are no simple screening tests for these disorders and, therefore, diagnostic investigations will generally be referred to specialist laboratories.

Consideration of the presenting clinical features will help in the selection of appropriate tests. Features commonly associated with lysosomal disorders are given in Table 7.3 and include an enlarged liver/spleen, progressive neurological deterioration, coarse facial features and bone deformities on X-ray examination, cloudy cornea or macular degeneration, and vacuolated lymphocytes or foamy histiocytes in bone marrow. Thus, following a clinical assessment it should be possible to select the types of investigation necessary to establish a diagnosis. For example, urinary screening for glycosaminoglycan excretion may point to a specific enzyme assay, but for the lipid storage disorders it may be necessary to screen for several of these by direct enzyme assay of blood leucocytes. For other disorders more specific tests will be indicated and these are discussed later in this chapter, following a brief description of different disorders.

Specimens used for diagnostic enzyme assays

Diagnostic enzyme assays are generally carried out on leucocytes, plasma or cultured cells. For mucopolysaccharide (MPS) disorders, preliminary glycosaminoglycan analyses should allow specific groups of enzyme assays to be selected. For the lipidoses and other degenerative disorders it is generally best to undertake a battery of enzyme assays. In many laboratories specializing in these investigations, it is convenient to assay some 12 or more lysosomal enzymes in leucocytes and plasma derived from 5 mL blood. Most lysosomal enzyme activities are relatively stable and it is therefore possible to post whole heparinized or ethylenediaminetetraacetate (EDTA) blood to a specialist laboratory. Leucocytes may be extracted either by differential sedimentation in dextran or following erythrocyte lysis with ammonium ions. The prepared cells may be stored frozen to await enzyme studies. The methods of enzyme assay have generally been optimized in individual laboratories (see Wenger & Williams, 1991) and it is important that each method has been verified on samples from known patients. In this respect fibroblast cultures have proved of great value.

Classification of disorders

Lysosomal disorders can be divided into different groups which reflect the nature of the accumulating storage material. It is convenient to consider these under the following headings: (a) lipidoses; (b) mucopolysaccharidoses; (c) glycoproteinoses; (d) mucolipidoses (ML); and (e) others.

Lipidoses (Table 7.4)

Most of these disorders are associated with the accumulation of glycosphingolipids or gangliosides. These are generally complex lipids, mostly associated with cell membranes, and in all cases contain a long-chain amino alcohol, known as the sphingoid base (usually sphingosine), to which various long-chain fatty acids are *N*-acylated. When sugars are attached to the hydroxyl group of ceramide through a glycosidic linkage, glycosphingolipids are formed, and when a negatively charged sialic acid (usually *N*-acetylneuraminic acid) is also attached, usually through a galactose moiety, the complex is known as a ganglioside. These may contain one (monosialoganglioside, e.g. GM_1 or GM_2) or more sialic acid residues (di-, tri-, tetra-sialogangliosides). Degradation of these complex glycosphingolipids follows a well-defined path (Fig. 7.5) in which specific lysosomal enzymes act at the non-reducing end of the molecule in a stepwise manner, as shown. Deficiency of any of these enzymes will result in a

Table 7.4 Lipidoses

Disorder (inheritance)*	Defect/enzyme deficiency	Storage materials	Diagnostic sample†
GM_1-Gangliosidosis (AR)	β-Galactosidase	GM_1-ganglioside, asialo-GM_1, galactose-rich oligosaccharides	leu, SF, AFC, CVS, (U)
GM_2-gangliosidosis			
Sandhoff (AR)	Total hexosaminidase	GM_2-ganglioside, asialo-GM_2, globoside, oligosaccharides	leu, pla, SF, AFC, CVS, U
Tay–Sachs (AR)	Hexosaminidase A	GM_2-ganglioside	leu, pla, SF, AFC, CVS
AB variant (AR)	GM_2-activator	GM_2-ganglioside	SF
Fabry (X-linked)	α-Galactosidase	Ceramide di- and trihexosides	leu, pla, SF, AFC, CVS, (U)
Gaucher's (AR)	β-Glucosidase	Glucocerebroside	leu, SF, AFC, CVS
Metachromatic leucodystrophy (AR)	Arylsulphatase A	Sulphatides	leu, SF, AFC, CVS, (U)
Krabbe (AR)	Galactocerebrosidase	Galactocerebroside, galactosyl-sphingosine	leu, SF, AFC, CVS
Niemann–Pick types A and B (AR)	Sphingomyelinase	Sphingomyelin	leu, SF, AFC, CVS
Niemann–Pick type C (AR)	Cholesterol esterification	Sphingomyelin and cholesterol	SF, AFC, CVC
Farber (AR)	Ceramidase	Ceramide	SF, AFC
Wolman's (cholesterol ester storage disease) (AR)	Acid lipase	Cholesterol esters and triglycerides	leu, SF, AFC, CVS

* Inheritance: autosomal recessive (AR) or X-linked recessive.
† leu (leucocytes), pla (plasma), SF (skin fibroblasts), AFC (cultured amniotic fluid cells), CVS (chorionic villus sample), CVC (cultured CVS), AF (amniotic fluid), (U) (urine useful but not confirmatory).
Prenatal diagnosis has been successfully carried out for those conditions showing AF, CVS or CVC.

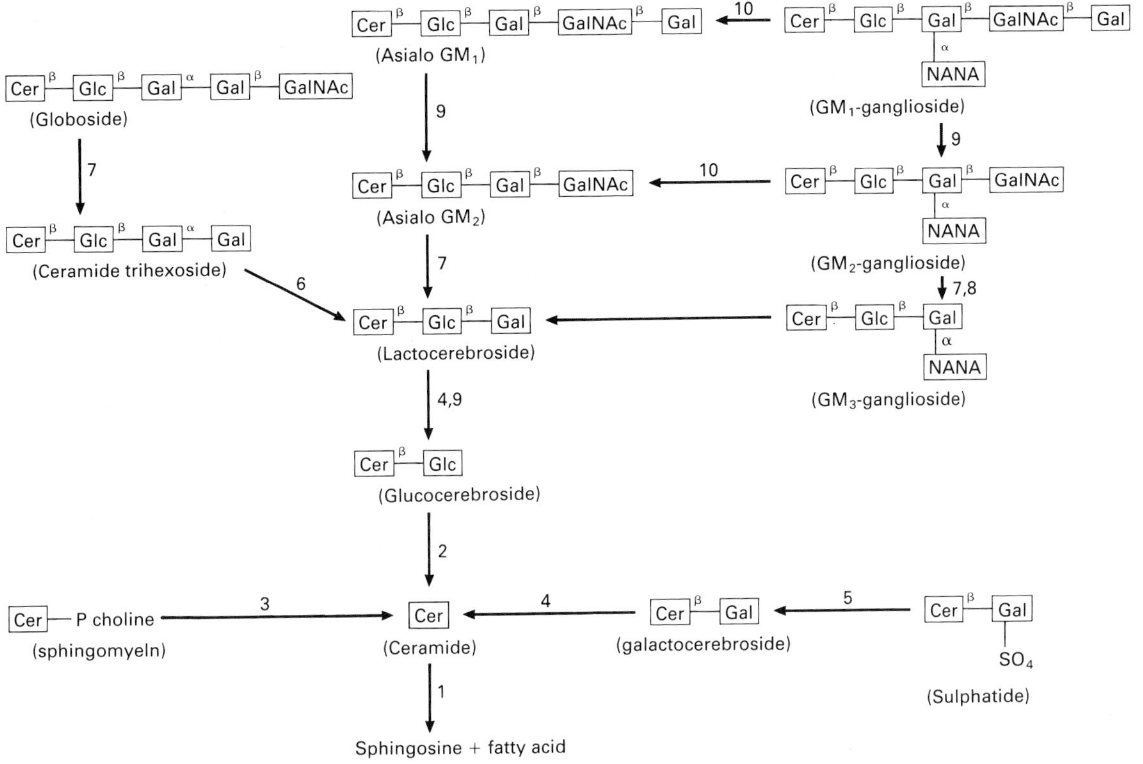

Fig. 7.5 Ganglioside and sphingolipid degradation, showing enzyme steps and metabolic blocks. Enzyme (disorder): 1, ceramidase (Farber); 2, β-glucosidase (Gaucher's); 3, sphingomyelinase (Niemann–Pick); 4, galactocerebrosidase (Krabbe); 5, arylsulphatase A (metachromatic leucodystrophy); 6, α-galactosidase (Fabry); 7, total hexosaminidase (Sandhoff); 8, hexosaminidase A (Tay–Sachs); 9, β-galactosidase (GM$_1$-gangliosidosis); 10, α-neuraminidase (mucolipidosis I). Abbreviations: Gal, galactose; GalNAc, *N*-acetylgalactosamine; Glc, glucose; NANA, *N*-acetylneuraminic acid; Cer, ceramide.

metabolic block with accumulation of undegraded substrate. A brief description of these disorders follows.

GM$_1$-gangliosidosis

The primary defect in this disorder is a deficiency of acid β-galactosidase activity. Since this enzyme acts on a number of substrates with a terminal-linked β-galactose, patients are found to accumulate GM$_1$-ganglioside (especially in the brain) as well as asialo-GM$_1$, galactosyl-oligosaccharides and keratan sulphate derivatives (in the viscera).

Patients with the severe infantile disease present in the first weeks of life and develop neurological deterioration, hepatosplenomegaly, often coarse features with bone involvement and oedema, cherry red spot in the eye (in about 50% of cases) with blindness, and a startle reflex. Vacuolated lymphocytes are present in peripheral blood and foam cells in the marrow. Galactose-rich oligosaccharides may be found on thin-layer chromatography of urine, and move a short distance from the origin. Patients with the milder juvenile type excrete less oligosaccharide, whereas the rarer adult type (more common in Japan) may involve apparently normal excretion. All types are deficient in β-galactosidase activity, which may be conveniently measured in leucocytes or cultured cells with the 4-methylumbelliferyl analogue. The enzyme normally exists as a high-molecular-weight aggregate with α-neuraminidase and a stabilizing protective protein.

Deficiency of this protein results in the combined deficiency of both enzyme activities, which should be measured in cultured fibroblasts. These patients have galactosialidosis and also excrete sialo-oligosaccharides. A different mutation affects the substrate specificity of β-galactosidase and leads to a Morquio (MPS IV) phenotype.

Prenatal diagnosis of these conditions is possible by β-galactosidase assay of chorionic villus samples (CVS) or cultured amniotic fluid cells. Enzyme activity should be measured in the presence of stabilizing chloride ions.

GM_2-gangliosidosis

Three main types of GM_2-gangliosidosis are recognized (Neufeld, 1989; Mahuran, 1991).

Sandhoff disease (O variant)

Patients with this condition are deficient in total hexosaminidase activity. The enzyme exists in two forms: hexosaminidase A (α and β heteropolymer) and hexosaminidase B (β polymer). Patients have a β-chain mutation. The enzymes together cleave *N*-acetylgalactosaminyl- and *N*-acetylglucosaminyl-residues from a variety of glycolipids, glycosamino-glycans and oligosaccharides. The enzymes are therefore generally referred to as hexosaminidases.

Patients with Sandhoff disease are almost identical to those with Tay–Sachs disease (see below), although the condition is generally panethnic and more likely to show organomegaly. Infantile and juvenile forms exist; diagnosis is straightforward and usually carried out on leucocytes or plasma. Residual activity is due to hexosaminidase S (α subunit only), which may give high activity with sulphated substrates. Sandhoff carriers show partial hexosaminidase deficiency but a raised percentage of the A component.

Tay–Sachs disease (B variant)

This results from an α-chain mutation leading to a deficiency of hexosaminidase A (and S) activity (Mahuran *et al.*, 1990). This enzyme is specific for negatively charged substrates such as GM_2-ganglioside and sulphated artificial substrates. Tay–Sachs disease is especially common in Ashkenazi Jews, where an incidence of 1 in 3900 births or a carrier frequency of 1 in 31 is predicted. Patients present at 3–5 months of age with mental and motor retardation, an exaggerated startle response and a characteristic 'cherry red' spot in the fundus of the eye. Carrier screening programmes have drastically reduced the incidence in certain populations. Diagnosis is by leucocyte or serum assay of hexosaminidase A activity, and DNA analysis for the two common Jewish mutations may assist in carrier detection.

Patients with the B1 variant have a different α-chain mutation that leads to deficient enzyme activity especially with the sulphated substrate.

AB variant

Patients with the AB variant lack the endogenous GM_2-ganglioside activator, and therefore retain normal hexosaminidase activities measured with artificial substrate. This is a rare variant which is difficult to diagnose.

Fabry disease

In young adults this X-linked disease generally presents with painful extremities, angiokeratoma, renal impairment and corneal opacities. Deficiency of α-galactosidase activity leads to an accumulation of glycosphingolipids with terminal α-linked galactose in tissues as well as in urinary sediment. Female carriers may manifest the disease due to Lyonization. Carrier detection has been by hair-root analysis as well as by the usual diagnostic blood samples. Two forms of α-galactosidase are measured with artificial substrates: the A form is deficient in Fabry disease, whereas the B form, better known as α-*N*-acetylgalactosaminidase, is deficient in a rare neurological condition known as Schindler disease.

Gaucher's disease

Three types of Gaucher's disease are recognized; all involve a deficiency in β-glucosidase activity and an accompanying accumulation of glucocerebroside.

The most common form, with an incidence between 1 in 600 and 1 in 2500 in Ashkenazi Jews, is the non-neuropathic type (type 1). Patients may present up to late adulthood with (hepato)splenomegaly, bone infiltration and anaemia. Haematological investigations may reveal characteristic Gaucher cells (storage macrophages) in bone marrow; these stain weakly periodic acid–Schiff (PAS) positive and have the appearance of crumpled tissue paper. Diagnosis by enzyme assay requires specific assay conditions to select for acid β-glucosidase activity, especially in leucocytes, and some patients may have significant residual activity. Deficiency is more marked in cultured fibroblasts, and carrier detection in Jewish families may be possible by DNA analysis (Beutler, 1993). The acute neuropathic form (type 2) is panethnic and presents in infancy with seizures, brainstem dysfunction, strabismus and feeding difficulty. Hepatosplenomegaly may not be so prominent. Gaucher's cells are present and diagnosis is as for type 1. An intermediate (type 3) form is also recognized.

Metachromatic leucodystrophy

This is a demyelinating condition in which deficiency of arylsulphatase A leads to an accumulation of sulphatides (important structural elements of the myelin sheath), especially in brain white matter and kidneys. Most patients present with late infantile disease at around 2 years of age; they have ataxia and mental deterioration. Myelin loss leads to hypodense areas on CT and MRI scans of the brain, and the nerve conduction velocity is slowed. Protein concentration in the CSF may be raised in this form as well as in the juvenile-onset type, but is not always raised in the adult cases which may present with dementia. Diagnosis is best made by enzyme assay of leucocytes using *p*-nitrocatechol sulphate as the substrate, but care must be taken to exclude other sulphatases (B and C), which may be present. A major complication in diagnosis is a common pseudodeficiency of arylsulphatase A activity, which results from a different genetic mutation (Gieselmann *et al.*, 1991). This mutation may be present in 15% of the general population and when it coexists with a metachromatic leucodystrophy mutation (heterozygote) may lead to activities of around 10% of normal; however, sufficient *in vivo* activity remains to prevent substrate accumulation or disease manifestation. DNA studies are necessary to evaluate such individuals.

Another diagnostic problem may arise at prenatal diagnosis when chorionic villus tissue is used, since this tissue has a high arylsulphatase C (steroid sulphatase) activity. Special steps are necessary to prevent this activity leading to a false negative diagnosis.

Krabbe disease

This is a more severe leucodystrophy and patients present at around 3–6 months of age, although cases of later onset are known. Patients generally show marked irritability, mental and motor deterioration, blindness and initially hypertonicity with scissoring. Later, patients become hypotonic. There is a loss of white matter on a brain scan, low nerve conduction velocity and a raised protein concentration in CSF. Patients are deficient in a specific β-galactosidase which acts primarily on galactocerebroside, an essential component of myelin. This enzyme, often referred to as galactocerebrosidase, also acts on the deacylated (i.e. less fatty acid group) derivative, psychosine, which is highly toxic and may account for much of the pathology found. The accumulating glycolipids are found in PAS-positive multinucleate globoid cells in the white matter. These characteristic cells are responsible for the term globoid-cell leucodystrophy. Assay of galactocerebrosidase activity is best carried out with radioactive natural substrate in the presence of taurocholate and oleic acid. Only under these conditions can a diagnosis of Krabbe disease be confidently made.

Niemann–Pick disease

Two major types of this condition are recognized. In type A and type B disease there is a deficiency of sphingomyelinase activity, leading to an accumulation of the phospholipid substrate in lysosomes of the reticuloendothelial system. Patients have marked hepatosplenomegaly and characteristic foam cells, often sea-blue histiocytes, in the bone marrow. Patients with the severe type A phenotype present in the first months with psychomotor retardation, wasting, abdominal distension and cherry red spot in 50% of cases. This acute/neuronopathic type is more common in Jewish populations. The chronic non-neuropathic variant type B may present in the first years of life or even into late adulthood. Infiltration of storage cells into the lung fields is commonly observed on X-ray. Sphingomyelinase deficiency may be less marked in type B patients, especially when assayed on leucocytes.

Patients with the type C (and type D) phenotype may also have an enlarged liver and spleen with characteristic foam cells or sea-blue histiocytes in the marrow. Storage of sphingomyelin is accompanied by other lipids including cholesterol. Recent evidence points to a defect in cholesterol esterification (Vanier *et al.*, 1988). Early neonatal jaundice and hepatosplenomegaly may give way in childhood to neurological signs including ataxia and vertical ophthalmoplegia. Sphingomyelinase activity is normal in leucocytes and tissues but may be partially reduced in cultured fibroblasts. Diagnosis is based on esterification defects in cultured cells.

Farber disease

This is a very rare sphingolipidosis where ceramidase deficiency leads to an accumulation of ceramide. Painful and swollen joints, subcutaneous nodules, respiratory difficulty and hoarseness are all hallmarks. The enzyme assay is performed in only a few laboratories, although loading studies with radiolabelled precursors in skin fibroblasts may assist in the diagnosis.

Wolman's disease

Deficiency of lysosomal acid lipase (or esterase) leads to the accumulation of cholesterol esters and triglycerides in this condition as well as the milder cholesterol ester storage disease. Patients with Wolman's disease may present with similar symptoms to those with Niemann–Pick disease. There is a marked failure to thrive, hepatosplenomegaly, diarrhoea and vomiting; death usually occurs by 6 months of age. Adrenal calcification is considered pathognomonic. Vacuolated lymphocytes are common in peripheral blood and bone marrow. Enzyme diagnosis may be carried out using artificial substrates but conditions should be optimized with samples from known patients.

Mucopolysaccharidoses (Table 7.5)

These are a group of disorders where defective degradation of glycosaminoglycans (formerly called mucopolysaccharides) leads to an accumulation, particularly in lysosomes, of the mesenchymal and parenchymal tissues (Kresse *et al.*, 1981). Skeletal deformities, growth and psychomotor retardation are common consequences of this pathology. Partially degraded glycosaminoglycans are excreted in excessive amounts in the urine, and recognition of this provides a useful first-line screening test.

Structurally, glycosaminoglycans are sulphated polysaccharides made up of repeating disaccharide units which, in their natural state, are attached through a neutral sugar linkage to a protein core. Four glycosaminoglycans are involved in MPS disorders: (a) chondroitin sulphate (repeating units of glucuronic acid and *N*-acetylgalactosamine which may be 4- or 6-sulphated); (b) dermatan sulphate (similar structure but with glucuronic or iduronic acids – the latter may be 2-sulphated); (c) heparan sulphate (contains α-linked glucosamine, which may be *N*-acetylated or *N*- or *O*-sulphated, and the uronic acid, either glucuronic or iduronic, which may be sulphated); and (d) keratan sulphate (contains *N*-acetylglucosamine, which may be 6-sulphated, and galactose, which may also be 6-sulphated). The structures of these compounds have been reviewed (Kresse *et al.*, 1981). Degradation of glycosaminoglycans is achieved in a sequential manner from the non-reducing end by a series of exoglucosidases and exosulphatases (Fig. 7.6). Deficiency of any of these leads to a mucopolysaccharidosis, brief details of which are given below.

Hurler–Scheie disease (MPS I)

Deficiency of α-iduronidase activity leads to an accumulation of dermatan and heparan sulphates. In its severe form, Hurler disease manifests with all

Table 7.5 Mucopolysaccharidoses

Disorder (inheritance)*	Enzyme deficiency	Storage materials	Diagnostic sample*
Hurler or Scheie (MPS I) (AR)	α-Iduronidase	Heparan and dermatan sulphates	leu, SF, AFC, CVS, (U), (AF)
Hunter (MPS II) (X-linked)	Iduronate-2-sulphatase	Heparan and dermatan sulphates	leu, pla, SF, AFC, CVS, (U), AF
Sanfilippo (MPS III)			
MPS IIIA (AR)	Heparan *N*-sulphatase	Heparan sulphate	leu, SF, AFC, CVS, (U), (AF)
MPS IIIB (AR)	α-Acetyl glucosaminidase	Heparan sulphate	leu, pla, SF, AFC, CVS, (U), (AF)
MPS IIIC (AR)	Acetyl CoA : α-glucosaminide *N*-acetyltransferase	Heparan sulphate	leu, SF, SFC, CVS, (U), (AF)
MPS IIID (AR)	*N*-Acetylglucosamine-6-sulphatase	Heparan sulphate	SF, AFC, (U), (AF)
Morquio (MPS IV)			
MPS IVA (AR)	*N*-Acetylgalactosamine-6-sulphatase	Keratan sulphate	leu, SF, AFC, CVS, (U), (AF)
MPS IVB (AR)	β-Galactosidase	Keratan sulphate	leu, SF, AFC, CVS, (U), (AF)
Maroteaux–Lamy (MPS VI) (AR)	Arylsulphatase B	Dermatan sulphate	leu, SF, AFC, CVS, (U), (AF)
Sly (MPS VII) (AR)	β-Glucuronidase	Dermatan and heparan sulphates	leu, SF, AFC, CVS, (U), (AF)
Multiple sulphatase deficiency (AR)	Various sulphatases	Various glycosaminoglycans, esp. heparan sulphate, plus sulphatides and steroid sulphate	leu, SF, AFC, CVS, (U), (AF)

* See Table 7.4.

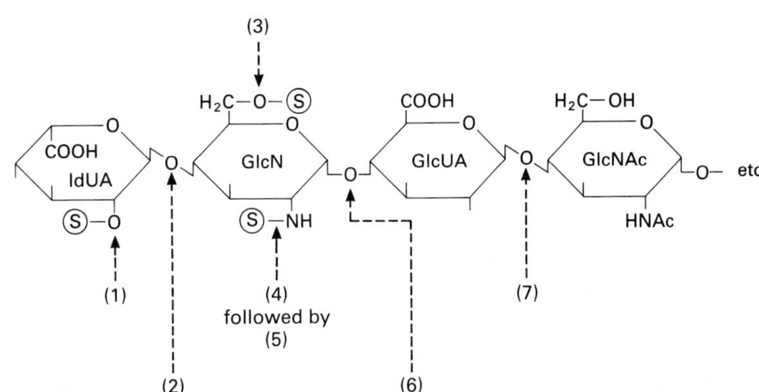

Fig. 7.6 Stepwise degradation of heparan sulphate-saccharide units: iduronic acid (IdUA), glucosamine (GlcN), glucuronic acid (GlcUA), *N*-acetylglucosamine (GlcNAc). Enzymes involved: (1) iduronate-2-sulphatase (MPS II); (2) α-iduronidase (MPS I); (3) *N*-acetylglucosamine-6-sulphatase (MPS IIID); (4) heparan *N*-sulphatase (MPS IIIA); (5) *N*-acetyltransferase (MPS IIIC) (transfer of an acetyl group is necessary for further degradation); (6) α-*N*-acetylglucosaminidase (MPS IIIB); (7) β-glucuronidase (MPS VII).

the clinical symptoms of MPS disorders, including psychomotor retardation, growth retardation, bone involvement (dysostosis multiplex), hepatomegaly and corneal opacities. The urinary excretion pattern is generally characteristic, and enzyme diagnosis usually straightforward. However, a low α-iduronidase activity in CVS has, in the past, led to problematic interpretation at prenatal diagnosis.

Hunter disease (MPS II)

Deficiency of iduronate-2-sulphatase activity leads to an accumulation of heparan and dermatan sulphates. The disorder is inherited in an X-linked manner and patients with the severe type may show similar symptoms to Hurler patients, apart from the lack of corneal clouding. Milder patients may have little or no mental retardation. Enzyme activity may be measured with a heparan-derived disulphated disaccharide which has been tritium labelled. The monosulphated product is separated by ion exchange chromatography or electrophoresis. Diagnosis may be conveniently carried out on serum/plasma, which may also help in carrier detection. However, activity increases during pregnancy – unless the fetus is affected with MPS II. Approximately 15% of patients have a significant deletion (Wilson *et al.*, 1991) in the iduronate sulphatase gene; in these families DNA analysis may assist in carrier detection.

Sanfilippo disease (MPS III)

Four different enzyme deficiencies give rise to MPS III. The most common in the UK is MPS IIIA, which is due to heparan *N*-sulphatase deficiency. All patients are clinically similar with relatively mild somatic changes and no corneal clouding, but they have marked mental retardation with behavioural disturbances. Excessive excretion of heparan sulphate should indicate the appropriate enzyme assays. MPS IIID has only rarely been reported.

Morquio disease (MPS IV)

These patients may have marked skeletal deformities and growth retardation. Excessive excretion of keratan sulphate often points to the diagnosis. Most patients are deficient in *N*-acetylgalactosamine-6-sulphatase activity, which can now conveniently be measured with a simple fluorogenic substrate (Van Diggelen *et al.*, 1990). A few patients are deficient in another enzyme, β-galactosidase, which presumably has a different mutation from that in GM_1-gangliosidosis.

Maroteaux–Lamy disease (MPS VI)

Patients with this rare disorder have mainly skeletal abnormalities, similar to those found in Hurler disease, but are generally spared the neurological deficit. There is a deficiency of *N*-acetylgalactosamine-4-sulphatase activity, which is normally assayed as arylsulphatase B activity. Excessive excretion of dermatan sulphate is characteristic of the condition.

Sly disease (MPS VII)

Sly disease is the least common of the mucopolysaccharidoses. Patients have a wide spectrum of phenotypes, from hydrops fetalis to mildly affected adults. The excretion pattern is variable but urine generally contains dermatan and heparan sulphates, and the deficient enzyme β-glucuronidase is easily demonstrated.

Glycoproteinoses (Table 7.6)

These are rare lysosomal storage disorders affecting enzymes involved with the degradation of oligosaccharides normally attached through an asparagine residue to the protein core of a glycoprotein. A typical structure of such an oligosaccharide is shown in Fig. 7.7.

Sialidosis

Deficiency of α-neuraminidase activity leads to an accumulation of sialic acid-rich oligosaccharides. The disorder is rare and may present with myotonic seizures, cherry red spot in the eye, mental retardation and vacuolated lymphocytes. The enzyme is not stable and should be assayed in freshly prepared cultured fibroblasts. The possibility of a combined

Table 7.6 Glycoproteinoses

Disorder (inheritance)*	Defect/enzyme deficiency	Storage materials	Diagnostic sample*
Sialidosis (AR)	α-Neuraminidase	Sialyloligosaccharides	SF, AFC, (U)
Galactosialidosis (AR)	Protective protein (β-galactosidase/ α-neuraminidase)	Sialyloligosaccharides	SF, AFC, (U)
Mannosidosis (AR)	α-Mannosidase	Mannose-rich oligosaccharides	leu, SF, AFC, CVS, (U)
Fucosidosis (AR)	α-Fucosidase	Fucose-rich oligosaccharides, glycoproteins and glycolipids	leu, SF, AFC, CVS, (U)
β-Mannosidosis (AR)	β-Mannosidase	Mannose-*N*-acetylglucosamine disaccharide	leu, pla, SF, (U)
N-Acetylglucosaminuria (AR)	Aspartylglucosaminidase	Aspartylglucosamine	leu, SF, AFC, CVS, (U)
Schindler (AR)	α-*N*-Acetylgalactosaminidase	*N*-Acetylgalactosamine-containing oligosaccharides	leu, pla, SF, (U)

* See Table 7.4.

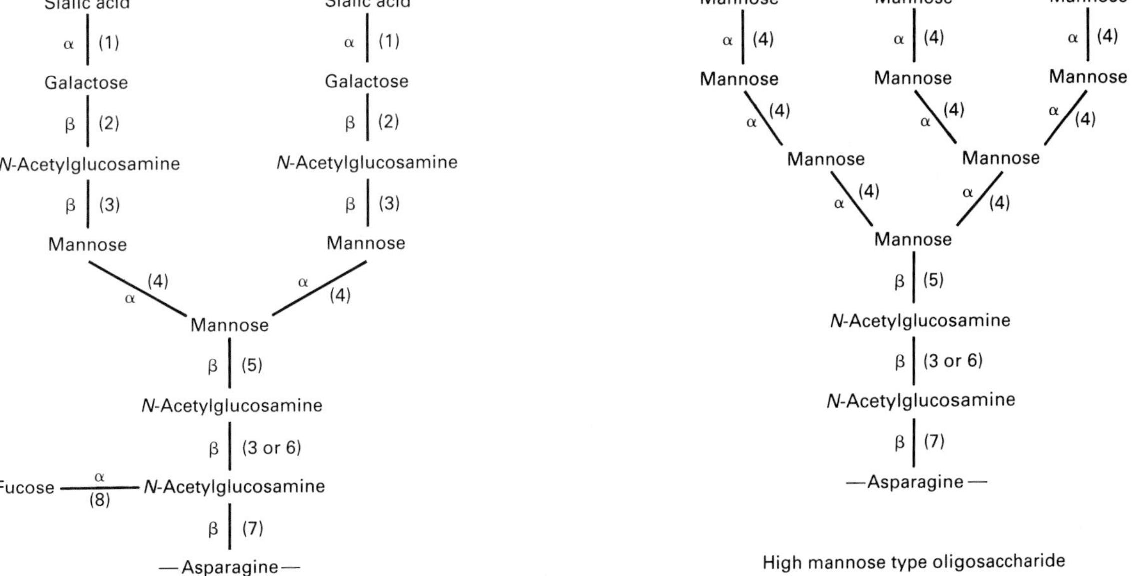

Fig. 7.7 Structures of two oligosaccharide chains as found on glycoproteins. α and β linkages are indicated and the specific degradative enzyme steps are numbered as follows: 1, α-neuraminidase; 2, β-galactosidase; 3, β-hexosaminidase; 4, α-mannosidase; 5, β-mannosidase; 6, endo-β-hexosaminidase; 7, aspartylglucosaminidase; 8, α-fucosidase.

Table 7.7 Mucolipidoses

Disorder (inheritance)*	Defect/enzyme deficiency	Storage materials	Diagnostic sample*
Mucolipidosis II (AR) (I cell disease)	*N*-Acetylglucosamine-1-phosphotransferase	Various glycosaminoglycans and oligosaccharides	pla, SF, AFC, CVC, (U), (AF)
Mucolipidosis III (AR) (pseudo-Hurler polydystrophy)	Same as ML II	Same as ML II	Same as ML II
Mucolipidosis IV (AR)	Not known	Phospholipids and gangliosides	SF, CVC

* See Table 7.4.

deficiency with β-galactosidase should always be considered.

Mannosidosis

Patients deficient in α-mannosidase activity excrete a number of mannose-rich oligosaccharides. The condition is rare in the UK and patients present with symptoms similar to those of MPS patients, with mental retardation, corneal opacities and deafness.

Fucosidosis

This is rare in the UK but more common in Italy. Patients may present in a manner similar to those with a mild Hurler phenotype, and in the longer-surviving type II patients there may be angiokeratoma. Enzyme diagnosis should be made on leucocytes since a common polymorphism may lead to a low activity in plasma from normal individuals.

β-Mannosidosis

A rare deficiency of β-mannosidase activity has been described in six families but with no clear phenotypic pattern. The mannose-rich disaccharide excreted is difficult to identify by thin-layer chromatography.

Aspartylglucosaminuria

Another rare condition, although first described in the UK but more commonly found in Finland, is that of aspartylglucosaminuria. Storage products in urine stain positively with ninhydrin as well as orcinol. Diagnosis should be confirmed by assay of aspartylglucosaminidase activity, an assay rarely performed in the UK.

Mucolipidoses (Table 7.7)

Four types of ML have been recognized. Patients with α-neuraminidase deficiency were formally classified as having ML I, although the term sialidosis is more generally used now. ML II (I cell disease) and ML III are biochemically similar but with different phenotypic expressions. In the former, patients present often in infancy with Hurler-like features but without excessive glycosaminoglycan excretion. Gum hyperplasia is common. The term 'I cell' is derived from the observation of large numbers of cellular inclusions (engorged lysosomes) in cultured cells and within connective tissue from these patients. Patients with ML III have a milder phenotype and stiff joints are a prominent feature. The underlying defect in both ML II and ML III is a failure to transfer the mannose-6-phosphate marker onto lysosomal enzymes during processing in the Golgi apparatus. As a result, most lysosomal hydrolases fail to be taken into the lysosome and therefore incompletely processed enzymes, which are fully active, are rerouted to the extracellular space. Consequently, high lysosomal enzyme activities are found in plasma, amniotic fluid and medium taken from ML II or ML III cell cultures.

In fibroblasts and other cell cultures from patients,

Table 7.8 Other conditions

Disorder (inheritance)*	Defect/enzyme deficiency	Storage materials	Diagnostic sample*
Pompe's (AR) (GSD II)	Acid α-glucosidase	Glycogen	Lymphocytes, SF, AFC, CVS
Cystinosis (AR)	Cystine transporter	Cystine	leu, SF, AFC, CVS, (U)
Salla (AR) and infantile sialic acid storage disease (AR)	Sialic acid transporter	Sialic acid	SF, U, AFC, CVS
Methylmalonic aciduria type *CblF* (AR)	Cobalamin transporter	Cobalamin (vitamin B_{12})	SF, (U)
Neuronal ceroid lipofuscinosis (Batten's disease) (AR)	Unknown	ATP synthase subunit C, ceroid pigment and lipofuscin	Lymphocytes, skin biopsy

* See Table 7.4.
GSD, glycogen storage disease.

most lysosomal hydrolase activities are low, the exceptions being β-glucosidase and acid phosphatase, which appear to be processed differently. It should be noted that normal hydrolase activities are generally found in ML II and ML III leucocytes and chorionic villus tissue.

ML IV is a rare condition with a complex lipid storage pattern which may be identified in cultured cells. A deficiency of ganglioside sialidase has not been confirmed.

Other conditions (Table 7.8)

Pompe's disease

The first lysosomal storage disorder in which the enzyme defect was identified in 1963 was Pompe's disease, in which deficiency of acid α-glucosidase (acid maltase) leads to glycogen storage disease type II. Until this time, the concept of lysosomal involvement in cellular metabolism had not been fully appreciated. Patients have marked hypotonia, cardiomegaly, hepatomegaly and generally succumb to cardiorespiratory failure. Vacuolated lymphocytes, which stain positive with PAS, are commonly found in peripheral blood. Leucocytes should be used with caution for enzyme diagnosis due to the presence of the 'renal' enzyme. Lymphocytes or fibroblasts may be preferred. The same enzyme deficiency may also lead to an adult myopathic disorder. (See also Chapter 16.)

Cystinosis

Cystine, taken into the lysosome directly or as a result of proteolysis, requires a specific transport protein for export. A defect of this transport system leads to the accumulation of free cystine within the lysosomes. Due to its low solubility, crystal formation occurs, leading to tissue damage. In the kidney, damage to the glomeruli leads to a typical Fanconi syndrome, whereas cystine crystals in the cornea leads to photophobia. Diagnosis may be confirmed by the demonstration of high levels of free cystine in leucocytes or by retention of [^{35}S]-cystine in cell cultures or chorionic villus tissue.

Salla disease

This condition was first reported in a group of patients in psychiatric institutions in a region (Salla) of northern Finland. Patients may be somewhat coarse featured with marked psychomotor retardation but have an almost normal life span. Most patients have prominent vacuolated lymphocytes and there is a marked increase in the urinary and tissue free sialic acid concentration, as a result of a defective lysosomal transport protein. A severe infantile form of the disease (infantile sialic acid storage disease) is also due to the defective transporter. Infantile patients tend to have higher levels (up to 100 × normal) of free sialic acid in the urine and tissues and may present at birth with ascites, hepatosplenomegaly, facial deformity, punctate calcifications and pale wispy hair. The presentation may be similar to Zellweger syndrome. Diagnosis should be established by free sialic acid measurement in urine and cultured cells.

Vitamin B_{12} transport defect

A deficiency of a lysosomal transport protein is the presumed defect in patients with the rare *Cbl*F mutation (Vassiliadis *et al.*, 1991). These patients excrete excessive amounts of methylmalonic acid and would be expected to have a raised plasma homocystine concentration.

Neuronal ceroid lipofuscinoses

This group of diseases was formally classified with the lysosomal storage disorders under the heading of familial amaurotic idiocy. The clinical presentation is similar, with progressive neurological deterioration, seizures and retinal blindness. There is a marked lysosomal accumulation of storage material in the neurones and other cells. On electron microscopy, characteristic curvilinear/fingerprint patterns may assist with diagnosis, especially of blood lymphocytes or skin biopsy. Recent studies have identified subunit C of the mitochondrial ATP synthase as the major storage product in late-infantile, juvenile and adult forms of the disease (Hall *et al.*, 1991). However, the primary biochemical lesion remains unknown.

Diagnostic tests

There are unfortunately no simple diagnostic screening tests for lysosomal disorders. Identification of characteristic vacuolated lymphocytes or foam cells in the bone marrow will help in the selection of diagnostic enzyme assays.

Microscopy of renal epithelial cells has been used in the past, for example for the metachromasia observed in metachromatic leucodystrophy or the 'Maltese cross' configuration in Fabry disease. However, microscopy is only reliable when fresh urine is used by an experienced person. Lipid analysis of urinary sediment may be useful when unusual variants are to be studied; for example, to confirm sulphatide excretion in individuals deficient in arylsulphatase A activity.

Urinary oligosaccharide analysis has been recommended for the investigation of patients with defects in glycoprotein, glycolipid and glycogen degradation (Sewell, 1980). One-dimensional thin-layer chromatograms stained with orcinol for sugars and with resorcinol for sialic acid-containing oligosaccharides have proved particularly useful for some conditions. Identification of abnormal bands may prove difficult in samples from neonates, especially when they are being fed breast milk. Oligosaccharide analysis is a useful adjunct to enzyme diagnosis but certain limitations should be recognized. Patterns from known patients should be established in individual laboratories.

A number of approaches have been used for the identification of abnormal glycosaminoglycan excretion in MPS disorders (see Whiteman (1984)). Spot tests, used in the past, are no longer considered reliable (Brimble *et al.*, 1990), and measurement of total glycosaminoglycan levels should take into account the reduction with age of the patient. The formation of complexes with cationic detergents, such as cetyl pyridinium chloride, or cationic dyes (Alcian blue 8GX) has been widely used but these may not help in the differential diagnosis. Uronic acid measurement is also a useful quantitative test, except for keratan sulphates (in Morquio disease). A recently described spectrophotometric test based on a reaction with dimethylmethylene blue (de Jong *et al.*, 1992) may provide a useful screening test.

In view of the limitations of quantitative analyses, particularly with regard to the identification of the type of MPS disorder, qualitative analysis is strongly recommended. Both one- and two-dimensional electrophoresis as well as thin layer chromatography have been used. The isolation of glycosaminoglycans from urine (or amniotic fluid) by complexing with Alcian blue 8GX, followed by two-dimensional electrophoresis and subsequent staining of the separated components, has proved one of the best approaches for diagnosis (Whiteman & Young, 1977).

Carrier detection

Theoretically, carriers of recessive disorders will have half the gene dosage, and for enzyme-deficient states they would be expected to have 50% of the normal activity. Unfortunately, in practice, this may not be easy to identify, in view of the range of activities normally measured in the general population. Studies within a family and the use of enzyme activity ratios may help. For autosomal recessive disorders it is generally felt that some 75% of carriers can be identified by enzyme assay alone. Screening heterozygotes for Tay–Sachs disease has proved to be a simple and inexpensive procedure when applied to Jewish populations at risk, particularly those in North America and Israel. Most carriers can be identified by assay of hexosaminidase A in leucocytes or serum. By combining enzyme studies with DNA analysis for the two common mutations in Jewish Tay–Sachs disease, the detection rate should now approach 100%.

For X-linked recessive disorders, the degree of Lyonization will significantly affect the activity which is measured. Hair-root analysis has, nevertheless, been used for carriers of Fabry and Hunter (MPS II) diseases. Since each hair root is almost of clonal origin, two populations, one with normal activity and one with deficient activity, may coexist in carriers.

With the development of DNA analyses and the recognition of specific mutations, it should soon be possible to offer more accurate methods of carrier detection, particularly for those families that are fully informative.

Prenatal diagnosis

Prenatal diagnosis is possible for almost all lysosomal storage disorders (Besley, 1992). The diagnostic test, usually an enzyme assay, must always be verified on samples (often cultured fibroblasts) from the proband. For most disorders, direct enzyme assay of chorionic villus tissue, taken at around 10 weeks of gestation, provides the most satisfactory approach but, alternatively, cultured amniotic fluid cells may be used. Such studies should only be undertaken by those laboratories with extensive experience in handling villus tissue and a thorough working knowledge of the normal enzyme activities in this tissue. Problems with maternal contamination, interfering activities and unexpectedly low activities have led to recognized difficulties with interpretation in the past. For some disorders, e.g. I cell disease, sialidosis and Niemann–Pick disease type C, direct studies on the villus are inappropriate but cultured villus cells may be used.

A few disorders (I cell disease, MPS II) may be diagnosed enzymatically on amniotic fluid, possibly in the first trimester. For MPS disorders, the analysis of glycosaminoglycans in second-trimester amniotic fluid should accompany an enzyme assay of cultured amniotic fluid cells. However, first-trimester fluid may not show significantly abnormal patterns in affected fetuses.

Treatment

There is at present no fully effective treatment for lysosomal storage disorders. Enzyme replacement therapy (infusion of purified enzyme) did not prove successful in most trials. However, removal of terminal oligosaccharide units to expose mannose-terminated chains has provided an effective means of targeting β-glucosidase to storage cells in patients with Gaucher's disease (Barton *et al.*, 1991). Such an approach may prove useful for non-neuropathic disorders, but those with neurological involvement will prove more difficult to treat. Firstly, the blood–brain barrier prevents delivery of the enzyme to the brain and, secondly, the pathology is already established during fetal development and this will prove difficult to reverse.

Bone marrow transplantation as a means of delivery of the deficient enzyme has been partially successful for some disorders. Frequently, stabilization of the disease process is all that can be achieved, and problems of graft-versus-host reaction have to be weighed against the expected benefits.

For cystinosis, renal transplant has alleviated the problems of renal failure and the use of specific cystine-depleting agents, such as β-mercapto-ethylamine (cysteamine), has proved effective in lowering tissue levels of cystine.

Since most therapeutic approaches have proved disappointing, it is not surprising that thoughts of gene therapy are now being seriously considered. The ability to express retrovirally delivered genes in a variety of cell types (Fink *et al.*, 1990; Anderson, 1992) opens the door to the serious ethical considerations which must now be addressed, and it is likely that somatic gene therapy will be undertaken for some disorders in the not-too-distant future.

References

Allard, J.C., Tilak, S. & Carter, A.P. (1988) CT and MR of MELAS syndrome. *Am J Neurorad* **9**, 1234–1238.

Anderson, S., Bankier, A.T., Barrell, B.G. *et al.* (1981) Sequence and organisation of the human mitochondrial genome. *Nature* **290**, 457–465.

Anderson, W.F. (1992) Human gene therapy. *Science* **256**, 808–813.

Argov, Z., Bank, W.J., Maris, J. *et al.* (1986) Treatment of mitochondrial myopathy due to complex III deficiency with vitamins K_3 and C: a ^{31}P NMR follow-up study. *Ann Neurol* **19**, 598–602.

Arnold, D.L., Taylor, D.J. & Radda, G.K. (1985) Investigation of human mitochondrial myopathies by phosphorus magnetic resonance spectroscopy. *Ann Neurol* **18**, 189–196.

Aubourg, P., Blanche, S., Jambagne, I. *et al.* (1990) Reversal of early neurologic and neuroradiologic manifestations of X-linked adrenoleukodystrophy by bone marrow transplantation. *N Engl J Med* **322**, 1860–1866.

Aynsley-Green, A., Weindling, A.M., Soltesz, G., Ross, B. & Jenkins, P.A. (1984) Dichloroacetate in the treatment of congenital lactic acidosis. *J Inher Metab Dis* **7**, 26.

Barth, P.G., Wanders, R.J.A., Schutgens, R.B.H., Bleeker-Wagemakers, E.M. & Van Heemstra, D. (1990) Peroxisomal Ā-oxidation defect with detectable peroxisomes: a case with neonatal onset and progressive course. *Eur J Pediatr* **149**, 722–726.

Barton, N.W., Brady, R.O., Darnbrosia, J.M. *et al.* (1991) Replacement therapy for inherited enzyme deficiency – macrophage targeted glucocerebrosidase for Gaucher disease. *N Engl J Med* **324**, 1464–1470.

Bauman, N., Frederico, A. & Suzuki, K. (1991) Late onset neurometabolic genetic disorders. From clinical to molecular aspects of lysosomal and peroxisomal disease. *Dev Neurosci* **13**, 181–376.

Bendahan, D., Desnuelle, C., Vanuxem, D. *et al.* (1992) ^{31}P NMR spectroscopy and ergometer exercise test as evidence for muscle oxidative performance improvement with coenzyme Q in mitochondrial myopathies. *Neurology* **42**, 1203–1208.

Benke, P.J., Parker, J.C., Lubs, M.-L., Benkendorf, J. & Feuer, A.E. (1982) X-linked Leigh's syndrome. *Hum Genet* **62**, 52–59.

Bennett, M.J., Pollitt, R.J., Goodman, S.I., Hale, D.E. & Vamecq, J. (1991) Atypical riboflavin-responsive glutaric aciduria and deficient peroxisomal glutaryl-CoA oxidase activity: a new peroxisomal disorder. *J Inher Metab Dis* **14**, 165–173.

Benson, P.F. & Fensom, A.H. (1985) *Genetic Biochemical Disorders. Oxford Monographs on Medical Genetics*, No. 12. Oxford University Press, Oxford.

Berenberg, R.A., Pellock, J.M., DiMauro, S. *et al.* (1977) Lumping or splitting? 'Ophthalmoplegia-Plus' or Kearns–Sayre syndrome? *Ann Neurol* **1**, 37–54.

Besley, G.T.N. (1992) Enzyme analysis. In: *Prenatal Diagnosis and Screening* (eds D.J.H. Brock, C.H. Rodeck & M.A. Ferguson-Smith), pp. 127–145. Churchill Livingstone, Edinburgh.

Beutler, E. (1993) Gaucher disease as a paradigm of current issues regarding single gene mutations of humans. *Proc Natl Acad Sci (USA)* **90**, 5384–5390.

Boustany, R.N., Aprille, J.R., Halperin, J., Levy, H. & DeLong, G.R. (1983) Mitochondrial cytochrome c oxidase deficiency presenting as a myopathy with hypotonia, external ophthalmoplegia, and lactic acidosis in an infant and as a fatal hepatopathy in a second cousin. *Ann Neurol* **14**, 462–470.

Bowen, P., Lee, C.S.M., Zellweger, H. & Lindenberg, R. (1964) A familial syndrome of multiple congenital defects. *Bull John Hopkins Hosp* **114**, 402–414.

Braybrooke, J., Lloyd, B., Nattrass, M. & Alberti, I.G.M.M. (1975) Blood sampling techniques for lactate and pyruvate estimation: a reappraisal. *Ann Clin Biochem* **12**, 252–254.

Bresolin, N., Doriguzzi, C., Ponzetto, C. *et al.* (1990) Ubidecarenone in the treatment of mitochondrial myopathies: a multi-center double-blind trial. *J Neurol Sci* **100**, 70–78.

Bresolin, N., Zeviani, M., Bonilla, E. *et al.* (1985) Fatal infantile cytochrome c oxidase deficiency: decrease of immunologically detectable enzyme in muscle. *Neurology* **35**, 802–812.

Brimble, A., Pennock, C. & Stone, J. (1990) Results of a quality assurance exercise for urinary glycosaminoglycan excretion. *Ann Clin Biochem* **27**, 133–138.

Brown, F.R., McAdams, A.J., Cummins, J.W. *et al.* (1982) Cerebro-hepato-renal (Zellweger) syndrome and neonatal adrenoleukodystrophy: similarities in phenotype and accumulation of very long chain fatty acids.

John Hopkins Med J **151**, 344–361.

Brown, G.K., Brown, R.M., Scholem, R.D., Kirby, D.M. & Dahl, H.H. (1989) The clinical and biochemical spectrum of human pyruvate dehydrogenase complex deficiency. *Ann N Y Acad Sci* **573**, 360–368.

Brown, G.K., Haan, E.A., Kirby, D.M. *et al.* (1988) 'Cerebral' lactic acidosis: defects in pyruvate metabolism with profound brain damage and minimal systemic acidosis. *Eur J Pediatr* **147**, 10–14.

Brown, R.M., Dahl, H.-H.M. & Brown, G.K. (1989) X-chromosome localisation of the functional gene for the E1α subunit of the human pyruvate dehydrogenase complex. *Genomics* **4**, 174–181.

Byrne, E. & Trounce, I. (1985) Oxygen electrode studies with human skeletal muscle mitochondria in vitro: a re-appraisal. *J Neurol Sci* **69**, 319–333.

Byrne, E., Dennett, X. & Trounce, I. (1985) Partial cytochrome oxidase (aa3) deficiency in chronic progressive external ophthalmoplegia. Histochemical and biochemical studies. *J Neurol Sci* **71**, 257–271.

Chow, C.W., Anderson, R.M. & Kenny, G.C.T. (1987) Neuropathology in cerebral lactic acidosis. *Acta Neuropathol (Berlin)* **74**, 393–396.

Christensen, E.J., Van Eldere, J., Brandt, N.J. *et al.* (1990) A new peroxisomal disorder: di- and trihydroxycholestanoic acidaemia due to a presumed trihydroxycholestanoyl-CoA oxidase deficiency. *J Inher Metab Dis* **13**, 363–366.

Clayton, P.T., Lake, B.D., Hjelm, M. *et al.* (1988) Bile acid analyses in pseudo-Zellweger syndrome: clues to the defect in peroxisomal β-oxidation. *J Inher Metab Dis* **11**, 165–168.

Dahl, H.-H.M., Brown, G.K., Brown, R.M. *et al.* (1993) Mutations and polymorphisms in the pyruvate dehydrogenase E1α gene. *Hum Mutat* **1**, 97–102.

Danks, D.M.P., Tippett, P., Adams, C. & Campbell, P. (1975) Cerebro-hepato-renal syndrome of Zellweger. A report of eight cases with comments upon the incidence, the liver lesion and a fault in pipecolic acid metabolism. *J Pediatr* **86**, 382–387.

Danpure, C.J. (1989) Recent advances in the understanding, diagnosis and treatment of primary hyperoxaluria type I. *J Inher Metab Dis* **12**, 210–224.

De Duve, C. & Baudhuin, P. (1966) Peroxisomes (microbodies and related particles). *Physiol Rev* **46**, 323–357.

Detre, J.A., Wang, Z., Bogdan, A.K. *et al.* (1991) Regional variation in brain lactate in Leigh syndrome by localised nuclear magnetic resonance spectroscopy. *Ann Neurol* **29**, 218–221.

DiMauro, S., Bonilla, E., Zeviani, M., Nakagawa, M. & deVivo, D.C. (1985) Mitochondrial myopathies. *Ann Neurol* **17**, 521–538.

DiMauro, S., Mendell, J.R., Sahenk, Z. *et al.* (1980) Fatal infantile mitochondrial myopathy and renal dysfunction due to cytochrome c oxidase deficiency. *Neurology* **30**, 795–804.

DiMauro, S., Nicholson, J.F., Hayes, A.P. *et al.* (1983) Benign infantile mitochondrial myopathy due to reversible cytochrome c oxidase deficiency. *Ann Neurol* **14**, 226–234.

Driscoll, P.F., Larsen, P.D. & Gruber, A.B. (1987) MELAS syndrome involving a mother and two children. *Arch Neurol* **44**, 971–973.

Dunger, D.B. & Leonard, J.V. (1984) An evaluation of urine lactate for detection of inborn errors of metabolism. *J Inher Metab Dis* **7** (Suppl. 2), 111–112.

Elef, S., Kennaway, N.G., Buist, N.R.M. *et al.* (1984) ^{31}P NMR study of improvement in oxidative phosphorylation by vitamins K_3 and C in a patient with a defect in electron transport at complex III in skeletal muscle. *Proc Natl Acad Sci USA* **81**, 3529–3533.

Eross, J., Silink, M. & Dorman, D. (1981) Cerebrospinal fluid lactic acidosis in bacterial meningitis. *Arch Dis Child* **56**, 692–698.

Fink, J.K., Correll, P.H., Perry, L.K., Brady, R.O. & Karlsson, S. (1990) Correction of glucocerebrosidase deficiency after retroviral-mediated gene transfer into haematopoietic progenitor cells from patients with Gaucher disease. *Proc Natl Acad Sci USA* **87**, 2334–2338.

Gieselmann, V., Fluherty, A.L., Tonnesen, T. & von Figura, K. (1991) Mutations in the arylsulphatase A pseudodeficiency allele causing metachromatic leucodystrophy. *Am J Hum Genet* **49**, 407–413.

Giles, R.E., Blanc, H., Cann, H.M. & Wallace, D.C. (1980) Maternal inheritance of human mitochondrial DNA. *Proc Natl Acad Sci USA* **77**, 6715–6719.

Goldfischer, S., Moore, C.L., Johnson, A.B. *et al.* (1973) Peroxisomal and mitochondrial defects in the cerebro-hepato-renal syndrome. *Science* **182**, 62–64.

Goto, Y.-I., Nonaka, I. & Horai, S. (1990) A mutation in the $tRNA^{leu(UUR)}$ gene associated with the MELAS subgroup of mitochondrial encephalomyopathies. *Nature* **348**, 651–653.

Hall, N.A., Lake, B.D., Dewji, N.N. & Patrick, A.D. (1991) Lysosomal storage of subunit C of mitochondrial ATP synthase in Batten's disease (ceroid lipofuscinosis). *Biochem J* **275**, 269–272.

Hansen, L.L., Brown, G.K., Kirby, D.M. & Dahl, H.-H.M. (1991) Characterisation of the mutations in three patients with pyruvate dehydrogenase E1α deficiency. *J Inher Metab Dis* **14**, 140–151.

Hanson, R.F., Szcepanik-van Leeuwen, P., Williams, G.C., Grabowski, G. & Sharp, H.L. (1979) Defects of bile acid synthesis in Zellweger's syndrome. *Science* **203**, 1107–1108.

Harding, A.E., Holt, I.J., Sweeney, M.G., Brockington, M. & Davis, M.B. (1992) Prenatal diagnosis of mitochondrial DNA$^{8993T-G}$ disease. *Am J Hum Genet* **50**, 629–633.

Hatefi, Y. (1985) The mitochondrial electron transport and oxidative phosphorylation system. *Ann Rev Biochem* **54**, 1015–1069.

Hayashi, J.-I., Ohta, S., Kikuchi, A. *et al.* (1991) Introduction of disease-related mitochondrial DNA deletions into HeLa cells lacking mitochondrial DNA results in mitochondrial dysfunction. *Proc Natl Acad Sci USA* **88**, 10614–10618.

Heymans, H.S.A., Schutgens, R.B.H., Tan, R., Van den Bosch, H. & Borst, P. (1983) Severe plasmalogen deficiency in tissues of infants without peroxisomes (Zellweger syndrome). *Nature (London)* **306**, 69–70.

Heymans, H.S.A., Van den Bosch, H., Schutgens, R.B.H. *et al.* (1984) Deficiency of plasmalogens in the cerebro-hepato-renal (Zellweger) syndrome. *Eur J Pediatr* **142**, 10–15.

Hillman, R.E. (1989) Primary hyperoxalurias. In: *The Metabolic Basis of Inherited Disease*, Vol. 2 (eds C.R. Scriver, A.L. Beaudet, W.S. Sly & D. Valle), 6th edn, pp. 933–944. McGraw-Hill, New York.

Ho, L., Wexler, I.D., Kerr, D.S. & Patel, M.S. (1989) Genetic defects in human pyruvate dehydrogenase. *Ann NY Acad Sci* **573**, 347–359.

Holt, I.J., Harding, A.E., Cooper, J.M. *et al.* (1989) Mitochondrial myopathy: clinical and biochemical features of 30 patients with major deletions of muscle mitochondrial DNA. *Ann Neurol* **26**, 699–708.

Holt, I.J., Harding, A.E. & Morgan-Hughes, J.A. (1988) Deletions in muscle mitochondrial DNA in patients with mitochondrial myopathies. *Nature* **331**, 717–719.

Holt, I.J., Harding, A.E., Petty, R.K.H. & Morgan-Hughes, J.A. (1990) A new mitochondrial disease associated with mitochondrial DNA heteroplasmy. *Am J Hum Genet* **46**, 428–433.

Howell, N. (1983) Origin, cellular expression, and hybrid transmission of mitochondrial CAP-R, PYR-IND and OLI-R mutant phenotypes. *Somat Cell Mol Genet* **9**, 1–24.

de Jong, J.G.N., Wevers, R.A. & Liebrand-van Sambeck, R. (1992) Measuring urinary glycosaminoglycans in the presence of protein: an improved screening procedure for mucopolysaccharidoses based on dimethylmethylene blue. *Clin Chem* **38**, 803–807.

Karpati, G., Arnold, D., Matthews, P.M. *et al.* (1991) Correlative multidisciplinary approach to the study of mitochondrial encephalomyopathies. *Rev Neurol (Paris)* **146**, 455–461.

Kearns, T.P. & Sayre, G.P. (1958) Retinitis pigmentosa, external ophthalmoplegia and complete heart block. *Arch Ophthalmol* **60**, 280–289.

Kelley, R.I., Datta, N.S., Dobijns, W.S. *et al.* (1986) Neonatal adrenoleukodystrophy: new cases, biochemical studies and differentiation from Zellweger and related peroxisomal polydystrophy syndromes. *Am J Med Genet* **23**, 869–901.

Kelly, D.P., Whelan, A.J., Ogden, M.L. *et al.* (1990) Molecular characterisation of inherited medium-chain acyl-CoA dehydrogenase deficiency. *Proc Natl Acad Sci USA* **87**, 9236–9240.

Kornfeld, S. & Mellman, I. (1989) The biogenesis of lysosomes. *Ann Rev Cell Biol* **5**, 483–525.

Kresse, H., Cantz, M., von Figura, K., Glossl, J. & Paschke, E. (1981) The mucopolysaccharidoses: biochemistry and clinical symptoms. *Klin Weschenschr* **59**, 867–876.

Kretzschmar, H.A., DeArmond, S.J., Koch, T.K. *et al.* (1987) Pyruvate dehydrogenase complex deficiency as a cause of subacute necrotising encephalopathy (Leigh disease). *Pediatrics* **79**, 370–373.

Latta, K. & Brodehl, J. (1990) Primary hyperoxaluria type I. *Eur J Pediatr* **149**, 518–522.

Lazarow, P.B. & Moser, H.W. (1989) Disorders of peroxisome biogenesis. In: *The Metabolic Basis of Inherited Disease*, Vol. 2 (eds C.R. Scriver, A.L. Beaudet, W.S. Sly & D. Valle), 6th edn, 1479–1509. McGraw-Hill, New York.

Lenz, H., Sluga, E., Bernheimer, H., Molzer, B. & Purgyi, W. (1979) Refsum Krankheit und ihr Verlauf bei diätetischen Behandlung durch 2 Jahre. *Nervenartzt* **33**, 237–244.

Linn, T.C., Pettit, F.H. & Reed, L.J. (1969) α-Keto acid dehydrogenase complexes X. Regulation of the activity of the pyruvate dehydrogenase complex from beef kidney mitochondria by phosphorylation and dephosphorylation. *Proc Natl Acad Sci USA* **62**, 234–241.

Luft, R., Ikkos, D., Palmieri, G., Ernster, L. & Afzelius, B. (1962) A case of severe hypermetabolism of nonthyroid origin with a defect in the maintenance of mitochondrial respiratory control; a correlated clinical, biochemical and morphological study. *J Clin Invest* **41**, 1776–1804.

McCormack, J.G., Halestrap, A.P. & Denton, R.M. (1990) Role of calcium ions in the regulation of mammalian intra-mitochondrial metabolism. *Physiol Rev* **70**, 391–425.

Machin, G.A. (1989) Hydrops revisited: literature review of 1414 cases published in the 1980s. *Am J Med Genet* **34**, 366–390.

McShane, M.A., Hammans, S.R., Sweeney, M. *et al.* (1990) Pearson syndrome and mitochondrial encephalomyopathy in a patient with a deletion of mitochondrial DNA. *Am J Hum Genet* **48**, 39–42.

Mahuran, D.J. (1991) The biochemistry of HEX A and HEX B gene mutations cause Gm2-gangliosidosis. *Biochem Biophys Acta* **1096**, 87–94.

Mahuran, D.J., Triggs-Raine, B.L., Feigenbaum, A.J. & Gravel, R.A. (1990) The molecular basis of Tay–Sachs disease: mutation identification and diagnosis. *Clin Biochem* **23**, 409–415.

Mannaerts, G.P. & van Veldhoven, P.P. (1992) Role of

peroxisomes in mammalian metabolism. *Cell Biochem Funct* **10**, 141–151.

Martinez, M. (1989) Polyunsaturated fatty acid changes suggesting a new enzymatic defect in Zellweger syndrome. *Lipids* **24**, 261–265.

Martinez, M. (1990) Severe deficiency of docosahexaenoic acid in peroxisomal disorders: an effect of Δ4-desaturation? *Neurology* **40**, 1292–1298.

Matthews, P.M., Allaire, C., Shoubridge, E.A. *et al.* (1990) In vivo muscle magnetic resonance spectroscopy in the clinical investigation of mitochondrial disease. *Neurology* **41**, 114–120.

Matthews, P.M., Berkovic, S.F., Tampieri, D. *et al.* (1991) Magnetic resonance imaging shows specific abnormalities in the MELAS syndrome. *Neurology* **41**, 1043–1046.

Matthews, P.M., Ford, B., Dandurand, R.J. *et al.* (1993) Coenzyme Q10 with multiple vitamins is therapeutically ineffective in mitochondrial disease. *Neurology* **43**, 884–890.

Miyabayashi, S., Ito, T., Abukawa, D. *et al.* (1987) Immunochemical study in three patients with cytochrome c oxidase deficiency presenting Leigh's encephalomyelopathy. *J Inher Metab Dis* **10**, 289–292.

Monnens, L.A.H. & Heymans, H.S.A. (1987) Peroxisomal disorders: clinical characterization. *J Inher Metab Dis* **10**, 23–32.

Montpettit, V.J.A., Anderman, F., Carpenter, S. *et al.* (1971) Subacute necrotising encephalomyelopathy. A review and a study of two families. *Brain* **94**, 1–30.

Moraes, C.T., DiMauro, S., Zeviani, M. *et al.* (1989) Mitochondrial DNA deletions in progressive external ophthalmoplegia and Kearns–Sayre syndrome. *N Engl J Med* **320**, 1293–1299.

Moraes, C.T., Shanske, S., Tritschler, H.-J. *et al.* (1991) mtDNA depletion with variable tissue expression: a novel genetic abnormality in mitochondrial disease. *Am J Hum Genet* **48**, 492–501.

Moser, A.B., Borel, J., Odone, A. *et al.* (1987) A new dietary therapy for adrenoleukodystrophy: biochemical and preliminary clinical results in 36 patients. *Ann Neurol* **21**, 240–249.

Moser, H.W. & Moser, A.B. (1991) Measurement of saturated very long chain fatty acids in plasma. In: *Techniques in Diagnostic Human Biochemical Genetics* (ed. F.A. Hommes), pp. 177–191. Wiley-Liss, New York.

Naidu, S., Hoefler, G., Watkins, P.A. *et al.* (1988) Neonatal seizures and retardation in girl with biochemical features of X-linked adrenoleukodystrophy. *Neurology* **38**, 1100–1107.

Neufeld, E.F. (1989) Natural history and inherited disorders of a lysosomal enzyme, β-hexosaminidase. *J Biol Chem* **264**, 10927–10930.

Neufeld, E.F. (1991) Lysosomal storage diseases. *Ann Rev Biochem* **60**, 257–280.

Nikoskelainen, E.K., Savontaus, M.-L., Wanne, O.P., Katila, M.J. & Nummelin, K.U. (1987) Leber's hereditary optic neuroretinopathy, a maternally inherited disease: a genealogic study in four pedigrees. *Arch Ophthalmol* **105**, 665–671.

Onkenhout, W., Van der Poel, P.F.H. & Van den Heuvel, M.P.M. (1989) Improved determination of very-long-chain fatty acids in plasma and cultured skin fibroblasts: application to the diagnosis of peroxisomal disorders. *J Chromat* **494**, 31–41.

Osumi, T., Tsukamoto, T., Hata, S. *et al.* (1991) Amino-terminal presequence of the precursor of peroxisomal 3-ketoacyl-CoA thiolase is a cleavable signal peptide for peroxisomal targeting. *Biochem Biophys Res Commun* **181**, 947–954.

Passarge, E. & McAdams, A.J. (1967) Cerebro-hepato-renal syndrome. A newly recognized hereditary disorder of multiple congenital defects including sudanophilic leukodystrophy, cirrhosis of liver and polycystic kidneys. *J Pediatr* **71**, 691–702.

Patel, M.S. & Roche, T.E. (1990) Molecular biology and biochemistry of pyruvate dehydrogenase complexes. *FASEB J* **4** 3224–3233.

Pavlakis, S.G., Phillips, P.C., DiMauro, S., DeVivo, D.C. & Rowland, L.P. (1984) Mitochondrial myopathy, encephalopathy, lactic acidosis, and stroke-like episodes: a distinctive clinical syndrome. *Ann Neurol* **16**, 481–488.

Pellock, J.M., Behrens, M. & Lewis, L. (1978) Kearns–Sayre syndrome and hypoparathyroidism. *Ann Neurol* **3**, 455–458.

Poll-The, B.T., Roels, F., Ogier, H. *et al.* (1988) A new peroxisomal disorder with enlarged peroxisomes and a specific deficiency of acyl-CoA oxidase (pseudo-neonatal adrenoleukodystrophy). *Am J Hum Genet* **42**, 422–434.

Poll-The, B.T., Saudubray, J.M., Ogier, H.A.M. *et al.* (1987) Infantile Refsum disease: an inherited peroxisomal disorder; comparison with Zellweger syndrome and neonatal adrenoleukodystrophy. *Eur J Pediatr* **146**, 477–483.

Poulton, J., Deadman, M.E. & Gardiner, R.M. (1989) Duplications of mitochondrial DNA in mitochondrial myopathy. *Lancet* **i**, 236–240.

Przyrembel, H., Wanders, R.J.A., van Roermund, C.W.T. *et al.* (1990) Di- and trihydroxycholestanoic acidaemia with hepatic failure. *J Inher Metab Dis* **13**, 367–370.

Rizzo, W.B., Leshner, R.T., Odone, A. *et al.* (1989) Dietary erucic acid therapy for X-linked adrenoleukodystrophy. *Neurology* **39**, 1415–1422.

Rizzo, W.B., Phillips, M.W., Dammann, A.L. *et al.* (1987) Adrenoleukodystrophy: dietary oleic acid lowers hexacosanoate levels. *Ann Neurol* **21**, 232–239.

Rizzo, W.B., Watkins, P.A., Phillips, M.W. *et al.* (1986) Adrenoleukodystrophy: oleic acid lowers fibroblast saturated C22–C26 fatty acids. *Neurology* **26**, 357–361.

Robinson, B.H. & Sherwood, W.G. (1984) Lactic acidaemia. *J Inher Metab Dis* **7** (Suppl. 1), 69–73.

Robinson, B.H., Chun, K., Mackay N. *et al.* (1989) Isolated and combined deficiencies of the alpha-keto acid dehydrogenase complexes. *Ann NY Acad Sci* **573**, 337–346.

Robinson, B.H., Glerum, D.M., Chow, W. *et al.* (1990) The use of skin fibroblast cultures in the detection of respiratory chain defects in patients with lacticacidemia. *Pediatr Res* **28**, 549–555.

Robinson, B.H., Taylor, J. & Sherwood, W.G. (1980) The genetic heterogeneity of lactic acidosis: occurrence of recognisable inborn errors of metabolism in a pediatric population with lactic acidosis. *Pediatr Res* **14**, 956–962.

Roesel, R.A., Carroll, J.E., Rizzo, W.B., Van der Zalm, T. & Hahn, D.A. (1991) Dyggve–Elchior–Clausen syndrome with increased pipecolic acid in plasma and urine. *J Inher Metab Dis* **14**, 876–880.

Rosing, H.S., Hopkins, L.C., Wallace, D.C., Epstein, C.M. & Weidenheim, K. (1985) Maternally inherited mitochondrial myopathy and myoclonic epilepsy. *Ann Neurol* **17**, 228–237.

Rotig, A., Cormier, V., Blanche, S. *et al.* (1990) Pearson's marrow–pancreas syndrome: a multisystem mitochondrial disorder in infancy. *J Clin Invest* **86**, 1601–1608.

Sakuta, R. & Nonaka, I. (1989) Vascular involvement in mitochondrial myopathy. *Ann Neurol* **25**, 594–601.

Santos, M.J., Imanaka, T., Shio, H., Small, G.M. & Lazarow, P.B. (1988) Peroxisomal membrane ghosts in Zellweger syndrome: aberrant organelle assembly. *Science* **239**, 1536–1538.

Schapira, A.H.V., Cooper, J.M., Morgan-Hughes, J.A. *et al.* (1988) Molecular basis of mitochondrial myopathies: polypeptide analysis in complex-I deficiency. *Lancet* **i**, 500–503.

Schrakamp, G., Schalkwijk, C.G., Schutgens, R.B.H. *et al.* (1988) Plasmalogen biosynthesis in peroxisomal disorders. *J Lipid Res* **29**, 325–334.

Schram, A.W., Goldfischer, S., Van Roermund, C.T.W. *et al.* (1987) Human peroxisomal 3-oxoacyl-coenzyme A thiolase deficiency. *Proc Natl Acad Sci USA* **84**, 2494–2496.

Schram, A.W., Strijland, A., Hashimoto, T. *et al.* (1986) Biosynthesis and maturation of peroxisomal β-oxidation enzymes in fibroblasts in relation to the Zellweger syndrome and infantile Refsum disease. *Proc Natl Acad Sci USA* **83**, 6156–6158.

Schutgens, R.B.H., Schrakamp, G., Wanders, R.J.A. *et al.* (1989) Pre- and perinatal diagnosis of peroxisomal disorders. *J Inher Metab Dis* **12**, 118–134.

Scriver, C.R., Beaudet, A.L., Sly, W.S. & Valle, D. (eds) (1989) *The Metabolic Basis of Inherited Disease*, 6th edn. McGraw-Hill, New York.

Sewell, A.C. (1980) Urinary oligosaccharide excretion in disorders of glycolipid, glycoprotein and glycogen metabolism. A review of screening for differential diagnosis. *Eur J Pediatr* **134**, 183–194.

Shapiro, Y., Cederbaum, S.D., Cancilla, P.A., Nielson, D. & Lippe, B.M. (1975) Familial poliodystrophy, mitochondrial myopathy and lactate acidemia. *Neurology* **25**, 614–621.

Shimozawa, N., Suzuki, Y., Orii, T., Tsukamoto, T. & Fujiki, Y. (1992) Molecular basis of Zellweger syndrome. *Science* **255**, 1132–1134.

Shoffner, J.M., Lott, M.T., Lezza, A.M. *et al.* (1990) Myoclonic epilepsy and ragged-red fiber disease (MERRF) is associated with a mitochondrial DNA $tRNA^{(Lys)}$ mutation. *Cell* **61**, 931–937.

Smeitink, J.A.M., Beemer, F.A., Espeel, M. *et al.* (1992) Bone dysplasia associated with phytanic acid accumulation and deficient plasmalogen biosynthesis: a peroxisomal entity amenable to plasmaphoresis. *J Inher Metab Dis* **15**, 377–380.

Stanley, C.A., Hale, D.E., Coates, P.M. *et al.* (1983) Medium-chain acyl-CoA dehydrogenase deficiency in children with non-ketotic hypoglycemia and low carnitine levels. *Pediatr Res* **17**, 877–884.

Steinberg, D. (1989) Refsum disease. In: *The Metabolic Basis of Inherited Disease* (eds C.R. Scriver, A.L. Beaudet, W.S. Sly & D. Valle), pp. 1533–1550. McGraw-Hill, New York.

Subramani, S. (1992) Targeting of proteins into the peroxisomal matrix. *J Mem Biol* **125**, 99–106.

Swinkels, B.W., Gould, S.J., Bodnar, A.G., Rachubinski, R.A. & Subramani, S. (1991) A novel, cleavable peroxisomal targeting signal at the amino-terminus of the rat 3-ketoacyl-CoA thiolase. *EMBO J* **10**, 3255–3262.

Tatuch, Y., Christodoulou, J., Feigenbaum, A. *et al.* (1992) Heteroplasmic mtDNA mutation (T-G) at 8993 can cause Leigh disease when the percentage of abnormal mtDNA is high. *Am J Hum Genet* **50**, 852–859.

Theill, A.C., Schutgens, R.B.H., Wanders, R.J.A. & Heymans, H.S.A. (1992) Clinical recognition of patients affected by a peroxisomal disorder: a retrospective study. *Eur J Pediatr* **151**, 117–120.

Toffaletti, J.G. (1991) Blood lactate: biochemistry, laboratory methods, and interpretation. *Crit Rev Clin Lab Sci* **28**, 252–268.

Tritschler, H.J., Andreetta, F., Moraes, C.T. *et al.* (1992) Mitochondrial myopathy of childhood associated with depletion of mitochondrial DNA. *Neurology* **42**, 209–217.

Tritschler, H.J., Bonilla, E., Lombes, A. *et al.* (1991) Differential diagnosis of fatal and benign cytochrome c oxidase-deficient myopathies of infancy: an immunochemical approach. *Neurology* **41**, 300–305.

Tsukamoto, T., Miura, S. & Fujiki, Y. (1991) Restoration by a 35kDa membrane protein of peroxisome assembly in a peroxisome deficient mammalian cell mutant. *Nature* **350**, 77–81.

Uziel, G., Bertini, E., Bardelli, P., Rimoldi, M. & Gambetti, M. (1991) Experience on therapy of adrenoleukodystrophy and adrenomyeloneuropathy. *Dev Neurosci* **13**, 274–279.

Van Costa, R.V., Lombes, A., DeVivo, D.C. *et al.* (1991) Cytochrome c oxidase-associated Leigh syndrome: phenotypic features and pathogenetic speculations. *J Neurol Sci* **104**, 97–111.

Van den Bosch, H., Schutgens, R.B.H., Wanders, R.J.A. & Tager, J.M. (1992) Biochemistry of peroxisomes. *Ann Rev Biochem* **61**, 137–147.

Van Diggelen, O.P., Zhao, H., Kleijer, W.J. *et al.* (1990) A fluorimetric enzyme assay for the diagnosis of Morquio disease type A (MPS IVA). *Clin Chim Acta* **187**, 131–140.

Vanier, M.T., Wenger, D.A., Comley, M.E. *et al.* (1988) Niemann–Pick disease group C: clinical variability and diagnosis based on defective cholesterol esterification: a collaborative study on 70 patients. *Clin Genet* **33**, 331–348.

Vassiliadis, A., Rosenblatt, D.S., Cooper, B.A. & Bergeron, J.J.M. (1991) Lysosomal cobalamin accumulation in fibroblasts from a patient with an inborn error of cobalamin metabolism (Cbl F complementation group): visualization by electron microscope radioautography. *Exp Cell Res* **195**, 295–302.

Wallace, D.C. (1986) Mitotic segregation of mitochondrial DNAs in human cell hybrids and expression of chloramphenicol resistance. *Somat Cell Mol Genet* **12**, 41–49.

Wallace, D.C., Singh, G., Lott, M.T. *et al.* (1988a) Mitochondrial DNA mutation associated with Leber's hereditary optic neuropathy. *Science* **242**, 1427–1430.

Wallace, D.C., Zheng, X., Lott, M.T. *et al.* (1988b) Familial mitochondrial encephalomyopathy (MERRF): genetic, pathophysiological, and biochemical characterization of a mitochondrial DNA disease. *Cell* **55**, 601–610.

Wanders, R.J.A., Heymans, H.S.A., Schutgens, R.B.H. *et al.* (1988) Peroxisomal disorders in neurology. *J Neurol Sci* **88**, 1–39.

Wanders, R.J.A., van Roermund, C.W.T., Brul, S., Schutgens, R.B.H. & Tager, J.M. (1992a) Bifunctional protein deficiency: identification of a new type of peroxisomal disorder in a patient with an impairment in peroxisomal β-oxidation of unknown etiology by means of complementation analysis. *J Inher Metab Dis* **15**, 385–388.

Wanders, R.J.A., van Roermund, C.W.T., Lageweg, W. *et al.* (1992b) X-linked adrenoleukodystrophy: biochemical diagnosis and enzyme defect. *J Inher Metab Dis* **15**, 634–644.

Wanders, R.J.A., van Roermund, C.W.T., Schelen, A. *et al.* (1991) Di- and trihydroxycholestanaemia in twin sisters. *J Inher Metab Dis* **14**, 357–360.

Wanders, R.J.A., van Roermund, C.W.T., Schutgens, R.B.H. *et al.* (1990a) The inborn errors of peroxisomal A-oxidation: a review. *J Inher Metab Dis* **13**, 4–36.

Wanders, R.J.A., Schelen, A., Feller, N. *et al.* (1990b) First prenatal diagnosis of acyl-CoA oxidase deficiency. *J Inher Metab Dis* **13**, 371–374.

Wanders, R.J.A., Schumacher, H., Heikoop, J., Schutgens, R.B.H. & Tager, J.M. (1992c) Human dihydroxyacetone-phosphate acyltransferase deficiency: a new peroxisomal disorder. *J Inher Metab Dis* **15**, 389–391.

Watkins, P.A., Chen, W.W., Harris, C.J. *et al.* (1989) Peroxisomal bifunctional enzyme deficiency. *J Clin Invest* **83**, 771–777.

Watts, R.W.E., Danpure, C.J., De Pauw, L. & Toussaint, C. (1991) Combined liver–kidney and isolated liver transplantations for primary hyperoxaluria type 1: the European experience. *Nephrol Dial Transplant* **6**, 502–511.

Wenger, D.A. & Williams, C. (1991) Screening for lysosomal disorders. In: *Techniques in Diagnostic Human Biochemical Genetics: A Laboratory Manual* (ed. F.A. Hommes), pp. 587–617. Wiley-Liss, New York.

Whitehouse, S., Cooper, R.H. & Randle, P.J. (1974) Mechanism of activation of pyruvate dehydrogenase by dichloroacetate and other halogenated carboxylic acids. *Biochem J* **141**, 761–774.

Whiteman, P.D. (1984) Lysosomal storage disorders. In: *Chemical Pathology and the Sick Child* (eds B.E. Clayton & J.M. Round), 1st edn, pp. 443–462. Blackwell Scientific Publications, Oxford.

Whiteman, P. & Young, E. (1977) The laboratory diagnosis of Sanfilippo disease. *Clin Chim Acta* **76**, 139–147.

Wicking, C.A., Scholem, R.D., Hunt, S.M. & Brown, G.K. (1986) Immunochemical analysis of normal and mutant forms of human pyruvate dehydrogenase. *Biochem J* **239**, 89–96.

Wilson, P.J., Suthers, G.K., Callen, D.F. *et al.* (1991) Frequent deletions at Xq 28 indicate heterogeneity in Hunter syndrome. *Hum Genet* **86**, 205–206.

Zeviani, M., Amati, P., Bresolin, N. *et al.* (1991) Rapid detection of the A-G (5344) mutation of mtDNA in Italian families with myoclonus epilepsy and ragged-red fibres (MERRF). *Am J Hum Genet* **48**, 203–211.

Zeviani, M., Peterson, P., Servidei, S., Bonilla, E. & DiMauro, S. (1987) Benign reversible muscle cytochrome c oxidase deficiency: a second case. *Neurology* **37**, 64–67.

Zeviani, M., Servidei, S., Gellera, C. *et al.* (1989) An autosomal dominant disorder with multiple deletions of mitochondrial DNA starting at the D-loop region. *Nature* **339**, 309–311.

Zeviani, M., Van Dyke, D.H., Servidei, S. *et al.* (1986) Myopathy and fatal cardiopathy due to cytochrome c oxidase deficiency. *Arch Neurol* **43**, 1198–1202.

Zinn, A.B., Kerr, D.S. & Hoppel, C.L. (1986) Fumarase deficiency: a new cause of mitochondrial encephalopathy. *N Engl J Med* **315**, 469.

8: Routine Screening Programmes

S.C. DAVIES, I.B. SARDHARWALLA & M. CLEARY

(a) Neonatal screening for the haemoglobinopathies

S.C. DAVIES

Definition

. . . the testing, not of carefully selected population samples, but of apparently healthy volunteers from the general population for the purpose of separating them into groups with high and low probabilities for a given disorder. (Sackett & Holland, 1975.)

Introduction

The haemoglobinopathies are inherited in a Mendelian recessive manner. The inherited globin-chain abnormalities are structural variants of either the β- or α-globin chains involved in making up the haemoglobin molecule (e.g. β^s in sickle haemoglobin), or there is an absence or reduction in production of normal β- or α-globin chains, known as thalassaemias and named for the chain that is deficient (e.g. β-thalassaemia). Occasional mutations result in a structural variant (e.g. haemoglobin E) with a thalassaemic phenotype. The clinically significant diseases, or major haemoglobinopathies, result from the inheritance of two severe β-chain abnormalities or loss of all α-chain production (i.e. hydrops fetalis).

Haemoglobinopathies are the most common genetic disorder worldwide with major haemoglobinopathies affecting some 250 000 births annually. Epidemiologically, it appears that sickle cell trait and both α- and β-thalassaemia trait confer some resistance against malaria (Pasvol & Wilson, 1982). They are very rare in people of pure northern European origin and the world distribution is shown in Fig. 8.1. The haemoglobinopathies have come to Britain, northern Europe and the United States as a result of immigration. The carrier rates for the different haemoglobinopathy traits in the ethnic groups commonly encountered in Britain are shown in Table 8.1.

β-Thalassaemia major results from the inheritance of two severe β-thalassaemia genes. There are over 100 described β-thalassaemia mutations and deletions resulting in the β-thalassaemia phenotype (Kazazian & Boehm, 1988). β-Thalassaemia major is a lethal disease early in the first decade unless the anaemia is relieved by regular blood transfusions. This, in turn, leads to death from iron overload unless iron chelation therapy, generally with parenteral desferrioxamine, is used (Weatherall & Clegg, 1981; Fosburg & Nathan, 1990). A small number of people inherit two β-thalassaemia genes but their disease state is ameliorated, either by the coinheritance of α-thalassaemia trait or because the mutations are mild, resulting in some β-chain production, so that the children grow and require only occasional blood transfusions. These cases are known as thalassaemia intermedia.

The family of sickle cell diseases (SCD) is made up of major haemoglobinopathies giving rise to clinical problems, having in common the inheritance of βS, including the homozygous SS state, also known as sickle cell anaemia, and the interaction of sickle haemoglobin with other β-globin abnormalities (e.g. haemoglobin (Hb) SC disease, Hb SO^{Arab}, Hb SD and sickle β-thalassaemia, both $S\beta^0$ and $S\beta^+$). SCD gives rise to a variety of clinical problems, pre-

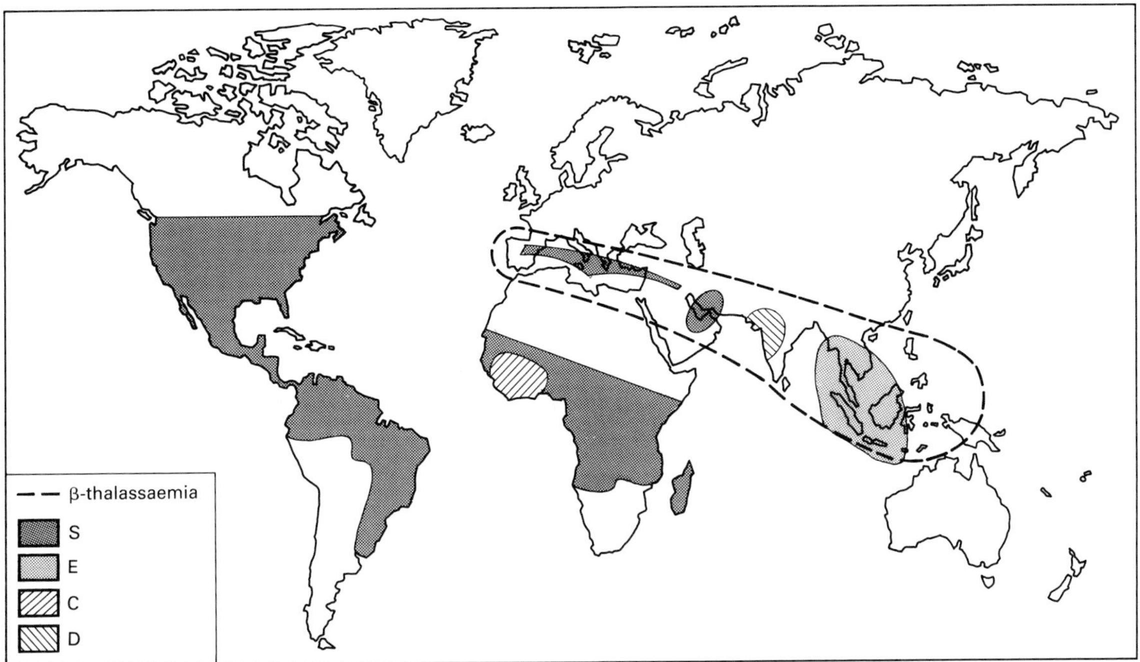

Fig. 8.1 The world distribution of haemoglobinopathies.

Table 8.1 Estimated prevalence of haemoglobinopathy traits and birth rate of affected children, by ethnic group

	Thalassaemia traits			Abnormal Hbs				
Origin of population	β	Hb E	α	S	C	D	Total	Affected births/1000
Italy	2–14	–	–	0–10	–	+	2–14	0.1–4.9
Greece	8	–	+	0–20	+	–	9.0	1.6
Portugal	1–2	–	–	0–2	+	+	1.5	0.06
Turkey	2–6	+	+	0–12	–	–	2.0	0.4*†
Cyprus	16	–	1	1	–	–	18.0	7.0
Middle East	1–3	–		1–2	–	–	3.5	0.31†
India	1–17	+	–	0–30	–	+	~3.5	0.3–1.2
Pakistan	6.5	+	–	+	–	+	6.5	2.0†
Bangladesh	1–3	1–3	–	–	–	+	4.0	1.2
S. China	3	+	6	–	–	–	9.0	1.1
Vietnam	1–6	2–10	+	–	–	–	3.0	0.23
Kampuchea	1–10	5–40	+	–	–	–	10.0	2.5
N. Africa	2–3	–	–	1–3	0–1	–	4.0	0.8†
Subsaharan Africa	1–3	–	–	15–25	1–20	–	22.0	12.0
Caribbean	1–2	–	–	6–12	1–3	–	14.0	5.0
Cabo Verde	1–2	–	–	2–9	3–12	+/–	5–12	0.6–3.6

* Uneven distribution is estimated to give a higher birth rate of homozygotes than would be calculated from the mean heterozygote prevalence.
† A tradition of consanguineous marriage increases the birth rate of homozygotes relative to the heterozygote frequency.

dominantly related to the occlusion of small blood vessels by sickled red cells (Davies & Wonke, 1991; Serjeant, 1992). The most frequent cause of hospital admissions is the painful vaso-occlusive crisis, but other unpleasant and life-threatening sequelae can arise, including priapism, aplastic crisis, the sickle chest syndrome and stroke. Sickling in the spleen gives rise to autoinfarction and diminished or absent immunological function. As a result, the children are at risk of serious infection, and even death, from encapsulated organisms, in particular *Streptococcus pneumoniae, Haemophilus influenza* and *Salmonella* species. Even with good care, patients with SCD have only a probäbility of living to the age of 20 years of 85% for those with SS and 89% for all types of SCD considered together (Leiken *et al.*, 1989). Bone marrow transplant is now accepted for patients with β-thalassaemia major (Lucarelli & Weatherall, 1991) and is under study for those with sickle cell anaemia (Davies, 1993; Vermylen & Cornu, 1993).

Screening children and adults for the haemoglobinopathies is relatively easy and cheap, based on the automated full blood count (FBC) and haemoglobin electrophoresis to detect variant haemoglobins with estimation, generally by column chromatography, of the haemoglobin A_2 level, which is raised in most cases of β-thalassaemia trait. Screening programmes can therefore be set in place to function opportunistically or on agreed sections of society. The rationale behind different screening programmes can be quite different, as shown in Table 8.2.

Table 8.2 The rationale of screening programmes for the haemoglobinopathies

Type of screening programme	Rationale
Symptomatic patients	To offer treatment
Family of a patient	Treatment and genetic advice*
Before general anaesthesia	To ensure safety during induction of anaesthesia
'At risk' community	To find mild cases and offer genetic advice
Schoolchildren	To find mild cases and offer genetic advice
Premarital	Genetic advice
Well-women and -men clinics	Genetic advice
Antenatal clinic	Genetic advice
Neonatal	To prevent morbidity and mortality (family genetic advice) and assist health planning

* Genetic advice includes non-directive discussion of prenatal diagnosis.

The objective of neonatal screening for the haemoglobinopathies is to commence therapy early in order to reduce morbidity and mortality. This is particularly important in SCD infants who need to start their routine prophylaxis with penicillin against pneumococcal infection as soon as possible, and at least by 10 to 12 weeks of age (Gaston *et al.*, 1986). The longest running neonatal haemoglobinopathy screening programme is that in New York State, where screening was mandated under the public health laws in 1975. Only one out of 159 infants, followed as part of the programme from 1981 to 1986, died as a result of overwhelming sepsis. This compares with an anticipated mortality rate at that time of 10% (Brown *et al.*, 1989). Studies have been performed to review the cost-effectiveness of such programmes and showed that, in 1991, screening and then treating affected black infants for SCD cost only $3100 more per life saved than not screening. Screening non-black populations with a high prevalence of haemoglobin S genes would cost $1.4 million per life saved, and screening low-prevalence populations would cost $450 million per life saved (Tsevat *et al.*, 1991). In Britain, there is an active debate as to the level, within the population of ethnic minorities, at which a neonatal screening programme for the haemoglobinopathies becomes sensible and cost-effective. The generally accepted figures are all in the range 10–15%.

There have been problems with establishing these programmes, in both North America and Britain, predominantly relating to inadequate or incorrect knowledge about SCD, a lack of understanding of the need for penicillin prophylaxis (Cummins *et al.*, 1991), poor knowledge in primary care about sickle cell trait (Shickle & May, 1989), poor communication between the screening laboratory and the infant's physician (Sternberg *et al.*, 1982) and 'the identification of prevention as the goal of genetic counselling' (Whitten, 1973). The procedure of neonatal detection of SCD also poses possible hazards to the family, such as misdiagnosis, exposure of non-paternity and the development of grief, guilt and anxiety. However, early detection remains a legitimate endeavour (Scott & Harrison, 1982).

Techniques of diagnosis

At birth the fetus is still producing significant amounts of γ-globin chains, which combine with the α chains to form fetal haemoglobin, as shown in Fig. 8.2. Very premature babies, prior to 25 weeks gestation, usually have no demonstrable haemoglobin A. Fetal haemoglobin interferes with gelation of the sickle molecules so that, even in SS, the sickle test will be negative until at least 6 months of life, and generally unreliable until 1 year of age. Most screening programmes are based on the detection of the abnormal globin protein or, in the case of β-thalassaemia major, the absence of Hb A in term babies. The only reason for screening within the first week or two of life is to ensure that coverage of the population is good, by linking this programme to neonatal screening for other diseases, in particular phenylketonuria and hypothyroidism, as the haemoglobin pattern develops with age. The maturing pattern of globin-chain production follows a genetic clock and is unaffected by the infant's health, feeding or any treatment.

Blood samples may be taken from the umbilical cord after delivery of the placenta (Pearson *et al.*, 1974; Henthorn *et al.*, 1984). Using this technique, there is a risk of contamination of the sample with maternal blood, which may alter the result and will not be detected unless looked for either by a Kleihauer procedure or by reviewing the distribution of mean cell volumes of the erythrocytes using an automated blood counter. We have shown that 1.7% of samples were contaminated with maternal blood. It is therefore better to sample directly from the neonate by using a heel prick. The blood from the heel-prick sample can be transported in heparinized capillary tubes (Evans & Blair, 1976; Griffiths *et al.*, 1988) or spotted onto filter paper (Garrick *et al.*, 1973). Dried blood spots can be sent through the post to a central laboratory for analysis for screening programmes.

The initial technology employed was haemoglobin electrophoresis on lysates eluted from the filter paper. This methodology is still utilized in New York State where the average daily workload is 1200 specimens (Schedlbauer & Pass, 1989). However,

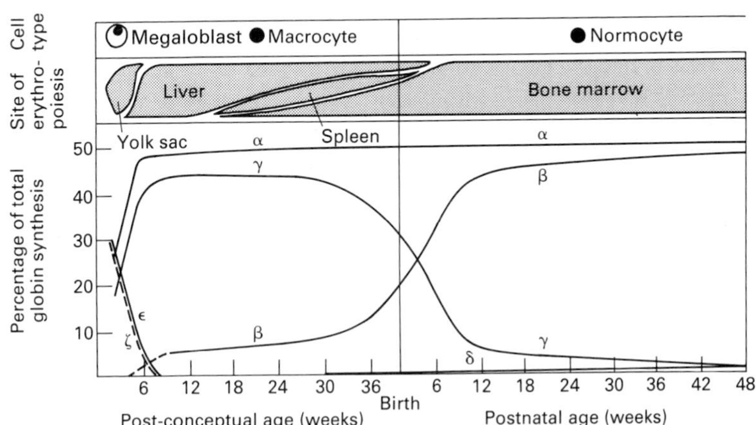

Fig. 8.2 Globin-chain production changes with development. (Modified by kind permission of Prof. Sir David Weatherall.)

prolonged delay before testing, as can occur with slow postage, results in poor electrophoretic resolution of the haemoglobin bands (Schmidt *et al.*, 1976). Undoubtedly, the separation of haemoglobin bands is better from heparinized blood samples than from these eluted samples. There is also a reduced ability to detect fast bands (e.g. haemoglobin Barts, which is diagnostic of α-thalassaemia) (Kinney *et al.*, 1989). Following haemoglobin electrophoresis, variant haemoglobin bands must always be confirmed, although this is a more laborious procedure; this iis because the preliminary screen is generally performed on cellulose acetate at an alkaline pH, when haemoglobins S and D run together, as do haemoglobins C and E. Bands running in either of these two positions need to be further checked by electrophoresis on citrate agar at acid pH, as shown in Fig. 8.3. Approximately 15% of samples tested by cellulose acetate electrophoresis require repeat analysis for final identification (Kleman *et al.*, 1989).

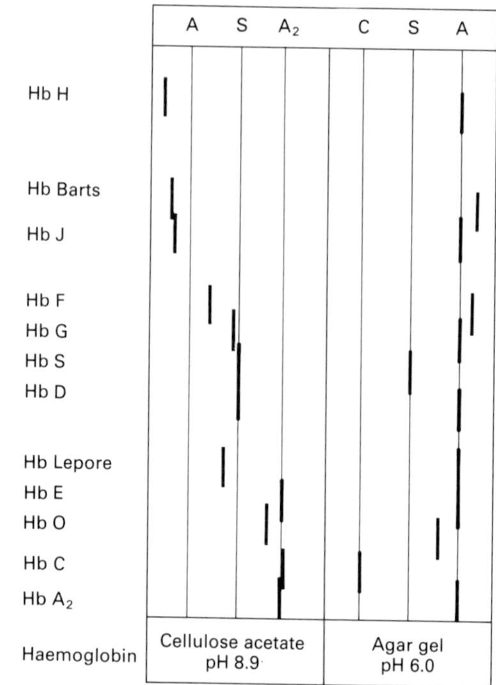

Fig. 8.3 The relative mobilities of the common haemoglobins on electrophoresis.

Isoelectric focusing as a method of screening newborns for abnormal haemoglobins was first described in 1977 (Altland, 1977). This was followed by the characterization of 70 haemoglobin variants by this technique and the publication of results of a neonatal screening programme from Paris (Basset *et al.*, 1978; Galacteros *et al.*, 1980). The advantage of isoelectric focusing is its high resolution, because the haemoglobins separate according to their isoelectric point in a stable pH gradient. With this technique, haemoglobins S and D, and C and E separate from each other, as shown in Fig. 8.4.

Chromatography as a method of neonatal screening for haemoglobinopathies was first reported in 1975 and involved the use of microcolumns (Powars *et al.*, 1975). With microcolumns the presence of presumptive haemoglobin S or C is always apparent but, as with electrophoresis, the presence of an abnormal band in a specific zone, while strongly presumptive, is not absolutely definitive for a diagnosis. The use of high performance liquid chromatography (HPLC) for neonatal screening was described in 1983 (Wilson *et al.*, 1983). This is the definitive method, not only for the quantitation of Hb F and Hb A_2 (except in the presence of Hb E), but also it provides easy differentiation between simple heterozygotes for a specific abnormality and persons with the same abnormality together with an additional β-thalassaemia heterozygosity. Probably the most important application of HPLC is in the quantitation of Hb A and the β-chain variants S, C, O^{Arab} and E in neonatal samples, thereby facilitating the diagnosis of conditions such as AE, EE, AS, SS, $S\beta^+$ thal, AC, CC, $C\beta^+$ thal, SC and SO^{Arab} in the newborn. The disadvantage of HPLC is its relatively high cost compared with the cost of isoelectric focusing and electrophoresis. It does, however, have an important role to play in the confirmation of the typing of variant bands found by other technologies and in the distinction of rare haemoglobin variants.

Even newer technologies are now becoming available for use in specialist laboratories. We use a monoclonal antibody against sickle haemoglobin to confirm that variant bands running as S are indeed

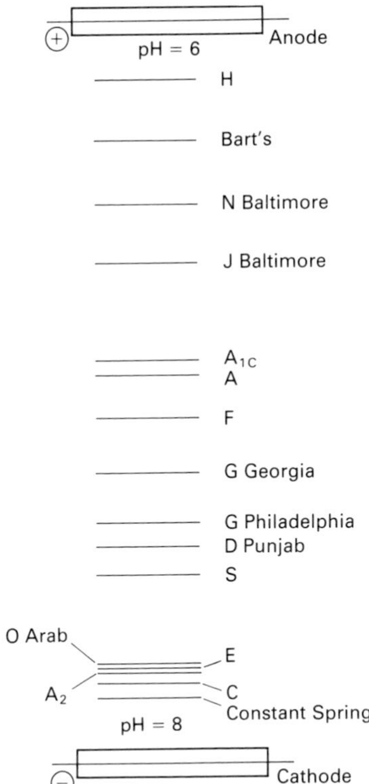

Fig. 8.4 The position of haemoglobin variants on isoelectric focusing.

sickle haemoglobin. Other antihaemoglobin variant monoclonal antibodies have been, and are being, made but are not yet available commercially in Britain (Tan-Wilson *et al.*, 1976; Garver *et al.*, 1984). Both deoxyribonucleic acid (DNA) and ribonucleic acid (RNA), extracted from dried blood spots on filter paper, have been amplified using the polymerase chain reaction with primers for the β-globin region, showing that diagnosis of variant β genes can be made using these technologies (Jinks *et al.*, 1989; Skogerboe *et al.*, 1991; Zhang & McCabe, 1992).

It is advisable to perform a repeat confirmatory test on a further specimen taken from the infant at 6 weeks or more of age, regardless of the technology used. This ensures that any variant band seen on the preliminary screen does indeed belong to that baby (or, in very rare cases, is a γ-chain fetal haemoglobin variant) and allows time for β-chain production to develop, and thus exclude prematurity and a few cases of sickle β^+-thalassaemia who develop their haemoglobin A late. In difficult cases, family studies or even gene analysis may be required. Neonatal screening, performed as outlined above, will detect all clinically significant/ major haemoglobinopathies. Table 8.3 gives the haemoglobin types that are, or may be, missed; the most important is β-thalassaemia trait. In order to diagnose β-thalassaemia trait it is best to repeat the screen at 1 year of age, or older. Fast-running haemoglobin Barts will be present in cases of α-thalassaemia trait, although, because of deterioration during storage, it may not be detected.

Table 8.3 The haemoglobin types not detected by neonatal screening

Always	Sometimes
	$S\beta^+$-Thalassaemia
	β-Thalassaemia intermedia
β-Thalassaemia trait*	α-Thalassaemia traits

* Recent work suggests that some β-thalassaemia-trait neonates may be predicted by haemoglobin quantitation, expressed as the ratio of Hb A : Hb F, taking into consideration birth weight, gestational age and sex.

Dissemination of results

A careful system is needed to ensure that, at a minimum, all positive results are communicated to the relevant health professionals, while respecting confidentiality, and to the family in a manner that does not cause unwarranted upset or anxiety. Any children with a major haemoglobinopathy should be enrolled as soon as possible into a comprehensive programme of care, arranged by a paediatrician or haematologist. A defined structure with nominated people holding responsibility for different aspects of the programme is essential in order to ensure full implementation, or a correct diagnosis may be made in the laboratory without the parents being made aware of the result and it's significance (Listernick *et al.*, 1992).

A person's haemoglobin type is inherited, and while the mature state may not be evident in the first year of life, it does not change. Ideally, therefore, all people who are tested should be made aware of their results and the significance of any abnormality found, and they should be offered haemoglobinopathy cards, stating their haemoglobin type, to carry in case of emergency. Central registers of babies and people screened can be helpful but many sickle crises present outside of routine working hours, when access to a central register may not be feasible (Losek, 1991).

Selection of neonates for screening

If an antenatal screening programme is in place, then it should be possible to predict all babies who are at risk of inheriting a major haemoglobinopathy. A feasible approach would then be to test only the infants of mothers with a haemoglobinopathy trait. In practice, this leads to a number of babies with sickle cell disease being missed (Adjaye *et al.*, 1989). In areas where less than 10% of mothers giving birth are of non-northern European origin, consideration, for cost reasons only, can be given to selective screening. In reality, all infants who have one parent not of pure northern European origin should be screened. It can be very difficult for the nurse, generally a midwife, taking the neonatal sample to enquire sensitively into the ethnic origins of the parents, leading to both errors and non-sampling as the easiest way out. It is not difficult to make the rules for a selective programme but it appears to be very much more difficult for the person responsible to remember the sampling rules and to apply them. As a result, a number of babies will be missed and there will be a consequent risk of increased morbidity and mortality.

Acknowledgments

I thank Dr Bernadette Modell for permission to use the table of the estimated prevalence of haemoglobinopathies in ethnic groups, Professor Sir David Weatherall for permission to use the diagram of globin-chain production changes, Joan Henthorn for constructive criticism, and Brian Dugan for typing the manuscript.

(b) Mass screening for inherited metabolic and other disorders

I.B. SARDHARWALLA & M. CLEARY

Introduction

Screening can be defined as a systematic search within a defined population for individuals with an abnormal genotype who are at risk of developing serious illness as a result of their genetic disease. The concept implies that appropriate action based on the screening result can improve the prognosis for individuals or their families.

In this section the principles of mass screening and the value of screening for individual diseases will be considered. Emphasis will be placed on mass screening of the newborn population. In addition, arguments for screening selected adult populations will be discussed where this has relevance to paediatrics.

Mass screening

There is a general consensus about the criteria which a disease must satisfy before being accepted as suitable for mass screening, but there is considerable debate about whether or not individual disorders satisfy the accepted criteria. Mass screening is usually confined to diseases which can be treated (e.g. phenylketonuria (PKU), hypothyroidism) or diseases in which there is the possibility of reducing morbidity or mortality in early life (e.g. the haemoglobinopathies; see Section (a)). If screening and diagnosis will not alter the clinical course, mass screening is unusual as it raises ethical problems and has the potential to cause much distress. Not everyone agrees with this view (see p. 207 for further comments).

The disorder must occur with sufficient frequency to justify screening. It is difficult to place an exact figure on what amounts to 'sufficient frequency' as this will vary, depending on local resources. More affluent communities may be more willing to utilize resources in screening for relatively rare disorders.

Administration

The main aim of a smooth administrative process for the screening programme is to ensure that a rapid diagnostic service and prompt clinical follow-up of affected children are provided for the community. These needs are best met when good communication exists between the hospital unit, family, general practitioner (GP) and health visitor. This ideal concept is realized when the screening programme, laboratory service and clinical input are concentrated together in a regional unit.

Analytical methods

The key element in methodology for mass screening of newborns has been the development of simple laboratory methods using microsamples of blood. The chosen test must be technically uncomplicated and provide an unequivocal result with no false negatives and few false positives. The use of urine as the screening sample is felt to be generally unsatisfactory as it is difficult to collect and the laboratory methods necessary to ensure sensitivity can be expensive and time consuming. For these reasons, methods using microsamples of blood collected by heel prick are employed by most groups for newborn screening.

Cost–benefit analysis

Many groups have presented data on cost–benefit analysis, particularly with regard to screening for PKU. The cost of providing special care of untreated persons is generally balanced against the costs of screening, diagnosis, treatment and normal schooling. There are other less easily quantifiable costs, such as the benefit of alleviating the patient's suffering and that of the families, as well as considering the positive contributions made to the social and economic life of the community.

Screening tests are available for a number of disorders. The arguments for or against the incorporation of these disorders in national screening programmes will be discussed.

Phenylketonuria

The first genetic disease to be sought by mass screening was PKU and in the UK the programme has existed for almost 25 years (for a review see Smith *et al.* (1991)). Screening using a dried blood spot and a bacterial inhibition assay (Guthrie & Susi, 1963) was adopted widely when screening on blood was introduced. Since then many laboratories have used fluorometry or thin layer or paper chromatography. The authors' laboratory uses one-dimensional paper chromatography (Scriver *et al.*, 1964), which has the advantage of detecting other amino acid disorders. All the children giving false negative results in the UK programme were screened with the bacterial inhibition assay.

The timing of the test is important. In the UK, sampling is performed between days 6 and 14 of life, and is dictated by the use of the health visitor as the community link. Testing too early incurs the risk of giving a false negative result (Starfield & Holtzman, 1975). There is now evidence that treatment should begin with minimal delay and certainly by the age of 20 days (Snyderman, 1986; Medical Research Council Working Party on Phenylketonuria, 1993b). Although early screening, diagnosis and treatment have led to great improvement in intellectual status, it has become apparent that the status is not as good as was originally thought (Medical Research Council Working Party on Phenylketonuria, 1993a,b). In addition, neurological deterioration has been observed in young adults (Thompson *et al.*, 1990).

A small number of infants have mutations affecting the metabolism of the hydroxylase cofactor, tetrahydrobiopterin. Such infants do not respond to dietary treatment. Routine secondary screening for biopterin defects should be performed on any infant giving a positive result at the first screening, even if the increase in phenylalanine in the blood sample is very small. Dried blood spots can be used for the measurement of dihydropteridine reductase and total biopterin concentrations (Smith *et al.*, 1991); alternatively, but less commonly, urine may be used for the determination of pterins.

Congenital hypothyroidism

Congenital hypothyroidism fulfils the recommended screening criteria and has been successfully added to the newborn screening programme in Britain since around 1980. Both thyroxine and thyroid-stimulating hormone (TSH) assays can be employed but in the UK the screening test estimates TSH using a radioimmunoreactive technique. The incidence of this condition varies throughout Europe but is around 1 in 3000–4000 (Pharaoh & Mappon, 1992). The outcome of hypothyroid screening has been prospectively studied in Toronto (Rovet *et al.*, 1986) and in the north Thames area of England (Fuggle *et al.*, 1991). The conclusions were that the intellectual development in treated children was normal in most aspects, although some had specific learning deficits. The north Thames study identified deficits in motor skills in hypothyroid children. In both studies, neurological abnormality was related more to the severity of hypothyroidism at birth and less to the inadequacy of treatment. Hypothyroid screening has been successful in avoiding the severe mental retardation associated with the untreated case.

Galactosaemia

Screening for galactosaemia can be undertaken either by measuring levels of blood galactose (Paigen *et al.*, 1982) or by detecting deficient galactose-1-phosphate uridyl transferase enzyme activity (Beutler & Baluda, 1966). Both methods are accurate but detection of blood galactose is dependent on the child having ingested food containing galactose (i.e. lactose). Although mass screening for galactosaemia is carried out in many countries, much controversy still remains about its role in newborn screening. Reasons for these reservations are as follows. It is a rare condition with a varying incidence from around 1 in 26 000 in Ireland to 1 in 1 million in Japan (Kawamura, 1987; Ng *et al.*, 1987). In the UK, screening is carried out in Scotland where the incidence is 1 in 70 000. There is no reliable evidence that early diagnosis via screening confers any long-term improvement in outcome. The outcome in treated galactosaemic patients, even in those in whom treatment is started on day 1 (as a result of sibling diagnosis) is disappointing, and IQ is in the range 70–90 (Sardharwalla & Wraith, 1987; Holton, 1990). It is argued that early treatment prevents the acute life-threatening presentation of liver failure and encephalopathy. Although this may be true, galactosaemia should in any case always be considered a possibility in the sick newborn, and urgent investigation should be undertaken, based on clinical presentation, before the situation worsens. The condition may present even before the time of screening. Paediatricians should have access to a skilled regional centre, experienced in the diagnosis of inborn errors of metabolism, where galactose-1-phosphate uridyl transferase estimation can be quickly carried out in the sick newborn. With this provision it is difficult to make a case for galactosaemia screening. Whilst awaiting the biochemical result, the use of a lactose-free synthetic milk will not harm an infant if the diagnosis of galactosaemia is not confirmed.

Biotinidase deficiency

Biotinidase deficiency can be detected in the newborn by use of a semi-quantitative colorimetric analysis of the Guthrie dried blood spot (Heard *et al.*, 1984). Children treated with pharmacological doses of biotin do not develop the neurological and skin problems associated with the disorder (Wolf *et al.*, 1985). Even if these signs have developed prior to the diagnosis being made, they are generally reversible with biotin. There have been reports of incomplete resolution of sensorineural deafness and optic atrophy after treatment (Wastell *et al.*, 1988). Since the method depends on the dried blood spot already in use, it is relatively cheap and administratively feasible. Reservations regarding biotinidase screening relate to the low incidence of this disorder of 1 in 100 000 to 1 in 200 000, the classical presentation leading to diagnosis, the generally good results from treatment and the detection of an apparently benign condition of partial biotinidase deficiency (Lawler *et al.*, 1992). National experience of screening in Scotland, where a pilot scheme was initiated in 1988, did not detect any cases and the scheme was discontinued in 1989 (Kennedy *et al.*, 1989; Wolf & Heard, 1990). Balancing these arguments, biotinidase deficiency does not seem to merit priority consideration in screening.

Congenital adrenal hyperplasia

Congenital adrenal hyperplasia (CAH) refers to a group of inherited diseases caused by a deficiency of one of the enzymes required for normal steroid synthesis (see Chapter 10). They are all autosomal recessive disorders. At least eight types of CAH have been described, but all except three are exceedingly rare. Over 90% result from a deficiency of 21-hydroxylase. In this type, the presenting signs are virilization (in female infants) and salt-wasting. The relatively high prevalence of CAH found in retrospective studies led to an interest in newborn screening for this disease and a method has been developed which entails assaying elevated levels of 17-α-hydroxyprogesterone (Pang *et al.*, 1977). The overall incidence worldwide is around 1 in 14 000 (Pang *et al.*, 1988). Taking this incidence for the UK, the screening programme would detect a maximum of four cases of CAH per year in the North-Western Region, where the birth rate is between 55 000 and 60 000 per year. Two of these cases would be female and would be expected to present with ambiguous genitalia at birth. Of the two males, one could be a salt-waster, although recent information is that the incidence of salt-wasting CAH may be three times higher and in this case the chances are high that the diagnosis would be made on clinical grounds. Therefore, the cost of detecting one case of the non-salt-losing variety of CAH which does not carry any long-term financial burden to the community would not be justified.

α_1-Antitrypsin deficiency

The prevalence of the clinically important form of α_1-antitrypsin deficiency (PI ZZ) is about 1 in 1500 (Orfanos *et al.*, 1982). It is estimated that around 60% of such individuals develop emphysematous lung disease in the third or fourth decade. Although there is no specific cure, preventative measures, including avoiding smoking, both passive and active, is felt to lessen the severity of the lung disease (Mittman *et al.*, 1973). α_1-Antitrypsin disease is also associated with liver disease in childhood in 10% of cases. The only therapy that can be offered for the relentless liver failure that ensues in some children is liver transplantation. However, as there is some form of curative treatment and the disease is relatively common compared with many genetic disorders, some argue that screening is indicated. The newborn screening test (Orfanos *et al.*, 1982) already in use is relatively cheap. The main argument against screening is the lack of evidence that identification of this genotype actually moves people to alter their smoking habits. Those who develop liver disease will be identified clinically, and since there are no preventative measures that can be taken to avoid liver problems, their outcome would not be improved by early identification through screening. Overall, there is not a strong argument for screening for this disorder.

Maple syrup urine disease

To screen effectively for maple syrup urine disease (MSUD) the blood sampling would need to be performed prior to the onset of neurological symptoms. As these frequently arise at around 4–7 days of life, such screening would have major implications for the mass screening programme already in service in the UK; this is scheduled around 6–14 days. In some parts of the USA, where screening is performed before 4 days of age and prior to hospital discharge, MSUD screening is feasible and is carried out in some states. On the whole, despite improved dietary treatment and intensive-care facilities, the long-term outcome in MSUD is not as satisfactory as in PKU (Danner & Elsas, 1989). Its place lies with screening of the sick newborn (see Chapter 4).

Cystic fibrosis

Although this is the most common serious genetic disease in the Caucasian population, the proposals for newborn screening for cystic fibrosis (CF) remain controversial. At present, the suggested method of newborn screening for CF utilizes a biochemical technique: the estimation of plasma immunoreactive trypsin performed on the dried blood spot. This method affords a sensitivity of 99.5% and a specificity of 96.3% (Wilcken & Brown, 1987).

Although this is an acceptable test, national screening has not been implemented in the UK. The main reason preventing screening is the concern regarding outcome of this condition, despite early diagnosis. Treatment regimes have improved over the past few years and the quality of life is undoubtedly better. However, there is no evidence to indicate that early diagnosis actually increases overall survival, although hospital admissions are reduced in the first few years of life (Wilcken & Chalmers, 1985; Dauphinais *et al.*, 1986; Chatfield *et al.*, 1991). With the recently acquired knowledge of the gene locus for CF (Riordan *et al.*, 1989; Rommens *et al.*, 1989), together with the identification of the abnormal transport protein produced by the mutated gene, the future for therapy for CF seems more promising. Currently, research aimed at treatment is progressing along two routes: pharmacological manipulation of the chloride channel pathways, and replacement via gene therapy. Presently, the role of mass screening for CF remains controversial. It seems probable that attitudes will change. The proposals for ascertaining the carrier status are discussed later.

In the first study on the demographic and social characteristics of adults with CF in the UK (Walters *et al.*, 1993), it was found that a high proportion of adults with the condition were living full and productive lives. The authors pointed out that this is contrary to an image of chronic ill-health and disability. The median age of the adults was 22 years and the maximum age was 67 for men and 56 for women.

Duchenne muscular dystrophy

In 1989, Plauchu *et al.* strongly advocated neonatal screening for Duchenne muscular dystrophy (DMD). The test detects elevated creatine kinase in a filter-paper blood spot, and in France the frequency was 1 in 5330. More recently, in Wales (Bradley *et al.*, 1993) 77 728 infants were screened, particular attention being paid to the emotional aspects. Justification for the programme included: the possibility of choice in a future pregnancy, the ability to plan for the future with a disabled child in the family, the avoidance of the experience of delayed diagnosis, and the identification of a presymptomatic cohort who might benefit from future treatments. There is promising research directed at replacement therapy for DMD, by monitoring the effect of injecting dystrophin into muscle or implanting normal myoblasts (Worton, 1992). Successful tests have been carried out in mice, and trials with humans are proceeding. The possibility of mass screening for DMD should be kept under review.

Sickle cell disease

See Chapter 8.

Glucose-6-phosphate dehydrogenase deficiency

The rationale for screening for this X-linked disorder relates to the impact that screening can have on morbidity. Although there is no treatment for glucose-6-phosphate dehydrogenase deficiency, environmental avoidance of trigger factors can markedly reduce morbidity in affected individuals. Since 1977, newborn screening has been undertaken in Greece, where the incidence is 1 in 22 males and 1 in 54 females who were either homozygous or heterozygous (Missiou-Tsagaraki, 1991). The results of screening in Greece suggest that the technique is successful. There are insufficient data regarding costing to be emphatic about the cost-effectiveness of this undertaking. The benefits were clear; a fourfold reduction in the hospital admission of patients for the treatment of haemolytic crises.

Neuroblastoma

See Chapter 21.

Relevant screening of adults

Maternal phenylketonuria

In screening for maternal PKU the underlying philosophy is different, and the precepts of screening as applied to the newborn may not apply. Nevertheless, maternal PKU has recently emerged as a major issue for public health and warrants special consideration.

Maternal PKU carries a risk of brain damage and congenital malformation to the fetus in addition to an increased incidence of spontaneous abortion (MacCready & Levy, 1972). Microcephaly occurs in over 90% of the children born to untreated mothers with PKU. Their children, who are carriers for PKU and do not require any dietary treatment, achieve IQs in the range 60–65 and require special education. The incidence of congenital malformation is 20–25%. All organs may be involved but congenital heart disease is the commonest anomaly seen in these patients (Levy & Waisbren, 1983).

These complications can be avoided, and a satisfactory outcome expected, if a really strict low-phenylalanine diet is started before conception, so that the mother-to-be has phenylalanine concentrations close to the normal range (Smith *et al.*, 1990). Good biochemical control must be maintained throughout the pregnancy (Lenke & Levy, 1980). Smith *et al.* (1990) showed that dietary treatment begun after conception did not appear to be of any benefit.

It is recognized that a small but significant number of untreated women with PKU would escape severe brain damage, attain intelligence in the subnormal range or in the lower range of normal, and have children (Hanley *et al.*, 1990). In the North-Western Region between 1970 and 1975 six such mothers were identified with PKU because their children were microcephalic and mentally retarded. The finding of these mothers prompted the introduction of a regional maternal PKU screening programme. The aim was to identify those mothers who were born before the national newborn screening programme was introduced in 1969. For reasons of economy and smooth administration it was decided to link the programme to the antenatal clinic, where blood is normally taken for a number of routine investigations. Since 1977, when the screening was first introduced, eight mothers with classical PKU and eight mothers with moderately elevated plasma phenylalanine concentrations have been identified out of about 500 000 mothers screened. Similar concerns have prompted action in other countries. In the USA, the problem was approached in a different way by routine estimation of phenylalanine in cord blood after delivery. In those mothers in whom PKU or hyperphenylalaninaemia was diagnosed, there was an association between phenylalanine levels in cord blood and the resultant IQ of the child (Levy & Waisbren, 1983).

The maternal PKU screening programme has a natural life span. By the end of the century, all women of child-bearing age will have been screened as newborns, regularly followed up by physicians and counselled about the risks of hyperphenylalaninaemia before conception and during pregnancy.

Carrier testing for inherited disorders

The identification of carriers of recessive traits in an at-risk family, where the primary defect is known, has been achieved by several methods, including the measurement of enzyme levels or documenting abnormal accumulation of metabolites following loading tests. In recent years, the rapid expansion of molecular genetic techniques has made it possible to identify (either directly or indirectly) carriers of an increasing number of mutant alleles. The methods demand highly skilled staff and sophisticated equipment. The aim in determining carrier status is to identify at-risk couples and provide genetic counselling and prenatal diagnosis.

Mass screening for the carrier state for inherited disorders has so far not been widely employed. Access to a whole adult population is very difficult, as there is no clinical indication for the individual to undergo the test. Consequently, the uptake tends to be low. Population screening for carriers has been undertaken for two disorders: cystic fibrosis and Tay–Sachs disease.

Cystic fibrosis

The incidence of cystic fibrosis is high (1 in 2500), the treatment is palliative but the quality and ultimate prognosis for life are improving significantly (Walters *et al.*, 1993). The recent discovery of the common CF gene mutation (△F508), which occurs at a frequency of 80% in CF carriers (Watson *et al.*, 1990), makes it possible to screen for this population. There have been several suggestions concerning ways of implementing such a service. Pilot screening programmes have been tested in two population groups: community-based persons and those attending antenatal clinics.

In the programme piloted by Watson *et al.* (1992) in the primary-care setting, individuals are screened before they have to make reproductive decisions. The advantage of this type of programme is that time is available for contemplation before having to make immediate decisions about a pregnancy. However, in Watson's study of opportunistic screening, the uptake was 66% at the GP clinic and 82% at the family planning clinic. Only 57% of those identified as carriers suggested to their partners that they should also be tested. In addition to poor uptake, the disadvantage is that the information is given outside the context of reproduction and the full meaning may be forgotten when decisions have to be made. Although the majority of people screened shared the information with their partners, about 52% did not disclose information to relatives outside the nuclear family. Administratively, this system would be fraught with difficulties. This model aimed to capture both opportunistically and by invitation; however, invitations were largely ignored.

In an attempt to assess the desire for population carrier screening, medically informed workers in a genetics laboratory were offered the screening test (Flinter *et al.*, 1992). Only 23 out of 110 workers took the opportunity. This report concluded that education alone is not the most important factor in determining the response to population screening.

An alternative screening programme based on primary care may target pregnant women when the first diagnosis of pregnancy is established by the GP. This approach has been piloted in a south Manchester practice in conjunction with the regional genetics unit (Harris *et al.*, 1992). Although this system has been reported to be successful in the practice concerned, it is difficult to imagine the extension of this plan nationally. The programme could not be developed within the present structure of screening and the extra administrative cost would be immense.

The second model (Mennie *et al.*, 1992) refers to the screening of pregnant women attending the antenatal clinic. The administrative arrangement for collecting blood and despatching it to the laboratory would be simple. As a captive and easily accessible population, all women could be offered DNA testing for the common mutation of the CF gene. When a woman is identified as heterozygotic for this mutation she can be privately counselled and testing offered to her partner. If the father is also found to be a carrier, the couple are counselled about the risk to the pregnancy and told that prenatal diagnosis by amniocentesis is available. The disadvantage is that with current obstetric practice, whereby the majority of women are not seen in the antenatal clinic before 10–12 weeks of pregnancy, it would be difficult to avoid second-trimester prenatal diagnosis. In the North-Western Region, where the birth rate is 55 000–60 000, if all expectant mothers were to accept the offer of the test, 16 out of 24 pregnancies with an affected fetus would be identified. Other disadvantages of this model of screening are that the couple are under a pressure of time to make decisions about the pregnancy, and the system prejudices against single mothers where the father cannot be traced.

Tay–Sachs disease

In almost all biochemical genetic disorders, the measurement in the plasma or white cells of enzyme activity alone fails to distinguish reliably a carrier from a normal subject. An exception is Tay-Sachs disease, in which of the two isoenzymes of hexosaminidase, A and B, A is absent. In carriers, isoenzyme A is reduced. The percentage of isoenzyme A of the total hexosaminidase in the plasma or white cells allows the detection of the carrier state, with the white cell assay affording the greater accuracy (Ludman *et al.*, 1986). Automated analytical methods to process large numbers of samples have been developed.

The screening test has been successfully applied for carrier testing in the Ashkenazi Jewish community, in whom the incidence of Tay–Sachs disease is about 1 in 2500 (Goodman, 1979). This gives a carrier rate in this group of 1 in 25 and since partnerships within the culture are common, carrier couples occur in about 1 in every 625 couples screened. As there is no treatment for this aggressive neurodegenerative illness, the aim of screening is to identify at-risk couples with the emphasis being on the prevention of the birth of affected children or the avoidance of planned marriage.

In the Willink unit, the Tay–Sachs screening programme of the Jewish community has been in progress for 2 years and covers the north of England. This privately funded programme employs a physician to: (a) educate the local Jewish population about the risks of disease and alert them to the screening facility on offer; (b) coordinate the sampling of blood specimens, as required; (c) keep records of the results; and (d) counsel couples where necessary. Despite these efforts, only about 2000 individuals have come forward for screening. Detection of the carrier state in selected subpopulations is dependent upon voluntary submission for testing and in such circumstances uptake is generally poor.

Anonymous screening of newborns

Where the risks to public health outweigh individual concerns, it is sometimes considered appropriate for the state to authorize screening without informing those screened. Such anonymous screening is designed to assess the incidence of a particular disease in the community and is presently practised in the UK to establish the frequency of human immunodeficiency virus (HIV) in the newborn population, thereby indirectly determining the incidence in mothers (Ades *et al.*, 1991; Tappin *et al.*, 1991). Since there is no treatment to be offered to those who are HIV positive, it is felt justified not to inform any positive cases and the tests are performed using numbering, rather than naming, systems to preserve this anonymity. Some may feel this is contrary to the rights of the individual, whereas others will espouse the argument of the greater need of the health of the society.

Parental attitudes to screening

Well-meaning and enthusiastic physicians may have a desire to extend screening programmes in the best interests of the health of the society they serve. It is important, however, to consider the wishes of the society. Many of the possibilities made available by the advent of molecular genetics challenge conventional morality and may be disturbing to the public. Several groups have attempted to analyse parental attitudes to screening.

A growing concern arising from the screening of newborns is the effect of false positive results. Since many programmes are designed to minimize the risk of missing a child who may develop a severe neurological disease, there is often an inherent false positive rate. Aside from the costs of reinvestigating these children, there may be a cost to be considered in relation to the damage to the parent–child relationship and the child's subsequent psychological development (Fyro & Bodegard, 1987).

A community-based study involving a questionnaire attempted to evaluate attitudes to CF carrier testing in several regions of England (Williamson *et al.*, 1989). The majority of participants agreed in principle to CF carrier testing. The authors acknowledge that such attitudes expressed may not be borne out by later actions. To assess the actions of people in practice, the uptake of screening services offered in pilot studies can be measured.

In an assessment of carrier screening for sickle cell disease in pregnant women, Rowley *et al.* (1986) found that in response to being identified as a carrier, 65% of women returned for specific counselling, 59% of fathers were tested, 59% of couples at risk opted for amniocentesis, and the experience did not provoke undue anxiety in the majority of mothers.

The psychological impact of being tested for CF heterozygote status was measured in an antenatal screening pilot study in Edinburgh (Mennie *et al.*,

1992). Anxiety was assessed using a questionnaire. Stress was apparent at the time of testing in those positively identified as CF heterozygotes. It disappeared once the partner was identified as negative, and did not return during the remainder of the pregnancy. Among a group positively detected in the community to be CF carriers, outside the domain of pregnancy, 81% were glad they had been tested and 6% were not (Watson *et al.*, 1992).

There are few studies that try to evaluate parental attitudes to newborn screening programmes. Neonatal screening for Duchenne/Becker muscular dystrophies was felt to be appropriate in 88.3% of mothers of healthy newborns and in 80% of parents of affected children (Firth & Wilkinson, 1983). These figures may include a bias since only mothers of healthy newborns were questioned, whereas both the mother and father of affected children replied. Parents rarely refuse to have their children tested in newborn screening programmes (Faden *et al.*, 1982). Nearly all parents whose children suffer from severe genetic disorders wish the diagnosis had been made sooner. Do parents want their children to be screened for disorders such as Alzheimer's or Huntington's disease, which do not manifest themselves until adulthood? We do not know the answer to this question. Although there seems to be no advantage to the individuals concerned to learn that in years to come they will suffer from a disease for which there is no treatment or cure at present, a recent report and accompanying editorial (Hayes, 1992; Wiggins *et al.*, 1992) suggest that there is psychological benefit from either a positive or a negative screening result in families at risk for Huntington's disease.

Conclusion

The majority of screening programmes already in use are designed to measure accumulating metabolites or enzyme activities rather than the primary genetic defect. Recent developments in the field of recombinant DNA technology have generated interest in adopting techniques based on DNA analysis for mass screening. It is at present inconceivable that DNA methodology could replace current biochemical techniques for widespread screening; the expense alone would be prohibitive. In addition, the phenotype of many disorders is recognized to arise from a range of genotypes; for example, over 70 mutations are recognized in PKU. The introduction of DNA screening techniques will depend upon the development of simple methods of scanning the gene for several mutations simultaneously. If such methods became available, then there would be advantages to DNA screening. The tests could be performed on cord blood, allowing the early commencement of treatment, and the risks of false positive and false negative results would be virtually eliminated. Indeed, the other major consideration for the not-too-distant future is the introduction of gene therapy, which, if successful, may greatly alter the number of treatable disorders and therefore have a bearing on the suitability of screening.

The growth in our knowledge and understanding of genetic and environmental disease has enabled the development of screening programmes for the early detection of some disorders. These programmes have, on the whole, been highly successful and enhanced the health of society. At the present time the case for national newborn screening is clear for PKU and congenital hypothyroidism. A case can be made for the introduction of antenatal screening for the carrier status in CF. The evidence for screening in the other diseases discussed remains controversial and must be viewed in the context of the priorities for health in a given community.

We endorse the feelings of the Committee of the National Academy of Sciences (1975) who stated that, 'The limits of screening for inherited disease should be set by the usefulness of the knowledge gained, the costs of obtaining it and its impact on persons tested rather than on the number of tests that can be devised'.

References

Ades, A.E., Parker, S., Berry, T. *et al.* (1991) Prevalence of maternal HIV-1 infection in Thames regions: results from anonymous unlinked neonatal testing. *Lancet* **337**, 1562–1565.

Adjaye, N., Bain, B.J. & Steer, P. (1989) Prediction and diagnosis of sickling disorders in neonates. *Arch Dis Child* **64**, 39–43.

Altland, K. (1977) Screening for abnormal hemoglobins in the newborn: a highly economic procedure using isoelectric focusing. In: *Electrofocusing and Isofachophoresis* (eds B. Radola & D. Grasslin), p. 295. Walter de Gruyter, Berlin.

Basset, P., Beuzard, Y., Garel, M.C. & Rosa, J. (1978) Isoelectric focusing of human hemoglobin: its application to screening, to the characterisation of 70 variants, and to the study of modified fractions of normal hemoglobins. *Blood* **51**, 971–982.

Beutler, E. & Baluda, M.C. (1966) A simple spot screening test for galactosaemia. *J Lab Clin Med* **68**, 137–141.

Bradley, D.M., Parsons, E.P. & Clarke, A.J. (1993) Experience with screening newborns for Duchenne muscular dystrophy in Wales. *Br Med J* **306**, 357–360.

Brown, A.K., Miller, S.T. & Agatisa, P. (1989) Care of infants with disease: the ultimate objective of newborn screening. *Pediatrics* **83** (Suppl.), 897–900.

Chatfield, S., Owen, G., Ryley, H.C. *et al.* (1991) Neonatal screening for cystic fibrosis in Wales and the West Midlands: clinical assessment after five years of screening. *Arch Dis Child* **66**, 29–33.

Committee for the Study of Inborn Errors of Metabolism (1975) *Genetic Screening: Programs, Principles and Research.* National Academy of Sciences, Washington DC.

Cummins, D., Heuschkel, P. & Davies, S.C. (1991) Penicillin prophylaxis in children with sickle cell disease in Brent. *Br Med J* **302**, 989–990.

Danner, D.J. & Elsas, L.J. (1989) Disorders of branched chain amino acid metabolism. In: *The Metabolic Basis of Inherited Metabolic Disease* (eds C.R. Scriver, A.L. Beaudet, W.S. Sly & D. Valle), 6th edn, pp. 671–692. McGraw-Hill, New York.

Dauphinais, R.M., Ganeshananthan, M. & Jezyk, P. (1986) Cystic fibrosis: early detection and clinical course. In: *Genetic Disease: Screening and Management* (eds T.P. Carter & A.M. Willey), pp. 65–79. Alan R. Liss, New York.

Davies, S.C. (1993) Bone marrow transplant for sickle cell disease – the dilemma. *Blood Rev* **7**, 4–19.

Davies, S.C. & Wonke, B. (1991) The management of haemoglobinopathies. In: *Clinical Haematology* (ed. I. Hann), **4**(2), p. 361. Baillière Tindall, London.

Evans, D.I.K. & Blair, V.M. (1976) Neonatal screening for haemoglobinopathy. Results in 7691 Manchester newborns. *Arch Dis Child* **51**, 127–130.

Faden, R.R., Chevalow, A.J., Holtzman, N.A. & Horn, S.D. (1982) A survey to evaluate parental consent as public policy for neonatal screening. *Am J Publ Hlth* **72**, 1347–1352.

Firth, M.A. & Wilkinson, E.J. (1983) Screening the newborn for Duchenne muscular dystrophy: parent's view. *Br Med J* **286**(1), 1933–1934.

Flinter, F.A., Silver, A., Mathew, C.G. & Bobrow, M. (1992) Population screening for cystic fibrosis. *Lancet* **339**, 1539–1540.

Fosburg, M.T. & Nathan, D.G. (1990) Treatment of Cooley's anemia. *Blood*, **76**, 435–444.

Fuggle, P.W., Grant, D.B., Smith, I. & Murphy, G. (1991) Intelligence, motor skills and behaviour at 5 years in early-treated congenital hypothyroidism. *Eur J Pediatr* **150**, 570–574.

Fyro, K. & Bodegard, G. (1987) Four-year follow-up of psychological reactions to false positive screening tests for congenital hypothyroidism. *Acta Paed Scand* **76**, 107–114.

Galacteros, F., Kleman, K., Caburi-Martin, J. *et al.* (1980) Cord blood screening for hemoglobin abnormalities by thin layer isoelectric focusing. *Blood*, **56**, 1068–1071.

Garrick, M.D., Dembure, P. & Guthrie, R. (1973) Sickle cell anemia and other hemoglobinopathies: procedures and strategies for screening employing spots of blood on filter paper as specimens. *N Engl J Med* **288**, 1265–1268.

Garver, F.A., Singh, H., Moscoso, H. *et al.* (1984) Identification of normal and variant haemoglobins after electrophoretic separation and transfer to nitrocellulose membranes. *Haemoglobin*, **8**, 105–115.

Gaston, M.H., Verter, J.I., Woods, G. *et al.* (1986) National co-operative study: penicillin. Prophylaxis with oral penicillin in children with sickle cell anemia. *N Engl J Med* **314**, 1593–1599.

Goodman, R.M. (1979) *Genetic Disorders Among the Jewish People.* Johns Hopkins University Press, Baltimore.

Griffiths, P.D., Mann, J.R., Darbyshire, P.J. & Green, A. (1988) Evaluation of eight and a half years of neonatal screening for haemoglobinopathies in Birmingham. *Br Med J* **296**, 1583–1585.

Guthrie, R. & Susi, A. (1963) A simple phenylalanine method for detecting phenylketonuria in large populations of newborn infants. *Pediatrics* **32**, 338–343.

Hanley, W.B., Clarke, J.T.R. & Schoonheyt, W.E. (1990) Undiagnosed phenylketonuria in adult women: a hidden public health problem. *Can Med Assoc J* **143**(6), 513–516.

Harris, H.J., Scotcher, D., Craufurd, D., Wallace, A. & Harris, R. (1992) Cystic fibrosis carrier screening at first diagnosis of pregnancy in general practice. *Lancet* **339**, 1539.

Hayes, C.V. (1992) Genetic testing for Huntington's disease: a family issue. *N Engl J Med* **327**, 1449–1451.

Heard, G.S., Secor McVoy, J.R. & Wolf, B. (1984) A screening method for biotinidase deficiency in newborns. *Clin Chem* **30**, 125–127.

Henthorn, J., Anionwu, E. & Brozovic, M. (1984) Screening cord blood for sickle haemoglobinopathies in Brent. *Br Med J* **289**, 479–480.

Holton, J.B. (1990) Galactose disorders: an overview. *J Inher Metab Dis* **13**(4), 476–486.

Jinks, D.C., Minter, M., Tarver, D.A. *et al.* (1989) Molecular genetic diagnosis of sickle cell disease using dried blood spot specimen on blotters used for newborn screening. *Hum Genet* **81**, 363–366.

Kawamura, M. (1987) Neonatal screening for galactosaemia in Japan. In: *Advances in Neonatal Screening* (ed. B.L. Therrell, Jr.), pp. 227–230. Elsevier, Amsterdam.

Kazazian, H. & Boehm, C. (1988) Molecular basis and prenatal diagnosis of beta thalassaemia. *Blood* **72**, 1107–1116.

Kennedy, R., Girdwood, R.W. & King, M.D. (1989) Neonatal screening for biotinidase deficiency. A pilot study in Scotland. *J Inher Metab Dis* **12**(3), 344–345.

Kinney, T.R., Sawtschenko, M., Whorton, M. *et al.* (1989) Techniques, comparison and report of the North Carolina experience. *Pediatrics* **83** (Suppl.), 843–848.

Kleman, K.M., Vichinsky, E. & Lubin, B.H. (1989) Experience with newborn screening using isoelectric focusing. *Pediatrics* **83** (Suppl.), 852–854.

Lawler, M.G., Frederick, D.L., Rodriguez-Anza, S., Wolf, B. & Levy, H.L. (1992) Newborn screening for biotinidase deficiency: pilot study and follow-up of identified cases. *Screening* **1**, 37–47.

Leiken, S.L., Gallagher, D., Kinney, T.R. *et al.* (1989) Co-operative Study of Sickle Cell Disease. Mortality in children and adolescents with sickle cell disease. *J Pediatr* **84**, 500–508.

Lenke, R.R. & Levy, H.L. (1980) Maternal phenylketonuria and hyperphenylalaninaemia: an international survey of the outcome of untreated and treated pregnancies. *N Engl J Med* **303**, 1202–1208.

Levy, H.L. & Waisbren, S.E. (1983) Effects of untreated maternal phenylketonuria and hyperphenylalaninaemia on the fetus. *N Engl J Med* **309**, 1269–1274.

Listernick, R., Frisone, L. & Silverman, B.L. (1992) Delayed diagnosis of infants with abnormal neonatal screens. *J Am Med Assoc* **267**, 1095–1099.

Losek, J.D. (1991) Sickle cell screening practice in paediatric emergency departments. *Pediatr Emer Care* **7**, 278–280.

Lucarelli, G. & Weatherall, D.J. (1991) Bone marrow transplantation for severe thalassaemia. *Br J Haem* **78**, 300–303.

Ludman, M.D., Grabowski, G.A., Goldberg, J.D. & Desnick, R.J. (1986) Heterozygote detection and prenatal diagnosis for Tay–Sachs disease. In: *Genetic Disease: Screening and Management* (eds T.P. Carter & A.M. Willey), pp. 19–48. Alan R. Liss, New York.

MacCready, R.A. & Levy, H.L. (1972) The problem of maternal phenylketonuria. *Am J Obstet Gynecol* **113**, 121–128.

Medical Research Council Working Party on Phenylketonuria (1993a) Phenylketonuria due to phenylalanine hydroxylase deficiency: an unfolding story. *Br Med J* **306**, 115–119.

Medical Research Council Working Party on Phenylketonuria (1993b) Recommendations on the dietary management of phenylketonuria. *Arch Dis Child* **68**, 426–427.

Mennie, M.E., Gilfillan, A., Compton, M. *et al.* (1992) Prenatal screening for cystic fibrosis. *Lancet* **340**, 214–216.

Missiou-Tsagaraki, S. (1991) Screening for glucose-6-phosphate dehydrogenase deficiency as a preventive measure: prevalence among 1 286 000 Greek newborn infants. *J Pediatr* **119**, 293–299.

Mittman, C.H., Barbella, T. & Lieberman, J. (1973) Antitrypsin deficiency and abnormal protease inhibitor phenotypes. *Arch Environ Hlth* **27**, 201–206.

Ng, W.G., Kawamura, M. & Donnell, G.N. (1987) Galactosaemia screening: methodology and outcome from worldwide data collection. In: *Advances in Neonatal Screening* (ed. B.L. Therrell, Jr.), pp. 243–249. Elsevier, Amsterdam.

Orfanos, A.P., Naylor, E.W. & Guthrie, R. (1982) Screening test for α-1-antitrypsin in dried blood specimens. *Clin Chem* **28**(4), 615–617.

Paigen, K., Pacholec, F. & Levy, H.L. (1982) A new method of screening for inherited disorders of galactose metabolism. *J Lab Clin Med* **99**, 895–907.

Pang, S., Hotchkiss, J., Drash, A.L. *et al.* (1977) Microfilter paper method for 17-hydroxyprogesterone radioimmunoassay: its application for rapid screening for congenital adrenal hyperplasia. *J Clin Endocrinol Metab* **45**, 1003–1008.

Pang, S., Wallace, M.A., Hofman, I. *et al.* (1988) Worldwide experience in newborn screening for classical congenital adrenal hyperplasia due to 21-hydroxylase deficiency. *Pediatrics* **81**, 866–874.

Pasvol, G. & Wilson, R.J.M. (1982) Red cells and malaria. *Br Med Bull* **38**, 133–140.

Pearson, H.A., O'Brien, R.T., McIntosh, S., Aspnes, G.T. & Yang, M.-M. (1974) Routine screening of umbilical cord blood for sickle cell diseases. *J Am Med Assoc* **227**, 420–421.

Pharaoh, P.O.D. & Mappon, M.P. (1992) Audit of screening for congenital hypothyroidism. *Arch Dis Child* **67**, 1073–1076.

Plauchu, H., Dorche, C., Cordier, M.P., Guibaud, P. & Robert, J.M. (1989) Duchenne muscular dystrophy: neonatal screening and prenatal diagnosis. *Lancet* **1**, 669.

Powars, D., Schroeder, W.A. & White, L. (1975) Rapid diagnosis of sickle cell disease at birth by microcolumn chromatography. *Pediatrics* **55**, 630–635.

Riordan, J.R., Rommens, J.M., Kerem, B.-S. *et al.* (1989) Identification of the cystic fibrosis gene: cloning and characterisation of complementary DNA. *Science* **245**, 1066–1072.

Rommens, J.M., Iannuzi, M.C., Kerem, B.-S. *et al.* (1989) Identification of the cystic fibrosis gene: chromosome walking and jumping. *Science* **245**, 1059–1065.

Rovet, J.F., Sorbara, D.-L. & Ehrlich, R.M. (1986) The

intellectual and behavioural characteristics of children with congenital hypothyroidism identified by neonatal screening in Ontario. The Toronto prospective study. In: *Genetic Disease: Screening and Management* (eds T.P. Carter & A.M. Willey), pp. 281–315. Alan R. Liss, New York.

Rowley, P.T., Loader, S. & Walden, M. (1986) Response of pregnant women to haemoglobinopathy carrier identification. In: *Genetic Disease: Screening and Management* (eds T.P. Carter & A.M. Willey), pp. 151–172. Alan R. Liss, New York.

Sackett, D.L. & Holland, W.W. (1975) Controversy in the detection of disease. *Lancet* **ii**, 357–359.

Sardharwalla, I.B. & Wraith, J.E. (1987) Galactosaemia. *Nutr Hlth* **5**, 175–188.

Schedlbauer, L.M. & Pass, K.H. (1989) Cellulose acetate/citrate agar electrophoresis of filter paper haemolysates from heel stick. *Pediatrics* **83** (Suppl.), 839–842.

Schmidt, R.M., Brosious, E.M., Holland, S. *et al.* (1976) Use of blood specimens collected on filter paper in screening for abnormal haemoglobins. *Clin Chem* **22**, 685–687.

Scott, R.B. & Harrison, D.L. (1982) Screening of the umbilical cord blood for sickle cell disease. *Am J Pediatr Hem Onc* **4**, 202–205.

Scriver, C.R., Davies, E. & Cullen, A.M. (1964) Application of a simple micromethod to the screening of plasma for a variety of aminoacidopathies. *Lancet* **ii**, 230–232.

Serjeant, G.R. (1992) *Sickle Cell Disease*, 2nd edn. Oxford University Press, Oxford.

Shickle, D. & May, A. (1989) Knowledge and perceptions of haemoglobinopathy carrier screening among general practitioners in Cardiff. *J Med Genet* **26**, 109–112.

Skogerboe, K.J., West, S.F., Murillo, M.D. *et al.* (1991) Genetic screening of newborns for sickle cell disease: correlation of DNA analysis with haemoglobin electrophoresis. *Clin Chem* **37**, 454–458.

Smith, I., Cook, B. & Beasley, M. (1991) Review of neonatal screening programme for phenylketonuria. *Br Med J* **303**, 333–335.

Smith, I., Glossop, J. & Beasley, M. (1990) Fetal damage due to maternal phenylketonuria: effects of dietary treatment and maternal phenylalanine concentrations around the time of conception. *J Inher Metab Dis* **13**, 651–657.

Snyderman, S.E. (1986) Newborn metabolic screening: follow-up and treatment results. In: *Genetic Disease: Screening and Management* (eds T.P. Carter & A.M. Willey), pp. 195–209. Alan R. Liss, New York.

Starfield, B. & Holtzman, N.A. (1975) A comparison of effectiveness of screening for phenylketonuria in the United States, United Kingdom and Ireland. *N Engl J Med* **293**, 118–121.

Sternberg, W.N., Carter, T.P., Humbert, J.R. & Rowley, P.T. (1982) Newborn screening for haemoglobinopathies in New York State: experience of physicians and parents of affected children. *J Pediatr* **100**, 373–377.

Tan-Wilson, A.L., Reichlin, M. & Noble, R.W. (1976) Isolation and characterisation of low and high affinity goat antibodies directed to single antigenic sites on human haemoglobin. *Immunochemistry* **12**, 921–927.

Tappin, D.M., Girdwood, R.W.A., Follett, E.A.C. *et al.* (1991) Prevalence of maternal HIV infection in Scotland based on unlinked anonymous testing of newborn babies. *Lancet* **337**, 1565–1567.

Thompson, A.L., Smith, I., Brenton, D. *et al.* (1990) Neurological deterioration in young adults with phenylketonuria. *Lancet* **336**, 602–605.

Tsevat, J., Wong, J.B., Pauker, S.G. & Steinberg, M.H. (1991) Neonatal screening for sickle cell disease: a cost-effectiveness analysis. *J Pediatr* **118**, 546–554.

Vermylen, C. & Cornu, G. (1993) Bone marrow transplantation in sickle cell anaemia. *Blood Rev* **7**, 1–3.

Walters, S., Britton, J. & Hodson, M.E. (1993) Demographic and social characteristics of adults with cystic fibrosis in the United Kingdom. *Br Med J* **306**, 549–552.

Wastell, H.J., Bartlett, K., Dale, G. & Shein, A. (1988) Biotinidase deficiency: a survey of 10 cases. *Arch Dis Child* **63**, 1244–1249.

Watson, E.K., Mayall, E.S., Lamb, J., Chapple, J. & Williamson, R. (1992) Psychological and social consequences of community carrier screening programme for cystic fibrosis. *Lancet* **340**, 217–220.

Watson, E.K., Mayall, E.S., Simova, L. *et al.* (1990) The incidence of delta F508 CF mutation, and associated haplotypes, in a sample of English CF families. *Hum Genet* **85**, 435–436.

Weatherall, D.J. & Clegg, J.B. (1981) *The Thalassaemia Syndromes*, 3rd edn. Blackwell Scientific Publications, Oxford.

Whitten, C.F. (1973) Sickle cell programming – an imperiled promise. *N Engl J Med* **288**, 318–319.

Wiggins, S., Whyte, P., Huggins, M. *et al.* (1992) The psychological consequences of predictive testing for Huntington's disease. *N Engl J Med* **327**, 1401–1405.

Wilcken, B. & Brown, A.R.D. (1987) Screening for cystic fibrosis in New South Wales, Australia: evaluation of the results of screening 400 000 babies. In: *Advances in Neonatal Screening* (ed. B.L. Therrell, Jr.), pp. 385–390. Elsevier, Amsterdam.

Wilcken, B. & Chalmers, G. (1985) Reduced morbidity in patients with cystic fibrosis detected by newborn screening. *Lancet* **2**, 1319–1321.

Williamson, R., Allison, M.E.D., Bentley, T.J. *et al.* (1989) Community attitudes to cystic fibrosis carrier testing in England: a pilot study. *Prenat Diag* **9**, 727–734.

Wilson, J.B., Headlee, M.E. & Huisman, T.H.J. (1983) A new, high performance liquid chromatographic procedure for the separation and quantitation of various hemoglobin variants in adults and newborn babies. *J Lab Clin Med* **102**, 174–186.

Wolf, B. & Heard, G.S. (1990) Screening for biotinidase deficiency in newborns: worldwide experience. *Pediatrics* **85**, 512–517.

Wolf, B. Heard, G.S., Weissbecker, K.A. *et al.* (1985) Biotinidase deficiency: initial clinical features and rapid diagnosis. *Ann Neurol* **18**, 614–617.

Worton, R.G. (1992) Duchenne muscular dystrophy: gene and gene product; mechanism of mutation in the gene. *J Inher Metab Dis* **15**(4), 539–550.

Zhang, Y.H. & McCabe, E.R. (1992) RNA analysis from newborn screening dried blood specimens. *Hum Genet* **89**, 311–314.

9: Growth and Its Problems

C. G. D. BROOK

Introduction

The tripartite nature of growth is clearly shown in Fig. 9.1. There is a rapid and rapidly decelerating phase of growth in infancy, a steady and slowly decelerating phase of growth in childhood, interrupted by the mid-childhood growth spurt, and an adolescent growth spurt. Since these phases of growth have different endocrine control mechanisms, the approach to the investigation of a child with growth problems must take them into account.

The endocrine control of growth

Consideration of the presentation of patients with intrauterine growth retardation, of infants born to a diabetic mother, of infants starved in the first year of life and of infants becoming obese during the first year of life will confirm the primacy of nutrition in the control of infantile growth. Children with congenital hypothyroidism or anencephaly have a virtually normal length at birth and, although severe panhypopituitarism may interfere with growth during the first year of life, growth hormone (GH) insufficiency is mainly a condition seen in childhood.

As soon as GH receptors become observable in the human infant (around 8 months of age), the rate at which such a child grows is determined by the amplitude of pulsatile GH secretion. Children who are destined to become tall grow more quickly than smaller children and have correspondingly greater amplitude of GH pulses. Children who are small but normal will have less GH, and GH insufficiency cannot be defined by absolute criteria because of the continuous nature of both GH secretion and growth rate.

The mid-childhood growth spurt is difficult to define in individual children without continuous observation. It results from the secretion of adrenal androgens from the zona reticularis of the adrenal cortex, which is, in turn, under the control of corticotrophin-releasing hormone (CRH) – adrenocorticotrophic hormone (ACTH). There is no evidence for an adrenal androgen-stimulating hormone that is separate from ACTH and adrenal androgens are clearly ACTH stimulable and dexamethasone suppressible.

Puberty is the outward and visible sign of sex steroid secretion secondary to pulsatile gonadotrophin secretion. In the newborn period there is exuberant gonadotrophin secretion; for reasons which are not at all clear, the amplitude of gonadotrophin secretion is not maintained during early childhood. What inhibits it remains to be demonstrated but, in children aged 5 to 7 years, the hypothalamo-pituitary-gonadal axis is relatively quiescent in the sense that there are occasional spikes of luteinizing hormone (LH) and follicle-stimulating hormone (FSH), demonstrable on some nights in a week, but such peaks are of small amplitude and few in nature. During the years 7 to 10, pulsatile gonadotrophin secretion at night becomes more evident and, as it becomes regular, it results in increasing the steroid secretion from the ovaries and testes. When oestradiol is present in amounts sufficient to bring about breast development, it amplifies the pulsatile GH secretion and brings about the increase in growth velocity seen in early female puberty. The concentration of testosterone needs to be higher to have a similar effect and at such a level testosterone is itself a powerful anabolic agent. Thus, although the signs of gonadal secretion in terms of pubertal

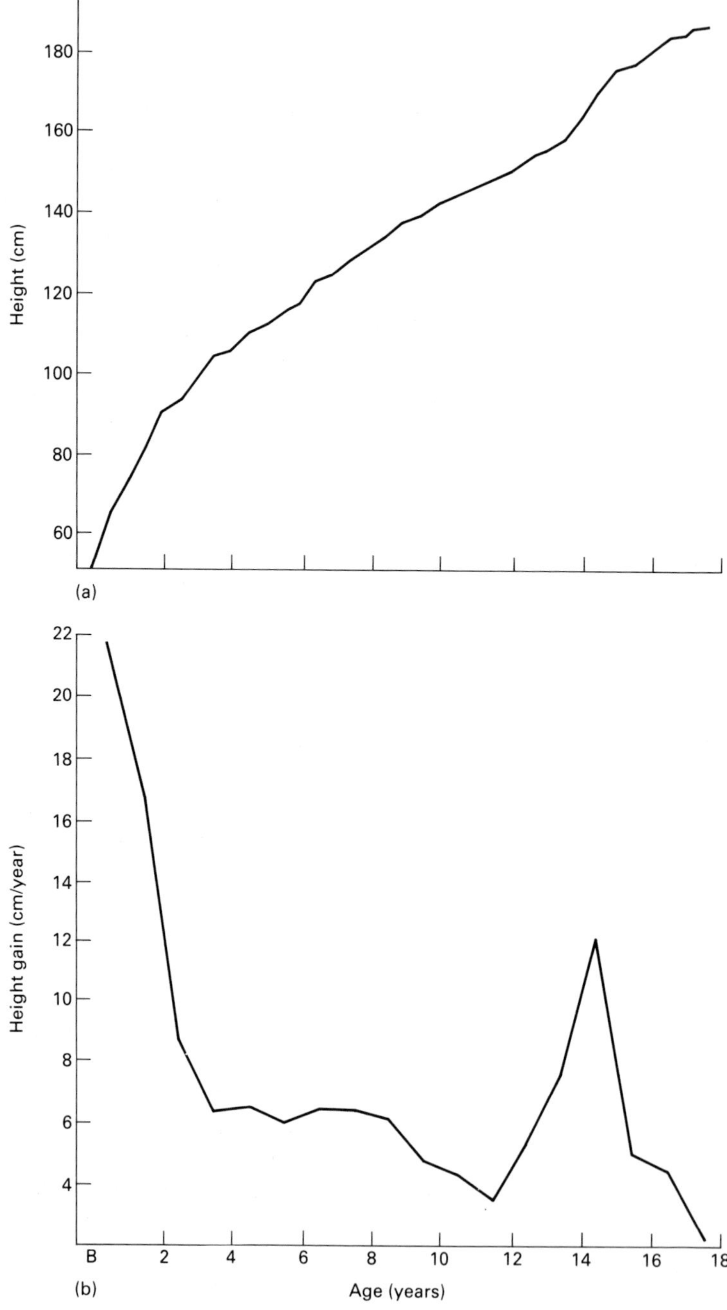

Fig. 9.1 The human growth curve. (a) The height, year on year, is plotted against chronological age. (b) The growth rate is plotted against chronological age. From this the three phases of growth in infancy, childhood (with the mid-childhood growth spurt) and puberty can be clearly seen.

development are contemporaneous in boys and girls, the growth spurt in boys does not occur until later in the sequence of the events of puberty, and thus men generally become taller than women.

These different control mechanisms define the approach to the application of chemical pathology to growth disorders in children.

Investigating growth disorders in infants

Hypopituitarism in infancy will generally present as hypoglycaemia (due to ACTH or GH deficiency), hyperbilirubinaemia (unconjugated due to thyroid-stimulating hormone (TSH) deficiency, conjugated due to ACTH deficiency) or micropenis (due to LH deficiency). It is very unlikely that a low growth velocity at this age is attributable to causes other than nutritional. Therefore, in a child who is failing to thrive tests of gastrointestinal function take precedence over tests of endocrine function.

The concept of 'failure to thrive' misleads many clinicians. In order to qualify for an abnormality which is diagnosable and treatable, the process has to be continuing. Thus, the child born inappropriately small for gestational age but who grows normally during the early weeks of life will remain extremely small but does not warrant investigation because the damage was done some time in the past. Clinicians should be warned that only a low growth velocity is appropriate for investigation because treatments can only affect growth velocity. In other words, failing to thrive is a cause for concern but past failure to thrive is not an appropriate indication for investigation.

Over-nutrition in infancy leads to an accelerated growth rate and this will not be recognizable by any of the conventional tests. It is not worth investigating GH secretion in an infant growing excessively quickly – a more likely cause is to be found in activation of the hypothalamo-pituitary-gonadal axis, that is in precocious puberty.

Investigating growth disorders in children

The child who is growing normally needs no further investigation. There may be an explanation for the tall or short stature. There may be a question of discussing how final stature can be manipulated but a normal child is a normal child and should not be assaulted. Investigating normal children is an assault – and an expensive one at that.

When a child is growing excessively quickly, the probability is that he or she has early puberty. Hyperthyroidism should be excluded clinically or by the measurement of thyroxine and TSH concentrations. The biochemical investigation of early puberty is less important than imaging (ultrasound and possibly magnetic resonance imaging (MRI)): the important thing from the point of view of chemical pathology is to define whether early puberty and an accelerated growth spurt are gonadotrophin dependent or independent. In the latter case, gonadotrophin levels will always be low, whereas in the former, they will be measurable. Excessive secretion of GH from an eosinophil adenoma is very rare: the biochemical hallmark is the finding of measurable concentrations of GH at all times, including after a glucose load.

For children who are becoming increasingly short, the whole gamut of paediatric medicine is potentially responsible. Endocrine disorders classically associated with poor growth (hypothyroidism, Cushing's syndrome, disorders of calcium metabolism, Turner's syndrome) can easily be excluded. The clinician has also to be careful to exclude cardiovascular, renal, respiratory, gastrointestinal, neurological, psychological and psychosocial causes of failure to grow. If all these have been excluded, with or without the help of special tests, the probability is that a slow-growing child who is otherwise normal has a problem with GH secretion.

Tests of GH secretion (see below) are difficult to perform and more difficult to interpret. They are not required to predict the effects of treatment with GH so their use has to be highly selective. If the clinician is stringent in the choice of criteria for pursuing investigation of a child growing slowly, the number of tests of GH secretion which will be required in an individual paediatric practice will be extremely few, so few that neither the laboratory nor the clinical staff will have sufficient experience to undertake them properly. Such tests should therefore be performed in centres specialized in their employment, i.e. in departments of paediatric endocrinology.

Investigating growth disorders in puberty

In a pubertal child growing excessively quickly, Klinefelter syndrome (and its variants) and GH excess leading to pituitary gigantism need to be excluded. The latter is difficult to diagnose without performing a 24-h GH profile for which the indication would be a persistently raised level of plasma GH concentration which was not suppressible by glucose administration. This is not to say that an exuberant normal secretion of GH may not require manipulation, but the important diagnosis to exclude is incipient acromegaly.

Children who are small in puberty may be growing at a rate which may or may not be appropriate for their stage of puberty. This must be established before further investigation is undertaken. A boy of 15 years with signs of early puberty may be expected to grow extremely slowly: investigation will almost certainly reveal an 'abnormality' in GH secretion but such a patient does not need GH. He needs sex steroids. There are some who would advocate the administration of sex steroids before the performance of a GH stimulation test (a primed test) but it is rare that such a test is actually required. It seems much more sensible to apply a small dose of sex steroids and measure the increment in growth velocity, which will be what the patient requires, rather than to investigate the hormonal situation on paper. If the administration of sex steroids fails to elicit an increase in growth rate, the patient requires comprehensive hypothalamo-pituitary evaluation.

For a girl who is small at pubertal age, the important diagnosis to exclude is Turner's syndrome and, in terms of chemical pathology, this is best done by measuring the gonadotrophin concentrations. Gonadotrophin and sex steroid concentrations should be at a very low level in both sexes in early puberty. If the gonadotrophins are elevated, the possibility of gonadal dysgenesis is correspondingly increased.

Tests used in the diagnosis of growth problems

The hypothalamo-pituitary-target gland axes are shown in Fig. 9.2. Assays are available for almost all the hormones mentioned but only some are used in routine clinical practice.

Hypothalamic hormones

Hypothalamic hormones cannot be measured because of the discrete nature of the communication between the hypothalamus and the pituitary. They may, however, be used as stimuli of the pituitary hormones and can be extremely useful in this respect.

Pituitary hormones

Assays of pituitary hormones are common and a sampling protocol in routine use for testing pituitary function is shown in Table 9.1. The administration of thyrotrophin-releasing hormone (TRH) to clarify hypothyroid disorders is useful (Fig. 9.3).

The use of CRH to amplify ACTH response has not been very helpful in children. It has been employed in the diagnosis of Cushing's syndrome in adults to localize the side of a pituitary microadenoma but this has not been profitable in children because of the vascular nature of the communication from side to side in the pituitary gland.

The use of gonadotrophin-releasing hormone (GnRH) to amplify gonadotrophin secretion is useful in disorders of puberty to separate gonadotrophin-dependent from gonadotrophin-independent precocious puberty and to identify those children in whom a gonadotrophin response cannot be elicited owing to gonadotrophin deficiency. It does not usefully differentiate constitutional delay of puberty from hypogonadotrophic hypogonadism.

Tests of GH secretion (Table 9.2) have their problems but their use is likely to be quite closely restricted.

The measurement of prolactin is an important guide to pituitary pathology. Where there is functional disconnection between the hypothalamus and the pituitary, the prolactin level is often moderately elevated. Where it is extremely elevated, a prolactinoma is a possibility. Where it is undetectable and unstimulable by TRH, a serious possibility of a pituitary lesion is raised.

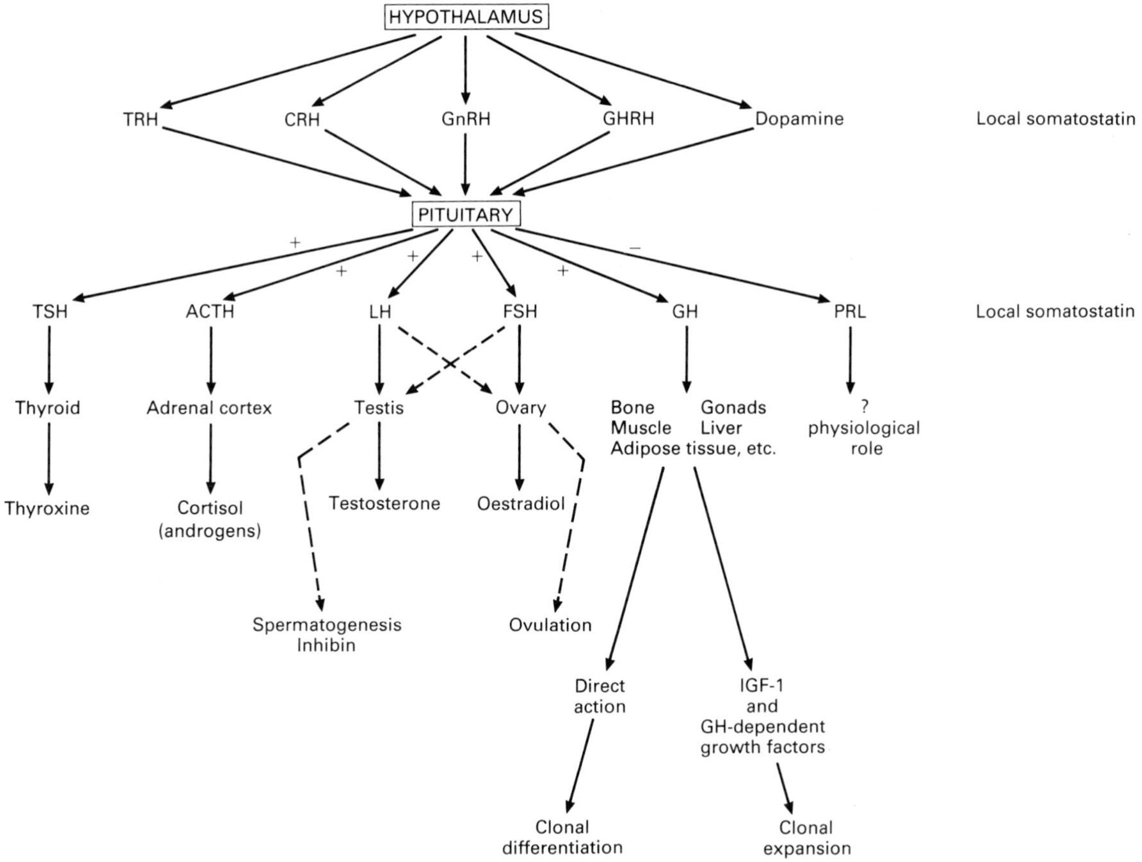

HYPOTHALAMUS
TRH
CRH
GnRH
GHRH
Dopamine
Local somatostatin
PITUITARY
+
+
+
+
+
−
TSH
ACTH
LH
FSH
GH
PRL
Local somatostatin
Thyroid
Adrenal cortex
Testis
Ovary
Bone
Gonads
Muscle
Liver
Adipose tissue, etc.
?
physiological
role
Thyroxine
Cortisol
(androgens)
Testosterone
Oestradiol
Spermatogenesis
Inhibin
Ovulation
Direct
action
IGF-1
and
GH-dependent
growth factors
Clonal
differentiation
Clonal
expansion

Fig. 9.2 Hypothalamo-pituitary-target gland axes. TRH, thyrotrophin-releasing hormone; CRH, corticotrophin-releasing hormone; GnRH, gonadotrophin-releasing hormone; GHRH, growth-hormone-releasing hormone; TSH, thyroid-stimulating hormone; ACTH, adrenocorticotrophic hormone; LH, luteinizing hormone; FSH, follicle-stimulating hormone; GH, growth hormone; PRL, prolactin; IGF, insulin-like growth-factor.

Serum TSH
Normal
range
0
20
60
Time after injection (min)
Injection of TRH

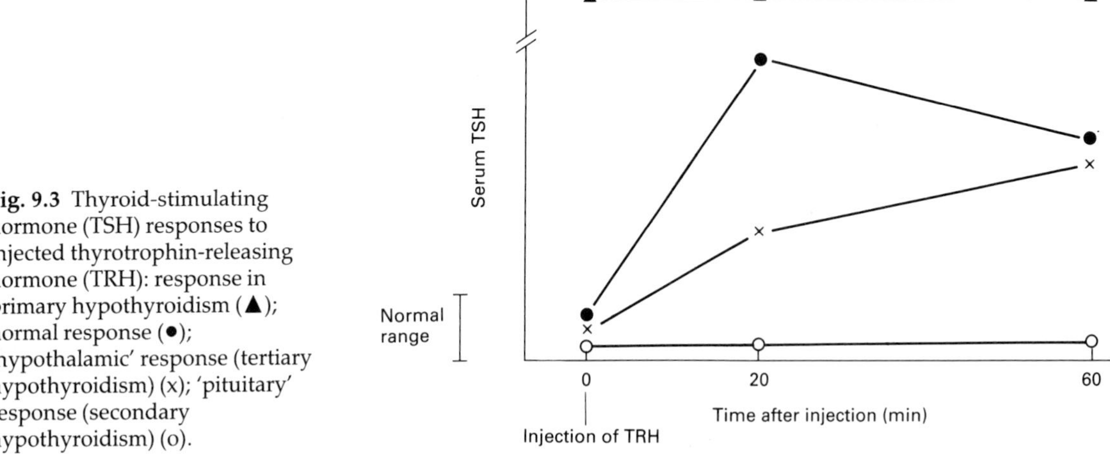

Fig. 9.3 Thyroid-stimulating hormone (TSH) responses to injected thyrotrophin-releasing hormone (TRH): response in primary hypothyroidism (▲); normal response (●); 'hypothalamic' response (tertiary hypothyroidism) (x); 'pituitary' response (secondary hypothyroidism) (o).

Table 9.1 Sampling protocol for the assessment of hypothalamo-pituitary function. After time 0 inject: soluble insulin (0.15 iu/kg) (use 0.1 iu/kg in cases of hypopituitarism and after cranial surgery or radiotherapy, which includes total body irradiation). Gonadotrophin-releasing hormone (GnRH) (2.5 μg/kg) up to a maximum dose of 100 μg. Thyrotrophin-releasing hormone (TRH) (7 μg/kg) up to a maximum dose of 200 μg

	Fluoride oxalate samples	Clotted samples							
Time (min)	Glucose	GH	Cortisol	TSH	LH	FSH	Prolactin	T_4	OE_2/testo.
0	+	+	+	+	+	+	+	+	+
20	+			+	+	+	+		
30	+	+	+						
60	+	+	+	+	+	+	+		
90	+	+	+						
120	+	+	+						

GH, growth hormone; TSH, thyroid-stimulating hormone; LH, luteinizing hormone; FSH, follicle-stimulating hormone; OE_2 oestradiol; testo., testosterone.

Table 9.2 Tests of growth hormone secretion

Type of test	Common tests available	Comments
Screening	IGF 1 IGF-binding proteins Exercise Urinary GH Serum or urine collagen markers Continuous GH withdrawal sampling	Can all detect excessively high or low levels of GH; no useful discriminant function in clinically doubtful cases
Physiological	24-h profile or sleep with intermittent sampling	Complex and expensive to perform and/or analyse results
Pharmacological	Insulin-induced hypoglycaemia Arginine Glucagon Glucose L-Dopa Clonidine Prostaglandin E_2 Bombesin Galanin GHRH	Wide within and between subject variability; all stimuli test readily releasable pool of GH (except possibly high-dose GHRH) so high false positive and negative results, depending on cut-off values; best sensitivity in detecting true positive results and specificity in excluding true negative results not more than 80%

IGF, insulin-like growth factor; GH, growth hormone; GHRH, growth-hormone-releasing hormone.

Thyroid function

In paediatric practice, basal total thyroxine concentration is the only required measurement. If the value is low or high, sensitive measurements of TSH concentration should reveal whether the patient is suffering from thyrotoxicosis or hypothyroidism.

Adrenal function

Because of the pulsatile nature of ACTH secretion and its diurnal variation, basal concentrations of serum cortisol are not helpful unless they are very high (above 600 mmol/L) or very low (less than 100 mmol/L), in which case simultaneous measurement of plasma ACTH concentration is required for the diagnosis of Cushing's syndrome or Addison's disease. These will rarely present simply as growth problems.

Gonadal steroids

Values of oestradiol and testosterone are useless in pubertal children unless targeted assays are used. If they are employed, they should be compared with simultaneous measurements of plasma gonadotrophin concentration.

Insulin-like growth factor (IGF) 1 and IGF-binding proteins

The measurement of these can reflect GH secretion but they are not as relevant as the measurement of growth velocity. IGF-binding protein 3 is GH dependent but IGF 1 and the other IGF-binding proteins are also regulated by nutritional status. They are not useful for the elucidation of growth problems in children.

10: Endocrine Disorders

J.W. HONOUR

Introduction

This chapter covers the contribution of the laboratory to the investigations of children with adrenal, gonadal and thyroid disorders. Some recent textbooks of paediatric endocrinology are recommended for more detailed information (Brook, 1989; Collu *et al.*, 1989).

Adrenal and gonadal disorders

Introduction

Diagnostic investigations of adrenocortical and gonadal function are usually based upon a series of blood hormone measurements. Steroid assays based on chemical reactions with functional groups, although still used, have in many laboratories been removed from the repertoire of tests (Rudd, 1983). Specific assays based on immunological methods (radioimmunoassay, RIA) for individual steroids are now widely accepted in endocrine practice. The pituitary-gonadal and pituitary-adrenal axes are assessed by measurement of the plasma levels of trophic and target gland hormones. The relevance and limitations of basal tests and confirmatory tests will be considered in relation to the diagnosis of clinical problems in paediatrics. Tests based on glandular stimulation and suppression are part of the clinical armamentarium for investigating patients with endocrine disorders. There is now little indication for the use *in vivo* of radioactive tracer steroids for the determination of secretion rates. Abnormalities of steroid receptors (Armanini *et al.*, 1985; Lipsett *et al.*, 1985; Lubahn *et al.*, 1989) and of the binding proteins (Barragry *et al.*, 1980; Ahrensten *et al.*, 1982; Baelen *et al.*, 1982; Roitman *et al.*, 1984) can be added to the disorders of steroid production.

A chemical pathology laboratory can meet most of the clinical requests for steroid hormones in sick children by providing, or having access to, assays for cortisol, testosterone, oestradiol, dehydroepiandrosterone sulphate (DHAS) and aldosterone. Assays for testosterone and oestradiol should be sensitive to 0.1 nmol/L and 10 pmol/L, respectively. A laboratory serving a large paediatric practice will need to provide assays for 17α-hydroxyprogesterone (17-OH-P), dihydrotestosterone (DHT), androstenedione (A_4) as well as certain peptide hormones (luteinizing hormone (LH), follicle-stimulating hormone (FSH), adrenocorticotrophin and plasma renin activity), possibly using regional, national or private facilities, particularly if results for cortisol precursors are required to support the diagnosis of one of the rare inborn errors of steroid metabolism in a newborn child with ambiguous genitalia. The chosen laboratory should have evaluated the method for samples from newborn infants and should have reference ranges in relation to gestational age, birth weight and age after birth. A number of excellent review articles cover important features of several of the hormone assays (Ratcliffe *et al.*, 1982, 1988; Wood *et al.*, 1985; Ismail *et al.*, 1986a,b; Beastall *et al.*, 1987; Diver & Nisbet, 1987; Masters & Hahnel, 1989; Seth *et al.*, 1989). Detailed analysis, by using capillary column gas chromatography (GC) (profile analysis), of all the major steroids in urine has a number of advantages in the diagnosis of problems, particularly in childhood, of steroid production or metabolism, but does require the experience of a specialist centre. Biochemists, chemical pathologists

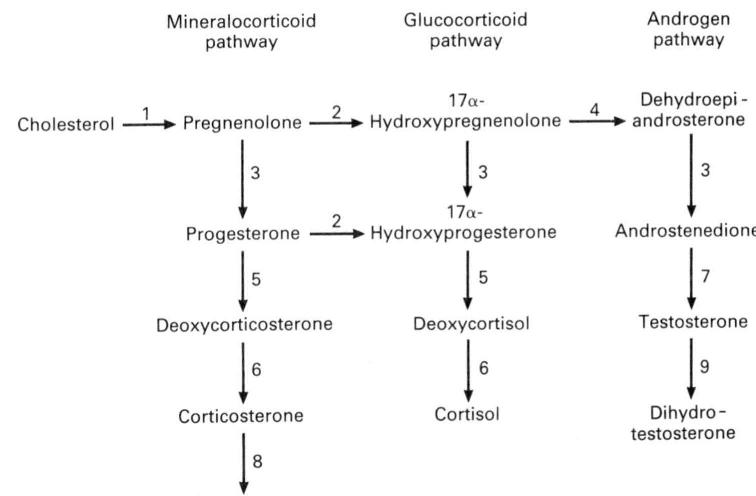

Fig. 10.1 Pathways of adrenal steroid biosynthesis. This is divided according to function. Enzymes: 1, 20,22-desmolase; 2, 17α-hydroxylase; 3, 3β-hydroxysteroid dehydrogenase, 4–5 isomerase; 4, 17,20-desmolase; 5, 21-hydroxylase; 6, 11β-hydroxylase; 7, 17β-hydroxysteroid dehydrogenase; 8, 18-oxidase; 9, 5α-reductase.

and paediatric endocrinologists should be aware of the nearest centres providing the relevant analyses.

Steroid hormones are synthesized from cholesterol through a series of enzymatic steps which reduce the carbon number from 27 to 21 (progestogens, glucocorticoids and mineralocorticoids), then to 19 (androgens) and 18 (oestrogens), the side-chain cleavage being performed by 20,22-desmolase and 17,20-lyase. Oestrogen formation is dependent on aromatase acting on C-19 steroids. In addition, hydroxylations occur at carbons 17, 21, 11 and 18. A 3β-hydroxysteroid dehydrogenase and 4,5-isomerase generate the 3-keto-4-ene structure of most of the active steroids.

In the adrenal cortex, the outer zona glomerulosa serves to synthesize aldosterone under the regulation of renin–angiotensin. The inner zona fasciculata principally makes cortisol under the control of adrenocorticotrophic hormone (ACTH). The pathways for the formation of the principal active steroids are shown in Fig. 10.1.

The gonads are the main source of sex steroids under gonadotrophic control. The synthesis of testosterone and oestradiol requires the action of 17-ketosteroid reductase. If, however, there are defects in the adrenal cortex of 3β-hydroxysteroid dehydrogenase (3β-OH-S-DH), 11β-hydroxylase or 21-hydroxylase, normal intermediates in the glucocorticoid and mineralocorticoid pathways are not metabolized further through the normal route, but become substrates for the formation of excess androgens. 17α-Hydroxylase is somewhat different in this regard because the enzyme is a product of the same gene as for 17,20-lyase and if the enzyme is deficient there is failure to produce both cortisol and sex steroids. In any of the adrenal defects the absence of cortisol causes increased secretion of ACTH and the adrenal cortex hypertrophies. Some of the intermediates, when produced in excess, have biological action in their own right, for example deoxycorticosterone (DOC) is a weak mineralocorticoid.

Collection of samples

Serum or plasma for steroid measurements should be separated from the blood cells as soon as possible after collection (Hilborn & Krahn, 1987). Hormone concentrations will change with prolonged contact with blood cells or the gels used in barrier tubes. ACTH, as measured by RIA, appears to be unstable at room temperature and plasma should be frozen as soon as possible. ACTH is destroyed by repeated freezing and thawing. Protein is adsorbed onto glass so all the tubes for ACTH monitoring should be plastic and kept cold. A stressful venepuncture can cause a marked increase in ACTH output within 30 min. Serial blood samples should be collected via a reliable intravenous catheter, inserted 90 min before the start of sampling. The tubing can be maintained patent with a heparin–saline solution.

It is important that the collection of blood for the measurement of aldosterone and renin is made under strictly controlled conditions. The patient should have a normal dietary intake of sodium and have been given potassium supplements. Blood is ideally taken after the patient has been lying down overnight, or after every effort has been made to keep the child lying down quietly for 2 h before the test. The blood must be collected into chilled heparinized tubes, centrifuged and plasma stored frozen until the assay. The electrolyte and acid–base status should be confirmed.

Urine for steroid analysis can be collected without preservative or with the addition of boric acid (10 mL of 10% aqueous solution) to the collection bottle.

Reference values

Serum hormones

Many of the individual hormone RIAs can be performed on 50–100 μL of serum or plasma. In addition to measuring the concentrations of a steroid hormone in plasma it is often useful to know the status of the regulatory peptides; problems may then arise with sample volumes if results for several analytes are required from the same sample. The major contributory factor to hormone variation is the circadian and pulsatile pattern of pituitary hormone release (Van Cauter, 1989). ACTH is secreted mainly at night in a series of pulses. Low values are found around midnight and highest values near 08.00 h. Gonadotrophins are secreted in pulses of varying amplitude and frequency. The patterns of LH and FSH secretion over a 24-h period are attenuated by age, sex steroids and disease states. A proper assessment of hormone secretion requires repeated blood sampling. The frequency of multiple samplings depends upon the half-life of the particular hormone being assessed (Veldhuis *et al.*, 1984). Some laboratories pool aliquots from multiple serum samples before performing a single hormone measurement on the pool in order to integrate variable results owing to the pulsatile pattern. In paediatrics, it may not be worthwhile to pool samples since pulses of low amplitude may become diluted with the lower hormone levels in a greater number of other aliquots.

Cortisol and adrenocorticotrophic hormone

ACTH is secreted by the pituitary in irregular bursts throughout the day and night, and plasma cortisol concentrations tend to rise and fall in consort. The ACTH pulses are more frequent in the early morning and less frequent in the evening (Krieger *et al.*, 1971). ACTH levels measured by RIA are normally less than 80 ng/L. The values determined by immunoradiometric assay (IRMA) are lower with peak concentrations of 10–20 ng/L, which reflect both improved sensitivity and specificity of IRMA (Fig. 10.2) (Horrocks *et al.*, 1990). Apart from in the newborn (Price *et al.*, 1983), there is a circadian rhythm of the hypothalamic-pituitary-adrenal axis in children (Wallace *et al.*, 1991). The lowest cortisol values (less than 140 nmol/L) are found around midnight and, following a number of peaks in cortisol and ACTH secretion during the night, maximal cortisol values at around 08.00 h are 200–700 nmol/L.

Aldosterone and renin

Plasma renin activity (PRA), and hence aldosterone concentrations, are influenced by many factors including age, sodium balance, body posture and activity. Renin release is normally stimulated on standing and by volume changes of the vascular compartment and by sodium depletion. In the first weeks of life, normal aldosterone concentrations may be up to 5000 pmol/L. PRA in the normal newborn can be up to 20 pmol/mL per h. PRA and aldosterone then decline over the first 18 months of life and from then on are close to adult ranges. Reference ranges for PRA and aldosterone with age are shown in Fig. 10.3 (Fiselier *et al.*, 1983). In adults who have been lying down for at least 30 min, the serum aldosterone is 50 to 400 pmol/L. In subjects who have been standing for 4 h this usually increases to between 200 and 700 pmol/L. In children this response to posture is less easily organized and is therefore not usually documented.

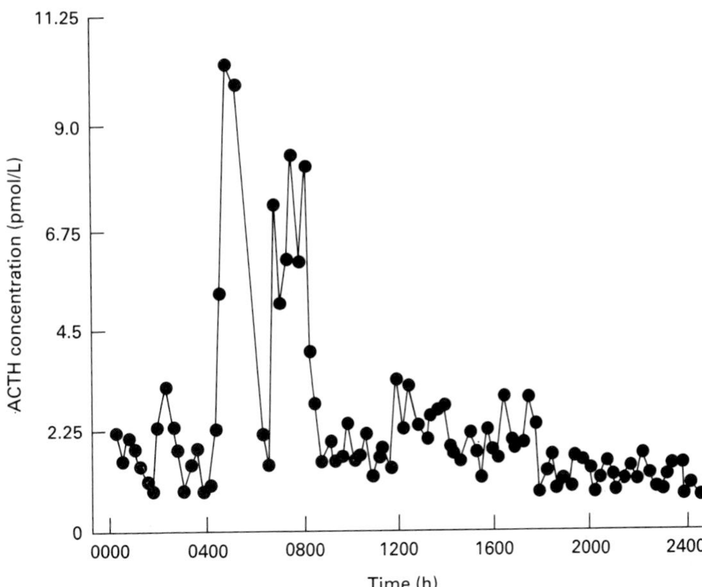

Fig. 10.2 Pattern of adrenocorticotrophic hormone (ACTH) pulsatility over 24 h in a normal female.

Sex steroids and gonadotrophins

Males. In the first week of life newborn males have serum testosterone levels (1–9 nmol/L) which approach those of adult males (10–30 nmol/L). This reflects stimulation of the testes by gonadotrophins and human chorionic gonadotrophin (HCG) remaining in circulation after detachment of the neonate from the placenta. The serum testosterone declines over the first week of life, but in response to an increase in gonadotrophin secretion there is a rise again to concentrations of 2–10 nmol/L and this continues for up to 5 months after birth (Fig. 10.4) (Forest *et al.*, 1980). From age 6 months to 8 years concentrations of serum testosterone remain below 1 nmol/L and are mostly of adrenal origin (Bidlingmaier *et al.*, 1986).

In early childhood there are occasional peaks of gonadotrophins of low amplitude. At the onset of puberty there is regular pulsatile LH secretion during the night with peak concentrations of 2–5 U/L (Wu *et al.*, 1990). A rise in serum testosterone occurs within 60–90 min of the initial pulse of LH. The highest testosterone concentrations near daybreak increase progressively to be around 10 nmol/L (Wu *et al.*, 1993). This drops to half that level by 09.00 h, and throughout the day may be below 2 nmol/L. The measurement of a basal serum testosterone during the day therefore has little value in boys during early puberty. During the day gonadotrophins may be 1–4 U/L and at night there are peak concentrations of around 10 U/L (Oerter *et al.*, 1990). At night, FSH exhibits less obvious pulsatility. As puberty progresses there is a rise in the amplitude of gonadotrophin pulses. The pulse frequency is approximately 2-hourly and the duration extends from the night to a pattern throughout the 24-h day. Between 9 and 17 years the testes increase in size, from less than 3 mL before puberty to the adult range of 12–25 mL. At this time there is a pulsatility in LH throughout the day and night. When 10-mL testes are achieved, plasma testosterone levels rise to the normal adult range. It is not helpful to have reference ranges for testosterone concentrations with age. Each boy progresses through puberty at his own pace (Fig. 10.5). Only when testosterone remains low after the age of 16–18 years is an abnormality worth investigating.

At the end of puberty, testosterone concentrations are 10 to 30 nmol/L and the morning testosterone concentrations may be 20–40% higher than in the same subject in the evening. Androstene-

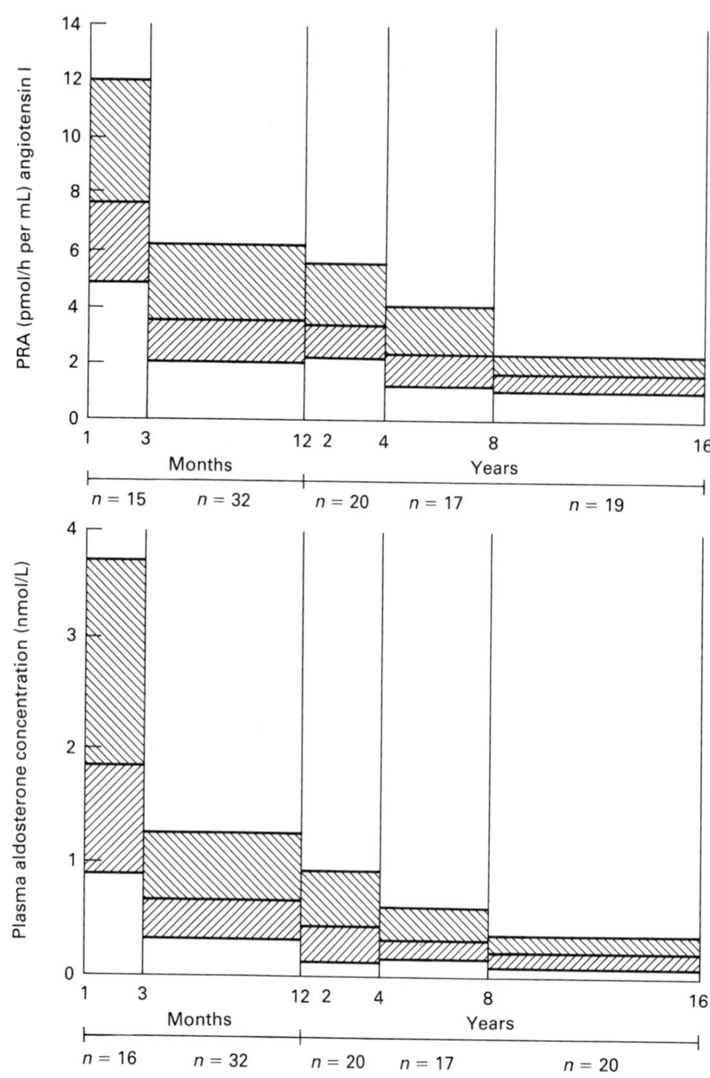

Fig. 10.3 Plasma renin activity (PRA) and plasma aldosterone concentrations during childhood. (Shading represents median and one standard deviation range.)

dione concentrations in the serum are 2–8 nmol/L in the morning. There is a diurnal variation of androstenedione and this is synchronous with cortisol.

Females. In girls under 12 months of age, serum oestradiol concentrations are less than 300 pmol/L. Thereafter, until the onset of puberty, oestradiol concentrations are less than 60 pmol/L. Basal gonadotrophin levels in prepubertal girls are almost always low (less than 5 U/L). In early puberty, a nocturnal release of gonadotrophins (FSH more so than LH) (Oerter *et al.*, 1990) leads to a slow rise in serum oestradiol concentrations, which peak at around midday at 500 pmol/L. In the prepubertal girl the ovary appears on an ultrasound scan to have a number of cysts less than 4 mm in diameter. Even when the gonadotrophins pulse throughout the 24-h period, ovulation cannot occur until an LH surge is sustained for 36 h. This relates to the gradual increase in sex steroid secretion by the ovaries in response to gonadotrophins.

Once menses begin there are marked changes in the frequencies of gonadotrophin secretion and in

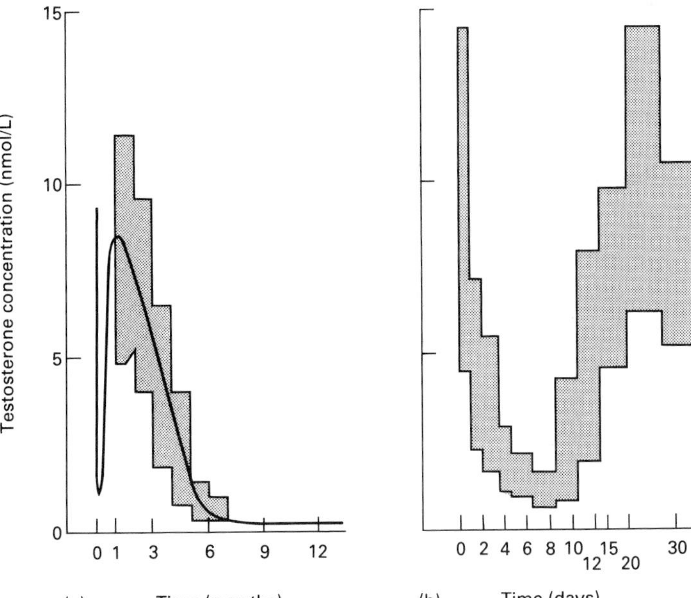

Fig. 10.4 (a) Testosterone in male newborns. Mean from birth to 12 months, ±2 SD ranges from 1 month shaded area. (b) Enlargement of first month (30 days only) testosterone levels from (a) showing ±2 SD ranges. (After Forest *et al.*, 1980.)

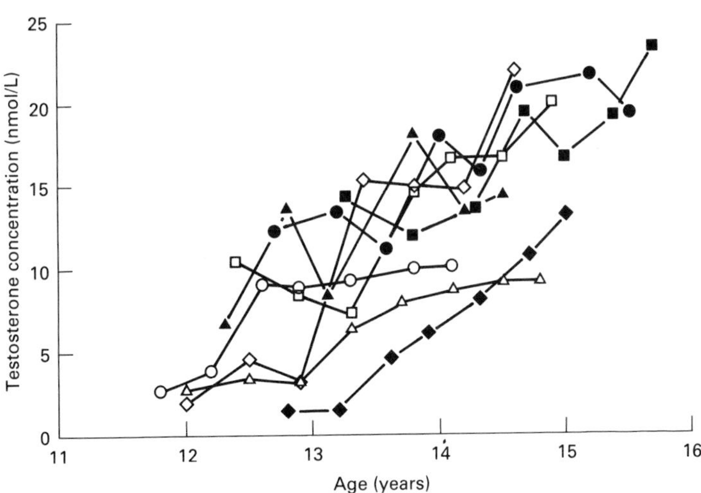

Fig. 10.5 Testosterone levels in eight boys passing through puberty.

the ratio of LH: FSH. In the first half of the normal cycle, LH pulses occur at 1- to 2-hourly intervals but slow to 4-hourly intervals in the mid and late luteal phases. In more mature girls with menstrual cycles the secretion of oestrogen is episodic, but fluctuations in serum oestradiol concentrations over a 24-h period are only discernible in the periovulatory period. Serum oestradiol may be moderately raised in patients with obesity, hyperthyroidism and liver disease, reflecting increased peripheral production of androgens from circulating androgens.

In normal menstruating girls serum testosterone is between 0.5 and 2.5 nmol/L with a small circadian variation and a modest increase in mid-cycle. Androstenedione concentrations rise during the menstrual cycle to 2–10 nmol/L at the time of ovulation.

Table 10.1 Plasma concentrations (μmol/L) of dehydroepiandrosterone sulphate in the neonate and throughout childhood

Age	Female	Male
Cord blood	0.4–7.6	0.6–6.9
1 day	0.5–10.7	1.0–13.5
1 month	0.1–1.2	0.2–2.6
1–6 months	0–0.3	0–0.8
5–12 months	0–0.3	0–0.3
1–5 years	0–0.3	0–0.3
6–8 years	0.1–0.6	0.1–0.6
8–10 years	0.3–1.6	0.2–2.8
10–12 years	0.8–3.2	0.9–3.8
12–14 years	1–5	1.3–4

Dehydroepiandrosterone sulphate

In children under 6 years of age, serum DHAS concentrations are in the range 0.2–0.4 μmol/L. The adrenal cortex then grows and develops a zona reticularis which secretes androgens. DHAS is thus produced in increasing amounts between age 7 and 15 years and leads to an increase in the circulating levels (Table 10.1). Serum DHAS concentrations respond to a lesser extent than cortisol to changes in ACTH production, with lower levels during the day and a nadir around midnight (Rosenfeld *et al.*, 1971; Reiter *et al.*, 1977; Korth-Schutz *et al.*, 1976). Adults have DHAS concentrations of about 2–12 μmol/L during the day.

Progesterone and 17α-hydroxyprogesterone

Serum progesterone concentrations are less than 2 nmol/L in young girls and in pubertal girls during the menstrual cycle prior to the LH surge. Although the ovarian follicles contain much progesterone, most of the circulating progesterone during the follicular phase is adrenal in origin. Progesterone rises in the serum with a peak at 30–80 nmol/L at 7–9 days after the LH surge in ovulatory cycles. This profile of progesterone concentration in the serum reflects the output of the hormone by the corpus luteum.

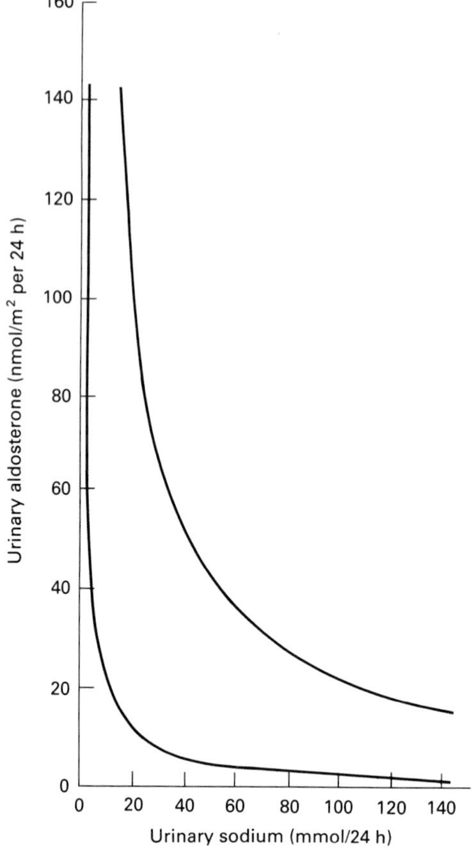

Fig. 10.6 Ranges observed in urine excretion rates of aldosterone-18-glucuronide in relation to sodium excretion.

In the normal newborn, 17-OH-P results will vary with age and gestation length. In unstressed neonates older than 3 days of age 17-OH-P will be less than 15 nmol/L, but if the infant is stressed less than 40 nmol/L is considered normal. Once menses start, 17-OH-P shows a similar cyclical pattern to progesterone with peak concentrations of less than 15 nmol/L in the luteal phase and levels of 1–3 nmol/L through most of the cycle.

Urine steroids

Free cortisol

Cortisol in the urine is normally excreted at the rate of

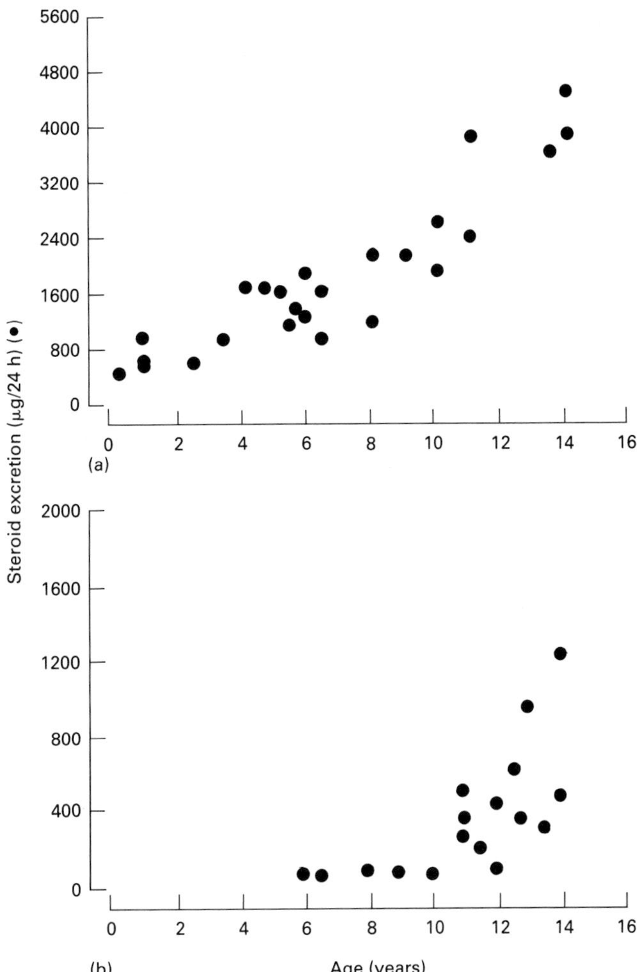

Fig. 10.7 Urine steroid excretion rates during childhood: (a) summed cortisol metabolites; (b) androsterone.

150–400 nmol/24 h (15–35 nmol/mmol creatinine). The variation within individual children may be less than 10%. Several 24-h urine collections may need to be obtained after admission of the patient to hospital to avoid stress effects, but the collection of even one 24-h urine sample is difficult in children.

Aldosterone-18-glucuronide

Aldosterone-18-glucuronide is a metabolite of aldosterone which represents about 10% of the total hormone production and between 10 and 60 nmol/24 h are normally excreted in urine. The urine aldosterone level is inversely related to urinary sodium excretion (Fig. 10.6) (New *et al.*, 1976). Although diuretics and purgatives are not often prescribed for children, it is desirable that such drugs are discontinued for at least 3 weeks before samples for aldosterone estimation are collected.

Steroid metabolites (profile analysis)

Daily cortisol production throughout life is maintained at 12 mg/m^2 (surface area). This reflects the control of the adrenal cortex by ACTH. In absolute terms, excretion rates of steroid metabolites increase throughout childhood to reach stable adult values (Fig. 10.7a). In contrast, there is little androgen excretion in the urine of children under 7 years of age, after which there is a dramatic rise, reflecting increased adrenal androgen secretion, later with superimposed gonadal function (Kelnar & Brook, 1983) (Fig. 10.7b).

Basal and confirmatory tests

The time of blood sampling must be recorded if hormone results are to be interpretable. The lower adrenal activity around midnight compared with that near 08.00 h is a useful combination of basal tests of the hypothalamic-pituitary-adrenal axis. The determination of basal ACTH and gonadotrophin concentrations assists in distinguishing primary adrenal and gonadal defects from hypothalamic and pituitary defects. Stimulation and suppression tests will usually give more exact information that allows decisions to be made with regard to the appropriate diagnosis and treatment.

Hypothalamic-pituitary-adrenal axis

The integrity of the pituitary-adrenal axis in children should be tested only when essential and always in an appropriate clinical setting and with caution. A number of provocative tests are available (Streeten *et al.*, 1984) and these assess the response of the pituitary to the negative cortisol feedback and stress as follows.

1 An increased signal from the hypothalamus following a stress stimulus, such as through insulin-induced hypoglycaemia.

2 The action of high glucocorticoid levels (dexameththasone) on ACTH release.

3 Lowering of serum cortisol concentrations through the action of metyrapone on adrenal synthesis; this is an unpleasant procedure and is not recommended for use in children.

Insulin-induced hypoglycaemia

Hypoglycaemia induced with insulin causes the release of ACTH and a subsequent increase in plasma cortisol, which can be measured by RIA. *This is a hazardous procedure which should be undertaken only with medical supervision.* Glucose (10% solution) and hydrocortisone must be to hand, ready for injection in the event of an emergency.

A suitable protocol is: a fasting blood sample is taken from the child using an indwelling cannula; insulin is injected at a dose which is appropriate to cause the blood glucose to fall below 2.2 mmol/L. In normal children this is 0.15 U insulin/kg body weight (0.1 U/kg in patients with suspected adrenal insufficiency). The ensuing hypoglycaemia usually causes some discomfort to the patient (sweating, tremor, disorientation, coma). The blood glucose should be checked using Clinistiks and blood spots should be taken at intervals during the test. The blood sugar can usually be brought back to normal by giving the patient a very sweet drink (e.g. Lucozade) and a high-carbohydrate snack (bowl of sweet cereal or toast with jam). In children, less than 5 g glucose is needed to relieve the symptoms of neuroglycopenia and bring glucose concentrations near to normal. If the glucose does not rise promptly after going below 2.2 mmol/L at 30 min then 10% glucose should be administered to restore normal concentrations. This glucose should be given at 0.7 mL/min per kg over 3 min, then at 0.1 mL/min per kg until the plasma glucose is kept at 5–8 mmol/L (Shah *et al.*, 1992).

During the insulin stress test, blood specimens for hormones and glucose are taken before and at 20, 30, 45, 60 and 90 min after insulin, and are assayed for cortisol and glucose. The laboratory will require heparin or clotted samples for hormones *and* samples in fluoride for glucose. Provided that hypoglycaemia can be confirmed, an increase in cortisol concentration of 200 nmol/L above the basal value, or a rise in concentration to at least 550 nmol/L, is interpreted as a normal response. The basal cortisol in the morning should usually be at least 200 nmol/L but in an anxious child may be above the normal range and the response may then be less than usual.

Pituitary suppression with dexamethasone

For children, the difficulties in collecting 24-h urine samples and the unpopularity of assays for 17-

hydroxycorticosteroids (17-OHCS) has heralded a number of changes in protocol for the dexamethasone test.

The dexamethasone suppression test serves two purposes (Liddle, 1960). The first is a low dose to exclude stress as a factor contributing to a high cortisol output. The second is a high dose to differentiate the causes of true hypersecretion of cortisol. Only patients with Cushing's syndrome fail to suppress cortisol secretion during the night after an oral tablet of dexamethasone at low dose. Measurement of serum dexamethasone concentrations at 08.00 h should be considered in patients who fail to suppress cortisol at that time after the low dose (Meikle, 1982). Serum dexamethasone should be between 2 and 20 nmol/L.

If serum cortisol concentrations are high, both at midnight and at 08.00 h, a further suppression test is justified to exclude Cushing's syndrome. For children dexamethasone should be given orally at 0.3 mg/m^2 (Hindmarsh & Brook, 1985). Since dexamethasone is prescribed in scored 0.5-mg tablets, the pharmacy will need to provide a suitable formulation for this test in children. After this low dose of dexamethasone has been administered orally at 23.00 h, the 08.00-h cortisol will be suppressed in normal children.

For the high-dose dexamethasone test in children, 1 mg dexamethasone should be taken every 6 h. The serum cortisol at 07.00 to 09.00 h should generally be suppressed below 100 nmol/L (Cronin *et al.*, 1990), except in patients with ectopic ACTH-secreting tumours or an adrenal tumour.

Adrenal stimulation

The adrenocorticotrophin stimulation test is useful in cases of suspected adrenal failure but there is a risk of anaphylaxis and the procedure should only be performed on a ward where resuscitation equipment is available. No child with any allergic condition should be tested. In children the ACTH dose should be 250 µg/m^2. This is still a pharmacological dose which stimulates the adrenal cortex for many hours. Blood is taken for cortisol measurement before and then at 30 and 60 min after the ACTH injection. If a young patient has been taking immunosuppressant or anti-inflammatory steroids for a long time (e.g. asthmatics), prolonged adrenal stimulation may be required, such as with an injected depot ACTH preparation (1 mg daily intramuscularly) for 3 days (Lindholm & Kehlet, 1987). In any patient being treated with steroids (particularly prednisolone) the cross-reaction of the steroid in the cortisol assay should be considered. It will be useful for the request form to show information about recent drug treatment. The ACTH test is also used to reveal minor defects of the steroidogenic enzymes in patients with late-onset forms of congenital adrenal hyperplasia (CAH) (Hague *et al.*, 1989). Intermediates in the biosynthetic pathway, e.g. 17-OH-P, dehydroepiandrosterone (DHEA) and 17-hydroxypregnenolone, will then have to be measured, probably by a specialist laboratory.

Hypothalamic-pituitary-gonadal axis

The gonadotrophin-releasing hormone (GnRH) and HCG stimulation tests provide limited and mostly indirect information about the hypothalamic-pituitary-gonadal (H-P-G) axis.

Gonadotrophin-releasing hormone stimulation test

When it is difficult to decide whether LH and FSH concentrations in pubertal children are normal or low, a GnRH test may be helpful. LH and FSH are determined before and at 30-min intervals for 1–2 h after the administration of a single intravenous injection of GnRH (25 µg in children). This test assesses pituitary function but interpretation of the results is not always straightforward. A two- to threefold rise in serum gonadotrophins following GnRH administration is characteristic of a normal pituitary. A failure of gonadotrophin concentrations to increase from the low levels seen in hypogonadotrophic hypogonadism suggests a hypothalamic defect. If the result is negative, however, injections of GnRH (2.5 µg/kg) should be repeated after 90 min. The first stimulus primes the pituitary and this may normalize gonadotrophin release in response to the second dose of GnRH in patients with hypogonadotrophic hypogonadism or patients with hypothalamic but not pituitary defects. The response to GnRH is variable and some consider this test to be of limited value (Abdulwahid *et al.*, 1985; Goji & Tanikaze, 1992). If testosterone levels are raised but the LH and FSH levels remain low, a diagnosis of gonadotrophin-independent precocious puberty can be supported.

Human chorionic gonadotrophin

A failure of gonadal endocrine function is supported by high gonadotrophin concentrations. In some patients, further insight into the gonadal problem is gained from gonadotrophin stimulation. The test can be helpful in establishing the presence of testicular tissue in children in whom cryptorchidism or intersex is being investigated. In male babies (46XY) with poor development of the genitalia this test may be used to reveal defects in the synthesis of testosterone and dihydrotestosterone. HCG (1000 U given on alternate days intramuscularly over 7 days, which is probably maximal stimulation) will normally evoke a rise of testosterone production, and hence a 5- to 10-fold increase in serum concentration over the basal level. A satisfactory response to this test will usually also predict a response of the child to HCG therapy.

Clinical tests

In addition to endocrine tests based on measurements of circulating hormones and clinical observation, the response of the reproductive tract of a teenage girl is amenable to direct clinical assessment by ultrasound examination of the abdomen. This will reveal in the ovary the number of follicles and their development, the size of the uterus and the thickness of the endometrium. There is a direct correlation between uterine dimensions and serum oestradiol concentrations.

Diseases of steroid hormone production

Cortisol and ACTH

Cortisol hypersecretion

Growth retardation and weight gain are the more likely presenting features in a child with Cushing's syndrome. High cortisol production may lead to a combination of more typical signs: moon-shaped face, buffalo hump and striae. The high production of other steroids (more typical of adrenal tumours) accounts for acne, hirsutism and, in some cases, hypertension, although it is difficult to exclude the direct effects of cortisol on blood pressure. Causes of cortisol excess are classified under the generic name of Cushing's syndrome. This can be the result of autonomous secretion of cortisol from an adrenal neoplasm or attributed to excess cortisol secretion from the adrenals, stimulated by ACTH from a number of sources (Fig. 10.8).

Testing for Cushing's syndrome first requires the demonstration of excess cortisol production and then the establishment of the degree of autonomy of this production (McHardy, 1984; Carpenter, 1988). A raised 24-h urine excretion rate of free cortisol is a good indicator for this condition. This is commonly requested in the investigation of an obese or short child, but a high cortisol excretion rate may reflect stress on the day of collection. In a patient with a high clinical suspicion of Cushing's syndrome but with a normal urine free cortisol excretion rate on the first occasion, repeated 24-h urine collections over many weeks might reveal a cyclical form of the disease, which has been characterized in adults with peak frequency over weeks to months (Atkinson *et al.*, 1985). In practice, this would be difficult to document in a child.

Even if the urine cortisol excretion rate is clearly elevated (typically above 400 nmol/24 h), in the majority of paediatric patients who are tested, the urine free cortisol excretion will be suppressed to the lower region of the normal range after a low dose of dexamethasone is given over 2 days. Correction of the urine excretion value for body surface area or in relation to the creatinine clearance can eliminate a false positive diagnosis in most cases of obesity. Serum cortisol concentrations can be measured at 08.00–09.00 h after the administration of 0.3 mg dexamethasone/m^2 at 23.00 h as an alternative to this urine test. Results for plasma cortisol above 100 nmol/L are abnormal. Accelerated dexamethasone metabolism is, however, seen in patients taking certain medications, particularly anticonvulsants. In this regard the simultaneous

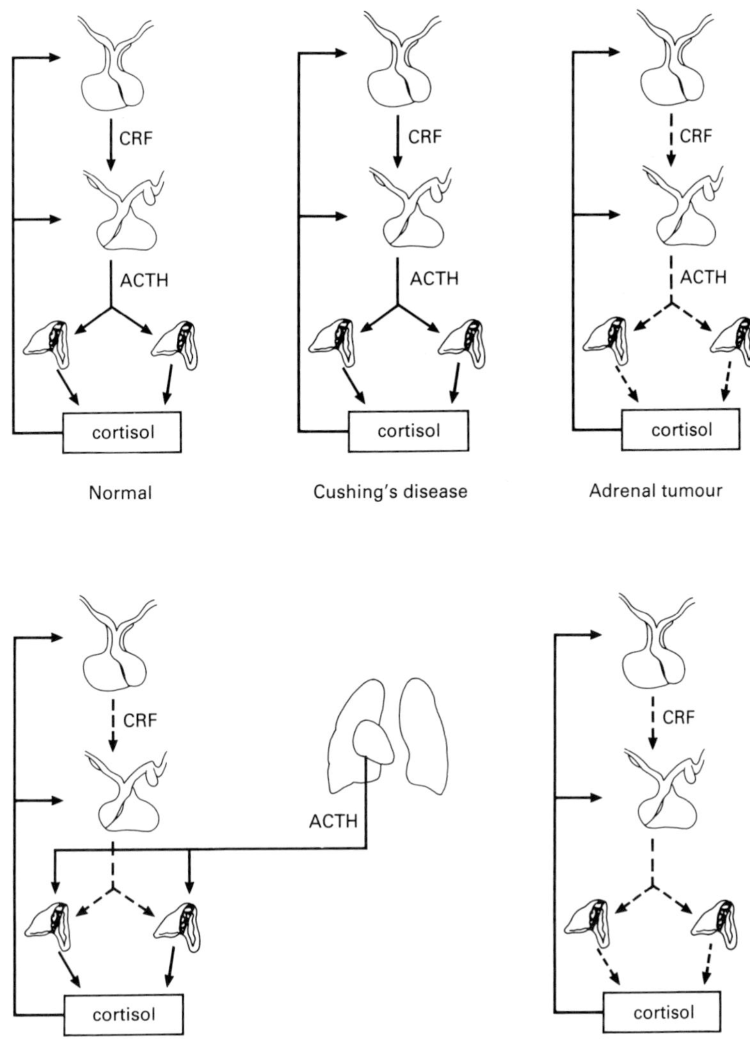

Fig. 10.8 Adrenal steroid excess. CRF, corticotrophic-releasing factor; ACTH, adrenocorticotrophic hormone.

measurement of cortisol and dexamethasone has improved the diagnostic precision of this test in adults (Meikle, 1982), but such data in children may not be available.

The pattern of cortisol secretion in Cushing's syndrome is such that there is a loss of the circadian rhythm of the serum concentrations of cortisol, with notably high levels in samples taken at midnight. Further biochemical tests and imaging procedures are justified at this stage. The protocol will vary between hospitals, depending upon the laboratory services, clinical practice and experience, and the availability of specialized tests, e.g. ACTH assay and computerized tomography (CT) scanning.

The diagnosis of pituitary-dependent Cushing's disease is confirmed when previously high cortisol production is substantially suppressed by high doses of dexamethasone (1 mg 6-hourly for 2 days in children). Plasma cortisol, ACTH and urine free cortisol (or better still, total cortisol metabolites) should all fall to less than 50% of their basal values in the majority of affected patients. Ectopic ACTH-secreting tumours are exceptionally rare in children (Himsworth *et al.*, 1977). Hypokalaemia is a common association with ectopic ACTH-secreting tumours (Howlett *et al.*, 1986). Other ectopic tumours, such as carcinoid, may not always be associated with hyperkalaemia and ACTH may be only moderately raised.

Measurements of the plasma ACTH concentrations at midnight and 08.00 h are invaluable in the distinction of Cushing's syndrome. Circulating ACTH is low in children with an adrenal tumour, and high or in the upper part of the normal range in patients with adrenal hyperplasia. In the ectopic syndrome, the serum ACTH, measured by RIA, is often very high (100–10 000 ng/L). In considering assays for ACTH, the specificity of the RIA or IRMA should be known with respect to the measurement of normal ACTH and of so-called big-ACTH, typical of ectopic tumours which may be poorly detected by IRMA (White *et al.*, 1987; Crosby *et al.*, 1988), and should show ACTH values between 100 and 1000 ng/L.

Dynamic tests can be used to gain further indications for the pathology of a patient with high cortisol production. An adrenal tumour will not usually respond with an increased steroid output to an ACTH stimulation test, but a hypersecreting adrenal will respond with a significant rise in serum cortisol concentrations. In contrast to the ACTH-stimulated adrenal, which secretes largely cortisol, many adrenal tumours produce a spectrum of steroids, such as dehydroepiandrosterone (DHA) (Yamaji *et al.*, 1984) or 17-hydroxypregnenolone (McKenna *et al.*, 1977) or other steroids, and measurements of these specifically or generally (steroid profile analysis) should be arranged through specialized centres. Analysis, using capillary column GC with mass spectrometry, of steroids in urine at a specialist laboratory may thus be the simplest way to reveal complex patterns of steroid production associated with tumours (Shackleton *et al.*, 1980b; Phillipou, 1982). CT scanning may localize an adrenal tumour or display adrenal hyperplasia.

The rare failure to distinguish the pituitary-dependent from the ectopic ACTH-secreting tumour and fear of the consequences of incorrect surgery are the reasons for the interest in finding further tests. If a pituitary adenoma is suspected, a corticotrophic-releasing factor (CRF) test may be informative in some cases. In Cushing's disease there is often an increase in cortisol above the normal response to 100 μg CRF (Grossman *et al.*, 1988). Indeed, those authors found that in adult patients the combination of high-dose dexamethasone and the CRF test with measurement of serum cortisol is superior to either test alone in the differential diagnosis of Cushing's syndrome.

Sampling of blood from the inferior petrosal sinuses for simultaneous assessment of ACTH concentrations is now used in some specialized centres to differentiate Cushing's disease from the syndrome and for lateralization of microadenoma. An intriguing observation from these studies has been the parallel increased secretion of prolactin (Schulte *et al.*, 1988), growth hormone, thyroid-stimulating hormone (TSH) and glycoprotein α subunit (Crock *et al.*, 1988; Zovickian *et al.*, 1988). These findings may reflect changes in the vasculature or a paracrine effect of β-endorphin from the tumour on adjacent tissue.

If a rare ectopic ACTH-secreting tumour is suspected, the pancreas and chest should be scanned by CT but the primary tumour may still not always be found. Venous catheter studies for the localization of tumours producing ACTH may be valuable, and the measurement of tumour markers, for example calcitonin, should also be considered (Howlett *et al.*, 1986).

Adrenal cortical insufficiency

Although, in 1855, Addison described tuberculous destruction of the whole adrenal glands, today autoimmune adrenalitis, which spares the adrenal medulla, is more common in children. Adrenal function is impaired after long-term treatment with steroids, such as hydrocortisone or prednisolone for rheumatoid arthritis and beclomethasone for asthma. This is an important clinical issue requiring tests (Law *et al.*, 1986). In newborns other rare causes of adrenal insufficiency include infarction or haemorrhage of the adrenals and adrenal hypoplasia. Adrenal cysts are rare and regress spontaneously. Adrenal cortical insufficiency may only be apparent when a child suffers a physical stress, such as trauma or surgery, and presents with tiredness, weakness, lethargy, anorexia, nausea, weight loss, dizziness and hypoglycaemia. Adrenal destruction may also be due to influenza, pneumococcal or haemolytic streptococcal infection or leukaemia. The differential diagnosis of hypoglycaemia in childhood should include adrenal cortical insufficiency, which can be due to congenital adrenal hypoplasia (Laverty *et al.*, 1973).

Adrenal hypoplasia may be the cause of adrenal insufficiency. This can take three forms, distinguished by the appearance of the adrenal cortex on histological examination, although recognizable prior to death and post mortem by biochemistry compatible with the morbid findings. The common pattern lacks a fetal adrenal zone and is called an anencephalic form because of the resemblance in appearance of the adrenal with the brainless state. The next most common pattern seen is small glands, which nevertheless have a normal ratio of fetal to definitive tissue. This pattern is called the miniature form and may be transmitted in an autosomal recessive pattern. A cytomegalic form is the rarest, but at autopsy the adrenal cortex is filled with giant cells and lacks the normal architecture into distinct zones. This form may be X-linked.

Adrenoleucodystrophy is adrenal insufficiency with progressive demyelinization of the cerebellum and cerebrum. Children may present first with the adrenal insufficiency. This is an X-linked lipid storage disease. Elevated levels of very long-chain fatty acids (particularly C24) are found in tissues and red blood cells (Moser *et al.*, 1991) and are worth measuring in a child with adrenal insufficiency.

A child with primary adrenal insufficiency may show pigmentation as a smokey brown coloration which affects the buccal mucosa (inside cheeks, gums and lips), on skin creases, scars, genitalia and areolae. This is a reflection of melanocyte-stimulating hormone action, which is a product of the pro-opiomelanocortin secreted by the pituitary when making ACTH. The child with adrenal insufficiency will probably have postural hypotension (Burke, 1985; Grant *et al.*, 1985) with high renin activity (PRA). Hyponatraemia due to the absence of mineralocorticoid steroids is a frequent finding in primary adrenal insufficiency. The low plasma sodium is associated with hyperkalaemia and a raised plasma urea. In the urine there will be low potassium loss.

Plasma cortisol concentrations, monitored between 07.00 and 09.00 h, which are repeatedly less than 170 nmol/L (Hagg *et al.*, 1987) are suggestive of adrenal cortical insufficiency but may not be seen until the disease is advanced. In a child older than 5 years, a short Synacthen (Ciba, Horsham, UK) test might be performed (with appropriate caution) to assess the adrenal reserve. Blood for serum cortisol is taken before and at 30 and 60 min after an intravenous injection of soluble Synacthen (250 $\mu g/m^2$). A normal response is characterized by an increment in the cortisol concentration of at least 200 nmol/L or a rise to levels above 500 nmol/L. If an assay for cortisol is specific enough to exclude cross-reaction with prednisone, this synthetic steroid can be given immediately after the basal blood has been taken, so as to afford glucocorticoid cover without affecting the adrenal response to exogenous ACTH. If there is pigmentation the child has primary adrenal failure and plasma ACTH measurements will be raised. ACTH will be normal or low in secondary adrenal failure. A metyrapone test can also be helpful in distinguishing primary from secondary adrenal insufficiency (Dolman *et al.*, 1979), but should be conducted only under medical supervision. Circulating antibodies to the adrenal cortex suggest an autoimmune process and other endocrine tests may be required to look for an extension of the autoimmune process to other hormonal tissues. Cortisol should be replaced in a child at a rate of 15 mg/m^2 per day and it is useful to check cortisol concentrations throughout the day in plasma samples taken at 30-min intervals over 2 h after a morning dose of hydrocortisone has been administered, and then at 2- to 3-h intervals throughout the day. A peak in cortisol concentrations above 700 nmol/L reflects adequate replacement therapy. Some of these patients may also need mineralocorticoid replacement.

Secondary adrenal insufficiency occurs when the adrenal cortex is deprived of ACTH stimulation. The commonest cause in children is after steroid therapy but an expanding pituitary tumour, which leads to impairment of pituitary function, can affect the pituitary hormone secretions in a sequential manner over a period of time. Gonadotrophin secretion fails first, followed by growth hormone, TSH and ACTH.

In some cases the adrenal insufficiency may be secondary to more general hypopituitarism, although the full expression of this is rare in young children. If hypocortisolism is due to ACTH deficiency resulting from pituitary or hypothalamic disease, such as tumours, infarction or trauma, there are usually signs of deficiency of other hormones (e.g. loss of body hair in older children). There will be no pigmentation. In the absence of ACTH, the production of aldosterone under the stimulus of the renin–angiotensin system will be unaffected and the blood pressure will probably be normal.

Hyponatraemia in secondary adrenal insufficiency is a result of dilution of the plasma because the cortisol deficiency prevents clearance of a water load. The low sodium is therefore associated with low plasma potassium and low blood urea. Urine sodium excretion is normal.

Mineralocorticoids and renin

Mineralocorticoid excess

Primary aldosteronism. Autonomous aldosterone secretion from the adrenal gland is incredibly rare in children and less than 30 cases has been reported in the literature. Bilateral adrenal hyperplasia or aldosterone-producing tumour has been described. Increased aldosterone secretion may be suspected in a child who has complained of muscle weakness and headaches and is shown to have hypokalaemia and hypertension. Before the laboratory begins investigations of the renin–angiotensin –aldosterone system, abnormalities of electrolytes and water balance need to be confirmed. Antihypertensive drugs (thiazide and loop diuretics, β-receptor blockers, calcium channel blockers) affect plasma renin and aldosterone and these should ideally be stopped 2 weeks before meaningful tests can be carried out.

Patients with primary hyperaldosteronism have repeated plasma potassium values below 3.7 mmol/L. Since the distal tubules are the prominent site of sodium reabsorption under the influence of aldosterone, patients on low-sodium diets have less available sodium for exchange because of the relative increase in sodium reabsorbed in the proximal segments. These patients should be retested after increasing the dietary sodium intake for 5 days. The urine potassium excretion rate should then exceed 30 mmol/day if a patient is to be investigated further. If potassium loss is less than 30 mmol/day the creatinine excretion should be checked to assess reliability of the daily urine collections. If the potassium depletion is not ascribed to renal losses, then other causes of potassium wasting should be investigated. Other causes for potassium depletion (diarrhoea, purgatives, liquorice ingestion, diuretics) must be excluded.

Tests for the diagnosis of primary aldosteronism may need to be modified for a child (Melby, 1985; Young & Klee, 1988), and some tests will need to be sent away to specialist laboratories. On a normal-salt diet, a urine aldosterone-18-glucuronide level above 50 nmol/day is a good screening test for primary aldosteronism. Since aldosterone synthesis is suppressed by hypokalaemia, an effort must be made to correct the hypokalaemia by prescribing potassium supplements before more detailed assessment of the renin–angiotensin–aldosterone system is undertaken. Normal children will show suppression of urinary aldosterone (below 30 nmol/day) on a high-salt diet. Patients with hyperaldosteronism, however, will fail to show this suppression. PRA will be suppressed (below 3 pmol/mL per h) in patients with an aldosterone-secreting adenoma.

The diagnosis of primary hyperaldosteronism rests on the finding that the plasma aldosterone concentrations are high for the corresponding PRA. In about 60–70% of adult patients with hyperaldosteronism, an adenoma can be visualized on a CT scan and then removed by surgery. In children, adrenal hyperplasia is more likely and this is difficult to confirm (Dillon, 1989). It would be advisable to seek the advice of specialists. Catheterization of the adrenal veins is, in some centres, a successful procedure but the venous drainage of the adrenal glands is often complex and even an experienced radiographer is not guaranteed success. In a patient with an aldosterone-secreting adenoma, the adrenal venous serum from the affected side usually shows a higher concentration of aldosterone compared with either the contralateral adrenal vein or a peripheral vein sampled simultaneously. Aldosterone and cortisol should be measured in all samples to check the authenticity of the site from which the blood is taken. The cortisol concentrations in the adrenal vein samples may be of the order of 6000 nmol/L and in most assays will have to be measured after dilution of the plasma. The secretion of 18-hydroxycorticosterone (18-OH-B) is raised in patients with adrenal adenoma such that serum levels are above 3000 pmol/L (Biglieri & Schambelan, 1979). A ratio of 18-OH-B:cortisol of more than 3 or an aldosterone:cortisol ratio above 2.2 is diagnostic of aldosterone-secreting tumours. These tests are not generally available.

Glucocorticoid remediable hyperaldosteronism is a rare familial cause of hypertension in which the biochemical features of hyperaldosteronism and the hypertension respond to glucocorticoid (dexamethasone) treatment. The hyperaldosteronism is responsive to ACTH but not to angiotensin II. Recently, it has been reported that this condition is associated with increased excretion of 18-oxocortisol and this may arise by 18-hydroxylation of cortisol in the adrenal zona fasciculata (Gomez-Sanchez *et al.*, 1988). An explanation for this has been proposed on the basis of a chimeric gene from two normal 11-hydroxylase genes (Lifton *et al.*, 1992) which separately determine cortisol and aldosterone production under ACTH and angiotensin control, respectively. The chimeric gene influences aldosterone production in response to stimulation by ACTH. Estimation of 18-oxocortisol is likely to prove useful in screening for this disorder (Davis *et al.*, 1988) but this assay has limited availability.

Bartter's syndrome. This syndrome describes a rare condition in childhood of growth and mental retardation together with hypokalaemic alkalosis. High aldosterone and high PRA are seen in Bartter's syndrome. Hypokalaemia with alkalosis owing to increased urinary losses of potassium and chloride are characteristics of this syndrome. PRA is high due to juxtaglomerular hyperplasia. The primary cause for this disorder may still not have been found but there is resistance of the vasculature to the pressor action of angiotensin II. Hypokalaemia without hypertension is the result of increased aldosterone but can also be due to diuretic or laxative abuse or psychogenic vomiting.

True and apparent mineralocorticoid excess. Low PRA may be encountered when mineralocorticoids other than aldosterone are in excess. This finding can be due to CAH owing to 11β-hydroxylase deficiency or 17α-hydroxylase deficiency (DOC excess) or to mineralocorticoid-secreting tumours. All are rare.

In the syndrome of apparent mineralocorticoid excess (AME) there is hypertension, hypokalaemia and reduced secretion of all steroids (Shackleton *et al.*, 1980a; Monder & Shackleton, 1984). The improvement brought about by treatment with an aldosterone antagonist, such as spironolactone, or a potassium-sparing diuretic, e.g. triamterine, suggests the presence of an unidentified mineralocorticoid, although no conclusive evidence for such a compound has been found so far. Currently, the disease is attributed to cortisol acting as both glucocorticoid and mineralocorticoid with a prolonged half-life owing to low activity of 11β-hydroxysteroid dehydrogenase, which normally oxidizes cortisol to inactive cortisone. This defect is most easily detected by a urinary steroid profile, which clearly displays a high excretion of cortisol metabolites relative to cortisone (Fig. 10.9). The disease is often fatal in childhood (Honour *et al.*, 1983); a single adult case has been described (Stewart *et al.*, 1988). Liquorice will inhibit the enzyme system 11β-hydroxysteroid dehydrogenase (Stewart *et al.*, 1987) and produce clinical effects like those seen in AME, so children should be questioned about the sweets they have eaten.

Mineralocorticoid deficiency

Primary hypoaldosteronism with hyperkalaemia, increased PRA and low aldosterone concentrations suggests Addison's disease, congenital adrenal hypoplasia or defects of aldosterone synthesis.

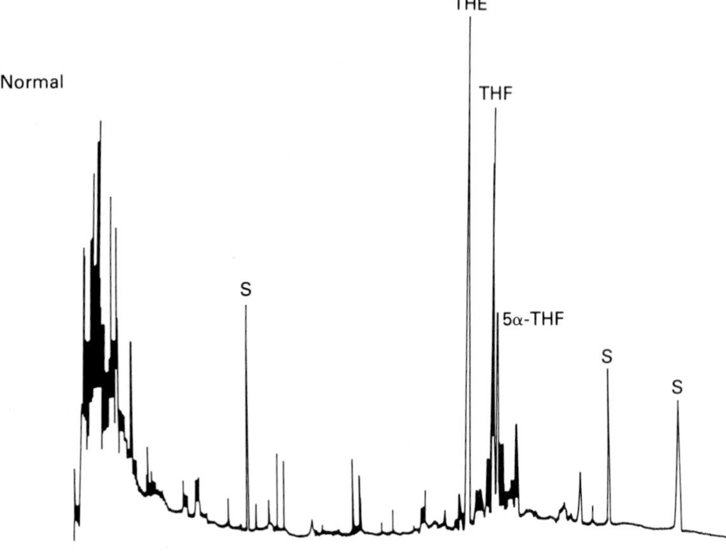

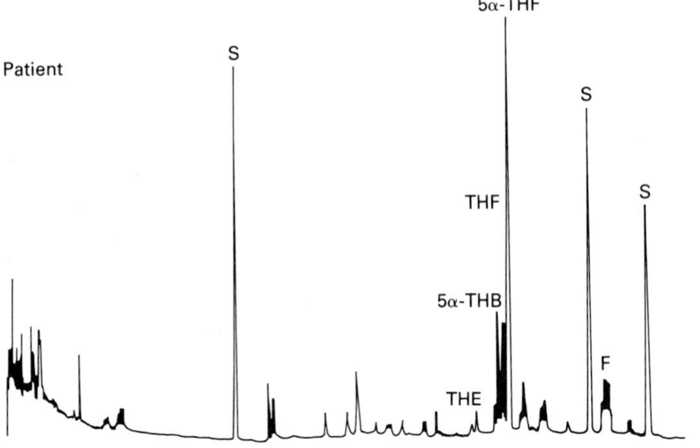

Fig. 10.9 The gas chromatography profiles of urine steroids from one child with hypertension due to 11β-hydroxysteroid dehydrogenase deficiency and from a normal child of the same age. Note the reduced tetrahydrocortisone (THE) in the patient and the change in ratios of 5α- to 5β-tetrahydrocortisol (5α-THF to THF). It has been suggested that this reflected an increase in 5α-reductase activity which if also included in the mineralocorticoid hormones would account for the hypertension.

The commonest cause in boys is CAH with the accompanying aldosterone deficiency. PRA and aldosterone are rarely subnormal but this secondary hypoaldosteronism is found in association with diabetes and chronic renal disease, and as an isolated idiopathic occurrence.

Salt-losing crisis in neonates. Hyponatraemia in the neonatal period is an urgent diagnostic problem; it is essential to consider whether the sodium intake is adequate (less than 4 mmol/kg per day in term babies, up to 12 mmol/kg per day in preterm infants), the water intake is high or there is sodium loss from the gastrointestinal tract or kidneys. Renal salt loss can be due to anatomical abnormalities and obstructive or renal tubular disorders, which can include failure to respond to aldosterone. Low production of mineralocorticoid owing to adrenal disease is a common cause of salt loss with hyperkalaemia in newborns.

Any male baby who collapses during the second week of life should be tested for CAH due to 21-hydroxylase deficiency. Some females with mild forms of CAH and only weak signs of virilization may also be affected. Blood should be taken for 17-OH-P measurement and a result above 100 nmol/L will support the defect. Of all children with this defect 60% show a salt-losing variant of the disease. The enzyme deficiency in cortisol production extends to the synthesis of aldosterone. PRA will be elevated and aldosterone will be inappropriately low. Since PRA is higher in all newborns than in adults, it is important to check the activity against a normal range for the age of the infant (Sassard *et al.*, 1975; Dillon *et al.*, 1976; Fiselier *et al.*, 1983).

If cortisol levels are normal the production of aldosterone needs to be evaluated. Defects of aldosterone production and action have been described (Ulick, 1976). In both cases, PRA will be elevated; the defects are distinguished by the serum concentrations of aldosterone or urine excretion rates of the metabolites. A defect in the late steps of aldosterone biosynthesis will be confirmed when 18-OH-B production is shown to be elevated. In disorders of aldosterone receptors, both aldosterone and 18-OH-B are elevated in the blood and urine. The diagnosis is difficult to make unless the child is sodium depleted (Honour *et al.*, 1982), but when the plasma sodium is low the pattern of steroids in a urine steroid profile will reflect the excess hormone production. The effects of the receptor defect seem to be partially correctable by increasing the dietary salt intake to satisfy the salt craving (Yasuda *et al.*, 1986). Aldosterone receptor resistance can be demonstrated in specialist laboratories with an assay for electrolyte transfer by mononuclear leucocytes (Armanini *et al.*, 1985).

Hyponatraemia is often seen in the first weeks of life in preterm infants (<30 weeks gestation) (Honour *et al.*, 1977), reflecting immaturity of renal function as well as immaturity in the adrenal production of aldosterone and the diuretic effect of increased vasopressin production (Al-Dahhan *et al.*, 1983; Rees *et al.*, 1984).

Sex hormones and gonadotrophins

Androgen excess

The investigations of androgen excess have different objectives, depending on the age and genetic sex of the patient. Newborn girls who present with ambiguous genitalia usually have an inherited metabolic disease (CAH) of the adrenal cortex. During childhood, boys with precocious puberty may have space-occupying lesions of the brain, CAH or, very rarely, tumours of the adrenals or gonads. A few girls with hirsutism and/or acne, with or without menstrual disturbance, may have late-onset CAH or, very rarely, excess secretion of androgens by ovarian or adrenal tumours. Polycystic ovaries commonly cause mild hyperandrogenization.

Newborn girls with ambiguous genitalia. The cause of ambiguous genitalia in a newborn infant must be diagnosed as soon as possible so that the best sex of rearing can be advised and the parents counselled. The commonest cause of ambiguous genitalia in the newborn female is an enzyme defect of cortisol synthesis such that intermediates are diverted to androgen production (New & Josso, 1988). The female genotype should be confirmed by chromosome analysis; blood should be taken in lithium heparin and sent without delay to an appropriate laboratory that offers a speedy return of results. An ultrasound of the genital tract by a specialist will also be helpful if it reveals a uterus and vagina. Other external causes of a virilized female are attributed to maternal ingestion of progestogens or androgens or to maternal production of androgens by an adrenal or ovarian tumour. In these cases the child will have normal endocrinology after birth but may need corrective surgery on the external genitalia.

A reduction in steroid 21-hydroxylase activity or an absence of 11β-hydroxylase or 3β-OH-S-DH can cause CAH. Steroid 21-hydroxylase deficiency is by far the commonest cause of CAH (90% of cases), and in Europe 60% of all patients with steroid 21-hydroxylase deficiency will present in the newborn period with a salt-losing crisis. Replacement treatment with steroids will be needed as soon as possible. An increase in serum potassium may be observed prior to a more pronounced fall in body weight than is seen in the normal baby, and occurs before hyponatraemia is detected.

17-OH-P is a precursor of cortisol and in patients with a deficiency of 21-hydroxylase the production of 17-OH-P increases and serum levels are elevated. The 21-hydroxylase enzyme is not absent but the reduced capacity to produce cortisol leads to high ACTH levels which, in turn, cause adrenal hyperplasia. There is also a high adrenal secretion of androgens, leading to virilization.

21-Hydroxylase deficiency is diagnosed by measuring 17-OH-P in serum, plasma or blood spots and the timing of sample collection is important. In all newborns, 17-OH-P concentrations in the serum are high (above 100 nmol/L) on the first day of life and the levels fall to less than 15 nmol/L during the first week. After day 3, 17-OH-P levels in affected cases are usually clearly raised (100–800 nmol/L) compared with the values in normal infants. Differences in results obtained with a direct method and an extraction method would suggest that steroids in the blood (probably steroid sulphates from the fetal adrenal) cross-react in the RIA, giving higher results in the direct assay (Wallace *et al.*, 1987; Makela & Ellis, 1988; Wong *et al.*, 1992). The suggestion that steroid sulphates affect the quality of the assay is supported by the observation that there is a greater discrepancy between the results obtained by the two methods in premature and low birth-weight babies, both of whom will have fetal adrenal activity. The difference is sustained after birth for many months longer than is usual in a normal infant.

A 17-OH-P assay which involves solvent extraction before the RIA is essential. Now that specific urinary metabolites of 17-OH-P have been recognized, diagnosis of the condition by GC analysis of steroids in the urine is reliable (Honour, 1986). Pregnanetriol is an unsatisfactory marker for the disorder in the newborn. A characteristic steroid pattern can be recognized, of which the most informative steroid is 17α-hydroxypregnanolone. It is essential that a laboratory offering this analysis should provide a rapid service. The pattern of steroids is complex and the identity of the steroids in the GC analysis must be confirmed by a further analysis with GC coupled to a mass spectrometer.

Once CAH due to 21-hydroxylase deficiency has been confirmed and lifelong cortisol treatment has commenced, compliance is best assessed in children by following their growth (Appan *et al.*, 1989). Parents must be advised to bring the children to hospital so that height and weight can be followed at 3-monthly intervals in the first 2 years of life, then at 6-monthly intervals. Bone age is checked yearly from X-rays of the wrists and hands. The replacement therapy should be adjusted according to body size. Hydrocortisone is usually prescribed at 20–25 mg/m^2 per day with two-thirds of the dose being taken in the morning and one-third in the evening. If fludrocortisone is given to control salt loss, it must be remembered that this steroid is a potent glucocorticoid itself and the dose should not exceed 0.15 mg/m^2. Electrolytes can be measured periodically but for long-term assessment of mineralocorticoid replacement the measurement of PRA is advisable. The measurement of 17-OH-P and androgens in blood (or saliva) taken at regular intervals will define the adrenal steroid output in relation to treatment, but in practice has little effect on compliance.

In rare cases of CAH the defect is a result of low activity of the 11β-hydroxylase enzyme. This defect is best identified by a raised serum concentration of 11-deoxycortisol (Perry *et al.*, 1982) or by a urine steroid profile. As with assays for 17-OH-P in the newborn period, there may be problems with measuring 11-deoxycortisol in the plasma of young children. The plasma assay must include solvent extraction of free steroids so as to reduce the possible cross-reaction of steroid sulphates from the fetal adrenal zone. Owing to other changes in steroid metabolism, a high excretion of 6-hydroxy-tetrahydro-11-deoxycortisol (6-hydroxy-THS) in urine is a better marker of the defect in the newborn than is THS, which is elevated in older patients but not so clearly raised in the newborn (Hughes *et al.*, 1986); this again emphasizes the need to involve specialist laboratories with relevant experience (Ratcliffe *et al.*, 1982).

Androstenedione measurements may be helpful in the management of CAH owing to 21-hydroxylase and 11β-hydroxylase defects. Measurement of PRA can be used to monitor the efficacy of mineralocorticoid treatment in CAH. Patients with 21-hydroxylase and 3β-OH-S-DH defects manifest an elevated PRA, while in the defects with mineralocorticoid excess (17α-hydroxylase and 11β-hydroxylase) PRA is suppressed. PRA is normalized with effective treatment

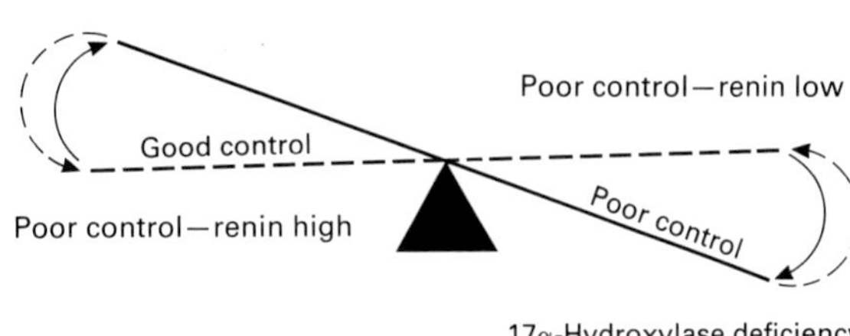

Fig. 10.10 Plasma renin activity in patients treated for congenital adrenal hyperplasia.

(Fig. 10.10). In the case of 21-hydroxylase deficiency and 3β-OH-S-DH defects, treatment is improved by the addition of fludrocortisone.

Precocious puberty. Normal puberty is the acquisition of secondary sexual characteristics between the ages of 8 and 18 years. The process is initiated by an increase in pulsatile secretion of GnRH at night, although the mechanism and timing of this initiation are still not understood. When sexual maturation appears before 8 years in girls and before 9 years in boys puberty is considered precocious.

When early puberty is being investigated in a girl, an ovarian tumour may be palpated as a pelvic mass or seen on an ultrasound scan of the abdomen, but such tumours are rare. Central precocious puberty (gonadotrophin-dependent precocious puberty) reflects early activation of the gonadotrophic drive to increased gonadal function, which is more common in girls than in boys. Among cases of central precocious puberty 75–95% are termed idiopathic as they have no detectable biochemical cause. A child presenting clinically with precocious puberty, pubertal gonadotrophin levels and augmented nocturnal gonadotrophin secretion has central precocious puberty (Stanhope *et al.*, 1986). In girls, a pelvic ultrasound scan is useful for the assessment of central precocious puberty (Stanhope *et al.*, 1985). A cerebral tumour is relatively more common in boys with central precocious puberty than in girls. Hamartomas, third ventricular cysts, astrocytomas or gliomas need to be excluded by radiological examination. Some dysgerminomas and hepatoblastomas secrete HCG, and sometimes α-fetoprotein (AFP), and the concentrations of these peptides in the blood should be measured.

Gonadotrophin-independent precocious puberty is due to inappropriate production of gonadal or adrenal hormones which affect secondary sexual characteristics. Children may have acne and behavioural problems and may become taller than their peers. There is early epiphyseal fusion as a result of rapid skeletal maturation, and resultant short stature in adulthood. Hypothyroidism should be excluded as increased TSH is associated with a concomitant increase in FSH and prolactin.

McCune–Albright syndrome is the eponym for precocious puberty associated with brown-pigmented irregular-edged skin marks and polyostotic fibrous dysplasia. Plasma gonadotrophin concentrations are often pubertal, but may in some cases be prepubertal (2–6 U/L). Symptoms may wax and wane, and on an ultrasound scan this can be correlated with the appearance and regression of unilateral ovarian cysts.

Pseudo-precocious puberty can be the result of exposure to exogenous sex steroids. Abnormal sex steroid secretion from adrenal or gonadal tumours is another cause. The most common adrenal tumour reported in the literature secretes DHAS, which is reflected in the high excretion of 17-oxosteroids. In general, the tumours have been quite large, probably because they are secreting weaker androgens than some other tumours and reach a later stage before the clinical signs are investigated. A number of cases of other patients with tumours secreting other androgens have been reported, and using urine steroid profile analysis the secretion of 11β-hydroxyandrostenedione has been defined on the basis of high excretion of 11β-hydroxyandrosterone (Honour *et al.*, 1984). In the reported cases 17-oxosteroid excretion was not grossly elevated and without scanning of the adrenals might have been dismissed as premature adrenarche. FSH and LH are suppressed to within prepubertal ranges (less than 5 U/L). The secretion of androgens by adrenal tumours is not suppressed by giving dexamethasone. In a boy, a testicular mass with grossly elevated 17-OH-P usually indicates non-salt-losing CAH with adrenal rests in the testes. Leydig cell tumours may, however, produce elevated 17-OH-P, but in contrast to CAH this is not suppressed with dexamethasone. 21-Deoxycortisol is not raised in these patients, unlike those with CAH (Solish *et al.*, 1989).

Premature adrenarche. The growth of pubic hair before breast development in girls or testicular enlargement in boys may be the outcome of increased secretion of DHAS from the adrenal cortex owing to early differentiation of the zona reticularis (premature adrenarche), a benign disorder that does not require treatment (Sklar *et al.*, 1980). DHAS concentrations in the serum should be interpreted against reference ranges for the age. Testosterone and androstenedione may be slightly elevated for the age owing to peripheral conversion of the adrenal DHAS. If 24-h urine excretion of androsterone, aetiocholanolone and cortisol metabolites is above the normal range for age and body size, advanced adrenal growth, a benign condition, is probable.

Acne, hirsutism and menstrual disturbance. In recent years non-classical forms of CAH have been recognized. Children with hirsutism and/or acne have been found to have a mild defect of the steroid 21-hydroxylase. In many such patients the basal 17-OH-P concentrations are more than 5 nmol/L, which is the upper limit of the reference range. An injection of 250 μg Synacthen/m^2 will lead to an increase in serum 17-OH-P above 10 nmol/L at 30 and 60 min after the trophic hormone injection; this is the increment seen in normal subjects (Hague *et al.*, 1989). A late-onset form of 3β-OH-S-DH deficiency has been described on the basis of an exaggerated DHA or 17-hydroxypregnenolone response to ACTH (Cravioto *et al.*, 1986).

Isolated breast development (premature thelarche) without growth acceleration. Increased oestrogen production in girls with premature thelarche is a benign condition that is not to be confused with central precocious puberty. The latter is more serious and causes progressive breast development, growth of pubic hair, accelerated growth rate and bone maturation, and early epiphyseal fusion. The two conditions are resolved by ultrasound and repeated blood sampling at 15-min intervals throughout the night (Stanhope *et al.*, 1986). At the time of breast enlargement a pelvic ultrasound scan will show a large isolated cyst in a patient with thelarche, whereas a patient with precocious puberty will almost certainly have a multicystic ovarian appearance on the ultrasound scan. In patients with thelarche, the FSH concentration is higher (2–7 u/L) than the LH level (1–3 u/L); this contrasts with precocious puberty where LH secretion predominates. A GnRH stimulation test may sometimes distinguish the two disorders by revealing differences in the dominant gonadotrophin (Pescovitz *et al.*, 1988).

Androgen deficiency

Newborn infants. Incomplete virilization of a 46XY fetus is the result of a failure in production of the sex steroids, or of resistance to these hormones because of receptor defects (Saenger, 1984; New & Josso, 1988). Males with defects in androgen action may be considered for rearing as females. Most laboratories will need the help of specialized centres to resolve such cases, although a number of investigations can be undertaken locally.

Following confirmation of the male karyotype, the serum concentrations of cortisol and testosterone should be checked. From these results some clear decisions can be made. Defects of sex steroid production can be separated into: (a) those affecting cortisol and androgens; and (b) those relating to androgens alone.

Low cortisol production in a poorly virilized male

Table 10.2 Steroid markers for adrenal diseases

Defect	Excess enzyme substrates	Excess urine metabolites
3β-Hydroxysteroid dehydrogenase	DHA 17-Hydroxypregnenolone	DHA Pregnanetriol
17α-Hydroxylase	Progesterone DOC B	Pregnanediol THDOC THB
21-Hydroxylase	17-Hydroxyprogesterone	17-Hydroxypregnanolone Pregnanetriol
11β-Hydroxylase	S	THS

DHA, dehydroepiandrosterone sulphate; (TH)DOC, (tetrahydro)-11-deoxycorticosterone; (TH)B, (tetrahydro)-corticosterone; (TH)S, (tetrahydro)-11-deoxycortisol.

can be attributed to defects of cholesterol 20,22-desmolase (side-chain cleavage), 17α-hydroxylase deficiency or 3β-OH-S-DH deficiency. These enzymes affect the production of all the important adrenal and gonadal steroid hormones and defective production is often fatal for the child, so that few cases are documented in the literature. 20,22-Desmolase deficiency was initially called lipoid adrenal hyperplasia, on account of the histological appearance of the adrenals at post mortem, reflecting the ACTH-stimulated adrenal which cannot process cholesterol to produce cortisol (Prader & Gurtner, 1955).

17α-Hydroxylase and 3β-OH-S-DH defects can be identified biochemically by demonstrating high serum levels and the presence in urine of metabolites of the respective enzyme substrates (Table 10.2). Experience with urine steroid profiles in such cases in the newborn period is limited. In one reported case, later confirmed to have 17α-hydroxylase deficiency, the pattern of steroids in the urine of a neonate showed a high excretion rate only of 16α-hydroxypregnanolone. At 15 months of age the child excreted corticosterone metabolites with a pattern similar to that of metabolites excreted in the urine of adults with this disease (Honour *et al.*, 1978; Dean *et al.*, 1984).

3β-OH-S-DH deficiency is, in practice, difficult to confirm in a newborn because, owing to normal inactivity of this enzyme in the fetal adrenal cortex, the markers for the defect (DHAS and pregnenolone) are the usual products of the adrenal in the newborn. A 24-h urine collection with determination of the excretion rates of all steroids is therefore useful. Since these infants are usually very sick, the collection of urine may be difficult. Plasma measurements of DHAS and ACTH may be better tests before treatment, but in the face of an adrenal crisis replacement therapy may be needed before the necessary tests are completed. To overcome this problem dexamethasone is the preferred treatment. If a child suspected of having 3β-OH-S-DH deficiency is treated with dexamethasone and fludrocortisone and then given daily injections of depot ACTH (Synacthen), the markers for the defect may be displayed in the urinary steroid profile without interference from dexamethasone. This approach has been used successfully in confirming 3β-OH-S-DH deficiency in one child maintained on hydrocortisone for several years, although ACTH was needed for several days before the suppressed adrenal secreted sufficient steroid to be detected in the assay (Taylor *et al.*, 1979). Males with defects in cortisol and androgen production may be reared as boys but will require adrenal steroid replacement with the addition of androgens at puberty.

If cortisol production is normal the further investigation of males with poorly developed genitalia is necessary to determine other defects in androgen metabolism, and action will depend upon the initial levels found for testosterone and gonadotrophins. A low basal testosterone with elevated gonadotrophins and a poor androgen response to HCG suggest either Leydig cell hypoplasia or an androgen biosynthetic defect. A ratio of androstenedione: testosterone above 2 in the basal state, exaggerated by HCG stimulation (ratio greater than 5), suggests a defect of 17-ketosteroid reductase. A newborn male with this defect can be reared as a boy but will need testosterone treatment. The usual presentation, however, is around puberty when a phenotypic girl will show signs of virilization owing to excess androstenedione production from the testes, which were undescended until this time.

A normal basal testosterone level with a rise following HCG suggests that the problem is the result of impaired action of testosterone, which can be due to a failure of target tissue 5α-reductase or to receptor defects. Specialist help will be needed to resolve this problem. A high testosterone:5α-DHT ratio (greater than 15) after HCG stimulation supports a diagnosis of 5α-reductase deficiency. In a urinary steroid profile there is also evidence for this disorder in the distribution of cortisol metabolites (5α-THF to THF) (Imperato-McGinley *et al.*, 1986). In newborns, however, the majority of cortisol metabolites have been oxidized to cortisone metabolites, leaving very little of the cortisol metabolites to facilitate effective confirmation of this diagnosis.

To confirm the number and stability of androgen receptors, genital skin has to be taken and sent to a specialist laboratory. The cell number will have to be increased through several courses of tissue culture before the investigations can be undertaken.

Delayed puberty. Among boys and girls 97% will show some signs of puberty by the 16th year. Complete absence of clinical signs by the age of 15 years warrants investigation. The causes of delayed puberty are many. An algorithm for the diagnosis of delayed puberty has been published elsewhere (Chapman, 1989) and in view of the rarity of this condition will not be discussed in detail here. The most common problem is isolated gonadotrophin deficiency. After confirming low levels of gonadotrophins and prolactin, the combined pituitary function test is the most useful laboratory investigation.

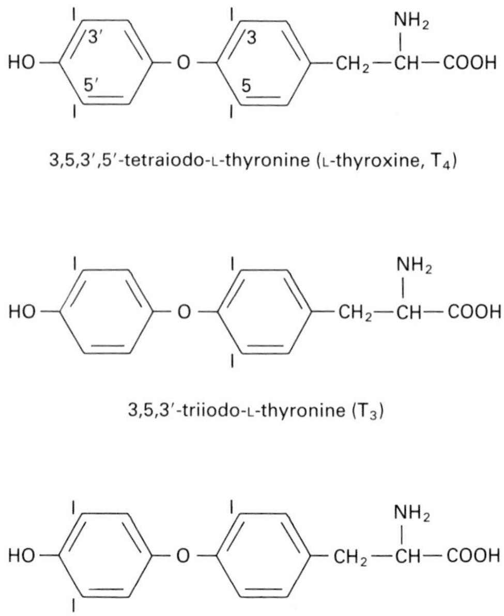

3,5,3′,5′-tetraiodo-L-thyronine (L-thyroxine, T_4)

3,5,3′-triiodo-L-thyronine (T_3)

3,3′,5′-triiodo-L-thyronine (reverse T_3)

Fig. 10.11 Structures of thyroid hormones.

Thyroid function

Introduction

The thyroid gland secretes two major hormones, tyrosine derivatives, which differ from each other in the number and arrangement of iodine atoms; in order of potency, these are designated T_3 and T_4 (triiodo- and tetraiodothyronine or thyroxine) (Fig. 10.11). Thyroxine is metabolized by 5′-deiodination to T_3. A third form of the hormone, reverse T_3 (rT_3), is a structural isomer of T_3 of even lower biological activity. In the fetus and newborn infant, rT_3 is relatively more important but this significance declines rapidly over the first month of life.

The thyroid glands require iodine for hormone synthesis and the need is met by a highly efficient mechanism for concentrating the halogen in the glands. Transport of iodine from the plasma against a concentration gradient requires energy, which is generated through the activity of a sodium pump. Iodine is oxidized by a membrane-bound peroxidase system, a haemoprotein which also catalyses the incorporation of iodine into the tyrosine molecules of thyroglobulin. Thyroglobulin is the major component of colloid, a fluid filling the follicles. Two iodinated tyrosine residues become coupled by an ether link between the amino acids, which are still within the thyroglobulin. A serine residue is left in place of one of the tyrosyls. The secretion of thyroid hormones involves the hydrolysis of thyroglobulin in order to dissociate T_4 from its peptide linkage with the protein. The thyroid is unique among the endocrine glands in storing hormone.

The trapping of iodine, the peroxidase activity, the coupling reaction and the release of thyroid hormones are all stimulated by TSH, secreted by the anterior pituitary. Production of TSH is itself under the neuroendocrine control of the hypothalamic thyrotrophin-releasing hormone (TRH). The hypothalamic-pituitary-thyroid system ensures that blood levels of the thyroid hormones are maintained constant by a negative feedback mechanism.

The active iodothyronine hormones in plasma are strongly bound to proteins; only about 0.015% of the total T_4 and 0.3% of the T_3 are present in the serum as free hormones. The strength of binding is greater to thyroxine-binding globulin (TBG) than to thyroxine-binding pre-albumin (TBPA), which has superior avidity than to serum albumin.

Thyroid hormones have general biological effects on oxygen consumption by tissues (calorigenic action) as well as being essential for normal growth and development. Many of the actions of thyroid hormones are mediated by T_3, which is secreted by the gland but also produced by deiodination of T_4. Most tissues in the body have a requirement for the thyroid hormones to a greater or lesser degree. Administration of thyroid hormones leads to an increase in metabolic rate. Thyroid hormones restore normal growth in thyroid-deficient children, in addition to improving energy and temperature control. The association of impaired neurological development (sporadic cretinism) in infants with atrophy of the thyroid was recognized more than a century ago. Myxoedema is the apparent oedema or subcutaneous mucus deposit that causes fluid retention in children with hypothyroidism.

The thyroid glands are well developed in the human fetus by 10–12 weeks of gestation, although they are functionally not capable of trapping iodine until mid-gestation (Fisher, 1981). The pituitary TSH content increases at mid-gestation and the serum concentrations of TSH rise progressively from 22 weeks gestation to term. There are, likewise, progressive increases in T_4 and free T_4 (fT_4) concentrations.

Collection of samples

Investigations of thyroid function are largely based on measurements of circulating hormone levels. The timing of the sampling during the day is less crucial for interpretation than is the case for adrenal hormones. The age of the child and the gestation age are important factors which influence the interpretation of results. Blood spots dried on filter papers are the basis of screening programmes to detect congenital hypothyroidism in neonates. There is much debate about the relative merits of separate and combinations of thyroid assays (T_4, T_3, total hormone, free hormones, TSH) for diagnostic purposes, and of the values of combinations of these analytes when monitoring treatment of thyroid disorders. For other diagnostic purposes it is usual to measure TSH or T_4 as a single screening test and then to perform further measurements if the initial result is equivocal. Increases or decreases in total serum thyroid hormone concentrations cannot be said to reflect an alteration in thyroid function until both thyroid-hormone-binding capacity and the levels of free hormones have been estimated. TSH can be undetectable in hospitalized children with non-thyroidal illness, both organic and psychiatric. Treatment with drugs, including corticosteroids, can also cause TSH to be undetectable and information about the administration of any drugs should be recorded with the request for the analysis.

Serum samples may also be needed for the measurement of thyroid autoantibodies when attempting to elucidate the aetiology of certain thyroid diseases. The assays are technically demanding and should be referred to a specialist centre.

Assays and reference values

Total T_4 assays using specific T_4 antibodies can adopt isotopic or non-isotopic methodologies. Measurements of the total T_4 and T_3 in serum are achieved by RIA after displacing hormones from their binding proteins with salicylic acid or anilino-naphthalene sulphonic acid (ANSA). The new sensitive TSH assays (to less than 0.01 mu/L), based on two antisera (immunometric or sandwich assays), are important thyroid function tests (Nicoloff & Spencer, 1990).

Free hormone assays should ideally be based on dialysis methods but many commercial kits use indirect methods, such as those based on labelled thyroid analogues of T_3 and T_4; these are said not to displace endogenous bound thyroid hormones from the binding proteins. By an immunometric procedure, the analogue competing with fT_4 in the sample for antibody binding offers an assessment of free hormone concentrations. The assays are known to be affected by other clinical circumstances and can give low results owing to abnormalities in serum T_4 binding (severe non-thyroidal illness) and high results with hereditary dysalbuminaemias. Other indirect assays for free thyroid hormones are two-step methods which involve two incubation and two washing steps. In the first step the serum is incubated with solid-phase coupled T_4 antibody in order to remove a percentage of the total T_4. After decanting and washing away the unbound serum constituents, the remaining unbound antibody-binding sites are measured by use of a labelled tracer; this is followed by a second decanting and washing step. This provides a value which is proportional to the absolute fT_4 or fT_3 value of the serum.

Most laboratories will strike a compromise between what is best and what is practical to operate in order to expedite the workload. Thus, analogue methods for free thyroid hormones will be widely used because the dialysis methods are too tedious for routine use (Hay *et al.*, 1991), but the diagnostic accuracy may be unknown.

Immediately after birth and probably as the result of a lowering of body temperature, there is a rapid increase in 5′-deiodination activity such that serum T_3 levels rise from less than 1.5 nmol/L to lie between 1.5 and 5 nmol/L, whilst rT_3 levels remain stable between 1 and 7 nmol/L. The serum T_4 rises above the reference ranges seen in young children (90–200 nmol/L) because of a sudden release of TSH to levels (70–100 mu/L) markedly above the adult range. Thereafter, TSH is less than 20 mu/L. TBG levels are essentially unchanged in the neonatal period so the fT_4 and fT_3 concentrations rise. The high neonatal levels of rT_3 decrease rapidly over 4 weeks to approach values seen throughout childhood (0.2–1.5 nmol/L).

Premature babies have a number of clinical problems, e.g. birth trauma, hypoxia, hypoglycaemia, hypocalcaemia and infection, which inhibit conversion of T_4 to T_3 and perpetuate a low T_3 state. All premature babies have, for 4–8 weeks after birth, lower thyroid hormone concentrations than do term babies. Thus, 50% of preterm babies delivered before 30 weeks have T_4 below 85 nmol/L. High thyroglobulin concentrations in a preterm baby also mean that fT_4 levels are very low. Serum T_3 may remain below 0.2 nmol/L for up to 2 months after birth.

During the first year of life after delivery at term, T_4 concentrations are 100–200 nmol/L and T_3 concentrations are 1.5–5 nmol/L. These values decrease progressively during the first decade to 80–160 nmol/L and 1.5–3.5 nmol/L, respectively. TBG levels in children are higher than in adults. The ranges for TBG narrow gradually from 10 to 60 nmol/L at birth to 18 to 33 nmol/L in puberty.

Tests of thyroid function

Following an initial investigation of circulating TSH levels, and perhaps total or free thyroid hormone concentrations, some further test procedures may be needed to address specific clinical questions regarding the aetiology of thyroid disease. The strategy will vary between laboratories. With the introduction of sensitive TSH assays, the TRH test is less useful because the TSH will almost certainly respond to the TRH stimulus if the basal TSH is detectable. The TRH test (3–5 μg/kg intravenously in children) allows differentiation between suppressed TSH in some cases of hypothalamic and pituitary disease, because in the former case TSH will rise to a peak after 20 min when the absent TRH is replaced.

Radioiodine uptake procedures to assess hyperfunction of the thyroid glands are rarely used in children. Imaging techniques may be necessary to confirm ectopic tissue.

Hypothyroidism

Congenital hypothyroidism is potentially the main cause of mental impairment in childhood and can affect about 1 in 4000 births. In many countries screening programmes for the condition have been in operation since 1982–1983 (Letarte & Garagorri, 1989). Early detection and early treatment are known to be important because the longer the delay, the greater the irreversibility of the mental retardation. Hypothyroidism is seldom apparent clinically at birth, except in a few cases where there is a goitre (enlarged thyroid gland). Suggestive signs of hypothyroidism are high birth weight, enlargement of the posterior fontanelles, persistence of neonatal jaundice and hypothermia. The age at which symptoms of hypothyroidism become recognizable will depend upon the degree of impairment of thyroid function in the infant. If unrecognized in the newborn, hypothyroidism leads to cretinism – children are dwarf, mentally retarded and have large protruding tongues and potbellies. Other signs include slow feeding, lethargy and delayed bone maturation.

A number of structural or functional abnormalities of the thyroid glands may lead to deficient production of thyroid hormones. Worldwide, the most common cause of hypothyroidism in newborns is maternal iodine deficiency (Glinoer *et al.*, 1992). The complete absence of thyroid tissue or failure of the thyroid gland to locate properly during fetal development is present in around 1 in 5000 newborns, and most of the cases are sporadic. Of all hypothyroid infants about 10–20% have T_4 levels in the normal range, possibly owing to the activity of some ectopic thyroid tissues. Thyroid dysgenesis is twice as common in girls as in boys. Autoimmune thyroiditis is the third most common cause of hypothyroidism in children but is later in onset, usually occurring in mid-childhood. About 15% of hypothyroid infants have defects of thyroid hormone synthesis. Congenital hypothyroidism due to biochemical defects in hormone synthesis (Table 10.3) presents in a similar way to embryological defects and is very rare (Lever *et al.*, 1983); distinction is aided by the time of onset of goitre and further tests based on uptake of radioactive iodine into the thyroid glands under different circumstances (Table 10.4). Rare instances of thyroid dysgenesis have been recorded consequent upon maternal autoimmune thyroiditis.

Table 10.3 Autosomal recessive biochemical defects of thyroid hormone production

Decreased response to thyroid-stimulating hormone (TSH) owing to abnormal coupling of TSH receptor to adenyl cyclase
Failure to concentrate iodide (Wolff, 1983)

Iodine organification defects
  Absent peroxidase
  With hearing loss

Iodotyrosine deiodinase defect
Defect in thyroglobulin synthesis or transport
Thyroid hormone resistance

If hypothyroidism is suspected in a newborn, TSH and T_4 can be assayed in cord blood or later. For screening purposes blood spots taken between days 3 and 14 are dried onto filter paper, as is done for screening tests for phenylketonuria (PKU). In Europe, TSH is assayed in the blood spots, whereas in the United States it is usual to measure T_4 first. Many mothers and children are discharged within 3 days of delivery. Early measurements of TSH may give a false positive result owing to a physiological neonatal TSH surge and must be interpreted against reference ranges for the age (Table 10.5). A TSH value above 20 mu/L makes the diagnosis of primary hypothyroidism almost certain. If the levels of T_3 and thyroglobulin are then found to be normal or near-normal this is evidence against the absence of at least some thyroid tissue. Screening based on TSH alone gives a few false positive results. Spuriously elevated TSH can rarely be due to the presence of maternal antibodies transferred across the placenta. TSH screening may miss some cases of secondary hypothyroidism.

In North America, T_4 concentrations are first determined in the screening for hypothyroidism.

Table 10.4 Characteristics of thyroid hormone synthetic defects

T_4	TSH	Presentation	Diagnostic aids
Low	High	Small thyroid	Normal radioiodide uptake
Low	High	Enlarged thyroid	Low radioiodide uptake
Low	Normal	Goitre, cretinism	Partial or rapid loss of iodine on administration of perchlorate or thiocyanate
Low	High	Goitre at birth	Loss of iodine in urine
Low	High	Goitre, familial	High levels of iodinated albumin in serum and urine
High	Detectable		

T_4, thyroxine; TSH, thyroid-stimulating hormone.

Table 10.5 Interpretation of results in thyroid screening

	T_4 (nmol/L)	TSH (miu/L)	
Cord serum	<75	>80	Primary hypothyroidism
Serum (2–3 days)	<90	>10	Primary hypothyroidism
Serum (>7 days)	<75	>50	Primary hypothyroidism
Serum (>7 days)	<75	20–50	Suspect hypothyroidism

T_4, thyroxine; TSH, thyroid-stimulating hormone.

All the samples which give values below 130–140 nmol/L (tenth percentile of the normal range) are then reassayed for TSH. An elevated TSH then confirms primary hypothyroidism, while a follow-up normal TSH indicates secondary hypothyroidism. Low TSH levels are recorded in children with pituitary or hypothalamic hypothyroidism, which can affect 1 in 60 000 to 140 000 births. Hypothalamic hypothyroidism can be distinguished from pituitary hypothyroidism by the presence in the former of a rise in plasma TSH following a test dose of TRH. The TSH response is usually normal in hypothalamic hypothyroidism, while it is increased in hypothyroidism due to thyroid disease and decreased in hyperthyroidism due to feedback of thyroid hormones on the pituitary gland. TSH deficiency may be isolated or associated with other pituitary hormone deficiency. Tests of growth hormone and ACTH production may be appropriate. An occasional infant may escape detection in newborn screening based on T_4 and TSH.

The diagnosis of hypothyroidism, indicated by screening, should be confirmed by quantitative assessment of T_4 and TSH on plasma samples. The reference ranges for thyroid hormones and related serum constituents are relatively stable between 2 and 6 weeks after birth (Table 10.6). The results of the screening are usually available before clinical evidence of hypothyroidism – slow feeding, lethargy and delayed bone maturation. Treatment should be started as soon as possible, even before the second test results are known. If the second test is not conducted in the newborn period, the replacement treatment should be continued for the next 3 years regardless of the test outcome. The therapy can then be withdrawn for 1 month at an

Table 10.6 Reference values for serum thyroid hormone concentrations in babies aged 2–6 weeks. After Fisher (1991)

Serum constituent	Concentration
T_4	20–200 nmol/L
T_3	1.5–4.5 nmol/L
Free T_4	10–30 pmol/L
TSH	1.5–9 mu/L
TBG	150–750 nmol/L
Thyroglobulin	15–400 pmol/L

T_4, thyroxine; T_3, triiodothyronine; TSH, thyroid-stimulating hormone; TBG, thyroxine-binding globulin.

appropriate time to confirm the low thyroid gland activity.

In some cases neonatal hypothyroidism is transient. Treatment should be stopped to re-evaluate the situation. Transient neonatal hypothyroidism usually reflects the circumstances in the newborn after maternal ingestion of substances which reach the fetus by placental transfer and interfere with thyroid hormone synthesis (goitrogenic substances). Iodine, prescribed in expectorants for the alleviation of asthma, may be the cause of transient hypothyroidism in some young infants. In children who have untreated primary hypothyroidism there may be poor growth and abnormal sexual maturation as a result of widespread disturbance of pituitary function (Buchanan *et al.*, 1988; Pringle *et al.*, 1988).

Acquired hypothyroidism

Iodine deficiency is the most important factor in endemic goitre and affects some 300 million people worldwide, particularly in Germany, Switzerland, Austria, Italy, Greece, Lebanon and Iraq. Dietary and environmental factors may influence the onset of hypothyroidism. The most important sign of acquired hypothyroidism is growth failure accompanied by skin changes (myxoedema – dry skin owing to decreased turnover of skin protein) and delayed sexual development.

Hypothyroidism may develop in childhood, often after 6 years of age in previously normal individuals. Around 2% of children may acquire antithyroid antibodies and autoimmune thyroiditis (Hashimoto disease). There is usually a strong family history of thyroid disease. The thyroid glands become irregularly enlarged and this may affect swallowing. Hashimoto thyroiditis is sometimes associated with diabetes mellitus either with or without the further complication of adrenal insufficiency. All children with diabetes mellitus should be screened for antithyroid antibodies. Serum TSH will be raised in children with acquired hypothyroidism. Gonadotrophins (particularly FSH) may be raised and form the basis of sexual precocity in girls and boys. Secondary menstrual irregularities are frequent complaints of postmenarcheal girls with hypothyroidism.

Hyperthyroidism

Thyrotoxicosis is characterized by nervousness, tachycardia, weight loss and heat intolerance owing to an increased metabolic rate. The skin feels warm and moist. The most common form of hyperthyroidism in childhood is Grave's disease. The eyeballs protrude, a condition called exophthalmos. The thyroid is diffusely enlarged and hyperplastic. Children may show rapid onset of unusual and erratic behaviour with poor school progress. High levels of circulating T_3 and T_4 suppress the secretion of TSH through the negative feedback effect, and TSH levels will be low. Only rarely is it necessary to confirm the diagnosis by estimating plasma TSH after an injection of TRH. Release of TSH is absent in hyperthyroidism, while a rise is considered to exclude the diagnosis. Circulating antibodies against TSH receptors in the thyroid activate, through adenyl cyclase, the receptor mechanism in the thyroid gland and cause thyroid stimulation. Hyperthyroidism caused by a functioning thyroid adenoma or adenomatous goitre is rare in the paediatric age range.

Neonatal hyperthyroidism is due to maternal Grave's disease and is caused by the passage of thyroid-stimulating immunoglobulins from the mother to the fetus. Plasma T_4 and T_3 levels will be above those levels normally expected in the newborn period.

References

Abdulwahid, N.A., Armar, N.A., Morris, D.V., Adams, J. & Jacobs, H.S. (1985) Diagnostic tests with luteinising hormone releasing hormone should be abandoned. *Br Med J* **291**, 1471–1472.

Ahrensten, O.D., Jensen, J.K. & Johnsen, S.G. (1982) Sex hormone binding globulin deficiency. *Lancet* **2**, 377.

Al-Dahhan, J., Haycock, G.B., Chantler, C. & Stimmler, L. (1983) Sodium homeostasis in term and preterm infants. I. Renal aspects. *Arch Dis Child* **58**, 335–342.

Appan, S., Hindmarsh, P.C. & Brook, C.G.D. (1989) Monitoring treatment in congenital adrenal hyperplasia. *Arch Dis Child* **64**, 1235–1239.

Armanini, D., Kuhnle, U., Strasser, T. *et al.* (1985) Aldosterone receptor deficiency in pseudohypoaldosteronism. *N Engl J Med* **313**, 1178–1181.

Atkinson, A.B., Kennedy, A.L., Carson, D.J. *et al.* (1985) Five cases of cyclical Cushings syndrome. *Br Med J* **291**, 1453–1457.

Baelen, H.V., Brepoels, R. & de Moor, P. (1982) Transcortin Leuven: a variant of human corticosteroid-binding globulin with decreased cortisol-binding affinity. *J Biol Chem* **257**, 3397–3400.

Barragry, J.M., Mason, A.S., Seamark, D.A., Trafford, D.J.H. & Makin, H.L.J. (1980) Defective cortisol binding globulin affinity in association with adrenal hyperfunction: a case report. *Acta Endocrinol* **95**, 194–197.

Beastall, G.H., Ferguson, K.M., O'Reilly, D.St J., Seth, J. & Sheridan, B. (1987) Assays for follicle stimulating and luteinising hormone; guidelines for the provision of a clinical biochemistry service. *Ann Clin Biochem* **24**, 246–262.

Bidlingmaier, F., Dorr, H.G., Eisenbenger, W., Kuhnle, U. & Knorr, D. (1986) Contribution of the adrenal gland to the production of androstenedione and testosterone during the first two years of life. *J Clin Endocrinol Metab* **62**, 331–334.

Biglieri, E.G. & Schambelan, M. (1979) The significance of elevated levels of plasma 18-hydroxycorticosterone in patients with primary aldosteronism. *J Clin Endocrinol Metab* **48**, 87.

Brook, C.G.D. (ed.) (1989) *Clinical Paediatric Endocrinology*. Blackwell Scientific Publications, Oxford.

Buchanan, C.R., Stanhope, R., Adlard, P. *et al.* (1988) Gonadotrophin, growth hormone and prolactin secretion in children with primary hypothyroidism. *Clin Endocrinol* **29**, 427–436.

Burke, C.W. (1985) Adrenocortical insufficiency. *Clinics Endocrinol Metab* **14**, 947–976.

Carpenter, P.C. (1988) Diagnostic evaluation of Cushing's syndrome. *Endocrinol Metab Clinics N Am* **17**, 445–472.

Chapman, A.J. (1989) Delayed puberty. In: *Endocrine System – Clinical Algorithms*, pp. 37–40. Cambridge University Press, Cambridge.

Collu, R., Ducharme, J.R. & Guyda, H.J. (1989) *Pediatric Endocrinology*, 2nd edn. Raven Press, New York.

Cravioto, M.A. del C., Ulloa-Aguirre, A., Bermudez, J.A. *et al.* (1986) A new variant of the 3β-hydroxysteroid dehydrogenase-isomerase deficiency syndrome: evidence for the existence of two isoenzymes. *J Clin Endocrinol Metab* **62**, 360–367.

Crock, P.A., Pestell, R.G., Calenti, A.J. *et al.* (1988) Multiple pituitary hormone gradients from inferior petrosal sinus sampling in Cushing's disease. *Acta Endocrinol* **119**, 75–80.

Cronin, C., Igoe, D., Duffy, M.J., Cunningham, S.K. & McKenna, T.J. (1990) The overnight dexamethasone test is a worthwhile screening procedure. *Clin Endocrinol* **33**, 27–33.

Crosby, S.R., Stewart, M.F., Ratcliffe, J.G. & White, A. (1988) Direct measurement of the precursors of adrenocorticotropin in human plasma by two-site immunoradiometric assay. *J Clin Endocrinol Metab* **67**, 1272–1277.

Davis, J.R.E., Burt, D., Corrie, J.E.T., Edwards, C.R.W. & Sheppard, M.C. (1988) Dexamethasone-suppressible hyperaldosteronism: studies on overproduction of 18-hydroxycortisol in three affected family members. *Clin Endocrinol* **29**, 297–308.

Dean, H.J., Shackleton, C.H.L. & Winter, J.S.D. (1984) Diagnosis and natural history of 17α-hydroxylase deficiency in a newborn male. *J Clin Endocrinol Metab* **59**, 513–520.

Dillon, M.J. (1989) Salt and water balance: sodium-losing states and endocrine hypertension. In: *Clinical Paediatric Endocrinology* (ed. C.G.D. Brook), pp. 463–484. Blackwell Scientific Publications, Oxford.

Dillon, M.J., Gillin, M.E.A., Ryness, J.M. & de Swiet, M. (1976) Plasma renin activity and aldosterone concentration in the human newborn. *Arch Dis Child* **51**, 537–540.

Diver, M.J. & Nisbet, J.A. (1987) Warning on plasma oestradiol measurement. *Lancet* **ii**, 1097.

Dolman, L.I., Nolan, G. & Jubiz, W. (1979) Metyrapone test with ACTH levels: separating primary from secondary adrenal insufficiency. *J Am Med Assoc* **241**, 1243–1251.

Fiselier, T., Lijnen, P., Monnens, L. *et al.* (1983) Levels of renin, angiotensin I and II, angiotensin-converting enzyme and aldosterone in infancy and childhood. *Eur J Pediatr* **141**, 3–7.

Fisher, D.A. (1981) Thyroid development and disorders of thyroid function in the newborn. *N Engl J Med* **304**, 702–712.

Fisher, D.A. (1989) The thyroid gland. In: *Clinical Paediatric*

Endocrinology (ed. C.G.D. Brook), pp. 309–337. Blackwell Scientific Publications, Oxford.

Fisher, D.A. (1991) Management of congenital hypothyroidism. *J Clin Endocrinol Metab* **72**, 523–529.

Forest, M.G., de Peretti, E. & Bertrand, J. (1980) Testicular and adrenal androgens and their binding proteins in the perinatal period: developmental patterns of plasma testosterone, 4-androstenedione, dehydroepiandrosterone and its sulphate in premature and small for dates infants as compared with that of full-term infants. *J Steroid Biochem* **12**, 25–36.

Glinoer, D., Delange, F., Laboureur, I. *et al.* (1992) Maternal and neonatal thyroid function at birth in an area of marginally low iodine intake. *J Clin Endocrinol Metab* **75**, 800–805.

Goji, K. & Tanikaze, S. (1992) Comparison between spontaneous gonadotropin concentration profiles and gonadotropin responses to low-dose GnRH in prepubertal and pubertal boys and patients with hypogonadotropic-hypogonadism: assessment by using time resolved immunofluorimetric assays. *Pediat Res* **31**, 535–539.

Gomez-Sanchez, C.E., Gill, J.R., Ganguly, A. & Gordon, R.D. (1988) Glucocorticoid-suppressible aldosteronism: a disorder of the adrenal transitional zone. *J Clin Endocrinol Metab* **67**, 444–448.

Grant, D.B., Barnes, N.D., Moncrieff, M.W. & Savage, M.O. (1985) Clinical presentation, growth and development in Addison's disease. *Arch Dis Child* **60**, 925–928.

Grossman, A.B., Howlett, T.A., Perry, L. *et al.* (1988) CRF in the differential diagnosis of Cushing's syndrome: a comparison with the dexamethasone suppression test. *Clin Endocrinol* **29**, 167–178.

Hagg, E., Asplund, K. & Lithner, F. (1987) Value of plasma cortisol assays in the assessment of pituitary-adrenal insufficiency. *Clin Endocrinol* **26**, 221–226.

Hague, W.M., Honour, J.W., Adams, J., Vecsei, P. & Jacobs, H.S. (1989) Steroid responses to ACTH in women with polycystic ovaries. *Clin Endocrinol* **30**, 355–365.

Hay, I.D., Bayer, M.F., Kaplan, M.M. *et al.* (1991) American Thyroid Association assessment of current free thyroid hormone and thyrotropin measurements and guidelines for future clinical assays. *Clin Chem* **37**, 2002–2008.

Hilborn, S. & Krahn, J. (1987) Effect of time of exposure of serum to gel-barrier tubes on results for progesterone and some other endocrine tests. *Clin Chem* **33**, 203–204.

Himsworth, R.L., Bloomfield, G.A., Coombes, R.C. *et al.* (1977) Big ACTH and calcitonin in an ectopic hormone secreting tumour of the liver. *Clin Endocrinol* **7**, 45–62.

Hindmarsh, P.C. & Brook, C.G.D. (1985) Single dose dexamethasone suppression test in children: dose relationship to body size. *Clin Endocrinol* **23**, 67–70.

Honour, J.W. (1986) Biochemical aspects of congenital adrenal hyperplasia. *J Inher Metab Dis* **9** (Suppl. 1), 124–134.

Honour, J.W., Dillon, M.J., Levin, M. & Shah, V. (1983) Fatal, low renin hypertension associated with a disturbance of cortisol metabolism. *Arch Dis Child* **58**, 1018–1020.

Honour, J.W., Dillon, M.J. & Shackleton, C.H.L. (1982) Analysis of steroids in urine for differentiation of pseudohypoaldosteronism and aldosterone biosynthetic defects. *J Clin Endocrinol Metab* **54**, 325–331.

Honour, J.W., Price, D.A., Taylor, N.F., Marsden, H.B. & Grant, D.B. (1984) Steroid biochemistry of virilising adrenal tumours in childhood. *Eur J Pediatr* **142**, 165–169.

Honour, J.W., Tourniaire, J., Biglieri, E.G. & Shackleton, C.H.L. (1978) Urinary steroid excretion in 17α-hydroxylase deficiency. *J Steroid Biochem* **9**, 495–505.

Honour, J.W., Valman, H.B. & Shackleton, C.H.L. (1977) Aldosterone and sodium homeostasis in preterm infants. *Acta Paed Scand* **66**, 103–109.

Horrocks, P.M., Jones, A.F., Ratcliffe, W.A. *et al.* (1990) Patterns of ACTH pulsatility over twenty-four hours in normal males and females. *Clin Endocrinol* **32**, 127–134.

Howlett, T.A., Drury, P.L., Perry, L. *et al.* (1986) Diagnosis and management of ACTH-dependent Cushing's syndrome: comparison of the features in ectopic and pituitary ACTH production. *Clin Endocrinol* **24**, 699–713.

Hughes, I.A., Arisaka, O., Perry, L.A. & Honour, J.W. (1986) Early diagnosis of 11β-hydroxylase deficiency in two siblings confirmed by analysis of a novel steroid metabolite in newborn urine. *Acta Endocrinol* **111**, 349–354.

Imperato-McGinley, J., Gautier, T., Pichardo, M. & Shackleton, C. (1986) The diagnosis of 5α-reductase deficiency in infancy. *J Clin Endocrinol Metab* **63**, 1313–1318.

Ismail, A.A.A., Astley, P., Burr, W.A. *et al.* (1986a) The role of testosterone measurement in the investigation of androgen disorders. *Ann Clin Biochem* **23**, 113–134.

Ismail, A.A.A., Astley, P., Cawood, M. *et al.* (1986b) Testosterone assays; guidelines for the provision of a clinical biochemistry service. *Ann Clin Biochem* **23**, 135–145.

Kelnar, C.J.H. & Brook, C.G.D. (1983) A mixed longitudinal study of adrenal excretion in childhood and the mechanism of adrenarche. *Clin Endocrinol* **19**, 117–129.

Korth-Schutz, S., Levine, L.S. & New, M.I. (1976) Dehydroepiandrosterone sulphate levels – a rapid test of abnormal adrenal androgen secretion. *J Clin Endocrinol Metab* **42**, 1005–1013.

Krieger, D.T., Allen, W., Rizzo, F. & Krieger, H.P. (1971) Characterisation of the normal temporal pattern of plasma corticosteroid levels. *J Clin Endocrinol* **32**, 266–284.

Laverty, C.R., Fortune, D.W. & Beischer, N.A. (1973) Congenital adrenal idiopathic hypoplasia. *Obstet Gynecol* **41**, 655–664.

Law, C.M., Marchant, J.L., Honour, J.W., Preece, M.A. & Warner, J.O. (1986) Nocturnal adrenal suppression in asthmatic children taking inhaled beclomethasone. *Lancet* **i**, 942–944.

Letarte, J. & Garagorri, J.M. (1989) Congenital hypothyroidism: laboratory and clinical investigations of early detected infants. In: *Paediatric Endocrinology* (eds R. Collu, J.R. Ducharme & H.J. Guyda), 2nd edn, pp. 449–471. Raven Press, New York.

Lever, E.G., Medeiro-Neto, G.A. & de Groot, L.J. (1983) Inherited disorders of thyroid metabolism. *Endocrine Rev* **4**, 213–239.

Liddle, G.W. (1960) Tests of pituitary adrenal suppressibility in the diagnosis of Cushing's syndrome. *J Clin Endocrinol Metab* **12**, 1539–1560.

Lifton, R.P., Dluhy, R.G., Powers, M. *et al.* (1992) A chimaeric 11β-hydroxylase/aldosterone synthase gene causes glucocorticoid-remediable aldosteronism and human hypertension. *Nature* **355**, 262–265.

Lindholm, J. & Kehlet, H. (1987) Re-evaluation of the clinical value of the 30-min ACTH test in assessing the hypothalamic-pituitary-adrenocortical function. *Clin Endocrinol* **26**, 53–59.

Lipsett, M.B., Chrousos, G.P., Tomita, M., Brandon, D.D. & Loriaux, D.L. (1985) The defective glucocorticoid receptor in man and non-human primates. *Rec Prog Hormone Res* **41**, 199–241.

Lubahn, D.B., Brown, T.R., Simenthal, J.A. *et al.* (1989) Sequence of the intron/exon junctions of the coding region of the human androgen receptor gene and identification of a point mutation in a family with complete androgen insensitivity. *Proc Natl Acad Sci USA* **86**, 9534–9538.

McHardy, K.C. (1984) Clinical algorithms: suspected Cushing's syndrome. *Br Med J* **289**, 1519–1521.

McKenna, T.J., Miller, R. & Liddle, G. (1977) Plasma pregnenolone and 17-hydroxypregnenolone in patients with adrenal tumours, ACTH excess or idiopathic hirsutism. *J Clin Endocrinol* **44**, 231.

Makela, S.K. & Ellis, G. (1988) Nonspecificity of a direct 17α-hydroxyprogesterone radioimmunoassay kit when used with samples from neonates. *Clin Chem* **34**, 2070–2075.

Masters, A.M. & Hahnel, R. (1989) Investigation of sex hormone binding globulin interference in direct radioimmunoassays for testosterone and estradiol. *Clin Chem* **35**, 979–984.

Meikle, A.W. (1982) Dexamethasone suppression tests: usefulness of simultaneous measurement of plasma cortisol and dexamethasone. *Clin Endocrinol* **16**, 401–408.

Melby, J.C. (1985) Diagnosis and treatment of primary aldosteronism and isolated hypoaldosteronism. *Clinics Endocrinol Metab* **14**, 977–1005.

Monder, C. & Shackleton, C.H.L. (1984) 11β-Hydroxysteroid dehydrogenase: a fact or fancy. *Steroids* **44**, 383–417.

Moore, A., Aitken, R., Burke, C. *et al.* (1985) Cortisol assays: guidelines for the provision of a clinical biochemistry service. *Ann Clin Biochem* **22**, 435–454.

Moser, H.W., Bergin, A., Naidu, S. & Ladenson, P.W. (1991) Adrenoleukodystrophy. *Endocrinol Metab Clinics N Am* **20**, 297–318.

New, M.I. & Josso, N. (1988) Disorders of gonadal differentiation and congenital adrenal hyperplasia. *Endocrinol Metab Clinics N Am* **17**, 339–366.

New, M.I., Baum, C.J. & Levine, L.S. (1976) Nomograms relating aldosterone excretion to urinary sodium and potassium in a pediatric population and their application to the study of childhood hypertension. *Am J Cardiol* **37**, 658–666.

Nicoloff, J.T. & Spencer, C.A. (1990) The use and misuse of the sensitive thyrotropin assays. *J Clin Endocrinol Metab* **71**, 553–558.

Oerter, K.E., Uriarte, M.M., Rose, S.R., Barnes, K.M. & Cutler, G.B. (1990) Gonadotropin secretory dynamics during puberty in normal girls and boys. *J Clin Endocrinol Metab* **71**, 1251–1258.

Perry, L.A., Al-Dujaili, E.A.S. & Edwards, C.R.W. (1982) A direct radioimmunoassay for 11-deoxycortisol. *Steroids* **9**, 115–128.

Pescovitz, O.H., Hench, K.D., Barnes, K.M., Loriaux, D.L. & Cutler, G.B. (1988) Premature thelarche and central precocious puberty: the relationship between clinical presentation and the gonadotrophin response to luteinising hormone releasing hormone. *J Clin Endocrinol Metab* **67**, 474–479.

Phillipou, G. (1982) Investigation of urinary steroid profiles as a diagnostic method in Cushing's syndrome. *Clin Endocrinol* **16**, 433–439.

Prader, A. & Gurtner, H.P. (1955) Das Syndrom pseudohermaphroditus masculinus bei kongenitaler Nebernierenrinden – Hyperplasia ohne Androgen über Produktion. *Helv Paediatr Acta* **10**, 397–412.

Price, D.A., Close, G.C. & Fielding, B.A. (1983) Age of appearance of circadian rhythm in salivary cortisol in infancy. *Arch Dis Child* **58**, 454–456.

Pringle, P.J., Stanhope, R., Hindmarsh, P. & Brook, C.G.D. (1988) *Clin Endocrinol* **28**, 479–486.

Ratcliffe, W.A. (1982) Evaluation of four methods for the direct assay of progesterone in unextracted serum. *Ann Clin Biochem* **19**, 362–367.

Ratcliffe, W.A., Carter, G.D., Dowsett, M. *et al.* (1988) Oestradiol assays: applications and guidelines for the provision of a clinical biochemistry service. *Ann Clin Biochem* **25**, 466–483.

Ratcliffe, W.A., McClure, J.P., Auld, W.H. *et al.* (1982) Precocious pseudopuberty due to a rare form of congenital adrenal hyperplasia. *Ann Clin Biochem* **19**, 145–150.

Rees, L., Shaw, J.C.L., Brook, C.G.D. & Forsling, M.L. (1984) Hyponatraemia in the first week of life in preterm infants. II. Sodium and water balance. *Arch Dis Child* **59**, 423–429.

Reiter, E.O., Fuldauer, V.G. & Root, A.W. (1977) Secretion of the adrenal androgen, dehydroepiandrosterone sulphate during normal infancy, childhood and adolescence in sick infants and in children with endocrinologic abnormalities. *J Pediatr* **90**, 766–770.

Roitman, A., Bruchis, S., Bauman, B., Kaufman, H. & Laron, Z. (1984) Total deficiency of corticosteroid-binding globulin. *Clin Endocrinol* **21**, 541–548.

Rosenfeld, R.S., Hellman, L., Roffwarg, A. *et al.* (1971) Dehydroepiandrosterone is secreted episodically and synchronously with cortisol by normal man. *J Clin Endocrinol* **33**, 87–92.

Rudd, B.T. (1983) Urinary 17-oxogenic and 17-oxosteroids. A case for deletion from the clinical chemistry repertoire. *Ann Clin Biochem* **20**, 65–71.

Saenger, P. (1984) Abnormal sexual differentiation. *J Pediatr* **104**, 1–17.

Sassard, J., Sann, L., Vincent, M., Francois, R. & Cier, J.F. (1975) Plasma renin activity in normal subjects from infancy to puberty. *J Clin Endocrinol Metab* **40**, 524–525.

Schulte, H.M., Allolio, B., Gunther, R.W. *et al.* (1988) Selective bilateral and simultaneous catheterisation of the inferior petrosal sinus: CRF stimulates prolactin secretion from ACTH-producing microadenomas in Cushing's disease. *Clin Endocrinol* **28**, 289–295.

Seth, J., Hanning, I., Bacon, R.R.A. & Hunter, W.M. (1989) Progress and problems in immunoassays for serum pituitary gonadotrophins: evidence from the UK external quality assessment schemes (EQAS) 1980–1988. *Clin Chim Acta* **186**, 67–82.

Shackleton, C.H.L. (1986) Profiling steroid hormones and urinary steroids. *J Chromat* **379**, 91–156.

Shackleton, C.H.L., Honour, J.W., Dillon, M.J., Chantler, C. & Jones, R.W.A. (1980a) Hypertension in a four-year-old child; gas chromatographic and mass spectrometric evidence for deficient hepatic metabolism of steroids. *J Clin Endocrinol Metab* **50**, 786–792.

Shackleton, C.H.L., Taylor, N.F. & Honour, J.W. (1980b) *An Atlas of Gas Chromatographic Profiles of Neutral Steroids in Health and Disease*. Packard-Becker DV, Delft.

Shah, A., Stanhope, R. & Matthew, D. (1992) Hazards of pharmacological tests of growth hormone secretion in childhood. *Br Med J* **304**, 173–174.

Sklar, C.A., Kaplan, S.L. & Grumbach, M.M. (1980) Evidence for dissociation between adrenarche and gonadarche: studies in patients with idiopathic precocious puberty, gonadal dysgenesis, isolated gonadotrophin deficiency and constitutional delayed growth at adolescence. *J Clin Endocrinol Metab* **51**, 548–556.

Solish, S.B., Goldsmith, M.A., Voutilainen, R. & Miller, W.L. (1989) Molecular characterisation of a Leydig cell tumor presenting as congenital adrenal hyperplasia. *J Clin Endocrinol Metab* **68**, 1148–1152.

Stanhope, R., Abdulwahid, N.A., Adams, J. & Brook, C.G.D. (1986) Studies of gonadotrophin pulsatility and pelvic ultrasound examinations distinguish between isolated premature thelarche and central precocious puberty. *Eur J Pediatr* **145**, 190–194.

Stanhope, R., Adams, J., Jacobs, H.S. & Brook, C.G.D. (1985) Ovarian ultrasound assessment in normal children idiopathic precocious puberty and during low-dose pulsatile GnRH therapy of hypogonadotrophic hypogonadism. *Arch Dis Child* **60**, 116–119.

Stewart, P.M., Corrie, J.E.T., Shackleton, C.H.L. & Edwards, C.R.E. (1988) Syndrome of apparent mineralocorticoid excess. A defect in the cortisol-cortisone shuttle. *J Clin Invest* **82**, 340–349.

Stewart, P.M., Wallace, A.M., Valentino, R. *et al.* (1987) Mineralocorticoid activity of liquorice: 11β-hydroxysteroid dehydrogenase deficiency comes of age. *Lancet* **2**, 821–824.

Streeten, D.H.P., Anderson, G.H., Dalakos, T.G. *et al.* (1984) Normal and abnormal function of the hypothalamic-pituitary-adrenocortical system in man. *Endocrine Rev* **5**, 371–394.

Taylor, N.F., Clymo, A.B. & Shackleton, C.H.L. (1979) Steroid excretion by an infant with 3β-hydroxysteroid dehydrogenase deficiency during conventional replacement therapy and following corticotrophin stimulation. *J Endocrinol* **80**, 62P.

Ulick, S. (1976) Diagnosis and nomenclature of the disorders of the terminal portion of the aldosterone biosynthetic pathway. *J Clin Endocrinol Metab* **43**, 92–96.

Van Cauter, E. (1989) Physiology and pathology of circadian rhythms. *Rec Adv Endocrinol Metab* **3**, 109–134.

Veldhuis, J.D., Evans, W.S., Rogol, A.D. *et al.* (1984) Intensified rates of venous sampling unmask the presence of spontaneous, high frequency pulsations of luteinising hormone in man. *J Clin Endocrinol Metab* **59**, 96–102.

Wallace, A.M., Beesley, J., Thomson, M. *et al.* (1987) Adrenal status during the first month of life in mature and immature infants. *J Endocrinol* **112**, 473–480.

Wallace, W.H.M., Crowne, E.C., Shalet, S.M. *et al.* (1991) Episodic ACTH and cortisol secretion in normal children. *Clin Endocrinol* **34**, 215–221.

White, A., Smith, H., Hoadley, M., Dobson, S.H. & Ratcliffe, J.G. (1987) Clinical evaluation of a two-site immunoradiometric assay for adrenocorticotrophin in unextracted human plasma using monoclonal antibodies. *Clin Endocrinol* **26**, 41–52.

Wolff, J. (1983) Congenital goitre with defective iodine transport. *Endocrine Rev* **4**, 240–254.

Wong, T., Shackleton, C.H.L., Covey, T.R. & Ellis, G. (1992) Identification of the steroids in neonatal plasma that interfere with 17α-hydroxyprogesterone radioimmunoassays. *Clin Chem* **38**, 1830–1837.

Wood, P., Groom, G., Moore, A., Ratcliffe, W. & Selby, C. (1985) Progesterone assays: guidelines for the provision of a clinical biochemistry service. *Ann Clin Biochem* **22**, 1–24.

Wu, F.C.W., Brown, D.C., Butler, G.E., Stirling, H.F. & Kelnar, C.J.H. (1993) Early morning plasma testosterone is an accurate predictor of imminent pubertal development in prepubertal boys. *J Clin Endocrinol Metab* **76**, 26–31.

Wu, F.C.W., Butler, G.E., Kelnar, C.J.H. & Sellar, R.E. (1990) Patterns of pulsatile luteinizing hormone secretion before and during the onset of puberty in boys: a study using an immunoradiometric assay. *J Clin Endocrinol Metab* **70**, 629–637.

Yamaji, J., Ishibashi, M., Sekihara, H., Itabashi, A. & Yanihara, T. (1984) Serum DHAS in Cushing's syndrome. *J Clin Endocrinol Metab* **59**, 1164.

Yasuda, T., Noda-Cho, H., Nishioka, T. *et al.* (1986) Pseudohypoaldosteronism: decreased aldosterone levels with age without significant changes in urinary sodium excretion. *Clin Endocrinol* **24**, 311–318.

Young, W.F. & Klee, G.G. (1988) Primary aldosteronism; diagnostic evaluation. *Endocrinol Metab Clinics N Am* **17**, 367.

Zovickian, J., Oldfield, E.H., Doppman, J.L., Cutler, G.B. & Loriaux, D.L. (1988) Usefulness of inferior petrosal sinus venous endocrine markers in Cushing's disease. *J Neurosurg* **68**, 205–210.

11: Renal Disorders

J.T. BROCKLEBANK

Introduction

More than a century ago Claude Bernard pointed out that the interstitial fluid bathing individual cells is the true environment of the body. The intravascular fluid and the interstitial fluid together comprise the extracellular fluid water (ECW), the biochemical equilibrium of which is maintained largely by the kidneys. When expressed as a percentage of total body weight, the ECW changes with maturity. In the fetus the ECW averages 60% of body weight, falling to 43% at birth, 25% at 1 year and 20% by the age of 13 years (Friis-Hansen, 1961). Similar changes are observed in total body water (TBW), which averages 94% of body weight in the fetus, 78% at birth and 60% at 1 year.

Sodium is the major cation in ECW and chloride and bicarbonate are the major anions. Their concentrations are under the influence of the kidney. Assuming a glomerular filtration rate (GFR) of 100 mL/min, and a plasma sodium concentration of 140 mmol/L, then over 20 000 mmol sodium is filtered by the kidney each day, yet only a fraction of this appears in the final urine. The major energy consumption by the kidney is concerned with reclaiming the filtered sodium. The reabsorption and excretion of many other solutes, such as potassium, hydrogen ions, glucose and amino acids, are linked to sodium transport. Glomerular filtration and sodium homeostasis are therefore the principal functions of the kidney and will be discussed in this chapter.

Glomerular clearance

The theory of glomerular clearance depends upon the hypothesis that the rate of removal of a solute from the plasma must equal its appearance in the urine, providing the kidney is the only route for its disappearance from the plasma. From this it follows that the rate of clearance of the solute from a unit volume of plasma (C) multiplied by its plasma concentration (P) must equal the solute's rate of appearance in the urine (urine concentration $U \times$ urine flow rate V):

$$C \times P = U \times V$$

$$\text{or} \quad C = \frac{UV}{P}$$

The concept of clearance can be applied to any solute excreted by the kidney, but it will only represent the true GFR when the solute is totally filtered by the glomerulus, not bound to plasma proteins and neither excreted nor reabsorbed by the renal tubules. Inulin is generally regarded as meeting all these conditions and its clearance is considered to be the gold standard for the measurement of GFR, against which other solutes are compared. It follows from this that if a solute has a clearance rate greater than inulin, then it is excreted by the renal tubules, and solutes with clearance rates less than inulin are reabsorbed. In routine clinical practice, creatinine is usually substituted for inulin in the determination of clearance rate.

Fractional excretion

The concept of fractional excretion (FE) of a solute is derived from these observations. Thus, the clearance of sodium (Na) is $U^{Na}V^{Na}/P^{Na}$, where U^{Na} is the urine sodium concentration, V^{Na} urine volume and P^{Na} plasma sodium concentration. This can then be expressed as a fraction of the clearance of inulin, a solute which is known to be 100% excreted:

$$FE_{Na} = \frac{U^{Na}V^{Na}}{P^{Na}} \div \frac{U^{I}V^{I}}{P^{I}}$$

If the same urine sample is used for each, then the term V will be cancelled and the equation becomes:

$$FE_{Na} = \frac{U^{Na}}{P^{Na}} \times \frac{P^{I}}{U^{I}}$$

This is more frequently expressed as a percentage:

$$FE_{Na}(\%) = \frac{U^{Na}}{P^{Na}} \times \frac{P^{I}}{U^{I}} \times 100$$

Conversely, the fractional reabsorption of a solute can be examined in a similar way. Thus, the fractional tubular reabsorption (TR) of phosphorus is expressed as:

$$TR_{P}(\%) = 100 - FE^{P}$$

$$= 100\left(1 - \frac{U^{P}}{P^{P}} \times \frac{P^{I}}{U^{I}}\right)$$

The FE of a substance can be used to examine renal tubular function. Thus, FE_{Na} is low in salt depletion and hypovolaemia, when there is active tubular reabsorption of sodium, and increased in salt-losing states and volume overload. The tubular reabsorption of phosphorus is reduced in hyperparathyroidism, Fanconi syndrome and familial hypophosphataemic rickets.

The TR_P is not a sensitive measure of renal phosphate handling. The tubular maximum (Tm) reabsorption of phosphate corrected for creatinine clearance is better ($TmPO_4$/GFR). It is a measure of the theoretical maximum reabsorption rate of phosphate per unit mass of the kidney. The $TmPO_4$/GFR can be calculated either from a nomogram, constructed by Walton and Bijvoet (1975), with knowledge of the serum phosphate and TR_P or from one of two equations derived from this nomogram (Kenny, 1973). The mean normal value in children is 1.67 mmol/L. It is not affected by age and there is an inverse relationship between $TmPO_4$/GFR and the plasma parathyroid hormone concentration (Shaw *et al.*, 1990).

Glomerular filtration rate

GFR is the most widely used test of renal function. Its value is related to the number of functioning nephrons. The measurement of GFR is helpful to determine the presence of renal disease, its extent and progression. The GFR is not constant during a 24-h period; there is a diurnal variation with higher values being measured during the day than at night. Diet and exercise also have significant effects and there is a marked effect of age. For these reasons it is important to standardize the conditions under which GFR is measured to obtain reproducible results.

Creatinine clearance

Creatinine is formed in the muscles by conversion from creatine. Its rate of production has a direct relationship to muscle mass and therefore also to body weight. The production rate remains approximately the same from day to day for an individual, unless there are changes in the muscle mass. As the muscle mass increases with growth, so plasma concentrations would be expected to increase with age. Creatinine is metabolically inert and is freely filtered by the glomerulus. It is not reabsorbed by the renal tubules under normal circumstances. A significant amount is actively secreted by the proximal renal tubule and this increases with increasing plasma creatinine concentration. Creatinine clearance may be up to 30% greater than inulin clearance when true creatinine is measured. Using methods which measure total chromogens, the plasma creatinine concentration is higher and the P value in the clearance formula is artificially high, making the calculated clearance value lower. It is fortuitous that creatinine clearance then approximates more closely the true GFR obtained by inulin clearance.

The measurement of creatinine clearance needs an accurately timed urine collection. There are major technical difficulties in obtaining a complete collection of urine from children. It has been calculated that creatinine clearance measurements must differ by 30% to be confident of a change (Chantler & Barratt, 1972). For these reasons the measurement of GFR by creatinine clearance has largely been abandoned in children and increasing use is being made of the plasma creatinine measurements alone.

Estimation of GFR from plasma creatinine

There are many factors which influence the measurement of plasma creatinine (Payne, 1986). Its concentration is not constant during 24 h and increases significantly after a protein meal and strenuous exercise. The measurement of creatinine may be spuriously elevated by ketone bodies and by drugs which block its tubular secretion, e.g. salicylate and trimethoprim. It may be misleadingly low in patients with jaundice, because bilirubin interferes with the measurement of creatinine, and in patients with muscle wasting, because of decreased synthesis. The cephalosporin antibiotics have variable effects on plasma creatinine by interfering with its analysis. Even when these variables are taken into account, the coefficient of variation of plasma creatinine measurement at a concentration of 200 μmol/L means that it must differ by at least 10% to be confident of a change. In children the plasma concentrations are normally much less than 100 μmol/L and a larger percentage change is probably required.

In clinical practice an accurate measurement of GFR is not generally required. The usual questions to be asked are whether the renal function is normal and has it changed between measurements. These can be answered by careful examination of the plasma creatinine concentration. Values of plasma creatinine at different ages are shown in Table 11.1. Although in adults there is an exponential relationship between plasma creatinine and GFR, so that plasma creatinine roughly doubles as the GFR falls by 50%, the relationship is less helpful in children because plasma creatinine increases with age. It has been shown that the ratio of height (cm) : plasma creatinine (μmol/L) has a constant linear relationship with GFR when corrected to 1.73 m^2 body surface area (Morris *et al.*, 1982). Furthermore, when the ratio is multiplied by 40 the result approximates to the measured GFR. For example, in a child 100 cm tall with a plasma creatinine of 80 μmol/L, the height : creatinine index would be 1.25 and the GFR $1.25 \times 40 = 50$ mL/min per 1.73 m^2. Using the formula height (cm) : plasma creatinine (μmol/L) × 40 = GFR (mL/min per 1.73 m^2) has proved to be a clinically useful method of following the progression of renal disease in children, but it cannot be applied to infants and the newborn, in whom it is better to follow plasma creatinine concentration. In the newborn the plasma creatinine approximates to the maternal concentration and progressively falls with age during the first few weeks. The level and rate of fall are dependent upon the gestational age of the neonate (Fig. 11.1) (Rudd *et al.*, 1983).

The relationship between GFR and plasma creatinine only holds in steady state conditions such as chronic renal disease. In acute renal failure, when there is a rapid and severe fall in GFR, the plasma creatinine may not rise for at least 24 h after the onset.

Table 11.1 Reference ranges for serum creatinine (μmol/L). From Savory (1990)

	Males		Females	
Age	Mean	Range	Mean	Range
1–3 months	43	19–67	44	29–59
3–6 months	45	28–62	43	29–57
6–12 months	44	23–65	43	20–66
1 year	48	22–74	45	24–66
2 years	46	26–66	46	21–71
3 years	49	32–66	49	26–72
4 years	51	28–74	51	35–67
5 years	53	34–72	50	32–68
6 years	55	34–76	51	36–66
7 years	56	39–73	54	36–72
8 years	57	38–76	58	43–73
9 years	60	38–82	57	45–69
10 years	64	40–88	60	47–73
11 years	63	47–79	63	44–82
12 years	66	46–86	60	51–69
13 years	69	44–94	66	53–79
14 years	70	52–88	69	48–90
15 years	77	53–101	69	54–84
16 years	80	52–108	72	51–93
17 years	81	62–100	76	59–93
18 years	86	64–108	76	59–95
19 years	89	68–110	77	59–95
20 years	85	63–107	75	53–97

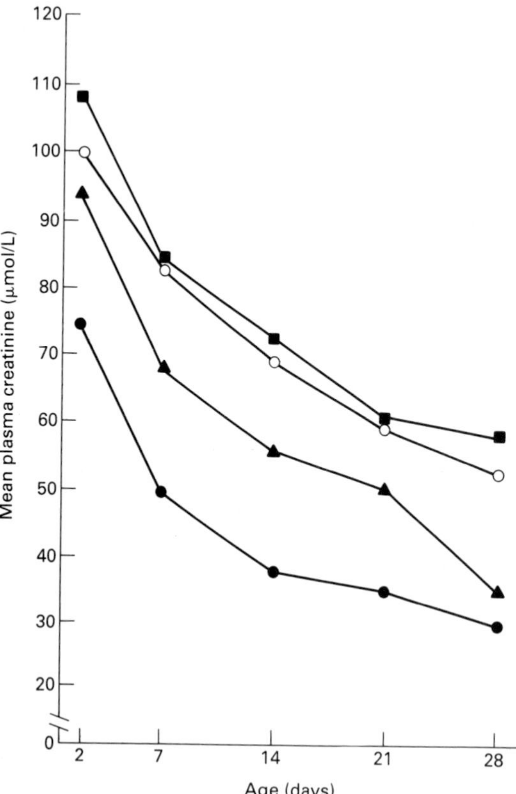

Fig. 11.1 Relationship between mean plasma creatinine level and postnatal age in different gestational age groups: less than 28 weeks (■); 29–32 weeks (○); 33–36 weeks (▲); and 37–42 weeks (●). (From Rudd *et al.*, 1983.)

Slope disappearance techniques

These methods were developed to avoid the need for urine collection or intravenous infusions and for these reasons are particularly suited to investigations in children. The techniques depend upon observing the rate of disappearance from the plasma of an intravenous bolus of a solute which is exclusively excreted by glomerular filtration. The GFR can be calculated either from the terminal slope of the plasma disappearance curve or from the area under the curve when plasma concentrations of the solute are examined over a period of 2–4 h. At least three blood samples are required to obtain a good fit to a straight line. [^{125}I]-Iothalamate, [^{51}Cr]-ethylenediaminetetraacetic acid ([^{51}Cr]-EDTA) and [^{99m}Tc]-diethylenetriaminepentacetic acid ([^{99m}Tc]-DTPA) are most frequently used. Iothalamate is excreted by the renal tubules and overestimates GFR. Both EDTA and DTPA are protein bound and underestimate GFR by 10% when compared with inulin clearance. DTPA has the advantage of having a shorter half-life (6 h) than EDTA (27 days); this reduces the exposure to radiation and provides the added convenience of being able to measure GFR at the time of renal imaging studies using the same radiopharmaceutical.

The technique tends to overestimate renal function when the GFR is low and in the presence of moderate oedema, and underestimates renal function when the GFR is high.

Plasma urea levels

Urea, the major product of protein catabolism, is formed in the liver, filtered by the glomerulus and both excreted and reabsorbed by the renal tubules. Its concentration in the plasma is dependent upon dietary protein intake and catabolism. Its excretion by the kidney is dependent upon the patient's state of hydration, as dehydration and reduced urine flow rates enhance renal tubular reabsorption of urea and lead to a rise in its plasma concentration. Plasma urea is not a useful marker of GFR. Its measurement in acute and chronic renal failure often reflects nutritional needs rather than GFR. A disproportionately high level suggests increased protein breakdown and the need to increase energy supplementation. Urea kinetic modelling in patients receiving dialysis can be used as a measure of the efficiency of dialysis and protein catabolic rate (Evans *et al.*, 1992).

Plasma β_2-microglobulin

β_2-Microglobulin plasma concentration has a logarithmic relationship with GFR and has been investigated as an alternative to plasma creatinine (Van-Acker *et al.*, 1984). In children over 3 years it has the advantage of not varying significantly with age. Its value as an index of GFR is limited in neoplastic and inflammatory diseases when its production is increased and plasma concentrations do not accurately reflect GFR. The measurement of β_2-microglobulin is expensive and it has no major advantage over plasma creatinine.

Effects of maturation on GFR

Glomerular filtration is low at birth, averaging 3 mL/min in the term neonate. There is a rapid increase during the first 6 months of life and thereafter a steady increment to reach normal adult values by the age of 14 years. In order to make valid comparisons of GFR between children of different ages, some correction is generally made for body size. It is conventional to use the body surface area (SA) correction and standardize GFR to that of a standard male (1.73 m^2). SA is computed from height (cm) and weight (kg) (Haycock *et al.*, 1978). When correction is made for differences in body SA the creatinine clearance of a term newborn averages 50 mL/min per 1.73 m^2 (Schwartz *et al.*, 1984), and increases rapidly to reach normal adult values of 120 mL/min per 1.73 m^2 by the age of 2 years. The plasma creatinine does not fall in association with the increasing GFR because of the concomitant increase in muscle mass and creatinine production due to growth. Thereafter, the plasma creatinine slowly increases as the increase in muscle mass accelerates faster than GFR, and at puberty there is a further increase in normal plasma creatinine (Table 11.1).

In the newborn there are strong arguments for standardizing GFR for body weight rather than SA (Coulthard & Hey, 1984). This is not only because weight is easy to measure but also because there is less variability when this correction is used. Using the weight correction, GFR averages 0.59 mL/min per kg at 26 weeks after conception and increases logarithmically to 1.14 mL/min per kg at 33 weeks (Wilkins, 1992).

It has been suggested that it would be more logical to correct the GFR measurement for the size of the fluid compartment over which the kidney has a major effect, either extracellular fluid volume or total body water (McCance & Widdowson, 1952). Both of these can be estimated from body weight but the proportion of water to body weight falls with increasing age, thereby exaggerating the increase in GFR observed as a result of maturation and increasing body size (Friis-Hansen, 1961). Direct measurement of fluid spaces is not practical and simple methods based upon bioelectrical impedance have not yet proved to be clinically useful in the neonate, although they are helpful in older children. It is generally better to correct GFR for body weight in the newborn and for SA in older children.

Proteinuria

The determination of urinary protein excretion is one of the most useful measurements in patients with renal disease. It is helpful not just in diagnosis, but also in assessing the response to treatment and the progression of a disorder. There are only a few progressive renal disorders which may not be associated with increased urinary protein excretion. These include familial nephronophthisis, uncomplicated obstructive uropathy, interstitial nephritis and the nephropathy associated with hypercalcaemia. In normal children the daily urine protein excretion averages 100 mg/m^2 per 24 h, but there are age and sex differences in addition to diurnal variations of protein excretion (Davies *et al.*, 1984). Significant proteinuria can be defined as follows.
Normal – less than 4 mg/m^2 per h.
Abnormal – greater than 4–40 mg/m^2 per h.
Nephrotic – greater than 40 mg/m^2 per h.

Measurement of proteinuria

In clinical practice the simplest and most widely used methods of measuring proteinuria are semi-quantitative tests performed on random urine samples. Their accuracy and reproducibility can be increased by ensuring that the urine examined is the first voided sample in the morning. For more accurate measurement, timed urine collections are required.

Semi-quantitative tests of proteinuria

Dipstix method

This method is based upon the fact that proteins change the colour of an indicator dye, tetrabromophenol, when maintained at constant pH 3.0. The dipstix test uses a paper impregnated with tetrabromophenol and a buffer to maintain the pH. The colour change is proportional to the amount of protein present, yellow representing the minimal and blue the maximal protein concentration. The relationship between protein concentration and colour change is only approximate; dipstix methods correlate in about two-thirds of cases with quantitative protein measurements (Rennie & Keen, 1967). There is also considerable interobserver error. It is, however, a specific test of protein, provided that extremes of pH are avoided. The major drawback is its insensitivity to protein concentrations of less than 30 mg/dL, when significant proteinuria may not be detected in dilute urines.

Precipitation methods

These techniques depend upon the characteristic of most proteins to precipitate when heated or in the presence of strong acids. It can be performed by adding 5% sulphasalicylic acid to an aliquot of urine or by heating urine and adding glacial acetic acid. The amount of precipitation is graded 0 to 4, depending upon the degree of flocculation. Protein concentrations as low as 10 mg/dL may be detected by these techniques. Precipitation methods are, however, more difficult to carry out and there are potential hazards associated with keeping strong acids in clinical areas. In addition, radiocontrast materials and some drugs, particularly penicillin, may give false positive reactions.

Quantitative tests of proteinuria

The standard quantitative measurement of proteinuria requires a 24-h collection of urine. This is often a difficult procedure to undertake in children; a timed urine collection can be substituted and the amount of protein extrapolated to provide a total for 24 h.

Urine protein : creatinine ratio

The 24-h urine protein excretion can be estimated from the measurement of both protein (mg/L) and creatinine (mmol/L) in the same random sample of urine. The ratio of protein to creatinine provides a reliable measure of the degree of proteinuria. In paediatric practice it has the added advantage of not requiring a timed collection of urine. Although it makes the assumption that the excretion of protein and creatinine is constant throughout a 24-h period, it provides a quantitative measure of proteinuria which is reasonably accurate for clinical practice. In addition, there is a good correlation between the ratio and 24-h urine protein excretion (Houser, 1984). Some laboratories multiply the protein : creatinine ratio by 1000 to produce a more manageable number and this is referred to as the protein : creatinine index. The normal ratio is less than 0.2 and the index less than 200.

Qualitative tests of proteinuria

From the electrophoretic patterns and immunochemical analysis of the urinary proteins present in those with renal disease, several categories of proteinuria emerge. The first, glomerular proteinuria, is the pattern most commonly seen in renal disease and comprises mainly albumin and proteins of higher molecular weight. In the second, tubular proteinuria, urine loss of low-molecular-weight proteins predominates and is seen in renal tubular damage. A third pattern, of mixed glomerular and tubular proteinuria, is found in advanced renal disease. A fourth type of proteinuria, referred to as overflow proteinuria, is characterized by the overproduction and urinary excretion of one specific protein or fragments of proteins and is rare in children. These patterns of proteinuria can be detected readily by the application of electrophoretic techniques to urine (Brocklebank *et al.*, 1991). A representation of the principal proteins detected in the urine by sodium dodecyl sulphate electrophoresis is shown in Fig. 11.2. Alternatively, specific proteins that are representative of glomerular or tubular proteins can be measured by radioimmunoassay.

Glomerular proteinuria

Albumin (molecular mass 69 000 Da) comprises 60–90% of the urinary protein excreted in glomerular

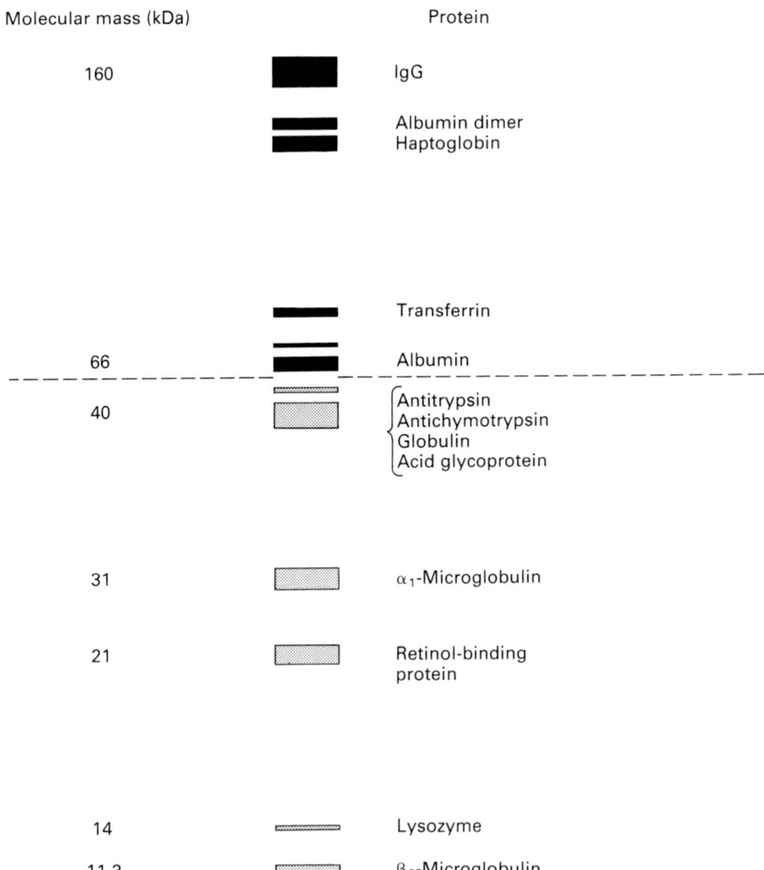

Fig. 11.2 Major urinary protein bands identified by sodium dodecyl sulphate polyacrylamide gel electrophoresis. (From Brocklebank *et al.*, 1991.)

disease. The remaining proteins are made up largely of haptoglobin, transferrin and immunoglobulins in varying amounts. The excretion of low-molecular-weight proteins is absent or minimal. Specific proteins may be measured by radioimmunoassay techniques. These methods are specific for the individual proteins and are sensitive enough to measure small quantities in normal urine.

Albumin. In urine, albumin is readily measured by radioimmunoassay and increased urinary excretion of it may be used as a marker for glomerular disease (Barratt *et al.*, 1970). The mean excretion rate in normal boys is 6.6 mg/24 h per 1.73 m^2 and in girls is 8.3 mg/24 h per 1.73 m^2. Excretion during the day is significantly greater than at night. The measurement of the ratio of albumin (mg/L) : creatinine (mmol/L) on a random urine sample is more practical to carry out than on a 24-h urine sample.

Small increases in urinary albumin excretion may be detected in the early stages of insulin-dependent diabetics in up to 40% of patients before other overt evidence of renal disease is apparent. This phenomenon is referred to as micro-albuminuria and may be predictive of diabetic nephropathy.

Urine protein selectivity. Glomerular proteinuria can be subdivided into selective and non-selective proteinuria, depending upon the comparison of the glomerular clearance of albumin with that of a larger molecular-weight protein. This makes the assumption that the glomerulus behaves as a simple sieve. There are, however, many factors other than molecular size which influence a protein glomerular clearance; these include shape, deformability and electrical charge on the one hand, and pore size and electrical charge of the glomerular sieve on the other hand. Despite these reservations, protein selectivity is still widely used in clinical practice. The clearance of IgG is usually the marker of a high-molecular-weight protein, and either albumin or transferrin (Cameron, 1966) as the low-molecular-weight protein. If the same samples of blood and urine are used for each measurement, then timed urine collections are not necessary.

$$\text{Selectivity } (S) = \frac{U_g P_t}{P_g U_t}$$

where U_g and P_g are the urine and plasma concentrations respectively of IgG and U_t and P_t denote the urine and plasma concentrations respectively of transferrin. Patients with selectivity of 0.1 or less are considered to have a modest increase in glomerular permeability and in nephrotic syndrome suggests corticosteroid responsive disease. A selectivity greater than 0.2 suggests that steroid resistance is likely. There is, however, considerable variability and the interpretation of selectivity can be misleading.

Tubular proteinuria

A pattern of proteinuria in which proteins of a lower molecular weight than albumin predominate is found in a group of kidney diseases characterized by damage to the proximal renal tubule. These include the primary Fanconi syndrome, or tubular damage secondary to cystinosis, tyrosinaemia, Wilson's disease, drugs and other poisons. Although many tubular proteins have been identified, those generally measured are lysozyme (molecular weight (MW) 14 000), β_2-microglobulin (MW 11 300), retinol-binding protein (MW 21 000) and α_1-microglobulin (MW 31 000). All are freely filtered, yet in healthy individuals little appears in the final urine, suggesting extensive tubular reabsorption and catabolism. Lysozyme (MW 14 000) is produced by neutrophils and upon their destruction it is released into the extracellular fluid space, where it is filtered, reabsorbed and catabolized by the proximal tubular cells. Its clinical usefulness is limited because plasma concentrations are also raised in chronic renal failure and leukaemia (Osserman & Lawlor, 1966). Some patients with tubular disease do not have lysozymuria (Harrison *et al.*, 1968). In children the upper normal limit for the lysozyme : creatinine ratio is 1.1 mg/μmol.

β_2-Microglobulin is produced by the lymphopoietic system, filtered by the glomerulus, reabsorbed and degraded by the proximal tubules. Its excretion is increased in renal tubular diseases, transplant rejection and nephrotoxicity due to drugs. The measurement is by enzyme-linked immunosorbent assay. However, it is rapidly degraded in acid urine and this limits its clinical usefulness. False positive results may also be seen in diseases of the lymphopoietic system and systemic infections (Cassuto *et al.*, 1978).

Retinol-binding protein and α_1-microglobulin are more stable proteins (Barnard & Lauwerys, 1981). Furthermore, the rates of synthesis of α_1-microglobulin are not affected by other diseases, except hepatic failure (Itoh *et al.*, 1983). The presence of low-molecular-weight proteins in the urine indicates proximal renal tubular damage but their concentration has not been shown to have prognostic significance.

The enzyme *N*-acetylaminoglucosaminidase is an intracellular enzyme released from the lysozomes on cell damage. Although present in most cells its molecular size is such that it is not normally filtered. Its presence in the urine suggests damage to the renal tubular cells. Increased urine excretion is found in transplant rejection, drug toxicity and interstitial nephritis.

Overflow proteinuria

Excretion of low-molecular-weight proteins produced in excess occurs mainly in the monoclonal gammopathies, including multiple myeloma and macroglobulinaemia. These are rare in childhood. Lysozymuria in patients with leukaemia (Osserman & Lawlor, 1966) and increased excretion of β_2-microglobulin in lymphoproliferative diseases (Cassuto *et al.*, 1978) are examples of conditions that occur in childhood. Myoglobulinuria in severe rhabdomyolysis is another example.

Tam–Horsfall protein

Tam–Horsfall protein is a large glycoprotein (MW 7×10^6) which is found in the cells of the thick ascending limb of the loop of Henlé. It is present in normal urine and is the major constituent of urinary casts. It probably protects the kidney from ascending infections of the urinary tract (Hoyer & Seiler, 1979).

Clinical categories of proteinuria

Intermittent proteinuria

There is often a transient increase in urinary protein excretion following strenuous exercise or fever in the absence of significant renal disease. There is no evidence that it is associated with significant renal disease, although minor renal histological abnormalities have been described in some patients (Muth, 1965). Postural proteinuria is a type of intermittent proteinuria which is present only in the upright posture. The amount of proteinuria rarely exceeds 20 mg/m^2 per 24 h. It can be detected by comparing the urinary protein content of an early morning urine sample with that of a sample obtained at the end of the day. The early morning protein: creatinine ratio should not be greater than 0.2. It is a phenomenon found commonly in adolescents, particularly boys (Ryland & Spreiter, 1981). Postural proteinuria may be found in the healing phase of acute glomerular nephritis, but more commonly it is not associated with renal pathology. The prognosis is excellent.

Persistent proteinuria

An increase in urinary protein excretion of between 4 and 40 mg/m^2 per 24 h, persisting despite changes in posture, suggests renal disease, especially when associated with an abnormal urinary sediment. Isolated proteinuria without other evidence of renal disease and a normal urinary sediment is found in up to 3% of schoolchildren (Vehaskari & Rapola, 1982). The majority have no significant renal pathology but important glomerular disease, particularly focal segmental glomerular sclerosis, may be found (Yoshikawa *et al.*, 1991).

Nephrotic syndrome

A urine protein excretion rate of more than 40 mg/m^2 per 24 h is associated with hypoalbuminaemia (plasma albumin less than 20 g/L), oedema and the clinical picture of nephrotic syndrome. In children it is due to steroid-responsive nephrotic syndrome in 80% of cases. Other glomerular pathology can only be identified with certainty by renal biopsy, but it is usually worth a trial of steroid treatment before subjecting children to this procedure (ISKDC, 1981).

Salt and water haemostasis

Serum sodium concentrations are normally maintained within the range 135–145 mmol/L. An increase in plasma sodium causes the release of antidiuretic hormone (ADH) as a result of stimulation of the osmoreceptors situated in the hypothalamus. ADH release is also under the control of volume receptors in the left atrium. Hypovolaemia causes ADH release and hypervolaemia would normally suppress it. In addition, the sensation of thirst is stimulated by increased plasma osmolality; under normal circumstances this should increase dietary water intake, although in young children who rely upon others to provide the water, this may not be the case. Finally, the dietary intake of sodium will influence plasma sodium concentration. The normal mature kidney is able to conserve sodium maximally at less than 1% of its filtered load and excrete up to 10%. Thus, a child with a plasma sodium concentration of 140 mmol/L, a GFR of 100 mL/min and filtering 14 mmol Na/min would be expected to be able to vary the urinary sodium loss between 0.14 and 1.4 mmol/min, depending upon the sodium intake. The normal urinary sodium excretion in children aged between 4 and 6 years has been shown to average 64 mmol/24 h and the urinary sodium: creatinine ratio 39 (DeCorcey *et al.*, 1986). It is therefore difficult to overwhelm the ability of the normal mature kidney to maintain sodium homeostasis. In contrast, the ability of the newborn to conserve and excrete a sodium load is immature compared with that of older children.

Hypernatraemia

Hypernatraemia is defined as a serum sodium concentration greater than 150 mmol/L. It may result from excessive intake of sodium, inadequate water intake or from a disproportionate loss of water relative to sodium. Sodium excess is rare and generally implies inadvertent administration during enthusiastic resuscitation or deliberately as a manifestation of child abuse (Meadow, 1993). It is important to measure the urinary sodium concentration; in salt poisoning this is often extremely high, reflecting the renal response to the increased excretion.

Water depletion usually results from increased urinary loss from either central diabetes insipidus, a lack of ADH or nephrogenic diabetes insipidus (NDI). NDI is inherited as a sex-linked recessive disorder but sporadic cases have been reported in both sexes and most female carriers suffer a partial defect. This can be explained by the Lyon hypothesis of random X-chromosome inactivation. Plasma vasopressin concentrations are appropriately normal and the basic defect is the failure of the renal collecting tubule to respond to ADH. Recent studies have suggested that there are two types of NDI; these depend upon the presence or absence of the response of urinary cyclic adenosine monophosphate to vasopressin (Zimmerman & Green, 1975).

Primary water deficit may result from a lack of water intake, occurring as a form of child abuse, the result of damage to the thirst centre in the hydrocephalus or following head injury (Hays *et al.*, 1963).

NDI may be secondary to other renal diseases, including obstructive nephropathy, renal dysplasia and medullary cystic and polycystic kidney disease. It has also been described in patients with sickle cell disease.

In both NDI and central diabetes insipidus, hypernatraemia is aggravated when access to water is limited, as is the case during infancy. Generally, the urine sodium concentration is low (less than 20 mmol/L), because of hypovolaemia, and the urinary specific gravity is constantly less than 1010. The diagnosis of NDI is confirmed by the failure of the urine flow rate or osmolality to change in response to pitressin injection or the nasal administration of arginine vasopressin (Aronson & Svenningsen, 1974).

Hypernatraemia may also occur when water loss exceeds the loss of sodium. This usually results from diarrhoea (sodium concentration 30–60 mmol/L) in association with vomiting. Hypernatraemic dehydration is now an uncommon consequence of diarrhoea and vomiting, but was relatively common when babies were fed formula milks based upon cows' milk and with a sodium concentration higher than that of present-day milks. Typically, the urine sodium excretion is low in hypernatraemic dehydration (less than 20 mmol/L) as the hypovolaemia takes precedence over the hypernatraemia. Urinary excretion of sodium does not increase until the hypovolaemia has been corrected with appropriate plasma expansion. Correction of the hypernatraemia should be carried out slowly over a period of at least 48 h to avoid neurological complications. There is often associated hypocalcaemia.

Hyponatraemia

Hyponatraemia is defined as a serum sodium less than 130 mmol/L. It is a relatively common problem, but does not usually become clinically significant until the serum sodium concentration is less than 120 mmol/L, when seizures and coma may occur owing to cerebral oedema. It may occur as a consequence of sodium loss or water excess.

Sodium loss

The most common cause of hyponatraemic dehydration is gastrointestinal loss of sodium, usually from gastroenteritis. Rarely, excessive sodium loss from sweat is seen in children with cystic fibrosis. The situation is aggravated when the replacement fluid used is a hypotonic solution. The associated hypovolaemia stimulates ADH release and reduces the GFR. The reduced GFR results in increased proximal tubular reabsorption of sodium and the decreased delivery of sodium to the distal nephron. Both these factors have the effect of reducing the excretion of free water, which aggravates the hyponatraemia. In this situation the urinary sodium concentration is less than 20 mmol/L.

If hyponatraemia is due to excessive renal loss of sodium, the urinary sodium is greater than 20 mmol/L, even in the presence of hypovolaemia. This occurs in Fanconi syndrome, familial nephronephthisis, obstructive uropathy, polycystic disease and the recovery phase of acute renal failure. Renal salt wasting is also found in adrenogenital syndrome, where the absence of mineralocorticoid impairs the distal tubular reabsorption of sodium in exchange for hydrogen ions and potassium.

Hyponatraemia is found in 40% of premature babies and probably occurs as a result of the inability of the immature kidney to conserve sodium and prevent excess urinary sodium loss (Wilkins, 1992).

Water excess

Hyponatraemia may occur rarely from excessive intake of water as a consequence of drowning in fresh water or having a tap-water enema. More commonly, it is the result of water retention from the inappropriate secretion of ADH. The syndrome of inappropriate ADH (SIADH) has many causes including drugs, such as vincristine and morphine, and infection, particularly of the central nervous system and lungs. The characteristic features of SIADH are a reduced urine flow rate with a high urine sodium concentration in the presence of a normal GFR. The treatment is by fluid restriction. Water overload may also occur in acute or chronic renal failure when excess water is administered and the reduced or absent glomerular filtration impairs its excretion.

ADH secretion in response to the hypovolaemia associated with the hypoalbuminaemia in patients with nephrotic syndrome and congestive cardiac failure may also result in hyponatraemia.

Treatment

Treatment of hyponatraemia requires fluid restriction or salt supplementation, depending upon the cause. There is some controversy surrounding the optimum rate of correction of chronic hyponatraemia because of the potential to develop osmotic encephalopathy syndrome, although this is rare in children (Valsanis *et al.*, 1971).

A plasma sodium concentration less than 120 mmol/L requires prompt correction, irrespective of the cause and particularly if it has occurred acutely and is causing neurological symptoms. Correction is achieved using a concentrated sodium solution calculated to provide sodium replacement so that the serum sodium is increased to a concentration of at least 125 mmol/L.

Factitious hyponatraemia

Plasma consists of approximately 95% water and 5% solids. High levels of plasma lipids, which typically occur in nephrotic syndrome, reduce the plasma water. Thus, 1 L plasma in a nephrotic child with hyperlipidaemia may, for the sake of argument, contain 900 mL water and 10% solids. As sodium is present in water, the measured sodium concentration per litre of plasma will be artificially low and does not represent true hyponatraemia.

Renal acidosis

The kidney regulates the acid–base balance by maintaining the plasma bicarbonate within its normal physiological concentration range. This is accomplished by the proximal tubule reclaiming the filtered bicarbonate and the distal tubule excreting hydrogen ions equal to the amount of non-volatile acid produced by metabolism. This usually amounts to 1–2 mmol/kg per 24 h. At both sites hydrogen is excreted in exchange for sodium.

The proximal tubule reclaims 85–90% of the filtered bicarbonate as sodium bicarbonate, in exchange for hydrogen ions which are excreted as carbonic acid (H_2CO_3). The breakdown of carbonic acid to CO_2 and H_2O is catalysed by the enzyme carbonic anhydrase, which is located in the brush border of the proximal tubular cells. This reaction reduces the concentration of carbonic acid in the tubular fluid and facilitates further secretion of hydrogen ions. The proximal tubule is the major pathway for hydrogen ion excretion by the kidney.

The distal tubule accounts for the remaining 10–15% of bicarbonate reabsorption. Here, the hydrogen ions are secreted and titrated by the urinary buffers Na_2HPO_4 and NH_3 to NaH_2PO_4 and NH_4^+

respectively. Titratable acid refers to the hydrogen ions excreted as NaH_2PO_4 and NH_4^+ and can be directly measured. The combined excretion rates of ammonium and titratable acid minus the excretion rate of bicarbonate is referred to as the net acid excretion. Because the distal tubule contributes a relatively small amount to the secretion of hydrogen ions by the kidney, it can be overwhelmed by relatively small amounts of bicarbonate rejected by the proximal tubule.

Renal acidosis may be caused by a reduction in the number of functioning nephrons or by renal tubular acidosis (RTA) where there is little or no evidence of renal insufficiency.

Uraemic acidosis tends to occur when the GFR is reduced to 25 mL/min per 1.73 m^2 or less, and is the result of both impaired proximal tubular reabsorption of bicarbonate and impaired ammonia production. The impaired bicarbonate reabsorption is not uncommon in patients with renal insufficiency; frank bicarbonate wastage may occur and require relatively large amounts of bicarbonate to correct the acidosis. The mechanisms of bicarbonate wastage in renal insufficiency are not clearly defined, but may be due to reduced carbonic anhydrase activity or to the increased filtered load of bicarbonate which overwhelms the remaining functioning nephrons. There is also evidence to suggest that impaired bicarbonate reabsorption in chronic renal insufficiency may result from secondary hyperparathyroidism because an increased circulatory concentration of parathyroid hormone can reduce bicarbonate reabsorption. In contrast, suppression of parathyroid hormone secretion by calcium infusion has been shown to reduce urinary bicarbonate wastage in patients with chronic renal failure. The rate of secretion of ammonium is also reduced in chronic renal insufficiency, probably owing to the reduction in the functioning nephron mass.

Renal tubular acidosis

In RTA there is little or no evidence of renal insufficiency, hyperchloraemic acidosis and an inappropriately high urine pH. At least four types of RTA have been described (McSherry, 1981). The classical or distal RTA (RTA type I) is inherited as a Mendelian dominant characteristic. It may also occur as a secondary phenomenon in association with other renal disorders (Table 11.2).

Table 11.2 Causes of type I renal tubular acidosis (distal)

Primary	Inherited (Mendelian dominant) Spontaneous
Secondary	
Autoimmune	Sjogren's syndrome Systemic lupus erythematosus
Genetic disease	Sickle cell disease Ehlers–Danlos syndrome Marfan's syndrome
Renal disease	Renal transplantation Obstructive uropathy Pyelonephritis Medullary sponge kidney Nephrocalcinosis
Drugs	Amphotericin B Lithium

The most common presentation is a failure to thrive, but a small percentage of children may present as an acute emergency with severe acidosis, shock and hypokalaemia, often associated with muscle weakness. The urine pH is high (greater than 5.5) and in contrast to the proximal RTA remains alkaline, even in the presence of severe metabolic acidosis (arterial pH less than 7.2). This is due to the inability of the distal nephron to generate and maintain a hydrogen ion gradient between the tubular fluid and the cell. In the typical distal RTA the acidosis can be corrected by bicarbonate supplementation sufficient to buffer the endogenous acid production, usually 2–3 mmol/kg per day. In some children with apparently classical distal RTA, bicarbonate supplements of more than twice this amount may be necessary to maintain normal plasma bicarbonate concentrations, suggesting that there is also a degree of bicarbonate wastage. This association has been classified as type III RTA.

Type II RTA results from a defect of bicarbonate reabsorption by the proximal tubule. It is typically found in association with reduced reabsorption of amino acids, phosphate, calcium and glucose and with tubular proteinuria. These features of Fanconi syndrome are associated with other systemic dis-

Table 11.3 Causes of renal tubular acidosis type II (proximal)

Primary	Idiopathic transient (infants)
Secondary	
With Fanconi syndrome	Idiopathic
	Cystinosis
	Tyrosinaemia
	Galactosaemia
	Fructose intolerance
	Lowe's syndrome
Toxins	Lead
	Outdated tetracycline
	Cisplatin

eases (Table 11.3). Here, the primary disease is the dominant clinical feature. It may occur as a primary defect of bicarbonate reabsorption in the newborn, usually as a transient abnormality. The urine pH is inappropriately alkaline but an acid urine (pH less than 5.5) can be produced in the presence of severe systemic acidosis (arterial pH less than 7.2) when the plasma bicarbonate is so low that the filtered bicarbonate falls. The primary defect is the result of increased bicarbonate wastage, which may exceed 15% of the filtered load. The bicarbonate wastage makes it impossible to maintain a normal bicarbonate concentration, even when large supplements of up to 20 mmol/kg per day are administered. The bicarbonaturia and secondary hyperaldosteronism both increase urinary potassium secretion by the distal tubule and it is usually necessary to replace the alkali with both potassium and sodium citrate.

Type IV RTA is now being recognized as the commonest type in children. It is characterized by hyperchloraemic acidosis and, in contrast to the other types of RTA, hyperkalaemia. The defect in urinary acidification, acidosis and potassium retention have been attributed to either a deficiency of or a lack of tubular response to circulatory aldosterone. A further subclassification into five subtypes has been proposed.

The diagnosis and distinction between proximal and distal RTA require the evaluation of bicarbonate and hydrogen ion excretion. Bicarbonate reabsorption can be investigated during a bicarbonate infusion (Rodriguez-Soriano *et al.*, 1967) or, more simply, by examining the fractional excretion of bicarbonate, which should not normally exceed 15% of the filtered load. The diagnosis of distal RTA is established by examining the ability of the kidneys to excrete an acid load; the urine pH, blood pH and plasma bicarbonate are monitored during systemic acidosis. Acidosis can be induced by an acid load of either ammonium chloride (100 mmol/m^2 orally) (Edelmann *et al.*, 1967) or intravenous arginine hydrochloride (100–150 mmol/m^2) (Loney *et al.*, 1982). Ammonium chloride is not tolerated well by children and arginine hydrochloride is preferred. Neither should be given in the presence of a persisting acidosis.

Renal alkalosis

Metabolic alkalosis may occur as a result of excess administration of bicarbonate or a loss of acid. The former is relatively rare; it may occur during resuscitation or occasionally from the deliberate administration of sodium bicarbonate as part of the syndrome of child abuse. The most common cause of alkalosis in paediatric practice is the loss of acid which accompanies the excessive vomiting of pyloric stenosis. Here, there is a loss of hydrochloric acid and sodium and potassium chloride. The consequences are metabolic alkalosis, potassium deficiency and contraction of the extracellular space, which causes activation of the renin–angiotensin aldosterone system. As a consequence, the kidney excretes large amounts of bicarbonate with sodium and potassium; this aggravates the hypokalaemia and alkalosis. The urine pH is high. Potassium deficiency is an important factor in alkalosis. Metabolic alkalosis itself increases urinary potassium loss because the kidney can only excrete bicarbonate in association with sodium or potassium. The renal loss of potassium is greater than that from the stomach in pyloric stenosis. In addition, sodium loss results in contraction of the extracellular space and stimulation of the renin–angiotensin aldosterone system, and aldosterone causes increased distal tubular reabsorption of sodium in exchange for increased excretion of potassium and hydrogen ions in the urine. Correction of the alkalosis of pyloric stenosis is dependent upon adequate sodium and potassium replacement to allow the kidney to excrete the excess alkali.

The kidney can generate an alkalosis when there is stimulation by aldosterone of the distal tubular sodium–hydrogen ion exchange process. Hyperaldosteronism has been recognized in a variety of clinical syndromes including Cushing's syndrome and primary hyperaldosteronism. The sodium retention results in hypertension. The hypokalaemia is associated with muscle wasting and weakness. More frequently, hyperaldosteronism occurs as a secondary phenomenon in association with extracellular fluid volume contraction, owing to loss of sodium, or in oedematous states such as the nephrotic syndrome. The blood pressure is usually normal.

In Bartter's syndrome hypokalaemic alkalosis is associated with hyperreninaemia, hyperaldosteronism and hyperplasia of the juxtaglomerular apparatus. Despite the hyperreninaemia, the blood pressure is normal and hypertension excludes Bartter's syndrome as a diagnosis. There is a marked loss of urinary potassium, resistance to the pressure effects of angiotensin II and norepinephrine, and the urinary excretion of prostaglandin E is increased. Less commonly, there may also be hyperuricaemia, hypercalcaemia and hypomagnesaemia. Children usually present with a failure to thrive and intermittent weakness due to the hypokalaemia. Progressive renal failure may develop as a consequence of interstitial nephritis.

The pathophysiology of Bartter's syndrome has been extensively investigated. The over-production of prostaglandin E has been well documented, but it is not clear whether this is the primary abnormality or an epiphenomenon resulting from an over-production of prostaglandin in response to hypokalaemia.

Treatment of Bartter's syndrome with potassium supplements is usually unsuccessful. Despite the hyperaldosteronism, the aldosterone antagonist spironolactone is generally unhelpful. Treatment with prostaglandin synthetase inhibitors does not completely correct the hypokalaemia but does reduce its severity and increase the growth velocity of children. Potassium-sparing diuretics, such as amiloride, and the renin–angiotensin antagonist captopril have also been used with some benefits.

Recently, a number of hereditary and congenital forms of urinary potassium wasting have been described and these differ from the classical picture of Bartter's syndrome. It is not clear whether they represent a spectrum of the same disorder or are different entities. They all involve hypokalaemic alkalosis, normal blood pressure and high plasma renin activity. Renal biopsies have shown evidence of interstitial nephritis rather than the juxtaglomerular hyperplasia which is typical of Bartter's syndrome.

Renal stone disease

Renal stone disease affects between one and two children per million of the population. Affected children normally present with abdominal pain, haematuria or urinary tract infections.

Stones develop when there is precipitation of crystals in the urine. Normally, there is a balance between factors which encourage and those which inhibit precipitation. The most important inhibitors of crystal growth are mucopolysaccharides, citrate and pyrophosphate. Stone formation is encouraged when there is an increased concentration of an insoluble substance or a change in the urine pH, and this results in its solubility product being exceeded. Precipitation is encouraged in the urine by the presence of cellular debris and other particles which provide a nidus for secondary participation and stone growth.

Infective stones

Urinary tract infections account for over 80% of renal stones in children. The usual infecting organism is *Proteus mirabilis*, which contains the enzyme urease. Its action on the urea present in the urine produces ammonia. This in turn increases the urine pH and the solubility product for calcium is exceeded, resulting in stone formation. Infective stones are either struvite (ammonium magnesium phosphate) or apatite (calcium hydrogen phosphate). These stones are more likely to develop when there is stasis of the urine, as occurs in hydronephrosis or other anatomical abnormalities of the urinary tract.

Hypercalciuria

Hypercalciuria is defined as a urinary calcium excretion in excess of 0.1 mmol/kg per 24 h or a calcium: creatinine ratio on the second voided urine sample of the day greater than 0.7 (Ghazali & Barralt, 1974). There is considerable variation in the amount of calcium excreted throughout the day. It varies with diet, plasma calcium concentration, GFR and the tubular handling of calcium. The calcium excretion index (CaE) is a way of correcting the urinary excretion of calcium for GFR changes. It is calculated using the following formula:

$$\mathrm{CaE} = U_{Ca}V \times \frac{P_{Cr}}{U_{Cr}V} = \frac{U_{Ca} \times P_{Cr}}{U_{Cr}}$$

where U_{Ca} is the urine calcium concentration, P_{Cr} the plasma creatinine concentration, U_{Cr} the urine creatinine concentration and V the urine volume. The upper limit of normal for CaE is 35 µmol/L.

The tubular handling of calcium can be assessed by examining its tubular maximum reabsorption, corrected for GFR by use of the formula derived by Marshall (1976)

$$\mathrm{T_mCa} = \frac{\mathrm{UFCa} - \mathrm{CaE}}{1 - 0.08\ \log_e(\mathrm{UFCa/CaE})}$$

where UFCa is the ultrafiltratable plasma calcium concentration. The normal range for T_mCa in children is 1.87–3.39 mmol/L (Shaw *et al.*, 1992).

Hypercalciuria may occur as a result of enhanced gut absorption of calcium or increased renal excretion. These may be distinguished by a calcium loading test (Stapleton, 1990). In the absorptive type, increased absorption by the gastrointestinal tract increases the plasma calcium, thus increasing the glomerular filtration of calcium. The increased plasma calcium also suppresses parathyroid hormone with a resultant reduction in the T_mCa, both of which lead to hypercalciuria. The basic defect is not understood but has been attributed to excessive production of 1,25-dihydroxy vitamin D.

Table 11.4 Causes of calcium stones

Hypercalciuria (normal plasma calcium)
Idiopathic hypercalciuria (absorptive type)
Idiopathic hypercalciuria (renal type)
Distal renal tubular acidosis
Hypercalcaemic hypercalciuria
Bone resorption
Primary hyperparathyroidism
Immobilization
Corticosteroid excess
Increased intestinal absorption
Vitamin D toxicity
Idiopathic hypercalcaemia of infancy

The primary defect in renal hypercalciuria is an impairment of renal tubular reabsorption of calcium. The resultant hypocalcaemia stimulates parathyroid hormone secretion, which in turn results in increased bone resorption, increased gut absorption of calcium and renal synthesis of 1,25-dihydroxy vitamin D.

Idiopathic hypercalciuria is a familial autosomal dominant disorder which is thought to be the result of an abnormality of enzyme transport across cell membranes. Other causes of hypercalciuria are shown in Table 11.4.

Hypercalciuria has been found in one-third of children who present with isolated haematuria, and it also represents a risk factor for future renal stone disease (Stapleton, 1990). Treatment requires an increased fluid intake, but in refractory cases the addition of chlorothiazide diuretics will reduce urine calcium excretion, largely by reducing its filtered load. A moderate reduction in dietary sodium may also be helpful.

Cystinuria

Cystinuria results from defective proximal renal tubular absorption of the dibasic amino acids cystine, arginine, lysine and ornithine; there is a resultant increased excretion of these in the urine. There may also be a similar absorptive defect in the gut. Cystine is insoluble when the urine pH is below 7.5.

Cystinuria is inherited as an autosomal recessive trait and at least three variants have been recognized

(Scriver *et al.*, 1985). It can be detected clinically by the presence of flat hexagonal cystine crystals in an acid urine and can be confirmed by quantitative urine amino acid analysis. Cystine stones are radio-opaque.

Treatment requires a high fluid intake. Alkalinization of the urine to pH 7.5 or greater will increase the solubility of cystine but runs the risk of inducing calcium stones. D-Penicillamine produces a penicillamine–cysteine complex, which is much more soluble than cystine, but treatment may cause proteinuria, which can progress to nephrotic syndrome.

Oxalate stones

Oxalate is a not uncommon constituent of renal stones but primary hyperoxaluria is rare. It may be due to one of two inborn errors of metabolism: a deficiency of the enzyme alanine: glyoxylate aminotransferase, located in the peroxisomes of the hepatocytes (type 1), or L-glycericaciduria (type 2).

Hyperoxaluria type 1 may present in infancy with renal failure, but more characteristically children present with recurrent stone formation and nephrocalcinosis, leading to renal failure in late childhood. Oxalate may precipitate in many other tissues including the joints, where it causes severe pain, the heart, causing arrhythmias, and in the retina.

Hyperoxaluria type 2 is a less severe disease. The urine contains increased amounts of oxalate and L-glyceric acid. Renal calculi are formed but renal failure occurs less frequently.

In children the normal daily excretion rate of oxalate is between 0.2 and 0.5 mmol/24 h per 1.73 m^2 (Barratt *et al.*, 1991). In renal failure the urinary excretion of oxalate is reduced, and the diagnosis of primary hyperoxaluria becomes more difficult in the presence of renal insufficiency. Further oxalate retention occurs in advanced renal failure, irrespective of the cause, and oxalate crystals may be precipitated in the tissues. The diagnosis can then only be established by measurement of the activity of the enzyme alanine: glyoxylate aminotransferase in a liver biopsy.

Pyridoxine is an important cofactor in the enzymatic conversion of glycine to glyoxylate, and treatment with high doses (100–200 mg/day) may reduce endogenous oxalate production in some patients. Enzyme replacement by liver transplant has been used successfully in some children with type 1 disease.

Hyperoxaluria may occur secondary to inflammatory bowel disease or malabsorption. It has been suggested that malabsorption results in increased fatty acids in the gut; these bind to calcium and reduce its intestinal concentration. The concentration of insoluble calcium oxalate is therefore reduced, leaving the more soluble compounds of oxalate to be absorbed and excreted in the urine, where there is the potential for stone formation.

Purine stones

Inborn errors of purine metabolism are rare causes of renal stones in children. Nevertheless, they are important because they are inherited and specific treatments are available (Simmonds *et al.*, 1989). There are three defects of purine salvage enzymes which result in renal stones: hypoxanthine guanine phosphoribosyltransferase (HGPRT), adenine phosphoribosyltransferase (APRT) and xanthine oxidase (XO). These are respectively associated with increased urinary excretion of the insoluble metabolites uric acid, dihydroxyadenine and xanthine. The resulting crystalluria may cause renal stones and interstitial nephritis. Acute renal failure may occur in up to one-third of cases.

A disorder of purine metabolism may be suspected if the renal calculi are radiolucent and can only be detected by renal ultrasound. The diagnosis is made by measurement of the specific metabolites of purine metabolism in the urine. The plasma urate level is usually very high in HGPRT deficiency. Although the urate may be elevated in renal failure, it is unusual for it to be more than twice the normal value. A complete deficiency of HGPRT gives rise to Lesch–Nyhan syndrome with the typical neurological picture, but partial defects cause gout and renal stones without the neurological features. In XO deficiency the plasma urate level is very low. Specific treatment with the XO inhibitor allopurinol is effective in the long-term management of HGPRT and APRT deficiency but should be carefully monitored to avoid complications. XO deficiency is usually asymptomatic and rarely causes renal calculi in children.

Tumour lysis syndrome

The rapid destruction of tumour cells by cytotoxic chemotherapy is associated with an increased metabolic load of uric acid, potassium and phosphate. The high plasma phosphate concentration in turn leads to hypocalcaemia. The syndrome occurs in tumours with a high growth fraction that are extremely sensitive to chemotherapy, such as acute lymphoblastic leukaemia with a white blood cell count greater than 100×10^9 per litre. The precipitation of uric acid in the renal tubules may cause acute renal failure or interstitial nephritis. Since uric acid is relatively insoluble in an alkaline urine, pretreatment with a sodium bicarbonate infusion to achieve a urine pH greater than 7 is advisable before chemotherapy is started. The XO inhibitor allopurinol will reduce the urine uric acid load and is also recommended.

References

Aronson, A.S. & Svenningsen, D. (1974) DDAVP test for estimation of renal concentrating capacity in infants and children. *Arch Dis Child* **49**, 654–659.

Barnard, A.M. & Lauwerys, R.R. (1981) Retinol binding protein in urine: a more practical index than urinary β_2 microglobulin for the routine screening of renal tubular function. *Clin Chem* **27**, 1781–1782.

Barratt, T.M., Kasidas, G.P., Muradoch, I. & Rosi, G.A. (1991) Urinary oxalate and glycolate excretion and plasma oxalate concentrations. *Arch Dis Child* **66**, 501–503.

Barratt, T.M., McLaine, P.M. & Soothill, J.F. (1970) Albumin excretion as a measure of glomerular dysfunction in children. *Arch Dis Child* **45**, 496–501.

Brocklebank, J.T., Cooper, E.H. & Richmond, K. (1991) Sodium dodecyl sulphate polyacrylamide gel electrophoresis patterns of proteinuria in various renal diseases of childhood. *Pediatr Nephrol* **5**, 371–375.

Cameron, J.S. & Blandford, G. (1966) The simple assessment of permeability in heavy proteinuria. *Lancet* **ii**, 242–245.

Cassuto, J.P., Krebb, B.J., Viot, A., Dujardin, P. & Masseyeff, R. (1978) β_2-Microglobulin, a tumour marker of lymph proliferative disorders. *Lancet* **4**, 108–109.

Chantler, C. & Barratt, T.M. (1972) Estimation of GFR from the plasma clearance of 51 chromium edetic acid. *Arch Dis Child* **47**, 613–617.

Coulthard, H.M. & Hey, E.N. (1984) Weight as the best standard for glomerular filtration in the newborn. *Arch Dis Child* **59**(4), 373–375.

Davies, A.G., Postlethwaite, R.J., Price, D.A. *et al.* (1984) Urinary albumin secretion in school children. *Arch Dis Child* **59**, 625–630.

DeCorcey, S., Mitchell, H., Simmonds, D. & MacGregor, G. (1986) Urinary sodium excretion in 4–6-year-old children: a cause for concern? *Br Med J* **292**, 1428–1429.

Edelmann, C.H. Jr, Biochis, H., Rodriguez-Soriano, J. & Stark, H. (1967) The renal response of children to acute ammonium chloride acidosis. *Pediatr Res* **1**, 452.

Evans, J.H.C., Smye, S.W. & Brocklebank, J.T. (1992) Mathematical modelling of haemodialysis in children. *Pediatr Nephrol* **6**, 349–353.

Friis-Hansen, B. (1961) Body weight compartments in children: changes during growth and related changes in body composition. *Pediatrics* **28**(2), 169–181.

Ghazali, S. & Barralt, T.M. (1974) Urinary excretion of calcium and magnesium. *Arch Dis Child* **49**, 97–101.

Harrison, J.F., Lunt, G.S., Scott, P. & Blainey, T.D. (1968) Urinary lysozyme, ribonurease and low-molecular-weight protein in renal disease. *Lancet* **i**, 371–374.

Haycock, G., Schwartz, G.J. & Wisotsky, D.H. (1978) Geometric method for measuring body surface area: a height–weight formula validated in infants, children and adults. *J Pediatr* **93**(1), 62–66.

Hays, R.M., McHugh, P.R. & Williams, H.E. (1963) Absence of thirst in association with hydrocephalus. *N Engl J Med* **269**, 227–231.

Houser, M. (1984) Assessment of proteinuria using random urine samples. *J Pediatr* **105**, 845–848.

Hoyer, J.R. & Seiler, M.W. (1979) Pathophysiology of Tam–Horsfall protein. *Kid Int* **16**, 279–289.

ISKDC (1981) The primary nephrotic syndrome in children. Identification of patients with minimal change nephrotic syndrome from initial response to prednisolone. *J Pediatr* **98**, 561–564.

Itoh, Y., Enomoto, H., Takagi, K., Kawai, T. & Yamanaka, T. (1983) Human $\alpha 1$ microglobulin in various hepatic disorders. *Digestion* **27**, 75–80.

Kenny, A.P. (1973) Tests of phosphate reabsorption. *Lancet* **II**, 158.

Loney, L.C., Norling, L.L. & Robson, A.M. (1982) The use of arginine hydrochloride infusion to assess urinary acidification. *J Pediatr* **100**(1), 95–97.

McCance, R.A. & Widdowson, E.M. (1952) The correct physiological basis on which to compare infant and adult renal function. *Lancet* **ii**, 860–862.

McSherry, E. (1981) Renal tubular acidosis in childhood. *Kid Int* **20**, 799–809.

Marshall, D.H. (1976) Calcium and phosphate kinetics. In: *Calcium, Phosphate and Magnesium Metabolism* (ed. B.E.J. Nordin), pp. 257–297. Churchill Livingstone, Edinburgh.

Meadow, S.R. (1993) Non-accidental salt poisoning. *Arch Dis Child* **68**, 448–452.

Morris, M.C., Allanby, C.W., Toseland, P., Haycock, G.B. & Chantler, C. (1982) Evaluation of a height/plasma creatinine formula in the measurement of glomerular filtration rate. *Arch Dis Child* **57**, 611–615.

Muth, R.G. (1965) Asymptomatic mild intermittent proteinuria. *Arch Int Med* **115**, 569–574.

Osserman, E.F. & Lawlor, D.P. (1966) Serum and urinary lysozyme (neuraminidase) in monocytic and myelomonocytic leukaemia. *J Exp Med* **124**, 921–951.

Payne, R.B. (1986) Creatinine clearance: a redundant clinical investigation. *Ann Clin Biochem* **23**, 243–250.

Rennie, I.D.B. & Keen, H. (1967) Evaluation of clinical methods of detecting proteinuria. *Lancet* **2**, 489.

Rodriguez-Soriano, J., Boichis, H. & Edelmann, C.M.J. (1967) Bicarbonate reabsorption and hydrogen ion excretion in children with renal tubular acidosis. *J Pediatr* **71**, 802–813.

Rudd, P.T., Hughes, E.A. & Placzek, M.M. (1983) Reference ranges for plasma creatinine during the first month of life. *Arch Dis Child* **58**, 212–215.

Ryland, D.A. & Spreiter, S. (1981) Prognosis in postural (orthostatic) proteinuria. Forty to fifty year follow-up of six patients after diagnosis by Thomas Addis. *N Engl J Med* **305**, 618–621.

Savory, D.J. (1990) Reference ranges for serum creatinine in infants, children and adolescents. *Ann Clin Biochem* **27**, 99–101.

Schwartz, G.J., Feld, L.G. & Langford, D.J. (1984) A simple estimate of glomerular filtration rate in full-term infants during the first year of life. *J Pediatr* **104**, 849–854.

Scriver, C.R., Claw, C.L., Reade, T.M. *et al.* (1985) Ontogeny modifies manifestations of cystinuria genes: implications for counseling. *J Pediatr* **106**, 411–416.

Shaw, N.J., Wheeldon, J. & Brocklebank, J.T. (1990) Indices of intact serum parathyroid hormone and renal excretion of calcium, phosphate and magnesium. *Arch Dis Child* **65**, 1208–1211.

Shaw, N.J., Wheeldon, J. & Brocklebank, J.T. (1992) The tubular maximum of calcium reabsorption: a normal range in children. *Clin Endocrinol* **36**, 193–195.

Simmonds, H.A., Cameron, J.S., Barratt, J.M. *et al.* (1989) Parine enzyme defects as a cause of acute renal failure in childhood. *Pediatr Nephrol* **3**, 433–437.

Stapleton, F.B. (1990) Idiopathic hypercalcinuria: associated with isolated haematuria and risk for urolithiasis in children. A report of the South West Pediatric Nephrology Study Group. *Kid Int* **37**, 807–811.

Valsamis, M., Peress, N.S. & Wright, L.D. (1971) Central Pontine myelinosis in childhood. *Arch Neurol* **25**, 307.

Van-Acker, K.J., Vlietinck, R.F. & Stech, D.H. (1984) Estimation of glomerular filtration rate from β_2-microglobulin serum levels in children. *Int J Pediatr Nephrol* **5**, 59–62.

Vehaskari, V.M. & Rapola, J. (1982) Isolated proteinuria: analysis of a school-age population. *J Pediatr* **101**, 661–668.

Walton, R.J. & Bijvoet, O.C.M. (1975) Nomogram for derivation of renal threshold phosphate concentrations. *Lancet* **II**, 509–510.

Wilkins, B. (1992) Renal function in sick very low birth weight infants: glomerular filtration rate. *Arch Dis Child* **56**(10), 1140–1145.

Yoshikawa, N., Kitagawa, K., Ohta, K., Tanak, R. & Nakamura, H. (1991) Asymptomatic isolated proteinuria in children. *J Pediatr* **119**, 375–379.

Zimmerman, D. & Green, C.O. (1975) Nephrogenic diabetes type II defect distal to the adenylate cyclase step (abstract). *Pediatr Res* **9**, 381.

12: Disorders of Bone Mineral Metabolism

J.M. GERTNER

Introduction

Calcium and phosphorus are the principal elements of the inorganic component of bone. Their importance in physiology and pathology extends far beyond the bounds of skeletal physiology, affecting almost all aspects of cellular reproduction and function. Although magnesium is not as important a constituent of bone mineral as calcium, it is present in body fluids as a divalent cation and its role in physiology and paediatric pathology will be considered in this chapter.

The physiology of calcium homeostasis

Calcium, in its free or ionized form, provides an essential link between stimulus and response in cellular systems ranging from endocrine secretion to myocardial contractility. Maintenance of the extracellular Ca^{2+} concentration within narrow limits is essential if this stimulus–response coupling is to operate reliably. Accordingly, a homeostatic system exerting tight control over the extracellular Ca^{2+} concentration has evolved. This system is governed by the parathyroid glands but also involves the vitamin D endocrine system. In addition, calcitonin, the parathyroid hormone-related peptide (PTHrP), growth hormone, insulin-like growth factor-1 (IGF-1 or somatomedin-C) and insulin may play a part in the hormonal control of calcium homeostasis and bone mineral metabolism.

Parathyroid hormone

The parathyroid glands are formed from cells of the third and fourth pharyngeal pouches at weeks 6–7 of gestation, and stain positively for parathyroid hormone (PTH) by the 12th week. The chief cells serve both as calcium sensors and as effectors of the control system. When extracellular fluid ionized calcium (Ca^{2+}) concentrations fall, transcription of the preproparathyroid gene is increased, proPTH is cleaved from preproPTH and stored in secretory granules, and the release of PTH from these granules is enhanced. PTH, an 84-amino-acid peptide, enters the circulation and elevates ionized calcium. The 34 N-terminal amino acids of PTH are those actually required for PTH to exert its biological effect.

PTH affects its target cells by interaction with a specific membrane-bound receptor whose function is to activate adenyl cyclase. The receptor is arranged as a tripartite system (Fig. 12.1) consisting of a hormone-binding domain, accessible from outside the cell, a stimulatory G-protein (G_s), which is activated by the binding of PTH to the receptor, and a catalytic unit, which is brought into function by the activated G-protein. The catalytic unit promotes the formation of cyclic adenosine monophosphate (cAMP) from adenosine triphosphate (ATP), but the ways in which cAMP then promotes new metabolic activity in the cell are not well understood. In the proximal renal tubule PTH promotes a decrease in phosphate reabsorption from the tubular fluid, i.e. it is phosphaturic. In bone it appears most likely that PTH activates bone-forming osteoblasts and that these produce some factor that leads directly to the recruitment and activation of bone-resorbing osteoclasts.

The parathyroid glands constitute a homeostatic system exerting tight control over the extracellular Ca^{2+} concentration (Fig. 12.2). The figure illustrates the two loops of this feedback system, both depen-

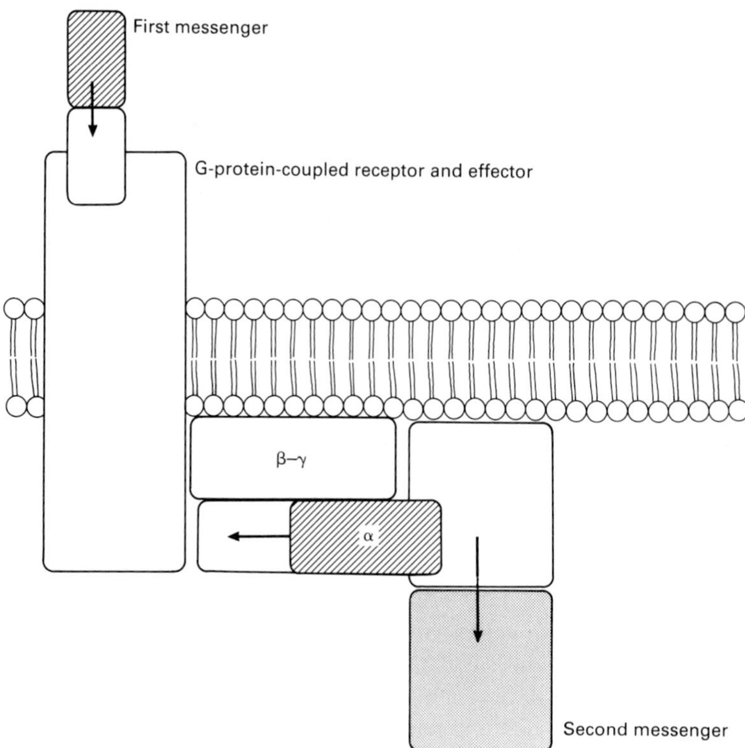

Fig. 12.1 The tripartite arrangement of a typical guanine nucleotide-associated cell-surface receptor such as the parathyroid hormone receptor. Note the central role of the α subunit.

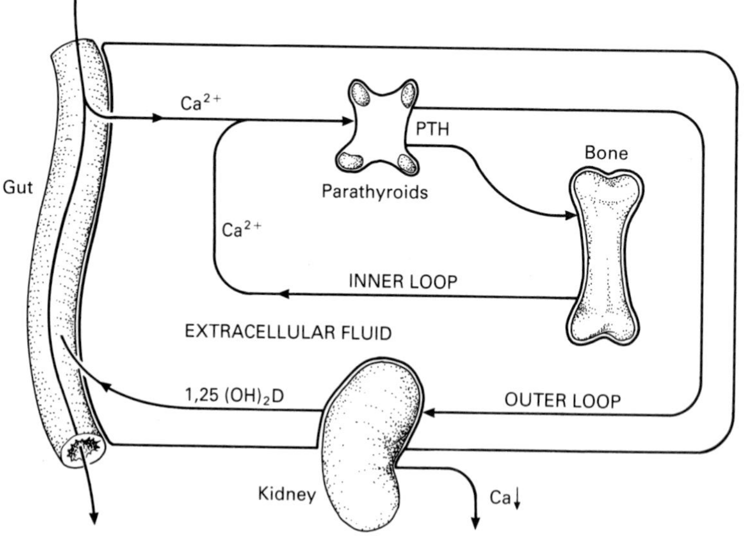

Fig. 12.2 The central role of the parathyroid glands as sensors and effectors in the control of the extracellular ionized calcium concentration. PTH, parathyroid hormone; 1,25-$(OH)_2D$, 1,25-dihydroxyvitamin D.

dent on the parathyroid glands to sense Ca^{2+} concentration and to release PTH to control it. The inner loop responds rapidly to changes in Ca^{2+}, using PTH to modulate the flux of calcium between bone and the extracellular fluid. PTH also serves as a trophic hormone stimulating the activation of calcidiol (25-hydroxycalciferol) to calcitriol (1,25-dihydroxycholecalciferol). This function of PTH is exploited in the slower response outer loop of PTH-mediated calcium homeostasis. Here, the exchange of calcium between the extracellular fluid and the external environment is modified by adjusting the rate of calcium absorption from the gut through control of the availability of calcitriol. Calcitriol also appears capable of directly regulating PTH secretion, but the significance of this interaction for normal physiology is not known.

Assays for PTH

Among the variety of immunoassays for PTH, assays for the N-terminal peptide or for the intact 1–84 molecule are regarded as most relevant to the physiological state of parathyroid activity at the time the measurement is made. Currently, the most sensitive technique appears to be a two-stage immunoradiometric assay (IRMA) (Nussbaum *et al.*, 1987) which depends on the presence of two separate epitopes and so measures quantities of the intact molecule rather than fragments of the C-terminal (inactive) or N-terminal (active) ends of the molecule. Reference values for PTH vary with the assay technique and age of the subject. The reference values given by the Nichols Institute for their two-stage IRMA are 10–65 pg/mL.

The vitamin D endocrine system

(Reichel *et al.*, 1989)

Vitamin D forms the starting material for the second major system controlling calcium homeostasis. Structurally, the vitamin and its metabolites are seco-steroids, all of which are lipid soluble. Vitamin D can be obtained from animal or plant sources (yeast). The vitamins from these sources differ with regard to the side structure at the C17 position: the animal form, vitamin D3 or cholecalciferol, has one methyl group more than the plant form, vitamin D2 or ergocalciferol. Specific assays for the D2 and D3 forms of the various vitamin D metabolites exist but the distinction is not necessary in clinical practice since the two forms of the vitamin are believed to function almost identically in humans.

Vitamin D is formed from cholesterol in the skin by photocatalysis of the reaction from 7-dehydrocholesterol to provitamin D. Provitamin D then undergoes non-enzymatic conversion to vitamin D, which is transported to storage sites bound to a protein known as the vitamin D-binding protein (DBP). Vitamin D is hydroxylated in the 25-position by a microsomal P-450 hydroxylase in the liver, a largely unregulated step. The final activation of the vitamin D endocrine system occurs in the cells of the proximal renal tubules, where a mitochondrial P-450 hydroxylase, 25-hydroxyvitamin D 1α-hydroxylase, catalyses the formation of calcitriol (1,25-dihydroxyvitamin D). This step is highly regulated, being promoted by PTH, calcium depletion and hypophosphataemia. Hypoparathyroidism, hypercalcaemia, hyperphosphataemia and growth hormone deficiency are among the factors which inhibit the 1α-hydroxylation reaction. The metabolic interconversions of vitamin D are shown in Fig. 12.3.

25-Hydroxyvitamin D, here referred to as calcidiol, is the most abundant vitamin D metabolite in the plasma. Like vitamin D it is bound to circulating DBP. The circulating concentration of calcidiol reflects the vitamin D nutritional status of the subject, with values varying according to diet and sunshine exposure (Poskitt *et al.*, 1979). The major function of calcitriol, the hormonally active form of vitamin D, is the promotion of intestinal calcium absorption (Norman, 1990), although clearcut actions on bone (Aksnes *et al.*, 1988), on the immune system (Rigby, 1988) and on other organ systems such as the skin (Holick, 1988) have been described. Calcitriol interacts with a cytoplasmic receptor of the same family as the glucocorticoid, sex hormone and thyroid hormone receptors. The hormone–receptor complex is translocated to the nucleus, where the complex alters genetic transcription in tissue-specific ways.

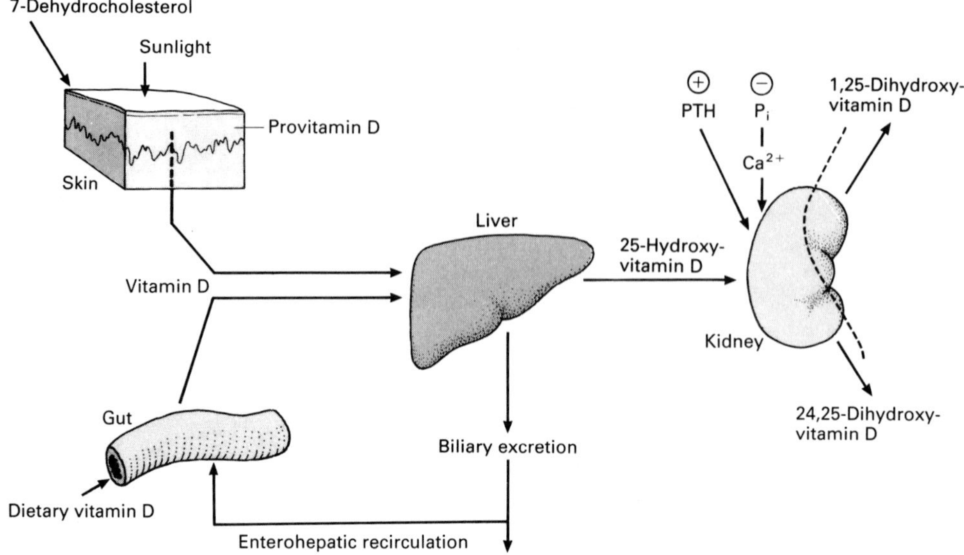

Fig. 12.3 A schema of the activating metabolism of vitamin D. ⊕ indicates a stimulatory effect and ⊖ an inhibitory effect. PTH, parathyroid hormone; P_i, inorganic phosphate.

Assays for vitamin D metabolites

Circulating levels of native vitamin D are rarely measured in clinical practice since calcidiol levels can give an accurate picture of vitamin D nutritional status. Circulating vitamin D is best measured directly by high performance liquid chromatography (HPLC) of lipid extracts of serum (Chen *et al.*, 1990). Calcidiol assays are used to assess vitamin D nutritional status. Assays on serum without preparative chromatography are generally adequate. Protein-binding assays usually employ diluted rat serum as a source of DBP (Haddad & Chyu, 1971), but recently a radioimmunoassay has become commercially available.*

Reference values for calcidiol vary with sunshine exposure and thus depend on latitude and season. They may also be higher in light-skinned than in black people. Values generally given for children are 20–50 ng/mL. It should be noted that for many populations a calcidiol concentration at the lower limit of normal (i.e. 2 SD below the mean) is still considerably higher than that normally observed in cases of clinically significant vitamin D deficiency.

A convenient assay for calcitriol is the radioreceptor assay of Reinhardt *et al.* (1984), which has been modified so that after lipid extraction of plasma, a single C_{18} SepPak chromatography step is sufficient to prepare the sample for assay. In the author's laboratory use of this assay gives the normal values for adults as from 22 to 47 pg/mL (mean ± 2 SD). Values in normal children are said to be a little higher than for adults (Taylor & Norman, 1988).

Calcitonin

Calcitonin, a 22-amino-acid peptide hormone, is secreted by the C-cells of the thyroid, which are derived from ultimobranchial (fifth pharyngeal pouch) tissue and are unrelated in origin or function to the cells which produce thyroxine. Acutely, calcitonin lowers the extracellular ionized calcium concentration by inhibiting bone resorption. However, its precise physiological role is unclear since humans of all ages maintain normal calcium homeostasis

*INCSTAR Corporation (1990 Industrial Blvd, PO Box 285, Stillwater, MN 55082-0285, USA).

and bone turnover in the absence of calcitonin (athyreosis) and in the presence of large excesses of the hormone (medullary carcinoma of the thyroid). With our present state of knowledge, information on calcitonin levels is chiefly important for the detection of medullary carcinoma of the thyroid, either in symptomatic individuals or in persons at risk for familial forms of this malignancy (Decker *et al.*, 1990).

Parathyroid hormone-related peptide

In recent years the role of another substance that may affect pre- and perinatal bone mineral physiology has come under investigation. PTHrP is a 141-amino-acid peptide that was first characterized in malignant cells derived from patients with the syndrome of humoral hypercalcaemia of malignancy. PTHrP shares with PTH a region of homology encompassing the 13 N-terminal amino acids, and it shares with PTH the power to bind to PTH receptors, activating bone resorption and renal tubular phosphate reabsorption. Once the gene for PTHrP had been cloned, a search for mammalian tissues expressing the ribonucleic acid (RNA) message for the peptide was conducted. Rather surprisingly, active peptide was discovered in the lactating breast, with considerable quantities secreted into the milk (Budayr *et al.*, 1989), and in placenta. It is not known what, if any, role is served by PTHrP to which the fetus might be exposed from the placenta (Barlet *et al.*, 1990) and which the young infant ingests orally from breast milk.

Growth hormone, IGF-1 and insulin

Growth hormone, a 191-amino-acid peptide made in the anterior pituitary, exerts trophic effects on various aspects of mineral metabolism. It is not certain which, if any, of these actions are direct effects of growth hormone and which result from changes in the availability of IGF-1 produced in response to growth hormone. In adults, excess growth hormone concentrations in acromegaly are associated with phosphate retention, hypercalciuria and increased calcitriol levels. The effects of growth hormone replacement on calcium and vitamin D metabolism in growth hormone-deficient children are probably not large (Gertner *et al.*, 1979). However, both in experimental animals and in children, growth hormone deficiency inhibits the normal increase in calcitriol during phosphate deprivation, an effect that can be reversed by the administration of growth hormone (Harbison & Gertner, 1990). Recently, IGF-1 synthesized from recombinant deoxyribonucleic acid (DNA) has become available for clinical trials. It appears that IGF-1 may well mediate the antiphosphaturic effect of growth hormone (Walker *et al.*, 1991). However, there is not much information about any direct effect of IGF-1 on vitamin D metabolism.

Disordered control of serum calcium

Serum calcium levels for neonates and older children are given in Table 12.1. Levels tend to be marginally higher (by about 0.2 mg) in growing children compared with adults (Krabbe *et al.*, 1982). The physiologically active component of serum calcium is the ionized calcium (Ca^{2+}) (normal concentration between 3.8 and 4.2 mg/dL), the remainder being largely protein (albumin) bound. Ca^{2+} can be measured directly with ion-sensitive electrodes, but values can also be calculated from the total calcium and serum albumin concentration using a formula such as that originally described by McLean and Hastings (1935) and later modified by Payne *et al.* (1974) and others. Such formulas have nnot been validated for the neonatal period, when different serum protein patterns may render inaccurate calculations of free calcium based on calcium–protein relationships derived from adult data.

Table 12.1 Normal range of total serum calcium in children. After Burritt *et al.* (1990); newborn data after Mayne and Kovar (1991)

Age category	Total calcium (mg/dL)
Newborn (1–3 weeks, fasting)	7.6–11.4
Boys 1–15 and girls 1–12 years	9.6–10.6
Boys 15–17 years	9.5–10.5
Boys 17–19 and girls 12–15 years	9.5–10.4
Girls 15–19 years	9.1–10.3

Hypocalcaemia

Clinically important categories of hypocalcaemia and certain features associated with the failure of calcium control are listed in Table 12.2. The symptoms and signs of hypocalcaemia are largely explained by the disturbance in neuromuscular excitability attributable to a reduction in extracellular fluid Ca^{2+} concentration. Muscular tetany and distressing paraesthesiae are more commonly seen in older children and adults, while hypocalcaemia often presents with epileptic seizures in babies and young children. In critically ill patients, hypocalcaemia may be associated with a reduction in cardiac output and mean arterial pressure. These physiological changes are reversed upon correction of the hypocalcaemia.

Hypocalcaemia can occur at any age but is much more common in the neonatal period than at any other time of life. It is convenient to cover the early and late forms of neonatal hypocalcaemia separately and then to discuss hypocalcaemia in older children.

Early neonatal hypocalcaemia

The clinical features are as follows: in the neonate, 'jittery' movements, convulsions and, occasionally, apnoea (Gershanick *et al.*, 1972) or myocardial dysfunction (Troughton & Singh, 1972) can all represent the clinical consequences of hypocalcaemia. If clinically indicated, a direct measurement of ionized calcium concentration is theoretically more useful than a total serum calcium level because no correction need be made for protein concentration or blood pH (Forman & Lorenzo, 1991). In practice, however, measurements of total serum calcium are used much more commonly because of the complexity and costs associated with the electrodes needed to measure Ca^{2+} directly. A Ca^{2+} concentration of less than 2.5 mg/dL can lead to clinical symptoms. Direct measurements of Ca^{2+} are not universally available, but a simple analysis of the electrocardiogram may provide a useful clue to the existence of a subnormal Ca^{2+} level.

This disorder commonly affects low birth-weight and sick infants between 1 and 4 days of age. It may be considered an exaggeration of the physiological drop in Ca^{2+} seen at this age (Fig. 12.4) (Mayne & Kovar, 1991).

During pregnancy the placenta acts as a 'pump' to transfer calcium actively from mother to fetus. The calcium concentration in the fetal circulation is higher than that normally seen in older children and adults, with consequent suppression of fetal parathyroid activity. At birth, the transplacental passage of calcium ceases and a transient inability of the parathyroid glands to mobilize bone calcium may permit

Table 12.2 Biochemical and hormonal changes in childhood hypocalcaemia

	Phosphorus	25-OHD	1,25-$(OH)_2$D	PTH
Neonate				
Early hypocalcaemia	N	N	N or L	N or L
Late hypocalcaemia				
Hypoparathyroidism	H	N	N or L	L
Uraemia	H	N	L	H
Older child				
Critical illness	N	N	N	L
Hypoparathyroidism	H	N	L	H
Pseudohypoparathyroidism	H	N	L	H
Vitamin D deficiency	L, N or H	L	L, N or H	H
Vitamin D dependency	L	N	H	H
Acute phosphorus overload	H	N	?	H

L, low; N, normal; H, high; 25-OHD, 25-hydroxyvitamin D; 1,25-$(OH)_2$D, 1,25-dihydroxyvitamin D; PTH, parathyroid hormone.

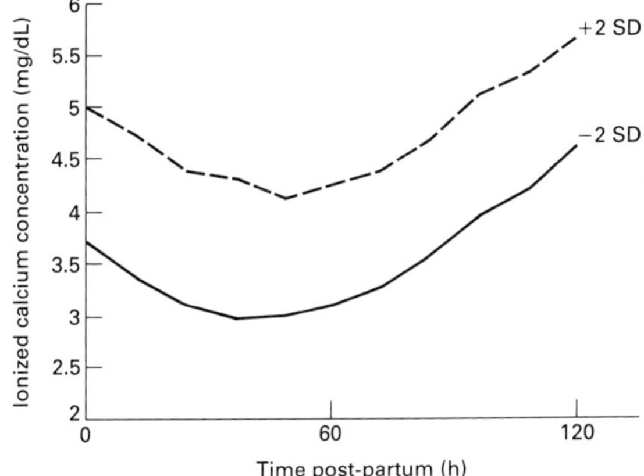

Fig. 12.4 Changes in serum calcium during the first 5 days of life. (After Mayne and Kovar, 1991.)

a dip in extracellular fluid Ca^{2+} levels. Calcitriol levels in newborn babies are not lower than in older individuals, but we do not know whether the 1α-hydroxylase system is capable of responding fully to the challenge of neonatal hypocalcaemia.

In the premature, small or sick neonate the defence of the Ca^{2+} concentration is further jeopardized by the lack of substantial food (and hence calcium) intake by mouth. It is also possible that stress hormones, such as calcitonin and cortisol, may act to 'stabilize' bone, diminishing PTH-induced calcium release. In addition to these 'natural' causes of hypocalcaemia, the transfusion of large volumes of citrate- or phosphate-containing whole blood may lower ionized calcium (Ca^{2+}) by the formation of non-ionizable calcium phosphates or citrate. The over-rapid correction of an acidosis with bicarbonate or by hyperventilation may also lead to a rapid fall in Ca^{2+} by increasing the fraction of circulating calcium which is bound to protein.

Hypomagnesaemia predisposes to neonatal hypocalcaemia and may exacerbate its symptoms. The reasons for these effects of low magnesium levels are not well understood.

Late neonatal hypocalcaemia

This term describes hypocalcaemia which presents clinically at 5–10 days of age in full-term, and apparently healthy, neonates. This type of hypocalcaemia is invariably associated with elevations in serum phosphorus levels. Hyperphosphataemia tends to shift the equilibrium in calcium flow between bone and the extracellular fluid towards bone. It also diminishes calcitriol synthesis. Both these changes operate to reduce the plasma calcium level.

Hypoparathyroidism

Hypoparathyroidism will be discussed below. It is very unusual for this condition to present in the neonatal period, but a fascinating exception is the temporary hypoparathyroidism associated with maternal hypercalcaemia. The mother's calcium should always be checked in cases where infantile hypocalcaemia is of late onset or prolonged duration, particularly if accompanied by hyperphosphataemia.

Phosphorus overload

Cows' milk contains six times as much phosphorus as human milk (950 vs 162 mg/L). Ingestion of a calorically adequate volume of cows' milk overwhelms the capacity of the neonatal kidney to excrete phosphate, with consequent hyperphosphataemia. This phenomenon was also observed with early infant-feeding formulas which still had a much higher phosphorus content than human milk. Even in modern 'humanized' cows' milk-based formulas the Ca : P ratio is lower than in breast milk. Recent work has shown that, compared with breast-fed babies, infants taking these formulas have slightly, but significantly, lower serum ionized calcium (Ca^{2+}) levels during the first 2 weeks of life (Specker *et al.*, 1991).

Uraemia

Acute or acute-on-chronic renal failure may present in the first week or two of life with seizures or tetany. These signs may be due to hypocalcaemia secondary to renal phosphate retention. The blood urea and creatinine should be measured in all cases of neonatal hypocalcaemia with hyperphosphataemia.

Hypocalcaemia beyond the newborn period

Older children and infants are able to maintain a normal serum calcium with much less difficulty than neonates. Nevertheless, hypocalcaemia may occur at times of critical illness and as a consequence of hypoparathyroidism (and pseudohypoparathyroidism, PsHP), disturbances of vitamin D nutrition or metabolism, and acute increases in serum phosphorus. Disorders of vitamin D metabolism will be dealt with in the section on Metabolic bone disease (p. 291).

Critical illness

Hypocalcaemia has been long known to occur in sick neonates. More recently a number of reports have described hypocalcaemia in critically ill non-neonatal children (Broner *et al.*, 1990). The occurrence of such hypocalcaemia is confined to extremely sick patients, often after cardiac surgery or major injury. There are conflicting views as to whether hypocalcaemia in these cases is due to a relative hypoparathyroidism or to some other factor such as hypercalcitoninaemia (Zaloga, 1992).

Hypoparathyroidism

The causes of parathyroid failure and resistance to PTH, together with some of the clinical features associated with these syndromes, are listed in Table 12.3. As one might expect, congenital disorders tend to present in young children and infants, while acquired disease is seen later, as an accompaniment to pathological or iatrogenic parathyroid destruction.

Hypoparathyroidism is generally diagnosed upon investigation of symptomatic hypocalcaemia. Occasionally, however, hypoparathyroidism or PsHP will be revealed by tests undertaken because of the presence of some of the 'associated features' listed in Table 12.3. Sensitive modern PTH assays, especially two-stage IRMA, for the intact PTH molecule have facilitated the diagnosis of hypoparathyroidism in suspicious cases.

Table 12.3 A classification of childhood hypoparathyroidism

Type	Parathyroid pathology	Age at presentation	Associated features
Transient neonatal	Physiological suppression	2–10 days	Maternal hypercalcaemia
Isolated congenital	Absence of glands or genetic defects in PTH biosynthesis	Prenatal	Autosomal recessive, dominant and X-linked forms
Di George syndrome	Absence	0–1 month	Athymia or reduced thymic function, congenital heart disease, del22q
Candida endocrinopathy	Autoimmune destruction	3+ years	Addison's disease, candidiasis, gonadal failure, alopecia, hypothyroidism
Haemosiderosis	Iron deposition	15+ years	Thalassaemia
Post-surgical	Ischaemic necrosis	10+ years	Post-thyroidectomy
Hypomagnesaemia	Reduced PTH secretion and resistance to PTH	Any	Diarrhoea, inherited gut or renal disorders
Pseudohypoparathyroidism	Peripheral resistance to PTH	1–5+ years	Short fourth metacarpals, mental retardation

PTH, parathyroid hormone.

Pseudohypoparathyroidism

In this rare condition tissues show a variable degree of resistance to PTH. PsHP, originally described by Albright in patients showing the phenotype of brachydactyly, round facies, short stature and mental retardation (Albright's hereditary osteodystrophy, AHO), is now believed to comprise a number of distinct entities. In some instances the physical appearance may be completely normal. Generally, PsHP with the somatic changes of AHO is referred to as PsHP type Ia, while PsHP without AHO is classified as PsHP type Ib.

The underlying biochemical lesion in PsHP type Ia is expressed as a failure of tissues normally containing a PTH-sensitive adenyl cyclase to generate cAMP upon exposure to PTH. The disorder of cAMP production is due to reduced function of the G_s component of the PTH receptor, which is, in turn, secondary to mutations affecting the gene coding for the α subunit of G_s. This reduction in activity is demonstrable in tissues, such as red blood cells, which are not normally responsive to PTH (Levine *et al.*, 1980). The mutation may be present in family members of patients with PsHP who themselves bear the physical stigmata of AHO but are biochemically normal (Patten *et al.*, 1990). Thus, the mutations described to date are necessary but not sufficient for the development of full-blown type I PsHP.

Biochemically, the critical distinction between hypoparathyroidism and PsHP is the level of circulating immunoreactive PTH. As expected, this is low in hypoparathyroidism and high in PsHP. Since modern assays for PTH readily permit this distinction, routine dynamic tests of the biochemical response to PTH are no longer needed.

Hypercalcaemia

Hypercalcaemia is less common than hypocalcaemia in childhood. Because of the potential clinical significance of slight elevations of serum calcium, it is important that this be measured accurately, taking alterations in serum albumin concentration into account. Some of the causes of elevated serum calcium levels in childhood and the associated biochemical findings are summarized in Table 12.4. In general, the reaction of the parathyroids and the vitamin D endocrine system to hypercalcaemia is towards suppression of PTH and calcitriol production. Such suppression is not seen in primary hyperparathyroidism, where an elevated PTH is itself the cause of hypercalcaemia and increased calcitriol production is a direct consequence of the trophic effect of PTH on the vitamin D 1α-hydroxylase.

Clinical features

The symptoms of hypercalcaemia vary little with age. They include general malaise and a failure to thrive, polyuria and thirst, vomiting, constipation and abdominal pain. Hypercalciuria, sometimes leading to renal stone formation, frequently accompanies chronic hypercalcaemia. Echogenic areas indicating areas of intrarenal calcification may be

Table 12.4 Biochemical and clinical features of childhood hypercalcaemia

	Serum					Urine	
Hypercalcaemia type	Ca	Phos	Calcidiol	Calcitriol	PTH	Ca	Notes
Idiopathic infantile (Williams)	H	N	N	?N	N	H	Elfin facies, aortic stenosis, mental retardation
Vitamin D toxicity	H	N	H	N	L	H	
Hyperparathyroidism	H	L	N	H	H	H	Often familial, MEN 1 and 2
Familial hypocalciuric	H	N or L	N	N	N or H	L	Dominant inheritance
Malignancy	H	N or L	N	L	L	H	Usually bone metastases, often painful
Immobilization	H	N	N	L	L	H	Acute quadriplegia

L, low; N, normal; H, high; PTH, parathyroid hormone; MEN, multiple endocrine neoplasia.

noted on renal sonography, and in more advanced cases renal function may be impaired. In addition to the general symptoms of hypercalcaemia, affected children may manifest the signs of the condition underlying the elevation in calcium.

Specific hypercalcaemic syndromes

Idiopathic infantile hypercalcaemia

In this non-familial syndrome, often given the eponym Williams' syndrome, congenital facial and cardiovascular anomalies and mental retardation are often accompanied by transient infantile hypercalcaemia. The facies are characteristic, and the cardiovascular anomalies, supravalvular aortic stenosis and peripheral pulmonary stenosis also form an easily recognizable combination. The hypercalcaemia may be severe and is exacerbated by dietary vitamin D supplementation.

The causes of the syndrome, and of the characteristic hypercalcaemia, remain mysterious. The recent finding that children with Williams' syndrome are heterozygous for a deletion involving the elastin gene on chromosome 7 (Ewart *et al.*, 1993) probably explains the observed vascular abnormalities. It is possible that Williams' syndrome is a 'contiguous gene syndrome' in which the hypercalcaemia and mental retardation are due to deletions in unidentified genes adjacent to that of elastin. Disorders of vitamin D metabolism (Garabedian *et al.*, 1985) have been proposed, as has deficient calcitonin release (Culler *et al.*, 1985). However, studies of the calcitonin gene in affected individuals have not shown any abnormality (Russo *et al.*, 1991). The mental retardation does not appear to be a consequence of high serum calcium levels in infancy. Elevated PTHrP levels have been described in a few patients with infantile hypercalcaemia who lacked the cardiovascular and mental manifestations of Williams' syndrome (Langman, 1992). This intriguing observation has yet to be confirmed.

Familial hypocalciuric hypercalcaemia

Familial hypocalciuric hypercalcaemia (FHH) is a dominantly inherited syndrome, linked, in some families, to the occurrence of severe neonatal hyperparathyroidism (Marx *et al.*, 1985). It is possible that the hyperparathyroid individuals are homozygotes for a gene which causes FHH in heterozygotes. FHH has also been called 'familial benign hypercalcaemia', a name that accurately reflects the usual clinical picture. Apart from instances of constipation and poor weight gain, affected children appear generally healthy. They are often diagnosed after routine biochemical screening or because asymptomatic hypercalcaemia has been found in a close relative.

By comparison with equivalently hypercalcaemic individuals with primary hyperparathyroidism, subjects with FHH have higher creatinine clearances and serum magnesium levels, and lower values for PTH and nephrogenous cAMP excretion. A characteristic of FHH is the lower-than-expected (for the degree of hypercalcaemia) urinary calcium excretion. In one series, no FHH patient excreted more than 0.2 mg/100 mL of glomerular filtrate over 24 h. Although the precise aetiology is not known, it appears that various cells, including renal tubule and, possibly, parathyroid, share a defect in divalent cation (Ca^{2+} and Mg^{2+}) transport.

Hyperparathyroidism

Hyperparathyroidism is very rare in childhood. Affected children are hypercalcaemic because of increased bone resorption and increased calcium absorption, the latter being a consequence of enhanced PTH-mediated calcitriol synthesis. Serum phosphorus is low because PTH decreases the renal tubular reabsorption of phosphate. PTH and calcitriol levels are elevated. In primary hyperparathyroidism the secretion of PTH is inappropriately high, while in secondary hyperparathyroidism overactivity of the parathyroids is an appropriate response to hypocalcaemia. Secondary hyperparathyroidism is seen in vitamin D deficiency rickets and in uraemia and is discussed in those sections. 'Tertiary hyperparathyroidism' is the phrase used to describe the development of autonomous hypersecretion in glands subject to prolonged stimulation. It is very rare in childhood but may occur in uraemia or during prolonged therapy for familial hypophosphataemic rickets.

Primary hyperparathyroidism may arise from

hyperplasia of all the parathyroid glands or from the development of tumours in one or more of the glands. In adults, the ratio of hyperplasia to tumour is 1:6, but in childhood the ratio is tilted towards hyperplasia with cases often associated with one of three familial syndromes: multiple endocrine neoplasia (MEN) types 1 and 2 and severe neonatal hyperparathyroidism.

Immobilization

Prolonged immobilization may lead to considerable loss of skeletal mineral. When immobilized, children and young adults are particularly susceptible to rapid bone loss, resulting in hypercalciuria and often hypercalcaemia. Hypercalcaemia may be observed after immobilization following burns or the application of extensive lower body casts. Most commonly, however, hypercalcaemia follows high spinal-cord injuries resulting in traumatic quadriplegia.

The hypercalcaemia may be severe, with impairment of renal glomerular function. The hypercalcaemia appears to be almost entirely attributable to increased bone resorption. Parathyroid function and the production of calcitriol are suppressed (Stewart *et al.*, 1982).

Vitamin D intoxication

Vitamin D intoxication is uncommon but can arise from inadvertent chronic overdosage of the vitamin given for therapeutic purposes or as a result of the ill-informed self (or parental) administration of vitamin preparations. Immobilization and lack of full parathyroid suppressibility (as seen in chronic renal failure) predispose to the development of vitamin D toxicity. All the available vitamin D preparations are toxic in excess, and in each case hypercalcaemia, often with hypercalciuria with nephrocalcinosis or stone formation, is the principal manifestation of toxicity.

The physiology of phosphate homeostasis

Phosphorus exists in the body largely as phosphate ions bound to organic molecules to form organophosphates. In the serum about two-thirds of the circulating phosphate forms part of such organic molecules bound to protein, but one-third is unbound and after separation from the bound form is measured as 'inorganic phosphate' (P_i). Concentrations of phosphate in the serum are given in molar units or in mass units, which refer to the quantity of elemental phosphorus within molecules of inorganic phosphate.

The systems which control phosphate homeostasis tend to regulate the total body phosphate content rather than, as in the case of calcium, absolute ionic concentration in the extracellular fluid. This difference accords with the different extracellular roles of calcium and phosphate. Tight control of extracellular Ca^{2+} is essential for the correct function of calcium-dependent cellular activation processes. In contrast, the function of extraskeletal phosphate is mostly as intracellular organophosphates providing intermediaries in energy metabolism, signalling systems and the genetic code. In accord with these observations is the fact that no well-characterized hormonal system operates primarily to regulate the serum phosphate concentration. Instead, it is total body phosphate that is sensed and regulated.

The intestinal absorption of dietary phosphate is largely unregulated, with about two-thirds of the phosphate content of digested foodstuffs being absorbed. The regulation of phosphate balance lies almost entirely at the level of the proximal renal tubule, which can vary the rate at which filtered P_i is reabsorbed from the tubular fluid. This reabsorption, which normally accounts for 85–95% of filtered P_i, is facilitated by a saturable sodium-dependent transporter located on the brush border membrane. The causes of disorders of phosphate homeostasis are attributable to intrinsic disorders of this transporter system or to dysregulation of its function.

Dietary regulation

The renal tubular reabsorption of P_i is extraordinarily sensitive to the dietary intake of phosphorus, specifically to the relationship of the intake of phosphate to that of total calories. In classic studies defining the phosphate-deprived state, Lotz *et al.* (1968) found a sharp reduction in urinary phosphate excretion, increased calciuria and symptoms such as muscle weakness. The reduction in urinary phosphate excretion is rapid, occurring before there is time for the serum phosphate to fall significantly, thus leading to the view that there exists a diet-sensitive hormonal system regulating urinary phosphate excretion. The mechanism whereby such a system might influence renal tubular phosphate transport is not known.

Parathyroid hormone and PTHrP

Although primarily functioning as a regulator of Ca^{2+}, PTH also affects renal phosphate control by inducing phosphaturia. Patients with hyperparathyroidism are hypophosphataemic and, conversely, hypoparathyroidism is associated with hyperphosphataemia. PTHrP, circulating in excess in patients with humoral hypercalcaemia of malignancy, leads to hypophosphataemia, presumably because of the interaction of PTHrP with PTH receptors on proximal tubular cells.

The vitamin D endocrine system

The adequacy of phosphate nutrition and the level of P_i in the serum are major regulators of the renal 1α-hydroxylation of calcidiol to calcitriol. Calcitriol levels are generally high during phosphate deprivation and in the face of hypophosphataemia, and are depressed by phosphate excess. Although phosphate regulates the activation of vitamin D, the converse, a vitamin D effect on phosphate homeostasis, appears to play only a minor part in normal physiology.

Insulin, growth hormone and IGF-1

Generally, the injection of insulin is accompanied by a profound fall in serum phosphate as phosphate moves into the intracellular compartment. The direct effect of insulin on renal phosphate control was shown to be antiphosphaturic in clamp studies in which glucose was held constant during insulin infusion (DeFronzo *et al.*, 1975). In diabetic ketoacidosis renal phosphate losses tend to be high because of the osmotic diuresis and a degree of competition between the proximal tubular transport systems for phosphate and glucose. The shift of phosphate into the intracellular compartment upon treatment with insulin can lead to profound hypophosphataemia in these phosphate-depleted patients.

Growth hormone has long been known to have an antiphosphaturic effect (Henneman *et al.*, 1960). The effect is seen during growth hormone treatment of deficient children and in the disease acromegaly, which is due to a growth hormone-secreting pituitary tumour. It is not known whether the effect of growth hormone upon the renal tubule is direct or mediated by the growth hormone-dependent protein IGF-1. Recent evidence suggests that exogenously administered IGF-1 does have an antiphosphaturic action in humans (Walker *et al.*, 1991), but this does not exclude a parallel direct action of growth hormone.

Somatic growth is itself associated with increased serum phosphate levels, almost certainly due to an increase in the threshold for tubular phosphate reabsorption during phases of rapid growth. Normative values for serum phosphate are substantially higher in growing children than in adults. In a longitudinal study of growing children, serum phosphate was found to correlate with growth velocity as children passed through the pubertal growth spurt (Round *et al.*, 1979). It is not known whether the changes in phosphate threshold at puberty are mediated by IGF-1 or growth hormone. Serum phosphate is also high during the rapid growth period of early infancy but growth hormone, and especially IGF-1 levels, are quite low in this age group compared with the levels in older children.

Disordered control of serum phosphate

Since the control of serum phosphate resides primarily in the proximal renal tubules, it follows that disorders of phosphate control are the result of primary tubular dysfunction or are secondary to processes which impact upon tubular function. They can be divided broadly into phosphate-retaining and phosphate-wasting states.

Hyperphosphataemia

Physiological states leading to hyperphosphataemia, such as those accompanying rapid growth, have been discussed, as have the pathological states relating to hypoparathyroidism and PTH resistance. Perhaps the most important phosphate-retaining disorder is uraemia, which leads to a complex bone disease known as uraemic osteodystrophy.

Dietary phosphate excess

The ingestion of unmodified cows' milk is no longer seen as a practical problem in developed countries. Cows' milk contains six times as much phosphorus as human milk (950 vs 162 mg/L). Ingestion of a calorically adequate volume of cows' milk overwhelms the capacity of the neonatal kidney to excrete phosphate, consequently causing hyperphosphataemia which shifts the equilibrium in calcium flow between bone and the extracellular fluid towards bone. It also diminishes calcitriol synthesis. Both these changes operate to reduce the plasma calcium level. Phosphate accumulation is also a feature of chronic glomerular renal failure, with hypocalcaemic tetany as a presenting feature in severely affected babies.

Acute endogenous phosphate release

Rhabdomyolysis

The term rhabdomyolysis refers to breakdown of skeletal muscle cells and liberation of their contents into the circulation, as evidenced invariably by elevated creatine phosphokinase and sometimes also by muscle weakness. Possible causes include a wide range of chronic and acute disorders, among them metabolic myopathies, drugs, toxins such as snake and insect venom, infections or infestations including influenza and Coxsackie virus, autoimmune inflammation, and physical trauma or vascular compromise (Brumback *et al.*, 1992).

Tumour lysis syndrome

The tumour lysis syndrome is defined by hyperuricaemia, hyperkalaemia, hyperphosphataemia and hypocalcaemia and appears shortly after administration of therapy for neoplasm (Zusman *et al.*, 1973). It is most often associated with rapidly dividing haematopoietic malignancies such as acute leukaemias and high-grade lymphomas. It has been suggested that the pathophysiology of the tumour lysis syndrome may resemble that of rhabdomyolysis: abrupt massive release of intracellular contents into the circulation. However, hypercalcaemia, while frequent in the recovery phase of rhabdomyolysis, has not been reported in tumour lysis syndrome (Dunlay *et al.*, 1989).

Acute exogenous phosphate poisoning

Hyperphosphataemia may result from enemas given to neonates and small children, both normal children and, more particularly, those with Hirschsprung's disease, imperforate anus, rectal atresia or renal insufficiency. A few cases of enema-induced hyperphosphataemia in adults have been reported, several of them associated with underlying renal insufficiency (Biberstein & Parker, 1985). Phosphate-containing laxatives may also induce hyperphosphataemia in adults (Wiberg *et al.*, 1978).

Uraemic osteodystrophy

Considered more fully in the section on disorders of the kidney, uraemic osteodystrophy (UOD) is a cause of growth failure and limb deformity in children with renal glomerular failure. These children's bones are affected by high levels of PTH, secreted in response to ionized calcium levels depressed by the high ambient phosphate concentration. At the same time, calcitriol levels are suppressed because of the limited availability of renal tissue, the only site within which significant 1α-hydroxylation of calcidiol can take place, and the suppression of 1α-hydroxylation due to high ambient phosphate ion concentrations.

Tumoral calcinosis

The name of this recessively inherited disorder refers to the 'lumps' or tumours of ectopic amorphously calcified material found in the soft tissues of affected persons, especially around the joints (Fig. 12.5) and in subcutaneous tissues. Pathophysiologically, the disease appears to be due to an inappropriately high 'set-point' for renal tubular phosphate reabsorption. Serum phosphate is very high but calcium, calcitriol and PTH levels are normal and respond in the expected way to phosphate deprivation and phosphate

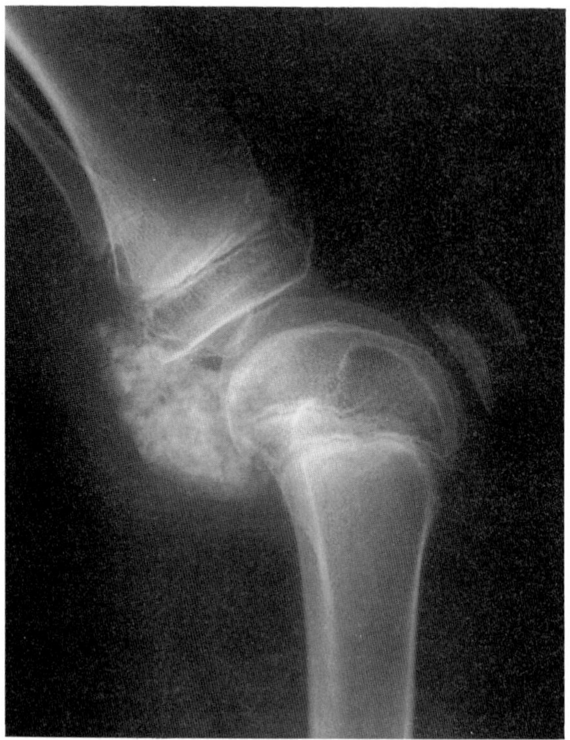

Fig. 12.5 Femoral calcinosis. The calcification in the popliteal area is due to the high ambient phosphate level.

loading (Lyles *et al.*, 1988). The ectopic calcification is presumably owing to the precipitation of calcium phosphates in a milieu which has a much higher calcium phosphate ion product than does healthy extracellular fluid. Tumoral calcinosis is very rare but is of considerable theoretical importance, not least because its pathophysiology is a mirror image of that of the more common phosphate-wasting condition, familial hypophosphataemic rickets (see below).

Phosphate-wasting states

The phosphate-wasting states are listed in Table 12.5. Because they lead to hypophosphataemia, their chief manifestations, when chronic, are rickets and osteomalacia, the general features of which are discussed under Metabolic bone disease.

Primary and secondary hypoparathyroidism are respectively discussed in the sections Hypercalcaemia (above) and Vitamin D deficiency (below).

Specific renal disorders

Isolated phosphaturia

Familial hypophosphataemic rickets (FHR). A condition dominated by proximal renal tubular phosphate wasting, FHR is by far the commonest cause of this type of phosphaturic rickets.

FHR is usually inherited as an X-linked dominant trait but there are quite credible reports of similar phenotypes inherited as autosomal dominant and recessive traits. A good animal model for FHR (the *hyp* mouse) exists but the aetiology is still unknown in both the mouse and human disorders. Although the mutation responsible for FHR has been mapped to the Xp 20 region of the X chromosome (Rowe *et al.*, 1992), the actual gene and its product remain to be identified. The manifestations are generally worse in male hemizygotes than in female heterozygotes. Rickets appears in the first year of life. The biochemical findings (see Table 12.7) are dominated

Table 12.5 A classification of renal phosphate wasting in childhood

Type of phosphaturia	Notes
Primary	
Isolated phosphate wasting	
Familial hypophosphataemic rickets	X-linked dominant
Hypercalciuric rickets	Autosomal recessive
Tumour rickets	Benign mesenchymal tumours
McCune–Albright syndrome (polyostotic fibrous dysplasia)	? Activation of PTH receptors
Combined tubular failure	
Fanconi syndrome	Renal
Cystinosis	Glycosuria
Other metabolic disorders	Amino aciduria
Toxic agents	Bicarbonaturia (acidosis)
Hormonal	
Primary hyperparathyroidism	Usually hyperplasia, often familial (autosomal dominant as in MEN 1 and 2a)
Secondary hyperparathyroidism	Vitamin D deficiency

PTH, parathyroid hormone; MEN, multiple endocrine neoplasia.

by hypophosphataemia with normocalcaemia. The renal tubular phosphate threshold (T_mP/GFR) (Walton & Bijvoet, 1975) is always subnormal in hypophosphataemic rickets. The normal calcitriol concentration in the face of hypophosphataemia (which generally stimulates calcitriol formation) suggests that the fundamental defect involves vitamin D metabolism as well as renal tubular phosphate transport. Recent evidence derived from work on the *hyp* mouse suggests that the renal tubular phosphate leak might be secondary to the action of humoral factors on the kidney rather than to an intrinsic defect of the kidney itself (Nesbit *et al.*, 1992).

Although the primary renal lesion consists of isolated proximal tubular phosphate wasting, renal glycosuria may also be seen in some older patients. Untreated, the disease results in severe growth retardation and deformity in males. The course is more variable in females; some are seriously affected, while in others short stature or the biochemical finding of a reduced renal tubular resorptive capacity for phosphorus may be the only manifestations.

Hypercalciuric rickets. This is a rare autosomal recessively inherited condition in which renal phosphate wasting leads to the 'expected' elevation of calcitriol concentrations. The condition stands in contrast to FHR, in which calcitriol levels are low despite phosphate wasting and hypophosphataemia. As a consequence of elevated calcitriol levels affected subjects may hyperabsorb calcium and excrete excessive quantities of calcium into the urine, with resulting nephrolithiasis (Tieder *et al.*, 1987). Most cases have been described in Middle Eastern and North African populations but the condition is also seen in persons with other ethnic backgrounds.

Tumour (or oncogenous) rickets (Weidner, 1991). Hypophosphataemic rickets is occasionally found in children and adults bearing a small mesenchymal tumour, usually benign. The tumour is often classified as a haemangiopericytoma. Resection of tumours from affected individuals has resulted in complete and rapid resolution of the hypophosphataemia, leading to the inescapable conclusion that the tumour, often very small, must be secreting an intensely phosphaturic substance. The hypophosphataemia is accompanied by inappropriately low levels of calcitriol, and the degree of osteomalacia or rickets may be severe. The mechanism of phosphaturia and the nature of any phosphaturic substance in these cases remain unknown, but the condition is clearly of great theoretical importance.

McCune–Albright syndrome. Also called polyostotic fibrous dysplasia, this is a pervasive multisystem disease of childhood onset that appears to be nonfamilial. Patients show a progressive fibrosis and deformity of bone affecting non-contiguous parts of the skeleton, while completely normal bony architecture is preserved elsewhere. The fibrous dysplasia may lead to pain, weakness and fractures with associated disturbances of gait. Children with fibrous dysplasia often manifest hyperactivity of one or more endocrine systems, particularly precocious puberty, thyrotoxicosis and acromegaly. They also show large, irregularly shaped pigmented skin naevi, which are characteristic of the disease. Many of these children have a phosphaturic state which can lead to rickets and osteomalacia with worsening of their bone pain and deformity.

Recent work, recognizing that most of the manifestations of the disease can be explained by hyper-responsiveness of cAMP-dependent hormone receptors, has demonstrated that McCune–Albright syndrome is caused by somatic mutations in the G-protein gene that controls the regulation of such receptors (Weinstein *et al.*, 1991). The receptors are thus constitutively activated, even though the target organs are not exposed to high levels of the hormonal or other stimulus that would normally be required for activation. On this model, the hypophosphataemia would presumably be due to activation of renal tubular PTH receptors in the absence of excessive PTH concentrations.

Generalized tubulopathies

This group of disorders is characterized by excessive urinary losses of all the factors primarily reabsorbed from the proximal tubular fluid. These include glucose, phosphate, bicarbonate and amino acids. Genetic causes are generally responsible for proximal tubular failure in children.

Cystinosis. Severe Fanconi syndrome is seen in cystinosis, an autosomal recessive lysosomal storage disease in which cystine accumulates intracellularly. The onset is in infancy and manifests as a failure to thrive and rickets. Affected children are hypophosphataemic and acidotic, and go on to develop progressive renal glomerular failure.

Other causes of Fanconi syndrome. Genetic conditions in which the impact of the associated Fanconi syndrome is relatively minor include the X-linked Lowe's syndrome (cause of tubular failure unknown), and Wilson's disease and type I glycogen storage disease, recessive conditions of copper and carbohydrate metabolism respectively.

Non-renal disorders

Dietary phosphate insufficiency including metabolic bone disease of prematurity

Dietary phosphate deficiency is very uncommon in adults and older children and is usually seen in association with the abuse of aluminium-containing antacid gels which bind phosphorus in the gastrointestinal tract. In the premature infant, however, a bone disease which encompasses a spectrum of disturbances, resulting in rickets, osteomalacia and osteoporosis, may occur as a result of an oral or parenteral intake insufficient in phosphorus. Milder cases may show only biochemical changes. The frequency of metabolic bone disease of prematurity (MBDP) varies depending on the diagnostic criteria. In one prospective radiological survey, fractures were detected in 20% of newborns with a birth weight of less than 1500 g and gestational age less than 34 weeks. Subclinical disease may be more common, as suggested by the prevalence of elevated serum alkaline phosphatase (ALP) and decreased bone mineral content.

The primary cause of MBDP is a deficiency of phosphate and calcium as a result of decreased intake of these. Initial reports of MBDP involved infants fed human milk and parenterally fed sick neonates, often with respiratory disease. Human milk is relatively low in calcium and phosphate. Infants on parenteral alimentation are at risk because of the low mineral concentration in the infusion fluid.

Of calcium and phosphate accretion for the fetal skeleton 80% occurs during the third trimester, by active transport across the placenta. After premature delivery an infant taking human milk (150–200 mL/kg per day) receives each day only 25–50% of the phosphorus and 35–70% of the calcium accumulated daily *in utero* during the third trimester. A standard formula provides only 55–90% of the phosphorus and 45–100% of the calcium. Adequate mineral delivery via total parenteral nutrition (TPN) is confounded by the difficulty of maintaining the solubility of concentrated mixtures of calcium and phosphorus.

Most preterm infants, including those with bone

disease, appear to have normal levels of calcidiol and calcitriol and are able to absorb efficiently calcium and phosphorus from the intestine. In some cases with MBDP, calcitriol levels are elevated, probably because of the stimulation by hypophosphataemia (Koo *et al.*, 1989).

Biochemical tests may detect early MBDP. An alkaline phosphatase level that is more than five times the maximum adult level may suggest MBDP even in an infant who is not on TPN and has no radiological bone disease.

Diabetic ketoacidosis

Diabetic ketoacidosis may lead to severe phosphate depletion with hypophosphataemia. During the diuretic phase of the disease phosphate is lost in the urine, reducing the body phosphate stores. When treatment with insulin is started, there is large-scale movement of phosphate into the cells. The combination of these two factors often leads to very low serum phosphorus levels during the treatment of ketoacidosis.

Metabolic bone disease

Measures of bone metabolic activity

Osteocalcin

Osteocalcin is a protein formed in bone by osteoblasts. The regulation of its formation is uncertain, but the gene is known to be highly responsive to calcitriol. Osteocalcin is an abundant protein in bone (10% of non-collagenous protein). Its alternative name, bone γ-glutamic acid (GLA) protein, refers to osteocalcin's three γ-carboxyglutamic acid residues, formed by post-translational carboxylation of glutamic acid, a vitamin K-dependent process. Since the identification of osteocalcin and the development of radio-immunoassays for the protein, a large amount of data has been published on circulating osteocalcin levels in healthy and diseased children. Osteocalcin levels are higher in growing children than in adults, while stimuli to bone growth or mineralization, such as the administration of growth hormone or vitamin D to children with the respective deficiency states, cause osteocalcin levels to rise. Nevertheless, the measurement of osteocalcin in the evaluation of skeletal activity in childhood is largely a research activity and not considered part of the clinical routine. This arises chiefly from the fact that osteocalcin measurements fail to provide much more, if any, information on bone turnover than does the far less elaborate measurement of ALP (Deftos, 1991). The inconsistency of osteocalcin measurements in various clinical situations may be partly due to changes induced when serum samples are frozen and thawed. Recently described new assays for osteocalcin may eliminate this problem and also reduce the cost and complexity associated with the current assays. Published values for serum osteocalcin in adults and growing children are shown in Fig. 12.6 (Hauschka *et al.*, 1989).

Alkaline phosphatase

The ALPs are a family of enzymes which hydrolyse organic phosphates at alkaline pH. Recent advances permit a clearer understanding of the relationship between the ALPs and provide some insight into the reasons why ALP levels are elevated or depressed in certain disease states (Harris, 1990). Four members of the ALP family are recognized: tissue nonspecific (TNSALP), placental, placental-like and intestinal. The TNSALP form is further subdivided into bone, liver and renal forms, which appear to be coded by a gene localized to 1p36.1–p34; the other forms have been mapped to 2q34–q37. The differences between forms arising from the same gene (e.g. bone and liver) are due to tissue-dependent post-translational modifications. Although assays for the ALPs are among the oldest established and most performed of clinical chemistry measurements, the lack of enzymatic specificity limits the value of such assays in evaluating skeletal disorders. Specificity for bone ALP can be enhanced by heat inactivation of the enzyme prior to assay, since bone ALP is much less stable than the liver enzyme at 65°C. The development of radioimmunoassays specific for bone ALP has further improved the specificity of assay for bone ALP (Duda *et al.*, 1988). Serum levels of ALP are expressed in units based on the rate of hydrolysis of an artificial substrate. Levels are higher

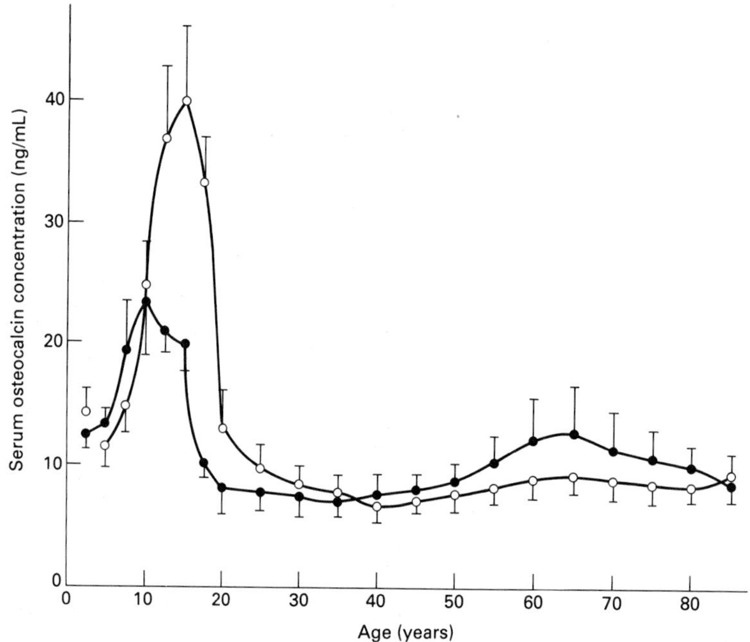

Fig. 12.6 Age dependency of serum osteocalcin levels, in females (filled circles) and males (open circles). (After Hauschka *et al.*, 1989.)

in growing children than in adults and have a generally accepted upper limit of three times the upper margin for adults.

ALP is an 'ectoenzyme' because of its localization to the outer cell surface and its ability to catalyse the hydrolysis of extracellular substrates. Bone ALP is necessary for adequate calcification, possibly operating by providing P_i from the hydrolysis of pyrophosphate.

Hypophosphatasia (Caswell *et al.*, 1991)

This term is used for a group of inherited diseases characterized by variable defects in bone mineralization, absent or subnormal serum ALP activity, and an excess of natural ALP substrates in the blood or urine. The disease is due to homozygosity or compound heterozygosity for deletions or mutations in the TNSALP gene. The severity is inversely related to the age of onset, which ranges from perinatal to adult. In addition to rickets-like bone deformity, affected children frequently show early loss of teeth. Serum ALP levels are low, but levels may fluctuate and may not correlate closely with clinical severity. Urinary excretion of phosphoethanolamine and inorganic pyrophosphate is increased, and in severe forms serum pyridoxal 5′-phosphate levels are also high. All these appear to be naturally occurring substrates for TNSALP.

Hyperphosphatasia

Abnormal elevations of ALP are observed quite regularly in large clinical chemistry departments dealing with specimens from paediatric patients. Most commonly, the high values are attributable to liver disease or a disorder of bone mineralization, such as rickets. Higher values may be seen in a rare familial condition called familial hyperphosphatasaemia, in which the biochemical abnormality is a marker for an extensive Paget's-like bone disease. Occasionally one sees isolated increases in ALP, often to levels that are 20 times or more higher than the upper adult limit, in the absence of any detectable bone disease. Heat stability tests suggest that bone ALP is responsible for these elevations, but this has not been confirmed by immunoassay. In some cases the onset of the elevation appears to coincide with a viral illness, while in other cases sensitivity to a drug such as co-trimoxazole may be responsible. In most cases no cause is ever found and the high level gradually fades after some weeks or months.

Urinary hydroxyproline

Hydroxyproline (OHP) is formed only in collagen, by the post-translational hydroxylation of proline residues in procollagen. Its urinary excretion represents contributions from dietary collagen, newly formed collagen fragments and the breakdown of collagenous tissues. The latter source generally predominates and, since most type I collagen is in the skeleton, urinary OHP excretion in a subject on a low-collagen diet can be used to estimate the rate of bone breakdown. As expected from the known increased bone turnover in childhood, urinary OHP : creatinine ratios, and even values for absolute OHP excretion, are higher in children than in adults (Schrier & Gottschalk, 1993).

Procollagen peptides

As mature collagen forms from procollagen, a terminal portion of each of the chains that will constitute part of the collagen triple helix is cleaved and liberated into the circulation. The C-terminal peptide of type I collagen, often referred to as procollagen peptide (PCP), is cleaved from newly formed type I collagen before the assembly of collagen molecules into mature collagen fibres occurs. The serum concentration of this soluble peptide can therefore be used as a guide to the overall formation rate of type I collagen and, since the bulk of type I collagen is in bone, as a marker for bone formation. Specific radioimmunoassays for the type I PCP have recently been devised. Using such an assay Trivedi *et al.* (1991) found the concentrations of type I PCP to be highest (220 ± 35 μg/dL) in young infants, falling to 33 ± 13 μg/dL in children aged 4–16 years and to still lower levels in adults. During growth hormone therapy significant increases in PCP were seen by these authors and by Allen and Goldberg (1992), who also found that children on glucocorticoid therapy had reduced serum PCP concentrations by comparison with untreated controls. Measurements of serum PCP may come to assume importance as a relatively simple non-invasive way of assessing physiological bone formation rates and the effects of various pathological states.

Pyridinium cross-links

The amino acids lysine and hydroxylysine participate in the cross-linking between peptide chains in mature collagen. Respectively, they form part of the compounds lysyl pyridinoline (LP) and hydroxylysyl pyridinoline (HP), both of which are classified as pyridinium compounds. Both these compounds are found in the body only as a result of the degradation of extracellular collagen. They are not absorbed from the intestine, and they are excreted unchanged in the urine without further catabolism in the body (Eyre, 1992). The pyridinium cross-linking compounds found in the urine come overwhelmingly from bone, and thus appear to constitute a sensitive marker for bone resorption. In particular, these assays will probably prove to be more specific measures of bone resorption than the quantification of urinary hydroxyproline.

The compounds are assayed fluorimetrically after HPLC (Black *et al.*, 1988), and considerable effort is currently being devoted to standardize the assays between laboratories. Early experience suggests that 24-h excretion levels in children are much higher than in normal adults but there are, as yet, no published standards for healthy children. A recent study by Branca *et al.* (1992) has shown that bone turnover rates in children, as gauged from the urinary excretion of pyridinium cross-links, are dependent on nutritional status. This study lends support to the view that pyridinium cross-link excretion will become a valuable tool for the assessment of skeletal status in childhood.

Disorders of matrix formation

Osteogenesis imperfecta

The term osteogenesis imperfecta (OI) describes a group of heritable diseases defined by excessive fragility of bones with increased liability to fracture

Table 12.6 Sillence's classification of osteogenesis imperfecta (OI)

OI type	Fragility	Sclerae	Dental involvement	Inheritance	Comments
IA	Present	Blue	Yes	Autosomal dominant	Relatively common
IB	Present	Blue	No	Autosomal dominant	Variable severity
II	Extreme	Blue	—	? Dominant (germ cell)	Perinatal
III	Severe	Normal	No	? Dominant (germ cell)	Skeletal deformity
IVA	Present	Normal	Yes	Autosomal dominant	Uncommon
IVB	Present	Normal	No	Autosomal dominant	Variable severity

(Gertner & Root, 1990). OI is a genetic disease caused by mutations in the genes coding for one of the two protein chains (COL1A1 and COL1A2) which combine to form type I collagen (Byers & Steiner, 1992). These affect either the structure of one or other of the protein chains or the relative rate at which one or other of the chains is synthesized. A range of secondary symptoms includes those due either directly or indirectly to osteopenia, dental defects and defects of non-skeletal connective tissue, such as ligament and sclera. The symptoms are found in various constellations and with variable severity in the four main clinically defined types of OI; individuals presumed to have inherited identical genetic defects may also differ in severity and expression of the disease.

At present no ideal taxonomy exists. Constant advances are being made in the association of clinical and genetic features with the underlying molecular pathology. The classification of Sillence *et al.* (1979) is the most useful interim solution to the clinical classification (Table 12.6). Definition of a molecular genetic basis for OI is difficult, as the synthesis of collagen is extraordinarily complex, and it is easy to appreciate that errors can occur at many points in the process. A summary of the currently recognized genetic lesions, as related to the classification of Sillence *et al.*, has been published recently by Byers (1988). The patterns of inheritance most commonly seen in the various types of OI are given in Table 12.6. In types II and III cases may occur in siblings with unaffected parents; this is the typical pattern for recessive inheritance. Recent work, however, has suggested that some (or all) of these cases are due to dominant mutations in a parental germline.

The natural history of OI depends on the type of disease and the ultimate outcome ranges from stillbirth to normal longevity. In severe cases, fractures *in utero* may result in short-limbed dwarfism at birth. Severely affected persons may also experience early spinal osteoporosis, increasing fixed deformity, confinement to a wheelchair, and risk of early death and morbidity as a result of pneumonia and respiratory failure, resulting from mechanical failure of the chest wall. Moderate to mild cases have an excellent chance of ambulation; the frequency of fractures will depend on the type, age and activity level of the patient.

Extraosseous involvement includes dentinogenesis imperfecta and obliteration of pulp space (Lukinmaa *et al.*, 1987); blueness of the sclerae (after infancy, when blueness is normal), resulting from qualitative or quantitative defects in scleral collagen; and ligamentous laxity. Both conductive and, for unknown reasons, sensorineural hearing are often affected. Mental retardation almost never occurs in OI.

The diagnosis of OI is still largely clinical, based on the findings described and on the family history. Laboratory approaches to the diagnosis include observation of the pattern of peptide fragments after protease digestion of collagen from patients' cells, often obtained from skin fibroblasts in culture. In principle, mutations known to be responsible for cases of OI could be sought in the *COL1A1* and *COL1A2* genes. However, at present this is scarcely practicable since such a large number of mutations have already been described in various forms of OI that searching for specific mutations could be tedious and lead to falsely reassuring results in a case of OI due to a 'new' mutation.

Disorders of mineralization

Rickets

Rickets results from the under-mineralization of the cartilaginous epiphyseal growth plate. It follows that rickets is confined to childhood, when the skeleton is growing, and affects only the growth plate. The equivalent in adults is a generalized softening of the skeleton due to a reduction in mineralization, known as osteomalacia. Children with rickets often suffer from osteomalacia too, so that the shafts of the long bones as well as the growth plates are affected. Although some characteristics of rickets vary with the cause, the widening and flaring of the epiphysis are seen with all forms. Frontal bossing occurs when the onset is in infancy. The ribs (costochondral junctions), ankles, wrists and knees are other sites commonly affected in growing children. The severity of involvement of particular epiphyses depends on the relative growth rate, which varies with age.

Deformity due to uneven epiphyseal growth and softening of the long bones is common in severe cases. The nature of the deformity is age dependent. The arms may be affected when they are used for weight-bearing in babies. The typical lower limb deformity is *genu varum* when the age of onset is under 3 or 4 years, and *genu valgum* when rickets starts in school-age children. These differences are possibly due to the changing relative rates of growth at various epiphyses.

Nomenclature

As the underlying causes of the varieties of rickets are discovered, more rational names can be applied to individual rachitic syndromes. The names 'vitamin D resistant rickets' and 'renal rickets' should be abandoned because they may be applied to a number of conditions that, respectively, do not resolve with nutritionally adequate doses of vitamin D or cannot be attributed to a variety of defects in renal function. International consultations designed to develop a logical and uniformly acceptable nomenclature are currently being held. In this chapter, the names of the various kinds of rickets have been selected as being the most unambiguous and informative with regard to aetiology.

Rickets can be caused by a number of distinct abnormalities of vitamin D intake and metabolism, and of mineral homeostasis. Some of these causes have been mentioned already under the headings of the various metabolic disturbances that can lead to rickets. The sources and metabolism of vitamin D are outlined in Fig. 12.3. A summary of the salient biochemical findings is given in Table 12.7.

Vitamin D deficiency

Vitamin D and its metabolites are essential components in the process of bone mineralization. The sterols appear not only to play a role in the maintenance of adequate extracellular fluid and calcium and phosphorus concentrations but also to act directly on skeletal tissues in order to aid mineralization. Children with vitamin D deficiency rickets often have bone pain and myopathy, the combination of which may lead to a loss of the skill of walking in affected toddlers. Hypocalcaemia is common and may lead to tetany and even convulsions.

Since vitamin D is obtainable from the diet, or is synthesized in the skin on exposure to sunlight, most affected children suffer from a combination of inadequate nutrition and a lack of exposure to the sun. In recent years some cases in northern Europe and North America have been reported in children who have received prolonged breast feeding, often from vegetarian mothers in whom vitamin D nutrition was also marginal.

In developed countries, vitamin D deficiency rickets usually follows gastrointestinal disease such as biliary atresia or coeliac disease. The occurrence of vitamin D deficiency in such children, despite adequate exposure to solar irradiation, gives credence to the notion of an enterohepatic circulation of vitamin D, the interruption of which can deplete the body of skin-derived, as well as dietary, vitamin D.

Vitamin D deficiency rickets is also seen in children on long-term anticonvulsant therapy, possibly because the drugs increase the demand for vitamin D by accelerating its hepatic elimination. Phenobarbitone and diphenylhydantoin have both been incriminated in the pathogenesis of anticonvulsant

Table 12.7 Biochemical and hormonal changes in rickets

	Serum					
Type of rickets	Ca	Phos	HCO_3^-	Calcidiol	Calcitriol	PTH
Vitamin D deficient	L	Variable	N	L	Variable	H
Calcitriol deficient*	L	L	N	N	L	H
Calcitriol resistant†	L	L	N	N	H	H
Fanconi syndrome	N	L	L	N	N or L	N or H
X-linked hypophosphataemia	N	L	N	N	N or L	N or H
Hypercalciuric	N	L	N	N	H	L
Tumour (oncogenous)	N	L	N	N	L	N
McCune–Albright syndrome	N	N	N	N	?	N
Prematurity‡	N	L	N	N	H	L

L, low; N, normal; H, high; PTH, parathyroid hormone.
* 25-Hydroxyvitamin D 1α-hydroxylase deficiency or 'pseudodeficiency vitamin D-dependent rickets, type I'.
† Formerly called 'pseudodeficiency rickets type II'.
‡ Usually due to phosphate deficiency.

rickets. The doses and duration of therapy that may lead to the development of rickets have not been defined clearly. However, it is evident that only modest supplementation with vitamin D can prevent anticonvulsant rickets and that this condition affects epileptic children who are already receiving inadequate vitamin D from the diet and inadequate exposure to sunlight.

Calcitriol deficiency (vitamin D-dependent rickets, type I)

This term is used to describe a very rare form of familial rickets. First described by Prader as 'pseudodeficiency rickets', the biochemical findings of this autosomal recessively inherited disease are very similar to those in nutritional rickets. Calcidiol levels are normal, while calcitriol levels are very low. The disease is due to diminished or absent renal 25-hydroxyvitamin D 1α-hydroxylase activity. The symptoms are not improved by vitamin D but respond readily to physiological doses of calcitriol (1–3 μg/day).

Calcitriol resistance (vitamin D-dependent rickets, type II)

In this recessively inherited condition, severe alopecia often accompanies rickets as a major clinical finding. The biochemical features resemble those of calcitriol deficiency, except that calcitriol levels are high. The patient's condition resists the administration of calcitriol, although some patients respond to doses as high as 12.5–20 μg/day. Resistance to calcitriol has been demonstrated at the cellular level with poor calcitriol binding to cytosolic receptors or diminished nuclear uptake of the sterol–receptor complex. In a few kindreds the responsible mutation has been precisely delineated and found to involve the DNA-binding 'zinc finger' region of the calcitriol receptor (Hughes *et al.*, 1988).

References

Aksnes, L., Rodland, O. & Aarskog, D. (1988) Serum levels of vitamin D3 and 25-hydroxyvitamin D in elderly and young adults. *Bone Min* **3**, 351–357.

Allen, D.B. & Goldberg, B.D. (1992) Stimulation of collagen synthesis and linear growth by growth hormone in glucocorticoid-treated children. *Pediatrics* **89**, 416–421.

Barlet, J.P., Davicco, M.J. & Coxam, V. (1990) Synthetic parathyroid hormone-related peptide (1–34) fragment stimulates placental calcium transfer in ewes. *J Endocrinol* **127**, 33–37.

Biberstien, M. & Parker, B.A. (1985) Enema-induced hyperphosphatemia. *Am J Med* **79**, 645–646.

Black, D., Duncan, A. & Robins, S.P. (1988) Quantitative analysis of the pyridinium crosslinks of collagen in urine using ion-paired reverse phase high-performance liquid chromatography. *Anal Biochem* **169**, 197–203.

Branca, F., Robins, S.P., Ferro Luzzi, A. & Golden, M.H. (1992) Bone turnover in malnourished children. *Lancet* **340**, 1493–1496.

Broner, C.W., Stidham, G.L., Westenkirchner, D.F. *et al.* (1990) Hypermagnesemia and hypocalcemia as predictors of high mortality in critically ill pediatric patients. *Crit Care Med* **18**, 921–929.

Brumback, R.A., Feeback, D.L. & Leech, R.W. (1992) Rhabdomyolysis in childhood: a primer on normal muscle function and selected metabolic myopathies characterized by disordered energy production. *Pediatr Clin N Am* **39**, 821–858.

Budayr, A.A., Halloran, B.P., King, J.C. *et al.* (1989) High levels of a parathyroid hormone-like protein in milk. *Proc Natl Acad Sci USA* **86**, 7183–7185.

Burritt, M.F., Slockbower, J.M., Forsman, R.W. *et al.* (1990) Pediatric reference intervals for 19 biologic variables in healthy children. *Mayo Clin Proc* **65**, 329–336.

Byers, P.H. (1988) Osteogenesis imperfecta: an update. *Growth Genet Horm* **4**, 1–9.

Byers, P.H. & Steiner, R.D. (1992) Osteogenesis imperfecta. *Ann Rev Med* **43**, 269–282.

Caswell, A.M., Whyte, M.P. & Russell, R.G. (1991) Hypophosphatasia and the extracellular metabolism of inorganic pyrophosphate: clinical and laboratory aspects. *Crit Rev Clin Lab Sci* **28**, 175–232.

Chen, T.C., Turner, A.K. & Holick, M.F. (1990) A method for the determination of the circulating concentration of vitamin D. *J Nutr Biochem* **1**, 272–276.

Culler, F.L., Jones, K.L. & Deftos, L.J. (1985) Impaired calcitonin secretion in patients with Williams' syndrome. *J Pediatr* **107**, 720–723.

Decker, R.A., Toyama, W.M., O'Neill, L.W., Telander, R.L. & Wells, S.A. (1990) Evaluation of children with multiple endocrine neoplasm type IIB. *J Pediatr Surg* **25**(9), 939–943.

DeFronzo, R.A., Cooke, C.R., Andres, R., Faloona, G.R. & Davis, P.J. (1975) The effect of insulin on renal handling of sodium, potassium, calcium, and phosphate in man. *J Clin Invest* **55**, 845–855.

Deftos, L.J. (1991) Bone protein and peptide assays in the diagnosis and management of skeletal disease. *Clin Chem* **37**, 1143–1148.

Duda, R.J., O'Brien, J.F., Katzmann, J.A. *et al.* (1988) Concurrent assays of circulating bone Gla-protein and bone alkaline phosphatase: effects of sex, age, and metabolic bone disease. *J Clin Endocrinol Metab* **66**, 951–957.

Dunlay, R.W., Camp, M.A., Allon, M. *et al.* (1989) Calcitriol in prolonged hypocalcemia due to the tumor lysis syndrome. *Ann Intern Med* **110**, 162–164.

Ewart, A.K., Morris, C.A., Atkinson, D. *et al.* (1993) Hemizygosity at the elastin locus in a developmental disorder, Williams syndrome. *Nature Genet* **5**, 11–16.

Eyre, D. (1992) Editorial: new biomarkers of bone resorption. *J Clin Endocrinol Metab* **74**, 470A–470C.

Forman, D.T. & Lorenzo, L. (1991) Ionized calcium: its significance and clinical usefulness. *Ann Clin Lab Sci* **21**, 297–304.

Garabedian, M., Jacqz, E., Guillozo, H. *et al.* (1985) Elevated plasma 1,25-dihydroxyvitamin D concentrations in infants with hypercalcemia and an elfin facies. *N Engl J Med* **312**, 948–952.

Gershanick, J.J., Levkoff, A.H. & Duncan, R. (1972) The association of hypocalcemia and recurrent apnea in premature infants. *Am J Dis Child* **113**, 646–652.

Gertner, J.M. & Root, L. (1990) Osteogenesis imperfecta. In: Pathologic Fractures in Metabolic Bone Disease. *Orthop Clin N Am* **21**(1), 151–162.

Gertner, J.M., Horst, R., Genel, M. & Rasmussen, H. (1979) The effect of human growth hormone replacement on parathyroid function and vitamin D metabolism. *J Clin Endocrinol Metab* **49**, 185–188.

Haddad, J.G. & Chyu, K.J. (1971) Competitive protein binding radioassay for 25-hydroxycholecalciferol. *J Clin Endocrinol Metab* **33**, 992–996.

Harbison, M.D. & Gertner, J.M. (1990) Permissive action of growth hormone on the renal response to dietary phosphorus deprivation. *J Clin Endocrinol Metab* **70**, 1035–1040.

Harris, H. (1990) The human alkaline phosphatases: what we know and what we don't know. *Clin Chim Acta* **186**, 133–150.

Hauschka, P.V., Lian, J.B., Cole, D.E. & Gundberg, C.M. (1989) Osteocalcin and matrix Gla protein: vitamin K-dependent proteins in bone. *Physiol Rev* **69**, 990–1047.

Henneman, P.H., Forbes, A.P., Moldawer, M., Dempsey, E.F. & Carroll, E.L. (1960) Effects of human growth hormone in man. *J Clin Invest* **39**, 1223–1238.

Holick, M.F. (1988) Skin: site of the synthesis of vitamin D and a target tissue for the active form, 1,25-dihydroxyvitamin D. *Ann NY Acad Sci* **548**, 14–26.

Hughes, M.R., Malloy, P.J., Kieback, D.G. *et al.* (1988) Point mutations in the human vitamin D receptor gene associated with hypocalcemic rickets. *Science* **242**, 1702–1705.

Koo, W.W., Sherman, R., Succop, P. *et al.* (1989) Serum vitamin D metabolites in very low birth weight infants with and without rickets and fractures. *J Pediatr* **114**, 1017–1022.

Krabbe, S., Transbol, I. & Christiansen, C. (1982) Bone mineral homeostasis, bone growth, and mineralisation during years of pubertal growth: a unifying concept. *Arch Dis Child* **57**, 359–363.

Langman, C.B., Budayr, A.A., Saler, D.E. & Strewler, G.J. (1992) Nonmalignant elevation of parathyroid hormone-related peptide as a cause of idiopathic infantile hypercalcemia. *J Bone Min Res* **7**, S93 (Abstract 2).

Levine, M.A., Downs, R.W., Singer, M. *et al.* (1980) Deficient activity of guanine nucleotide regulatory protein in erythrocytes from patients with pseudohypoparathyroidism. *Biochem Biophys Res Comm* **94**, 1319–1324.

Lotz, M., Zisman, E. & Bartter, F.C. (1968) Evidence for a phosphorus-depletion syndrome in man. *N Engl J Med* **278**, 409–415.

Lukinmaa, J.-P., Ranta, H., Ranta, K. *et al.* (1987) Dental findings in osteogenesis imperfecta. 1. Occurrence and expression of type I dentinogenesis imperfecta. *J Craniofac Genet Devel Biol* **7**, 115–125.

Lyles, K.W., Halsey, D.L., Friedman, N.E. & Lobaugh, B. (1988) Correlations of serum concentrations of 1,25-dihydroxyvitamin D, phosphorus, and parathyroid hormone in tumoral calcinosis. *J Clin Endocrinol Metab* **67**, 88–92.

McLean, F.C. & Hastings, A.B. (1935) Clinical estimation and significance of calcium ion concentration in the blood. *Am J Med Sci* **189**, 601–613.

Marx, S.J., Fraser, D. & Rapoport, A. (1985) Familial hypocalciuric hypercalcemia. Mild expression of the gene. *Am J Med* **78**, 15–22.

Mayne, P.D. & Kovar, I.Z. (1991) Calcium and phosphorus metabolism in the premature infant. *Ann Clin Biochem* **28**, 131–142.

Nesbit, T., Coffman, T.M., Griffiths, R. & Drezner, M.K. (1992) Crosstransplantation of kidneys in normal and *hyp* mice. Evidence that the *hyp* mouse phenotype is unrelated to an intrinsic renal defect. *J Clin Invest* **89**, 1453–1459.

Norman, A.W. (1990) Intestinal calcium absorption: a vitamin D hormone-mediated adaptive response. *Am J Clin Nutr* **51**, 290–300.

Nussbaum, S.R., Zahradnik, R.J., Lavigne, J.R. *et al.* (1987) Highly sensitive two-site immunoradiometric assay of parathyrin, and its clinical utility in evaluating patients with hypercalcemia. *Clin Chem* **33**, 1364–1367.

Patten, J.L., Johns, D.R., Valle, D. *et al.* (1990) Mutation in the gene encoding the stimulatory G protein of adenylate cyclase in Albright's hereditary osteodystrophy. *N Engl J Med* **322**, 1412–1419.

Payne, R.B., Little, A.J., Williams, R.B. & Milner, J.R. (1974) Letter: Correction of plasma calcium measurements. *Br Med J* **1**(904), 393.

Poskitt, E.M.E., Cole, T.J. & Lawson, D.E.M. (1979) Diet, sunlight, and 25-hydroxyvitamin D in healthy children and adults. *Br Med J* **1**, 221–223.

Reichel, H., Koeffler, H.P. & Norman, A.P. (1989) The role of the vitamin D endocrine system in health and disease. *N Engl J Med* **320**, 980–991.

Reinhardt, T.A., Horst, R.L., Orf, J.W. & Hollis, B.W. (1984) A microassay for 1,25-dihydroxyvitamin D not requiring high performance liquid chromatography: application to clinical studies. *J Clin Endocrinol Metab* **58**, 91–98.

Rigby, W.F.C. (1988) The immunobiology of vitamin D. *Immunol Today* **9**, 54–58.

Round, J.M., Butcher, S. & Steele, R. (1979) Changes in plasma inorganic phosphorus and alkaline phosphatase activity during the adolescent growth spurt. *Ann Hum Biol* **6**, 129.

Rowe, P.S., Read, A.P., Mountford, R. *et al.* (1992) Three DNA markers for hypophosphataemic rickets. *Hum Genet* **89**, 539–542.

Russo, A.F., Chamany, K., Klemish, S.W., Hall, T.M. & Murray, J.C. (1991) Characterization of the calcitonin/CGRP gene in Williams' syndrome. *Am J Med Genet* **39**, 28–33.

Schrier, R.W. & Gottschalk, C.W. (1993) *Diseases of the Kidney*, 5th edn. Little, Brown, Boston.

Sillence, D.O., Senn, A.S. & Danks, D.M. (1979) Genetic heterogeneity in osteogenesis imperfecta. *J Med Genet* **16**, 101–116.

Specker, B.L., Tsang, T.C., Ho, M.L., Landi, T.M. & Gratton, T.L. (1991) Low serum calcium and high parathyroid hormone levels in neonates fed 'humanized' cow's milk-based formula. *Am J Dis Child* **145**, 941–945.

Stewart, A.F., Adler, M., Byers, C.M., Segre, G.V. & Broadus, A.E. (1982) Calcium homeostasis in immobilization: an example of resorptive hypercalciuria. *N Engl J Med* **306**, 1136–1140.

Taylor, A. & Norman, M.E. (1988) Vitamin D metabolite profiles in moderate renal insufficiency of childhood. *Pediatr Nephrol* **2**, 453–459.

Tieder, M., Modai, D., Shaked, U. *et al.* (1987) 'Idiopathic' hypercalciuria and hereditary hypophosphatemic rickets. Two phenotypical expressions of a common genetic defect. *N Engl J Med* **316**, 125–129.

Trivedi, P., Risteli, J., Risteli, L. *et al.* (1991) Serum concentrations of the type I and III procollagen propeptides as biochemical markers of growth velocity in healthy infants and children and in children with growth disorders. *Pediatr Res* **30**, 276–280.

Troughton, O. & Singh, S.P. (1972) Heart failure and neonatalhypocalcemia. *Br Med J* **4**, 76–79.

Walker, J.L., Ginalska-Malinowska, M., Romer, T.E., Pucilowska, J.B. & Underwood, L.E. (1991) Effects of the infusion of insulin-like growth factor I in a child with growth hormone insensitivity syndrome (Laron dwarfism). *N Engl J Med* **324**, 1483–1488.

Walton, R.J. & Bijvoet, O.L.M. (1975) Nomogram for derivation of renal threshold phosphate concentration. *Lancet* **2**, 309–310.

Weidner, N. (1991) Review and update: oncogenic osteomalacia-rickets. *Ultrastruct Pathol* **15**, 317–333.

Weinstein, L.S., Shenker, A., Gejman, P.V. *et al.* (1991) Activating mutations of the stimulatory G protein in the McCune – Albright syndrome. *N Engl J Med* **325**, 1688–1695.

Wiberg, J.J., Turner, G.G. & Nuttall, F.Q. (1978) Effect of phosphate or magnesium cathartics on serum calcium. *Arch Int Med* **138**, 1114–1116.

Zaloga, G.P. (1992) Hypocalcemia in critically ill patients. *Crit Care Med* **20**, 251–262.

Zusman, J., Brown, D.M. & Nesbit, M.E. (1973) Hyperphosphatemia, hyperphosphaturia, and hypocalcemia in acute lymphoblastic leukemia. *N Engl J Med* **289**, 1335–1340.

13: Liver Disorders

A.P. MOWAT

Introduction

During much of childhood the liver has considerable reserve capacity for all its functions. There is a temporary hepatic inefficiency in the first 2 years of life and this is most marked in the neonatal period, particularly in infants born prematurely (Rosenthal *et al.*, 1986; Mowat, 1987; Lentze & Reichen, 1992). Brief consideration of some of the functions of the liver and of the cells and structures involved is essential to understand the pathophysiological changes associated with disease (Arias *et al.*, 1988; McIntyre *et al.*, 1991). In this chapter the value and limitations of available tests for hepatobiliary disorders are critically appraised and their role in the investigation of common syndromes of hepatobiliary disease is summarized.

The liver is a highly organized tissue of interacting specialized cells in intimate contact with a complex system of channels carrying blood and bile. The structural organization of the liver and its vascular elements is designed to allow it to occupy a central place in metabolism, while interposed as a guardian between the digestive tract (and spleen) and the rest of the body. An essential function of the liver is to maintain the concentration of a vast number of macromolecules and solutes in hepatic vein blood and in bile within a very narrow range, irrespective of dietary intake or the metabolic demands of other organs. The supply of carbohydrates, proteins, lipids and macromolecules to other tissues is carefully regulated in spite of wide variations in dietary intake and metabolic demands. The liver is the site of chemical transformation of many endogenous and exogenous substances, thereby changing their metabolic, therapeutic or toxic effects, and often facilitating biliary excretion. By the formation of bile it contributes to digestion and absorption. It stores carbohydrates, in the form of glycogen, and vitamins A, D, E, K, B_1 and B_2.

Although many of the metabolic processes which are necessary to perform these functions occur in hepatocytes, other specialized liver cells also play an important role. Hepatocytes, Kupffer cells and sinusoidal endothelial cells are in intimate contact with blood in a specialized network of capillaries termed *'sinusoids'*. These have a unique structure which facilitates transfer of metabolites between hepatocytes and the blood perfusing the liver. The sinusoids are lined by three types of highly specialized cells which play a critical role in maintaining homeostasis. The *sinusoidal endothelial cells* are large flat cells with fine processes, and form the barrier which gives the sinusoids their vascular structure. They have in their cytoplasm fenestrae of approximately 100 nm; these occur in small patches called sieve plates and provide channels between the plasma and the space of Disse through which fluid, metabolites and chylomicron remnants devoid of triglyceride can readily move. The diameter of these channels is changed by endogenous and exogenous factors. The endothelial cells, by the process of receptor-mediated phagocytosis, are important in the turnover and catabolism of many glycoproteins, lipoproteins and mucopolysaccharides in the plasma. They synthesize important mediators such as prostaglandin E_2 and prostacyclin. Cells of the second type, usually given the eponym *Kupffer cells* and comprising 10% of liver cells, are anchored to the sinusoidal aspect of endothelial cells by fine cytoplasmic processes. Their functions include phagocytosis of particulate matter, catabolism of endotoxin, lipids, and glycoproteins including many enzymes, antigen processing and secretion of mediators and cytotoxic agents. The Kupffer cells are also thought to have the ability to bulge into the lumen of the sinusoid and perhaps control sinusoidal blood flow. A third type of perisinusoidal cell, on the non-luminal side of the endothelial cells, is the *Ito cell* (also called a 'fat-storing cell'), a stellate cell reflecting two of its functions: vitamin A storage and metabolism and its ability to send out prolonged cytoplasmic projections which may control the diameter of sinusoids. During fibrogenesis these cells become myofibroblasts and secrete massive amount of extracellular matrix components.

Between these endothelial cells and the hepatocytes in health there is no basement membrane but a large extravascular space, the space of Disse, extending to the tight junction between adjacent hepatocytes. The space of Disse contains very scanty amounts of such elements of the extracellular matrix as collagens, laminin, fibronectin and proteoglycans which modify cell function via specific receptors on cell surfaces (Reif *et al.*, 1992).

The functional unit of the liver is an *acinus*, a three dimensional sphere of parenchyma dependent upon blood supply through a single the portal tract. Blood exits the acinus by two or more hepatic veins. These hepatic venules, called 'central veins' are at the periphery of the acinus. In this concept, parenchymal cells, arranged in concentric zones surrounding the terminal afferent vessel, are heterogeneous in the function. Zone 1 nearest the portal tract receives blood with a high content of oxygen, insulin and glucagon, has a high metabolic activity and is the last to undergo necrosis and the first to show signs of regeneration. It contains more Kupffer cells than the other zones. Zone 3, nearest the terminal hepatic veins, receives blood last and is described as being in the microcirculatory periphery of the acinus. This perivenular area is formed by the most peripheral portions of zone 3 of several adjacent acini. There are striking differences in morphology, biochemistry and function between hepatocytes in zones 1 and 3, with a continuous and gradual transition in characteristics in the 20 to 25 hepatocytes lying between the beginning of zone 1 and the end of zone 1. Sequential modification of blood and bile occurs as these traverse the acinus (Gumucio, 1989; Sokal *et al.*, 1989a,b). The *hepatocyte heterogeneity* and the considerable reserve capacity of hepatocytes to change biochemically in response to changes in the concentration of metabolites perfusing the sinusoids play a major role in ensuring that the concentrations of solutes in the hepatic venules and bile remain within closely defined limits. Bile is further modified by secretion and reabsorption in bile ducts, both inside and outside the liver (Suchy *et al.*, 1987; Reubner *et al.*, 1990). Only in the presence of severe liver dysfunction, hepatocellular necrosis and/or porto-systemic shunts does the liver fail to maintain homeostasis.

Specific hepatocellular functions: pathophysiological aspects

Carbohydrate metabolism

The monosaccharides glucose, galactose and fructose, absorbed from the intestinal tract, are avidly taken up by the liver, where they may be utilized for immediately required energy or glycogen formation. Hepatic glycogen is an invaluable source of carbohydrate which prevents hypoglycaemia by releasing glucose to other tissues during fasting. Gluconeogenesis in the liver increases with prolonged fasting and utilizes amino acids released from peripheral tissues.

Hypoglycaemia occurs in inborn errors of carbohydrate, amino acid and lipid metabolism, fulminant hepatic failure, Reye's syndrome and with poisons such as ethanol or hypoglycin (see also Chapter 5a).

Lipids: bile salts

The liver plays a key role in the synthesis, catabolism and biliary excretion of lipids, lipoproteins, phospholipids, sphingolipids, cholesterol-derived steroids and enzymes involved in their metabolism. Neutral fats absorbed from the small intestine are oxidized within the liver into glycerol and free fatty acids. Fatty acids may be further oxidized to acetyl-

CoA, which enters Krebs tricarboxylic acid cycle. Glycerol may be utilized to form other triglycerides or degraded via acetyl-CoA. Hepatic fat is usually triglyceride, synthesized from fatty acids and glycerol-3-phosphate. There is continuous recycling of fatty acids between the liver and adipose tissue. Hepatic uptake increases with serum fatty acid concentrations. The liver has a limited ability to oxidize fatty acids and secrete triglyceride as very low-density lipoproteins. Thus, fat accumulation in the liver may occur because of increased lipolysis in adipose tissue or because of hepatic malfunction limiting fatty acid oxidation or triglyceride secretion.

Cholesterol is synthesized in the liver, intestinal mucosa, adrenal cortex and arterial walls. It is excreted in the bile as a neutral steroid. As well as providing the basic structure for steroid hormones, cholesterol is the precursor of bile salts (Lentze & Reichen, 1992). These are synthesized only within the liver. The liver also has a key role in the uptake, reabsorption and conjugation of free bile acids arriving in the portal blood, having been reabsorbed in the ileum. These are rapidly conjugated with glycine or taurine prior to biliary secretion.

Mild hypertriglyceridaemia, decreased cholesterol esters and a reduced α- and increased β-lipoprotein band occur in all forms of hepatocellular injury and in cholestasis. With cholestasis there is also a striking rise in free cholesterol and unusual lipoproteins, particularly lipoprotein-X, with a deficiency of cholesterol acyltransferase (LCAT).

Protein and amino acid metabolism

Along with muscle the liver plays a key role in amino acid metabolism. Amino acids absorbed from the intestine are rapidly taken up by the liver. There they are utilized in protein synthesis, gluconeogenesis or ketogenesis, following deamination or transamination. Ammonia produced by the deamination of amino acids is rapidly converted to urea.

Most of the proteins in plasma, other than the immunoglobulins, are synthesized by the rough endoplasmic reticulum of the hepatocytes. Quantitatively, albumin is the most important of these, but haptoglobin, transferrin, caeruloplasmin, C-reactive protein, α_1-antitrypsin, α_2-globulin, ferritin and α- and β-lipoprotein are also formed in the liver.

The other important proteins are those involved in clotting mechanisms. Fibrinogen, prothrombin and factors V, VII, IX, X and, to some extent, VIII are formed within the liver. Factors II, VII, IX and X are dependent on the presence of vitamin K for their synthesis. Inhibitors of the coagulation system, notably antithrombin-III and components of the fibrinolytic system, for example plasminogen, are also synthesized within hepatocytes. The liver avidly binds desialylated glycoproteins, thereby clearing them from the circulation.

If liver function is impaired severely, serum concentrations of amino acids increase, except for branched-chain amino acids which are metabolized preferentially by muscle (Vilstrup *et al.*, 1990). Protein synthesis decreases, as evidenced clinically by low concentrations of clotting factors and serum albumin, although fluid retention may contribute to the latter.

During hepatic regeneration α-fetoprotein production increases. α-Fetoprotein production is prolonged and increased in neonatal obstructive liver diseases; it is also produced by 70–90% of hepatocellular tumours.

Drug, toxin and xenobiotic metabolism

The metabolism of drugs, toxins and xenobiotics within the liver occurs in two stages in most instances. The first step (phase I) is a biochemical transformation resulting in oxidation, reduction, hydrolysis, hydration or isomerization. This is followed (phase II) by conjugation (glucuronidation, glycosylation, sulphation, methylation, acetylation or amino acid, glutathione or fatty acid conjugation or condensation), which renders the compound more water soluble, making it more readily excreted in the urine or bile.

In hepatic insufficiency the effects on drug metabolism are complex and depend on the degree of biotransformation in other tissues, changes in binding protein concentrations in addition to changes in liver function. In general, the pharmacological half-life is prolonged.

Vitamins

Hepatocytes play a critical role in the uptake, storage, metabolism and transport of both water-soluble and fat-soluble vitamins and are frequently the main sites of synthesis of specific transport lipoproteins. Ito cells are involved in vitamin A metabolism. Thiamine, riboflavin, niacin, folic acid, biotin and pantothenic acid are metabolized in the liver and are essential for other aspects of hepatic metabolism. The liver produces essential active metabolites of vitamins D and E. Vitamin K is essential for the synthesis of clotting factors II, VII, IX and X, and proteins C and S. Normal bile production is essential for absorption of fat-soluble vitamins. In hepatic insufficiency the main effects are deficiency of fat-soluble vitamins, but other tissues suffer from deficiency of metabolically active vitamin products.

Laboratory assessment of hepatobiliary disease

As in all branches of medicine, the thoughtful appraisal of a full history and careful clinical examination provides the basis for the prudent use of tests on urine and blood to confirm the presence of significant liver disease and to identify the cause. Standard, routinely available, automated laboratory investigations for liver disease will usually verify or exclude the presence of clinically significant hepatobiliary disease. These investigations should include at least serum (or plasma) total bilirubin, an aminotransferase, alkaline phosphatase (ALP) and albumin. Some laboratories test for γ-glutamyl transpeptidase and bile acids (Gilmore, 1986). The results are of value in monitoring the progress of disease and in the assessment of possible hepatotoxic effects of new drugs. They are particularly important in identifying hepatic disease in the patient with no jaundice or other clinical features of liver disease, for example acute unexplained coma in childhood abnormalities might suggest a diagnosis of Reye's syndrome. In such circumstances a blood ammonia concentration can be invaluable in directing attention to disorders such as urea cycle defects.

It is unfortunate that none of the automated investigations are specific for liver disease. The direct bilirubin is specific but is less rigorously standardized than the tests mentioned above (Mair & Klempner, 1987). It is no more specific, and much more expensive, than the under-used urine test strips impregnated with diazo reagent, which detect as little as 1 μmol bilirubin/L (Mowat, 1990). Serum bile salt concentrations are also relatively specific but are no more sensitive than the batch of routinely available investigations (Matsui *et al.*, 1982) prompted by the clinician's request for 'liver function tests'. Tests for liver function are non-specific. They are frequently influenced by many non-hepatic factors. Elevated serum activities of single enzymes may occur in children without disease (Perrault *et al.*, 1990). Only a prolonged prothrombin time (international normalized ratio, INR) or reduced factor V concentration indicates the severity of acute hepatic damage. Too often these are forgotten. For a fuller assessment, imaging and a skilfully interpreted liver biopsy are essential. Psychometry, electroencephalography and other electrophysiological measurements may be required to detect hepatic encephalopathy.

Laboratory tests for liver disease

Aminotransferases

Aminotransferases (transaminases) are intracellular enzymes found in nearly all tissues and with greatest activity in liver, heart, skeletal muscle, adipose tissue, brain and kidney. Increases in serum activities of *aspartate aminotransferase* (AST) and *alanine aminotransferase* (ALT) are sensitive indicators of hepatocellular necrosis. Leakage from damaged tissues is thought to be the main reason for the rise, rather than impaired clearance by hepatic endothelial cells. ALT is more sensitive and liver specific than AST, which has advantages in epidemiological studies, but in paediatric clinical practice AST is as useful. Increases in serum values of both AST and ALT occur in all types of liver disease (Reichling & Marshall, 1988; Weidemann *et al.*, 1989; Lackman & Tollner, 1992). Their activities are a poor guide to the possible pathology

(Manolaki *et al.*, 1983). Levels may be normal in cirrhosis. Elevations may also occur in severe cardiac or skeletal muscle damage (see Chapter 16), pancreatitis, infarction of the kidney, brain or lung, and in haemolytic anaemia. Typically, normal paediatric values are about 2.5 times the adult value in the first month of life, falling gradually to adult values by the age of 2 years.

Serum alkaline phosphatase

Isoenzymes of ALP are found in the liver, bone, kidney, intestines and placenta. In the liver they are located mainly on canalicular microvilli but are also present on the sinusoidal border of the hepatocytes, where they increase in experimental bile duct ligation. Serum concentrations are increased in many types of liver disease, particularly if there is impaired bile production. The degree of elevation is of no value in distinguishing between intrahepatic or extrahepatic disease. An increased ALP with other tests normal may occur in malignant and benign infiltrations of the liver, including liver abscess, and in congenital hepatic fibrosis. Elevations occur in disorders with no liver involvement and as a familial trait.

The ALP level is raised throughout childhood, particularly in the first few months of life and during puberty. Considerable variations in reference ranges are seen, depending upon age and substrate used for the determination at 37°C (see Appendix A2).

γ-Glutamyl transpeptidase

γ-Glutamyl transpeptidase (γGT) is present throughout the hepatobiliary tree and in other organs such as the heart, kidney, lungs, pancreas and seminal vesicles. Enzymatic activity is increased by enzyme-inducing drugs in the absence of other evidence of liver disorder. The finding of elevated serum γGT activity is thus of limited usefulness. In patients with known hepatobiliary disorders it may serve to follow the course of these disorders or indicate recrudescence of disease in the anicteric patient when the AST and ALT are still normal. Conversely, values may become more abnormal for some months when all other features of liver disease are improving. In the individual patient the activity frequently correlates with that of ALP. The activity levels do not help in identifying the nature of the liver disorder (Manolaki *et al.*, 1983), except that a persistently normal value in the presence of other clinical and laboratory evidence of chronic liver disease is found in a proportion of infants with progressive intrahepatic cholestasis. Normal levels can occur in benign recurrent cholestasis (Maggiore *et al.*, 1991).

Enzymatic activity is less than 40 iu/L from the age of 1 year, but in the newborn period the upper limit of normal may be three times this value; the activity falls gradually during infancy (Knight & Hammond, 1981).

5′-Nucleotidase

5′-Nucleotidase is an enzyme which has a similar distribution within the hepatocytes as ALP, and may play a role in membrane-mediated transport. Although present in bone, its concentration does not change with growth. Its main role may be in distinguishing between hepatic and non-hepatic causes of a raised serum ALP. Reference values at all ages are less than 15 iu/L.

Serum protein determination

Serum albumin is formed by the liver. Hypoalbuminaemia is found in advanced chronic liver diseases. This may be due to decreased synthesis or increased degradation but may be the result of an increased plasma volume. It is a useful but imprecise index of severity of liver damage. It is of little value in predicting the patient's prognosis or response to surgery such as porto-systemic shunting. Temporary low levels are found in extrahepatic portal hypertension following alimentary bleeding. In acute disease its value is less striking, perhaps because of the long half-life of albumin (about 20 days). Serum γ-globulins are composed of immunoglobulins produced in the reticuloendothelial system. Elevations of serum immunoglobulins IgG, IgA and IgM may be found in chronic liver disease. IgM is raised in viral hepatitis (Aksu & Mietens, 1984). Very high levels of IgG are commonly found in autoimmune chronic hepatitis and in primary sclerosing cholangitis in children. Abnormally low levels of IgA and complement component C4 occur in some patients with autoimmune chronic active hepatitis. In Wilson's disease the serum immunoglobulins may all be elevated. The caeruloplasmin is low in 70–95% of cases.

Non-organ-specific autoantibodies

The presence in serum of non-organ-specific autoantibodies directed against elements in the subcellular organelles indicates an underlying genetic predisposition to autoimmune disorders. High levels of antinuclear factor and/or smooth muscle antibodies are found in children with primary sclerosing cholangitis, autoimmune chronic active hepatitis and less commonly in Wilson's disease. Liver–kidney microsomal antibodies alone are found in a category of autoimmune chronic active hepatitis (Mieli-Vergani *et al.*, 1989).

Prothrombin time

The liver is the site of synthesis of many coagulation factors. The half-life of these varies from a few hours to 4 days. Thus, a decrease in hepatic synthesis owing to acute liver dysfunction is rapidly reflected in a decrease in the plasma levels of these factors (I, II, V, VII, IX and X). All except factor I require vitamin K for synthesis and activation. The one-stage prothrombin time with standardized substrate and expressed as INR is a widely available test of these factors and fibrinogen.

An increased INR, which is rapidly corrected by intramuscular injection of vitamin K, indicates vitamin K malabsorption owing to reduced bile secretion without marked synthetic impairment. A vitamin K-resistant increase in INR indicates severe hepatic disease or disseminated intravascular coagulation (Bernuau *et al.*, 1986). Such increases are commonly found in fulminant hepatic failure, metabolic disorders, severe chronic active hepatitis and decompensated cirrhosis. An INR of 1.3 to 1.5 is present in the majority of patients with portal vein obstruction.

Serum bilirubin and its metabolism

Total serum bilirubin values are of little diagnostic or prognostic value. A fall to normal is a good omen. A gradual rise without obvious additional correctable or self-limiting cause in chronic liver disease is ominous. In patients of all ages, investigation of the causes of jaundice is simplified if it is classified into *unconjugated* or *conjugated* hyperbilirubinaemia. Laboratory estimation of *'direct'* bilirubin underestimates serum bilirubin glucuronide concentrations at normal serum bilirubin concentrations, overestimates it at total serum bilirubin concentrations of less than 70 to 80 μmol/L and underestimates it at higher levels (Blankaert & Heirwegh, 1986; Mair & Klempner, 1987). Classification of jaundice into 'direct' and 'indirect' may be misleading if interpreted without urine analysis and/or standard laboratory biochemical tests for the detection of liver disease. Accurate measurement of bilirubin mono- and di-conjugates is possible with alkaline methanolysis and high performance liquid chromatography. These show that 2–5% of bilirubin is conjugated in health. Using these methods Gilbert's syndrome may be differentiated from jaundice due to haemolysis. In Gilbert's syndrome less than 2% will be conjugated.

Urinary bilirubin

Bilirubin in the urine may be the first indication of hepatobiliary disease in patients who are not overtly jaundiced. In jaundiced patients it indicates the presence of conjugated bilirubin in the serum, a sign of hepatocellular parenchymal damage or bile duct obstruction (Mowat, 1990). The commercially prepared Icto-test* is the best available test, detecting as little as 1 μmol/L. Urine that has been standing at room temperature for several hours may give a false negative result. Beeturia and chlorpromazine metabolites in the urine may interfere with the test.

* Icto-test is available from Miles Inc., USA, and Bayer Diagnostics, Basingstoke, UK.

Serum cholesterol

Cholesterol is synthesized in the liver and intestinal wall. Esterification occurs in the liver. Low serum values of cholesterol and its esters are found in those with acute and chronic liver diseases. Very low values (<2 μmol/L) in chronic disease indicate a poor prognosis (Ballin, 1986). Very high serum cholesterol values (six times above normal) are found in patients with chronic biliary obstruction, particularly paucity of intrahepatic bile ducts.

For reference data for serum cholesterol see Chapter 14. Reference data may vary from community to community, depending on dietary habits and genetic background.

Ammonia

The concentration of blood ammonia is primarily regulated by the liver. Endogenous ammonia is produced in the colon by the action of bacterial urease, or synthesized in the liver, kidney or small intestine. Exogenous sources include dietary protein and amino acids.

Ammonia may be raised in acute and chronic hepatic encephalopathy. In these circumstances, diagnosis and monitoring of the patient's progress can usually be determined in other ways (Mowat, 1991, 1992). The main role of blood ammonia determination is in the investigation of unexplained neurological dysfunction. As a rough guide, concentrations exceeding 150 μmol/L in the newborn or 80 μmol/L in children should prompt full investigation to exclude inherited errors of metabolism, but values up to 300 μmol/L may occur in any severe illness (Clayton, 1991a) (see also Chapter 5b).

Serum bile salts

Primary bile salts are synthesized and conjugated in the liver and secreted in the bile. Over 95% are reabsorbed from the intestine and are efficiently extracted from the portal blood into the hepatocytes to be recirculated. As hepatic extraction becomes less efficient the plasma concentration of bile salts rises. Thus, a raised serum bile salt concentration is specific for hepatobiliary disease. The maximum values are found 2 h after a meal. The test is no more sensitive than a combination of the standard laboratory tests for liver damage. The lack of a reliable automated method of analysis has prevented the widespread introduction of this test. A further problem is that serum values rise in healthy babies in the first week of life and remain above normal adult values until 4 to 8 years of age. The normal bile salt levels found in genetic disorders of bilirubin metabolism, as opposed to raised values found with hyperbilirubinaemia caused by hepatobiliary disease, are the best example of the value of bile acids in specific diagnosis (Verling *et al.*, 1982).

Urinary bile salts

Bile salts in the urine reflect those in the serum. The absence of primary salts and the presence of abnormal C24 bile salt derivatives, detected by gas chromatography–mass spectrometry, are diagnostic of inherited abnormalities of bile salt metabolism (Clayton, 1991b).

Quantification of liver function

As already mentioned, albumin and the INR are the only commonly available tests of liver synthetic function. Tests of hepatic clearance of substances from the circulation (a measure of perfusion, uptake and/or metabolism), using dyes such as indocyanine green or sulphobromophthalein, have been largely replaced in clinical practice by radionucleotides. Their rate of uptake by hepatocytes, and their excretion in the biliary system, can be quantitatively monitored by gamma cameras. Sulphobromophthalein sodium may cause systemic allergic or anaphylactic, occasionally fatal, reactions (Bar-Meir *et al.*, 1986). Caffeine clearance from serum or saliva (Lewis & Rector, 1992) and lidocaine metabolite formation (Oellerich *et al.*, 1990) provide a practical means of estimating microsomal function. Both are being evaluated for their prognostic significance but have not yet been introduced as routine clinical tests.

Specific investigations

Caeruloplasmin and copper studies

Caeruloplasmin is a copper-binding protein which possesses copper oxidase enzymatic activity. In Wilson's disease caeruloplasmin may be absent or present in very low concentrations in 70–95% of cases. In fulminant hepatic failure or subacute hepatic necrosis low levels may be found, returning to normal if the liver recovers. In most other forms of liver disease the caeruloplasmin concentration is high. Low values are found in the nephrotic syndrome and protein-losing enteropathy. The normal range is from 1.25 to 2.5 μmol/L; much lower values are found in cord blood and in the first 2 months of life. For the diagnosis of Wilson's disease an increased 24-h urinary copper excretion, particularly with a penicillamine challenge (0.5 g at 0 and 12 h), is more specific than caeruloplasmin levels (da Costa *et al.*, 1992). Values above 25 μmol/24 h are highly specific.

α_1-Antitrypsin deficiency

The genetic variant of α_1-antitrypsin associated with liver disease, protease inhibitor (PI) genotype ZZ (PIZZ), is most easily identified by isoelectric focusing complemented by immunofixation procedures. Serum levels do not identify all PIZZ individuals. In advanced liver disease some PIZZ sera may be identified as SZ. Parental PI studies or deoxyribonucleic acid (DNA) techniques using genotype-specific oligonucleotide probes clarify the phenotype in these circumstances (Hussain *et al.*, 1991).

α-Fetoprotein

α-Fetoprotein is synthesized by embryonic or poorly differentiated liver cells. In intrauterine life it is the major serum protein. It reaches its highest levels in the serum between 13 and 20 weeks gestation, after which it declines gradually. It is detectable in cord blood but is not normally detectable after the age of 6 to 8 weeks.

High α-fetoprotein concentrations are found in infants with intrahepatic cholestatic syndromes in the first 10–20 weeks of life (Johnston *et al.*, 1976). Particularly high levels are found in tyrosinaemia. Very high levels of α-fetoprotein are useful markers of hepatoblastoma or hepatocellular carcinoma, and occur in the serum in up to 90% of cases. If positive, it is also a useful test for assessing whether surgical removal has been complete or the response to chemotherapy has been satisfactory.

Step-wise investigations

Infants with conjugated hyperbilirubinaemia

Investigations to be initiated immediately

Prothrombin time.
Bacterial culture of blood and urine.
Urine microscopy, 'Dipstick' analysis and Clinitest analysis for non-glucose-reducing substances.
Peripheral blood count including reticulocyte count if anaemia is present.
Blood group, saving serum for cross-matching.
Blood sugar and urea or creatinine.
Serum sodium, potassium, bicarbonate and calcium.
If acutely ill, save as much urine and serum as possible at −70°C for subsequent biochemical analysis.
Diagnostic paracentesis of abdomen if appropriate.

Investigations to be initiated as soon as the normal laboratory service is available

Standard biochemical tests of liver function, including total and direct bilirubin and γGT.
Serum cholesterol.
α_1-Antitrypsin phenotype.
Red blood cell galactose-1-phosphate uridyl transferase.
Sweat electrolyte determination or, preferably, genetic screen for cystic fibrosis.
Immunoreactive trypsin.
Serum and urinary amino acids followed by urinary succinyl acetone and γ-aminolaevulinic acid if suggestive of tyrosinaemia.
Urinary bile acid identification by gas chromatography–mass spectrometry.

Serum thyroxine, thyroid-stimulating hormone, cortisol (09.00 h).
Viral culture of urine.
Wasserman reaction (WR) or venereal disease research laboratory (VDRL).
Hepatitis B surface antigen (HbsAg).
Antibody tests for toxoplasmosis, cytomegalovirus, herpes simplex, rubella, hepatitis C virus and human immunodeficiency virus (HIV).
Ultrasound of abdomen to identify focal hepatic lesion or dilated ducts; if either are present, refer to a specialist centre.
X-ray chest, to exclude cardiac lesion, long bones, for evidence of rickets or intrauterine infection, and vertebral bodies, for failure of fusion.
Ophthalmological examination for embryotoxon, optic nerve hypoplasia, choroidoretinitis.

If stools are acholic, contact a specialist centre to arrange transfer of the patient to exclude biliary atresia.

Specialist centre

If no contraindication, percutaneous liver biopsy: tissue for histological, electron microscopic and enzymatic studies; viral and bacterial culture.

Laparotomy

If the tests performed so far exclude genetic disorders and endocrine deficiencies, and if stools lack green or yellow pigment and biopsy is compatible with biliary atresia: laparotomy, operative cholangiography or definitive surgery by an experienced paediatric hepatobiliary surgeon.

If biopsy equivocal and stool colour not convincingly green or yellow, ^{99m}Tc-tagged methylbromoiminodiacetic acid biliary excretion scan after 3 days of phenobarbitone may be useful to demonstrate bile duct patency.

If biliary atresia is a possibility but laparotomy can be deferred, carry out endoscopic cholangiography.

If no cause is identified and cholestasis is incomplete

Bone marrow aspiration for storage disorders.
White blood cell lysosomal enzyme concentration.
Urinary mass spectroscopy to exclude defects of fatty acid metabolism.
Fibroblast culture for enzyme studies.

Apparent acute hepatitis

Prothrombin time*.
Bilirubin, total and direct.
AST.
γGT.
Albumin*.
Total serum protein*.
*If abnormal:**
Serum immunoglobulins.
Tissue autoantibodies.
Caeruloplasmin.
Copper studies to exclude Wilson's disease.

Serological investigations for viral hepatitis

	1	2	3	4	5	6	7	8	9
Anti-HAV IgM	+	−	−	−	−	−	−	−	−
HBsAg	−	+	−	+	−	+	−	+	−
Anti-HBc IgM	−	+	+	+	+	−	−	+	−
Anti-HDV	−	−	−	+	+	+	−	−	−
Anti-HCV	−	−	−	−	−	−	+	+	−
Anti-HEV IgM	−	−	−	−	−	−	−	−	+

1, Acute hepatitis A virus (HAV) infection; 2, acute hepatitis B virus (HBV) infection or reactivation; 3, acute HBV infection with early seroconversion; 4, acute HBV/HDV coinfection; 5, acute hepatitis D virus (HDV) infection with HBV suppression; 6, acute HDV infection in HBV carrier; 7, acute HCV infection; 8, HBV/HCV coinfection; 9, acute hepatitis E virus (HEV) infection. HBC, hepatitis B core.

Serological tests for Epstein–Barr, cytomegaloviruses, leptospirosis and toxoplasmosis

If encephalopathy is present:
Blood glucose.
Ammonia.
Calcium.
Electrolytes.
Creatinine.

Uric acid.
Amylase.
Consider:
Investigations to exclude metabolic disorders.
Presenting with Reye's syndrome.
Computerized axial tomography (CAT) scan of brain to exclude other causes of coma.

Apparent chronic hepatitis or cirrhosis

Investigations as for acute hepatitis *plus*:
Ultrasonography of liver, to assess heterogeneity, venous and arterial blood-flow patterns, spleen size and evidence of porto-systemic collaterals; of biliary system, to detect dilatation; and of kidneys for features of polycystic disease.
α-Fetoprotein.
Cholesterol.
α_1-Antitrypsin phenotype.
Sweat electrolytes or, preferably, genetic screen for cystic fibrosis.
Urinary succinyl acetone, fatty acid metabolites and porphyrins.
Serum complement components C3 and C4.
Serum iron and iron-binding capacity.
Urinary bile acid identification by gas chromatography–mass spectrometry.
Blood gases.
Liver biopsy.
Consider:
Endoscopy for oesophageal varices.
Endoscopic or percutaneous cholangiography.
Cardiac and hepatic vein catheterization.
Bone marrow for storage materials.

References

Aksu, F. & Mietens, S.T. (1984) Serum immunoglobulin levels in acute viral hepatitis in childhood. *Klin Pediat* **196**, 83–86.

Arias, I.M., Jakoby, W.B., Popper, H., Schachter, D. & Shafritz, D.A. (eds) (1988) *The Liver – Biology and Pathobiology*, 2nd edn. Raven Press, New York.

Bailin, J.A. (1986) Lipid metabolism in relation to liver physiology and disease. In: *Liver Annual*, Vol. 5 (eds I.M. Arias, M. Frenkel & J.H.P. Wilson), p. 22. Elsevier, Amsterdam.

Bar-Meir, S., Bar-Tal, L., Papa, M.Z. & Peled, Y. (1986) Bromosulfophthalein clearance and aminopyrine test in patients with Gilbert's syndrome. *Israel J Med Sci* **22**(5), 376–379.

Bernuau, J., Rueff, B. & Benhamou, J.P. (1986) Fulminant and subfulminant liver failure: definition and causes. *Sem Liver Dis* **6**, 97–106.

Blankaert, N. & Heirwegh, K.P.M. (1986) Analysis and preparation of bilirubins and biliverdins. In: *Bile Pigments and Jaundice: Molecular, Metabolic and Medical Aspects* (ed. J.D. Ostrow), pp. 31–80. Marcel Dekker, New York.

Clayton, P.T. (1991a) Diagnosis of metabolic liver disease in infancy. *Curr Paediatr* **1**, 224–227.

Clayton, P.T. (1991b) Inborn errors of bile acid metabolism. *J Inher Metab Dis* **14**, 478–496.

da Costa, C.M., Baldwin, D., Portmann, B. *et al.* (1992) The value of urinary copper excretion after penicillamine challenge in the diagnosis of Wilson's disease. *Hepatology* **15**, 609–615.

Gilmore, I.T. (1986) Modern methods of diagnosis in liver disease. *J R Coll Phys London* **22**, 201–204.

Green, A. (1988) When and how should we measure plasma ammonia? *Ann Clin Biochem* **25**, 199–209.

Gumucio, J.J. (1989) Hepatocyte heterogeneity: the coming of age. From the description of a biological curiosity to the partial understanding of its physiological meaning and regulation. *Hepatology* **9**, 154–160.

Hussain, M., Mieli-Vergani, G. & Mowat, A.P. (1991) Alpha-1-antitrypsin deficiency and liver disease: clinical presentation, diagnosis and treatment. *J Inher Metab Dis* **14**, 497–511.

Johnston, D.I., Mowat, A.P., Orr, H. & Kohn, J. (1976) Serum alphafetoprotein levels in extrahepatic biliary atresia, idiopathic neonatal hepatitis and alpha-1 antitrypsin deficiency (PiZ). *Acta Paed Scand* **65**, 623–627.

Knight, J.A. & Hammond, R.E. (1981) Gammaglutamyl transphorase and alkaline phosphatase activities compared in serum of normal children and children with liver disease. *Clin Chem* **27**, 48–53.

Lackmann, G.M. & Tollner, U. (1992) Enzymaktivitaten im Serum gesunder Neugeborener. *Monatsschr Kinderheilkunde* **140**(3), 171–176.

Lentze, M.J. & Reichen, J. (1992) *Paediatric Cholestasis. Novel Approaches to Treatment.* Kluwer Academic, Dordrecht.

Lewis, F.W. & Rector, W.G. (1992) Caffeine clearance in cirrhosis: the value of simplified determinations of liver metabolic capacity. *J Hepatol* **14**, 157–162.

McIntyre, N., Benhamou, J.-P., Bircher, J., Rizzetto, M. & Rodes, J. (eds) (1991) *Oxford Textbook of Clinical Hepatology*. Oxford University Press, Oxford.

Maggiore, G., Bernard, O., Hadchouel, M., Lemonnier, A. & Alagille, D. (1991) Diagnostic value of serum gamma-glutamyl transpeptidase activity in liver disease in children. *J Ped Gastr Nutr* **12**(1), 21–26.

Mair, B. & Klempner, L.B. (1987) Abnormally high values in direct bilirubin in the serum of newborns as measured with the DuPont aca Analyser. *Am J Clin Pathol* **87**, 642–644.

Manolaki, N., Larcher, V.F., Mowat, A.P. *et al.* (1983) Prelaparotomy diagnosis of extrahepatic biliary atresia. *Arch Dis Child* **58**, 591–594.

Matsui, A., Psacharopoulos, H.T., Mowat, A.P. *et al.* (1982) Radioimmunoassay of serum glycocholic acid. Standard laboratory tests of liver function and liver biopsy findings: comparative study of children with liver disease. *J Clin Pathol* **35**, 1011.

Mieli-Vergani, G., Lobo-Yeo, A., McFarlane, B.M. *et al.* (1989) Different immune mechanisms leading to autoimmunity in primary sclerosing cholangitis and autoimmune chronic active hepatitis of childhood. *Hepatology* **9**, 198–203.

Mowat, A.P. (1987) *Liver Disorders in Childhood*, 2nd edn. Butterworth, London.

Mowat, A.P. (1990) Urine analysis in the assessment of jaundiced infants. In: *Clinical Urinalysis* (eds R.G. Newall & R. Howell). Ames Division, Miles Ltd.

Mowat, A.P. (1991) Acute liver failure. *Curr Paediatr* **1**, 218–223.

Mowat, A.P. (1993) Hepatic encephalopathy: acute and chronic. In: *Baillière's Clinical Paediatrics* (ed. J.A. Eyre). Saunders, London.

Oellerich, M., Brudelski, M., Lautz, H.-U. *et al.* (1990) Lidocaine metabolite formation as a measure of liver function in patients with cirrhosis. *Therap Drug Monit* **12**, 219–226.

Perrault, J., O'Brien, J.F. & Tremaine, W.J. (1990) Macrotransaminase of aspartate aminotransferase (AST): a benign cause of elevated AST activity. *J Pediatr* **117**(3), 444–445.

Reichling, J.J. & Marshall, M.K. (1988) Clinical use of serum enzymes in liver disease. *Digest Dis Sci* **33**(12), 1601–1614.

Reif, S., Sykes, D., Rossi, T. & Weiser, M.M. (1992) Changes in transcripts of basement components during rat liver development: increase in laminin messenger RNAs in the neonatal period. *Hepatology* **15**, 310–315.

Reubner, B.H., Blankenberg, T.A., Burrows, D.A., SooHoo, W. & Lund, J.K. (1990) Development and transformation of the ductal plate in the developing human liver. *Paediatr Pathol* **10**, 55–68.

Rosenthal, P., Blankaert, N., Kabra, P.M. & Thaler, M.M. (1986) Formation of bilirubin conjugates in human newborns. *Paediatr Res* **20**, 947–950.

Sokal, E., Trivedi, P., Cheeseman, P., Portmann, B. & Mowat, A.P. (1989a) The application of quantitative cytochemistry for the study of the acinar distribution of enzymatic activities in human liver biopsy sections. *J Hepatol* **9**, 42–48.

Sokal, E., Trivedi, P., Portmann, B. & Mowat, A.P. (1989b) Developmental changes in the intra-acinar distribution of succinate dehydrogenase, glutamate dehydrogenase, glucose 6 phosphatase and NADPH dehydrogenase in rat liver. *J Ped Gastr Nutr* **8**, 522–527.

Suchy, F.J., Bucuvalas, J.C. & Novak, D.A. (1987) Determinants of bile formation during development: ontogeny of hepatic bile acid metabolism and transport. *Sem Liver Dis* **7**, 77–84.

Vierling, J.M., Berk, P.D., Hofmann, A.F. *et al.* (1982) Normal fasting-state levels of serum cholyl-conjugated bile acids in Gilbert's syndrome: an aid to diagnosis. *Hepatology* **2**, 340–344.

Vilstrup, H., Gluud, C., Hardt, F. *et al.* (1990) Branched chain enriched amino acid versus glucose treatment of hepatic encephalopathy. A double-blind study of 65 patients with cirrhosis. *J Hepatol* **10**(3), 291–296.

Wiedemann, G., Armann, O., Reinhardt, M. & Biesenbach, R. (1989) Untersuchungen zur Ermittlung von Referenzbereichen der Serumenzyme Alaninaminotransferase (ALAT), Aspartataminotransferase (ASAT), Gamma-Glutamyltransferase (GGT), Lactatdehydrogenase (LDH) and Creatinkinase (CK) fur Kinder vom 2 bis 7 Lebensjahr. *Z Klin Med* **44**(12), 1059–1065.

14: Plasma Lipid Disorders

J.K. LLOYD

Introduction

Although requests for estimation of plasma lipids and lipoproteins in children are relatively uncommon and such analyses probably still form only a small part of the total workload of a Department of Chemical Pathology, greater awareness of the importance of hyperlipidaemia in childhood as a risk factor for coronary heart disease in adult life will result in such laboratories receiving increasing numbers of requests in the future, and biochemists will be asked to advise on appropriate investigations, interpret findings and sometimes advise on management. Advances in both the methodology of analysis and the understanding of the metabolic pathways of lipoproteins have been considerable and the data on which to base concepts of normality have become more secure. Whilst it is beyond the scope of this chapter to give a detailed review of the biochemistry and genetics of lipid metabolism, knowledge of this is clearly essential and a brief and simplified summary may be helpful. An excellent review (Havel & Kane, 1989) can be found in *The Metabolic Basis of Inherited Disease* (Scriver *et al.*, 1989); the most recent edition should be consulted.

Biochemistry and metabolism

Although lipoproteins were originally classified according to their mobility on electrophoretic separation, they are now defined in relation to their density, and terminology based on this is used. Table 14.1 gives the average composition of the major classes of lipoproteins. Within each class there is a spectrum of molecular size, and the fatty acids esterified to cholesterol, triglyceride and phospholipid vary greatly in respect of both chain length and degree of unsaturation.

Figure 14.1 illustrates the essential features of lipoprotein metabolism in over-simplified form. Ingested triglyceride is transported out of the intestinal absorptive cell in the form of chylomicrons which, in addition to the triglyceride, contain small amounts of cholesterol (both unesterified on the surface and esterified within the core), phospholipid and a mixture of apoproteins. Within the circulation chylomicrons, together with very low-density lipoproteins (VLDL) carrying endogenous triglyceride from the liver, are exposed to extrahepatic lipoprotein lipases at the endothelial surfaces of adipose tissue and muscle (and probably other tissues); these hydrolyse the triglyceride prior to uptake of fatty acids into the cell for energy or storage purposes. As a result, the chylomicrons and VLDL particles shrink in size and increase in density. They also lose some of their surface apoproteins, cholesterol and phospholipids by transfer to high-density lipoproteins (HDL). The remnants of chylomicrons are taken up by specific receptors in the liver and degraded. The small amount of dietary cholesterol entering by the chylomicron route is rapidly cleared from the circulation by this process and thus, for practical purposes, meals do not affect the plasma cholesterol concentration.

Hydrolysis of VLDL triglyceride results in a particle of intermediate density (IDL), whose triglyceride is further hydrolysed by a hepatic lipase. At the same time, interchange occurs between the cholesterol of IDL and HDL; the unesterified cholesterol in HDL is esterified by the action of the enzyme lecithin-cholesterol acyltransferase (LCAT) and incorporated into IDL. Removal of all apoprotein apart

Table 14.1 Composition of lipoproteins

	Density (g/mL)	Approximate composition (%)				Apoproteins*	
		Protein	Cholesterol†	Phospholipid	Triglyceride	Major	Minor
Chylomicrons	< 0.95	1–2	3–7	3–6	80–95	A B C	E
VLDL	< 1.006	6–10	20–30	15–20	45–65	C B	E
LDL	1.019–1.063	20–25	50–60	18–24	4–8	B	
HDL	1.063–1.21	40–50	18–25	26–32	2–7	A	C E

* Apoproteins may have several alleles, designated A-I, A-II, etc.
† Includes cholesterol esters.
VLDL, very low-density lipoprotein; LDL, low-density lipoprotein; HDL, high-density lipoprotein.

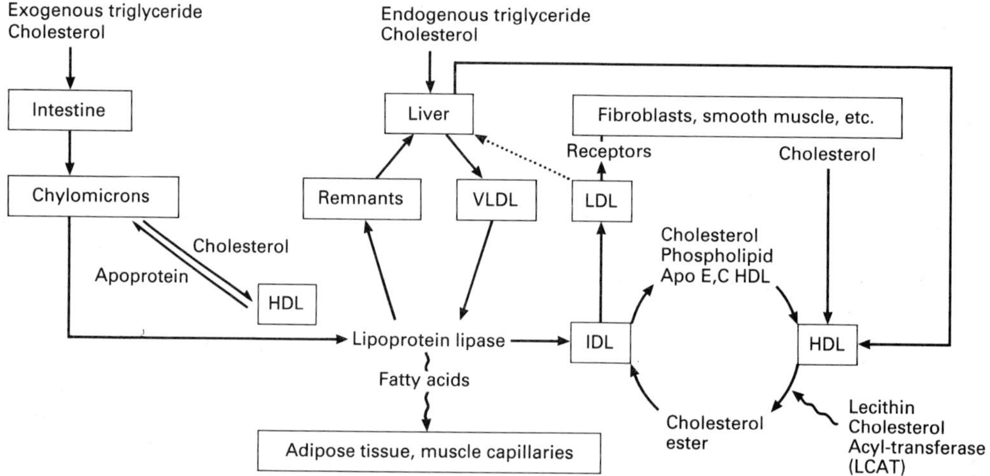

Fig. 14.1 Lipoprotein metabolism and inter-relationships. LDL, low-density lipoprotein; VLDL, very low-density lipoprotein; IDL, intermediate-density lipoprotein; HDL, high-density lipoprotein.

from apoprotein B (apoB), and the loss of most of its triglyceride, results in the conversion of IDL into low-density lipoprotein (LDL).

LDL, carrying about two-thirds of its cholesterol in unesterified form, is bound by specific receptors present on the surface of cells (Brown & Goldstein, 1986). Receptor-bound LDL is then taken into the cell where the cholesterol ester is hydrolysed by a lysosomal acid-lipase. Free cholesterol thereby liberated is used for membrane synthesis and also regulates both the amount of endogenous synthesis of cholesterol by the cell, and the formation of further surface receptors. HDL is synthesized in the liver and intestine and takes up the surplus cholesterol resulting from cell death and membrane turnover. It appears to function primarily as a reservoir for the exchange of free cholesterol and phospholipid between lipoproteins, and for interchange between certain apoproteins. Cholesterol ester transfer protein is responsible for the transfer and exchange of cholesterol ester and triglyceride between HDL and apolipoprotein B-containing lipoproteins in the plasma (Johnson *et al.*, 1991). The close inter-relationship that exists between the lipoprotein classes is such that in any disorder affecting lipoprotein metabolism some derangement can be expected in all lipoproteins. The molecular structure of, and genes coding for, most of the apoproteins and of the LDL receptor have now been described. It is probable that technological advances will soon enable more precise diagnosis at the genetic level and pave the way for therapy in some cases.

Sampling and interpretation

Many analyses, and certainly total cholesterol and triglyceride, can now be made on capillary blood, provided that the specimen is collected by an experienced operator. Venous blood is usually required for more detailed studies. Because plasma triglyceride concentrations fluctuate during lipid absorption, and because plasma cholesterol concentrations are slowly influenced by the total quantity and nature of dietary fat, samples for lipid analysis should ideally be obtained in the fasting state and when the subject has been known to have been eating a usual diet over the preceding 2 or 3 weeks. In practice however, a random sample is acceptable for plasma cholesterol estimation and, because the clearing of dietary fat in children is normally a very efficient process (the half-time for clearance of chylomicrons is only 4–5 min, but the absorptive process of course continues for much longer), it may also be possible to interpret the triglyceride value in such a sample. Nevertheless, if it is likely that the patient has a primary abnormality of lipid metabolism that requires precise diagnosis, blood should be obtained after a fast of not less than 8 h. Specific instructions are advisable: 'nothing to eat or drink, except water, after your last meal on the night before the test'. For milk-fed babies the last milk feed should be at midnight, with blood being taken at 08.00 h. If this degree of fasting is unacceptable, a fruit drink or other small amount of carbohydrate in the form of bread and jam is unlikely to interfere with the interpretation of the results.

Confirmation of abnormal results is a wise measure, especially if the values are in the 'borderline range'. In the primary lipoprotein disorders treatment is not likely to be required urgently (an exception is familial hyperchylomicronaemia with an acute abdominal crisis; see p. 313), follow-up will be lengthy, and two estimations before treatment is started will not only confirm the diagnosis but also provide a more secure baseline from which to judge the efficacy of therapy.

Lipoprotein studies by electrophoretic separation are now seldom carried out. Quantitation by estimating the amount of cholesterol present in HDL after precipitation of the LDL is a routine measure in many laboratories and enables LDL cholesterol to be calculated. This analysis, coupled with inspection of the plasma (if necessary stored in the refrigerator at 4°C overnight), knowledge of the timing of the sample in relation to the last meal, and an understanding of the composition of the lipoproteins (Table 14.1), should enable an estimation of the nature of any lipoprotein abnormality to be made. Table 14.2 indicates the deductions, regarding the lipoprotein status, that may be so derived. Only rarely will studies of fatty acid composition, apoprotein distribution, or associated enzyme concentrations be required for diagnosis and management; if necessary, these more specialized determinations are best carried out in centres with special interest and expertise in lipid metabolism. Identification of an abnormal lipoprotein pattern does not, of course, by itself provide a diagnosis; this can only be established after consideration of the clinical and family data. The typing system based on the electrophoretic appearance (Beaumont *et al.*, 1970) should preferably not be used when reporting results. A report of 'type II hyperlipidaemia' may be misinterpreted by the clinician as indicating that the patient has familial hypercholesterolaemia (FH), whereas it only indicates raised LDL without any increase in chylomicrons or VLDL. Nevertheless, for some disorders there is still no other convenient nomenclature, for example types III, IV and V (see Table 14.4).

Genetic analysis can be used to identify family members who have inherited defects in either the LDL receptor or *apoB100* genes. Plasma lipid levels and clinical characteristics of familial defec-

Table 14.2 Interpretation of some plasma lipid and lipoprotein analyses

Appearance		Total lipids			
Turbidity at room temp.	Cream layer at 4°C	C	T	HDL-C	Interpretation
0	0	N	N	N	Normal
±	±	N	±	N	Probably non-fasting sample
+	0	N	+	N(−)	Slight increase VLDL
++	0	N(+)	++	−	Increase VLDL
++	0	++	++	−	Increase IDL
+++	+++	++	+++	−	Pathological increase chylomicrons
0	0	++(+)	N(±)	N(−)	Increase LDL
0	0	−	N	N(−)	Decrease LDL
0	0	−−	−−	−−	Absent LDL, VLDL
++	(+)	−	+	−−	Absent HDL
0	0	+(+)	N	+(+)	Increase HDL

N, normal range; + or −, degree of increase or decrease; C, cholesterol; T, triglyceride; LDL, low-density lipoprotein; VLDL, very low-density lipoprotein; IDL, intermediate-density lipoprotein; HDL, high-density lipoprotein; HDL-C, high-density lipoprotein-cholesterol.

tive apoB100 are similar to those reported for heterozygous FH; for this reason it is important not to use a diagnostic procedure based on linkage analysis without first clarifying which particular gene is at fault (see Chapter 2).

A number of mutations are known to cause LDL-receptor defects. Deletions caused by recombination between the 'Alu' repeat sequences interspersed at regular intervals in the introns of the LDL-receptor gene are relatively common. Restriction fragment length polymorphisms provide a means of following the inheritance of the defective gene within families in whom the mutation has not been identified, thereby detecting individuals at risk of early cardiovascular disease (Humphries *et al.*, 1988).

A single common nucleotide change has been described in the *apoB100* gene and this leads to an amino acid substitution at position 3500 of apoB100 (Innerarity *et al.*, 1990). This change can be easily detected using the polymerase chain reaction and has been reported in 3% of cases of patients with FH, suggesting that the cause of FH in these individuals was not an LDL-receptor defect (Tybjaerg-Hansen *et al.*, 1990).

Normal values

Knowledge of what should be regarded as 'normal' for plasma cholesterol and triglyceride concentrations in children is based on an increasing number of epidemiological studies. Values vary between communities and are influenced by a combination of genetic and environmental factors. Changes occur with age, and those around puberty are still inadequately studied. The relationship between 'normal' values, defined statistically in relation to the values obtaining in a population, and the values desirable for health or indicative of disease, remain somewhat controversial. For example, the World Health Organization (WHO) has proposed an 'ideal mean' for total cholesterol concentration in children of 5–18 years of 2.85 mmol/L (SD 0.52 mmol/L) (WHO, 1982). This would result in well over half the child population in the UK (and many other countries) being regarded as 'abnormal'. The Bogalusa heart study (Berenson, 1980) in the USA provides ranges for total plasma cholesterol, triglyceride and lipoproteins from birth through childhood and adolescence which are probably applicable to children

Table 14.3 'Normal' values* for plasma lipids and lipoproteins (median, 5th and 95th centiles, mmol/L)

Age	Total cholesterol	Total triglyceride	HDL cholesterol
Newborn	1.8	0.4	0.9
(cord blood)	(1.1–2.6)	(0.1–0.9)	(0.3–1.5)
6 months	3.4	1.0	1.3
	(2.3–4.9)	(0.6–1.9)	(0.6–2.2)
1 year	3.9	0.9	1.3
	(2.5–4.9)	(0.5–1.8)	(0.6–2.2)
2–14 years	4.1	0.7	1.7
	(3.1–5.4)	(0.4–1.4)	(0.8–2.6)

* Values derived from Bogalusa heart study (USA) (Berenson, 1980).
HDL, high-density lipoprotein.

in industrialized countries (Table 14.3). There is no evidence in the USA that values have changed appreciably over the past decade. Levels of cholesterol in the UK and Australia may be somewhat higher; 8- to 15-year-old children in Australia have mean total cholesterol around 4.5–4.6 mmol/L (5th–95th centile, 3.5–5.8 mmol/L) (Boulton *et al.*, 1991).

When considering the significance of values in an individual against the background of the population data, the following points are relevant:

1 At birth, plasma cholesterol concentrations do not predict levels in later infancy or childhood. The reason for this may be due in part to the fact that a smaller proportion of total cholesterol is in the LDL fraction at birth than at any other age, even LDL values at birth lack correlation with those at 6 months of age. The type of feeding and the changes that occur as the composition of the diet alters may account for some of the variability.

2 During the first year of life, and especially during the first 6 months, concentrations of plasma lipid and lipoproteins rise rapidly. Thereafter, there is relatively little change with age up to puberty, such change being largely confined to a slight increase in total and HDL cholesterol.

3 After 1 year of age (and perhaps as early as 6 months) 'tracking' (staying within the same centile band) of plasma cholesterol can be demonstrated. The Muscatine study (USA) (Lauer *et al.*, 1988), reporting a 15-year follow-up of schoolchildren tested at 6–15 years, showed that for those children tested at the age of 7–8 years and followed to 20–25 years, the *r* value for total cholesterol was 0.56 for males and 0.64 for females (similar figures were obtained for LDL cholesterol but there was no significant correlation for HDL). Of those children with cholesterol in childhood greater than the 90th centile, 43% had levels above the 90th centile between ages 20 and 30 years. An Australian study of tracking from 3 months to 13 years showed a steady increase in the coefficient of correlation with an *r* value of 0.76 between 8 and 13 years (Boulton *et al.*, 1990). A greater proportion of children in the uppermost quintile remained in their track (47%) compared with those in the lowest quintile (26%). Tracking of plasma triglyceride is much less marked, with a correlation coefficient of only about 0.3.

4 Up to 5 years of age, lipid and lipoprotein levels do not differ significantly between the sexes. From 5 to 14 years girls tend to have higher triglyceride, LDL and VLDL levels and lower HDL levels than do boys.

5 During puberty, variations occur which may make interpretation in individuals difficult. Total cholesterol and LDL levels tend to fall and triglyceride and VLDL to rise. Sometime during puberty and in later adolescence LDL and VLDL concentrations in boys become higher and HDL levels become lower than in girls, thus approaching the adult pattern.

6 At all ages diet and other environmental factors may need to be considered (see Environmental hypercholesterolaemia and cholesterol screening, p. 318).

Hyperlipoproteinaemia

The most common abnormality found on analysis of plasma lipids is some form of hyperlipoproteinaemia. This may be a secondary manifestation in another disease state or may represent a primary genetically determined disturbance. Table 14.4 summarizes the hyperlipoproteinaemias which may present in childhood according to the predominant lipoprotein affected.

Table 14.4 Hyperlipoproteinaemias

Main lipoprotein abnormality	Lipoprotein type*	Plasma lipids: Cholesterol	Plasma lipids: Triglyceride	Primary disorders	Secondary disorders
Increased chylomicrons	I	++	++++	Familial lipoprotein lipase deficiency	Diabetes mellitus (rare)
Increased LDL	IIa	++(+)	N	Familial hypercholesterolaemia	Hypothyroidism, nephrotic syndrome, obstructive jaundice†, hepatic glycogenoses (rare), acute porphyria, cholesterol ester storage disease
Increased LDL	IIb	++(+)	+	Familial combined hyperlipidaemia	
Increased IDL	III	++	++	Familial type III hyperlipidaemia	Hypothyroidism (rare)
Increased VLDL	IV	N or +	++	Familial hypertriglyceridaemia	Diabetes mellitus, nephrotic syndrome, renal dialysis, hepatic glycogenoses, obstructive jaundice, lipodystrophy, idiopathic hypercalcaemia
Increased VLDL and chylomicrons	V	++(+)	++(+)	Familial type V hyperlipidaemia	Diabetes mellitus, obstructive jaundice
Increased HDL	–	+	N	Familial hyperalphalipoproteinaemia	

* WHO classification (Beaumont *et al.*, 1970).
† In obstructive jaundice there may be an accumulation of lipoprotein X.
LDL, low-density lipoprotein; IDL, intermediate-density lipoprotein, abnormal in composition and electrophoretic and ultracentrifugal characteristics; VLDL, very low-density lipoprotein; HDL, high-density lipoprotein; N, normal range; +, degree of increase.

Primary hyperlipoproteinaemias

All the primary disorders can occur in children but most will be uncommon findings in chemical pathology laboratories: some because the disease is itself rare (lipoprotein lipase deficiency, hyperchylomicronaemia); some because, although fairly common in adults, the disorder is infrequently expressed during childhood (hypertriglyceridaemia type IV, combined hyperlipidaemia, type III, type V); and some because children rarely have clinical abnormalities and therefore plasma lipids are not estimated (FH). However, with an increase in studies of the families of adults with primary hyperlipoproteinaemia and/or early coronary heart disease, it can be expected that more children will be found who have a genetic disorder, the most common of which is likely to be FH. Detailed accounts of all the disorders can be found in Scriver *et al.* (1989).

Familial hyperchylomicronaemia

The majority of patients presenting with familial hyperchylomicronaemia have deficiency of lipoprotein lipase. In some populations it is relatively common with a frequency of 1 in 5000 to 1 in 1000 (Gagne *et al.*, 1989). Some patients may have the rare disorder of apoCII deficiency, apoprotein CII being an essential cofactor for the activation of lipoprotein lipase. Both defects are inherited as an autosomal recessive trait and have similar clinical features and management (Santamara-Fojo & Brewer, 1991). ApoCII-deficient patients do, however, differ in their response to infusions of normal plasma.

Clinical presentation can occur at any time during childhood with eruptive xanthomata, unexplained hepatosplenomegaly or acute attacks of abdominal pain. The disorder may also first be recognized in the laboratory when plasma is received for investigation of some other condition. Impaired hydrolysis of chylomicrons and VLDL owing to a lack of lipoprotein lipase activity results in accumulation of these lipoproteins with gross turbidity of the plasma and an obvious cream layer on standing. The biochemical defect is fully expressed at birth and the lipoprotein abnormalities are evident as soon as the infant starts to ingest fat. It is not known why the clinical manifestations are so variable in expression between different children, and even in the same child at different ages.

Analysis of the plasma lipids shows greatly increased triglycerides, usually in the range 40–100 mmol/L. Cholesterol concentrations are also high because the degree of hyperchylomicronaemia is sufficient to make the small proportion of cholesterol in the molecule significant in terms of total plasma values. If the concentrations of triglyceride and cholesterol are expressed as mg/dL, the triglyceride : cholesterol ratio is of the order of 10 : 1; about 50% of the cholesterol is unesterified, reflecting the composition of chylomicrons. Estimations of LDL and HDL show these lipoproteins to be reduced in concentration and this feature becomes even more noticeable after treatment.

The diagnosis can usually be made on the plasma findings, together with rapid clearing of the hyperchylomicronaemia (generally within 4–5 days) on withdrawal of dietary fat, and the exclusion of any other disorder known to cause secondary hyperchylomicronaemia. Lipoprotein lipase activity can be estimated after heparin stimulation but if this is performed the test must be carried out before (or within a few days of) starting a low-fat diet; such investigations are best undertaken by a specialized laboratory.

Investigation of the parents (heterozygotes) may show mild lipoprotein abnormalities but is not often rewarding unless, of course, genetic studies are being undertaken. Siblings should be examined so that symptomless individuals with the disorder may be detected.

Treatment is by reduction of dietary fat of all types. In children with abdominal crises (acute pain, often with vomiting, which is thought to be due to pancreatitis and in which plasma amylase levels may be raised) all fat should be withdrawn. Plasma infusion will be helpful in cases that are the result of apoCII deficiency. Usually a fat-free diet is given until the plasma is optically clear, and initial treatment can be easily monitored by observing the decline in plasma turbidity. For subsequent management the degree of reduction in dietary fat required in the long term depends on the individual child and should be low enough to prevent symptoms. Medium-chain triglycerides may prove a helpful adjunct to the diet as they do not utilize the chylomicron route of absorption. Laboratory monitoring of plasma cholesterol and triglyceride generally shows that triglyceride levels remain moderately elevated, but cholesterol is within, or even below, the normal range. The disorder is not thought to be associated with premature atherosclerosis and attempts to maintain plasma triglyceride within the normal range are probably neither justified nor practicable.

Familial hypercholesterolaemia

This is the most common of the familial hyperlipoproteinaemias to be expressed during childhood. It may be due to defects in the LDL receptors for apoB100 at cell surfaces (Brown & Goldstein, 1986) or to an abnormality in apoB100 inhibiting its binding by receptors (Innerarity *et al.*, 1990; Myant *et al.*, 1991). Both defects are inherited as autosomal dominant traits and have a gene frequency perhaps as great as 1 in 500. Clinical manifestations are probably the same, although homozygotes for defective apoB100 do not yet appear to have been described.

The homozygous form

The homozygous form is very rare, presenting during early childhood with tendon and skin xanthomata, and causing premature atherosclerosis with clinical evidence of cardiac dysfunction towards the end of the first decade and death from myocardial infarction in either the second or third decade. Plasma cholesterol levels are extremely high (of the order of 17–26 mmol/L), triglycerides are normal and HDL is often low. The diagnosis must be confirmed by demonstrating the heterozygous state in both parents or, if one parent has already died, the disorder should be shown to be present in both parental families. In the event of a parent having normal cholesterol and LDL the rare situation of coinheritance of a factor suppressing hypercholesterolaemia in a heterozygote should be considered (Hobbs *et al.*, 1990).

A child who has tendon or skin xanthomata, plasma lipid and lipoprotein levels compatible with the homozygous state, no evidence of any other disease that might cause hyperlipidaemia, and parents who have entirely normal plasma lipids may have the even rarer condition of pseudohomozygous hypercholesterolaemia (Mishkel, 1976). This condition may in fact be due to sitosterolaemia (Low *et al.*, 1991) and plasma concentrations of plant sterols should therefore always be measured in such cases.

In both the homozygous and pseudohomozygous disorders, intermittent arthralgia may occur and the erythrocyte sedimentation rate may be raised. These findings, coupled with cardiac murmurs, have sometimes resulted in an erroneous diagnosis of rheumatic carditis.

The heterozygous form

The heterozygous form of FH is usually suspected in a child because a parent is known to have the disorder, or premature ischaemic heart disease has occurred in a first-degree relative. Clinical manifestations in childhood are rare but skin xanthomata and corneal arcus may occur. The lipoprotein abnormality is fully expressed in childhood and the diagnosis usually presents no problem, with plasma cholesterol being unequivocally high at around 8–12 mmol/L, triglycerides normal, HDL normal or low and LDL high. Difficulties may arise, however, in interpreting cholesterol concentrations at, or only a little above, the 'normal' range. There is no absolute cut-off point for either plasma cholesterol or LDL cholesterol above which the diagnosis is certain, and it is essential that family studies are undertaken and the disease demonstrated in a first-degree relative. Under these circumstances the studies of Leonard *et al.* (1977) showed that, at least in the UK and at that time, the plasma cholesterol concentration at which minimal misclassification occurred was 6.77 mmol/L. Even with values at or below this level uncertainties may still exist, especially if the child has already been receiving a modified fat diet. It is most important that in all cases in which there is doubt, and even in those in which the diagnosis appears clear, repeat estimations are undertaken before the definitive diagnosis is made. Diagnosis by genetic analysis is an attractive proposition but the number of allelic variations may make this difficult. It is anticipated that the use of DNA technology will expand in the near future and will increasingly be used to sort out the problems in which heterogeneity may be a component, e.g. in the aetiology of cardiovascular disease.

Children, examined as part of a family study, who are found to have unequivocally normal values can be reassured and do not require further follow-up. If one parent is known to be heterozygous the diagnosis of FH can probably be made at birth by examination of cord-blood plasma using LDL cholesterol values rather than total cholesterol (Kwiterovich *et al.*, 1973). There appears, however, to be no major therapeutic advantage in establishing the diagnosis at birth and it can be delayed to the end of the first year, by which time the child is usually taking a mixed diet. Screening the general population for hypercholesterolaemia at birth or in later childhood is not a satisfactory method for detecting FH.

Treatment of FH in children is undertaken in the hope that by lowering plasma cholesterol and LDL concentrations the premature development of atherosclerosis may be prevented, or at least delayed. There are as yet no hard data to show that treatment is effective in these respects, although the accumulated evidence from intervention studies in adults (Watts *et al.*, 1992), together with data derived from homozygous children (West *et al.*, 1985), supports the desirability of treatment.

Plasma cholesterol may be lowered by means of diet or drugs or a combination of both (Kwiterovich, 1990). Once treatment has been started regular estimations of plasma lipids should be made both to ensure the efficacy of the treatment procedure in lowering plasma cholesterol, and to check compliance. Initially, monthly measurements are desirable but after treatment has been well established 3- to 4-monthly checks may suffice. In general, compliance with treatment in the long term is not good (West *et al.*, 1980) and it is unwise to relax on follow-up estimations.

The drug therapy currently most likely to be used for children is an ion-exchange resin such as cholestyramine. Patients on long-term treatment with this drug should be receiving additional folic acid and should have their folate and vitamin D status and clotting function checked. The hydroxymethylglutaryl-CoA (HMGCoA) reductase inhibitors which appear to be so successful in adults are not yet licensed for use in young children, but the expectation is that they will become available and prove effective and more acceptable.

Homozygous FH

This generally responds poorly to diet and drug therapy and more radical measures are eventually required. Regular plasmapheresis is effective but burdensome (Thompson *et al.*, 1989). Liver transplantation is probably the treatment of choice (East *et al.*, 1986). Gene therapy is entirely experimental and likely to be some way in the future.

Familial combined hyperlipidaemia (HyperapoB)

This disorder, or group of disorders, thought to be inherited in an autosomal dominant manner, should probably now be designated by the term hyperapoB (Sniderman *et al.*, 1992). Most adult patients show moderate hypercholesterolaemia and modest hypertriglyceridaemia with elevation of both LDL and VLDL, and HDL somewhat depleted. Plasma lipid levels are, however, quite variable and the critical abnormality is an increase in the number of LDL particles owing to excess hepatic production of apoprotein B. The disorder is associated with accelerated atherosclerosis, and the children in affected families should therefore be investigated. However, only 10–20% of those with the gene defect will have hyperlipidaemia in childhood (most of these showing hypertriglyceridaemia) (Kane & Havel, 1989) and apoB should preferably be measured. Follow-up must be maintained into adult life.

Familial type III hyperlipoproteinaemia

In this condition, which is inherited in an autosomal dominant manner, IDL accumulates in the plasma and results in the elevation of both cholesterol and triglyceride. The basic abnormality resides in apoprotein E, a mutant form of which cannot bind normally to lipoprotein receptors. Neither clinical nor biochemical expression during childhood is common; clinical suspicion may be alerted by the presence of xanthomata in the palmar creases. The disorder is associated with premature atherosclerosis causing both ischaemic heart disease and peripheral vascular disease. Obesity is often present and the lipoprotein abnormality responds to both weight reduction and clofibrate.

Familial hypertriglyceridaemia (type IV)

This disorder, characterized by an increase in VLDL with variable, but often quite marked, hypertriglyceridaemia and normal, or only slightly elevated, plasma cholesterol with normal apoprotein B, is relatively common in adults and is often associated with carbohydrate-rich diets, the taking of contraceptive pills or alcohol ingestion. Inheritance is probably as an autosomal dominant trait but the lipoprotein abnormality is rarely found in children before puberty, and becomes more common during adolescence. Children of affected parents should therefore be examined both at the time of parental diagnosis or concern and, if the lipid pattern is normal, again during adolescence and early adult life.

Familial type V hyperlipoproteinaemia

In this condition, which almost certainly represents more than one basic disorder, the lipoprotein pattern is complex with an increase of chylomicrons and VLDL. There is therefore marked hypertriglyceridaemia and hypercholesterolaemia with excess chylomicrons and VLDL. The clinical features resemble those of primary lipoprotein lipase deficiency, and attacks of acute abdominal pain owing to pancreatitis are a particular feature. The primary disorder appears to be very rare during childhood and has to be distinguished from lipoprotein lipase deficiency. Post-heparin lipolytic activity (adipose tissue lipoprotein lipase) is, however, normal and as the condition is inherited in a dominant manner family studies should help to clarify the diagnosis. Treatment is essentially similar to that for lipoprotein lipase deficiency.

Hyperalphalipoproteinaemia

With the more widespread measurement of HDL cholesterol concentrations it has become clear that in some children hypercholesterolaemia is the result of an increase in HDL rather than LDL. The strong inverse correlation of HDL with coronary heart disease in adults highlights the importance of determining whether raised total plasma cholesterol concentrations are due to increased HDL, so that inappropriate treatment and unnecessary anxiety are avoided.

Criteria for the diagnosis of familial hyperalphalipoproteinaemia (Glueck *et al.*, 1978) include plasma HDL cholesterol levels greater than 1.8 mmol/L, absence of diseases to which hyperalphalipoproteinaemia might be secondary, and detection of a similar pattern in a first-degree relative. The condition appears to be associated with reduced coronary heart disease and longevity. Population studies indicate that about half the HDL cholesterol variability is genetic (polygenic) and the other half environmental (Breslow, 1989).

Secondary hyperlipoproteinaemia

Raised concentrations of plasma cholesterol and/or triglyceride associated with a variety of different lipoprotein patterns may occur as a secondary manifestation in a number of diseases (Durrington, 1990) (Table 14.4). This possibility must be excluded in any child with hyperlipoproteinaemia, although in most instances the underlying condition will have already been diagnosed. Occasionally, however, hyperlipidaemia may be the initial feature, and its presence may influence further investigations and management.

The disorders most commonly associated with secondary hyperlipoproteinaemia in childhood are poorly controlled diabetes mellitus, hypothyroidism, the nephrotic syndrome, the hepatic glycogenoses and obstructive liver disease.

Diabetes mellitus

The poorly controlled diabetic child may have hyperlipidaemia with almost any type of lipoprotein pattern, although elevation of VLDL and hyperchylomicronaemia are most usual. The degree and nature of the hyperlipoproteinaemia do not necessarily correlate with the degree of glycaemic control, although the most gross forms of hyperlipidaemia are generally associated with ketoacidotic states. An excellent review of the changes in plasma lipids and lipoproteins and the mechanisms involved has been made by Taskinen (1990). In the presence of gross accumulation of chylomicrons or VLDL plasma electrolyte results may be fallacious, with apparent very low levels (Frier *et al.*, 1980), and if treatment is based on these misleading figures serious consequences can result. In any markedly lipaemic plasma the electrolytes should only be measured in the aqueous phase, the lipids having been previously extracted with an appropriate solvent.

Plasma lipid estimations are of limited help in the assessment of overall diabetic control and there would appear to be no indication for performing either cholesterol or triglyceride measurements routinely in the management of childhood diabetes. Nevertheless, in some individuals the finding of abnormally turbid plasma or hypercholesterolaemia will prompt a reappraisal of the insulin control. Occasionally, diabetes mellitus and a primary hyperlipoproteinaemic state will coexist and family studies should be undertaken if hyperlipidaemia persists in spite of improved control. Whilst hyperlipoproteinaemia undoubtedly plays a part in the pathogenesis of large blood-vessel disease, its role in relation to the diabetic microvascular complications remains controversial.

Hypothyroidism

Hypercholesterolaemia due to raised LDL levels is common, although not invariable, in untreated hypothyroidism in children, but is rarely found in infancy. Occasionally, levels of IDL are raised, resulting in hypertriglyceridaemia in addition to hypercholesterolaemia, and the plasma is then turbid. The lipoprotein pattern reverts to normal with treatment; unusually low plasma cholesterol can be an indication of excess thyroxine administration.

Renal disease

Hypercholesterolaemia is a classic feature of the nephrotic syndrome. Plasma LDL is increased and VLDL levels may also be raised, with associated hypertriglyceridaemia. There is a strong negative correlation between plasma triglyceride and albumin. Plasma lipoproteins return to normal with control of proteinuria and restoration of plasma albumin levels. Measurement of plasma lipids does not contribute to diagnosis or management and need not be performed routinely.

Hyperlipoproteinaemia may occur in other forms of renal disease but is less well documented. An increase in VLDL with moderate hypertriglyceridaemia and some hypercholesterolaemia may be found in children undergoing regular haemodialysis. It is suggested that plasma cholesterol and triglyceride concentrations should be measured regularly in such children, and in those with chronic renal insufficiency and likely to require dialysis, in order that dietary manipulation aimed at reducing the hyperlipidaemia may be attempted. Premature atherosclerosis is known to be associated with chronic renal disease in adults and occurs in adults on long-term haemodialysis.

Hepatic glycogenoses

Raised levels of VLDL with turbidity of the plasma and hypertriglyceridaemia are commonly found in hepatic glycogen storage disease owing to glucose-6-phosphatase, debrancher or phosphorylase deficiencies. Hypercholesterolaemia and some increase in LDL may also occur. The finding of turbid plasma and hyperlipidaemia in a child with hepatomegaly but no evidence of obstructive liver disease should prompt investigations for glycogen storage disease. Hyperlipidaemia is not a feature of the lipid storage disorders, with the exception of cholesterol ester storage disease, in which hypercholesterolaemia is common.

Obstructive liver disease

Hypercholesterolaemia and hypertriglyceridaemia associated with raised levels of LDL and/or VLDL and reduced concentrations of HDL are common in children with obstructive liver disease. When liver failure supervenes, however, all lipoprotein levels decline.

In intrahepatic biliary atresia an unusual type of hyperlipoproteinaemia may be found as the result of an accumulation of so-called lipoprotein X. This abnormal lipoprotein contains high proportions of cholesterol and phospholipid, and most of its cholesterol is in the unesterified form. Investigations show gross hypercholesterolaemia and hyperphospholipidaemia. LCAT activity is reduced, as are levels of HDL. Xanthomata associated with this type of hyperlipoproteinaemia have a rather characteristic flat appearance and occur on the palms, soles and mucous membranes, as well as on other skin areas.

Environmental hypercholesterolaemia and cholesterol screening

In the populations of most industrialized countries plasma cholesterol and LDL concentrations are higher than in poorer societies and this is probably due, at least in part, to the higher consumption of both total fat and saturated fatty acids in the richer nations. With growing recognition and acceptance of the fact that the atherosclerotic process originates in early life and may already be present in a significant number of young children (Stary, 1987), the role of plasma cholesterol in childhood as a risk factor for adult coronary heart disease requires consideration. This is not a new topic; as long ago as 1972 the American epidemiologists Kannel and Dawber were recommending that paediatricians should be 'detecting and treating hyperlipidaemia' in children, but this has not in fact happened to any extent in relation to the general population of children. This is due in part to uncertainty that measures taken to lower cholesterol concentrations in childhood would actually delay or prevent adult coronary disease, and in part to the difficulty of defining an appropriate cut-off point for plasma cholesterol (or LDL), above which children would be labelled as 'at risk'. Although the WHO (1982) proposed an ideal mean of 2.85 mmol/L (SD 0.52 mmol/L) for total plasma cholesterol in children aged 5–18 years, they accepted a value of 3.62 mmol/L (SD 0.65 mmol/L) as a feasible mean. Even this would result in a large number of children being identified if a figure of 4.92 mmol/L (2 SDs above this mean) were adopted (WHO Expert Committee, 1990). The 95th centiles for plasma cholesterol in the USA (Kwiterovich, 1990) of 5.32 mmol/L for boys (5.29 mmol/L for girls) and in Australia (Boulton *et al.*, 1991) of 5.82 mmol/L for boys (5.90 mmol/L for girls) are considerably in excess of WHO recommendations.

Screening

Pressure undoubtedly exists to introduce plasma cholesterol screening programmes in childhood, and the increasing availability of rapid 'desk-top' cholesterol analysers means that this may be done in an uncontrolled manner. Whilst there is general agreement that targeted screening of children in families known to have hyperlipoproteinaemia and/or early-onset coronary heart disease is justified, population screening remains controversial (Newman *et al.*, 1990; Lloyd, 1991; Stuhldrener & Orchard, 1991). Before any population screening can be introduced, assurance is needed that the degree of tracking is sufficient, an appropriate cut-off point has been agreed and, most importantly, management and follow-up arrangements for those identified as at risk are in place. The formidable problems still outstanding are discussed in a number of recent publications (Boulton *et al.*, 1991; Lloyd, 1991; Arneson *et al.*, 1992; Lauer, 1992). The Expert Panel of the National Cholesterol Education Program (NCEP) (1992) in the USA has carefully considered the advantages and disadvantages of universal screening and decided not to recommend it; it is the opinion of this author that the criteria for total population screening are not met and its introduction should be resisted. Research on all aspects is needed. Recommendations to lower the plasma cholesterol levels of all children, and thereby move the population distribution curve to the left, are justified, although the effects of the dietary changes needed in order to do this should be monitored (American Academy of Pediatrics, Committee on Nutrition, 1986).

Hypolipoproteinaemia

As with hyperlipoproteinaemia, hypolipoproteinaemia may be the result of a primary genetic disorder or be a secondary manifestation of another disease. Hypolipoproteinaemia affecting the LDL species, and therefore presenting as hypocholesterolaemia on routine analysis, is common in most disorders of fat malabsorption in children; cholesterol concentrations may be reduced to the order of around 2.0–2.5 mmol/L and levels rise as the underlying abnormality is treated. Hypocholesterolaemia may occur in hyperthyroidism, in hepatic failure and in association with certain chronic haemolytic disorders (e.g. thalassaemia); the mechanism in the latter condition is not yet understood. Reduction in HDL is a fairly common non-specific finding in many disease states and very low levels may be found in liver disease.

Primary hypolipoproteinaemias

Although most of the primary disorders causing hypolipoproteinaemia are rare, some (for example, abetalipoproteinaemia) are important to diagnose because treatment is possible, and all give insight into the metabolism and function of lipoproteins.

Deficiency of betalipoprotein

Since an absence of betalipoprotein (abetalipoproteinaemia) was first reported in 1960, it has become clear that there are at least three (and probably more) distinct disorders in which synthesis of LDL, is impaired (Kane & Havel, 1989).

Abetalipoproteinaemia

The term abetalipoproteinaemia should be confined to the recessively inherited disorder that is characterized clinically by fat malabsorption and a failure to thrive, present from birth, and by a pigmentary retinopathy and ataxic neuropathy, which develop in later childhood and adolescence, and in which it can be shown that both parents have normal concentrations of lipoproteins. The genetic defect does not involve the synthesis of apoprotein B (Talmud *et al.*, 1988) and is due to absence of microsomal triglyceride transfer protein (Wetterau *et al.*, 1992).

Investigation of the patient reveals the total absence of LDL, VLDL and chylomicrons from the plasma, acanthocytosis of the red cells, and vacuolation of the intestinal epithelial cells on jejunal biopsy. Presentation to the clinical chemistry laboratory is likely to be either because no other cause has been detected for fat malabsorption, or acanthocytosis has been reported, or because of unusual ophthalmological or neurological features. Acanthocytosis, although always arousing the suspicion of abetalipoproteinaemia, is not pathognomonic for this disorder (Herbert *et al.*, 1983). The plasma is completely clear, even in the postabsorptive state, total cholesterol is very low (usually below 1.3 mmol/L) and triglycerides are virtually undetectable. Confirmation of the diagnosis should be obtained by immunochemical estimation of LDL, and family studies (parental lipoproteins) are necessary to exclude homozygous hypobetalipoproteinaemia (see below). Acanthocytes, if not already visualized, should be looked for in a fresh wet preparation of blood, and intestinal biopsy will show the typical appearances. Because of the fat malabsorption and lack of LDL, fat-soluble vitamins are depleted and estimations of plasma vitamins A, D, E and K (by prothrombin time) are required. Levels of vitamins A and E are likely to be very low and the prothrombin time is likely to be prolonged. Increased peroxide haemolysis of the red cells is a sensitive test of vitamin E function at the cellular level.

Treatment comprises reduction of fat intake and administration of large doses of vitamins A and E, with the probable need for additional vitamin K. As it is now established that the neurological sequelae can be alleviated, and even prevented, by prolonged therapy with vitamins A and E (Muller & Goss-Sampson, 1990), plasma vitamin levels should be monitored regularly, although levels of vitamin E will always be low because of the lack of the carrier protein (LDL). It should be remembered that a lack of vitamin E causing neuropathy may occur in any condition resulting in severe fat malabsorption but especially in chronic cholestasis (Guggenheim *et al.*, 1982; Goss-Sampson *et al.*, 1989), and that spinocerebellar disease due to vitamin E deficiency can occur without steatorrhoea (Sokol *et al.*, 1988). Clinical chemistry laboratories are therefore likely to be asked to estimate this vitamin more frequently.

Familial hypobetalipoproteinaemia

This is a dominantly inherited disorder which is characterized by decreased LDL and cholesterol levels in the plasma. The basic defect results in truncated forms of the *apoB* gene and a large number of different mutations have now been described (Young, 1990; Gabelli, 1992). The different classes of mutation give rise to variable clinical features but homozygous familial hypobetalipoproteinaemia can be indistinguishable, both clinically and biochemically, from abetalipoproteinaemia, and management should be the same.

In the heterozygous state LDL is reduced to about half its normal concentration; plasma cholesterol levels are of the order of 1.7–2.2 mmol/L. Many individuals may be asymptomatic with normal fat absorption, normal red cell morphology and no neuro-ophthalmological features. The diagnosis is then established from the plasma lipoprotein pattern, the finding of a similar abnormality in at least one first-degree relative and the absence of other disease to explain the hypolipoproteinaemia. In some patients, however, failure to synthesize chylomicrons results in malabsorption; in others, acanthocytosis or neurological abnormalities may be found.

Normotriglyceridaemic abetalipoproteinaemia

In 1981, Malloy and colleagues described a disorder in which plasma LDL and VLDL are absent but chylomicrons can be formed and fat absorption is normal. Acanthocytosis is not a feature. The finding of severe hypocholesterolaemia and absent LDL in the presence of turbid plasma after a meal containing fat, with chylomicrons visible as a cream layer on standing, is therefore an indication for more detailed studies of lipoprotein metabolism. It has been postulated that the molecular defect in this condition is selective deletion of the hepatic form of apoB (B100) with retention of the intestinal form (B48) (Malloy *et al.*, 1981). In 1991, Hardman *et al.* reported the case of a patient with a truncated variant (apoB50) associated with VLDL which is very rapidly removed from the circulation.

Chylomicron retention disease

This disorder appears to be inherited in an autosomal recessive manner. Chylomicrons cannot be formed and fat accumulates in the enterocytes, probably owing to a defect in the intestinal processing of apoprotein B48 (Levy *et al.*, 1987; Roy *et al.*, 1987).

Apoprotein B100 is synthesized normally but total LDL and apoB levels in the plasma are reduced by about 50%. Patients have severe steatorrhoea and growth failure; plasma vitamin E concentrations are either undetectable or markedly reduced, and neurological complications occur without treatment. Treatment should therefore include a low-fat diet and adequate amounts of vitamins A and E, as for abetalipoproteinaemia.

Deficiency of alphalipoprotein

A number of syndromes have now been described in which there is a genetically determined deficiency of HDL (Schmitz *et al.*, 1990). The clinical manifestations are variable and not all are associated with accelerated atherosclerosis.

Tangier disease

Tangier disease (Assman *et al.*, 1989) was the first disorder recognized to be due to deficiency of HDL. Inherited as an autosomal recessive trait, it is characterized clinically by the accumulation of cholesterol (mainly in esterified form) in the reticuloendothelial system (especially the tonsils), by corneal opacities, and by an unusual peripheral neuropathy. Any of these clinical features may prompt an examination of plasma lipids, which will show low levels of total cholesterol (around 2.0–2.5 mmol/L), modest hypertriglyceridaemia and a virtual absence of HDL. Ultracentrifugal and immunochemical studies should be undertaken to confirm the HDL deficiency, and family studies may reveal reduced HDL levels in the heterozygotes. Other causes of HDL deficiency must be excluded, especially LCAT deficiency (see below) and liver diseases.

Familial deficiency of LCAT

Familial deficiency of LCAT (McIntyre, 1988), another recessively inherited disorder, presents clinically with proteinuria, a mild normochromic anaemia and corneal opacities. The renal lesion progresses to renal failure and most patients present in adult life, although presumably the defect is detectable in childhood. Plasma levels of total cholesterol and triglyceride are variable and HDL is very low. Most of the cholesterol is in the unesterified form and if LCAT deficiency is suspected, the determination of unesterified cholesterol is an important initial investigation. Further studies include determination of the enzyme activity and characterization of the HDL abnormality.

Fish-eye disease

Fish-eye disease (McIntyre, 1988) is characterized by striking corneal opacities but without anaemia or renal anomalies. HDL is reduced and LCAT activity is abnormal owing to a point mutation in the gene.

References

American Academy of Pediatrics, Committee on Nutrition (1986) Prudent life style for children: dietary life style and cholesterol. *Pediatrics* **78**, 521–525.

Arneson, T., Leupker, R., Pirie, P. & Sinaiko, A. (1992) Cholesterol screening by primary care paediatricians. *Pediatrics* **89**, 502–505.

Assmann, G., Schmitz, G. & Brewer, H.B. Jr (1989) Familial high density lipoprotein deficiency: Tangier disease. In: *The Metabolic Basis of Inherited Disease* (eds C.R. Scriver, A.L. Beaudet, W.S. Sly & D. Valle), 6th edn, pp. 1267–1282. McGraw-Hill, New York.

Beaumont, J.L., Carlson, L.A., Cooper, G.R. *et al.* (1970) Classification of hyperlipidaemias and hyperlipoproteinaemias. *Bull WHO* **43**, 891–915.

Berenson, G.S. (1980) *Cardiovascular Risk Factors in Children*. Oxford University Press, Oxford.

Boulton, T.J.C., Magarey, A. & Cockington, R. (1990) Cholesterol from infancy to age thirteen: tracking and parent–child association. *Ann Nestlé* **48**, 70–76.

Boulton, T.J.C., Seal, J.A. & Magarey, A.M. (1991) Cholesterol in childhood: how high is OK? *Med J Aust* **154**, 847–850.

Breslow, J.L. (1989) Familial disorder of high density lipoprotein metabolism. In: *The Metabolic Basis of Inherited Disease* (eds C.R. Scriver, A.L. Beaudet, W.S. Sly & D. Valle), 6th edn, pp. 1251–1266. McGraw-Hill, New York.

Brown, M.S. & Goldstein, J.L. (1986) A receptor-mediated pathway for cholesterol homeostasis. *Science* **232**, 34–47.

Durrington, P.N. (1990) Secondary hyperlipidaemia. In: *Lipids and Cardiovascular Disease* (eds D.J. Galton & G.R. Thompson), *Br Med Bull* **46**, 1005–1024. Churchill Livingstone, Edinburgh.

East, C., Grundy, S.M. & Bilheimer, D.W. (1986) Normal cholesterol levels with lovastatin (Mevinolin) therapy in a child with homozygous familial hypercholesterolaemia following liver transplantation. *J Am Med Assoc* **256**, 2843–2848.

Frier, B.M., Steer, C.R., Baird, J.D. & Bloomfield, S. (1980) Misleading plasma electrolytes in diabetic children with severe hyperlipidaemia. *Arch Dis Child* **55**, 771–775.

Gabelli, C. (1992) The lipoprotein metabolism of apolipoprotein B mutants. *Curr Op Lipidol* **3**, 208–214.

Gagne, C., Brum, L.D., Julien, P. *et al.* (1989) Primary lipoprotein lipase deficiency: clinical investigation of a French Canadian population. *Can Med Assoc J* **140**, 405–411.

Glueck, C.J., Kelly, K., Mellies, M.J., Gartside, P.S. & Steiner, P.M. (1978) Hypercholesterolaemia and hyperalphalipoproteinaemia in school children. *Pediatrics* **62**, 478–487.

Goss-Sampson, M., Muller, D.P.R. & Lloyd, J.K. (1989) Clinical importance of vitamin E: a review. *J Hum Nutr Diet* **2**, 145–150.

Guggenheim, M.A., Ringel, S.P., Silverman, A. & Grabert, B.E. (1982) Progressive neuromuscular disease in children with chronic childhood cholestasis and vitamin E deficiency: diagnosis and treatment with alpha tocopherol. *J Pediatr* **100**, 51–58.

Hardman, D.A., Pullinger, C.R., Hamilton, R.L. *et al.* (1991) Molecular and metabolic basis for the metabolic disorder normotriglyceridaemic abetalipoproteinaemia. *J Clin Invest* **88**, 1729–1772.

Havel, R.J. & Kane, J.P. (1989) Structure and metabolism of plasma lipoproteins. In: *The Metabolic Basis of Inherited Disease* (eds C.R. Scriver, A.L. Beaudet, W.S. Sly & D. Valle), 6th edn, pp. 1129–1138. McGraw-Hill, New York.

Herbert, P.N., Assman, G., Gotto, A.M. & Fredrickson, D.S. (1983) Familial lipoprotein deficiency. In: *The Metabolic Basis of Inherited Disease* (eds J.B. Stanbury, J.B. Wyngaarden, D.S. Fredrickson, J.L. Goldstein & M.S. Brown), 5th edn, pp. 589–622. McGraw-Hill, New York.

Hobbs, H.H., Russell, D.W., Brown, M.S. & Goldstein, J.L. (1990) The LDL receptor locus in familial hypercholesterolaemia: mutational analysis of a membrane protein. *Ann Rev Genet* **24**, 133–170.

Humphries, S.E., Taylor, R. & Munroe, A. (1988) Resolution by DNA probes, of uncertain diagnosis of inheritance of hypercholesterolaemia. *Lancet* **ii**, 794–795.

Innerarity, T.L., Mahley, R.W., Weisgraber, K.H. *et al.* (1990) Familial defective apolipoprotein B-100: a mutation of apolipoprotein B that causes hypercholesterolaemia. *J Lipid Res* **31**, 1337–1349.

Johnson, W.J., Mahlberg, F.H., Rothblat, G.H. & Phillips, M.C. (1991) Cholesterol transport between cells and high density lipoproteins. *Biochem Biophys Acta* **1085**, 273–298.

Kane, J.P. & Havel, R.J. (1989) Disorders of the biogenesis and secretion of lipoproteins containing the B apolipoproteins. In: *The Metabolic Basis of Inherited Disease* (eds C.R. Scriver, A.L. Beaudet, W.S. Sly & D. Valle), 6th edn, pp. 1139–1164. McGraw-Hill, New York.

Kannel, W.B. & Dawber, T.R. (1972) Atherosclerosis as a pediatric problem. *J Pediatr* **80**, 544–554.

Kwiterovich, P.O. Jr (1990) Diagnosis and management of familial dyslipoproteinemia in children and adolescents: current issues in pediatric and adolescent endocrinology. *Pediatr Clin N Am* **37**, 1489–1523.

Kwiterovich, P.O., Levy, R.I. & Fredrickson, D.S. (1973) Neonatal diagnosis of familial type II hyperlipoproteinaemia. *Lancet* **i**, 118–122.

Lauer, R.M. (1992) Should children, parents and paediatricians worry about cholesterol? *Pediatrics* **89**, 509–510.

Lauer, R.M., Lee, J. & Clarke, W.R. (1988) Factors affecting the relationship between childhood and adult cholesterol: the Muscatine study. *Pediatrics* **82**, 309–318.

Leonard, J.V., Whitelaw, A.G.L., Wolff, O.H., Lloyd, J.K. & Slack, J. (1977) Diagnosing familial hypercholesterolaemia in childhood by measuring serum cholesterol. *Br Med J* **1**, 1566–1568.

Levy, E., Marcel, Y., Deckelbaum, R. *et al.* (1987) Intestinal apoB synthesis, lipids and lipoproteins in chylomicron retention disease. *J Lipid Res* **28**, 1263–1274.

Lloyd, J.K. (1991) Cholesterol: should we screen all children or change the diet of all children. *Acta Paed Scand* **373**, 66–72.

Low, L.C.K., Lau, K.S., Kung, A.W.C. & Yeung, C.Y. (1991) Phytosterolemia and pseudohomozygous type II hypercholesterolemia in two Chinese patients. *J Pediatr* **118**, 746–749.

McIntyre, N. (1988) Familial LCAT deficiency and fish eye disease. *J Inher Metab Dis* **11**, 45–56.

Malloy, M.J., Kane, J.P., Hardman, D.A., Hamilton, R.L. & Dalal, K.B. (1981) Normotriglyceridemic abetalipoproteinaemia: absence of the B-100 apolipoprotein. *J Clin Invest* **67**, 1441–1450.

Mishkel, A. (1976) Pseudohomozygous and pseudoheterozygous type II hyperlipoproteinemia. *Am J Dis Child* **130**, 991–993.

Muller, D.P.R. & Goss-Sampson, M.A. (1990) Neurochemical, neurophysiological, and neuropathological studies in vitamin E deficiency. *Crit Rev Neurobiol* **5**, 239–263.

Myant, N.B., Gallagher, J.J., Knight, B.L. *et al.* (1991) Clinical signs of familial hypercholesterolaemia in patients with familial defective apolipoprotein B-100 and normal low density lipoprotein receptor function. *Arteriosclero Thromb* **11**, 691–703.

National Cholesterol Education Program (NCEP) (1992) Expert panel on blood cholesterol levels in children and adolescents. *Pediatrics* **89**, 495–501; Part **2** (Suppl.), 525–577.

Newman, T.B., Bronner, W.S. & Hulley, S.B. (1990) The case against childhood cholesterol screening. *J Am Med Assoc* **264**, 3039–3043.

Roy, C.C., Levy, E., Green, P.H.R. *et al.* (1987) Malabsorption, hypocholesterolaemia, fat filled enterocytes with increased intestinal apoprotein B: chylomicron retention disease. *Gastroenterology* **92**, 390–399.

Santamara-Fojo, S. & Brewer, H.B. Jr (1991) The familial hyperchylomicronemia syndrome. New insights into underlying genetic defects. *J Am Med Assoc* **265**, 904–908.

Schmitz, G., Brüning, T., Williamson, E. & Nowicka, G. (1990) The role of HDL in reverse cholesterol transport and its disturbances in Tangier disease and HDL deficiency with xanthomas. *Eur Heart J* **11** (Suppl. E), 197–211.

Scriver, C.R., Beaudet, A.L., Sly, W.S. & Valle, D. (eds) (1989) *The Metabolic Basis of Inherited Disease*, 6th edn. McGraw-Hill, New York.

Sniderman, A., Brown, B.G., Stewart, B.F. & Cianflone, K. (1992) From familial combined hyperlipidemia to hyperapoB: unravelling the over production of hepatic apolipoprotein B. *Curr Op Lipidol* **3**, 137–142.

Sokol, R.J., Kayden, H.J., Bettis, D.B. *et al.* (1988) Isolated vitamin E deficiency in the absence of fat malabsorption – familial and sporadic cases. Characterisation and investigation of causes. *J Lab Clin Med* **III**, 548–559.

Stary, H.C. (1987) Macrophages, macrophage foam cells, and eccentric internal thickening in the coronary arteries of young children. *Atherosclerosis* **64**, 91–108.

Stuhldrener, W.L. & Orchard, T.J. (1991) Use of routine cholesterol testing in childhood to classify risk status. *Curr Op Pediatr* **3**, 681–687.

Talmud, P.J., Lloyd, J.K. & Muller, D.P.R. (1988) Genetic evidence that the apolipoprotein B gene is not involved in abetalipoproteinaemia. *J Clin Invest* **82**, 1803–1806.

Taskinen, M.R. (1990) Hyperlipidemia in diabetes. *Clin Endocrinol Metab* **4**, 743–775.

Thompson, G.R., Barbir, M., Okabayashi, K. *et al.* (1989) Plasmapheresis in familial hypercholesterolaemia. *Arteriosclerosis* **9** (Suppl. 1), 152–157.

Tybjaerg-Hansen, A., Gallagher, J., Vincent, J. *et al.* (1990) Familial defective apolipoprotein B-100: detection in the United Kingdom and Scandinavia, and clinical characteristics of 10 cases. *Atherosclerosis* **80**, 235–242.

Watts, G.F., Lewis, B., Brunt, J.N.H. *et al.* (1992) Effects on coronary artery disease of lipid-lowering diet, or diet plus cholestyramine. In: St Thomas's Atherosclerosis Regression Study (STARS). *Lancet* **339**, 563–569.

West, R.J., Gibson, P.J. & Lloyd, J.K. (1985) Treatment of homozygous familial hypercholesterolaemia: an informative sibship. *Br Med J* **2**, 1079–1080.

West, R.J., Lloyd, J.K. & Leonard, J.V. (1980) Long-term follow-up of children with familial hypercholesterolaemia treated with cholestyramine. *Lancet* **ii**, 873–875.

Wetterau, J.R., Aggerbeck, L.P., Bouma, M.-E. *et al.* (1992) Absence of microsomal triglyceride transfer protein in individuals with abetalipoproteinemia. *Science* **258**, 999–1000.

WHO (1982) Prevention of coronary heart disease. *WHO Technical Report Series* **678**. WHO, Geneva.

WHO Expert Committee (1990) Prevention in childhood and youth of adult cardiovascular diseases: time for action. *WHO Technical Report Series* **792**. WHO, Geneva.

Young, S.G. (1990) Recent progress in understanding apolipoprotein B. *Circulation* **82**, 1574–1594.

15: Diabetes Mellitus

D.J. BETTERIDGE

Introduction

The child with diabetes, the most common endocrine disorder of childhood, represents a considerable clinical challenge to the physician. In addition to alleviating symptoms, the physician, with the collaboration of parents and child, has to negotiate a particularly difficult ridge between 'good' metabolic control on the one side (to ensure normal growth and development, and to prevent, or at any rate delay, long-term diabetic complications) and on the other side the avoidance of severe hypoglycaemia, which children fear most and which marks them as different from their peers. This requires extensive child and parent education by the diabetes care team and the appropriate use of insulin, diet and exercise. These difficult clinical management considerations have to be pursued within the important constraint of ensuring as near normal a life as possible for the diabetic child. In many hospitals separate paediatric diabetic clinics have been established, enabling joint management by the specialist diabetologist and the paediatrician. The advent of the diabetic specialist nurse has revolutionized the education process of the parents and child and provides a point of contact for advice between clinic visits.

Assessment of biochemical parameters plays a central role in the everyday management of diabetes and during acute metabolic decompensation, particularly diabetic ketoacidosis. This involves the chemical pathology laboratory, but increasingly biochemical tests are performed at home and at the bedside and this has attendant quality control problems.

This chapter outlines the definition and classification of diabetes mellitus, the recent developments in understanding of the pathogenesis of the disorder, the clinical course of diabetes, the short- and long-term complications and the contribution of laboratory, bedside and home biochemical analyses in the overall management of this disease.

Definitions and classification

The term diabetes mellitus has been used extensively for many years but it remains difficult to provide a succinct definition that covers the protean features of this important condition. Furthermore, mystery still surrounds the aetiology and pathogenesis of the major forms of the disease. Various classifications have been suggested over the years and in the 1970s developing knowledge in relation to the association of insulin-dependent diabetes with certain human leucocyte antigen (HLA) types and the presence of islet cell antibodies pointing to a different pathology led to a reclassification by the National Diabetes Data Group (1979) in the USA; this was later adopted by the World Health Organization (WHO Expert Committee on Diabetes Mellitus, 1980; WHO Study Group, 1985). The WHO classifications of primary diabetes mellitus and glucose intolerance are shown in Table 15.1. Other types of diabetes are shown in Table 15.2.

The diagnosis of diabetes mellitus is often clear, with the presence of the classic symptoms of thirst, polyuria and weight loss with glycosuria and an elevated random blood glucose. Diabetes mellitus, particularly non-insulin-dependent diabetes mellitus (NIDDM), is often diagnosed incidentally when urine is tested routinely or a random blood glucose is performed as a routine procedure along with other investigations. The WHO has provided guide-

Table 15.1 Classification of diabetes mellitus and glucose intolerance

Clinical classes
Insulin-dependent diabetes mellitus (IDDM)
Non-insulin-dependent diabetes mellitus (NIDDM)
  Non-obese
  Obese
Non-insulin-dependent diabetes mellitus of youth
Malnutrition-related diabetes mellitus
Gestational diabetes mellitus
Other types of diabetes mellitus associated with specific syndromes and conditions (see Table 15.2)

Statistical risk classes
Previous abnormality of glucose tolerance, e.g. women with gestational diabetes but normal glucose tolerance post partum
Potential abnormality of glucose tolerance, e.g. individuals with islet cell antibodies or HLA identical siblings of diabetic patients

HLA, human leucocyte antigen.

lines for the interpretation of these random blood glucose tests (Table 15.3). In the majority of cases that fall in the 'diabetes likely' category, a formal oral glucose tolerance test (OGTT) should be performed. Similarly, caution should be exercised in the interpretation of fasting blood glucose tests as many diabetics have normal fasting blood glucose values at diagnosis (Harris *et al.*, 1987).

The oral glucose tolerance test

The OGTT is recommended for confirmation of the diagnosis of diabetes mellitus when this is not readily apparent. Urine testing for glucose and measurement of glycated proteins are not used for diagnostic purposes. In most cases of diabetes the diagnosis is readily apparent and a formal OGTT is not necessary. However, in borderline cases the OGTT is performed because the labelling of a patient as having diabetes mellitus has important consequences. The diagnostic criteria recommended by the WHO, based on the OGTT (75-g load), are shown in Table 15.4.

General guidelines for the performance of an OGTT include the following.

1 The test should be performed in the morning

Table 15.2 Other types of diabetes mellitus

Diabetes due to pancreatic disease
Pancreatitis
Haemochromatosis

Diabetes due to certain drugs
Glucocorticoids
Diuretics
Diazoxide
Catecholaminergic agents

Diabetes due to endocrine disease
Cushing's syndrome
Acromegaly
Thyrotoxicosis
Phaeochromocytoma
Glucagonoma

Diabetes associated with genetic syndromes
Didmoad syndrome
Myotonic dystrophy
Lipoatrophy
Cystic fibrosis
Glycogen storage disease type I

Diabetes due to insulin or insulin receptor abnormalities
Insulinopathies
Insulin receptor defects
Anti-insulin receptor antibodies

From *Diabetes Mellitus*. Report of a WHO Study Group, 1985.

Table 15.3 Interpretation of random glucose tests

Specimen	Blood type	Glucose (mmol/L)	Occurrence of diabetes
Whole blood	Venous	≥10	Likely
		4.4–10	Uncertain
		<4.4	Unlikely
	Capillary	≥11.1	Likely
		4.4–11.0	Uncertain
		<4.4	Unlikely
Plasma	Venous	≥11.1	Likely
		5.5–11.0	Uncertain
		<5.5	Unlikely
	Capillary	≥12.2	Likely
		5.5–12.1	Uncertain
		<5.5	Unlikely

From *Diabetes Mellitus*. Report of a WHO Study Group, 1985.

Table 15.4 Diagnostic criteria for the oral glucose tolerance test (OGTT)

(a) Diabetes mellitus

		Glucose (mmol/L)	
Specimen	Blood type	Fasting	2-h sample
Whole blood	Venous	≥6.7	≥10
	Capillary	≥6.7	≥11.1
Plasma	Venous	≥7.8	≥11
	Capillary	≥7.8	≥12.2

(b) Impaired glucose tolerance

		Glucose (mmol/L)	
Specimen	Blood type	Fasting	2-h sample
Whole blood	Venous	<6.7	6.7–10.0
	Capillary	<6.7	7.8–11.1
Plasma	Venous	<7.8	<7.8–11.1
	Capillary	7.8–11.1	8.9–12.2

From *Diabetes Mellitus*. Report of a WHO Study Group, 1985.

after at least 3 days of unrestricted diet (>150 g carbohydrate/day).

2 The patient should have fasted for at least 10 h and no more than 16 h.

3 The patient should remain seated and refrain from smoking during the test.

4 An appropriate loading dose of glucose for children should be administered as 1.75 g/kg body weight to a maximum of 75 g.

5 The glucose load should be consumed over 5 min.

6 Blood samples should be obtained fasting and at 2 h after the glucose load. Although not required for the diagnostic criteria, samples at intermediate times provide useful confirmatory data.

7 Samples should be collected into fluoride-oxalate tubes and sent to the laboratory for analysis. Bedside tests with dry chemistry strips should not be used.

8 Blood taken from a vein is preferable for diagnostic purposes.

Children rarely require an OGTT to be performed because they usually present with severe symptoms and marked hyperglycaemia. However, some children and adolescents with less severe symptoms may require an OGTT. The recommended diagnostic criteria are the same for children as for adults.

The great majority of children and adolescents with diabetes will have classical insulin-dependent diabetes mellitus (IDDM). A small minority will have NIDDM presenting at a young age – maturity-onset diabetes of the young (MODY). More rarely, hereditary insulin resistance syndromes will be encountered in childhood, and diabetes (often not requiring insulin therapy) may be associated with chromosomal defects, single gene defects and miscellaneous endocrine and other syndromes (see Table 15.2).

Epidemiology and aetiology of IDDM

For a variety of logistic reasons incidence and prevalence data available for IDDM are mainly based on children and young adults, the group in which the condition is most prevalent. There are wide variations in IDDM incidence; the countries of northern Europe have the highest incidence rates, particularly Finland (approximately 30 in 100 000). Japan, on the other hand, has a very low incidence (about 1 in 100 000). In some countries, including Scotland and Finland, the incidence of IDDM appears to be rising, whilst in other countries, such as the USA, there has been no change (Diabetes Epidemiology Research International, 1987). Two interesting features emerge from epidemiological studies of IDDM. There is a marked seasonal variation in presentation, the highest incidence being in the autumn and winter months and the lowest in the spring and summer. In addition, the incidence of IDDM varies with age, with the peak incidence at 11–13 years.

Leslie *et al.* (1989) have provided an up-to-date succinct account of the aetiology of IDDM. There is no doubt that genetic influences are important but not overwhelming in IDDM; in twin studies only approximately 36% of identical twins of diabetic individuals develop the disease (Leslie & Pyke, 1986). In addition, only 10% of siblings of patients become diabetic. Population studies have demonstrated that the HLA genes DR_3 and DR_4 in the class II region on the short arm of chromosome 6 with HLA *DQ* genes linked to the *DR* genes are associated with diabetes (Michelson & Lernmark, 1987). Those individuals with HLA $DR_{3/4}$ heterozygosity are at higher risk than those with DR_3 or DR_4 alone, whereas DR_2 protects against the disease, suggesting that several genes are involved in the susceptibility to IDDM. However, the HLA genes are not specific for IDDM as approximately 60% of the normal population has them, and the relative risk of developing IDDM associated with the HLA genes is small. Furthermore, the association between HLA genes and IDDM is not seen in all populations. It has been estimated that the HLA associations account for approximately 60% of the genetic predisposition to IDDM. Other possible non-HLA gene associations with IDDM that are being pursued are the hypervariable region of the insulin gene on chromosome 11 (Bell *et al.*, 1986) and the constant region of the T-cell receptor β-chain gene on chromosome 77 (Millward *et al.*, 1987).

The fact that the genetic associations identified with increased susceptibility to IDDM involve genes coding for proteins involved in the immune response points to a role for the immune system in the pathogenesis of IDDM. This is supported by many strands of evidence including pancreatic histology in individuals with IDDM. Pancreatic specimens examined from individuals with long-standing IDDM show insulin deficiency owing to the absence of B cells in the islets of Langerhans. Interestingly, pancreatic A cells (glucagon secreting), D cells (somatostatin secreting) and PP cells (pancreatic polypeptide secreting) are preserved with normal numbers and distribution of the cells within the islets. In pancreas obtained from individuals with recent-onset IDDM (<1 year) a different picture emerges. Many islets have no B cells but in others B cells are still present, and some show what has been termed insulitis (Gepts, 1965), characterized by infiltration with chronic inflammation cells, mostly cytotoxic/suppressor T cells and thought to represent autoimmune destruction. In addition to this pathological evidence of involvement of the immune process, circulating islet cell antibodies (ICA) are found in the majority of individuals with newly diagnosed IDDM (Bottazzo *et al.*, 1974; Lendrum *et al.*, 1975). The cytoplasmic antigens to which these antibodies are directed in both B cells and other cells of the islet are as yet unidentified. Autoantibodies to insulin (IAA) have also been demonstrated in newly diagnosed patients (Palmer, 1987). Both ICA and IAA may be present for many years prior to the clinical onset of IDDM, as shown by the Bart's Windsor prospective study, which suggests an ongoing autoimmune process damaging the B cells (Bottazzo *et al.*, 1986). Prospective family studies have also demonstrated that siblings of diabetic patients who share HLA haplotypes with the propositus develop ICA. Of those siblings with high titres of ICA or complement fixing ICA, approximately 75% will develop diabetes within 8 years (Tarn *et al.*, 1988).

It is thought that the major HLAs play a fundamental role in modulating the immune response. The class II antigens are important in the activation of T-helper lymphocytes and it is possible that DR_3 and DR_4, which are associated with an increased risk of IDDM, facilitate the presentation of B-cell autoantigens to T-helper lymphocytes, thereby promoting autoimmune damage. On the other hand, DR_2, which appears to protect against IDDM, may impede autoantigen presentation (Todd *et al.*, 1987).

There is considerable evidence that environmental factors, particularly viruses, play an important role in the development of IDDM in the genetically susceptible individual. It is possible that viruses or other environmental agents could lead to abnormal HLA class II expression on islet B cells, thereby exposing their cell surface antigen to T-helper lymphocytes. It is likely that cytokines are also involved in B-cell destruction.

Although much is known about the immune features of IDDM, much still needs to be learnt about the genetic determinants, environmental factors and cellular processes that ultimately lead to destruction of the islet B cells.

Epidemiology and aetiology of NIDDM

In the population as a whole, NIDDM is by far the most prevalent form of diabetes. In the United Kingdom the overall prevalence of NIDDM is approximately 1–2%. This figure increases to approximately 4% in those over 65 years. In Asian immigrants the corresponding figure is 16% (Mather & Keen, 1985). Other populations with high prevalence rates of NIDDM include several North American Indian tribes, particularly the Pima Indians of Arizona (30% of adults have the disease) (Bennett *et al.*, 1976), and certain populations in the Pacific region (Zimmet, 1979).

There is a strong genetic tendency to NIDDM, as demonstrated by twin and family studies. In identical twin studies the concordance rate for the disease is over 90% (Barnett *et al.*, 1981; Newman *et al.*, 1987). In some populations, for instance the Micronesian population of Nauru in the Pacific, there is evidence of autosomal dominant transmission (Kirk *et al.*, 1985). In some white and black families it has been possible to demonstrate an autosomal dominant pattern of inheritance wiiith affected individuals typically presenting at an early age (Tattersall, 1984). This has been termed MODY and will be described later.

Despite the strong genetic influences in NIDDM, which are likely to be polygenic in the majority of cases, the search for genetic markers has so far been unrewarding. It is likely that environmental factors, such as diet and obesity, do not initiate NIDDM but they are likely to lead to earlier clinical presentation. However, there is currently considerable interest in the relationship between central obesity, as demonstrated by measurement of the waist:hip ratio, and insulin resistance and hyperinsulinaemia. Dyslipidaemia, hypertension and increased vascular risk are often associated. Reaven (1988) has termed this cluster of disorders 'syndrome X' and he and others have suggested that insulin resistance is the central abnormality.

Pancreatic histology in NIDDM differs considerably from that previously described in IDDM. Most studies indicate that the volume of insulin-producing B cells is reduced to about 50% (Rahier *et al.*, 1983). On the other hand, the number of A cells (glucagon secreting) is approximately doubled (Stefan *et al.*, 1982). A possible breakthrough in the search for clues to the pathogenesis of NIDDM has come from the observation of amyloid deposition in the islets of over 50% of NIDDM individuals studied (Clark *et al.*, 1984). Furthermore, a novel peptide, amylin, has been identified; this is produced by islet B cells in both normal and NIDDM subjects but is the basis of amyloid deposition in NIDDM islets (Westermark *et al.*, 1987). This phenomenon may reflect altered cell function and give a clue to the nature of the B-cell abnormality in NIDDM.

There has been considerable argument over the years regarding the fundamental metabolic abnormality in NIDDM with regard to the relative importance of B-cell dysfunction versus impaired insulin sensitivity, particularly in muscle and adipose tissue but also in the liver. Although it is highly probable that NIDDM will prove to be heterogeneous, it is clear that abnormalities of B-cell function and impaired insulin sensitivity coexist in many individuals. As with B-cell function, the mechanism of reduced insulin sensitivity remains obscure. In terms of insulin secretion the most prominent abnormality in NIDDM is the loss or impairment of the first-phase insulin response. In most patients, however, this abnormality in insulin secretion dynamics in response to stimuli coexists with chronic hyperinsulinaemia.

Maturity-onset diabetes of the young

Attention was first drawn to this condition by Tattersall (1984), who noted atypical features in three women who had started insulin therapy before 1930. They had developed diabetes at the ages of 12, 14 and 22 years; it was possible to stop insulin therapy and substitute oral hypoglycaemic agents, and long-term diabetic complications were rare. Each individual had a strong family history, the family members showed a similar clinical picture, and the family trees suggested autosomal dominant inheritance. Since this original description, more families have been reported and, not surprisingly, more heterogeneity in the clinical features has been noted. According to Tattersall (1991) this condition is best defined as 'hyperglycaemia diagnosed before 25 years of age and treatable for at least 5 years without insulin in patients without immune or HLA markers of IDDM'. MODY appears to be more common in American black people (Winter *et al.*,

1987) and Asians, either in southern India (Mohan *et al.*, 1985) or living in South Africa (Asmal *et al.*, 1981), than in Caucasians. An interesting feature of the initial three families described was the rarity of diabetic complications. This has not been a consistent feature of subsequent reports.

Well-delineated families showing the clinical and genetic characteristics of MODY afford the opportunity for the identification of the genetic defect(s). No HLA associations have been described and no relationship to polymorphisms of the insulin gene has been identified. However, recent reports point to a possible association of MODY with a glucose transporter gene (Bell *et al.*, 1991; Vionnet *et al.*, 1992). These are very exciting findings and may provide clues to the genetic defects of NIDDM in general.

Chromosomal defects associated with diabetes mellitus

Some well-described syndromes characterized by chromosomal abnormalities associated with an increased risk of diabetes mellitus include Turner's syndrome, Klinefelter's syndrome and Down's syndrome.

In Turner's syndrome (about 1 in 2500 newborn females), which is characterized by a 45XO karyotype or 46XX/45XO mosaic pattern, bilateral 'streak' ovaries and other well-known clinical features, approximately 15% of young children have impaired glucose tolerance (Wilson *et al.*, 1987). In older individuals up to 60% are shown to have diabetes on OGTT (Nielson *et al.*, 1969a). Diabetes is often asymptomatic and does not require insulin therapy. It is associated with a diminished insulin response to a glucose challenge (van Companhout *et al.*, 1973).

In Klinefelter's syndrome, phenotypic males have a 47XXY karyotype or a 46XY/47XXY mosaic pattern. In addition to the well-known features, such as a eunuchoid appearance, gynaecomastia and small soft testes, about a quarter of individuals have diabetic OGTT (Nielson *et al.*, 1969b). The diabetes may be asymptomatic and is NIDDM in type. The increased insulin response to a glucose challenge tends to suggest insulin resistance in this condition, which is associated with decreased insulin binding to erythrocyte receptors (Breyer *et al.*, 1981).

In contrast to the mainly asymptomatic diabetes associated with Klinefelter's and Turner's syndromes, individuals with Down's syndrome (trisomy chromosome 21) often have IDDM. It has been estimated that the prevalence of diabetes in Down's syndrome children (<14 years) is 21 in 1000 (expected <1.3 in 1000) (Jeremiah *et al.*, 1973). Insulin therapy was required in about three-quarters of Down's syndrome patients with diabetes in one series of 88 patients below the age of 20 years (Milunski & Neurath, 1968). Several authors have pointed to the association of Down's syndrome with IDDM and primary hypothyroidism.

Cystic fibrosis

This common autosomal recessive condition (1 in 2000 live births) is associated with pancreatic exocrine failure in approximately 90% of cases. The pancreatic islets are not damaged until relatively late in the disease but 30% of patients develop glucose intolerance, with symptomatic diabetes in 1–2% rising to 13% in older patients (Schwachman, 1978).

Acute intermittent porphyria

In this autosomal dominant disorder of porphyrin metabolism, an impaired glucose tolerance and occasional frank diabetes mellitus may occur (Waxman *et al.*, 1967). In the majority of patients oral hypoglycaemic agents are not required, but if they are, glipizide and metformin are thought to be safe.

Insulin resistance syndromes

Several rare inherited disorders are associated with insulin resistance and hyperinsulinaemia. In the majority of cases this is relatively mild but in some there is severe insulin resistance. Those which present in childhood will be described briefly.

Prader–Willi syndrome

In this condition, which is characterized by obesity, hypogonadotrophic hypogonadism, mental retardation and muscular hypotonia, diabetes has been reported in up to 10% of cases. It is NIDDM in type with evidence of insulin resistance and decreased numbers of insulin receptors (Kousholt *et al.*, 1983). It is of interest to note that although the diabetes is usually 'mild' and responds to dietary measures and weight reduction, diabetic retinopathy has been described in an adolescent with this condition (Savir *et al.*, 1974).

Dystrophia myotonica

Impaired glucose tolerance associated with insulin resistance and hyperinsulinaemia is common (20–30%) in this autosomal dominant disorder, although, again, frank diabetes is rare (Hudson *et al.*, 1987). The typical clinical features, including myotonia, muscle wasting, cataracts and hypogonadism, point to the diagnosis.

Alström's syndrome

Impaired glucose tolerance and diabetes of the NIDDM type associated with hyperinsulinaemia and acanthosis nigricans is present in the majority of individuals (*c.* 90%) with this rare autosomal recessive disorder (Goldstein & Fialkow, 1973). Other clinical features include nerve deafness, retinitis pigmentosa and diabetes insipidus. Hypertriglyceridaemia and hyperuricaemia may occur.

Laurance–Moon–Biedl syndrome

Diabetes mellitus has been reported in about 6% of individuals with this rare autosomal recessive disorder, which is diagnosed by the constellation of mental retardation, obesity, abnormal digits, hypogonadism, retinitis pigmentosa and glomerular sclerosis.

Cockayne's syndrome

Impaired glucose tolerance and hyperinsulinaemia without frank diabetes mellitus have been described in this rare autosomal recessive disorder characterized by microcephaly, mental retardation, dwarfism, deafness, retinal degeneration, glomerulonephritis and skin atrophy. Hyperlipidaemia has also been reported (Fujimoto *et al.*, 1969).

Ataxia telangiectasia

Diabetes mellitus is more severe in this autosomal recessive condition with marked hyperglycaemia but without ketosis. Hyperinsulinaemia is present because of circulating antibodies which interfere with the binding of insulin to its receptor (Bar *et al.*, 1978). Acanthosis nigricans is also present, together with the characteristic telangiectasia, cerebellar ataxia and immune deficiency.

Leprechaunism

This rare congenital syndrome is characterized by an elfin-like facies, hirsutism, growth retardation, acanthosis nigricans, decreased subcutaneous fat and early death. Insulin resistance and hyperinsulinaemia are severe with hyperglycaemia in the postprandial state and paradoxically fasting hypoglycaemia. The pancreatic histology shows hyperplasia of the islets. No abnormality in the insulin molecule has been described but reduced insulin receptor affinity and number have been described together with decreased insulin-stimulated receptor kinase activity. These abnormalities have also been demonstrated in cell lines obtained from affected individuals and maintained in culture (Reddy *et al.*, 1988).

Rabson–Mendenhall syndrome

This very rare autosomal recessive condition is characterized by precocious puberty, pineal hyperplasia, hirsutism, dystrophic dentition and nails, and diabetes. Insulin resistance is very severe and hyperglycaemia and ketosis may not respond to huge doses of insulin (thousands of units) (West *et al.*, 1975). Decreased insulin receptor numbers and receptor affinity explain the resistance to insulin (Taylor *et al.*, 1983). It is of interest to note that there is an initial report showing that the administration of insulin-like growth factor 1 in this condition led to some reduction in the blood glucose level (Quin *et al.*, 1989).

Lipoatrophic diabetes

Insulin-resistant diabetes mellitus and the partial or complete absence of subcutaneous adipose tissue characterize this rare group of conditions.

Congenital generalized lipoatrophy can occur as a dominantly inherited disorder, Dunnigan syndrome, or as an autosomal recessive disorder, Seip–Berardinelli syndrome. Dunnigan reported two Scottish families with lipoatrophy affecting the trunk and limbs symmetrically but sparing the face. Other features included acanthosis nigricans and eruptive xanthomata typical of severe hypertriglyceridaemia (Dunnigan *et al.*, 1974). The autosomal recessive form is more common, and in addition to lipoatrophy other features include mental retardation, accelerated growth and acromegalic facies with thick skin and large hands and feet; the growth hormone levels are normal. The lipodystrophy is manifest in childhood and diabetes develops in the early teenage years. So far no explanation has been advanced for the insulin resistance. Other metabolic features include marked hypertriglyceridaemia. Very large doses of insulin are necessary to control the hyperglycaemia, although this may not be achieved.

A sporadic form of acquired generalized lipoatrophy has been described with onset in young children or teenagers and subsequent development of insulin-resistant diabetes – the Lawrence syndrome. The lipoatrophy is generalized and, again, severe hyperlipidaemia is a feature.

Progressive partial lipoatrophy affects subcutaneous adipose tissue on the face and trunk. It is usually manifest in childhood. Its development may be preceded by a viral illness. Some individuals have glomerulonephritis and features of systemic lupus erythematosus. Again, insulin-resistant diabetes is a feature.

Management of IDDM

Clinical presentation

The vast majority of children presenting with diabetes will have IDDM. However, the rare syndromes described above should be borne in mind together with the causes of secondary pancreatic damage. A minority of children (less than a third) will present with frank diabetic ketoacidosis, and this is described separately. Other children present with the classical symptoms of polyuria, polydipsia and weight loss. However, polydipsia may not be so readily appreciated, particularly in young children. Nocturia may present as enuresis in a child who has been previously toilet-trained. Polyphagia is sometimes present, the increased appetite being particularly for carbohydrate-rich foods. The presentation is usually subacute with symptoms appearing over days, and less frequently weeks. Occasionally, the onset is insidious and this may lead to a problem in diagnosis. In this situation the symptoms are non-specific with weakness, lethargy and weight loss. In the symptomatic child with a raised fasting or random glucose the diagnosis is clear. In addition, some degree of ketonuria is usually present. A formal OGTT is unnecessary.

With the development of community diabetes care with specialist diabetic nurses and health visitors it is usually possible in the child who is not ketotic to commence insulin therapy as an outpatient treatment. This approach has tremendous advantages in avoiding hospitalization. It is also more sensible to stabilize the blood glucose levels in the normal environment rather than in the artificial situation of a hospital ward. Education of the parents and child is a fundamental principle of treatment and again this is best performed away from the hospital environment.

Insulin treatment

By definition, IDDM patients require insulin therapy. There are many different types of insulin (Table 15.5) and insulin regimes, and these can be tailored to the individual. In the early stages of the disease there is often continuing β-cell function, as indicated by the presence of measurable C-peptide in the blood (Fig. 15.1), and in young children a single injection of an intermediate-acting insulin given before breakfast is often sufficient to achieve reasonable control. If post-breakfast blood glucose tests indicate a high postprandial excursion, a short-acting insulin can be added or, alternatively, a premixed insulin containing a proportion of soluble insulin can be given.

Table 15.5 Insulin preparations

Preparations	Timing of action (h) Onset	Peak	Duration
Short-acting			
Neutral soluble insulins: e.g. Actrapid, Humulin S, Velosulin, Hypurin Neutral, Pur-in-Neutral	0.25–1	2–4	5–8
Intermediate-acting			
Semilente, lente and isophane insulins:			
Semilente	0.5–1	4–6	8–12
Lente	2–4	6–10	12–24
Isophane	2–4	6–10	12–24
e.g. Insulatard, Semitard Protophane, Humulin I, Pur-in-Isophane, Hypurin Isophane, Monotard, Humulin Lente			
Long-acting			
Ultralente	3–4	14–20	24–36
Protamine zinc	3–4	14–20	24–36
e.g. Humal Ultratard, Hypurin Protamine Zinc			
Biphasic/premixed			
e.g. Penmix range, Humulin range, Human Mixtard, Initard and Actraphane, Pur-in-Mix range, Rapitard	The timing of action depends on constituent parts		

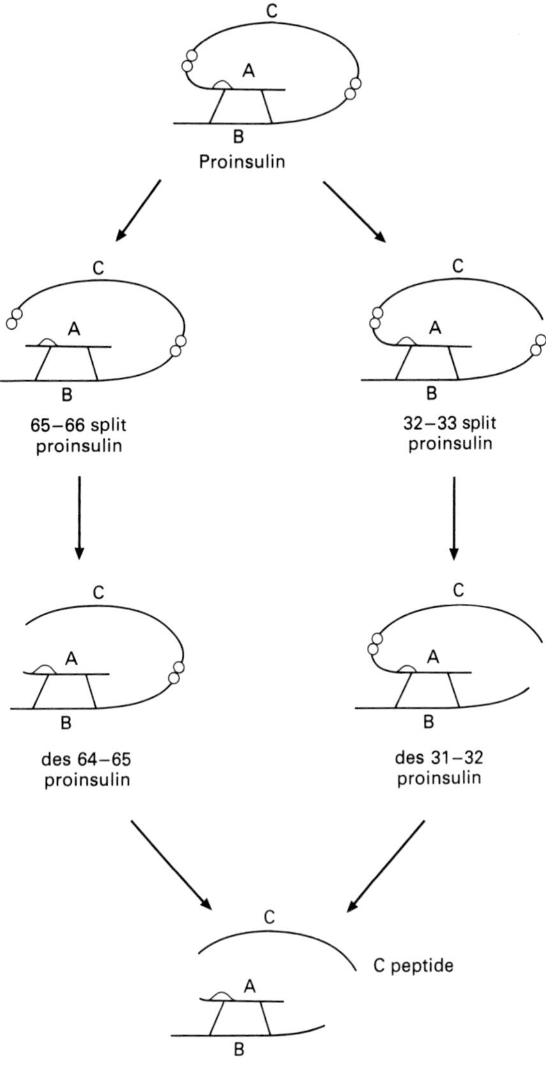

Fig. 15.1 The processing of proinsulin to insulin. (After Clark and Hales, 1991.)

In older children it is usual to start with two injections per day of a mixture of a short- and intermediate-acting insulin. The average insulin requirement ranges from 0.3 to 0.5 units insulin/kg body weight per day. Insulin treatment has certainly become more convenient over the last few years with the advent of pen injection devices. These devices, which use cartridges of insulin, remove the chore of drawing up insulin from the vial. In addition, there are now premixed insulins available and these can be given with the pen injection devices with a variable content of soluble insulin ranging from 10 to 50%.

A commonly used regime involves two injections per day of a mixture of regular- and intermediate-acting insulin given before breakfast and before the evening meal. This insulin regime theoretically gives four peaks of insulin activity; the soluble insulin component given before breakfast will cover the post-breakfast increase in glucose and the intermediate-acting insulin given at that time will begin to work around lunchtime and continue during the afternoon. Similarly, the quick-acting insulin given before the evening meal will cover the glucose excursion following this meal and the intermediate-acting insulin given at this time will cover the night-time period. In some individuals hypoglycaemia is a problem in the early part of the night. This can be overcome in many cases by splitting the evening injection and giving the soluble insulin before the evening meal and the intermediate-acting insulin at bedtime.

The pen injection devices have enabled multiple injections to be given more conveniently and there are now older children and adolescents taking what is known as the basal/prandial regime with an intermediateacting insulin at bedtime and then an injection of soluble insulin with each main meal: breakfast, lunch and dinner. This regime can give increased flexibility with regard to meal times and the variable lifestyle of adolescents.

It is very rare in children to have to resort to constant subcutaneous insulin infusion devices. These miniaturized insulin pumps, which deliver a constant insulin infusion and can be programmed to give additional boluses to cover meals, have been used more extensively in the United States but have not been popular in the United Kingdom. It is debatable whether they achieve better control than multiple injection regimes. Furthermore, they require more attention and a down-side is the rapid development of ketoacidosis in the event of pump failure.

Table 15.6 Dietary guidelines*

Carbohydrate	>55% of energy
Fat	<30–35% of energy (saturated fat <10%)
Protein	10–15% of energy
Salt	<6 g daily
Sucrose (added)	<25 g daily
Dietary fibre	>30 g daily

Energy intake should be tailored to achieve a BMI of 22.

$$BMI = \frac{\text{Wt in kg}}{(\text{Ht in m})^2}$$

* Data from Nutrition and Diabetes Group of the European Association for the Study of Diabetes (1988). Nutritional recommendations and principles for individuals with diabetes mellitus. *Diabet Nutr Metab* **1**, 145–149.

Diet

Insulin therapy must be balanced by appropriate dietary intake and at an early stage the child and parents need to have several discussions with a dietitian. It is important to make the dietary guidelines as simple and unrestricted as possible and broad guidelines should be given. The most important point to emphasize is the need for regular meals separated by mid-meal snacks. Refined sugar should be removed from the regular diet. The general dietary guidelines are given in Table 15.6.

Exercise

Exercise is the third main plank of management of IDDM and there is no doubt that regular exercise aids good glycaemic control. The child must be educated in how to cope with diabetes during exercise to avoid hypoglycaemia and occasionally the development of ketoacidosis. Planned exercise can be covered by taking additional carbohydrate in the form of readily assimilated glucose with or without a decrease in insulin dosage. In the child with poorly controlled diabetes, exercise can actually precipitate hyperglycaemia and occasionally diabetic ketoacidosis. Occasionally, hypoglycaemia occurs several hours after a period of exercise and it is important to educate the child and parents about the possibility of a delayed hypoglycaemic reaction. This can be avoided by taking additional carbohydrate for the main meal following the exercise, and occasionally an additional night-time snack is necessary.

Sick-day rules

An important part of the management of the diabetic child is to educate the child and parents in how to cope with diabetes during an intercurrent illness. Viral illnesses, particularly those affecting the upper respiratory tract or causing gastroenteritis, are frequent in children and cause hyperglycaemia. In these situations blood glucose monitoring should be increased and medical advice sought. The insulin dosage may have to be increased and certainly should not be reduced. A reduction of insulin during intercurrent illnesses is a common cause of diabetic ketoacidosis. Testing for ketones is useful in these situations. If anorexia and vomiting are a problem it is often possible to maintain carbohydrate intake by using glucose-containing drinks. However, hospitalization may be required for the child with persistent vomiting and diarrhoea.

Management during planned surgery

If surgical intervention is necessary in a child with IDDM it is usual to admit the child 24 h prior to the planned surgery. If possible, the child should be placed first on the morning list. On the day before surgery the usual insulin and meals can be taken. On the morning of surgery and at the time the child would normally have breakfast, an infusion of 5% dextrose is set up together with an infusion of soluble insulin. The infusion rate of insulin can be calculated from the child's usual insulin requirements. By these means the blood glucose can be kept constant during surgery and the postoperative period. If emergency surgery is required the combination of glucose and insulin infusion works very well.

Growth and development

Growth and development of secondary sexual characteristics proceed normally in the child whose diabetic control is reasonable. However, poor diabetic control can lead to growth retardation and delayed puberty, as can any other chronic illness of childhood. Fortunately, the full-blown Mauriac syndrome is now extremely rare and the sight of the stunted child with absent secondary sexual characteristics is hopefully a thing of the past. Obesity can be a problem in the adolescent child and weight reduction can be difficult. Careful attention to the insulin requirements is needed as obesity may be secondary to chronic overdosage of insulin. As mentioned previously, diet is an important part of therapy and the child is educated repeatedly in the appropriate diet. Because of this attention to diet it is perhaps not surprising that eating disorders, such as anorexia nervosa and bulimia, are seen more frequently in diabetic children.

Puberty

It is usual to see an increase in insulin requirements around the time of puberty. In addition, puberty in adolescents can lead to considerable management problems. This is often a troublesome time for parents and children and the adolescent process is not helped by the presence of diabetes, particularly if the diabetes has been present since infancy. There is no doubt that the summer camps organized by the British Diabetic Association for adolescent children help them in many ways in coping with the condition and learning from their peers. They are also a useful means of improving the adolescent's independence and in re-education.

Assessment of glycaemic control

There is a general consensus that the achievement of near normoglycaemia in diabetic patients will help to delay, and maybe prevent, the development of long-term diabetic complications. Although many diabetic individuals insist that they can 'feel' whether their blood glucose concentration is too high or too low, unfortunately this is not the case and objective measures of glycaemic control are necessary in patient management. Since the mid-1970s there has been a revolution in the approach to this problem and patients are encouraged in the self-monitoring of their blood glucose levels. In addition, most laboratories now provide effective measurements of more long-term glycaemic control.

Laboratory assessment of control

Glycated haemoglobin

The term glycated haemoglobin refers to haemoglobin that has been modified by the attachment of a glucose molecule involving a continuous slow non-enzymatic process. Several minor components – HbA1a, HbA1b and HbA1c – collectively known as HbA1 were first isolated in the 1950s using cation exchange chromatography (Allen *et al.*, 1958). The ionic charge of the haemoglobin is altered by the adduction of a molecule of glucose to the N-terminal valine residue of the β chain of haemoglo-

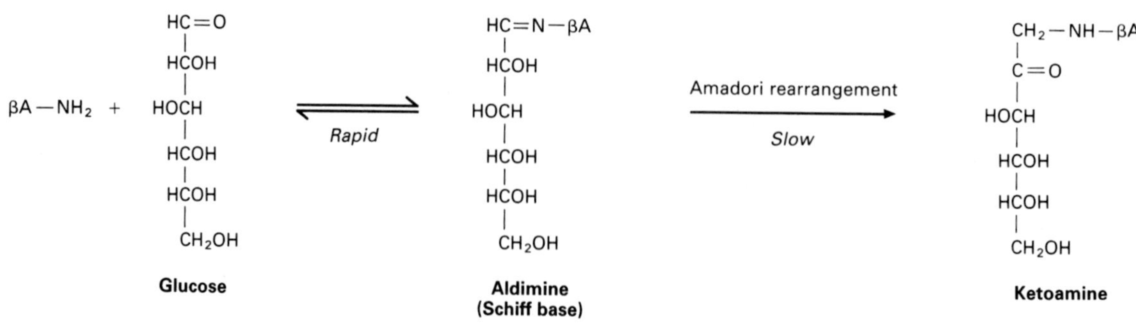

Fig. 15.2 Glycated haemoglobin. The addition of glucose to the N-terminal valine residue of the β chain of haemoglobin A to form haemoglobin $A1_c$.

bin through the formation of an aldimine (Schiff) base. This labile compound subsequently undergoes an Amadori rearrangement to form a stable ketoamine product (Fig. 15.2). In the mid-1970s glycated haemoglobin was synthesized by incubating whole blood with glucose (Flückiger & Winterhalter, 1976) and it was suggested that in diabetes the measurement of such glycated haemoglobin could reflect blood glucose concentrations over a period of 6–8 weeks, corresponding to the half-time of the red blood cell. Subsequent clinical studies confirmed this suggestion (Koenig *et al.*, 1976).

Several methods are available for the measurement of glycated haemoglobin; these include ion exchange chromatography, high performance liquid chromatography, agar gel electrophoresis, isoelectric focusing and affinity chromatography. Because of the wide variety of techniques available for the measurement of glycated haemoglobin, it is clear that individual diabetic clinics will need to familiarize themselves with the local laboratory's reference ranges.

Ion exchange chromatography, which depends on the separation of the haemoglobin by charge, was the first technique to be used in the measurement of HbA1 and now mini-column systems are available (Trivelli *et al.*, 1971; Welch & Boucher, 1978). As with all laboratory techniques, care has to be taken in this assay. Small changes in pH and temperature can affect the assay (Kortlandt *et al.*, 1985). Using this method the coefficient of variation is about 3%, while interassay variation is 4–5%.

High performance liquid chromatography is very precise and quick but is not available in most laboratories because of the expense of the equipment. Many laboratories use agar gel electrophoresis to separate HbA1 (Menard *et al.*, 1980); this method has somewhat higher intra-assay and interassay variations, up to 7%. Isoelectric focusing involves the application of the haemoglobin lysate to polyacrylamide gel containing ampholite and the haemoglobin components are quantified by microdensitometry (Spicer *et al.*, 1978; Simon & Cuan, 1982).

It is important to be aware of factors that can affect the measurement of glycated haemoglobin, giving either spuriously high or spuriously low levels. Measured HbA1 will be spuriously low if the red cell survival is reduced, such as will occur in blood loss or haemolysis (Horton & Huisman, 1965; Bernstein, 1980; Starkman *et al.*, 1983). In addition, if abnormal haemoglobins, such as haemoglobin S or haemoglobin C, are present HbA1 will be underestimated because these abnormal haemoglobins elute with the unglycated haemoglobin. On the other hand, the presence of haemoglobin F will lead to a spuriously high HbA1 as this coelutes with HbA1 (Eberentz-Lhomme *et al.*, 1984). If a particular clinic is dealing with a large proportion of patients who may well have haemoglobinopathies it would be an advantage to adopt an assay system such as affinity chromatography, which is unaffected by haemoglobin variants. Unfortunately, uraemia may be present in diabetic individuals as a consequence of diabetic nephropathy and this leads to modification of the haemoglobin by carbamylation. As this coelutes with HbA1 this can lead to spuriously high levels (Flückiger *et al.*, 1981).

It is the author's practice to perform a measurement of glycated haemoglobin at each clinic visit, i.e. every 3–6 months, in IDDM patients. This has the advantage of being objective and unaffected by the time of day and relationship to meals. It is useful for the test to be performed in the week prior to the clinic visit so that the result can be discussed at the time of the consultation. It is possible to develop collecting systems for the assay of glycated haemoglobin on finger-prick blood; this is particularly important when dealing with small children where routine venesection is difficult. The disadvantage of the measurement is that although it may indicate poor control, it is of no help in trying to improve the control. The test also remains relatively expensive and has a coefficient of variation higher than that of most other laboratory estimations. However, it has superseded the routine measurement of random blood glucose levels, which do not provide a useful assessment of control in the patient with IDDM. When postprandial glucose levels have been correlated to glycated haemoglobin measures, little relationship has been shown. A small but significant correlation is, however, obtained when the random glucose levels are averaged (Tchobroutsky *et al.*, 1980).

In NIDDM, rare in childhood and adolescence, a single blood glucose estimation does correlate well with overall control, as assessed by glycated haemoglobin, because in these individuals the glycaemic profiles show a similar pattern to those in the normal population, although they are set at a higher level with somewhat wider excursions in relation to meals (McCance *et al.*, 1988).

Fructosamine

Use has been made in some laboratories of the fact that other plasma proteins undergo glycation and, as these proteins have shorter half-times in the plasma, their glycation provides an overview of blood glucose concentrations over a much shorter time period (7–14 days) (Kennedy *et al.*, 1981). The fructosamine assay measures all glycosylated proteins but mainly reflects glycated albumin. In adult clinics it has been suggested that fructosamine may be useful in the management of diabetic pregnancy, where it is important to be able to monitor short-term changes. However, the fructosamine assay would seem to have little to recommend it in the management of childhood and adolescent diabetics. There is no doubt that the fructosamine assay, which is based on the ability of ketoamines to act as reducing agents in alkaline solution, is a cheap, reproducible and rapid assay with a low coefficient of variation of 2–3% (Johnson *et al.*, 1982). However, it has to be interpreted with care because of inherent variability (10–12%) owing to fluctuations in plasma proteins (McCance *et al.*, 1987).

Blood glucose estimation

A single isolated measurement of blood glucose yields little information about overall glycaemic control and has largely been abandoned in outpatient clinics. However, the ability of the patient or the patient's parents to measure blood glucose at home has revolutionized the day-to-day management of diabetes and has enabled day-to-day adjustments to glycaemic control.

Various techniques are available for the self-monitoring of blood glucose levels. Most diabetic patients use strip tests which are read visually against a colour chart. Well-known strip tests include the BM test 1-44 (Boehringer Mannheim (UK) Ltd, Lewes, East Sussex) and Glucostix (Bayer Diagnostics, Basingstoke, Hampshire, UK). There is no doubt that these have an acceptable accuracy and are more than adequate for day-to-day adjustments. However, it is not surprising that some patients will produce fictitious blood-test results for the doctor's benefit and this may be suspected if there is a discrepancy between the reported self-monitored blood glucose tests and the glycated haemoglobin.

For those patients who have difficulty with visual assessment of the strips small meters are available for use with various testing strips, which are read by either colour change or, more recently, electrochemical methods. Although there is no doubt that these tests are convenient for patients, whether or not they aid in improving diabetic control is a different matter. In themselves they probably do not, and they have to be combined with detailed education packages enabling the patient to act on the information obtained from self-monitoring.

Various meters have been used in medical and other wards in hospitals so that bedside tests can be performed. In this situation there is no doubt that mistakes can be made. Careful education of all staff in the performance of these tests is necessary and this is best organized by the chemical pathology department. In addition, frequent quality control assessments of the blood glucose machines are necessary.

Ketonuria

The tests for detecting a ketonuria, such as Acetest tablets, Ketostix (Bayer Diagnostics, Basingstoke, UK) or Keturtest (BM Pharmaceuticals, Lewes, UK), use the nitroprusside reaction, which does not detect β-hydroxybutyrate. This can be a problem in some cases of acidosis where β-hydroxybutyrate may be the ketone that is high. Some patients test for ketonuria to aid monitoring during intercurrent illness.

Urine testing

Testing urine for glucose is now more or less obsolete in the management of diabetic patients. The amount of glycosuria bears very little relationship to the prevailing blood glucose levels because of the very variable renal threshold for glucose. However, urine testing has a useful role in the initial screen for diabetes. The various ways for measuring urinary glucose are shown in Table 15.7. Clinitest is the most well established of these tests but is not specific for glucose and will detect other reducing agents. Newer tests are based on the specific glucose oxidase method for glucose. The tests for urinary glucose can be interfered with by several substances and these are shown in Table 15.8.

Diabetic ketoacidosis

Diabetic ketoacidosis (DKA) represents an urgent medical emergency in children and adolescents. Its onset can be extremely rapid and prompt diagnosis and management with correction of fluid and electrolyte disturbances and of the acidosis are necessary to prevent a potential life-threatening situation. A minority of children will present with DKA as

Table 15.7 Methods for testing urine for glucose

Method	Advantages	Disadvantages
Clinitest tablets (Bayer Diagnostics) Reduction of cupric sulphate to cuprous oxide by reducing agent such as glucose	Cheap	Inconvenient: test tubes and dropper needed Non specific: detects other reducing agents such as vitamin C, galactose, some drugs Insensitive: only reliable when urine glucose concentration >1 g/L Tablets are caustic and hygroscopic
Clinistix plastic reagent strip (Bayer Diagnostics) Glucose oxidase enzyme which converts glucose to glucuronic acid and hydrogen peroxide. The latter oxidizes a chromogen system in the presence of peroxidase	Specific for glucose	Not quantitative, gives 'low', 'medium' and 'high' readings
Diabur-Test 5000 plastic reagent strip (BM Pharmaceuticals) As above for Clinistix	Accurate and sensitive	Incubation time of two months
Diastix plastic reagent strip (Bayer Diagnostics) As above for Clinistix	Specific for glucose	Not easy to distinguish colours in poor light Colour development inhibited by high concentration of ketones

Table 15.8 Substances which may interfere with tests of glycosuria

Ascorbic acid
Cephalosporins
Nalidixic acid
Nitrofurantoin
Penicillin
Probenecid
Salicylates
X-ray contrast media

As a general rule, drugs produce false positives with Clinitest and false negatives with glucose oxidase test strips.

their first manifestation of diabetes mellitus. Unfortunately, a delay in diagnosis does sometimes occur in these children and often has tragic consequences. Management of DKA is a situation *par excellence* where close collaboration is required between the physician and the chemical pathology department in the assessment and monitoring of the condition, and is a prerequisite for the successful outcome of this severe metabolic emergency. DKA remains the most important cause of death in patients below the age of 20 years and accounted for 15% of deaths in diabetics below the age of 50 years (Tunbridge, 1981).

Definition

DKA is characterized by hyperglycaemia (glucose >20 mmol/L) associated with ketonaemia (>5 mmol/L), acidosis (pH < 7.3), bicarbonate <15 mmol/L, glycosuria, ketonuria and the classical clinical features. Differing degrees of impairment of consciousness can occur in patients with DKA, and diabetic precoma and coma are terms that have been used to some extent interchangeably with DKA. Certainly those individuals presenting with frank coma carry a poor prognosis.

Pathophysiology

The pathophysiology of DKA has been well reviewed recently (Alberti, 1989; Krentz & Nattrass, 1991). The severe metabolic disturbances are related to insulin deficiency, often in combination with increased secretion of the counter-regulatory hormones catecholamines, glucagon, cortisol and growth hormone, as a consequence of intercurrent illness which frequently precipitates DKA. The severe hyperglycaemia is due to increased hepatic glucose production as a consequence of gluconeogenesis and glycogenolysis combined with reduced glucose uptake in peripheral tissues. Glucagon and catecholamines will stimulate gluconeogenesis in the absence of insulin. The severe hyperglycaemia leads to glycosuria and consequent renal fluid loss and dehydration.

Increased ketogenesis is an important metabolic feature of DKA. Ketogenesis occurs in the mitochondria of liver cells and is accelerated in DKA as a consequence of the severe insulin deficiency and the increased levels of counter-regulatory hormones. As a result of these hormonal abnormalities there is an increased supply of free fatty acids to the liver owing to increased adipose tissue lipolysis. Hepatic metabolism shifts away from fatty acid esterification and triglyceride synthesis towards β-oxidation within mitochondria to yield ketone bodies. The rate of ketogenesis is such that the rate of utilization in peripheral tissues is exceeded. It is usual to find elevations of acetoacetate, β-hydroxybutyrate and acetone in DKA. β-Hydroxybutyrate and acetoacetate are strong organic acids and lead to profound acidosis. Occasionally, only β-hydroxybutyrate accumulates.

The fluid and electrolyte disturbance in DKA is profound. Along with this massive osmotic diuresis there are major deficits in the body sodium and potassium levels. Despite these massive deficits in total body electrolytes, serum concentrations are generally normal in DKA and, in fact, the serum potassium may be slightly raised.

Occasionally, severe hypertriglyceridaemia complicates insulin deficiency owing to decreased activity of the enzyme lipoprotein lipase, which depends on insulin for its activity. This can lead to problems with interpretation of serum electrolytes and gives spuriously low figures.

Hyperosmolality is an important consequence of the hyperglycaemia and correlates well with the

Plasma osmolality (mosmol/L)

= 2 × (plasma sodium + plasma K)

+ plasma glucose

+ plasma urea

Fig. 15.3 The calculation of plasma osmolality. All concentrations are in mmol/L.

degree of impairment of consciousness. It is unusual to find coma unless the plasma osmolality exceeds 340 mosmol/L. Osmolality can be readily calculated at the bedside using the formula shown in Fig. 15.3.

Precipitating factors

Apart from new-onset IDDM, which accounts for approximately 25% of cases, the major precipitating factors for DKA include too little insulin and intercurrent illness. Unfortunately, it is still quite common for inappropriate measures to be taken during an intercurrent illness such that the insulin dosage is reduced or even stopped because the patient is unable to take carbohydrate. This is a recipe for rapid metabolic decompensation. More difficult to untangle are the individuals who deliberately induce ketoacidosis by withholding their insulin.

Presenting symptoms

The major symptoms and signs include severe polyuria and polydipsia, often associated with vomiting and abdominal pain. It is important to point out that the abdominal pain may be quite severe and mimic an intra-abdominal emergency. The acidosis is manifest as Kussmaul's breathing and acetone can be smelt on the breath. The patient may be profoundly dehydrated with sunken eyes and a cold periphery. Severe dehydration with circulatory collapse may be present.

Investigations and management

Once the diagnosis is suspected, emergency investigations should be performed with an assessment

Table 15.9 Principles of management of diabetic ketoacidosis in children. After Greene (1991)

Urgent hospital admission

Fluid replacement
Volumes

Isotonic saline; dextrose–saline when blood glucose <10 mmol/L

Consider bicarbonate if arterial blood pH < 7.0

Consider plasma or plasma expander (25 mL/kg) initially if severe hypotension and coma

Potassium replacement
Generally 20 mmol/L intravenous fluid

Adjust according to plasma potassium (or ECG)

Insulin replacement
Continuous intravenous infusion, initially 0.1 U/kg/h

Adjust by blood glucose monitoring

Other measures
Blood glucose monitoring, hourly until stable

Fluid balance monitoring

Plasma urea and electrolytes monitoring, 3-hourly until stable

Arterial blood gases and pH monitoring if acidotic or hypoxaemic

Consider oxygen; review need for intubation and ventilation

ECG monitoring, if arrhythmias or electrolyte disturbances develop

Nasogastric intubation, if persistent vomiting or gastric stasis occurs

Urinary catheterization, if retention or apparent oliguria develops

If cerebral oedema suspected
(a) avoid fluid overload and use of hypotonic solutions
(b) consider intravenous mannitol or dexamethasone

Table 15.10 Calculation of fluid deficit. After Greene (1991)

Calculate fluid deficit from degree of dehydration:		
Symptomatic hyperglycaemia	= 5%	× body weight (kg) = fluid deficit (L)
Dry mouth, sunken eyes, cold peripheries	= 10%	
Ketotic breathing, coma	= 15%	
Impending circulatory collapse	= 20%	
Add maintenance daily fluid requirement:		
Age 1–2 y (weight 10–13 kg)	= 120 mL/kg	× body weight (kg) = maintenance fluid intake (mL)
Age 3–6 y (weight 14–21 kg)	= 100 mL/kg	
Age 7–9 y (weight 22–29 kg)	= 80 mL/kg	
Age 10–15 y (weight 30–55 kg)	= 60 mL/kg	
Immediate fluid requirement:		
Give 20% in hours 1–2		
20% in hours 3–6		
Consider adding fluid losses after 6 hours:		
Urine		
Vomit or gastric aspirate		
Diarrhoea		
Give remaining 60% over 20 hours		

of the acid–base status by arterial blood gas estimation together with glucose, urea and electrolyte concentrations. It is usual to send blood and urine for culture.

The principles of management include replacement of the large fluid deficit and the electrolyte disturbance, correction of the acidosis and appropriate insulin therapy (Table 15.9). Replacement of fluid is most critical and in severe dehydration and shock plasma or blood is given at a rate of 25 mL/kg in the first hour. Once the blood pressure is stable (>60 mmHg systolic), then the replacement fluid regime outlined in Table 15.10 may be followed. It is essential to repeat the blood glucose, urea and electrolyte measurements every 2 h. If acidosis is severe (pH < 7) the arterial blood gases should be monitored every 2 h. Although the initial potassium measurement may be normal or even slightly raised, it is essential to replace potassium at an early stage to prevent profound hypokalaemia as a result of the insulin and glucose therapy, which pushes potassium back into the intracellular compartment. Hyperglycaemia may cause an osmotic shift between the intracellular and extracellular compartments and lead to hyponatraemia. However, this is a pseudodilution of sodium and most patients have a normal to high plasma sodium.

The use of bicarbonate is controversial and in most instances the acidosis corrects itself following fluid replacement and insulin therapy, which inhibits ketogenesis and encourages the metabolism of ketones to bicarbonate. The principal reason for the avoidance of bicarbonate therapy, particularly with the 8.4% hypertonic solution, is the possible worsening of intracellular acidosis, particularly cerebral acidosis. This is a result of the bicarbonate ions combining with the hydrogen ions and dissociating to CO_2 and water. CO_2 diffuses freely across the blood–brain barrier and hence cerebral acidosis may be exacerbated; therefore, bicarbonate is reserved for only the most severe acidosis (pH < 7). In this situation isotonic bicarbonate solution (1.4%) should preferably be given by infusion rather than bolus injection. Insulin therapy is best administered by constant intravenous insulin infusion at a rate of 0.1 U/kg per h. This will need to be adjusted, depending on the blood glucose response.

Other supportive measures may include nasogastric aspiration in the vomiting child, and if gastric distension is present. It is important to monitor urine output and catheterization may be necessary in the comatosed child or if there is a large palpable bladder.

The bedside measurement of blood glucose has aided the treatment of DKA and in some high-dependency units bedside measurement of electrolytes is also possible. However, there is no doubt that mishaps have occurred owing to incorrect use of bedside diagnostic equipment. Another point to make with regard to bedside testing strips is that hyperketonaemia may be underestimated as patients sometimes have a very high 3-hydroxybutyrate: acetoacetate ratio and 3-hydroxybutyrate is not detected by the test strips.

DKA is often precipitated by infection but as leucocytosis is an almost invariable finding in DKA it is therefore unhelpful in diagnosing infection. Furthermore, mild hypothermia is common as a result of the acidosis producing peripheral vasodilatation. This again may lead to problems in the diagnosis of infection.

Although it is clear that the fluid deficit in DKA is of hypotonic fluid, it is usual to use isotonic physiological saline as replacement fluid so that a rapid fall in plasma osmolality is avoided. As the child's condition improves, small amounts of fluid can be introduced by mouth but it is usual to administer insulin and fluid intravenously until the child can tolerate an adequate carbohydrate intake without vomiting. Subcutaneous insulin is then administered.

It is clear that careful monitoring of the fluid and electrolyte balance, insulin therapy and supportive measures with frequent biochemical monitoring are necessary for the resolution of this severe metabolic emergency. If supreme care is not exercised complications can occur, including hypokalaemia, inhalation of vomit and fluid overload, but perhaps the most serious complications are respiratory distress syndrome and cerebral oedema. Fortunately, the respiratory distress syndrome is rare. It is characterized by severe dyspnoea and hypoxia, and diffuse pulmonary infiltrates can be seen on the chest X-ray. It carries a high mortality. The cause remains to be elucidated; however, many of the case reports involved patients who were rehydrated with hypotonic fluid or in whom sodium bicarbonate was administered. Cerebral oedema is another serious complication which may occur particularly in children. With the advent of computerized tomography (CT) scanning it is clear that asymptomatic cerebral oedema is common in children and adults early in DKA (Krane *et al.*, 1985). Again it carries a high mortality and appears to be related to use of hypotonic fluid and an over-rapid reduction of blood osmolality.

Prevention of DKA is an important part of overall management and the precipitating cause, if not an obvious infection, should be pursued. The cause may be faulty sick-day rules with inappropriate reduction or withdrawal of insulin. However, with repeated episodes of DKA it is likely that insulin is being omitted by the child in response to difficulties at home or at school.

Non-ketotic hyperosmolar coma

This is a rare situation in children but has been reported. It is characterized by a very high glucose level (usually >30 mmol/L) and a high serum osmolality with absence of ketones. Treatment is along the lines outlined for DKA but these patients are exquisitely more sensitive to insulin.

Hypoglycaemia

This is a most distressing complication of insulin therapy and children, as well as adults, dread a severe 'hypo' reaction. The symptoms and signs are shown in Table 15.11 and consist of those owing to the secretion of counter-regulatory hormones, particularly catecholamines, and those due to the cerebral glucopenia. An important part of the management of the diabetic child is the avoidance of hypoglycaemia, which, in addition to its psychological effect in marking that individual as abnormal, may provoke epileptic seizures. There is no doubt that an increased frequency of hypoglycaemic reactions is the down-side of modern therapy designed to achieve near normoglycaemia. In addition, the symptoms of hypoglycaemia may be less pro-

Table 15.11 Symptoms and signs of hypoglycaemia*

Secondary to release of catecholamines		
Nervousness	Pallor	Apprehension
Tremor	Sweating	Hunger
Palpitations	Tachycardia	Weakness
Secondary to neuroglucopenia		
Restlessness, mental instability and agitation		
Irritability and obstinacy		
Mental confusion, negativism and amnesia		
Psychopathic behaviour, incoherent speech and delirium		
Headache, diplopia and dysphasia		
Unsteady gait		
Hyperreflexia and extensor plantar responses		
Signs of paraplegia, hemiplegia or monoplegia		

* Symptoms of hypoglycaemia may be more difficult to recognize in children and may include frequent yawning, episodic staring, 'startlement jerk', bizarre behaviour and twitching.

nounced in individuals who are well controlled. There has been much publicity recently suggesting that human insulin treatment may be associated with fewer warning signs of hypoglycaemia. This situation remains to be resolved, but the author is not impressed with the evidence linking hypoglycaemic unawareness with human insulin therapy.

The major causes for hypoglycaemic reactions are delayed or missed meals and/or excess exercise. In the adolescent an important cause of severe hypoglycaemia may be an alcohol binge, particularly if this is taken as spirit. The adolescent may have had diabetes for over a decade and may have developed an autonomic neuropathy, which again will reduce the warning symptoms of hypoglycaemia. In small infants the initial symptoms of hypoglycaemia may not be recognized and the condition will not be obvious until severe neuroglycopenia develops.

Hypoglycaemia is confirmed by measuring the blood glucose concentration, carried out either by the individual in a mild attack or by family or friends who have been educated to spot the warning symptoms and signs of hypoglycaemia. Most diabetic children are educated to carry with them a readily assimilated form of carbohydrate such as Dextrasol tablets. A more recent innovation is a glucose-containing gel (Hypostop). In more severe cases the child's carers may administer an injection of glucagon (1 mg) which will mobilize hepatic glycogen stores. If this does not work within a few minutes in the semi-conscious or comatosed child, then urgent admission to hospital is required where intravenous glucose may be administered initially as a bolus of 50% dextrose followed by an infusion of 10% dextrose at a rate of 100 mL/h until the child recovers consciousness. Recovery is usually quick once intravenous glucose is administered, although the child is often left with a headache and occasionally abdominal pain. There are usually no apparent sequelae to the hypoglycaemic coma in terms of neurological deficit. However, fatalities have occurred during hypoglycaemia.

Long-term diabetic complications

It is beyond the scope of this chapter, which deals particularly with children with diabetes, to give an exhaustive description of the long-term diabetic complications, and hopefully in the majority of cases evidence of these complications will not be seen in the child or adolescent. However, the bulk of evidence suggests that long-term complications, such as neuropathy, retinopathy and nephropathy, are a product of time and poor glycaemic control and therefore the prevention of complications begins with the diagnosis of diabetes and the foundations for good metabolic control at an early stage.

It is now accepted policy to tell most older children with diabetes of the need for good diabetic control to delay or prevent the development of long-term complications. Obviously this information and education has to be handled very sensitively and the establishment of continuity of care is an important aspect of this evolving educational and information process. In many hospitals there are now joint paediatric diabetic clinics staffed by a paediatrician and a diabetologist. This would seem to be an optimum way to proceed and certainly makes the transfer from the paediatric clinic to the adult diabetic clinic much easier.

It is usual to check the fundi through dilated pupils on an annual basis, and certainly after 5 years of disease in those who are post-puberty. Similarly, evidence should be sought for the presence of neuropathy affecting either the legs or the autonomic system. Joint problems can also be seen in insulin-dependent diabetics and manifests with some stiffness and minor deformities, particularly of the hands, known as cheiroarthropathy. Fortunately, the vast majority of diabetic children will have a good life expectancy and remain free of serious morbidity. Therefore, it is important to be optimistic and to emphasize that short-term poor control is unlikely to have any serious long-term effects.

Many adult clinics are now introducing the microalbuminuria test and it is likely that these tests will also become routine in paediatric diabetic clinics. It is now quite clear that long before the development of Albustix (Bayer Diagnostics, Basingstoke, UK) positive proteinuria it is possible to detect increased albumin excretion rates using sensitive immunoassays for small amounts of albumin (Keen & Chlouverakis, 1963). These assays have been set up in many chemical pathology departments and now there are clinic tests (Microbumintest, Ames; Albusure, Cambridge Life Sciences), which are useful for screening in the clinic (Gatling *et al.*, 1985), also available to detect microalbuminuria. It is important to bear in mind that urinary albumin excretion has a high day-to-day variation (40–50%) and is influenced by diet, posture, exercise and urine flow (Walker & Viberti, 1991). Because of this variation it is important to make repeated measurements in the individual patient. Attempts to standardize the measurement of urinary albumin excretion have included the collection of timed overnight urine collections. However, in the clinic, as opposed to research protocols, the urinary albumin:creatinine ratio is a useful indicator of albumin excretion. A ratio greater than 2.0 mg/mmol in the first morning urine specimen has a specificity of 100% to detect an albumin excretion rate of >30 μg/min as the cut-off point for the presence of microalbuminuria (Gatling *et al.*, 1988). The importance of microalbuminuria is that it predicts the development of full-blown proteinuria and is the first sign of diabetic nephropathy (Viberti *et al.*, 1988). In addition, individual patients with proteinuria, whether it be microalbuminuria or Albustix-positive proteinuria, seem to be at a hugely increased risk of morbidity and mortality from cardiovascular disease (Borch-Johnsen & Kreiner, 1987). The possible link between the two is a current exciting area of research. In terms of clinical practice, the presence of microalbumin points to an at-risk individual and every attention should be given to try to improve glycaemic control, ensuring optimum blood pressure and blood lipids.

References

Alberti, K.G.M.M. (1989) Diabetic emergencies. *Br Med Bull* **45**, 242–263.

Allen, D.W., Schroeder, W.A. & Balog, J. (1958) Observations on the chromatographic heterogeneity of normal adult and fetal haemoglobin. *J Am Chem Soc* **80**, 1628–1634.

Asmal, A.C., Dayal, B., Jailal, I. *et al.* (1981) Non insulin dependent diabetes mellitus with early onset in Blacks and Indians. *SA Med J* **60**, 93–98.

Bar, R.S., Levis, W.R., Rechler, M.M. *et al.* (1978) Extreme insulin resistance in ataxia telangiectasia. *N Engl J Med* **298**, 1164–1171.

Barnett, A.H., Eff, C., Leslie, R.D.G. & Pyke, D.A. (1981) Diabetes in identical twins: a study of 200 pairs. *Diabetologia* **20**, 87–93.

Bell, G.I., Horito, S. & Karam, J.H. (1986) A polymorphic locus near the human insulin gene is associated with insulin dependent diabetes mellitus. *Diabetologia* **33**, 176–183.

Bell, G.I., Xiang, K.S., Newman, M.V. *et al.* (1991) Gene for non-insulin-dependent diabetes mellitus (maturity onset diabetes of the young subtype) is linked to DNA polymorphism on human chromosome 20q. *Proc Natl Acad Sci USA* **88**, 1484–1488.

Bennett, P.H., Rushforth, N.B., Miller, M. *et al.* (1976) Epidemiological studies of diabetes in the Pima Indians. *Rec Prog Hormone Res* **32**, 333–376.

Bernstein, R.E. (1980) Glycosylated haemoglobins: haematologic considerations determine which assay for glycohaemoglobin is advisable. *Clin Chem* **26**, 174–175.

Borch-Johnsen, K. & Kreiner, S. (1987) Proteinuria – predictor of cardiovascular mortality in insulin dependent diabetes mellitus. *Br Med J* **294**, 1651–1655.

Bottazzo, G.F., Florin-Christensen, A. & Doniach, D. (1974) Islet cell antibodies in diabetes mellitus with autoimmune polyendocrine deficiency. *Lancet* **ii**, 1279–1283.

Bottazzo, G.F., Pujol-Borrell, R. & Gale, E.A.M. (1986) Autoimmunity and diabetes: progress, consolidation and controversy. In: *The Diabetes Annual*, Vol. 2 (eds

K.G.M.M. Alberti & L.P. Krall), pp. 13–29. Elsevier Amsterdam.

Breyer, D., Cvitkovic, P., Zdenko, S., Pedersen, O. & Rocic, B. (1981) Decreased insulin binding to erythrocytes in subjects with Klinefelter's syndrome. *J Clin Endocrinol Metab* **53**, 654–655.

Clark, A., Holman, R.R., Mathews, D.R., Hockaday, T.D.R. & Turner, R.C. (1984) Non uniform distribution of amyloid in the pancreas of maturity onset diabetic patients. *Diabetologia* **27**, 527–528.

Clark, P.M.S. & Hales, C.N. (1991) Assay of insulin. In: *Textbook of Diabetes* (eds J. Pickup & G. Williams), pp. 335–347. Blackwell Scientific Publications, Oxford.

van Companhout, J., Antaki, A. & Rasio, E. (1973) Diabetes mellitus and thyroid autoimmunity in gonadal dysgenesis. *Fertil Steril* **24**, 1–9.

Diabetes Epidemiology Research International (1987) Preventing insulin dependent diabetes mellitus. *Br Med J* **195**, 279–481.

Dunnigan, M.G., Cochrane, M.A., Kelly, A. *et al.* (1974) Familial lipoatrophic diabetes with dominant transmission: a new syndrome. *Quart J Med* **43**, 33–48.

Eberentz-Lhomme, C., Ducrocq, R., Intrator, S. *et al.* (1984) Haemoglobinopathies: a pitfall in assessment of glycosylated haemoglobin HbA_1. *Diabetologia* **27**, 596–598.

Flückiger, R. & Winterhalter, K.H. (1976) In vitro synthesis of haemoglobin A_{1c}. *FEBS Lett* **71**, 356–360.

Flückiger, R., Marmon, W., Meier, W., Loo, S. & Gabbay, K.H. (1981) Haemoglobin carbamylation in uraemia. *N Engl J Med* **304**, 823–827.

Fujimoto, W.Y., Green, M.L. & Seegmiller, J.E. (1969) Cockayne's syndrome: report of a case with hyperlipoproteinaemia, hyperinsulinaemia, renal disease and normal growth hormone. *J Pediatr* **75**, 881–884.

Gatling, W., Knight, C. & Hill, R.D. (1985) Screening for early diabetic nephropathy: which sample to detect microalbuminuria. *Diabet Med* **2**, 451–455.

Gatling, W., Knight, C., Mullee, M.A. & Hill, R.D. (1988) Microalbuminuria in diabetes: a population study of the prevalence and an assessment of three screening tests. *Diabet Med* **5**, 343–397.

Gepts, W. (1965) Pathologic anatomy of the pancreas in juvenile diabetes mellitus. *Diabetes* **14**, 619–633.

Goldstein, J.L. & Fialkow, P.J. (1973) Alström syndrome. *Medicine* **52**, 53–71.

Greene, S.A. (1991) Diabetes mellitus in childhood and adolescence. In: *Textbook of Diabetes* (eds J. Pickup & G. Williams), pp. 866–883. Blackwell Scientific Publications, Oxford.

Harris, M.I., Hadden, W.C., Knowler, W.C. & Bennett, P.H. (1987) Prevalence of diabetes and impaired glucose tolerance and plasma glucose levels in US population aged 20–74 years. *Diabetes* **36**, 523–534.

Horton, B.F. & Huisman, T.H.J. (1965) Studies on the heterogeneity of haemoglobin: VII. Minor haemoglobin components in haematological diseases. *Br J Haem* **11**, 296–304.

Hudson, A.J., Huff, M.W., Wright, C.G. *et al.* (1987) The role of insulin resistance in the pathogenesis of myotonic muscular dystrophy. *Brain* **110**, 469–488.

Jeremiah, D.E., Leyshon, G.E., Rose, T., Francis, H.W.S. & Elliot, R.W. (1973) Down's syndrome and diabetes. *Psychol Med* **3**, 455–457.

Johnson, R.N., Metcalf, P.A. & Baker, J.R. (1982) Fructosamine: a new approach to the estimation of serum glycosyl protein. An index of diabetic control. *Clin Chim Acta* **127**, 87–95.

Keen, H. & Chlouverakis, C. (1963) An immunoassay for urinary albumin at low concentrations. *Lancet* **ii**, 913–916.

Kennedy, L., Mehl, T.D., Riley, W.J. & Merimee, T.J. (1981) Nonenzymatically glycosylated serum protein in diabetes mellitus: an index of short term glycaemia. *Diabetologia* **21**, 94–98.

Kirk, R.L., Serjeantson, S.W., King, H. & Zimmet, P. (1985) The genetic epidemiology of diabetes mellitus. In: *Disease of Complex Aetiology in Small Populations: Ethnic Differences and Research Approaches* (eds R. Chakraborty & E. Szathmary), pp. 119–146. Alan R. Liss, New York.

Koenig, R.J., Peterson, C.M., Jones, R.L. *et al.* (1976) Correlation of glucose regulation and haemoglobin A_{1c} in diabetes mellitus. *N Engl J Med* **295**, 417–420.

Kortlandt, W., Van Rijn, H.J.M., Hocke, J.O.O. & Thissen, J.H.H. (1985) Comparison of three different assay procedures for the determination of HbA_1 with special attention to the influence of pre HbA_{1c}, temperature and haemoglobin concentrations. *Ann Clin Biochem* **22**, 261–268.

Kousholt, A.M., Beck Neilson, H. & Lund, H.T. (1983) A reduced number of insulin receptors in patients with Prader–Willi syndrome. *Acta Endocrinol* **104**, 345–351.

Krane, E.J., Rockoff, M.A., Wallman, J.K. & Wolfsdorf, J.I. (1985) Subclinical brain swelling during treatment of diabetic ketoacidosis. *N Engl J Med* **312**, 1147–1151.

Krentz, A.J. & Nattrass, M. (1991) Diabetic ketoacidosis, non-ketotic hyperosmolar coma and lactic acidosis. In: *Textbook of Diabetes* (eds J. Pickup & G. Williams), pp. 479–494. Blackwell Scientific Publications, Oxford.

Lendrum, R., Walker, I.G. & Gamble, D.R. (1975) Islet cell antibodies in juvenile diabetes mellitus of recent onset. *Lancet* **i**, 880–882.

Leslie, R.D.G. & Pyke, D.A. (1986) The genetics of diabetes. In: *The Diabetes Annual*, Vol. 3 (eds K.G.M.M. Alberti & L.P. Krall), pp. 39–54. Elsevier, Amsterdam.

Leslie, R.D.G., Lazarus, N.R. & Vergani, D. (1989) Aetiology of insulin-dependent diabetes. *Br Med Bull* **45**, 58–72.

McCance, D.R., Coulter, D., Smye, M. & Kennedy, L. (1987) Effect of fluctuations in albumin on serum fructosamine. *Diabet Med* **4**, 434–436.

McCance, D.R., Ritchie, C.M. & Kennedy, L. (1988) Is HbA_1 measurement superfluous in NIDDM. *Diabet Care* **11**, 512–514.

Mather, H.M. & Keen, H. (1985) The Southall Diabetes Survey: prevalence of known diabetes in Asians and Europeans. *Br Med J* **291**, 1081–1084.

Menard, L., Dempsey, M.E., Blankstein, L.A. *et al.* (1980) Quantitative determination of glycosylated haemoglobin A_1 by agar gel electrophoresis. *Clin Chem* **26**, 1598–1602.

Michelson, B. & Lernmark, A. (1987) Molecular cloning of a polymorphic DNA endonuclease fragment associates insulin-dependent diabetes with HLA DQ. *J Clin Invest* **75**, 1144–1152.

Millward, B.A., Welsh, K.I., Leslie, R.D.G., Pyke, D.A. & Demaine, A.G. (1987) T cell receptor beta-chain polymorphisms are associated with insulin-dependent diabetes. *Clin Exp Immunol* **70**, 152–157.

Milunski, A. & Neurath, P.W. (1968) Diabetes in Down's syndrome. *Arch Environ Hlth* **17**, 372–376.

Mohan, V., Ramachandran, A., Snehalatha, C. *et al.* (1985) High prevalence of maturity onset diabetes of the young (MODY) among Indians. *Diabet Care* **8**, 371–374.

National Diabetes Data Group (1979) Classification and diagnosis of diabetes mellitus and other categories of glucose intolerance. *Diabetes* **28**, 1039–1057.

Newman, B., Selby, J.V., King M-C. *et al.* (1987) Concordance for type 2 (non-insulin-dependent) diabetes mellitus in male twins. *Diabetologia* **30**, 763–768.

Nielson, J., Johansen, K. & Yde, H. (1969a) The frequency of diabetes mellitus in patients with Turner's syndrome and pure gonadal dysgenesis. *Acta Endocrinol* **62**, 251–269.

Nielson, J., Johansen, K. & Yde, H. (1969b) Frequency of diabetes mellitus in patients with Klinefelter's syndrome of different chromosomal constitutions and the XYY syndrome. Plasma insulin and growth hormone level after a glucose load. *J Clin Endocrinol* **29**, 1062–1073.

Palmer, J.P. (1987) Insulin autoantibodies: their role in the pathogenesis of IDDM. *Diabet Metab Rev* **3**, 1005–1015.

Quin, J.D., Fisher, B.M., Paterson, K.R., Beastall, G.H. & MacCuish, A.C. (1989) Preliminary studies with subcutaneous and intravenous insulin-like growth factor-1 in Mendenhall's syndrome. *Diabet Med* **6** (Suppl. 2), 5A.

Rahier, J., Goebbels, R.M. & Henquin, J.C. (1983) Cellular composition of the human diabetic pancreas. *Diabetologia* **24**, 366–371.

Reaven, G.M. (1988) Banting lecture 1988: role of insulin resistance in human disease. *Diabetes* **37**, 1595–1607.

Reddy, S.S., Lauris, V. & Kahn, C.R. (1988) Insulin receptor function in fibroblasts from patients with leprechaunism, kinase activity and receptor-mediated internalization. *J Clin Invest* **82**, 1359–1365.

Savir, A., Dickerman, Z., Zarp, M. & Laron, Z. (1974) Diabetic retinopathy in an adolescent with Prader–Labhardt–Willi syndrome. *Arch Dis Child* **49**, 963–964.

Schwachman, H. (1978) Cystic fibrosis. *Curr Prob Pediatr* **8**, 1–16.

Simon, M. & Cuan, J. (1982) Haemoglobin A_{1c} by isoelectric focusing. *Clin Chem* **28**, 9–12.

Spicer, K.M., Allen, R.C. & Buse, M.G. (1978) A simplified assay of haemoglobin A_{1c} in diabetic patients by use of isoelectric focusing and quantitative microdensitometry. *Diabetes* **27**, 384–388.

Starkman, H.S., Wacks, M., Soeldner, S. & Kim, A. (1983) Effect of acute blood loss on glycosylated haemoglobin determinations in normal subjects. *Diabet Care* **6**, 291–294.

Stefan, Y., Orci, L., Malaisse-Lagae, F. *et al.* (1982) Quantitation of endocrine cell content in the pancreas of non-diabetic and diabetic humans. *Diabetes* **31**, 694–700.

Tarn, A.C., Thomas, J.M., Dean, B.M. *et al.* (1988) Predicting insulin-dependent diabetes. *Lancet* **i**, 845–850.

Tattersall, R.B. (1984) Mild familial diabetes with dominant inheritance. *Quart J Med* **43**, 339–357.

Tattersall, R.B. (1991) Maturity onset diabetes of the young (MODY). In: *Textbook of Diabetes* (eds J.C. Pickup & G. Williams), pp. 243–246. Blackwell Scientific Publications, Oxford.

Taylor, S.I., Underhill, L.H., Hedo, J.A. *et al.* (1983) Decreased insulin binding to cultured cells from a patient with the Rabson–Mendenhall syndrome: dichotomy between studies with cultured lymphocytes and cultured fibroblasts. *J Clin Endocrinol Metab* **56**, 856–861.

Tchobroutsky, G., Charibanski, D., Blouquit, Y. *et al.* (1980) Diabetic control in 102 insulin-treated patients. *Diabetologia* **18**, 447–452.

Todd, J.A., Bell, J.I. & McDevitt, H.O. (1987) HLA DQ beta gene contributes to susceptibility and resistance to insulin-dependent diabetes mellitus. *Nature* **329**, 599–604.

Trivelli, L.A., Ranney, H.M. & Lai, H.-T. (1971) Haemoglobin components in patients with diabetes mellitus. *N Engl J Med* **284**, 353–357.

Tunbridge, W.M.G. (1981) Factors contributing to deaths of diabetics under fifty years of age. *Lancet* **i**, 569–572.

Viberti, G.C., Wiseman, M.J. & Redmond, S. (1988) Microalbuminuria in diabetes: a population study of the prevalence and an assessment of three screening tests. *Diabet Med* **5**, 343–397.

Vionnet, N., Stoffel, M., Takeda, J. *et al.* (1992) Nonsense mutation in the glucokinase gene causes early onset non-insulin-dependent diabetes mellitus. *Nature* **356**, 721–722.

Walker, J.D. & Viberti, G.C. (1991) Aetiology and pathogenesis of diabetic nephropathy: clues from early functional abnormalities. In: *Textbook of Diabetes* (eds J. Pickup & G. Williams), pp. 657–670. Blackwell Scientific Publications, Oxford.

Waxman, A.D., Schalch, D.S., Odell, W.D. & Tschudy, D.P. (1967) Abnormalities of carbohydrate metabolism in acute intermittent porphyria. *J Clin Invest* **46** (Suppl. 1), 1129.

Welch, S.G. & Boucher, B.J. (1978) A rapid micro-scale method for the measurement of haemoglobin $A_{1(a+b+c)}$. *Diabetologia* **14**, 209–211.

West, R.J., Lloyd, J.K. & Turner, W.M.L. (1975) Familial insulin-resistant diabetes, multiple somatic anomalies and pineal hyperplasia. *Arch Dis Child* **50**, 703–708.

Westermark, P., Wernstedt, C., Wilander, E. *et al.* (1987) Amyloid fibrils in human insulinoma and islets of Langerhans of the diabetic cat are derived from a novel neuropeptide-like protein also present in normal islet cells. *Proc Natl Acad Sci USA* **84**, 3881–3885.

WHO Expert Committee on Diabetes Mellitus (1980) Second report. *WHO Technical Report Series* **646**. Geneva.

WHO Study Group (1985) Diabetes Mellitus – Report of a WHO Study Group. *WHO Technical Report Series* **727**. WHO, Geneva.

Wilson, D.M., Frane, J.W., Sherman, B. *et al.* (1987) Carbohydrate and lipid metabolism in Turner's syndrome: effect of therapy with growth hormone, oxandrolone and a combination of both. *J Pediatr* **112**, 210–217.

Winter, W.E., MacLaren, N.K., Riley, W.J. *et al.* (1987) Maturity onset diabetes of youth in black Americans. *N Engl J Med* **316**, 285–291.

Zimmet, P. (1979) Epidemiology of diabetes and its macrovascular manifestations in Pacific populations: the medical effects of social progress. *Diabet Care* **2**, 144–153.

16: Muscle Diseases

J.M. ROUND & M.J. JACKSON

Introduction

Muscle diseases present in many different ways. A child may be weak and miserable or show delay in achieving the normal motor milestones, a baby may present as a floppy infant, or an adolescent may complain of excessive fatigue and show a disinclination to take part in active sport. Patients can generally be divided into three groups: (a) those who are weak even when rested; (b) those who are of normal strength at rest but in whom even a small amount of exercise leads to premature or excessive fatigue; and (c) those who are not necessarily weak or especially easily fatigued but have some disturbance of function such as spasticity, inability to relax the muscle after contraction (myotonia), abnormally severe cramps or episodes of weakness. Weakness generally arises because either the child has lost functional muscle, or there has been some failure in muscular development. The primary cause of this loss may be a defect in the muscle itself (myopathy) or in the nervous system innervating that muscle or muscles (neuropathy).

The myopathies may be broadly subdivided into two categories: the atrophic myopathies, in which the muscle fibres shrink in size but do not decrease in total number, and the destructive myopathies, where fibres are destroyed and lost. This latter category, which is more common in childhood, can be further subdivided into the muscular dystrophies, which are of genetic origin, and the inflammatory myopathies, where the destruction is due to an autoimmune process.

In neuropathic conditions the muscle fibres become denervated as a result of degeneration of the motor nerve input, or fail to receive impulses owing to defective development of the motor neurones. In both categories there are disorders which are thought to be of genetic origin and those which are acquired.

Excessive fatiguability is a common complaint, but only in a few cases are there satisfactory explanations for the clinical symptoms. Defects in the glycolytic pathway or the electron transport chain severely limit exercise, but are comparatively rare (see also Chapter 7(b)). Myasthenia gravis, an autoimmune disorder, affects the postsynaptic membrane of the neuromuscular junction and leads to a rapid loss of force during sustained efforts.

Abnormalities of muscle function are also relatively rare; most are associated with changes in the electrical properties of the muscle fibre membrane which render the fibre either over- or under-excitable.

Before moving on to a more detailed discussion of muscle diseases in children, it will be useful to examine the contribution which biochemical measurements can make both in assisting diagnosis and in monitoring the progress of diseases and the effects of treatment.

Biochemical indicators of muscle disease

Definitive diagnoses of muscle diseases in children or adults can rarely be achieved through measurement of circulating biochemical parameters, but such measurements can prove an aid to diagnosis and to monitoring the progress of disease in a number of situations. The available techniques provide an index of ongoing muscle damage, of muscle protein loss or of muscle bulk.

Indices of muscle damage

One of the features of acute damage or degeneration of tissue is a loss of cytoplasmic components, such as proteins, into the circulation. This feature of muscle damage has been widely examined; a number of protein markers of muscle damage are used and the limitations in their interpretation are recognized.

Creatine kinase

Elevations of the total creatine kinase (CK) activity in the plasma or serum of patients with a variety of degenerative muscle diseases have been extensively reported, but such elevations also occur in normal subjects following unaccustomed or excessive exercise and in patients following myocardial infarction. Further refinements of the test have therefore been developed to help improve the specificity, particularly in terms of differentiating between CK derived from skeletal or heart muscle.

The functional CK protein consists of a dimer of two distinct monomers, M and B. The protein produced by skeletal muscle is in the M form and the functional protein is CK-MM, while brain produces CK-BB; cardiac muscle produces about 30% MB enzyme, the rest being MM (Van der Veen & Willebrands, 1966; Eppenberger *et al.*, 1967). Differential diagnosis between damage to heart and skeletal muscle can therefore be achieved by the detection of a rise in the circulating CK-MB activity. The problem of whether an elevated circulating CK-MM activity is due to exercise rather than muscle disease can usually be overcome by repeated sampling, since the rise following even the most extreme forms of exercise normalizes within a few days.

The recent recognition that the CK enzyme is post-translationally modified in the circulation to produce at least two modified forms of the CK-MM protein, which retain enzymatic activity, has extended the use of this measurement. It appears that a carboxypeptidase present in serum sequentially removes the C-terminal groups from each M subunit, producing two further isoforms with different isoelectric points. There is some confusion concerning the nomenclature of these isoforms, but the generally utilized form is to designate the pure gene product CK-MMI, which is modified to produce CK-MMII in plasma followed by CK-MMIII (Yasmineh *et al.*, 1981). The time course of this conversion is considerably shorter than that of the decline of total CK activity in plasma so that following an episode of muscle damage an initial rise in CK-MMI occurs and is followed by a rise in MMII and then MMIII. Thus, a high MMI:MMIII ratio is associated with the short time periods following an episode of muscle damage, and a low ratio is associated with a much later time course following damage, although the total CK-MM activity in plasma may be equivalent (Page *et al.*, 1989).

CK activities in the plasma or serum of patients with degenerative muscle diseases show considerable variability and it is therefore difficult to use this test as a precise means of assessing disease progression in, for example, patients with polymyositis or Duchenne muscular dystrophy. In such cases, analysis of CK isoforms may prove helpful since patients with Duchenne muscular dystrophy display a consistent elevation of the MMI:MMIII ratio, indicative of continuing muscle damage (Page *et al.*, 1989).

CK has been the most widely studied of the proteins released from damaged muscle. Studies of the processes of damage to muscle in isolated systems have revealed that these proteins are released by defined biochemical mechanisms, rather than by simple lysis of the cells, and a clear pattern of proteins is released (Jackson *et al.*, 1991). CK is only a minor component of this group of proteins, but is readily and sensitively measured in the circulation owing to its long half-life and the high specific activity of the protein in human blood. Nevertheless, these studies suggest that more specific and/or sensitive indicators of skeletal muscle damage than CK might be present and a number of candidate proteins have been examined.

Myoglobin

This haem protein is found in large quantities in both skeletal and cardiac muscle. Following an episode of 'rhabdomyolysis' the rise in plasma myoglobin occurs prior to the rise in plasma CK activity and deposition of myoglobin in the renal tubules after an episode of acute rhabdomyolysis can lead to acute renal failure. It is unusual for a rise in myoglobin to occur in isolation without a subsequent rise in CK, and the relative difficulties in measuring myoglobin, which requires the use of radioassay techniques (in comparison to CK where a relatively straightforward enzyme assay is used), mean that there is little advantage in routine analysis of this protein.

Other muscle-derived enzymes

Circulating activities of lactate dehydrogenase, pyruvate kinase, aldolase, the transaminases and carbonic anhydrase III may all be raised in muscle disease. All except carbonic anhydrase III are less specific to muscle and are less sensitive than CK as an index of muscle damage. Carbonic anhydrase III may offer significant advantages over CK in terms of the specificity of its location within skeletal muscle compared with cardiac muscle (Carter *et al.*, 1983), but, as for myoglobin, it is necessary to utilize immunoassay techniques to quantify the protein in the bloodstream and hence it is comparatively infrequently studied.

Indicators of muscle protein degradation

In conditions where atrophy of muscles occurs in the absence of acute muscle damage (e.g. the endocrine myopathies) the only abnormal biochemical parameters which can be monitored reflect increased loss of muscle protein. Traditionally this has involved monitoring the nitrogen balance of patients, although the balance of other major intracellular muscle ions, such as potassium or zinc, will also be negative during severe muscle wasting (Edwards *et al.*, 1979).

3-Methylhistidine; a specific excretion product appearing in the urine following the breakdown of actin has been widely used as an index of myofibrillar protein degradation in humans. The interpretation of the urine 3-methylhistidine excretion is, however, complicated when the muscle mass is reduced, since in these situations non-muscle sources provide a relatively greater contribution to the urinary 3-methylhistidine excretion. For example, if intestinal actin turnover was elevated in a wasted patient with inflammatory myopathy an elevated urinary 3-methylhistidine excretion would be found when compared with the muscle bulk (i.e. an elevated urinary 3-methylhistidine:creatinine ratio), even though muscle protein degradation might not be elevated. Caution is therefore required in the interpretation of urinary 3-methylhistidine excretion rates (Rennie & Millward, 1983).

Indicators of muscle bulk

The primary origin of the creatinine excreted in human urine is skeletal muscle, and hence urinary creatinine excretion over 24 h is related to the muscle mass. This can be of particular importance when muscles are severely wasted and infiltrated by fat and fibrous tissue, as occurs in young boys with Duchenne and other muscular dystrophies. Excessive weight gain is a potential problem in such patients and it is important to be able to assess their true muscle bulk in order to predict their ideal weight. Measurement of serial urinary creatinine excretion allows an estimate of muscle bulk under these conditions (Edwards *et al.*, 1983).

Muscle biopsy

Definitive diagnosis of many muscle diseases can only be made following muscle biopsy. In many cases diagnosis is achieved by histochemical or immunohistochemical techniques, but some diagnoses can only be made following biochemical analyses or by applying the techniques of molecular biology to biopsy samples. Such analyses include the estimation of the size and abundance of dystrophin as an aid to the differential diagnosis of Becker and Duchenne muscular dystrophies. The diagnosis of muscle phosphorylase deficiency (McArdle's disease) is made by revealing both the absence of enzyme activity and the nature of the underlying defect at the messenger ribonucleic acid (mRNA) level (McConchie *et al.*, 1989). Firm diagnosis of a number of defects in muscle mitochondrial oxidative metabolism, including carnitine palmitoyl transferase (CPT) deficiency (Caroll *et al.*, 1978) and defects in the mitochondrial electron transport chain (e.g. those at the level of complexes I, III and IV), can also be achieved using material from a muscle biopsy (see also Chapter 7(b)).

Inflammatory myopathies

Myositis may occur in childhood and it has been suggested that the term juvenile dermatomyositis (JDMS) should be used in the paediatric age group (Pachman & Maryjowski, 1984) where dermatomyositis is much more common than polymyositis. There is good evidence that JDMS is a separate disease entity and as with polymyositis there is a female to male predominance of 2:1. Dermatomyositis is almost certainly the result of an autoimmune process. Presentation is with weakness, malaise and misery. The inflammatory changes lead to a loss of muscle tissue and resultant weakness; there may be dysphagia, calcified deposits, gastrointestinal vasculitis and restrictive defects in the ventilatory capacity. There is, however, a relative absence of muscle pain. A myopathic electromyograph (EMG) is often, but not always, found. In addition to the muscle lesions, the skin is affected with a lilac discoloration of the upper eyelids and face and a patchy erythematous rash over the hands and sometimes the forearms.

Histological examination shows the major pathology to be damage to the capillaries, venules and small arteries; changes seen in a muscle biopsy also include muscle fibre degeneration and necrosis with inflammatory infiltration in perivascular, perimysial and endomysial areas. Atrophied fibres, particularly in perifascicular areas, and fibres with abnormal architecture may be seen. Regenerating fibres are often present in acute cases but are less often seen in chronic disease.

The plasma CK may be markedly elevated and values up to 100 times normal have been reported, but in some cases the CK is normal. In cases where the plasma CK is raised this test may be useful in monitoring progress. A raised CK is often associated with a high lactate dehydrogenase level, although the order of magnitude is less than the rise in CK.

In 50% of cases γ-globulin levels are also raised and the erythrocyte sedimentation rate (ESR) may be high, although in 40% of cases it is normal (Morrow & Isenberg, 1987).

Cardiac abnormalities have been reported and myoglobinuria owing to the release of soluble protein from the damaged muscles may, in severe cases, lead to renal failure (Askari & Huettner, 1982).

Treatment is with steroids (1 mg/kg body weight per day) and other anti-inflammatory agents are sometimes used. Careful monitoring of the steroid therapy is necessary to maximize effects on the immune system while minimizing the tendency of corticosteroids to put the child into a negative nitrogen balance, which will lead to a failure in growth and further loss of strength owing to muscle fibre atrophy, particularly of the type 2 (fast) fibres. With careful management the prognosis is good, and once the clinical response is sustained the steroid dose can be progressively reduced. The clinical response is a more useful guide to progress than the plasma CK, which may remain elevated during recovery.

Muscular dystrophies

The term *muscular dystrophy* includes all hereditary progressive disorders of muscles which result in fibre destruction and the relentless replacement of muscle cells with fat and fibrous tissue.

The most serious, although not the most common, type is *Duchenne muscular dystrophy* (DMD). Incidence figures for this disease vary from 13 to 33 per 100 000 live male births with a prevalence in the population of 1.9 to 3.4 per 100 000. It has a sex-linked recessive mode of inheritance so that boys are affected but girls act as carriers. If there is no family history the condition may not be noticed at first because there are few physical signs of the disease at birth (although the plasma CK can be as high as 10 000 iu/L), but by 3 to 4 years a delay in reaching the usual motor milestones becomes apparent. Many children present because of difficulties with running, jumping and climbing stairs; earlier presentation may be due to a delay in motor milestones accompanied by slow speech development, and in these cases delay is usually accompanied by intellectual retardation. Biochemical investigations reveal a grossly elevated plasma CK; a normal or only moderately raised level is against the diagnosis. Subsequently, the child has increasing difficulty standing and walking and by the beginning of the second decade is usually confined to a wheelchair. As the disease progresses, weakness of the respiratory muscles, often accentuated by scoliosis, becomes a problem and death usually occurs before the third decade. Life expectancy depends on the quality of nursing care available; major aims are to minimize contractures and prevent the development of the respiratory problems which lead to congestion, infections and often a fatal pneumonia.

The defective gene in DMD has now been assigned to a deletion on the short arm of the X chromosome at position Xp21. The gene is large, spanning a genomic region of about 2000 kb, and the protein encoded for by this region has been identified and named *dystrophin* (Hoffman *et al.*, 1987). Monoclonal antibodies raised against dystrophin have shown its location to be at the periphery of muscle fibres in association with the plasma membrane, and some evidence suggests that it has a structural role in the surface membrane. Isolation of deoxyribonucleic acid (DNA) segments from the Xp21 region has provided probes which are useful for the prenatal diagnosis of DMD in mothers known to be at risk (see Chapter 2). Nevertheless, even with the most efficient prenatal screening the incidence of DMD will still remain significant, because spontaneous mutations in this large gene are responsible for about one-third of all reported cases.

The clinically less severe form, *Becker muscular dystrophy*, is now known to be allelic with the more severe disease. In Becker dystrophy the production of dystrophin appears to be reduced and variable in amount (Kunkel *et al.*, 1986). The essential pathological features are the same and the course is progressive in both forms, but with a slower time course in Becker dystrophy, where patients may survive into the fourth or fifth decade. Differential diagnosis between these two forms of muscular dystrophy is now generally dependent upon the abundance and relative size of the dystrophin present in muscle biopsies: Duchenne dystrophy is associated with a virtual lack of dystrophin, whereas Becker dystrophy is associated with the presence of a protein of abnormal size or in reduced quantities.

Fascioscapulohumeral dystrophy is another well-recognized form of muscular dystrophy. Inheritance is dominant but there is a wide variety of gene expression within families. Patients usually have a normal life span. Distribution of the weakness is predominantly in the upper limb-girdle, but later in the disease the lower limbs may also be affected. Also dominantly inherited is *occulopharyngeal dystrophy*, in which abnormalities of mitochondrial structure and function and changes of immune tolerance have been reported. Other forms of dystrophy include the X-linked recessive *Emery–Dreifuss humeroperoneal* form, the autosomal recessive and relatively benign *limb-girdle dystrophy* and the *scapulohumeral* form, which also has an autosomal recessive inheritance and is characterized by cardiomyopathy and contractures. An absence of dystrophin does not appear to be associated with any of these dystrophies.

Myotonic dystrophy is dominantly inherited with variable penetrance; it is a multisystem disorder characterized primarily by myotonia and progressive muscle weakness. The incidence is calculated to be 13.5 per 100 000. The gene maps on chromosome 19 in the q13.2–13.3 region. The mutation has recently been identified as an unstable trinucleotide CTG repeat, which is present 50–130 times in the normal population but is amplified 2000 times in myotonic dystrophy patients. Weakness starts peripherally and only later are proximal muscle groups involved. Myotonia of the hands is often the initial symptom. This is a multisystem disease in which many organs, including the brain, heart, endocrine organs, skin and eyes, are affected. Although giving its name to the disorder, myotonia itself is not the major problem.

Before 1960, myotonic dystrophy was considered primarily as a disease of teenagers and adults but more recently a congenital form has been clearly defined; it is characterized by extreme hypotonia, muscular atrophy, neonatal respiratory distress and feeding difficulties. Transmission is almost exclusively maternal, and during pregnancy increased amounts of amniotic fluid and reduced fetal movements are frequent findings. Mental retardation and facial paralysis may also occur. Myotonia and cataracts, two prominent features of the adult disease, are absent in the congenital form.

Among the disorders of muscle which occur *in utero* are the autosomal recessive *congenital muscular dystrophies* in which infants are born with weakness, hypotonia and a 'dystrophic' muscle biopsy picture. A raised plasma CK is seen in about 50% of cases. The disease progresses little after birth so long as immobility is avoided.

Congenital myopathies

Congenital myopathies may manifest as a failure in the development of one particular fibre type, leading to a predominance of the other (usually type 1), or the presence within the muscle fibres of rod-like bodies (*nemeline myopathy*) or cores (*central core disease*). These are disorders of early childhood which remain static or progress slowly. For a full discussion of childhood dystrophies the reader is referred to the specialist text by Dubowitz (1985).

Atrophic myopathies

Atrophic myopathies are some of the most commonly encountered causes of weakness and muscle wasting in adults but are much less common in children. They are frequently secondary to a clinical condition affecting some other system. The muscle wasting is due to a reduction in the cross-sectional area of individual muscle fibres, often affecting one fibre type more than another. Most frequently it is the type 2 fibres that show the greatest atrophy and type 2 fibre atrophy is commonly seen in hypothyroidism, rickets, as a result of prolonged steroid therapy and in the wasting which occurs as a result of malignant disease or malnutrition. It is not known why the fast type 2 fibres are more often affected. One theory is that they atrophy as a result of the inactivity caused by the primary clinical problem, but limited studies on patients with atrophy after limb immobilization suggest that in this situation it is the type 1 fibres which are more severely affected (Edstrom, 1970). Alternatively, it is possible that the type 2 fibres represent a pool of protein which is utilized by the body during times of stress, fast fibres being less essential for survival than the fatigue-resistant type 1 fibres. The plasma CK is usually normal, except in thyroid myopathies; the raised CK seen in hypothyroid cases is well documented but its origin is obscure.

Preferential atrophy of type 1 fibres is less common; it has been reported in myotonic dystrophy, in some childhood myopathies and, as mentioned above, in the quadriceps muscles of young people who have been immobilized for some time after knee injury or lower limb fracture. A complete absence of type 1 fibres can be seen in biopsies from the quadriceps of paraplegic patients after traumatic transection of the spinal cord. The plasma CK in these patients remains normal.

In most cases of atrophic myopathy, if the underlying disorder can be successfully treated, the muscle strength and size usually recover, with the atrophic fibres growing back to their full size, although this can be a slow process that takes many months.

Childhood neuropathies

Muscle requires an intact, healthy and active nerve supply for its development and normal function, and damage to the nervous supply leads to pathological changes in the muscle. Three groups of neurogenic muscle disease can be identified according to the site of the lesion: (a) lesions in the anterior horn cells, as in the *spinal muscular atrophies*; (b) axonal lesions, as in *peripheral neuropathies*; and (c) lesions at the motor end-plate, as in *myasthenia gravis*.

Biochemical tests are of little value in the diagnosis of neuropathic conditions; indeed, a raised CK, for example in a floppy infant, would be a pointer towards a probable myopathic cause. The microscopic appearance of a muscle biopsy specimen and an EMG examination, including conduction velocity measurements, will usually provide a diagnosis.

Neuromuscular changes in human immunodeficiency virus infection

At the present time the incidence of human immunodeficiency virus (HIV)-associated myopathy in children is largely unknown, but during the past few years reports have appeared of neuromuscular diseases associated with HIV infections in adults. These include *Guillain–Barré syndrome* and other peripheral neuropathies, type 2 fibre atrophy, a necrotizing myopathy and a polymyositis-like syndrome with a raised CK and a myopathic EMG. Electron microscopic examination of muscle biopsies has also revealed mitochondrial abnormalities.

Some myopathic changes are thought to be related to use of the drug zidovudine (AZT), which is used in the treatment of HIV-positive individuals to protect against the development of acquired immunodeficiency syndrome (AIDS)-related complex. Plasma CK levels in some patients on zidovudine therapy may occasionally be found to be raised to values that are several hundred units above the top of the reference range, but levels in any one individual can fluctuate and the next measurement may yield a normal value. Muscle biopsy in such cases usually reveals only non-specific changes. Where an inflammatory myopathy is present, the plasma CK is usually much higher, reaching levels of several thousand units per litre. The type of neuropathy or myopathy encountered seems to be related to the specific stage of HIV infection and chronic asymptomatic HIV infection should be considered, particularly in high-risk individuals, in the differential diagnosis of certain acquired polyneuropathies and muscle disorders. As children who are HIV positive are now living longer it seems likely that in the near future HIV-related myopathies will begin to be recognized (for a review see Dalakas Illa, 1990).

The chronic fatigue syndrome (postviral fatigue syndrome; myalgic encephalomyelitis)

This syndrome has been reported in teenage children, particularly girls, presenting with an array of symptoms, but one of the main complaints is of excessive fatigue after mild exercise; delayed-onset muscle pain may also occur. Psychological problems, particularly depression, are also a common feature. Persistence of virus particles from an original precipitating infection, abnormal muscle membrane conductance, excessive intracellular acidosis during exercise, disturbance of the T4:T8 lymphocyte ratio (Lloyd *et al.*, 1989) and altered surface markers in T8 lymphocytes (Landay *et al.*, 1991) have all been reported. Opinions differ, sometimes sharply, as to whether the problem is a psychological disorder or one in which there is an abnormality of skeletal muscle function. Plasma CK and other biochemical tests are usually normal. Tests of muscle function are also inconclusive or normal but this may reflect the difficulty of devising tests that truly reflect the activity of daily life. Muscle biopsy in these cases generally shows only non-specific changes. The condition covers a wide variety of problems and overlaps with a similar set of symptoms, known as *effort syndrome*. The fact that the symptoms are generally similar to those of any healthy teenager who is very tired as a result of excess mental or physical effort raises the possibility that these young patients may have a heightened awareness of the normal aches and pains of everyday life. This condition has been the subject of several recent reviews (Shafran, 1991; James *et al.*, 1992).

Disorders of energy metabolism

There are a number of rare genetically determined conditions which give rise to defects in muscle energy metabolism. These include defects in the enzymes of the glycolytic pathway, mitochondrial enzymes concerned with both pyruvate and fatty acid metabolism, and the cytochrome components of the electron transport chain. All these pathways are important in the secondary supply of energy to muscles as the primary reserve of phosphocreatine becomes depleted.

In general, children with metabolic defects are of normal or near-normal strength, although in some cases weakness is also reported, but they are limited in their exercise endurance. When a defect occurs in the glycolytic pathway the exercised muscle tends to go into a painful contracture (an electrically silent contraction similar to rigor mortis). Children with mitochondrial disorders have a very limited exercise capacity, and mild exercise is associated with breathlessness, acidosis and a high blood lactate, as would be seen with a normal muscle exercising under hypoxic conditions.

Glycolytic disorders

Glycogen is a branched-chain polysaccharide which is present in most mammalian cells; the main sites of storage are skeletal muscle and the liver. The molecule consists of straight chains of glucose molecules which are branched at various points. The residues in the straight chains are joined by $\alpha(1-4)$ linkages and the branches are initiated by $\alpha(1-6)$ linkages. When required for energy metabolism, glycogen is broken down in a stepwise process which starts with the splitting of the accessible $\alpha(1-4)$ linkages by the enzyme phosphorylase; the $\alpha(1-6)$ linkages are then hydrolysed by the debrancher enzyme (amylo 1,6-glucosidase) to yield further straight chains that are then broken down in turn by phosphorylase to yield glucose-1-phosphate. This is transformed by the enzyme phosphoglucomutase into glucose-6-phosphate, which enters the glycolytic pathway to yield pyruvate; this is either further metabolized in the mitochondria or transformed into lactate under anaerobic conditions.

Research into metabolic disorders of skeletal muscle began with the discovery in 1951 of a myopathy where there was a defect in glycolysis (*McArdle's disease*). In this disease there is a specific lack of *myophosphorylase*, the first enzyme in the pathway utilizing muscle glycogen.

Children with myophosphorylase deficiency often present in their early teens when their inability to undertake normal amounts of exercise during the school day may lead to the acquisition of the label 'lazy' or even to the suggestion that some psychological disturbance may be limiting exercise performance. These patients usually have normal strength at rest but pain is experienced in the muscles during exercise and becomes severe if exercise is continued at the same intensity. There is rapid fatigue, and the muscle may go into a painful contracture if the subject is forced to continue with the exercise; in some cases continued exertion can lead to rhabdomyolysis with the risk of acute renal failure. A period of rest or more gentle exercise often leads to diminution of the pain (the *second wind* phenomenon) and the ability to continue exercise for a time. The absence of myophosphorylase means that the muscle is unable to utilize stored glycogen and this severely limits its ability to function under anaerobic conditions. As a result, there will be no rise in plasma lactate after exercise. The second wind phenomenon is a result of the utilization of blood-borne substrates, such as glucose and fatty acids, by the muscle as an alternative to muscle glycogen. Patients can be advised to eat carbohydrates to raise the blood glucose just before exercise. The specific enzyme defect can be demonstrated histochemically in sections from a muscle biopsy specimen; large amounts of glycogen are also seen in the muscle fibres. Small specimens of muscle can also be examined biochemically. There is increasing evidence for biochemical heterogeneity in McArdle's disease at both the translational and transcriptional levels. The gene, which has been partially sequenced, maps to chromosome 11. Although the majority of patients have no detectable muscle phosphorylase activity, a few cases with up to 30% of normal activity have been reported; patients with non-functional mRNA or inactive protein have also been described (McConchie *et al.*, 1989, 1991). It is becoming increasingly clear that the same phenotype is elicited by a range of genomic lesions.

McArdle's disease is categorized as type 5 in a group of glycogenoses, and a summary of these conditions is given in Table 16.1.

Mitochondrial defects

Mitochondrial myopathies are associated with profound exercise intolerance and a persistently raised blood lactate in young adults, and with muscle weakness and hypotonia in babies and young children (see also Chapter 7(b)). Microscopic examination of muscle from patients with mitochondrial myopathies shows characteristic *ragged red fibres;* this abnormal staining is due to subsarcolemmal aggregations of defective mitochondria.

A variety of defects, including the transport of pyruvate into mitochondria, lack of pyruvate dehydrogenase and defects in the electron transport chain, have been described; all these will reduce

Table 16.1 Glycogenoses: types 1–7

Type	Deficiency	Notes
1 (Von Gierke's disease)	Glucose-6-phosphatase	Hepatomegaly, growth retardation, hypoglycaemia, lactic acidosis Liver, kidney and skeletal muscles affected
2 (Pompe's disease; acid maltase deficiency)	α-1,4- and α-1,6-glucosidase	*Children*: all tissues affected; cardiomyopathy, severe weakness, large accumulations of glycogen in muscle; death usually within first year *Adults*: proximal myopathy of variable severity; weakness of respiratory muscles
3 (Cori–Forbes disease)	Debrancher enzyme	Production of limit dextrin because myophosphorylase removes from glycogen the glycosyl units attached by 1,4 linkages but cannot attack branch points; abnormal glycogen accumulates in muscle *Symptoms*: severe muscle weakness in children to asymptomatic adult forms
4 (Andersen's disease)	Brancher enzyme	Deposition of large amounts of abnormal glycogen in childhood; disease is rapidly progressive with cirrhosis and hepatosplenomegaly
5 (McArdle's disease)	Myophosphorylase	See text
6 (Her's disease)	Glycogen phosphorylase	Hepatomegaly, hypoglycaemia, lactic acidosis Skeletal muscles not affected
7 (Tauri's disease)	Phosphofructokinase (in skeletal muscle and red cells)	Clinical symptoms and exercise impairment similar to McArdle's disease Haemolytic anaemia may occur, often detected by raised reticulocyte count in peripheral blood

oxidative metabolism. The inheritance of mitochondrial disorders is unique since the mitochondrial DNA, which codes for 13 protein components of the electron transport chain, is transmitted exclusively by the mother (Harding, 1989). Poor oxidative metabolism has two consequences: a reduction in the rate of adenosine triphosphate (ATP) synthesis; and an accumulation of lactate and pyruvate, leading to metabolic acidosis. Even with an intact circulation the muscles in these patients are continually working under effectively hypoxic conditions, and the muscle function is like that of a normal subject working ischaemically.

A detailed investigation of mitochondrial function *in vitro* is required in order to diagnose fully these disorders; such tests are the province of a specialist laboratory.

Disorders of lipid metabolism

Disorders of fat metabolism should properly be classified under mitochondrial myopathies, but because defects in fatty acid metabolism do not have such a drastic effect on energy supply as do defects in the electron transport chain, they are considered separately. Young patients with disorders of fatty acid metabolism have symptoms of weakness, exercise intolerance, muscle stiffness and pain (sometimes accompanied by myoglobinuria). Symptoms are most evident at times when free fatty acids are the main substrates for energy metabolism, such as during prolonged submaximal exercise, particularly in the fasting state.

Carnitine and carnitine palmitoyl transferase

Carnitine is important in the metabolism of fats to provide muscle energy. The β-oxidation of fatty acids depends on the transport of free fatty acids into the mitochondria by a shuttle mechanism involving L-carnitine and the CPT enzymes. L-Carnitine (β-hydroxytrimethyl-γ-aminobutyric acid) can be obtained from the diet or synthesized in the liver from lysine. It is taken up by muscle from plasma and its concentration is higher in muscle than in blood. Low plasma and muscle carnitine and deficiencies of CPT have been described. Patients lacking in carnitine are weak, whereas those lacking CPT enzymes are of fairly normal strength at rest, but after fasting or exercise they may show evidence of muscle damage with a raised CK and occasionally myoglobinuria. Assays for carnitine in plasma and muscle are available but both these measurements and the assay for CPT activity are the province of a specialist laboratory.

Carnitine deficiencies may be myopathic, systemic or secondary.

Myopathic carnitine deficiency

This condition was first described by Engel and Sickert in 1972, and usually presents with lactic acidosis. Plasma L-carnitine is slightly low or normal but muscle carnitine is reduced. DL-Carnitine has been used successfully in treatment (Angelini *et al.*, 1970) and some benefit from corticosteroid therapy has been reported (Van Dyke *et al.*, 1975).

Systemic carnitine deficiency

Systemic carnitine deficiency was first reported by Karpati *et al.* (1975). Patients have a progressive neuromuscular disorder with nausea and vomiting. The condition may progress to coma and death. Plasma carnitine levels are greatly reduced to as little as 50% of normal, possibly due to impaired hepatic synthesis. Treatment with oral DL-carnitine has reportedly restored plasma levels to normal in some cases, although liver and muscle carnitine levels were not restored. If carnitine deficiency is suspected in a child, specimens of plasma and urine should be collected and stored frozen. Definitive diagnosis will require muscle and probably liver biopsies.

Myopathic and systemic carnitine deficiencies are both genetic in origin and siblings may be affected.

Secondary carnitine deficiency

Secondary carnitine deficiency has been reported in a number of conditions, including malnutrition, glutamicaciduria and Spanish oil poisoning syndrome, and in low birth-weight babies receiving total parenteral nutrition. Carnitine deficiency syndromes have recently been reviewed by Breningstall (1990).

CPT deficiency

The enzyme defect in this condition was established by DiMauro and DiMauro (1973). Serum and muscle carnitine are normal. The enzyme, which is located in both the outer and inner mitochondrial membranes, may represent two entities or may be the same enzyme in two locations.

Skeletal muscle CPT deficiency can present at any age and the youngest reported patient to date was 6 years old. There is rhabdomyolysis, often with renal failure, under conditions (e.g. prolonged heavy exercise) where there is a need for increased oxidation of fatty acids. Diagnosis is by the demonstration of the enzyme defect in fresh muscle biopsy specimens and is the province of a specialist laboratory. Treatment with a diet high in carbohydrates and low in fat, coupled with attention to the life style, has proved remarkably successful.

The periodic paralyses

The familial periodic paralyses are rare disorders characterized by periodic attacks of weakness. There are two main types, which give rise to similar muscular symptoms but have quite different causes.

Hypokalaemic periodic paralysis

Hypokalaemic periodic paralysis is a rare autosomal dominant condition which is about 100% penetrant in males but only 8% penetrant in females; it is sometimes associated with thyrotoxicosis. Presentation is most common during the second decade of life but cases have been reported from the age of 4 years. Attacks of weakness, which may progress to flaccid paralysis, occur, often during the night. Attacks are precipitated by carbohydrate meals, especially if these are taken after exercise. Giving insulin will also provoke an episode of paralysis. These attacks, regardless of the cause, can be aborted by 10–15 g oral potassium chloride or potassium citrate.

Characteristically, the weakness is accompanied by a fall in plasma potassium (values as low as 1.5 mmol/L have been reported). This potassium is not lost to the body but moves into the skeletal muscles. It is now generally accepted that the paralysis is accompanied by a large depolarization of the muscle membranes; evidence comes from recordings *in vivo* and *in vitro*. Relief of paralysis by oral KCl is accompanied by repolarization of the muscle fibres. The disorder is associated with some abnormality in the transport of K^+ and glucose into the muscle.

Hyperkalaemic periodic paralysis

It was found that treating some patients who had periodic paralysis with potassium salts made the condition worse. Investigation of this phenomenon led to the description of hyperkalaemic periodic paralysis. This condition, due to an autosomal dominant defect, generally presents in the first decade; attacks are provoked by potassium and can be relieved by glucose and insulin. In these respects the condition is the reverse of hypokalaemic periodic paralysis. During attacks, plasma potassium may rise as potassium moves out of the muscles and possibly also from other tissues.

Acetazolamide has proved effective in the long-term treatment of both types of periodic paralysis. It appears to act by reducing potassium fluxes across the muscle membranes.

Malignant hyperpyrexia

Malignant hyperpyrexia is a rare condition but one that presents considerable problems for anaesthetists. It was first described in Australia in the 1960s. Apparently healthy children and young people undergoing routine surgery can develop an alarming, and often fatal, hyperpyrexia. The disease is not confined to children; it can present at any age as the result of receiving anaesthesia for the first time. Typically, after anaesthesia has been induced, muscle stiffness is noted in the patient and is accompanied by a rapid rise in body temperature. Jaw rigidity during induction of anaesthesia is often the first indication that all is not well. There is a rapid increase in plasma potassium and lactate, and subsequent cardiac failure. Unless the signs are recognized at an early stage during anaesthesia and treatment is begun promptly, the prognosis is poor, death resulting in about 50% of cases. Myoglobinuria and renal damage are additional problems for patients who survive the acute episode. It is now clearly established that the precipitating factors are the use of *halothane* as an anaesthetic agent and/or *suxamethonium* as a muscle relaxant. These substances can potentiate the release of calcium in normal skeletal muscle, as does caffeine. In susceptible individuals this effect seems to be dangerously exaggerated. Treatment is with *dantrolene*, a substance that reduces calcium release from the sarcoplasmic reticulum.

Inheritance is autosomal dominant, although the reported incidence is twice as high in males as in females. The incidence has been variously reported as from 1 per 20 000 to 1 per 200 000 persons subjected to anaesthesia (figures for children alone are not available). This apparent difference in incidence probably represents differing anaesthetic procedures in the UK and other countries. The management of susceptible patients requires careful monitoring of body temperature and the avoidance of known precipitating agents and stress. If a hyperpyrexic attack develops, intravenous dantrolene should be administered and strenuous efforts made to limit the rise in body temperature. Reports by patients or their relatives regarding problems with anaesthesia experienced by other members of the family should never be dismissed as mere gossip; because there is at present no satisfactory screening test for this condition, such anecdotal facts may give a valuable guide to anaesthetists.

References

Angelini, C., Lucke, S. & Cantarutti, F. (1970) Carnitine deficiency of skeletal muscle: report of a treated case. *Neurology* **26**, 633–637.

Askari, A.D. & Huettner, T.L. (1982) Cardiac abnormalities in polymyositis/dermatomyositis. *Sem Arth Rheum* **12**, 208–219.

Breningstall, G.N. (1990) Carnitine deficiency syndromes. *Paediatr Neurol* **6**, 75–81.

Carroll, J.E., Brooke, M.H., De Vivo, D.C., Kaiser, K.K. & Hagberg, J.M. (1978) Biochemical and physiological consequences of carnitine palmityl transferase deficiency. *Muscle Nerve* **1**, 103–110.

Carter, N.D., Heath, R., Jeffrey, S. *et al.* (1983) Carbonic anhydrase III in Duchenne muscular dystrophy. *Clin Chim Acta* **133**, 201–208.

Dalakas, M. & Illa, I. (1990) HIV associated myopathies. In: *Paediatric AIDS: Challenge of HIV Infection in Infants, Children and Adolescents* (eds P.A. Pizzo & C. Wilford), pp. 420–429. Williams and Wilkins, Baltimore.

DiMauro, S. & DiMauro, P.M.M. (1973) Muscle carnitine palmityl transferase deficiency and myoglobinuria. *Science* **182**, 929–931.

Dubowitz, V. (1985) *Muscle Biopsy: A Practical Approach*, 2nd edn. Baillière Tindall, London.

Edstrom, L. (1970) Selective atrophy of red muscle fibres in the quadriceps of longstanding knee joint dysfunction injuries to the anterior cruciate ligament. *J Neurol Sci* **11**, 551–556.

Edwards, R.H.T., Round, J.M., Jackson, M.J., Griffiths, R.G. & Lilburn, M.F. (1983) Weight reduction in boys with muscular dystrophy. *Develop Med Child Neurol* **26**, 384–390.

Edwards, R.H.T., Wiles, C.M., Round, J.M., Jackson, M.J. & Young, A. (1979) Muscle breakdown and repair in polymyositis: a case study. *Muscle Nerve* **2**, 223–228.

Engel, A.G. & Sickert, R.G. (1972) A lipid storage myopathy responding to prednisone. *Arch Neurol* **127**, 174–181.

Eppenberger, H.M., Dawson, D.M. & Kaplan, N.O. (1967) The comparative enzymology of creatine kinases. *J Biol Chem* **242**, 204–209.

Harding, A.E. (1989) The mitochondrial genome—breaking the magic circle. *N Engl J Med* **320**, 1341–1343.

Hoffman, E.P., Monaco, A.P., Feener, C.C. & Kunkel, L.M. (1987) Conservation of the DMD gene in mice and humans. *Science* **238**, 348–350.

Jackson, M.J., Page, S. & Edwards, R.H.T. (1991) The nature of the proteins lost from skeletal muscle during experimental damage. *Clin Chim Acta* **197**, 1–8.

James, D.G., Brook, M.G. & Bannister, B.A. (1992) The chronic fatigue syndrome. *Postgrad Med J* **68**, 611–614.

Karpati, G., Carpenter, S., Engel, A.G. *et al.* (1975) The syndrome of systemic carnitine deficiency. *Neurology* **25**, 16–24.

Kunkel, L.M., Hejchancik, J.F., Caskey, C.T. *et al.* (1986) Analysis of deletions in DNA from patients with Becker and Duchenne muscular dystrophy. *Nature* **322**, 73–75.

Landay, A.L., Jessop, C., Lennette, E.T. & Levy, J.A. (1991) Chronic fatigue syndrome, clinical condition associated with immune activation. *Lancet* **338**, 707–712.

Lloyd, A., Wakefield, D. & Boughton, C. (1989) Immunological abnormalities in the chronic fatigue syndrome. *Med J Aust* **151**, 123–124.

McConchie, S.M., Benyon, R.J., Coakley, J. & Edwards, R.H.T. (1989) The expression of glycogen phosphorylase in McArdle's disease: a Southern, Northern and Western blot study. In: *Advances in Myochemistry*, Vol. 2 (ed. G. Benzi), pp. 222–223. John Libbey, Paris.

McConchie, S.M., Coakley, J. & Edwards, R.H.T. (1991) Molecular heterogeneity in McArdle's disease. *Biochem Biophys Acta* **1096**, 26–32.

Morrow, J. & Isenberg, D. (eds) (1987) Polymyositis. In: *Autoimmune Rheumatic Disease*, pp. 234–255. Blackwell Scientific Publications, Oxford.

Pachman, L.M. & Maryjowski, M.C. (1984) Juvenile dermatomyositis and polymyositis. *Clin Rheum Dis* **10**, 95–115.

Page, S., Jackson, M.J., Coakley, J. & Edwards, R.H.T. (1989) Isoforms of creatinine kinase MM in the study of skeletal muscle damage. *Eur J Clin Invest* **19**, 185–191.

Rennie, M.J. & Millward, D.J. (1983) 3-Methylhistidine excretion and the urinary 3-methylhistidine/creatinine ratio are poor indicators of skeletal muscle protein breakdown. *Clin Sci* **65**, 217–225.

Shafran, S. (1991) The chronic fatigue syndrome. *Am J Med* **90**(6), 730–739.

Van der Veen, K.J. & Willebrands, A.F. (1966) Isoenzymes of creatine phosphokinase in tissue extracts and in normal and pathological sera. *Clin Chim Acta* **13**, 312–316.

Van Dyke, D.H., Griggs, R.C., Markesbery, W. & Di Mauro, S. (1975) Hereditary carnitine deficiency of muscles. *Neurology* **25**, 154–159.

Yasmineh, W.G., Yamada, M.K. & Cohn, J.N. (1981) Postsynthetic variants of creatine kinase MM. *J Lab Clin Med* **98**, 109–118.

17: Acute Poisoning

A.T. PROUDFOOT

Introduction

Humans use many substances in attempts to alter internal and external environments to their benefit; however, most of those used to combat disease produce effects in addition to those for which the substance is primarily taken. Some of the unwanted effects are unpredictable and are usually said to be due to hypersensitivity or idiosyncrasy, while others are predictable and often directly related to the magnitude of the exposure, whether by inhalation, ingestion or skin application. When unwanted effects complicate the correct use of accepted therapeutic amounts of a drug they are usually termed adverse reactions and when they result from excessive doses, incorrect use or exposure to some substance not used therapeutically, the term poisoning is used. It will be obvious that there is no clear dividing line between most adverse reactions and poisoning. Peter Mere Latham (1789–1875) crystallized the situation by saying that 'poison and medicine are often-times the same substance given with different intents'.

Classification of childhood poisoning

The classification of poisoning is unsatisfactory. It is based on at least three factors, including the age at which poisoning occurs, the circumstances preceding it and whether it is self-inflicted or caused by another person. Five classes are commonly recognized.

Neonatal poisoning

This type of poisoning occurs very early during life. It is caused by the transfer of drugs from mother to child across the placenta before or during delivery. It is rarely, if ever, due to transmission through breast milk.

Therapeutic poisoning

This term is used when intoxication occurs during the therapeutic use of drugs and such poisoning is often due to miscalculation of doses; excessive amounts for the age and weight of the patient are given. Unfortunately, the response to drugs often varies widely from one individual to another and cannot be predicted with certainty. At one extreme a 'standard' dose may be subtherapeutic and ineffective, while at the other extreme features of overdosage rapidly develop. In some cases therapeutic intoxication is due to inherited factors, immaturity or to the limited capacity of some drug-metabolizing enzymes. Clearly, this type of poisoning cannot be confined to a specific age group, although it is more common in younger children and infants.

Accidental poisoning

Accidental poisoning is the commonest type of childhood poisoning and involves the child eating some substance which is potentially harmful. It is uncommon before the age of 7 months but after that age the incidence of accidental poisoning rises rapidly, reaching a peak in the second and third years of life (Fig. 17.1). This is simply because crawling and walking permit curious children, who explore the environment with their mouths as well as their hands and eyes, to have access to potential poisons in an unsupervised moment. By the fifth year of life the incidence has declined sharply to

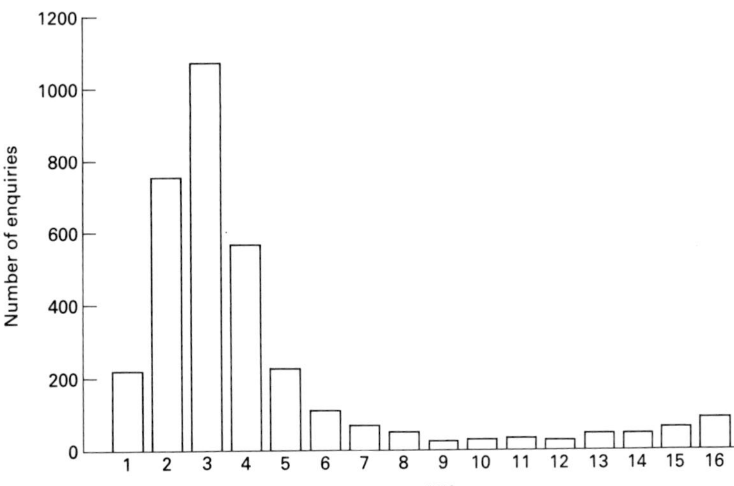

Fig. 17.1 Age distribution of toxic exposures in childhood (based on enquiries to the Scottish Poisons Information Bureau).

continue at a low level until adulthood is approached.

Occasionally, accidental poisoning results from a child being given a substance by another person, e.g. drugs proffered as sweets by another youngster who knows no better, or surgical spirit given in mistake for gripe water by a mother.

Self-poisoning

This term is used to describe the knowing, deliberate consumption of an overdose of a drug or the ingestion (very rarely inhalation) of some other substance. Such behaviour is the most common cause of poisoning in adults and is seldom encountered before the age of 10 years. However, it is well documented in children of younger ages and from the age of 8 years upwards any drug overdose should be regarded with suspicion as it may not be accidental. Failure to appreciate the true nature of the poisoning may prevent the identification and possible correction of the interpersonal, social and occasionally psychiatric factors which commonly precipitate or contribute to such acts. Self-poisoning is often the response to a disagreement with a close relative and in this context often appears unreasonable or excessive. However, to the individual no other solution or escape may be apparent at the time of crisis. Less frequently, self-poisoning is intended to punish another person or manipulate that person into doing something he or she would otherwise not do.

Non-accidental poisoning

Non-accidental poisoning is an unsatisfactory term used to describe the deliberate, covert poisoning of a child, usually by one parent. It is a variant of the syndrome sometimes referred to as Munchausen by proxy. This form of child abuse has only been recognized in recent years and presents considerable diagnostic challenges. The illness which results from non-accidental poisoning may be used by one spouse to distress the other or, by making the child the focus of attention, the parents may be able to cooperate more harmoniously for a period.

Epidemiology of childhood poisoning

Size of the problem

It is impossible to estimate with any certainty the number of children poisoned in a year. Many incidents probably never come to medical attention. Hospital admissions indicate the minimum size of the problem and analysis of 10% of samples for England and Wales showed that the number of admissions annually for poisoning in children under the age of 15 years more than doubled between 1964 and 1969, to reach 28 000. Thereafter, the number rose to 29 000 in 1976, only to fall to about 13 000 in 1985. The admission rate for children up to 4 years of age is about 10 times that for those aged 5–9 years and in both age groups boys predominate.

The poisons involved

The poisons involved vary noticeably from one age group to another, even in what might be termed childhood. They probably also vary from one society to another. Thus, children below the age of 5 years seem able to eat the most unpalatable and unlikely things. The range of substances taken varies widely (Table 17.1) but, regardless of age, drugs are responsible for at least 50% of incidents, and of them paracetamol in one formulation or another is by far the most important. Vitamin preparations, oral contraceptive agents, dental caries prophylaxis tablets, salbutamol, cough mixtures, inhalants and antibiotics are also common. Fortunately, more toxic drugs, such as tricyclic antidepressants, theophylline derivatives and other psychotropic agents, are less frequently involved. The non-drug poisons are drawn from a vast array of household products kept in cupboards below kitchen sinks and in bathrooms. Polishes, cosmetics, bleaches, disinfectants and turpentine-substitute, amongst other things, are consumed with monotonous regularity, as are house plants and bulbs. The appearance of serious toxins, such as mercury, sodium fluoride and methanol, in the list of those consumed frequently is alarming until it is realized that they can be explained on the basis of their presence (usually in non-toxic forms or in very low concentrations) in the bulbs of clinical thermometers, dental caries prophylaxis tablets and perfumes, respectively. Plants become a more common source of poisons in the 5- to 9-year-old age group, while alcohol (ethanol), butane, 'magic' mushrooms and typewriter correcting fluid reflect experimentation with 'recreational' substances by those aged 10–15 years.

Table 17.1 Comparison of exposure to groups of toxins in young children and adults (based on enquiries to the Scottish Poisons Information Bureau)

Group of toxins	Age <5 years (%) (n = 2972)	Age >15 years (%) (n = 2873)
Drugs	50.9	74.8
Household products	20.4	7.3
Toiletries	7.4	1.4
Petroleum distillates	5.7	1.5
Chemicals	6.2	10.1
Plants and fungi	5.3	1.5
Others	4.2	3.2

Morbidity from poisoning

Children admitted to hospital because of poisoning are unlikely to have a duration of stay of longer than 1 day and it is commonly less. This suggests that the morbidity from childhood poisoning is usually minimal and that admission is usually a precautionary measure. Many years ago, a detailed analysis of 178 episodes in children under the age of 15 years showed that only 22% had symptoms of poisoning, while only 1% suffered permanent injury. In a further 13% of cases there was good evidence that the child had taken some substance, although symptoms did not develop. The majority (65%), however, were regarded as 'scares' or 'non-toxic exposures', either because there was no evidence that the child had absorbed any poison or because it was inherently innocuous or taken in only small amounts. Even with psychotropic drugs serious intoxication is uncommon. There is no reason to think that the situation has anything but improved in the past decade or more, although the belief would be difficult to substantiate. Unfortunately, the difficulty of distinguishing between 'scares' and true poisoning will continue indefinitely since it can only be made with certainty in retrospect. 'Unnecessary' hospital admissions can never be completely avoided.

Mortality from poisoning

Deaths from poisoning in British children aged less than 10 years between 1958 and 1977 showed the peak numbers to occur in 1964 and 1965. That 81%

of 598 deaths were due to drugs highlights the importance of the latter, but the individual drugs or groups of drugs have changed over the years. Whereas salicylates were by far the most important in the early part of the period studied, their contribution declined while that of tricyclic antidepressants increased to the extent that they were the single most important cause of drug deaths at the end of the period studied. Recent surveys indicate that they remain important.

Corrosive substances, hydrocarbons and unusual chemicals are other important causes of non-drug deaths. In contrast, plants accounted for only three of the 598 deaths in 20 years.

Clinical biochemistry in acute poisoning

The importance of clinical chemistry in the management of acute poisoning should not be underestimated. Poisons tend to cause multiple organ damage and, of the 20% of children who have symptoms as a result of accidental poisoning, only a minority will be suitable for treatment with antidotes or techniques intended to enhance elimination of the poisons. In the majority of cases the clinician has no choice but to support vital functions until the toxin is metabolized and/or excreted and the patient recovers. Biochemical investigations are therefore essential to:

1 Detect acid–base and electrolyte disturbances.
2 Detect depression of function (e.g. respiration).
3 Assess the severity of organ damage (principally renal and hepatic).
4 Monitor the efficacy of treatment intended to correct these abnormalities.

Arterial blood gas analysis

Arterial blood gas analysis is probably the single most important and urgent biochemical investigation in severe acute poisoning. Elevation of the arterial carbon dioxide tension indicates inadequate ventilation, whether spontaneous or mechanically assisted, and a need for corrective measures. Hyperventilation, indicated by a low arterial carbon dioxide tension, is usually due to overdosage of salicylates and rarely to sympathomimetic drugs (mainly bronchodilators) or phenoxyacetate ('hormone') weedkillers.

Table 17.2 Poisons which may cause metabolic acidosis

Benzene hexachloride	Metaldehyde
Carbon monoxide	Methanol
Cyanide	Paraquat
Digoxin	Quinidine
Ethylene glycol	Salicylates
Fluoride	Solvent inhalation
Iron	Theophylline
Isoniazid	

Low arterial oxygen tensions may also be the result of ventilatory failure, such as may occur in deep coma owing to a variety of drugs in patients whose lungs are clinically and radiologically normal and in the absence of significant carbon dioxide retention. Mismatching of ventilation and perfusion is the most likely explanation. In other cases, hypoxia is secondary to pulmonary aspiration of gastric contents or pneumonia.

Arterial blood gas analysis is also essential to detect and assess the severity of any acid–base disturbance. Metabolic acidosis is a common finding in any severe poisoning, particularly with the drugs and chemicals listed in Table 17.2. In such cases the acid–base abnormalities are usually associated with clinical evidence of serious toxicity. While profound changes in arterial hydrogen concentration may be fatal, they may also enhance or diminish the toxicity of some poisons by altering their distribution throughout the body fluids. Thus, acidaemia (elevated arterial hydrogen ion concentration) facilitates the movement of weakly acidic drugs such as salicylates from the plasma into the brain, thereby increasing toxicity (see below). Conversely, the toxic effect on the central nervous system of phencyclidine, a basic drug widely abused in North America, is reduced by metabolic acidosis.

Plasma electrolyte concentrations

The alterations of plasma electrolyte concentrations which may occur in association with acute poisoning or its treatment are listed in Table 17.3.

Table 17.3 Alteration of plasma electrolyte concentrations in poisoning

	Plasma concentration	
Electrolyte	Raised	Reduced
Sodium	Saline emetics	
Potassium	Metabolic acidosis Cardiac glycoside poisoning Overdosage with KCl or potassium and diuretic combinations Rhabdomyolysis	Sympathomimetic poisoning Forced alkaline diuresis
Calcium	Chronic hypervitaminosis D	Ethylene glycol Oxalic acid Inorganic fluorides Forced alkaline diuresis
Bicarbonate		Metabolic acidosis Respiratory alkalosis

Hypernatraemia

Hypernatraemia is more likely to be the result of the over-enthusiastic use of salt solutions to induce vomiting rather than a feature of poisoning itself. It may be fatal in both adults and children. Syrup of ipecacuanha is now the emetic of choice.

Hypokalaemia and hyperkalaemia

Hyperkalaemia may result from the ingestion of potassium salts, when elimination is impaired secondary to acute renal failure and when there is release from damaged tissue, particularly skeletal muscle (rhabdomyolysis). The acute changes in plasma potassium concentrations found in association with poisoning with cardiac glycosides and sympathomimetics are caused by redistribution between the intracellular and extracellular fluids and are particularly important in the myocardium. Overdosage with these drugs is frequently associated with cardiac arrhythmias.

Digoxin, the most widely used cardiac glycoside, inhibits the membrane sodium–potassium pump, thereby permitting the leakage of potassium from the cells. Hyperkalaemia may result, and in adults at least has been used as an indication of the severity of poisoning. In contrast, ephedrine stimulates cyclic adenosine monophosphate (cAMP) and theophylline inhibits phosphodiesterase. Both these drugs therefore increase intracellular potassium at the expense of extracellular concentrations and the resulting hypokalaemia can be severe and develop with striking rapidity. Hypokalaemia has also been reported in association with salbutamol and terbutaline overdosage. Clearly, the more acute the change in potassium gradient across the myocardial cell membrane, the greater the potential for arrhythmias.

Hypocalcaemia and hypercalcaemia

Plasma calcium concentrations may be severely reduced after the ingestion of oxalic acid or inorganic fluorides. These compounds combine with calcium to form insoluble calcium oxalate and fluoride, respectively, which in turn may be deposited in tissues and excreted in the urine. Ethylene glycol is also important in this respect since it is widely used as an antifreeze and is metabolized by alcohol and aldehyde dehydrogenases to produce oxalic acid. Calcium also stabilizes the myocardium and frequent recurring ventricular fibrillation is a well-documented result of severe hypocalcaemia after ingestion of fluorides. Fortunately, the fluoride preparations available in the community (mainly for the prophylaxis of dental caries) contain very small quantities and ingestion of even 100 tablets or more is unlikely to produce serious toxicity.

Despite the importance of vitamin D in the control of plasma calcium concentrations, acute overdosage with this vitamin does not cause problems. Hypercalcaemia is only likely with chronic hypervitaminosis D.

Forced alkaline diuresis used to be the standard treatment for moderate and severe salicylate and phenobarbitone poisoning and produced hypocalcaemia and hypokalaemia. However, it is now used much less frequently for the former and has been superseded by repeated oral doses of activated charcoal for the latter.

Plasma urea and creatinine

The measurement of plasma urea and creatinine to monitor renal function is too well known to warrant comment. It should be remembered that in the presence of severe hepatic necrosis (as in paracetamol poisoning) the plasma urea is a poor indicator of renal function, and the plasma creatinine is much more reliable.

Liver function tests

Overdosage with paracetamol, iron salts or phenylbutazone and ingestion of some halogenated hydrocarbons may cause acute hepatocellular necrosis. Serial measurement of the plasma bilirubin and alanine aminotransferase and alkaline phosphatase activities is useful for following the severity and course of the liver damage, but for the clinician the prothrombin time (or the international normalized ratio, INR) is usually available more quickly and is more relevant. Severe hepatic damage soon causes hypoprothrombinaemia, which may lead to potentially lethal haemorrhage and require administration of fresh frozen plasma or clotting factor concentrates. The time course for abnormalities of the prothrombin time tends to parallel the biochemical tests of liver function.

Plasma glucose

Measurement of the plasma glucose concentration as an emergency is valuable in ascertaining whether or not undiagnosed coma is due to hypoglycaemia. It is also mandatory for the assessment of the effects of insulin and sulphonylureas and for monitoring the effects of treatment. Plasma glucose estimations may have to be made frequently over the course of several days when chlorpropamide has been ingested, since the plasma half-life of the drug is of the order of 48 h and hypoglycaemia may recur for up to 5 days after acute poisoning, despite adequate carbohydrate intake. Patients who have been severely hypoglycaemic for protracted periods may not recover consciousness after administration of intravenous dextrose and it may be important for the clinician to confirm that the treatment has restored normoglycaemia and that failure to improve is probably owing to underlying cerebral oedema. Regular measurement of the plasma glucose concentration is also indicated when a poison has produced severe hepatocellular necrosis, since hypoglycaemia may be one of the few treatable complications.

Hypoglycaemia may rarely result from ethanol and salicylate overdosage in childhood.

Hyperglycaemia and glycosuria have been reported after poisoning with isoniazid, organophosphate insecticides and theophylline and related sympathomimetic drugs.

Methaemoglobinaemia

Methaemoglobinaemia is an uncommon complication of acute poisoning, although it is well documented after the ingestion of chlorates, nitrates, aniline, sulphonamides, phenacetin, dapsone, some local anaesthetics and phenazopyridine. It has rarely been reported after the ingestion of a formulation of paraquat with monolinuron and with phenol. Severe methaemoglobinaemia is usually complicated by haemolysis with subsequent haemoglobinuria, hyperkalaemia and jaundice resulting from an increase in unconjugated bilirubin.

Other enzymes

Apart from aspartate and alanine aminotransferases, the plasma activity of various enzymes may be increased after acute poisoning. Lactate dehydrogenase activity and creatine kinase activity are often greatly increased in patients who have been unconscious, and may reflect skeletal muscle damage. Myoglobinuria and acute renal tubular necrosis with its usual biochemical features may follow.

Myopathy and rhabdomyolysis are uncommon and have usually been reported in adults after poisoning with substances such as carbon monoxide, isopropanol, phencyclidine, cocaine, theophylline, amphetamines, opioid analgesics, benzene hexachloride, colchicine and 2,4-dichlorophenoxyacetic acid.

Cholinesterase activity may be severely reduced after exposure to organophosphate and carbamate insecticides, and although the degree of reduction correlates poorly with clinical intoxication, measurement is important in confirming exposure and monitoring the response to antidotes.

Examination of the urine

The colour of the urine may change after the ingestion of some poisons. Bilirubinuria secondary to severe hepatocellular necrosis does not require comment. A green or blue discoloration of the urine usually indicates the presence of methylene blue. Rifampicin, phenindione and phenazopyridine make the urine turn orange/red as does ferrioxamine, the chelate resulting from the treatment of acute iron overdosage with desferrioxamine. Phenols and cresols produce shades varying from grey to black. The presence of cresols is readily confirmed by their characteristic odour. A brown-coloured urine that is not owing to bilirubinuria may also be found in association with myoglobinuria, haemoglobinuria and the presence of high concentrations of paracetamol metabolites.

Unusual crystalluria occasionally occurs in acute poisoning. Oxalate crystals may be present after ingestion of ethylene glycol, and following massive overdosage of the anticonvulsant primidone this may crystallize out of the urine as it cools.

Plasma drug concentrations

The majority of poisoned patients recover with supportive care alone and knowledge of the plasma concentration of the poison seldom alters clinical management. However, the demonstration of a poison in the plasma and measurement of its concentration are potentially useful in order to:

1 Confirm exposure to, and absorption of, significant amounts. This alone is usually of little value except in patients in whom non-accidental poisoning or brain death is suspected.

2 Assess the severity of poisoning and determine the need for treatment with specific antidotes (for example iron salts, paracetamol, methanol, ethylene glycol) or, rarely, measures to enhance elimination, such as haemodialysis (for salicylates, ethylene glycol, methanol and lithium) or haemoperfusion (for barbiturate hypnotics, paraquat and theophylline).

3 Monitor the efficacy of treatment. This is particularly important when techniques such as exchange transfusion, peritoneal dialysis, haemodialysis and haemoperfusion are used in attempts to increase elimination of poisons.

4 Help determine when to restart regular drug therapy. The usual clinical situation is an epileptic who has superimposed acute poisoning with anticonvulsants on long-term treatment.

5 Investigate the pharmacokinetics of serious poisons so that rational treatment can be devised. This indication for the measurement of plasma drug concentrations may appear superfluous for a routine hospital laboratory but maximum information from each episode is essential if progress is to be made. Paraquat poisoning is a pertinent illustration. In the knowledge that the mortality from paraquat ingestion is high, many patients were treated in blind desperation with protracted peritoneal dialysis, haemodialysis and haemoperfusion and suffered significant complications from the treatment. Later development of a satisfactory paraquat assay and analysis of samples collected at the time made possible retrospective assessment of the severity of poisoning and made clear that many patients who were not seriously poisoned had (with the best of intentions) been exposed unnecessarily to the hazards of a 'treatment' which was probably of no value.

Interpretation of plasma drug concentrations

The plasma concentration of a poison is determined by many factors including the dose, route of absorption, volume of distribution, rates of distribution, metabolism and excretion, and the time since exposure; age, tolerance, other drugs, biologically active metabolites and some diseases (particularly hepatic and renal) influence toxicity. With so many variables, it is hardly surprising that the correlation between plasma concentrations and toxicity is often poor. Equally, it will be clear that statements about 'fatal' plasma concentrations and concentrations above which certain treatments should be implemented must be interpreted with extreme caution.

Although the results of laboratory drug analyses are important, they can only be interpreted in the context of the overall clinical situation. In the great majority of cases it is important to treat the patient and not the plasma concentration of the poison.

It must be emphasized that at present there is considerable confusion concerning the most appropriate units for expression of the concentrations of drugs and other poisons in body fluids. Both molar and mass units are used, even though drugs are prescribed in mass units. This is a dangerous situation and can lead to confusion and delay when reporting results by phone to a clinician.

The time response for drug assays

The degree of urgency of a drug analysis should be determined by how much the result will influence clinical management. In Table 17.4 toxicological analyses have been divided into three groups. The first group requires an immediate response from the laboratory since the results largely determine whether specific antidotes are administered or steps are taken to increase the elimination of the poison. It must be remembered that most antidotes (particularly that for iron poisoning) are potentially toxic and their indiscriminate use cannot be condoned.

Since all but a very small number of patients poisoned with central nervous system depressants are managed by supportive measures alone, emergency measurement of the plasma concentrations of these drugs is rarely indicated. However, such requests are justified in cases of very severe poisoning if some special measure, such as haemodialysis or haemoperfusion, is being contemplated.

Investigation of new or unfamiliar drugs or poisons usually requires the development and validation of new assays, and even if urgent analysis were possible it is unlikely that the result would be helpful since the lack of data would make interpretation impossible or uncertain. Although a 'leisure' response is therefore sufficient in this situation, the importance of collecting as much information as possible should not be underestimated. Similarly, many of the drugs which have been in clinical use for many years have not been adequately investigated after an overdosage of them has been taken, and would bear scrutiny using present-day laboratory techniques.

Unfortunately, as with many laboratory investigations, clinicians frequently fail to ask themselves how the result of a drug analysis will alter treatment, and many requests (particularly regarding salicylates in unconscious patients) seem to originate at a spinal reflex level. The only mitigating factor in the approach is that the responsibility for the immediate management of acute poisoning frequently falls on the most junior and inexperienced members of medical teams who gain comfort and confidence (which is not always justified) from toxicological investigations.

Table 17.4 The laboratory response time for measurement of plasma drug concentration

Immediate response	Salicylate Paracetamol Iron Lithium Theophylline Paraquat
Response within 24 h	Carbamazepine Phenobarbitone Phenytoin
'Leisure' response	New drugs Serious poisoning with 'old' drugs

Analytical methods

The responsibility for the choice of analytical method lies with the laboratory staff and it is essential that they should use specific methods which measure the concentrations of active drugs only. Some years ago problems arose with paracetamol because some laboratories used assays which measured not only paracetamol but also its inactive metabolites, thus producing totally misleading results upon which physicians no doubt acted. Other commonly used assays, particularly for sedative drugs, may also be non-specific and of limited value. Highly specific methods now need not be more complex and time consuming. The passage of time has created a much greater awareness of such issues and similar problems are unlikely unless some new toxin appears on the clinical scene.

In some cases the effort spent trying to measure only the active drug concentration would not be justified because toxicity correlates poorly with plasma concentrations. In general, however, there is a need for laboratories which perform drug analysis only intermittently to review their methods so that valid results can be obtained within the restraints imposed by the need for speed and simplicity.

Drug screening

Clinicians frequently request drug 'screens' and the measurement of plasma salicylate concentrations in unconscious patients, including children, in whom there is no obvious focal neurological lesion. From a laboratory point of view, the fact that few young physicians today seem to appreciate that salicylates seldom cause impairment of consciousness is counterbalanced by their almost total unfamiliarity with barbiturates, and consequently a much reduced demand for urgent barbiturate levels. Drug 'screens' are tedious, expensive and time consuming and even the best laboratories are unable to identify the agent responsible in every case of confirmed poisoning. Inevitably, drug screening has to be based on the expectation that drugs commonly prescribed (mainly psychotropic drugs and analgesics) are likely to be consumed accidentally. A screen should not be undertaken without prior detailed discussion with the consultant looking after the child. It is essential not only to determine what drugs may have been available for poisoning, but also to obtain a list of those which may have been given during the course of resuscitation or subsequent treatment, lest the latter appear in the screen and create confusion. Drug screens may justifiably be requested when:

1 A patient is desperately ill and the clinician feels that it is necessary to know which poison has been taken, in case specific treatment or methods to enhance elimination are required.

2 A patient is ill with a presumed diagnosis of poisoning. In the absence of improvement or in the face of deterioration the clinician may wish to confirm that poison is present and, if so, which, in case treatment should be modified.

3 Non-accidental poisoning of children is suspected. Such cases present the ultimate challenge to paediatric chemical pathologists, and require analysis of samples contemporaneous with clinical events. The possible toxins are legion in number and the use of techniques involving mass spectrography will almost certainly be necessary.

Drug analysis should never become a substitute for careful clinical and social assessment, or a comforter for insecure physicians. However, not all doctors deal with acute poisoning sufficiently frequently to gain the confidence that allows them to dispense with requests for drug analysis except in the circumstances stated above. Careful discussion between the clinicians and the laboratory staff should be able to reduce the number of requests for drug 'screens' to those which are likely to be clinically profitable.

Specific drug intoxications

Paracetamol

In Britain, paracetamol is the most widely used non-prescription analgesic and suspected overdosage with it was the single most important substance in the age groups 1–4, 5–9 and 10–14 years who were the subjects of enquiries to the Scottish Poisons Information Bureau in 1992. Fortunately, serious childhood poisoning with paracetamol is uncommon and only a few cases of severe toxicity have been reported. The most likely explanation for this is that young children tend to take liquid paediatric formulations which contain such low concentrations that consumption of even a whole bottle is unlikely to lead to serious toxicity. Of possible relevance in children is that early induction of vomiting and qualitative differences in the metabolism of the drug may reduce conversion to the toxic metabolite, but the fetus is capable of metabolizing sufficient paracetamol after maternal overdosage to incur liver damage.

Mode of toxicity

Paracetamol is extensively metabolized in the liver to form sulphate and glucuronide conjugates. However, a small proportion is converted to a highly reactive intermediary metabolite, *N*-acetyl-*p*-benzoquinone imine (NAPQUI), by hepatic mixed-function oxidases and is normally inactivated by conjugation with hepatic glutathione before being excreted in the urine as cysteine and mercapturic acid conjugates. When a large amount has been taken the hepatic stores of glutathione are rapidly exhausted and the metabolite is then free to combine with the SH groups in cell macromolecules, causing hepatocellular necrosis. The same process occurs in the renal tubules. In adults, liver enzyme induction by anticonvulsant therapy increases the susceptibility to paracetamol overdosage, and the same is probably true in children.

Features of poisoning

Most children are unlikely to develop symptoms and, whatever the reasons, all the evidence indicates that serious toxicity in childhood is rare compared with that which develops in adults. In the early stages, symptoms include nothing more than nausea and vomiting, which resolve within a few hours. In exceptional cases features of hepatocellular damage may develop 12–24 h after ingestion with abdominal pain, hepatic tenderness and, later, mild jaundice. Proteinuria and haematuria may presage the onset of renal tubular necrosis.

Maximum abnormalities of liver function tests occur 3–4 days after ingestion and plasma alanine aminotransferase activities may be among the highest recorded. Values of up to 15 000 U/L are not uncommon in severe poisoning in adults, although unlikely to be attained in children. The plasma bilirubin is only moderately elevated and the alkaline phosphatase is usually normal. Patients with very severe poisoning may go into hepatic encephalopathy and die 3–7 days after ingestion, usually from cerebral oedema. It must be emphasized that such occurrences are rare in accidental childhood poisoning.

Plasma paracetamol concentrations

The severity of paracetamol poisoning cannot be reliably assessed from the dose alleged or thought to have been taken, or from the symptoms in the first few hours. As in adults, the amount ingested is often only a matter for speculation and there may well be considerable inter-individual variation in the response to a given quantity. Paracetamol overdosage was identified as a problem in adults long before it occurred in children and, inevitably, experience with older patients has had to be extrapolated to the young. The severity of liver damage is therefore predicted from the plasma paracetamol concentration related to the time interval since ingestion. The limited data available for children suggest that the same relationships pertain and the laboratory must therefore be prepared to measure plasma paracetamol concentrations as an emergency.

Laboratory investigations

Liver function tests should be monitored if the patient is at risk of liver damage, but not otherwise. Measurement of the plasma urea, creatinine and electrolytes is obviously required if there is any suggestion of renal failure.

Treatment

Most children who ingest excessive quantities of paracetamol require no more treatment than emptying the stomach to reduce further absorption. However, if the plasma concentration related to the time since ingestion indicates a risk of hepatic damage it would be appropriate to give specific antidotes such as *N*-acetylcysteine or, if this is not available, methionine, in doses appropriate to the weight of the patient. In adults, treatment with intravenous *N*-acetylcysteine within 8 h gives virtually complete protection against liver damage and may also have some effect up to 24 h. It is generally assumed that children will benefit similarly and there is little choice but to manage them as adults since the data necessary to develop more appropriate risk assessment and treatment are, for ethical reasons, unlikely ever to be forthcoming. The current expert view is that the treatment threshold should be reduced for those who are at additional risk of liver damage by virtue of enzyme induction. It is recommended that a plasma paracetamol concentration of half the value of that of the conventional value at any given time after ingestion is appropriate.

Salicylate poisoning

Salicylates are derived from salicylic acid, which was originally obtained from willow bark. Salicylic acid itself is now used only in corn solvents and ointment for removing the scales of skin diseases such as psoriasis. Aspirin (acetylsalicylic acid) has analgesic, anti-inflammatory and antipyretic properties, and although it is largely being replaced by paracetamol it is still commonly kept as a household remedy for a wide variety of common minor ailments. Paediatric preparations, some attractively flavoured and coloured, are also available and ease of access, particularly in bathroom medicine cabinets, may be partly responsible for the incidence of accidental salicylate poisoning in childhood. The relatively new and widespread therapeutic role for aspirin in coronary artery and cerebrovascular diseases in adults to some extent offsets its disuse for other purposes and may again increase its availability to young children. Methyl salicylate (oil of wintergreen) is toxic in relatively small amounts and is fortunately seldom encountered. Newer salicylates such as benorylate and diflunisal have not yet been shown to be important causes of poisoning in childhood. Salicylates have been implicated in Reye's syndrome (see Chapters 5 and 13).

Mode of action

Salicylates have a direct toxic action on the inner ear and renal tubules and stimulate the respiratory centre in the mid-brain. They also uncouple oxidative phosphorylation and have complex effects on intermediary metabolism.

Pharmacokinetic considerations

Salicylates are readily absorbed in the gastrointestinal tract, but absorption may be delayed if enteric-coated or delayed-release formulations have been ingested. Salicylic acid in ointments may be absorbed through the skin; this is particularly important in children where the thickness of the dermis is less and its surface area in relationship to body weight is greater than in adults. Salicylates are also contained in teething gels, excessive application of which may lead to poisoning.

Salicylates are extensively bound to plasma albumin and are metabolized in the liver to salicyluric acid (the major metabolite), salicyl acyl and phenolic glucuronides and gentisic acid. Salicyluric acid formation is readily saturated, even with therapeutic doses, and children are at great risk of therapeutic intoxication. Neonates also metabolize and eliminate salicylates more slowly than their seniors. Renal excretion becomes an important route of elimination as plasma concentrations increase. Salicylates are weak acids and their distribution in body fluids is profoundly influenced by changes in hydrogen ion concentration. Acidaemia reduces ionization and facilitates the transfer of salicylates from plasma into tissues, while the reverse occurs with alkalaemia.

Features of poisoning

In overdosage, salicylates commonly cause nausea, vomiting, ringing in the ears, deafness and some degree of dehydration owing to sweating and hyperventilation. The metabolic rate is increased as a result of the uncoupling of oxidative phosphorylation and this in turn causes warm, flushed extremities and bounding pulses. Hyperpyrexia may occur, especially in young children. In severe poisoning consciousness may be impaired with delirium and, rarely, convulsions, hypoglycaemia, pulmonary oedema or renal failure may develop.

The acid–base disturbance

The acid–base changes in acute salicylate poisoning are complex. Most patients have a combination of respiratory alkalosis, secondary to the hyperventilation produced by stimulation of the respiratory centre, and metabolic acidosis, owing to disturbances of intermediary metabolism as a result of uncoupling of oxidative phosphorylation. Increased urinary excretion of lactic acid and ketoacids has been demonstrated in poisoned children. The age of the patient is the most important factor in determining which component dominates the acid–base disturbance. Children under the age of 4 years commonly present with a dominant metabolic acidosis, while older children and the vast majority of adults have a dominant respiratory alkalosis. Since the arterial hydrogen ion concentration has such an important influence on the distribution of salicylate throughout the body fluids, interest in the acid–base disturbance is more than academic. Knowledge of the arterial hydrogen ion concentration in an ill patient is as essential to assessment as measurement of the plasma salicylate concentration. The longer the duration of poisoning the more likely a child is to have a dominant metabolic acidosis.

Treatment

Treatment of salicylate poisoning is directed towards preventing further absorption by emptying the stomach and replacing fluids and electrolytes lost by vomiting, sweating and hyperventilation. Activated charcoal may also be given.

The symptoms of salicylate poisoning can be alleviated more rapidly by enhancing elimination of the drug from the body. Fortunately, few children become so severely poisoned with aspirin that it is necessary to resort to haemodialysis, which is the treatment of choice for such cases. Until recently, forced alkaline diuresis was the standard treatment for mild and moderate poisoning but this is now falling into disuse since it is appreciated that alkalinization of the urine is more important than inducing rapid urine flow. Instead, repeated doses of oral activated charcoal (provided that they are tolerated – a far from certain hope) are the first measure to be taken, since by adsorbing salicylate in the gut near-zero concentrations are maintained, thereby creating a gradient between plasma in the capillaries of the intestinal villi and the lumen across which the drug may move. Potentially any child who has ingested aspirin is suitable for this form of treatment but it is especially suitable for those under 5 years of age who have plasma concentrations exceeding 350 mg/L (2.5 mmol/L) and older children with levels of 500 mg/L (3.6 mmol/L) or more, as in adults. The reasons for differentiating the two groups is that the acid–base disturbance may seriously increase toxicity in younger children.

Plasma salicylate concentrations

Plasma salicylate concentrations are widely used to assess the severity of salicylate poisoning, although in general they correlate less well with clinical severity than one might wish. This is hardly surprising since the commonly used assays measure free (active) and protein-bound (inactive) drug, the former only becoming the greater at concentrations encountered in severe acute poisoning. Cerebrospinal fluid (CSF) salicylate concentrations correlate much better, presumably because they reflect the concentrations of unbound drug in the plasma, but it is inconceivable that CSF salicylate concentrations will ever be accepted as the routine method of assessing the severity of poisoning.

Tricyclic antidepressants

Tricyclic antidepressants, such as amitriptyline, dothiepin, doxepin and imipramine, are widely available in the community and may be accidentally consumed by children.

Toxicity

The toxicity of tricyclic antidepressants is largely due to their ability to antagonize the effects of acetylcholine and to potentiate the effects of catecholamines at peripheral nerve endings, ganglia and in the brain. They also have a direct depressant effect on the activity of the myocardium.

Features of poisoning

The peripheral features of poisoning with this group of drugs include dilatation of skin blood vessels (producing flushed warm skin), absence of sweating, dilatation of the pupils and paralysis of accommodation, leading to blurred vision. These manifestations are neatly summarized as 'red as a beet, hot as a hare, dry as a bone and blind as a bat'. Urinary retention may also occur. Some patients can also be described as being 'mad as a wet hen' since the central nervous system anticholinergic effects include drowsiness, confusion and delirium with visual hallucinations.

Large overdoses cause coma, convulsions and cardiac effects, including hypotension, tachycardia, cardiac conduction abnormalities and arrhythmias. The latter are serious and occasionally fatal, although with the exception of sinus tachycardia they are uncommon.

Treatment

Treatment of tricyclic antidepressant poisoning is almost entirely a matter of supporting vital functions. There is no satisfactory antidote. A clear airway and adequate ventilation must be maintained and, depending on the circumstances, the stomach emptied. Hypotension and hypothermia are seldom serious problems and sinus tachycardia alone does not merit treatment. The management of more serious arrhythmias is unsatisfactory and a matter of clinical debate. Identification and correction of hypoxia and metabolic acidosis are more important than the administration of drugs.

Laboratory investigation

Since the management of tricyclic antidepressant poisoning is supportive, the main demand on the laboratory should be for blood gas analysis and electrolyte estimations.

Plasma concentrations of tricyclic antidepressants can be measured and are likely to be greater than 1 mg/L in patients with evidence of serious toxicity. In the author's opinion there is little merit in measuring such concentrations as an emergency, but it has been claimed that this may be indicated in a child with disturbed behaviour or coma without a history of possible ingestion.

Benzodiazepines

The group of substances known as benzodiazepines contains many members and must be one of the most widely prescribed groups of agents in recent decades. They have been extensively used as anxiolytics, sedatives, hypnotics and anticonvulsants, and, not surprisingly in view of their uses, are amongst the most common agents swallowed deliberately by adults. Their widespread availability puts them within reach of children.

Toxicity

In general, the benzodiazepines are of relatively low toxicity and certainly much safer in overdosage than the barbiturates which they have replaced. They depress the central nervous system and, potentially, all the body functions controlled by the brain. For the same reasons they potentiate the effects of other central nervous system depressants taken simultaneously.

Features of poisoning

Drowsiness with slurred speech and an unsteady gait are the early signs of poisoning. Coma may then develop, but respiratory depression, hypotension and hypothermia are uncommon. However, some individuals seem to react more severely than others and require more intensive management.

Treatment

Support of the vital functions is usually all that is required, although activated charcoal may be given if there is no risk of the child aspirating it into the respiratory tract. Complete recovery within a few hours is to be expected. In severe poisoning, administration of flumazenil, a specific benzodiazepine antagonist, will reverse (albeit temporarily) all the features of intoxication.

Laboratory investigations

No laboratory investigations are likely to be required with this form of poisoning. Rarely, blood gas analysis may be indicated.

Iron salts

Overdosage with iron salts, particularly ferrous sulphate, has long been one of the most serious forms of childhood poisoning. Despite this, the mortality from iron poisoning has probably been exaggerated by biased reporting of serious cases, as with other poisonings.

Features of poisoning

Iron salts have corrosive effects on the gastrointestinal tract and early features of poisoning include vomiting, abdominal pain and diarrhoea. Tablet residues may make the vomitus and stool dark grey or black and blood may also be present if ulceration occurs. Hypotension and impairment of consciousness may develop later and are the best clinical indicators of severe poisoning. Jaundice owing to hepatic necrosis may become apparent after 3–4 days and lead to hepatic encephalopathy and death in rare cases. Renal failure may also occur. Laboratory features include a polymorphonuclear leucocytosis, metabolic acidosis and hyperglycaemia. Weeks or months after ingestion children may present with vomiting and a metabolic alkalosis secondary to pyloric stricture subsequent to healing of mucosal ulceration by fibrosis.

Treatment

Absorption of iron from the gastrointestinal tract is reduced by emptying the stomach but theoretically attractive methods of chelating iron remaining in the stomach are now regarded as of no value. Desferrioxamine may be given by intramuscular or intravenous injection to chelate circulating iron. These measures are supplementary to supportive care including replacement of electrolytes, fluid and blood lost through the gut and, if necessary, conventional treatment for hepatic and renal failure.

Serum iron concentrations and other laboratory investigations

Reliable assessment of the severity of iron poisoning is seldom possible from the early features and since parenteral desferrioxamine has potentially serious adverse effects it should not be given indiscriminately. The laboratory must therefore assist with the measurement of serum iron concentrations as an emergency. There is no need, however, to measure the total iron-binding capacity, which may be falsely elevated and is of no value in deciding management.

The limited data available indicate that concentrations peak within 2–3 h of ingestion, sometimes attaining values such as 90 mg/L (1600 µmol/L); these then decline very rapidly as the iron is distributed to tissues. Chelation with desferrioxamine is most effective if it is given while plasma iron concentrations are high, and values in excess of 5 mg/L (90 µmol/L) are usually accepted as indications for desferrioxamine administration. Subsequent serum iron estimations may be necessary to monitor the efficacy of treatment and help decide when to stop the administration of the antidote. It should be noted that ferrioxamine (the iron–desferrioxamine complex) is excreted in the urine and colours it orange/red.

The child with serious iron poisoning also requires measurement of plasma electrolytes, to guide replacement therapy, together with arterial blood gas analysis, for correction of acid–base disturbance, urea and creatinine and liver function tests.

Other drugs

Other drugs which occasionally cause childhood deaths (mainly digoxin, quinine, anticholinergic drugs, theophylline and narcotic analgesics) have few implications for clinical laboratory investigation other than the biochemical investigation required for the appropriate supportive care. Childhood overdosage with thyroxine is not uncommon: measurement of the T_3 and T_4 concentrations in a sample taken 6–12 h after ingestion should be undertaken. Normal results make follow-up unnecessary.

Non-drug poisonings

Paraquat poisoning

Weedkillers containing paraquat are undoubtedly amongst the most toxic chemicals which are freely available in the community. Many of the early deaths from paraquat, including some of children, were due to accidental ingestion of liquid concentrations such as Gramoxone (20% paraquat), which had been transferred to soft-drink bottles from the original containers. Since then accidental poisoning has become rare and most cases are due to deliberate self-poisoning by adults. Children may still have access to paraquat in granular formulations, such as Weedol and Pathclear, which are available to the general public but the paraquat content of these is low (2.5%); it is difficult to swallow dry granules and the risk of poisoning is small, although the lethal dose obviously depends on the age and weight of the child.

The ingestion of fruit or vegetables which have recently been sprayed with properly diluted paraquat is exceedingly unlikely to lead to significant poisoning. Inhalation of paraquat aerosols does not cause systemic poisoning.

Toxicity

Solutions of paraquat are directly corrosive to the upper gastrointestinal tract and skin. It also causes renal failure and hepatic damage, but death is due to progressive respiratory failure secondary to pulmonary alveolar cell proliferation and fibrosis.

Pharmacokinetic considerations

Ingested paraquat appears to be very rapidly absorbed, peak plasma concentrations being attained within 2–3 h. They then decline rapidly as the paraquat is taken up by tissues, particularly the lungs and kidneys. Most of the absorbed paraquat is excreted in the urine.

Features of poisoning

While most cases of paraquat ingestion in childhood are likely to be uneventful, the potential for serious harm should not be underestimated. Paraquat causes nausea, vomiting, abdominal pain and diarrhoea. About 2 days after ingestion burns of the mouth, tongue, pharynx and oesophagus become apparent, making swallowing, coughing and speaking difficult. These features are a function of the concentration of the paraquat solution swallowed rather than the total amount. In severe poisoning acute renal tubular necrosis and renal failure may develop within 24–36 h and haemodialysis may be necessary. Mild hepatocellular damage also occurs but is less common. Death from paraquat poisoning is usually due to profound hypoxia secondary to progressive pulmonary fibrosis, but with massive ingestion deaths from cardiorespiratory failure may occur within a few hours.

Treatment

There is no effective treatment for paraquat poisoning. Measures to prevent further absorption by emptying the stomach, giving adsorbents, such as Fuller's earth or bentonite, and removing paraquat from the circulation by haemodialysis and haemoperfusion are of little or no value. Similarly, attempts to mitigate its cellular effects with corticosteroids, immunosuppressive drugs and superoxide dismutase have been remarkable only for their lack of success in preventing fatal outcomes when toxic amounts have been ingested.

Laboratory investigation

The toxicity of paraquat is known to many members of the public and, not surprisingly, any possibility of poisoning creates considerable alarm. The laboratory has a vital role in allaying these anxieties at an early stage and in protecting patients from potentially hazardous, unnecessary or ineffective treatment.

Absorption of paraquat can be readily confirmed or refuted by adding small quantities of sodium dithionite and sodium bicarbonate to 5 mL urine. If paraquat is present a colour, varying between pale green and dark blue, develops instantly. A negative test on urine passed within about 4 h after the alleged ingestion indicates that little or no paraquat has been absorbed and that there is no risk. If the urine test is positive, interpretation is more difficult because the severity of systemic poisoning cannot be inferred from the intensity of the blue colour developed on testing the urine. Strongly positive tests are compatible with both insignificant and significant poisoning, depending on the time interval between ingestion and passing the urine used for the test. Accurate assessment of severity requires measurement of the plasma paraquat concentration, which must then be related to the time interval since ingestion. Concentrations exceeding 2.0, 0.6, 0.2 and 0.1 mg/L at 4, 6, 13 and 24 h, respectively, after ingestion are likely to be fatal.

Volatile substance abuse (solvent inhalation, glue sniffing)

The practice of inhaling the volatile hydrocarbons contained in glues, dry-cleaning agents, paint thinners, typewriter correction fluids and petrol has increased greatly in recent years and is a cause for considerable concern. A variety of solvents are involved, including acetone, carbon tetrachloride, toluene, trichloroethane, trichloroethylene and tetrachloroethylene. Deliberate solvent inhalation is usually a group activity, predominantly among boys, the peak incidence being in those between the ages of 13 and 15 years, although some individuals are as young as 8 years. Adhesives are usually squeezed into a polythene or potato-crisp bag before being inhaled, while dry-cleaning agents are sniffed after being poured on to a piece of cloth.

Features

The solvents mentioned above are lipid soluble and their major early effects are on the brain, causing features similar to alcohol intoxication. There may be euphoria or mild excitement with dysarthria, ataxia and disturbed behaviour. Some children may experience hallucinations or become suicidal. As exposure continues, consciousness is depressed, leading to coma, convulsions and metabolic acidosis. Ulceration of the lips and nostrils may be present. Rarely, deaths occur, most frequently from asphyxia caused by the polythene bags used to contain the adhesive and less frequently from burns and the toxic effects of the solvents.

Recovery from solvent inhalation is usually complete but a variety of short- and long-term complications have been reported, including permanent brain damage (petrol and toluene), reversible renal damage (toluene and halogenated hydrocarbons), Fanconi's syndrome, renal tubular acidosis and peripheral neuropathy (hexane and methyl-isobutyl ketone).

Treatment

Removal from exposure is as much treatment as is required in most cases. Supportive measures are necessary if consciousness is impaired, and renal and hepatic damage are managed conventionally.

Laboratory investigations

Laboratory investigations are unnecessary in most cases of solvent inhalation. If exposure has been prolonged or heavy, renal and liver function tests may be required. In severe poisoning arterial blood gas analysis is indicated.

A diagnosis of solvent inhalation is not always possible from the clinical features alone. Toluene is present in most of the abused adhesives and its presence in the blood may be demonstrated by gas chromatography.

Carbon monoxide

Children are as much at risk as adults from poisoning by carbon monoxide, often as one of the major components of smoke in domestic fires or, less frequently, when home-heating appliances are incorrectly installed or flues become blocked. The features of toxicity are numerous and often so easily attributed to other causes that the correct diagnosis may not be considered unless circumstantial evidence is strong. Headache, nausea, vomiting and diarrhoea are amongst such minor complaints, while more serious intoxication causes progressive impairment of consciousness and all the other complications of depression of the central nervous system. Supportive care, particularly the administration of high inspired oxygen concentrations, is indicated.

Carboxyhaemoglobin concentrations

The laboratory might reasonably be asked to measure carboxyhaemoglobin concentrations if carbon monoxide poisoning is suspected, and to do so as a matter of urgency in some cases. Such requests stem from a belief (strongly held by a few clinicians and presented to others who have not considered the issues until they have to manage a crisis) that any victim of carbon monoxide exposure who has subsequently shown neurological features or, regardless of the features of poisoning, has a carboxyhaemoglobin concentration exceeding 40%, should be referred immediately for hyperbaric oxygen therapy. While such a course of action remains controversial, it undoubtedly imposes significant logistical problems which the laboratory can help resolve and will have to face until further evidence on the value of hyperbaric oxygen therapy is forthcoming.

Household products and toiletries

As indicated in Table 17.1, exposure to these everyday products is commonplace and seldom seems to cause serious harm. The most toxic components of toiletries are ethanol and methanol, the latter being present in such low concentrations that adverse effects after ingestion are improbable unless large amounts have been swallowed – something which packaging usually makes impossible. Important household toxins include drain cleaners (sodium hydroxide) and battery and descaling fluids (usually strong acids); these can cause severe corrosive injury to the upper alimentary tract. Supportive measures and surgical intervention may be required but there are no unusual laboratory implications. Petroleum distillates (long-chain hydrocarbons) are common constituents of many household products ranging from cosmetics and ointments to paraffin and petrol. Occasionally, ingestion of them causes significant respiratory problems but, again, there are no unusual implications for the laboratory.

Mercury

Children are frequently exposed to metallic mercury by crushing the bulbs of clinical thermometers placed in their mouths. This is not a cause for concern. Possibly of greater toxicological importance is exposure to mercury salts leaking from the small batteries used to power hearing aids, toys, calculators and a wide variety of common household appliances. The leak is usually due to the effects of gastric acid on the casing and is only a problem if the battery is retained in the body for several days. Leakage may not only cause local corrosive damage to the mucosa but significant absorption of mercury has been demonstrated. The precise magnitude of this risk is not clear but it has been suggested that blood mercury concentrations should be measured to identify children who should be treated with chelating agents. Fortunately, most batteries pass through the gut within 4 days of having been swallowed.

Further reading

General

Proudfoot, A.T. (1993) *Acute Poisoning*. Butterworth Heinemann, Oxford.

Epidemiology

Campbell, D. & Oates, K.R. (1992) Childhood poisoning – a changing profile with scope for prevention. *Med J Aust* **156**, 238–240.

Hsu, R., Coles, E.C. & Routledge, P.A. (1986) Childhood poisoning in Wales: experience of the Welsh National Poisons Unit. *Hum Toxicol* **5**, 373–376.

Lacroix, J., Gaudreault, P. & Gauthier, M. (1989) Admission to a pediatric intensive care unit for poisoning: a review of 105 cases. *Crit Care Med* **17**, 748–750.

Litovitz, T. & Manoguerra, A. (1992) Comparison of pediatric poisoning hazards: an analysis of 3.8 million exposure incidents. *Pediatrics* **89**, 999–1006.

Vale, J.A., Meredith, T.J. & Buckley, B.M. (1987) Acute pesticide poisoning in England and Wales. *Hlth Trends* **19**, 5–7.

Wiseman, H.M., Guest, K., Murray, V.S.G. & Volans, G.N. (1987) Accidental poisoning in childhood: a multicentre survey. 1. General epidemiology. *Hum Toxicol* **6**, 293–301.

Neonatal poisoning

Elhassani, S.B. (1986) Neonatal poisoning: causes, manifestations, prevention and management. *South Med J* **79**, 1535–1543.

Accidental poisoning

Ashton, M.R., Sutton, D. & Nielson, M. (1990) Severe magnesium toxicity after magnesium sulphate enema in a chronically constipated child. *Br Med J* **300**, 541.

Jakobson, A.M., Kreuger, A., Mortimer, O. *et al.* (1992) Cerebrospinal fluid exchange after intrathecal methotrexate overdose. A report of two cases. *Acta Pediatr* **81**, 359–361.

Jonville, A.P., Barbier, P., Blond, M.H. *et al.* (1990) Accidental lidocaine overdosage in an infant. *Clin Toxicol* **28**, 101–106.

Mucklow, E.S. (1988) Accidental feeding of a dilute antiseptic solution (chlorhexidine 0.05% with cetrimide 1%) to five babies. *Hum Toxicol* **7**, 567–569.

Springer, M., Olson, K.R. & Feaster, W. (1986) Acute massive digoxin overdose: survival without use of digitalis-specific antibodies. *Am J Emerg Med* **4**, 364–368.

Non-accidental poisoning

Alexander, R., Smith, W. & Stevenson, R. (1990) Serial Munchausen syndrome by proxy. *Pediatrics* **86**, 581–585.

Dockery, W.K. (1992) Fatal intentional salt poisoning associated with a radio-opaque mass. *Pediatrics* **89**, 964–965.

Hickson, G.B., Altemeier, W.A., Martin, E.D. & Campbell, P.W. (1989) Parental administration of chemical agents: a cause of apparent life-threatening events. *Pediatrics* **83**, 772–776.

Paracetamol

Haibach, H., Akhter, J.E., Muscato, M.S. *et al.* (1984) Acetaminophen overdose with fetal demise. *Am J Clin Pathol* **82**, 240–242.

Jones, A.F., Harvey, J.M. & Vale, J.A. (1989) Hypophosphataemia and phosphaturia in paracetamol poisoning. *Lancet* **ii**, 608–609.

Penna, A. & Buchanan, N. (1991) Paracetamol poisoning in children and hepatotoxicity. *Br J Clin Pharmacol* **32**, 143–149.

Salicylates

Proudfoot A.T. (1990) Salicylates and salicylamide. In: *Poisoning and Drug Overdose* (eds L.M. Haddad & J.F. Winchester), pp. 909–920. WB Saunders, Philadelphia.

Iron salts

Klein-Schwartz, W., Oderda, G.M., Gorman, R.L. *et al.* (1990) Assessment of management guidelines. Acute iron ingestion. *Clin Pediatr* **29**, 316–321.

Tenenbein, M. (1991) The total iron-binding capacity in iron poisoning. *Am J Dis Child* **145**, 437–439.

Theophylline and other bronchodilators

Jarvie, D.J., Thompson, A.M. & Dyson, E.H. (1987) Laboratory and clinical features of self-poisoning with salbutamol and terbutaline. *Clin Chim Acta* **168**, 313–322.

Sessler, C.N. (1990) Theophylline toxicity: clinical features of 116 cases. *Am J Med* **88**, 567–576.

Thyroxine

Lin, T.-H., Kirkland, R.T. & Kirkland, J.L. (1988) Clinical features and management of overdosage with thyroid drugs. *Med Toxicol* **3**, 264–272.

Laboratory

Wiley, J.F. (1991) Difficult diagnoses in toxicology. Poisons not detected by the comprehensive drug screen. *Pediatr Clin N Am* **38**, 725–737.

18: Non-Essential Trace Elements and Their Toxic Effects

H.T. DELVES

Introduction

Most of the 92 elements in the Periodic Table have no known essential function in humans and are present in body tissues and fluids as environmental contaminants. The body-fluid reference concentrations that are generally considered as acceptable are given in Table 18.1 for 13 of the more commonly encountered toxic trace elements. When these concentrations are exceeded there is a risk of toxicity; this needs to be assessed by further analyses and by biochemical and clinical investigations.

The non-essential trace elements exert their toxic effects by competing with chemically similar essential trace elements (e.g. Al and Ca, Pb and Zn, Hg and Se) for binding sites in enzymes, or in molecules with transport or structural roles. Three elements are chosen for detailed discussion: aluminium, lead and mercury. All are commonly encountered in the environment but have contrasting biochemical and clinical effects when accumulated in body tissues and fluids. In severe toxicity all three produce severe damage to the central nervous system. Aluminium and lead (but not mercury) can induce anaemias and although both are mainly deposited in bones, only aluminium causes bone disease. Lead and mercury (but not aluminium) cause renal tubular damage but only mercury causes glomerular nephritis.

Aluminium

Despite its ubiquity in foods and in the environment, aluminium has no known biological function in humans (Underwood, 1977) and is toxic when excessively accumulated in body stores. Exposure to aluminium can occur via oral intake from foods, water or pharmaceutical products; intravenous fluids; or inhalation of dusts and from skin contact. The major excretory route for absorbed aluminium is renal with little or no biliary elimination (Alfrey, 1986). Excessive accumulation of aluminium and its subsequent toxicity are usually associated with impaired renal function, together with either abnormally high oral intakes or direct input of aluminium into the circulatory system, e.g. from infusion or dialysis fluids. Infants might be at particular risk of aluminium toxicity because of the immaturity of their gastrointestinal tract, renal system and blood–brain barrier.

Sources

Dietary sources

The concentrations of aluminium in foods and beverages range from <1 μg/kg to >1 g/kg depending upon the type of food, the degree of contamination during manufacture/processing and the presence of permitted aluminium-containing additives (Delves *et al.*, 1988). Consequently, there is a wide range of dietary intakes of aluminium. Breast-fed infants receive about 3–5 μg/day, whereas infants fed specialized formulas can receive 500–2000 μg/day (McGraw *et al.*, 1986; Greger, 1992). The aluminium content of some commercial babyfoods can exceed 10 mg/kg and thus provide intakes of more than 2 mg/day, which is comparable with the adult daily intake of 2–4 mg/day (Suchak, 1992; Delves *et al.*, 1993).

Aluminium concentrations in drinking water are dependent upon the local geology, pH, etc., but are

Table 18.1 Concentrations of some toxic metals in body fluids. After Delves and Shuttler (1991); Walker (1992)

Element	Specimen	Units	Reference range
Al	Serum	μmol/L	<0.4
Ag	Serum	nmol/L	<1
Au	Serum	μmol/L	<0.4
Ba	Serum	nmol/L	<7
Be	Urine	nmol/L	<100
Bi	Blood	nmol/L	0.5–4.0
Cd	Blood	nmol/L	<15
Hg	Urine	nmol/L	<20
Ni	Serum	nmol/L	<30
Pb	Blood	μmol/L	<0.5
Sb	Blood	nmol/L	<8
Te	Blood	nmol/L	<2
Tl	Urine	nmol/L	<1

often less than 10 μg/L. The use of aluminium sulphate to clarify some water supplies produces elevated concentrations but these should be less than 200 μg/L (Water Supply Regulations, 1989).

Pharmaceutical products

Aluminium-containing phosphate-binding gels have long been used to treat hyperphosphataemia in patients with chronic renal disease. The high oral doses (100 mg/kg per day) needed to minimize the intestinal absorption of phosphate have often produced aluminium toxicity (Greger, 1992). Alternative preparations which do not contain aluminium, e.g. $MgCO_3$ or $CaCO_3$, are in use (see Schneider, 1986).

Dialysis fluids containing high concentrations of aluminium – either from the water supply or from other contamination – have produced severe toxicity in patients with renal disease (Kerr & Ward, 1986). The upper concentration limit recommended for these fluids is 30 μg/L (Berlin, 1986).

Aluminium contamination of intravenous pharmaceuticals, nutrients and blood products has been well documented (Fell *et al.*, 1986; Koo & Kaplan, 1988). Infants receiving intravenous feeding solutions with high aluminium levels have had elevated concentrations of aluminium in the plasma, urine and bone with a high risk of toxicity (Sedman *et al.*, 1985). A 3-month-old infant died with a brain-aluminium concentration of 40 μg/g (20 times normal) after receiving prolonged total parenteral nutrition with an aluminium input of 22 μg/kg per day (Bishop *et al.*, 1989). There are proposals to limit the aluminium load to infants from infusion solutions to below 0.074 μmol/kg per day (2 μg/kg per day), which is regarded as safe (ASCN/ASPEN, 1991).

Other sources

The concentration of aluminium in house-dust can be about 2% m/m (Delves *et al.*, 1988). Since children ingest approximately 100 mg/day of soil or dust from hand-to-mouth activity (Davies, 1987) this source could add about 2000 μg/day to a child's oral intake of aluminium.

Uptake and excretion

The intestinal absorption of dietary aluminium is low, only about 1% for adults receiving 1–5 mg/day, but at higher intakes (1–3 g/day) some 20–300 mg/day is retained in the body (Gorsky *et al.*, 1979). The absorption is enhanced by dietary constituents such as citrate and is manifested by increases (temporary) in plasma-aluminium concentrations (Slanina *et al.*, 1986; Sieniawska, 1993). It is possible that citrate could have caused increased absorption of aluminium from infant formulas to produce elevated plasma-aluminium concentrations (N. Hawkins *et al.*, in preparation).

Little is known about the retention of absorbed aluminium. However, an elegant study by Priest *et al.* (1991) on one adult who received ^{26}Al intravenously has shown that although most is excreted in the first day, some 7% is retained even after 1 year. Sedman *et al.* (1985) found that premature infants on total parenteral nutrition and with normal renal function retained about 75% of infused aluminium. This is similar to the 60–74% retention reported for adults with impaired renal function, and greater than the 40% retention reported for adults with normal renal function (Klein *et al.*, 1982; Maharaj *et al.*, 1987).

Toxicity

The manifestations of aluminium toxicity are: microcytic anaemia, osteomalacia, osteodystrophy and a progressive encephalopathy leading to dementia and death (Kerr & Ward, 1986). The microcytic anaemia does not involve iron deficiency and the mechanism by which it is induced by aluminium is not clear. Aluminium accumulates at the mineralization front of bone and reduces bone formation, bone mineralization and both activity and numbers of osteoblasts. Aluminium may also cause bone disorders involving a suppression of parathyroid hormone secretion and depressed levels of 1,25-dihydroxyvitamin D (Koo & Kaplan, 1988; ASCN/ASPEN, 1991).

The inhibitory effect of aluminium on dihydropteridine reductase, leading to reduced brain neurotransmitters (Leeming & Blair, 1979), and the effect of aluminium on cholinergic neurotransmission (Marquis, 1982) provide a strong basis for the neurotoxicity of aluminium. Reduced levels of biopterin and of dihydropteridine reductase have been observed in Alzheimer-type dementia (Young *et al.*, 1982), and epidemiological studies (Martyn *et al.*, 1989) have shown an increased incidence of Alzheimer's disease in some regions of the UK where aluminium is used in the treatment of drinking-water supplies. Aluminium has been identified as aluminosilicate in neurofibrillary tangles and in senile plaques of the brains of patients with Alzheimer's disease (Candy *et al.*, 1986). However, the role of aluminium as a causative agent in the development of this disease remains speculative and there is some doubt about the accumulation of aluminium in the brains of patients with Alzheimer's disease (Lovell *et al.*, 1993).

Indicators of toxicity

Serum

The concentration of aluminium in the serum or plasma is the most commonly used index of increased aluminium uptake and toxicity. The reference values given in Table 18.2 are a useful guide in patient management but are generally poor predictors of the pathologies associated with increased uptake of aluminium. Although most clinical manifestations of aluminium toxicity are associated with serum-aluminium levels in excess of 3.7 pmol/L, lower concentrations may not be regarded as 'safe' because of the excessive accumulation of aluminium in bone (Channon *et al.*, 1986). Ideally, the serum-aluminium concentration should be maintained as low as possible, preferably below 0.4 pmol/L.

Table 18.2 Reference ranges for aluminium in serum

Concentration of Al (µmol/L)	Notes
<0.4	Healthy subjects, normal renal function
0.5–2.1	Increased uptake of aluminium
≥2.2	Excessive accumulation of aluminium; risk of toxicity
≥3.7	Cause for concern; high risk of toxicity

Other indices

Chappuis *et al.* (1989) investigated the relationship between hair-, serum- and bone-aluminium in haemodialysis (adult) patients. They found that hair-aluminium could not replace bone- and serum-aluminium as a criterion of osteodystrophy in haemodialysed patients. Serum-aluminium correlated fairly well ($r = 0.67, p < 0.05$) with aluminium in cortical bone but not in trabecular bone. Accumulation of aluminium in bone may also be detected by aluminium staining of biopsy samples (McClure, 1986).

The level of aluminium in cerebrospinal fluid (CSF) has been claimed to be a better index of risk of encephalopathy (>5 µg/L vs 'normal' of <1 µg/L) than is the concentration of aluminium in plasma (Brancaccio *et al.*, 1986). However, neither this nor the other above-mentioned measures are suitable for routine patient management.

Treatment of aluminium poisoning

Intravenous administration of desferrioxamine (DFO) has been proven to be effective in removing aluminium from patients with renal disease (Ackrill, 1986). Dialysis dementia, aluminium osteodystrophy and the anaemia induced by excessive aluminium uptake have all improved with DFO treatment, at doses from 40 to 100 mg/kg (Ackrill, 1986; Weiss *et al.*, 1989). The latter authors compared the efficacy of haemodialysis, haemofiltration and haemoperfusion following DFO infusion. They showed that haemodialysis was effective in removing aluminium overload and was recommended for patients without severe symptoms. However, for patients with encephalopathy or painful bone disease subsequent treatment using either haemofiltration or a combination of haemodialysis and haemoperfusion with activated charcoal was more efficient and was preferred (Weiss *et al.*, 1989).

Lead

Reference concentrations in blood

There has been a general decline in blood-lead concentrations of children in the UK since the early 1970s. The upper limit of the reference range for healthy children, determined from biological monitoring studies in 1979–1981 as part of European-wide surveys, was 1.2 μmol/L (DoE, 1983). This was significantly lower than the previously accepted limit of 1.8 μmol/L established in the mid-1960s (Moncrieff *et al.*, 1964). Subsequently, further declines in blood lead were noted in the UK Environmental Lead Monitoring (ELM) Programme 1984–1987, which was designed to assess the effect of the reduction of petrol lead. These studies showed that 95% of children attending schools in heavily trafficked areas had blood-lead concentrations below 0.65 μmol/L (DoE, 1990). Furthermore, extrapolation of the fall in the average 95th percentile of all groups studied (almost 2500 people per year) predicted a current (1993) upper limit of about 0.50 μmol/L. This prediction is supported by recent (1991/1992) studies carried out in Southampton on 55 children living in a rural area, but in housing built on contaminated land. The highest blood-lead concentration found was 0.45 μmol/L. These data suggest that childhood blood-lead concentrations greater than 0.50 μmol/L indicate increased exposure to, and uptake of, lead and this requires investigation to identify and remove the source. This upper limit is comparable with that seen in Sweden (0.34 μmol/L) (Schütz *et al.*, 1989), and with the upper limit of 0.50 μmol/L proposed by the Centers for Disease Control in the USA (CDC, 1991). Interpretation of blood–lead concentrations in excess of this upper limit is given in Table 18.3.

Table 18.3 Interpretation of blood-lead concentrations

Blood-lead concentration (μmol/L)	Comment
≤0.5	Levels generally considered acceptable
0.6–0.75	Increased lead uptake; confirm result with repeat analysis; if necessary, identify and remove the source of lead
0.76–1.2	Identify and remove the source of lead
1.2–1.9	Indicates excessive exposure and the result must be considered in conjunction with the clinical findings; signs and symptoms unlikely
≥2.0	Clinical symptoms and signs may be apparent
≥4.0	Dangerous concentrations; risk of encephalopathy
~5.5	Lowest levels at which deaths from acute encephalopathy have been reported

Notes: If excessive exposure is found in a child, as indicated by a high blood-lead concentration, siblings in the family should have their blood-lead levels measured. In suspected severe/acute lead poisoning an X-ray examination of the abdomen is of value in identifying radio-opaque fragments of the ingested source of lead.

Sources

The provisional tolerable weekly intake of lead for one child is only 140 μg/week (0.68 μmol/week). Increases above this may occur from lead in drinking water containing more than 10 μg/L (0.05 μmol/L)

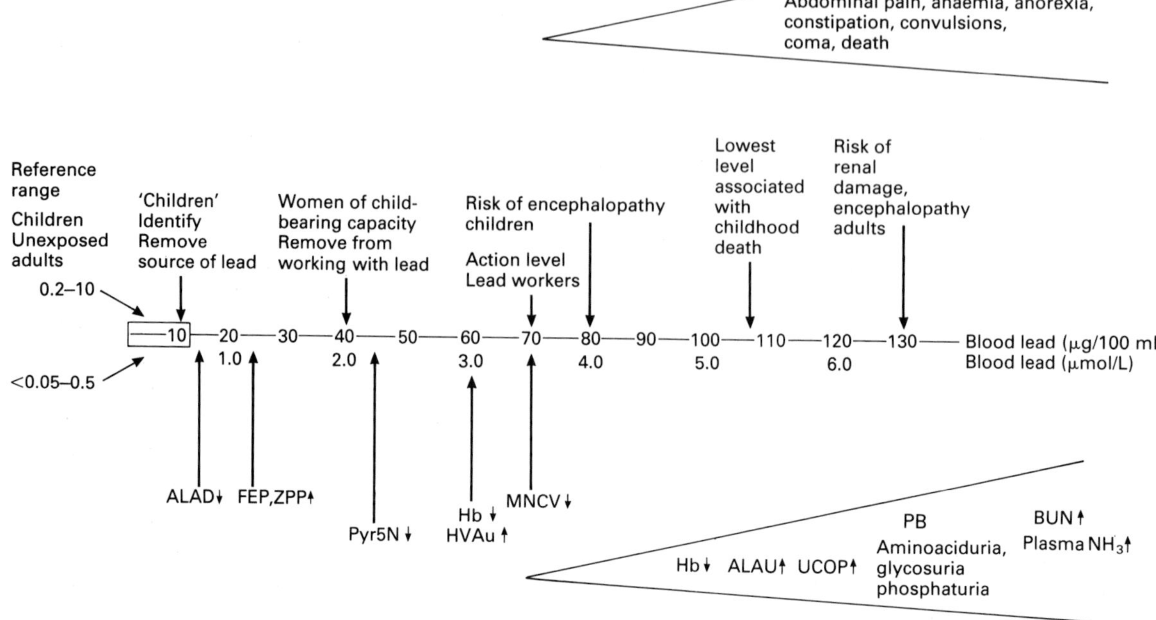

Fig. 18.1 Biochemical and clinical effects of increased lead absorption. ALAD, δ-aminolaevulinic acid dehydratase; FEP, free erythrocyte protoporphyrin; ZPP, zinc protoporphyrin; Pyr5N, pyrimidine 5′-nucleotidase; Hb, haemoglobin; HVAu, homovanillic aciduria; MNCV, mean nerve conduction velocity; ALAU, aminolaevulinic aciduria; UCOP, urinary coproporphyrins; BUN, blood urea nitrogen; PB, punctate basophilia.

and from involuntary ingestion of lead from household dust via hand-to-mouth activity (Davies, 1987). Children with pica may excessively ingest lead from commonly available sources, e.g. painted surfaces, dusts and soil.

The source which still most frequently causes acute (and chronic) lead poisoning in young children is old paint, usually located in old housing. Concentrations of lead in the dried paint film can exceed 5% m/m, so that ingestion of small amounts can produce toxicity in a young child. Ethnic medicines and cosmetics, e.g. 'surma', can also contain very high concentrations of lead and have caused lead poisoning.

It is possible to identify the source of lead poisoning from a number of suspected materials by measurement of stable lead isotopes in these materials and in the child's blood (Delves, 1988).

Indicators of lead exposure

The best index of increased exposure to inorganic lead is a blood-lead concentration which increases sharply after exposure. The biochemical and clinical effects associated with increased blood-lead concentrations are summarized in Fig. 18.1. The diverging sections in this figure imply a progressive increase in the severity of the manifestations encompassed within the lines.

Decreased δ-aminolaevulinic acid dehydratase activity in blood is a sensitive index of the blood-lead level, particularly for moderately increased exposure. Zinc protoporphyrin in blood is not sufficiently sensitive to detect increased exposure at moderately increased blood-lead levels of 0.6 to 1.2 μmol/L and is not specific for lead (e.g. Fe deficiency anaemia).

Effects of increased lead uptake

Acute poisoning

The initial symptoms of acute childhood lead poisoning may include vomiting, apathy, stupor and ataxia. An abdominal X-ray is of value in revealing radio-opaque fragments of the ingested source of lead. In addition to a high blood-lead concentration, other biochemical signs which may be found include increased protoporphyrin, amino acids – especially δ-aminolaevulinic acid – and sugars in the urine, plus a raised zinc protoporphyrin in the blood and punctate basophilia (see Fig. 18.1). The nature of the treatment is determined by the child's clinical condition and the blood-lead concentration.

Chronic poisoning

Anaemia and increased red-cell fragility are seen with prolonged increased exposure to lead and are a consequence of the effect of lead on haem biosynthesis. Although δ-aminolaevulinic acid dehydratase begins to be inhibited at blood-lead concentrations of <0.5 to 0.75 μmol/L the clinical signs are rarely apparent at blood-lead concentrations below 2.0 μmol/L.

The effects of blood-lead levels <1.2 μmol/L on children's intellect and behaviour are beyond the remit of this work and are excellently reviewed elsewhere (MRC, 1988).

Treatment of lead poisoning

Chelation therapy is usually only necessary with acute poisoning and when the blood-lead concentration is greater than 4.0 μmol/L. With elevated blood-lead levels (2.5–4.0 μmol/L) the decision whether or not to chelate is dependent upon the child's clinical condition. It is rarely necessary to use chelation therapy with blood-lead levels <2.5 μmol/L and an effective treatment is to separate the child from the source of lead.

Four chelators are in current use: $CaNa_2EDTA$ (ethylenediaminetetraacetate), BAL (dimercaptopropanol), penicillamine and dimercaptosuccinic acid (DMSA). The first two are given intramuscularly or intravenously and are established for the safe reduction of very high (>4 μmol/L) levels of lead in blood. Usually these drugs are used alternately in a treatment regime, typically 12.5 mg/kg per day of BAL in divided doses given 4-hourly and 40 mg/kg per day of $CaNa_2EDTA$ for 5–6 days (CDC, 1991).

Oral penicillamine (25–35 mg/kg per day divided into two or three doses up to a total of about 500 mg/day) is useful for reducing lower blood-lead levels of 2.5 to 3.5 μmol/L when there is no evidence of encephalopathy. It should be noted that D-penicillamine is mildly nephrotoxic and that the related compound *N*-acetyl-DL-penicillamine has fewer side effects. Although DMSA has been known to bind heavy metals, its use in chelation therapy is relatively new. It has a similar chemical structure to BAL but is water soluble and may be given orally. It has been shown that for adults 30 mg/kg per day of DMSA is more effective in reducing blood lead than is intravenous $CaNa_2EDTA$. More work is needed to establish the usefulness of DMSA in children and the extent of any potential adverse drug effects (Graziano, 1986).

The decrease in blood-lead concentration during treatment should be monitored every 2–3 days until it has fallen below 2.5 to 3.0 μmol/L, and thereafter at 5- to 10-day intervals until it is below 1.5 μmol/L. In the absence of chelation therapy the half-life of lead in blood is about 28 days (Chamberlain, 1985).

Mercury

Sources

Mercury is a non-essential element which occurs widely in the environment and which can cause toxicity following excessive uptake via ingestion, inhalation or skin absorption. Inorganic mercury has many industrial uses, particularly in electrical components, and is also used in dental amalgams. Organomercurials are used in some paints as preservatives and in agriculture as fungicides or pesticides. The clinical use of organomercurials as antimicrobial agents is rare nowadays, as is the use of inorganic mercury compounds such as calomel for teething powder and mercuric chloride solutions for antiseptic lavage. However, Lauwerys *et al.*

(1987) reported prenatal and early postnatal toxicity caused by the use of a mercuric iodide soap and mercury-containing cream during pregnancy and lactation.

Dietary intakes of mercury are only 1–2 μg/day (MAFF, 1987). Higher intakes (4–20 μg/day) may occur because of mercury vapour released intra-orally from amalgam fillings during chewing (WHO, 1991). Higher intakes are also caused by increased consumption of fish, which normally contains mercury at about 300 μg/kg, mostly (90%) as methyl mercury (MAFF, 1987). Fish from polluted waters can contain 100 times this level of methyl mercury (Baselt & Cravey, 1989). Inorganic mercury released into waterways can be converted by microflora and other aquatic species into methyl mercury, which is then concentrated by fish.

Indicators of increased uptake

The best indicator of chronic exposure to elemental mercury vapour or inorganic mercury salts is analysis of the mercury in urine (HSE, 1991; Goering *et al.*, 1992). Reference ranges are given in Table 18.4. However, for investigations of suspected acute inorganic mercury toxicity both whole blood and urine should be analysed. Recent investigations by the author of a 2-year-old boy who had swallowed a mercury battery revealed that 6 days after the event his whole-blood-mercury level was 394 nmol/L (reference range <20 nmol/L) and this eventually fell after 6 months to 2 nmol/L. The corresponding urinary mercury concentrations were 160 and 40 nmol/L.

Table 18.4 Interpretation of urinary mercury concentrations with reference to exposure to elemental or inorganic mercury

Concentration of mercury (nmol/L)	Notes
<20	Range considered acceptable; children with no/few amalgam fillings will be in the low region
20–250	Increased exposure to and uptake of mercury; identify and remove source; clinical/biochemical signs unlikely
>250	Risk of toxicity; changes in renal function likely

In the absence of additional sources, whole-blood-mercury concentrations are used to indicate exposure to organomercury compounds, e.g. methyl mercury. These are usually below 20 nmol/L but can exceed 200 nmol/L in people eating large amounts of fish (Singerman, 1984).

Effects of increased uptake

The sequence of biochemical and clinical effects of mercury toxicity differs for exposure to elemental mercury, inorganic mercury compounds and organomercury compounds. Inhalation of elemental mercury initially produces adverse effects on the central nervous system, followed by renal toxicity after oxidation to mercuric ions. Ingestion of inorganic mercury compounds produces gastroenteritis and both glomerular and tubular nephritis, whereas excessive uptake of organomercury compounds can produce severe central nervous system toxicity (Baselt & Cravey, 1989).

Elemental/inorganic mercury poisoning

About 70% of inhaled mercury vapour is absorbed into the blood where it is distributed in a 2:1 ratio between red cells and plasma (Underwood, 1977; Goering *et al.*, 1992). The mercury in the plasma is lipophilic and easily able to enter the brain and induce cognitive and neurological disturbances. These manifest as nervousness, lassitude and, in extreme cases, tremor and erethism (Baselt & Cravey, 1989). Placental barriers also present no hindrance to the passage of mercury in plasma (Lauwerys *et al.*, 1987).

The elemental mercury in the erythrocytes is oxidized to Hg^{2+} by the catalase–hydrogen peroxide complex and leaves the cells as a complex with glutathione. This is redistributed throughout the body with the kidneys as the main storage organ. Excessive accumulation in the kidneys produces both glomerular and tubular nephritis, mediated at least partly by the action of mercury on the immune system (WHO, 1991; Goering *et al.*, 1992). Glomerular damage is characterized by the urinary excretion of albumin and other high-molecular-weight proteins, e.g. transferrin. Additional renal damage is manifested by increased urinary excretion of γ-glutamyltranspeptidase, an enzyme found in the brush borders of renal tubular cells, and of the enzymes β-galactosidase and *N*-acetyl-β-glucosaminidase (WHO, 1991).

Other signs of inorganic mercury toxicity include neutropenia and acrodynia. However, 'Pink disease' is rarely encountered now that calomel teething powders are no longer used. The use of nappies treated with phenylmercury (Gotelli *et al.*, 1985) has also been responsible for 'Pink disease', presumably by production of Hg^{2+} from phenylmercury.

The major excretory route for ingested elemental mercury released from dental amalgam is via the faeces. Malmström (1992) showed that after just one small amalgam filling in an 11-year-old girl with no previous amalgam fillings, the faecal excretion of mercury increased from 23 to 3200 μg/kg. Although overt mercury toxicity is unlikely from a small number of fillings, it is known that the release of mercury can stimulate an increase in Hg-resistant and antibiotic-resistant bacteria in the intestinal flora (Summers *et al.*, 1993).

Organomercury poisonings

Mass poisonings have occurred from the ingestion of organic mercury in contaminated seafood (Minimata disease) or from mercury-treated seed grain in Iraq. Elevated concentrations of mercury were noted in blood (>1000 nmol/L), and the liver (up to 79 mg/kg) and of methyl mercury in hair (60 mg/kg). There were severe neurological consequences, including paraesthesia, impaired sensory functions, ataxia and dysarthria, and many fatalities (Magos *et al.*, 1976; Clarkson, 1977; Baselt & Cravey, 1989).

Treatment of mercury poisoning

Chelation therapy using BAL, *N*-acetylpenicillamine or penicillamine has been used to treat toxicity from all forms of mercury. DMSA has also proved effective for chelation therapy for increased mercury exposure (Graziano, 1986).

References

Ackrill, P. (1986) Aluminium removal by desferrioxamine, clinical practice. In: *Aluminium and Other Trace Elements in Renal Disease* (ed. A. Taylor), pp. 193–199. Baillière Tindall, London.

Alfrey, A.C. (1986) Aluminium metabolism. *Kid Int* **29** (Suppl. 18), S8–S11.

ASCN/ASPEN (1991) Working Group: Parenteral drugs containing aluminium as an ingredient or a contaminant: response to FDA notice of intent. *Am J Clin Nutr* **53**, 399–402.

Baselt, R.C. & Cravey, R.H. (1989) *Disposition of Toxic Drugs and Chemicals in Man*, pp. 500–505. Year Book Medical Publishers, Chicago.

Berlin, A. (1986) Prevention and monitoring of aluminium exposure during dialysis in the European community. In: *Aluminium and Other Trace Elements in Renal Disease* (ed. A. Taylor), pp. 167–170. Baillière Tindall, London.

Bishop, N.J., Robinson, M.J., Lendon, M. *et al.* (1989) Increased concentration of aluminium in the brain of a parenterally fed preterm infant. *Arch Dis Child* **64**, 1316–1317.

Brancaccio, D., Bugrami, O., Pacini, L. *et al.* (1986) Blood brain barrier derangement in aluminium intoxicated dialysis patients. In: *Aluminium and Other Trace Elements in Renal Disease* (ed. A. Taylor), pp. 19–23. Baillière Tindall, London.

Candy, J.M., Klinowski, J., Perry, R.H. *et al.* (1986) Aluminosilicates and senile plaque formation in Alzheimer's disease. *Lancet* **i**, 354–357.

CDC (1991) *Preventing Lead Poisoning in Young Children: A Statement by the Centers for Disease Control*. US Department of Health and Human Services, Atlanta.

Chamberlain, A.C. (1985) Prediction of response of blood lead to airborne and dietary lead from volunteer experiments with lead isotopes. *Proc R Soc London* **B224**, 149–182.

Channon, S.M., Osman, O.M., Boddy, I. *et al.* (1986) Serum aluminium measurement – a false sense of security? In: *Aluminium and Other Trace Elements in Renal Disease* (ed. A. Taylor), pp. 171–176. Baillière Tindall, London.

Chappuis, P., de Vernejoul, M.-C., Paolaggi, F. & Rousselet, F. (1989) Relationship between hair, serum and bone aluminium in hemodialyzed patients. *Clin Chim Acta* **179**, 271–278.

Clarkson, T.W. (1977) Mercury poisoning. In: *Clinical Chemistry and Chemical Toxicology of Metals* (ed. S.S. Brown), pp. 189–200. Elsevier, Amsterdam.

Davies, D.J.A. (1987) An assessment of the exposure of young children to lead in the home environment. In: *Lead in the Home Environment* (eds I. Thornton & E. Culbard), pp. 189–196. Science Reviews Ltd, Northwood.

Delves, H.T. (1988) Biomedical applications of ICP-MS. *Chem Brit* **24**, 1009–1012.

Delves, H.T. & Shuttler, I.L. (1991) Elemental analysis of body fluids and tissues by electrothermal atomization and atomic absorption spectrometry. In: *Atomic Absorption Spectrometry* (ed. S.J. Haswell), pp. 381–438. Elsevier, Amsterdam.

Delves, H.T., Sieniawska, C.E. & Suchak, B. (1993) Total and bioavailable aluminium in foods and beverages. *Analyt Proc* **30**, 358–360.

Delves, H.T., Suchak, B. & Fellows, C.S. (1988) Determination of aluminium in foods and in biological material. In: *Aluminium in Food and Environment* (eds R. Massey & D. Taylor), pp. 52–67. Royal Society of Chemistry, London.

DoE (1983) European Community screening programme for lead: United Kingdom results for 1981. *Pollution Report*, No. 18. Department of Environment, London.

DoE (1990) UK blood lead monitoring programme 1984–1987. *Pollution Report*, No. 28. Department of Environment, London.

Fell, G.S., Shenkin, A. & Halls, D.J. (1986) Aluminium contamination of intravenous pharmaceuticals, nutrients and blood products. *Lancet* **i**, 380.

Goering, P.L., Galloway, W.D., Clarkson, T.W. *et al.* (1992) Toxicity assessment of mercury from dental amalgams. *Fund Appl Toxicol* **19**, 319–329.

Gorsky, J., Dietz, A., Spencer, H. & Osis, D. (1979) Metabolic balance studies of aluminium in six men. *Clin Chem* **25**, 1739–1743.

Gotelli, C.A., Astolfi, E., Cox, C., Cernichiari, E. & Clarkson, T.W. (1985) Early biochemical effects of an organic mercury fungicide on infants. *Science* **227**, 638–640.

Graziano, J.H. (1986) Role of 2,3-dimercaptosuccinic acid in the treatment of heavy metal poisoning. *Med Toxicol* **1**, 155–162.

Greger, J.L. (1992) Dietary and other sources of aluminium intake. In: *Aluminium in Biology and Medicine. Ciba Foundation Symposium 169*, pp. 26–49. John Wiley & Sons, Chichester.

HSE (1991) *Guidance on Laboratory Techniques in Occupational Medicine*, 5th edn. Health and Safety Executive, London.

Kerr, D.N.S. & Ward, M.K. (1986) The history of aluminium related disease. In: *Aluminium and Other Trace Elements in Renal Disease* (ed. A. Taylor), pp. 1–14. Baillière Tindall, London.

Klein, G.L., Alfrey, A.C., Miller, N.L. *et al.* (1982) Aluminium loading during total parenteral nutrition. *Am J Clin Nutr* **35**, 1425–1429.

Koo, W.K.K. & Kaplan, L.A. (1988) Aluminium and bone disorders: with specific reference to aluminium contamination of infant nutrients. *J Am Coll Nutr* **7**, 199–214.

Lauwerys, R., Bonnier, C., Evrard, P., Gennart, J. & Bernard, A. (1987) Prenatal and early postnatal intoxication by inorganic mercury resulting from the maternal use of mercury-containing soap. *Hum Toxicol* **6**, 253–256.

Leeming, R.J. & Blair, J.A. (1979) Dialysis, dementia, aluminium and tetrahydrobiopterin metabolism. *Lancet* **i**, 556.

Lovell, M.A., Ehmann, W.D. & Markesbury, W.R. (1993) Laser microprobe analysis of brain aluminium in Alzheimer's disease. *Ann Neurol* **1**, 36–42.

McClure, J. (1986) The demonstration of aluminium in articular and hypertropic cartilage in a case of aluminium-related renal osteodystrophy. In: *Aluminium and Other Trace Elements in Renal Disease* (ed. A. Taylor), pp. 123–126. Baillière Tindall, London.

McGraw, M., Bishop, N., Jameson, R. *et al.* (1986) Aluminium content of milk formulae and intravenous fluids used in infants. *Lancet* **i**, 157.

MAFF (1987) Survey of mercury in food: second supplementary report. *Food Surveillance Paper*, No. 17. Ministry of Agriculture, Fisheries and Food. HMSO, London.

Magos, L., Bakir, F. & Clarkson, T.W. (1976) Tissue levels of mercury in autopsy specimens of liver and kidney. *Bull WHO* **53** (Suppl.), 93–97.

Maharaj, D., Fell, G.S., Boyce, B.F. *et al.* (1987) Aluminium bone disease in patients receiving plasma exchange with contaminated albumin. *Br Med J* **295**, 693–696.

Malmström, C. (1992) Amalgam-derived mercury in faeces. *J Trace Elem Exp Med* **5**, 123 (Abstract 122).

Marquis, J. (1982) Aluminium toxicity: an environmental perspective. *Bull Env Contam Tox* **29**, 43–49.

Martyn, C.N., Barker, D.J.P., Osmond, C. *et al.* (1989) Geographical relation between Alzheimer's disease and aluminium in drinking water. *Lancet* **i**, 59–62.

Moncrieff, A.A., Koumides, O.P., Clayton, B.E. *et al.* (1964) Lead poisoning in children. *Arch Dis Child* **39**, 1–13.

MRC (1988) The neuropsychological effects of lead in children: a review of the research 1984–1988. *Report from MRC Advisory Group on Lead and Neuropsychological Effects in Children*. Medical Research Council, London.

Priest, N.D., Newton, D., Talbot, R.J. *et al.* (1991) Metabolism of aluminium-26 and gallium-67 in a volunteer following their injection as citrates. *Report AEA-EE-0206*, pp. 1–57. Harwell Biomedical Research, Harwell, Oxon.

Schneider, H. (1986) Alternatives to aluminium-containing phosphate binders. In: *Aluminium and Other Trace Elements in Renal Disease* (ed. A. Taylor), pp. 127–135. Baillière Tindall, London.

Schütz, A., Attewell, R. & Skerfving, S. (1989) Decrease in

blood lead of Swedish children 1978–1988. *Arch Environ Hlth* **44**, 391–394.

Sedman, A.B., Klein, G.L., Merritt, R.J. *et al.* (1985) Evidence of aluminium loading in infants receiving intravenous therapy. *N Engl J Med* **312**, 1337–1343.

Sieniawska, C.E. (1993) *Bioavailability of Aluminium from Foods and Beverages*. MPhil thesis, University of Southampton.

Singerman, A. (1984) Exposure to toxic metals: biological effects and their monitoring. In: *Hazardous Metals in Human Toxicology* (ed. A. Vercruysse), pp. 17–94. Elsevier, Amsterdam.

Slanina, P., Frech, W., Ekstron, L. *et al.* (1986) Dietary citric acid enhances absorption of aluminium in antacids. *Clin Chem* **32**, 539–541.

Suchak, B. (1992) *Determination of Aluminium in Foods and Beverages*. MPhil thesis, University of Southampton.

Summers, A.O., Wireman, J., Vimy, M.J. *et al.* (1993) Mercury released from dental 'silver' fillings provokes an increase in mercury- and antibiotic-resistant bacteria in oral and intestinal floras of primates. *Antimicrob Agents Chemotherapy* **37**, 825–834.

Underwood, E.J. (1977) *Trace Elements in Human and Animal Nutrition*, 4th edn. Academic Press, New York.

Walker, A. (ed.) (1992) *Trace Element Analyses*, 2nd edn. Handbook of SAS Laboratories of NHS. Royal Surrey County and St Luke's Hospitals, Guildford, Surrey.

The Water Supply (Water Quality) Regulations (1989) No 1147. HMSO, London.

Weiss, L.G., Danielson, B.G., Fellström, B. & Wikstrom, B. (1989) Aluminium removal with hemodialysis, hemofiltration and charcoal hemoperfusion in uremic patients after desferrioxamine infusion: a comparison of efficiency. *Nephron* **51**, 325–329.

WHO (1991) IPCS. Inorganic mercury. In: *Environmental Health Criteria 118*. World Health Organisation, Geneva.

Young, J.H., Kelly, B. & Clayton, B.E. (1982) Reduced levels of biopterin and dihydropteridine reductase in Alzheimer-type dementia. *J Clin Exp Gerontol* **4**, 389–402.

19: Gastrointestinal Disorders

J.A. WALKER-SMITH

Introduction

The biochemical investigation of gastrointestinal function in children with gastrointestinal disease is based upon three types of investigation which may be direct or indirect.

1 Tests based on the measurement of absorption of a nutrient after an oral loading test.

2 Tests based on the measurement of non-absorbed substrate, either by direct examination of the stool (i.e. evidence of classical malabsorption or loss into the gut), or by indirect measurement of non-absorbed substrate by hydrogen in the breath.

3 Tests based on the measurement of secretions from the gastrointestinal tract.

Some of these procedures involve invasive techniques, e.g. venepuncture, small bowel biopsy or intubation, whereas others are non-invasive, such as stool examination or hydrogen breath tests. Clearly, non-invasive tests are particularly appropriate for children but some non-invasive tests, such as faecal fat, are no longer routinely carried out.

In this chapter a practical guide to the use of such investigations, largely based upon practice at the Queen Elizabeth Hospital for Children in London, will be given.

Tests of gastric acid secretion

Measurement of the pH of fasting gastric juice has long been used as a simple assessment of therapy; for example, when using hydrogen blocking agents a fasting pH of >4, compared with the normal pH of 1–2, indicates appropriate control of acid secretion.

Attachment of a pH electrode to a small-bowel biopsy tube is a simple way to assess gastric acidity and duodenal pH in children having a small intestinal biopsy. Gastric hypoacidity, especially in children with *Helicobacter pylori* infection, may be a risk factor for small intestinal enteropathy (Thomas *et al.*, 1992). Hypoacidity also occurs in malnutrition, Menetrier's disease (Frank & Kern, 1967) and the Verner–Morrison syndrome (Verner & Morrison, 1958).

Formal acid-output tests are now rarely performed in children. Such testing may be needed when the Zollinger–Ellison syndrome is a diagnostic possibility (Schwartz *et al.*, 1974).

Tests of small intestinal function

These include tests of absorption and oral loading tests.

Sugar

Sugar intolerance, especially lactose intolerance, is an important cause of chronic diarrhoea in infancy. It usually is a temporary syndrome occurring secondary to gastroenteritis (Table 19.1). In infancy it may be due to a rare syndrome of congenital enzyme deficiency. In the older child it may be the result of late-onset or genetic lactose intolerance.

The term intolerance implies a disorder requiring treatment, whereas sugar malabsorption is a term used to describe the situation where there is laboratory evidence of disordered sugar absorption, e.g. a flat lactose tolerance test or an abnormal hydrogen breath test may not necessarily indicate clinical intolerance.

Oral loading tests are of most value for making a

Table 19.1 Clinical disorders of carbohydrate absorption

Primary disorders
Congenital lactase deficiency
Sucrase–isomaltase deficiency
Congenital glucose–galactose malabsorption
Acquired disorder
Late-onset lactase deficiency
Secondary disorders
Disaccharides
Gastroenteritis
Cows' milk protein intolerance
Malnutrition
Coeliac disease
Tropical sprue
Surgery to gastrointestinal tract
Phototherapy
Immune deficiency syndromes
Inflammatory bowel disease
Giardiasis
Monosaccharides
Gastroenteritis
Malnutrition
Necrotizing enterocolitis

diagnosis in the older child or for assessing the persistence, or otherwise, of lactose intolerance in an infant already on a lactose-free diet. Such loading tests may be combined with a lactose-hydrogen or other breath test (see below).

The sugar to be investigated is given by mouth to the fasting child, usually at a dosage of 2 g/kg body weight in 200 mL, but never exceeding 12% concentration. Although in the past capillary blood samples were collected fasting and then at 30-min intervals for 2 h, this practice is seldom followed today. An increase in blood glucose of more than 1.7 mmol/L above the fasting level is recognized as normal; a rise of 1.1–1.7 mmol/L is doubtful; and a rise of less than 1.1 mmol/L is abnormal (Holzel, 1967). Rather more important than the rise in blood sugar levels in children is observation of the stools after the oral load and, in particular, the development of diarrhoea. Following an oral lactose load, the demonstration of a 'flat' lactose tolerance test, without diarrhoea and accompanied by excess stool reducing substances, is not clinically significant. It should not on its own be regarded as an indication that the child should be treated for lactose malabsorption. It could be due to delayed gastric emptying or indicate some disorder of sugar handling which is not severe enough to produce symptoms.

The xylose absorption test is no longer recommended as a routine diagnostic test for coeliac disease or other small intestinal enteropathy. Xylose is absorbed both by passive absorption and facilitated active transport, and so interpretation is difficult (Lambadusuriya *et al.*, 1975).

Fat

There was a time when oral fat loads were routinely used for diagnosis (Stone & Thorp, 1966; Robbards, 1973) but these tests are no longer used in practice.

Tests of intestinal permeability

In recent times great attention has been paid to the role of intestinal permeability as a measure of the integrity of function of the small intestinal mucosa. When permeability is abnormal it forms a clear indication for small bowel biopsy, and can also be used serially to follow the response to elimination therapy and as a guide to relapse after a food challenge, e.g. with gluten or milk. The clinical assessment of intestinal permeability *in vivo* to relatively small test molecules is now practicable. In fact, a variety of test molecular probes have been used. They include xylose, lactulose (Muller *et al.*, 1969), mannitol (Fordtran *et al.*, 1965), polyethylene glycols (Thomas *et al.*, 1990), a mixture of two sugars (e.g. lactulose and rhamnose (Ford *et al.*, 1985; Lelercq-Foucart *et al.*, 1987) or lactulose and mannitol (Weaver *et al.*, 1984a,b; 1985)) and [^{51}Cr]-ethylenediaminetetraacetate (EDTA) (Bjarnason & Peters, 1983; Forget *et al.*, 1985).

Differential absorption of two or more inert sugars: lactulose and rhamnose

After an overnight fast of at least 6 h, an isotonic load containing 3.5 g lactulose and 0.5 g rhamnose in 50 mL is given by mouth. All urine passed in the subsequent 5 h is collected, the volume recorded and an aliquot, preserved with merthiolate, is kept at −20°C until analysis. Stored in this way remain stable for many months. Urinary concentrations of each sugar are measured by quantitative thin layer chromatography. Results are expressed as a ratio of the percentages of the oral sugar loads excreted in the 5-h urine collection (Ford *et al.*, 1985). The lactulose:rhamnose ratio is normally less than 0.08. A major advantage of this technique is that by expressing the urinary excretion of the sugars as a ratio of recovery, the effects of the many variables influencing the individual sugar probes can be overcome. These variables include the complete ingestion of the oral load, gastric emptying time, intestinal transit time, dilution of the marker by intestinal secretions, renal clearance and the completeness of the urine collection. The disadvantages are the difficulty in collecting any urine in young children, particularly those with diarrhoea, and the complexity of the assay.

Using differential absorption, Noone *et al.* (1986) have shown increased intestinal permeability in adults and children with acute gastroenteritis. Ford *et al.* (1985) also found that children with acute gastroenteritis had a greatly increased urinary recovery of lactulose. When tested again, after recovery, normal permeability results were obtained. Children with chronic diarrhoea also had increased sugar permeability but significantly less than was found in the acute gastroenteritis group. An abnormal proximal small-bowel morphology was associated with increased permeability, and in particular a strong correlation was observed between crypt depth and intestinal permeability.

Permeability to lactulose and rhamnose is also altered in Crohn's disease of the small intestine but improves with an elemental diet (Sanderson *et al.*, 1987).

In general, abnormal intestinal permeability is associated with mucosal damage of the small intestine and this investigation can be used as a screen for mucosal damage. Occasionally, abnormal permeability can occur in the presence of a normal mucosa (Pearson *et al.*, 1982).

In a study using lactulose and mannitol, Sullivan *et al.* (1992) found that although in general there is a correlation between morphology and permeability, there was not a direct quantitative correlation between small intestinal mucosal damage, assessed by morphometry, and lactulose–mannitol absorption. Normally, the urinary lactulose:mannitol ratio should be less than 0.008. Also using lactulose and mannitol, Andre *et al.* (1987) have shown that in patients with food allergy there is a decrease in mannitol recovery and an increase in lactulose recovery after ingestion of the offending food allergen, thereby indicating the value of this test in the challenge situation. Oral sodium cromoglycate, administered before a food provocation test in patients with gastrointestinal food allergy, protected against the development of abnormal intestinal permeability.

Permeability of the small intestinal mucosa to lactulose and rhamnose has been shown to be altered in young children with iron deficiency (Berant *et al.*, 1992). There is a lower urinary recovery of rhamnose, which passes across the epithelium via the transcellular route, but normal recovery of lactulose, which passes through a paracellular route. Permeability returns to normal when a normal iron status is achieved. Thus, iron status must be taken into account when interpreting results of permeability studies.

These tests of differential sugar absorption therefore appear to have great potential in the investigation and diagnosis of small intestinal disease in childhood because they are non-invasive and should give an objective measurement of significant changes in the integrity of the intestinal mucosa; their precise role in diagnosis still awaits clarification. They cannot by themselves replace small intestinal biopsy, although in centres where this test is available and biopsy is not, sugar permeability provides a useful screen.

Combined approach

Another approach is to use a combination of non-invasive tests simultaneously. This involves giving an oral dose of several test substances and then making a total urine collection over 6 h and a breath sample analysis every 30 min for 4 h (Obinna *et al.*, in preparation) (Fig. 19.1). This technique permits

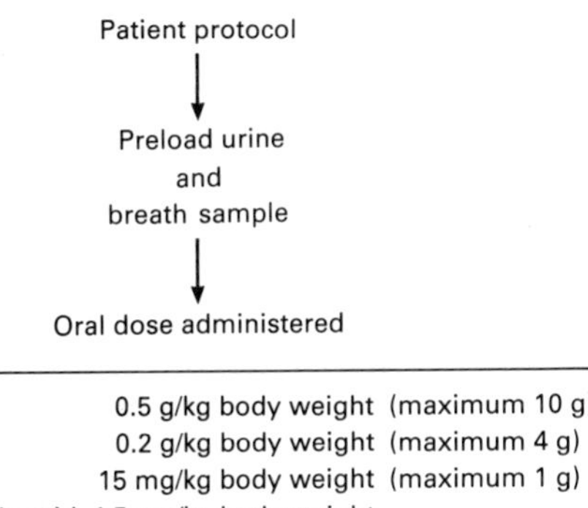

Lactulose	0.5 g/kg body weight (maximum 10 g)
Mannitol	0.2 g/kg body weight (maximum 4 g)
Bentiromide	15 mg/kg body weight (maximum 1 g)
p-Amino salicylic acid	4.5 mg/kg body weight

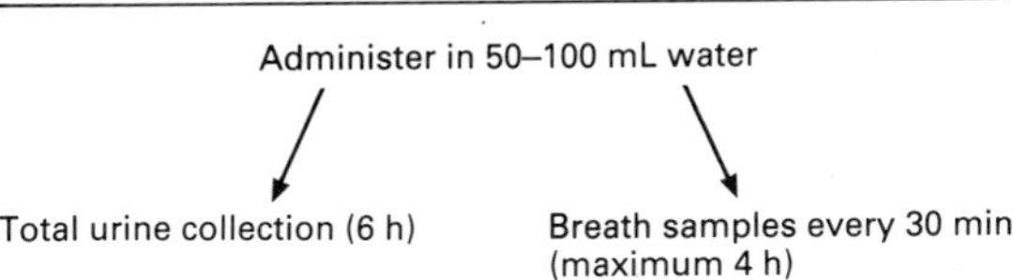

Fig. 19.1 Procedure for a combination of non-invasive tests.

Table 19.2 Effect of allowing stools to soak into a nappy

Time (min)	Nature of stool	Reducing substances (%)
Before soaking		
0	Yellow–green watery and curds	1
After soaking		
0	Yellow–green curds and little fluid	0.5–0.75
15	Yellow–green curds	0.25–0.5
30	Yellow–green curds	0.25–0.5
60	Yellow–green curds	0.25–0.5

simultaneous analysis and allows assessment of: (a) chymotrypsin activity via luminal bentiromide hydrolysis (see below); (b) sugar permeability of lactulose and mannitol; (c) bowel transit time, as assessed by the lactulose-hydrogen breath test; and (d) urinary oxalate excretion (increased excretion occurs in malabsorption). Highly sensitive and specific high performance liquid chromatography (HPLC) and mass spectrometric analysis, which is not widely available, is required for this approach.

Tests of non-absorbed substrate

Stool studies

Sugar

Stool reducing substances and stool chromatography for sugars

When the stools are loose the fluid part of the stool may be monitored for the presence of reducing substances (Kerry & Anderson, 1964); this is best done as a simple ward test. The fluid portion of the stool is tested by first diluting one part fluid stool with two parts water, placing 15 drops of the mixture into a test tube and adding a Clinitest tablet (Ames Company, Bayer Diagnostics, Basingstoke, UK). The amount of reducing substances may then be estimated according to the resultant colour and will range from 0 to 2%; a value of 1.0% or more is regarded as abnormal. Alternatively, the fluid part of the stool may be diluted with twice its volume of water, homogenized in the laboratory and tested as above. This refinement is recommended only when chromatography for sugars is planned. The homogenized suspension is centrifuged and the supernatant used for chromatography. Stool chromatography is only indicated when an excess of reducing substances is present.

The fluid stool for testing must be collected in such a way that the fluid does not soak into the infant's nappy and so largely be lost. This may be done by collecting the stool on a non-absorbable material, such as plastic, or straight into a container. Table 19.2 shows the effect of allowing the stool to soak into the nappy in a case of sugar intolerance; a significant result may turn into a non-significant one. There is a need for rapid analysis or rapid freezing of stools.

When paper chromatography is to be undertaken, the supernatant should be spotted on to Whatman No.1 chromatography paper (Soeparto *et al.*, 1972). The technique of Menzies (1973), based upon thin layer chromatography, is now often used instead.

Stool pH

A stool pH of less than 6 is regarded as abnormal and has been described as a characteristic finding in children with sugar intolerance (Holzel, 1967). It is very important to test fresh stools so the stool pH is usually tested at the bedside with pH indicator paper with a range 1–11. Clinically, such a method correlates satisfactorily with true pH meter readings (Soeparto *et al.*, 1972).

Fat

The need to monitor daily faecal fat over a 3- to 5-day period has declined considerably over recent years. The usual method used is that described by van de Kamer *et al.* (1949). The percentage of fat absorbed from dietary fat, i.e. the coefficient of fat absorption, is perhaps a more reliable measure than the absolute value of daily faecal fat estimations, but depends on a precise knowledge of the child's fat intake during the period of study; this may be difficult to obtain. Nonetheless, some attempt should be made to ensure an adequate fat intake during the period of stool collection as misleading results may occur when the child is anorexic and the fat intake is low. Coloured markers may be added.

Steatorrhoea is no longer regarded as a suitable screening test for coeliac disease. The main indications for this investigation in children are: (a) small intestinal disease: massive small intestinal resection, evidence of a stagnant loop syndrome and chronic diarrhoea of obscure origin when other investigations have proved negative; and (b) pancreatic disease: a more important indication is the response to treatment in patients with cystic fibrosis in assessing the correct dosage of pancreatic enzyme therapy (Shmerling *et al.*, 1970; Penny *et al.*, 1986).

Protein

Monitoring of faecal nitrogen is an investigation that is not often of much practical help in diagnosis. It is only important in very severe pancreatic dysfunction.

Diagnostic tests for enteric protein loss

Abnormal protein loss from the gastrointestinal tract may occur in a wide variety of disease states owing to abnormal permeability; these include mucosal ulceration and obstruction to lymph flow. Faecal α_1-antitrypsin has been used but its role is uncertain (Walker-Smith & Andrews, 1975; Sharp, 1976; Davidson & Robb, 1985; Durie, 1985; Magazzu *et al.*, 1985; Hill *et al.*, 1987; Quigle *et al.*, 1987).

Radioisotope studies

The diagnosis of protein-losing enteropathy may be definitively established by means of a radioisotope study. A simple test is to use intravenous $^{51}CrCl_3$, which binds to plasma protein *in vivo* (Walker-Smith *et al.*, 1967); $^{51}CrCl_3$ suspended in saline is given intravenously, usually as a 30-μCi (1.11 MBq) dose in adults. The stools are collected, homogenized, made up to a final volume of 500 mL and counted in a scintillation counter. The total radioactivity in a 5-day collection is expressed as a percentage of the injection dose. In controls, less than 1% is excreted (Walker-Smith *et al.*, 1967). In practice, it is not necessary to resort to this test when the clinical picture may be diagnostically clear, e.g. with clinical features of lymphangiectasia, but it is helpful when the diagnosis of hypoproteinaemia is unclear. The test does have the drawbacks of demanding complete faecal collection and separation of urine from the stools, because a substantial fraction of the radioisotope is excreted in the urine.

Indirect tests

Lactose-hydrogen breath test

Hydrogen in the breath is the result of bacterial fermentation, which breaks down within the intestines substrates that have been eaten. Normally after eating, no hydrogen appears in the breath for 1–3 h, i.e. until the ingested substrate reaches the bacteria of the colon. A simple technique using the hydrogen breath test as a measure of lactose absorption in childhood has been described. An oral load of lactose (2 g/kg in 100 mL water to a maximum of 20 g) is given and a small plastic tube is placed in a nostril and used to sample the child's breath before and every 30 min after the oral load for 3 h (Maffei *et al.*, 1977). A rise in hydrogen excretion of greater than 20 p.p.m. represents a positive peak.

Higher levels of hydrogen than those normally found in the breath occur when carbohydrates, malabsorbed by the small intestine, reach the colon, there to be fermented by colonic bacteria. Thus, in children who have carbohydrate malabsorption the peak in breath hydrogen occurs 1–3 h after an oral load of lactose. This investigation has been very useful in understanding pathophysiology, but the role of such a test in routine practice in infants has not yet been established (Tolbloom *et al.*, 1986); stool reducing substances remain the most useful test. In older children, as part of an oral load test it is a valuable investigation.

Other hydrogen breath tests

The lactulose-hydrogen breath test is a practical technique to measure the oral–caecal transit time. The interval between the ingestion of this non-absorbed sugar and a rise in breath hydrogen is a measure of the transit time (Solomons *et al.*, 1979). The normal gut transit time is 90–120 min from the start of the investigation.

Then there is the glucose-hydrogen breath test. Normally, most hydrogen produced locally is absorbed in the proximal small intestine and excreted by the lungs. When there is bacterial contamination of the small intestine, there is an early peak of hydrogen and there may be high fasting loads (Metz *et al.*, 1976). The peak may occur in less than 1 h. This is a useful test to diagnose bacterial colonization of the small intestine.

To summarize, sugar malabsorption in infants and younger children, while they are having feeds containing the offending sugar, can usually be adequately diagnosed by the demonstration of an abnormal amount of reducing substances in their stools, but in older children, and sometimes in infants, the lactose-hydrogen breath test may be used. The practical clinical proof that the diagnosis of sugar intolerance is correct in these children will be provided by a clinical response to the removal of the offending sugar from the child's diet.

Enzyme assay of small intestinal biopsies

The disaccharidase assay has only a small, but important, diagnostic role in routine gastroenterological practice. Although assays of other enzymes, such as dipeptidase, have also been investigated extensively in research studies, assay of these enzymes is not performed routinely, and therefore they will not be discussed here.

Disaccharidase assay

Because the intestinal biopsy specimen that is assayed for disaccharidase activity is so small, and because proximal small intestinal mucosal abnormalities may be patchy and of variable extent, there is often a lack of correlation between disaccharidase activity and other measures of sugar absorption, e.g. sugar tolerance tests and stool studies. An example of this is in coeliac disease where there may sometimes be no detectable lactase activity on assay of a biopsy specimen, and yet the child has no clinical evidence of lactose intolerance. This is because although lactase activity is absent in the biopsy specimen, the total amount of small intestinal lactase activity is adequate for the child to absorb lactose further down the small intestine.

The estimation of disaccharidase activity in a single biopsy specimen is of no specific diagnostic importance in the management of disorders where secondary disaccharidase deficiency occurs, such as in coeliac disease and post-gastroenteritis malabsorption. In these circumstances a depressed disaccharidase activity merely reflects the state of mucosal damage that is present. Lactase activity appears to be more susceptible to damage than is the activity of the other disaccharidases; the depression of lactase levels is usually more severe and takes longer to recover (Smith, 1985).

The principal indication for a disaccharidase assay in clinical practice is in the diagnosis of primary congenital deficiency syndromes, namely sucrase–isomaltase (Burgess *et al.*, 1965), and congenital alactasia, and serial estimation as a measure of

Table 19.3 Disaccharidase activity (units/g protein) in normal jejunal mucosa in children

	Lactase	Sucrase	Isomaltase	Maltase
Range	14–132	32–338	31–177	83–615
Mean	49	95	89	260

Table 19.4 Disaccharidase activity (units/g wet mucosa per min) in normal small intestinal mucosa

	Lactase	Sucrase	Maltase
Range	2–16	3–20	15–77
Mean	5.6	10.2	31.2

mucosal damage after a challenge, e.g. cows'-milk-sensitive enteropathy.

With abnormal results this may then reflect mucosal damage induced by the ingestion of cows' milk, although when evaluated critically, there may be a lack of correlation between the histology and disaccharidase activity (Harrison & Walker-Smith, 1977).

Disaccharidase activity may be expressed as per gram of protein in the biopsy specimen, or as per gram of wet mucosa (used at the Queen Elizabeth Hospital for Children) (Tables 19.3 & 19.4).

There is a much wider range of values in the Melbourne series (Burke *et al.*, 1965) than in the series reported from the Queen Elizabeth Hospital for Children. The reason for this is not clear but may be related to differing criteria for 'normality'. Levels for disaccharidase activity are somewhat lower in the duodenum than the values obtained in the proximal jejunum. There is a report that disaccharidase activity can be measured in jejunal fluid and shows a good correlation with biopsy levels (Aramayo *et al.*, 1983).

Tests of pancreatic function

A biochemical diagnosis of exocrine pancreatic insufficiency is required for the rare disorders of isolated pancreatic lipase deficiency and for the general disorders of cystic fibrosis and Shwachman's syndrome.

Direct tests of pancreatic exocrine function

The exocrine pancreas secretes fluids and ions, in response to the endogenous secretion of secretin, and also enzymes, in response to the endogenous secretion of cholecystokinin (CCK). These endogenous hormones are released from the small intestinal mucosa in response to exposure to nutrients or gastric acid.

The function of the exocrine pancreas may be assessed directly by stimulating it via intravenous hormones or by intestinal nutrients.

Ideally, for precise quantitative measures the distal duodenum should be occluded by a balloon, or else there should be continuous perfusion of a non-absorbable marker which permits correction for distal levels (Go *et al.*, 1970; Durie & Goldberg, 1986). In addition, gastric juice (i.e. pepsin and acid) should be excluded via gastric aspiration.

In practice, this is a cumbersome and time-consuming procedure, so a simpler quantitative approach is now often used for routine, as opposed to research, purposes. This is a technique of intubating the duodenum without a balloon, collecting a sample for basal enzyme levels and then infusing intravenously, in sequence, secretin and CCK at doses needed to achieve maximal pancreatic stimulation. A separate nasogastric tube enables aspiration of gastric juice and minimizes contamination of the duodenal contents by acid and pepsin. The results of such a procedure are indicated in Table 19.5.

In children, nutrient stimulation of the exocrine pancreas can be undertaken using the Lundh meal (McCollum *et al.*, 1977).

Indirect test of pancreatic function (bentiromide test)

A non-invasive tubeless test, the bentiromide test of pancreatic chymotrypsin pain activity, may be used to assess pancreatic function. A measured dose of synthetic substrate is given to the patient and pancreatic enzyme activity is estimated as an index of metabolic urinary products (Weizman *et al.*, 1985).

Bentiromide is a non-absorbable synthetic peptide (*N*-benzoyl-L-tyrosyl-*p*-aminobenzoic acid). It is specifically split by pancreatic chymotrypsin in the

Table 19.5 Direct test of pancreatic exocrine function

	Basal	Post-CCK	Difference	Change (%)	Post-secretin
Trypsin (mmol/mL per min)					
Median	29	75	38	160	53
Mean	32	87	–	–	38
5–95%	2–74	21–207	3–156	2–1150	0–96
Lipase (mmol/mL per min)					
Median	350	980	615	159	510
Mean	433	1172	–	–	595
5–95%	0–1200	200–2720	90–2040	20–2500	10–1590

CCK, cholecystokinin.
Values from 100 children (mean age 4.2 years), subsequently proven to have no pancreatic involvement.
Interpretation
An abnormal result is indicated when there is no response observed to CCK stimulation on top of minimal basal activity of the enzymes. Before an abnormal result is accepted, reference must be made to both the pH and volume of the collected juices to verify the correct positioning of the tube.
A small or negative increment in enzyme activity can be observed on top of a normal or raised basal value; because activity is present then some pancreatic function is present, although previous stimulation by food must be questioned. There is a wide range of stimulated values; however, careful interpretation of the lower values must be made.
Data reproduced by permission of Dr Andrew Johnson, Queen Elizabeth Hospital for Children, London.

upper small intestine, resulting in the release of *p*-aminobenzoic acid (PABA). PABA is rapidly absorbed, conjugated in the liver and excreted in the urine, where it serves as a marker of chymotrypsin activity. PABA can be measured in blood and urine by a colorimetric assay or HPLC.

The bentiromide test was introduced in 1972. The initial reports relied on a one-stage test with a urinary collection. The methods used involved collections over varying time periods and use of varying doses of substrate. In order to correct for potential defects of absorption, hepatic conjugation or excretion, a two-stage test was suggested with an equivalent dose of free *p*-aminosalicylic acid (PASA) administered subsequently as an internal standard; the urine was collected for an identical time period. This allows the urinary recovery of PABA after bentiromide to be corrected for the urinary recovery of equimolar free PASA, and the results are expressed as the PABA excretion index (PEI):

$$\text{PEI} = \frac{\text{PABA recovered after bentiromide (\%)}}{\text{PASA recovered after free PASA (\%)}}$$

Normally, this ratio should be greater than 0.6, more than 50% of the oral dose should be excreted in a 6-h urine collection and there should be more than 20 mmol PABA/mL serum at 150 min post-injection. Further refinement may improve the capability of this test to identify patients with lesser degrees of impairment of exocrine function. Thus, the bentiromide test may discriminate pancreatic steatorrhoea from other causes of steatorrhoea, and could potentially provide a method of monitoring the effect of pancreatic enzyme supplementation (Puntis *et al.*, 1988). However, for the present, direct testing of pancreatic exocrine function remains the method of choice.

Table 19.6 Features of osmotic and secretory diarrhoea

	Osmotic	Secretory
Fasting	Stops	Continues
Faecal osmolality	400	280
Na^+	30	100
K^+	30	40
(Na^+ plus K^+) × 2	120	280
Solute gap	280	0

Tests of stool electrolytes

The main functions of the large intestine are the storage of the intestinal contents prior to excretion, the absorption of water and electrolytes entering from the ileum and, in particular, the conservation of sodium ions and water.

The determination of stool electrolytes, in particular sodium, potassium and chloride ions, is a useful diagnostic test in patients with explosive watery stools. For example, stool electrolytes may be helpful in distinguishing osmotic from secretory diarrhoea (Krejs *et al.*, 1980). In a secretory state, stool electrolytes (Na^+, K^+ and their respective anions) account almost completely for the total faecal osmolarity, whereas in osmotically induced diarrhoea there is a large deficit (solute gap) which results from the presence of metabolic products arising from bacterial action on malabsorbed carbohydrate (see Table 19.6). A stool chloride concentration which exceeds the sum of the sodium and potassium concentration is pathognomonic for congenital chloridorrhoea (Launiala *et al.*, 1968), a primary defect of electrolyte absorption.

References

Andre, C., Andre, F., Colin, L. & Cavagna, S. (1987) Measurement of intestinal permeability to mannitol and lactulose as a means of diagnosing food allergy and evaluating therapeutic effectiveness of disodium cromoglycate. *Ann Aller* **59**, 127–130.

Aramayo, L.A., De Silva, D.G., Hughes, C.A., Brown, G.A. & McNeish, A.S. (1983) Disaccharide activities in jejunal fluid. *Arch Dis Child* **58**, 686–691.

Berant, M., Khourie, M. & Menzies, I.S. (1992) Effect of iron deficiency on small intestinal permeability in infants and young children. *J Ped Gastr Nutr* **14**, 17–20.

Bjarnason, I. & Peters, T.J. (1983) A persistent defect in intestinal permeability in coeliac disease demonstrated by a ^{51}Cr-labelled EDTA absorption test. *Lancet* **1**, 323–325.

Burgess, E.A., Levin, B., Mahalanabis, D. & Tonge, R.E. (1965) Hereditary sucrose intolerance levels of sucrase activity in jejunal mucosa. *Arch Dis Child* **39**, 431.

Burke, V., Kerry, K.R. & Anderson, C.M. (1965) The relationship of dietary lactose to refractory diarrhoea in infancy. *Aust Paediat J* **1**, 147–149.

Davidson, G.P.O. & Robb, T.A. (1985) Value of breath hydrogen analysis in management of diarrhoeal illness in childhood: comparison with duodenal biopsy. *J Ped Gastr Nutr* **4**, 381–387.

Durie, P.R. (1985) Intestinal protein loss and faecal antitrypsin. *J Ped Gastr Nutr* **4**, 345–347.

Durie, P.R. & Goldberg, D.M. (1986) Biochemical tests of pancreatic function in infancy and childhood. *Adv Clin Enzymol* **4**, 77–92.

Ford, R.P.K., Menzies, I.S., Phillips, A.D., Walker-Smith, J.A. & Turner, M.W. (1985) Intestinal sugar permeability: relationship to diarrhoeal disease and small bowel morphology. *J Ped Gastr Nutr* **4**, 568–574.

Fordtran, J.S., Rector, F.C., Ewton, M.F., Soter, N. & Kinney, J. (1965) Permeability characteristics of the human small intestine. *J Clin Invest* **44**, 1935–1944.

Forget, P., Sodoyez-Goffaux, F. & Zapitelli, A. (1985) Permeability of the small intestine to [^{51}Cr] EDTA in children with acute gastroenteritis of eczema. *J Ped Gastr Nutr* **4**, 393–396.

Frank, B.F. & Kern, F. (1967) Menetrier's disease. *Gastroenterology* **53**, 953–960.

Go, V.L.W., Hofmann, A.F. & Summerskill, W.H.J. (1970) Simultaneous measurements of total pancreatic, biliary and gastric outputs in man using a perfusion technique. *Gastroenterology* **58**, 321–328.

Harrison, M. & Walker-Smith, J.A. (1977) Re-investigation of lactose intolerant children: lack of correlation between continuing lactose intolerance and small intestinal morphology, disaccharidase activity and lactose tolerance test. *Gut* **18**, 48.

Hill, R.E., Hercz, A., Corey, M.L., Gilday, D.L. & Hamilton, J.R. (1981) Fecal clearance of CR 1-antitrypsin: a reliable measure of enteric protein loss in children. *J Paediatr* **99**, 416–419.

Holzel, A. (1967) Sugar malabsorption due to deficiencies of disaccharidase activities and of monosaccharide transport. *Arch Dis Child* **42**, 341.

van de Kamer, J.H., Ten Bokkel Huinink, H. & Weijers, H.A. (1949) Rapid method for the determination of fat in faeces. *J Biol Chem* **177**, 347.

Kerry, K.R. & Anderson, C.M. (1964) A ward test for sugar in the faeces. *Lancet* **1**, 981.

Krejs, G.J., Hendler, R.S. & Fordtran, J.S. (1980) Diagnostic and pathophysiologic studies in patients with chronic diarrhoea. In: *Secretory Diarrhoea, Clinical Physiology Series* (eds M. Field, J.S. Fortran & S.G. Schultz), pp. 67–83. American Physiological Society, Bethesda.

Lambadusuriya, S.P., Packer, S. & Harries, J.T. (1975) Limitations of xylose tolerance test as a screening procedure in childhood coelic disease. *Arch Dis Child* **50**, 34–39.

Launiala, K., Perheentupa, J., Pasternack, A. & Hallman, N. (1968) Familial chloride diarrhoea–chloride malabsorption. *Mod Prob Paediatr* **11**, 137–149.

Leelercq-Foucart, J., Forget, P.P. & Van Cutsem, J.L. (1987) Lactulose–rhamnose intestinal permeability in children with cystic fibrosis. *J Ped Gastr Nutr* **6**, 66–70.

McCollum, J.P.K., Muller, D.P.R. & Harries, J.T. (1977)

Test meal for assessing intraluminal phase of absorption in childhood. *Arch Dis Child* **52**, 887–889.

Maffei, H.V.L., Metz, G., Bampoe, V. *et al.* (1977) Lactose intolerance, detected by the hydrogen breath test, in infants and children with chronic diarrhoea. *Arch Dis Child* **52**, 766.

Magazzu, G., Jacono, I., Pasquale, G.D. *et al.* (1985) Reliability and usefulness of random fecal alpha-antitrypsin concentration: further simplification of the method. *J Ped Gastr Nutr* **4**, 402–408.

Menzies, I.S. (1973) Quantitative estimation of sugars in blood and urine by paper chromatography using direct densitometry. *J Chromat* **81**, 109.

Metz, G., Gassull, M.A., Seeds, A.R., Blendis, L.M. & Jenkins, D.J. (1976) A simple method of measuring breath hydrogen in carbohydrate malabsorption by end expiratory sampling. *Clin Sci Mol Med* **50**, 237.

Muller, M., Walker-Smith, J.A., Shmerling, D.H., Curtius, H.Ch. & Prader, A. (1969) Lactulose: a gas-liquid chromatography method of determination and evaluation of its use to assess intestinal mucosal damage. *Clin Chim Acta* **24**, 45–49.

Noone, C., Menzies, I.S., Banatvala, J.E. & Scopes, J.W. (1986) Intestinal permeability and lactose hydrolysis in human rotaviral gastroenteritis assessed simultaneously by non-invasive differential sugar permeation. *Eur J Clin Invest* **16**, 217–225.

Pearson, A.D.J., Eastham, E.J., Laker, M.F., Craft, A.W. & Nelson, R.H. (1982) Intestinal permeability in children with Crohn's disease and coeliac disease. *Br Med J* **285**, 20–21.

Penny, D.J., Ingall, C.B., Boulton, P., Walker-Smith, J.A. & Basheer, S.M. (1986) Intestinal malabsorption in cystic fibrosis. *Arch Dis Child* **61**, 1127–1128.

Puntis, J.W.L., Berg, J.D., Buckley, T.M., Booth, I.W. & McNeath, A.S. (1988) Simplified oral pancreatic function test. *Arch Dis Child* **63**, 780–784.

Quigle, E.M.M., Ross, I.N., Haeney, M.R., Holbrokke, I.B. & Marsh, M.N. (1987) Reassessment of faecal alpha 1-antitrypsin extraction for use as a screening test for intestinal protein loss. *J Clin Pathol* **40**, 61–66.

Robbards, M. (1973) Changes in plasma nephelometry after oral fat loading in children with normal and abnormal jejunal morphology. *Arch Dis Child* **48**, 656.

Sanderson, I.R., Boulton, P., Menzies, I. & Walker-Smith, J.A. (1987) Improvement of abnormal lactose/rhamnose permeability in active Crohn's disease of the small bowel by an elemental diet. *Gut* **28**, 1073–1076.

Schwartz, D.L., White, J.J., Saulsbury, F. & Haller, J.R. (1974) Gastrin response to calcium infusion: an aid to improved diagnosis of Zollinger–Ellison syndrome in children. *Pediatrics* **54**, 599–602.

Sharp, H.L. (1976) The current status of alpha 1-antitrypsin: a protease inhibitor in gastrointestinal disease. *Gastroenterology* **70**, 611–621.

Shmerling, D.H., Forrer, J.C.W. & Prader, A. (1970) Faecal fat and nitrogen in healthy children and in children with malabsorption or maldigestion. *Paediatrics* **46**, 690.

Smith, M.V. (1985) Expression of digestive and absorptive function in differentiating enterocytes. *Ann Rev Physiol* **47**, 247–260.

Soeparto, P., Stobo, E.A. & Walker-Smith, J.A. (1972) The role of chemical examination of the stool in the diagnosis of sugar malabsorption in children. *Arch Dis Child* **47**, 56.

Solomons, N.W., Garcia, R., Schneider, R., Viter, F.E. & Von Kaenel, A. (1979) H_2 breath test during diarrhoea. *Acta Paed Scand* **68**, 171–172.

Stone, M.C. & Thorp, J.M. (1966) A new technique for the investigation of the low density lipoproteins in health and disease. *Clin Chim Acta* **4**, 812.

Sullivan, P.B., Lunn, P.G., Northrop-Clewes, C. *et al.* (1992) Persistent diarrhoea and malnutrition – the impact of treatment on small bowel structure and permeability. *J Ped Gastr Nutr* **14**, 208–210.

Thomas, J.E., Gibson, G.R., Darboe, M.K., Dale, E. & Weaver, L.T. (1992) Isolation of *Helicobacter pylori* from human faeces. *Lancet* **340**, 1194–1195.

Thomas, Y., Hollander, D. & Krugliak Pokatz, K. (1990) PEG 400, a hydrophilic molecular probe for measuring intestinal permeability. *Gastroenterology* **98**, 39–46.

Tolbloom, J.J.M., Moteet, M., Kabir, H., Molatseli, P. & Fernandes, J. (1986) Incomplete lactose absorption from breast milk during acute gastroenteritis. *Acta Paed Scand* **75**, 151–155.

Verner, J.V. & Morrison, A.B. (1958) Islet cell tumour and a syndrome of refractory water diarrhoea and hypokalaemia. *Am J Med* **25**, 374–380.

Walker-Smith, J.A. & Andrews, J. (1975) Alpha 1-antitrypsin, autism and coeliac disease. *Lancet* **2**, 883–884.

Walker-Smith, J.A., Skyring, A.P. & Mistilis, S.M. (1967) The use of $^{51}CrCl_3$ in the diagnosis of protein-losing enteropathy. *Gut* **8**, 116.

Weaver, L.T., Chapman, P.D., Madeley, C.R., Laker, M.F. & Nelson, R. (1985) Intestinal permeability changes and the excretion of micro-organisms in the stools of infants with diarrhoea and vomiting. *Arch Dis Child* **60**, 326–330.

Weaver, L.T., Laker, M.F. & Nelson, R. (1984a) Intestinal permeability in the newborn. *Arch Dis Child* **49**, 236–241.

Weaver, L.T., Laker, M.F. & Nelson, R. (1984b) Enhanced intestinal permeability in preterm babies with bloody stools. *Arch Dis Child* **59**, 280–281.

Weizman, Z., Forstner, G.G., Gaskin, K.J. *et al.* (1985) Bentiromide test for assessing pancreatic dysfunction using analysis of para-aminobenzoic acid in plasma and urine: studies in cystic fibrosis and Shwachman's syndrome. *Gastroenterology* **89**, 596–604.

20: The Child with Learning Disability

J. HAMMOND & J. STERN

Introduction

The great majority of pregnancies end in the birth of a normal healthy child. However, amongst liveborn babies some 2% have major abnormalities and a further 3% have minor abnormalities. The central nervous system is affected in about 15 to 20% of these cases and frequently these infants will be mentally retarded.

The terminology surrounding mental retardation is in some confusion following attempts to provide accurate definitions and to avoid stigma. Mental retardation, mental handicap, learning difficulties and learning disability are all being used synonymously. The WHO (1980) describes impairment as a defect which may be inborn or acquired in early life, disability as the failure to perform on equal terms with one's peers, and handicap as the resulting limitation of functioning within society, often due to social disadvantage. Mental retardation is thus a disability with impaired maturation, learning and social adjustment, a disability that may be general or specific, mild, moderate or severe. In schools in the UK, children are classed as having moderate disability if they have an intelligence quotient (IQ) of 50–70 and severe disability if they have an IQ around 50 or below, although there is greater emphasis on actual performance in school than on IQ. The Department of Health has accepted the term 'learning disability'. In this chapter the terms are used interchangeably, but for the sake of simplicity we shall, in the main, refer to mental retardation.

Prevalence

The prevalence of severe retardation in school-age children remains at approximately 3 to 4 per 1000, and that of moderate retardation at 20 to 30 per 1000. At birth the incidence of many disorders with severe retardation is considerably higher, perhaps up to 1%, but many of these infants die during the first year of life, often before their learning defect has been ascertained. Adult prevalence is increasing owing to the improved longevity of people with disabilities, especially of those with Down's syndrome.

Aetiology

The diagnosis of learning disability in a young baby may be very difficult. Neuropsychiatric defects often present at a later age and the problem may only be realized during regular developmental surveillance, from a failure to achieve relevant milestones, or with the development of associated physical signs or symptoms, such as skin lesions, microcephaly or the onset of convulsions. The expectation of learning disability may sometimes be inferred from its association with dysmorphic features, for example in Down's syndrome. Other babies may be stillborn or die in infancy, when lesions of the nervous system, such as malformation, may be found upon autopsy.

The aetiological pattern of mental retardation is not static. For example, the contribution to prevalence by phenylketonuria, congenital hypothyroidism and rubella embryopathy has been, or is being, eliminated. On the other hand, with improved techniques in cytogenetics, enzymology and molecular biology new disorders are being discovered at an increasing rate, diminishing the proportion of unclassified cases. Sadly, industrial pollution and

Table 20.1 Causes of severe and mild mental retardation in Swedish patients. After Hagberg and Hagberg (1984)

	Severe mental retardation (n = 73)	Mild mental retardation (n = 91)	Comments
Prenatal			
Genetic			
Chromosomal	21	4	Higher incidence of MMR patients found by Göstason *et al.* (1991)
Mutations	4	1	Modest contribution to prevalence of handicap
Environmental			
Alcohol	–	7	Fetal alcohol syndrome
Infections	5	–	Predominantly TORCH
Other	1	–	Exposure to drug/poison
Unknown			
Specific syndromes	–	2	Syndromes of unknown aetiology, many described by Jones (1988)
Non-specific multiple anomalies	9	7	Antenatal abnormalities of the brain often found at autopsy
Perinatal			
Placental insufficiency, asphyxia, anoxia	9	15	With better antenatal care incidence is decreasing
Infections of the nervous system	2	1	Meningitis, encephalitis
Postnatal	8	2	Head injuries, battered baby syndrome, infection
Psychosis	1	2	
Unidentified			
Epilepsy, cerebral palsy	8	5	The size of this group depends on intensity of biomedical 'work-up' and criteria for classification
Without epilepsy or cerebral palsy	5	45	

MMR, mild mental retardation; TORCH, toxoplasma, rubella, cytomegalovirus and herpes.

iatrogenic disasters have occasionally added to the causes of mental handicap. The impact of some natural disasters, such as human immunodeficiency virus (HIV) infection, is not yet known. Table 20.1 shows data from Swedish studies.

Severe mental handicap may result from the effects of a single gene or a well-defined environmental hazard, such as meningitis or head injury. More often than not, the aetiology of the defect is multifactorial. For example, the very preterm baby, requiring intensive care for the first few months of life, may well suffer periods of hypoxia, is likely to experience a variety of metabolic stresses and is also at high risk from intraventricular or periventricular haemorrhage. However, the premature birth itself may be due to multiple antenatal factors, maternal or environmental. It then becomes extremely difficult to apportion weight to the many adverse factors contributing to the aetiology.

In general, the more severely retarded a child is, the earlier is presentation likely. Even then, parental concern is often not expressed until the child fails to achieve developmental milestones. Early diagnosis can only be achieved by comprehensive surveillance of all children and by screening, where relevant, for metabolic disorders. The shift of responsibility to primary care has the advantage that the family doctor will have a detailed knowledge of the family pathology, but also emphasizes the need for awareness of the range of normality and for continued alertness. A comprehensive assessment of abnormality will include a full medical, social and psychological evaluation. The paediatrician will decide, in consultation with the laboratory, which biochemical investigations are indicated.

An account of the many inborn errors and other metabolic disturbances which may affect the brain will be found in Chapters 4 to 8 and in the reference volumes by Scriver *et al.* (1989), Fernandes *et al.* (1990), Aicardi (1992) and Jaeken (1993). This chapter will be restricted to aspects directly relevant to mental retardation.

Clinical chemistry investigations for mentally retarded children

The success of biochemical investigations in all children, particularly those with learning problems, in whom collection of samples is especially difficult, depends in no small measure on cooperation between laboratory staff and those looking after the children. The importance of care in collecting samples and of prompt and careful transport to the appropriate laboratory cannot be overstressed. Children with severe retardation often react badly to outpatients and admissions; nevertheless, complex tests and sample collections may be more easily completed on a day ward. For all routine investigations physicians have access to the pathology services of the district in which the child resides. A case can be made for regular screening of children on long-term medication with anticonvulsants or psychoactive drugs in order to provide early warning of anticonvulsant rickets, liver damage or anaemia. Thyroid function is abnormal in a considerable proportion of mentally retarded adults, particularly those with Down's syndrome, but screening for T_4, free T_3 and sensitive thyroid-stimulating hormone (TSH) (not to be confused with neonatal screening for congenital hypothyroidism) is somewhat less productive of abnormal results in children. Monitoring of thyroid status and renal function is mandatory in patients on lithium therapy. Abnormal levels of sex hormones are seen in conditions associated with hypogonadism, such as Klinefelter's syndrome or the Laurance–Moon–Biedl syndrome; adrenal insufficiency may be the presenting sign of adrenoleucodystrophy. With community care, the problem in the laboratory of HBsAg-(Australia antigen)-positive patients, in the past common in long-stay hospitals and institutions, has been reduced but not eliminated. Impairment of mental function has not yet been a prominent feature of HIV-positive infants.

Mention must be made of the many non-specific biochemical abnormalities which have been reported in children with mental retardation, particularly when many were cared for in institutions (Crome & Stern, 1972). Some were the consequences of the frequent, and often chronic, infections to which such children are liable. Some degree of malnutrition is not uncommon in the severely slow child, often owing to an inability or unwillingness to eat appropriate food, to poor absorption or to vomiting – gastro-oesophageal reflux is a common finding. Severely slow children who cannot chew need to be offered food of the correct consistency as well as nutritional value. There is a wider scatter of hormonal and biochemical assay results than in normal children. An inappropriate response to stress and inadequate homeostasis are common. In general, mild abnormalities are more likely to be secondary to the child's life style than of significance in the aetiology of the disability. They are found less frequently in children living in the community than previously found in children resident in institutions.

Physical stigmata and neurological signs in the mentally retarded

Children with learning disabilities often look 'different', but genuine physical signs must be distinguished from the impression conveyed by delayed growth, immaturity or inappropriate behaviour. A number of retarded children have definite somatic abnormalities. These may be obvious from birth in some chromosomal disorders, such as Down's syndrome, or from a recognizable pattern of clinical signs, for example in the de Lange syndrome ('Amsterdam dwarfs') or Apert syndrome (acrocephaly syndactyly). An excellent monograph for the diagnostic use of physical signs (Jones, 1988) also notes useful laboratory findings.

Sometimes, symptoms of one system will suggest a more generalized disease that also affects the central nervous system. For example, the finding of a cardiac lesion following intrauterine exposure to the rubella virus will suggest the possibility of brain involvement as well. Infants with obvious physical defects involving the central nervous system, for example neural-tube defects, will often have, or develop, learning disability. The increased use of imaging in pregnancy leads to unexpected findings, the significance of which may be difficult to predict in terms of learning disability. For example, absence of the corpus callosum presents a new type of problem in interpreting findings to parents.

Ataxia, dyskinesia and dystonia

Ataxia must be distinguished from other forms of unsteadiness and the immature coordination commonly seen in retarded children. A number of metabolic disorders have been detected in patients presenting with ataxia, in some cases with a clinical picture of spinocerebellar degeneration. Abnormal movements (dyskinesia) and abnormal tone (dystonia), signs of extrapyramidal dysfunction, are also seen in many neurometabolic disorders (Table 20.2).

Muscle weakness and hypotonia

Of special interest is the infant with hypotonia (Dubowitz, 1980). Hypotonia and weakness may result from a disorder of the motor unit (the anterior horn cell, peripheral nerves, neuromuscular junction of muscle fibres), but may also be secondary to disorders affecting levels above the motor neurone. In the newborn in particular, hypotonia is often seen following trauma, hypoxia-ischaemia, intracranial infection or haemorrhage and in numerous metabolic disorders, giving way at a later date to cerebral palsy.

Investigations may include computerized tomography (CT) scanning, magnetic resonance imaging (MRI), myelography, ultrasound imaging, nerve conduction studies, electromyography, examination of the cerebrospinal fluid (CSF) and assay of serum creatine phosphokinase (CK), which has proved to be most organ specific (see Chapter 16). In a child with muscle weakness a normal CK does not exclude a myopathy; conversely, a very high CK does not necessarily imply a severe myopathy (Dubowitz, 1980). Examination of the CSF may be diagnostically useful in the polyneuropathies. Elevation of protein without an increase in cells, 'albumino cytological dissociation', is found, for example, in the Guillain–Barré syndrome and in Refsum's disease.

The nerve conduction velocity depends on the diameter and degree of myelination of the nerve. In the demyelinating neuropathies the nerve conduction velocity is markedly reduced. The results of nerve conduction studies help in the choice of biochemical investigations, particularly when the peripheral neuropathy is associated with involvement of the central nervous system, for example in the leucodystrophies. Muscle weakness and hypotonia in mental retardation are detailed in Table 20.2.

Sensory organs

Ocular findings are common in severely retarded children. Among 285 such children admitted to Queen Mary's Hospital for Children 20% were blind or partially sighted. Amongst common signs nystagmus, strabismus, optic atrophy and microphthalmia do not normally call for specific biochemical investigations, nor does retrolental fibroplasia, nowadays seen only rarely. Laboratory investigations have a diagnostic contribution to make in cases presenting with the signs listed in Table 20.3.

Hearing loss is probably as prevalent as blindness in the severely mentally retarded. It may be acquired prenatally, as in maternal rubella and other infections, perinatally, when it may be associated with cerebral palsy and other evidence of an insult to the nervous system, and postnatally, for example as sequelae to meningitis or the action of ototoxic drugs. Deafness is also a feature of a number of distinct hereditary syndromes, some with biochemical features (Table 20.4).

Some disorders associated with abnormalities of the skin and hair are shown in Table 20.5. A number of organic acidurias are associated with peculiar odours. However, there is wide variation amongst

Table 20.2 Some neurological signs in children with learning disability

Clinical sign and disorder	Comments
Ataxia	
Environmental aetiology	Drugs (anticonvulsants, psychoactive drugs, cytotoxic drugs), viral infections, lead intoxication, hypoxia, ante- or postnatal cerebellar lesions
Amino acidopathies	Late-onset urea cycle disorders (due to ammonia intoxication), Hartnup disease (most patients not retarded)
Organic acidurias	Intermittent branched-chain ketoaciduria, glutaric aciduria type I, pyruvate dehydrogenase deficiency
Respiratory chain disorders	Mitochondrial encephalomyopathies
Peroxisomal disorders	Refsum's disease, adrenoleucodystrophy
Lysosomal disorders	Juvenile variants of some lipidoses, mucolipidoses and leucodystrophies
Other metabolic disorders	Abetalipoproteinaemia (symptoms attributable to vitamin E deficiency); Lesch–Nyhan disease (particularly in partial HGPRT deficiency); ataxia telangiectasia (defective DNA repair); cerebrotentinous xanthomatosis; Wilson's disease
Dyskinetic–dystonic syndromes	Abnormal movements such as chorea and athetosis, and defective control of tone, signs of extrapyramidal dysfunction
Amino acidopathies	Tetrahydrobiopterin-deficient phenylketonuria; rare complication in homocystinuria
Organic acidurias	Glutaric aciduria type I; branched-chain organic acidurias; propionic aciduria
Lysosomal disorders	Krabbe's disease; metachromatic leucodystrophy; Gaucher's disease (acute neuropathic variant); late rare complication in some gangliosidoses
Other metabolic disorders	Lesch–Nyhan disease; Wilson's disease; Huntington's chorea (may present in childhood); Crigler–Najjar syndrome; xeroderma pigmentosum (some cases); hypoparathyroidism and pseudohypoparathyroidism (rare complication)
Hypotonia without significant weakness	
Non-specific mental retardation	Very common in unclassified mental retardation; often variable, it tends to get less severe with age
Down's syndrome	Treatment with 5-hydroxytryptophan of no benefit; some trisomies show increased tone
Prader–Willi syndrome	Adiposity, hypogenitalism, endocrine abnormalities; deletions on chromosome 15
Amino acidurias	Marked in non-ketotic hyperglycinaemia, also noted in hyperlysinaemia
Organic acidurias	Seen in a number of organic acidurias
Lysosomal disorders	Observed in some sphingolipidoses, mucopolysaccharidoses and leucodystrophies
Peroxisomal disorders	Zellweger syndrome, adrenoleucodystrophy
Other metabolic disorders	Carbohydrate-deficient glycoprotein syndrome, hypothyroidism, hypercalcaemia of infancy
Muscle weakness with hypotonia	
Congenital myopathies	Mitochondrial encephalomyopathies; Pompe disease

continued

Table 20.2 *Continued*

Clinical sign and disorder	Comments
Congenital myotonic dystrophy	Mental retardation common but non-progressive; serum CK and CSF protein normal; electromyogram shows 'dive bomber' effect
Congenital muscular dystrophy	Often associated with mental retardation; serum CK raised in early stages
Duchenne muscular dystrophy	Non-progressive mental retardation in some patients; dystrophin gene expressed in brain; serum CK very high in early stages
Peripheral neuropathies	Lower motor units affected in many lysosomal and peroxisomal disorders, abetalipoproteinaemia, familial dysautonomia, infectious and toxic polyneuropathies; CSF protein assays diagnostically useful; Werdnig–Hoffmann disease, Refsum's disease, myasthenia gravis not usually associated with mental handicap

For more extensive documentation refer to Dubowitz (1980), Adams and Lyon (1982), Fernandes *et al.* (1990) and Aicardi (1992).
HGPRT, hypoxanthine-guanine phosphoribosyl transferase; CK, creatine kinase; CSF, cerebrospinal fluid.

Table 20.3 Ocular findings and learning disability in some metabolic disorders

Sign*	Disorder and comments
Cataracts	Galactosaemia; galactosuria only occurs after intake of milk Lowe's syndrome; also glaucoma and buphthalmos, massive renal amino aciduria Pseudohypoparathyroidism; not always associated with mental retardation Dystrophia myotonica; dominant inheritance, gene on chromosome 19; variable genotype Galactokinase deficiency; galactosuria; no learning disability
Corneal opacities	Lysosomal disorders; a frequent finding, helpful in diagnosis Wilson's disease; Kayser–Fleischer rings Tyrosinaemia type II; some patients mentally retarded
Retinal degeneration	Abetalipoproteinaemia; low plasma cholesterol and vitamin E, acanthocytes in blood film; vitamin E therapy prevents neuropsychiatric deficits Hyperornithinaemia; ornithine transaminase deficiency; gyrate atrophy of choroid and retina; may respond to pyridoxine or proline therapy; not retarded Laurance–Moon–Biedl syndrome; sex hormones abnormal Lysosomal disorders; a frequent sign Peroxisomal disorders; prominent in Refsum's disease, many ocular changes in Zellweger syndrome and related disorders Respiratory chain disorders; seen in Kearns–Sayre syndrome and related disorders Sjögren–Larsson syndrome; fatty alcohol oxidoreductase deficiency; ichthyosis, spasticity
Dislocation of lens	Homocystinuria Sulphite oxidase deficiency; more severe course than homocystinuria; often linked to molybdenum cofactor deficiency
Conjunctiva	Ataxia telangiectasia; telangiectasiae of the conjunctiva; immune deficiencies, multisystem disorder; mild mental retardation in some patients

* Most prominent ocular sign; other parts of the eye may also be affected.

Table 20.4 Some hereditary disorders involving hearing loss

Disorder	Comments
Mucopolysaccharidoses I, II and III	Neurosensory and conduction mechanisms may be involved in hearing loss
Mucolipidoses I and II, sialidosis	Hearing loss is comparatively frequent in these disorders
Mannosidosis	Hearing loss occurs in both the α and β variants
Biotinidase deficiency	Vitamin therapy may not afford complete protection from hearing loss
Aspartoacylase deficiency (Canavan–van Bogaert disease)	Aspartoacylase active in some patients with Canavan–van Bogaert disease; hearing loss reported in some patients
Respiratory chain disorders	Sensorineural hearing loss occurs in some mitochondrial encephalomyopathies
Peroxisomal disorders	Hearing loss prominent in Refsum's disease, also seen in Zellweger syndrome, often in X-linked adrenoleucodystrophy
Pendred syndrome	Defect in thyroxine synthesis; sensorineural hearing loss; mental development normal in most patients

Table 20.5 Abnormalities of skin and hair in children with learning disability

Disorder	Abnormality	Comments
Menkes' disease	Twisted and brittle hair (pili torti and trichorrhexis nodosa)	Death in early infancy usual, a few milder cases have survived with mental retardation
Argininosuccinic aciduria	Brittle hair	Urinary amino acid chromatogram diagnostic; wide spectrum of severity
Homocystinuria	Sparse and brittle hair, malar flush	
Phenylketonuria (untreated)	Dry skin, sometimes with eczema; dilution of hair colour	Symptoms improve on low-phenylalanine diet
Tyrosinaemia type II	Hyperkeratotic plaques on palms, soles and elbows	Richner–Hanhart syndrome (cytoplasmic tyrosine transaminase deficiency)
Hartnup disease	Pellagra-like rash	Most patients not retarded, psychiatric symptoms intermittent; transport defect of some monoamino-monocarboxylic amino acids
Multiple carboxylase deficiency	Extensive skin rash, alopecia	Due to deficiency of biotinidase or holocarboxylase synthetase; wide range of symptoms, biotin supplements mandatory
Fabry's disease	Keratotic papules (angiokeratoma)	Most patients not mentally retarded
Farber's disease	Subcutaneous nodules over joints	Lipogranulomatosis (ceramide lipidosis); variable course
Cerebrotendinous xanthomatosis	Xanthoma of Achilles tendon	Storage of cholestanol in nervous system; plasma cholestanol level elevated, plasma cholesterol low

staff in how these odours are perceived and described. We recommend that the urine of any patient with an unusual smell that persists or appears during an episodic metabolic crisis should be checked for organic acid abnormalities.

Growth deficiency and endocrine disorders

Jones (1988) has pointed out that many patterns of malformation associated with growth retardation are the result of congenital hypoplasia in the skeletal system and in other organs, which will often include the brain. The disorders may be caused by teratogens, chromosomal abnormalities or mutant genes (Table 20.6). In many instances there is good correlation between the degrees of linear growth retardation and brain growth deficiency and mental defect. Postnatally, there is usually a lack of catch-up growth owing to irreversible antenatal impairment of brain and skeletal development. In mentally retarded children secondary growth retardation is also common, but in contrast to primary growth deficiency this can sometimes be prevented or reversed by appropriate treatment. Tall stature is mostly genetically determined but is a feature of a few mental retardation syndromes (Table 20.6).

In the mentally retarded child endocrine abnormalities are more often than not the consequence, rather than the cause, of impairment of the nervous system. Exceptions are untreated hypothyroidism (see Chapter 10) and hypoparathyroidism (see Chapter 12). Mental retardation has been found in several of the rare inborn errors of thyroid hormonogenesis, particularly the iodide peroxidase defect, but also in the dehalogenase defect and in some cases involving deficits in thyroglobulin synthesis or breakdown. Numerically most important as a cause of mental handicap was, until recently, congenital non-goitrous cretinism associated with thyroid dysgenesis. With an incidence of between 1 in 3000 and 1 in 4000 births this is probably the most common endocrine disorder in infants. Neonatal screening and prompt early treatment have greatly reduced the prevalence of mental handicap by this cause (see Chapter 8). However, evidence is now accumulating that not all the effects of fetal hypothyroidism are reversed by early treatment. The mean IQ in treated hypothyroid infants is lower than in controls, and those with low levels of thyroxine at diagnosis have a worse prognosis (Fuggle *et al.*, 1991). Complete prevention of neurological damage would presumably require antenatal thyroxine therapy. Currently, this is only practicable in selected cases, for example in a pregnancy with a family history of dyshormonogenesis when a fetal goitre is diagnosed on routine ultrasonography.

In hypoparathyroidism and pseudohypoparathyroidism mental retardation of varying severity is quite common. It is tempting to attribute the mental handicap to hypocalcaemia, a feature of both syndromes, in view of the key role of this cation not only in normal neuronal activity but also in epileptogenesis and 'epileptic brain damage'. However, in some late-onset cases of pseudohypoparathyroidism mental retardation and epilepsy precede the hypocalcaemia. In others, hypocalcaemia and the morphological features of the syndrome may be present without mental retardation. The cause of the mental defect remains to be established. The diagnosis of these disorders is described in Chapter 12, as is another disorder of calcium metabolism with mental retardation, hypercalcaemia of infancy. Psychological aspects of the syndrome are reviewed by Taylor (1991).

Self-injurious behaviour, behaviour disturbance and autism

Mild stereotyped self-injurious behaviour occurs in over 10% of mentally retarded patients; severe self-injurious behaviour resulting in irreversible mutilation, blindness or brain injury occurs in about 1 in 10 000 of the total mentally handicapped population (Corbett & Campbell, 1981). Biochemical factors are probably involved because certain agents known to influence neurotransmitters, such as the 5-hydroxytryptamine precursor 5-hydroxytryptophan, and the γ-aminobutyric acid (GABA) analogue Baclofen (lioresal), affect this behaviour at least temporarily. The rare Lesch–Nyhan syndrome (hypoxanthine-guanine phosphoribosyl transferase deficiency) is the best-known syndrome with severe self-mutilation. The diagnosis will be suggested by a raised urine uric acid:creatinine ratio. Definitive

Table 20.6 Growth disorders and learning disability*

Disorder	Comments
Primary growth deficiency	
Environmental	
Alcohol	Fetal alcohol syndrome, a major cause of congenital malformations and mental handicap
Anticonvulsants	As with all teratogenic agents, effects are neither regular nor uniform; risk of withholding anticonvulsants greater than risk of teratogenicity
Thalidomide	Fetal thalidomide syndrome usually not associated with mental retardation
Aminopterin	Similar effects produced by methotrexate, a cytotoxic methyl derivative of aminopterin
Rubella virus	Incidence of fetal rubella syndrome dramatically reduced by immunization; growth retardation also seen in other early intrauterine infections
Chromosomal imbalance	Growth retardation seen in autosomal aneuploidy
Mutant genes	The mutation may affect brain and skeletal growth in parallel, as in the mucopolysaccharidoses
Mechanism unknown	Growth retardation seen in many syndromes with congenital malformations and mental retardation of unknown aetiology; reduced brain weight is the commonest single abnormality in cases of mental retardation, cerebral palsy and epilepsy
Secondary growth deficiency	
Environmental	
Malnutrition	May be due to inappropriate diet
Radiotherapy	Risk of learning disability if CNS irradiated
Psychosocial deprivation	Formerly common, notably in large institutions
Metabolic	
Rickets	May be exacerbated by anticonvulsants
Renal disease	May form part of a syndrome or be the result of repeated urinary infections
Heart disease	Common complication, for example, in Down's syndrome
Respiratory disease	Common in retarded patients, often exacerbated by deficient cough reflex
Endocrine dysfunction	Hypothyroidism; pseudohypoparathyroidism; other endocrine dysfunction is usually non-specific, secondary to suboptimal function of the nervous system
Cerebral palsy	Common, mechanism of growth retardation unknown
Chronic severe infection	Mechanism of action unknown
Tall stature	
XXY and XYY karyotypes	The alleged aggressive behaviour of patients with the XYY karyotype has not been confirmed
Cerebral gigantism	Soto's syndrome; no consistent biochemical or endocrinological abnormality has been demonstrated
Homocystinuria	Patients with the clinically similar Marfan's syndrome are usually not mentally retarded
Beckwith–Wiedemann syndrome	Mild to moderate mental retardation; hypoglycaemia often caused by pancreatic hyperplasia
Megalencephaly	
Primary megalencephaly	Brains show structural changes, such as disorders of migration and organization, or spongy degeneration of the white matter

continued

Table 20.6 *Continued*

Disorder	Comments
Dysmorphic syndromes	These include cerebral gigantism and the Beckwith–Wiedemann syndrome
Neurocutaneous disorders	Seen, for example, in neurofibromatosis, tuberous sclerosis, the Sturge–Weber syndrome
Metabolic megalencephaly	Reported, for example, in Canavan–van Bogaert disease (aspartoacylase deficiency), some lysosomal disorders, glutaric aciduria type I

* For extensive documentation refer to Jones (1988).

diagnosis is by enzyme assay. This can be done on blood collected on filter paper as for infant screening tests. Severe self-mutilation is not confined to the Lesch–Nyhan syndrome; we have seen very similar behaviour in a child with Cornelia de Lange syndrome.

Behaviour problems are common in the retarded, but often they are only behaviour appropriate for the mental age of the patient. A point mutation in the monoamine oxidase A gene is associated with aggressive behaviour and mental retardation (Brunner *et al.*, 1993). We have seen severe behaviour problems in a patient with γ-glutamyl transferase deficiency (glutathionuria; see Scriver *et al.*, 1989), a condition detectable by the urinary nitroprusside test. Behaviour problems have been described in some, but not all, patients with this disorder. Retarded children as a group show no adverse behavioural or cognitive reactions to food additives, although a few individuals may do so (Rutter & Casaer, 1991).

Infantile autism as a specific syndrome was first described by Kanner nearly 50 years ago. Affected children fail to develop social relationships, exhibit ritualistic and repetitive behaviour and suffer from disturbance of language and perception, with onset in the first 3 years of life. The prevalence of the disorder appears to be rising as more children with moderate and severe retardation are included in the 'spectrum of autism'. Overall, more than three-quarters are retarded, over a third have non-specific brain dysfunction, and nearly half have an associated somatic condition, notably the fragile X syndrome and neurocutaneous disorders. Autistic features have been incidental findings in some children with inborn errors of metabolism, for example untreated phenylketonuria, homocystinuria and some lipidoses. A metabolic screen is therefore indicated in investigating children with autism. With so many underlying aetiologies it is not surprising that biochemical research has, on the whole, not been very rewarding. The significance of abnormalities of catecholamines and endorphins in the body fluids remains to be established. A comprehensive review of this topic has recently been published by Gillberg (1992).

Degenerative diseases of the nervous system with learning disability

A significant proportion of degenerative disorders are genetically determined and presumably therefore have a biochemical basis. Investigation of suspected cases is by no means straightforward. Neurological examination of severely retarded children is notoriously difficult. Observations over weeks or months and psychological, electrophysiological and imaging investigations may be needed to establish first slowing, then arrest of acquisition of skills, and finally regression. Regression is not the same as a drop in IQ. When slow progress is related to age it appears as a lowering of the IQ.

Sometimes, clinical signs such as the retinal cherry red spot in some lipidoses or the characteristic appearance of children with mucopolysaccharidoses or mucolipidoses will narrow the diagnostic possibilities. While regression is the hallmark of many hereditary metabolic disorders, it is also seen in many encephalopathies that are due to other causes (Table 20.7). In some lysosomal disorders (see Chapter 7(c)) pathological processes may be active before birth. In these patients perinatal compli-

Table 20.7 Some disorders leading to intellectual deterioration*

Tumours
Infections (subacute sclerosing panencephalitis)
Chronic poisoning (lead, organic mercury)
Autoimmune and postinfectious disorders (Schilder's disease)
Heredodegenerative disorders
Neurocutaneous disorders (tuberous sclerosis)
Spinal and spinocerebellar degenerations (Friedreich's ataxia)
Lysosomal storage disorders (see Chapter 7(c))
Leucodystrophies (Krabbe's, metachromatic, adrenoleucodystrophy)
Huntington's chorea
Wilson's disease
Childhood autism (Heller dementia infantilis)

* Examples are shown in parentheses. Conditions presenting in neonates have been excluded.

cations are common and their effects may mask the underlying disease process. An authoritative treatment of the diagnostic problems of these disorders has been published by Adams and Lyon (1982).

Definitive diagnosis of most heredodegenerative disorders involves assay of the enzyme or isoenzyme affected, or direct DNA analysis. These investigations are nearly always carried out in reference laboratories. Table 20.8 lists some exploratory investigations. Their availability in peripheral hospitals is conducive to judicious use of the reference centres.

Problems are posed by the child or adolescent who presents in the developmental clinic with sudden unexpected failure at school, often with psychiatric and sometimes with minor neurological symptoms. The biochemical findings can be ambiguous. For example, a low level of arylsulphatase A may suggest the diagnosis of metachromatic leucodystrophy which has an incidence of approximately 1 in 40 000 in the general population. However, 1 in 200 to 1 in 400 individuals are homozygous for the arylsulphatase A pseudodeficiency gene, also with low arylsulphatase levels but harmless, and about 1 in 1000 of the general population are compound heterozygotes for the deficiency and pseudodeficiency alleles, also with low enzyme levels and clinically well. Unrelated neurological symptoms in these individuals may lead to the wrong diagnosis. Fortunately, in most cases the

Table 20.8 Investigations of children with learning disability and convulsions or suspected degenerative disease of the nervous system

Investigation	Comments
Urea, creatinine, electrolytes	Abnormalities may be caused by vomiting, failure to drink or polydipsia
Blood gases, acid–base balance	Informative in non-ketotic hyperglycinaemia, hyperammonaemia and organic acidurias; note that patients with organic acidurias are not always acidotic
Liver function tests	Often abnormal due to drugs or infections; in Wilson's disease, children usually present with hepatic dysfunction rather than with neurological symptoms
Blood calcium, phosphate, alkaline phosphatase, magnesium	Onset of hypocalcaemia may be insidious in pseudohypoparathyroidism; mild rickets not uncommon in patients on anticonvulsants; magnesium assays have low priority after age 3 months
Blood copper and caeruloplasmin	Caeruloplasmin is not invariably low in Wilson's disease; blood copper low in Menkes' disease, copper transporting ATPases are involved in both disorders (Tanzi *et al.*, 1993)
Blood lead	Regression and convulsions only rarely due to lead poisoning; children with pica often have raised blood-lead levels, a consequence, not the cause, of their mental retardation
Blood glucose (fasting)	Often as low as 2.2 mmol/L, even in older children, seldom identifiable as the cause of fits or regression

continued

Table 20.8 *Continued*

Investigation	Comments
Blood lactate, pyruvate	Monitor 1 h after a meal; screen for disorders of carbohydrate metabolism and oxidative mechanisms; specificity improved if acetoacetate and 3-hydroxybutyrate are assayed concurrently
Blood ammonia	Monitor 1 h after a meal; elevated levels occur in late-presenting variants of urea cycle disorders, in female carriers of ornithine carbamyl transferase deficiency and in some patients with Rett's syndrome
Blood biotinidase	Treatment with biotin of affected patients should prevent, or at least halt, intellectual deterioration
Very long-chain fatty acids, bile acid metabolites	Screen for peroxisomal disorders
Blood and urine amino acids	Enhanced susceptibility to epilepsy in some amino acidurias; non-specific renal amino acidurias common in retarded children
Organic acidurias	Interpretation of the often highly complex excretion patterns of many organic acidurias is best left to reference laboratories
Urine screen	Cells and culture; pH, osmolality, protein; glucose, non-glucose reducing substances, ketones; odour; one purpose of the screen: to ensure sample is suitable for further tests
Urine uric acid : creatinine ratio	Screen for Lesch–Nyhan disease in patients with self-mutilation; preferable to plasma urate assay
Urinary oligosaccharides, mucopolysaccharides; intracellular metachromasia	Screen for lysosomal disorders
Blood and bone marrow films	Abnormal inclusions or vacuoles are found in lymphocytes and bone marrow cells in some lysosomal disorders
Enzyme assays and genotyping	For definitive diagnosis
Infection screen	Includes haematological and microbiological tests; under age 3 years tests for TORCH and measles antibody titre for the diagnosis of SSPE
EEG	Diagnostic in SSPE, some late-onset lipidoses, non-ketotic hyperglycinaemia, helpful in other disorders; up to 10% of normal children have some abnormalities in their EEG
Electrophysiology	Nerve conduction useful when peripheral neuropathy present; visual, auditory and somatosensory evoked potentials test integrity of respective pathways, electroretinography for retinal degeneration; electromyography
Imaging	CT scans, ultrasound imaging (while fontanelles are open) and MRI allow brain development and the evolution or resolution of lesions to be followed non-invasively; in a few centres positron emission computerized tomography and single photon emission computerized tomography have opened the way to studying brain function *in vivo*
Biopsies	Skin biopsies for the diagnosis of some lysosomal disorders and as a source of fibroblasts; muscle biopsies for the diagnosis of some encephalomyopathies; brain biopsies are hardly ever justified

TORCH, toxoplasma, rubella, cytomegalovirus and herpes; SSPE, subacute sclerosing panencephalitis; EEG, electroencephalogram; CT, computerized tomography; MRI, magnetic resonance imaging.

pseudodeficiency gene can now be identified by specific oligonucleotide primers and the polymerase chain reaction (Gieselmann & von Figura, 1990). A recent review of late-onset neurometabolic disorders has been edited by Baumann *et al.* (1991).

In X-linked adrenoleucodystrophy extreme phenotypic variability is found (Moser *et al.*, 1992). Patients may present over a wide age range with a rapidly progressive cerebral form, with Addison's disease without neurological signs, or with a late-presenting slowly progressive adrenomyelopathy affecting mainly the spinal cord, in about a third of cases with normal adrenal function. At least 16% of female heterozygotes have some neurological disability. This wide phenotypic variability may occur within one kindred or even one family (Moser *et al.*, 1992), and although reliable diagnostic tests are available this variability affects diagnosis, prognosis, carrier detection and genetic counselling.

Convulsions and learning disability

Convulsions are relatively common in children: about 40 per 1000 are affected, and nearly 70 per 1000 if febrile convulsions, breath-holding attacks and similar events are included. Fits are more frequent in those with mental retardation – about 10% of the mildly retarded and over one-third of the severely retarded may be affected (Aicardi, 1992). Epileptics as a group are somewhat less intelligent than non-epileptics, although some are not only normal but intellectually brilliant. In the mentally retarded, seizures may be the primary cause of the handicap, seizures and retardation may both be the result of an underlying disease process, or they may be iatrogenic, related to therapy.

Seizures causing retardation

The view that convulsions harm the brain is deep-rooted. Many neurologists and neuropathologists have maintained that fits themselves or their consequences, such as cerebral hypoxia, asphyxia or vascular disturbances and associated biochemical changes, may cause permanent injury. This may occur in a previously well or retarded child. Sometimes, a vicious circle is set up, the lesions caused by fits producing further fits. The acute illness responsible for the fits may be associated with acute metabolic disturbances. It must not be assumed without careful consideration of all the evidence that these disturbances are causally related to any pre-existing or ensuing retardation. Transient metabolic changes under stress are more common in mentally retarded than in normal children.

Seizures and retardation arising from a common cause

A disease process, perhaps hitherto unrecognized, may give rise to both fits and mental retardation. For example, infantile spasms have been observed in a number of hereditary disorders, and myoclonic epilepsy may be in the forefront of the clinical picture of mitochondrial and lysosomal disorders (Aicardi, 1992). While most often seen in infants and toddlers these disorders may present at any age.

Iatrogenic seizures

The sudden withdrawal of anticonvulsants or administration of other drugs which interfere with the metabolism of anticonvulsants can precipitate fits. The risk is increased in the mentally retarded child in whom both compliance with therapy and absorption of oral drugs can be unreliable.

Common investigations of children with fits are listed in Table 20.8. Depending on the severity, duration and nature of the fits the majority of, but probably not all, these investigations will need to be carried out. Clinical aspects of epilepsy have been reviewed by Laidlaw *et al.* (1993) and the morbid anatomy of epilepsy in the mentally retarded by Crome and Stern (1972).

Many anticonvulsants have enzyme-inducing effects and this commonly results in an asymptomatic rise in the level of serum liver enzymes. Hyperammonaemia, usually mild, occurs in some patients on valproate therapy. Isolated cases of severe hepatotoxicity caused by valproate have been reported, notably in mentally retarded children with enhanced vulnerability to this organic acid. Mild rickets, which sometimes develops in mentally retarded children on long-term anticonvulsants, responds readily to moderate supplements of vit-

Table 20.9 Reasons for monitoring drug levels in mentally retarded children with epilepsy

Therapeutic doses close to toxic doses (low margin of safety)
Wide individual variation in the rate of metabolism of the drug
Signs of toxicity difficult to recognize clinically
Presence of renal or liver disease
Drug interactions are suspected in children taking several drugs
Non-compliance is suspected
Seizure control is poor, or 'breakthrough seizures' occur in a previously well-controlled child

amin D (75 μg/week). It is well established that there is an increased incidence of folate deficiency causing macrocytosis in severely retarded treated epileptic children. This deficiency can be prevented by increasing the dietary folate intake.

Therapeutic monitoring

The indications for monitoring plasma drug levels are summarized in Tables 20.9 and 20.10. Routine monitoring is recommended for phenytoin and is often helpful for patients receiving carbamazepine, ethosuximide and phenobarbitone. In the case of

Table 20.10 Some drugs commonly used in the management of epilepsy in mentally retarded children. After Wallace (1990); Pedley and Meldrum (1992)

Drug	Mode of action	Half-life (h)	Optimum range (μmol/L)	Comments on plasma levels
Phenytoin	Blocks Na^+ channels	9–40	40–80	Steep relationship between dose and plasma level; monitoring of plasma levels recommended
Phenobarbitone	Enhances action of GABA	35–75	70–160	Toxic levels ill defined, wide variation in response to increased plasma levels
Primidone	Enhances action of GABA	35–75*	70–160*	The active metabolite of primidone is phenobarbitone
Carbamazepine	Blocks Na^+ channels	8–24	16–40	Monitoring may be useful in optimizing therapy
Ethosuximide	Blocks some Ca^{2+} channels	20–40	300–700	Monitoring useful in cases of unusual dose response
Valproate	Augments action of GABA	7–15	350–700	Drug action not reflected by serum level; monitoring only checks compliance
Diazepam	Enhances action of GABA	10–20	–	Tolerance develops rapidly; monitoring virtually of no clinical value
Clonazapam	Enhances action of GABA	20–30	–	Tolerance develops rapidly; monitoring virtually of no clinical value
Clobazam	Enhances action of GABA	10–30	–	Tolerance develops rapidly; monitoring virtually of no clinical value
Vigabatrin	GABA aminotransferase inhibitor	5–7	–	Drug action not related to plasma level
Lamotrigine	Blocks Na^+ channels	13–35	4–12	Monitoring may be useful in optimizing plasma levels
Gabapentin	GABA mimetic?	5–7	12–90	Levels found in clinical trials

* As phenobarbitone.
GABA, γ-aminobutyric acid.

sodium valproate the therapeutic effect appears to correlate better with the dose per kilogram body weight than with the actual plasma level. The 'optimum range' is only a guide to therapy. Many patients are well controlled below the stated ranges; a few tolerate levels above the range without toxic symptoms. Monitoring of anticonvulsant levels is valuable in the severely retarded because eliciting neurological signs of intoxication is difficult in these patients. Recent reviews of this topic are by Wallace (1990) and Meijer (1991).

Table 20.11 Factors pointing to a biochemical contribution to the aetiology of mild mental retardation

Intelligence, while not grossly impaired, is significantly lower than that of parents and rest of family
Consanguinity in parents
More than one slow child in family
Behaviour disorders, autistic features or other psychiatric disturbance
Neurological or physical signs
Deteriorating performance at school
Backwardness not explained by birth history or environmental deprivation
Possibility of exposure to drugs or poisons

The child with mild learning disability

The aetiology of mild mental retardation has been the subject of some controversy. For a long time the so-called high-grade defectives were thought to constitute a 'subcultural' group. Their mental handicap was ascribed to 'normal biological variation', largely conditioned by the operation of polygenes. This approach does not commend itself to the pathologist. Where the brains of mildly retarded patients have been examined they have mostly shown changes similar to, but less severe than, those of severely retarded patients (Crome & Stern, 1972). The handicap associated with monogenic disorders, such as tuberous sclerosis, some chromosomal disorders or those caused by environmental agents, such as the rubella virus, may span the full range from normal to the most gross forms of mental disability.

It is a daunting task to identify aetiologically significant biochemical factors in the mildly retarded. Sometimes it is like looking for a needle in a haystack – in the dark and with no guarantee that the needle is actually there! Nevertheless, there are some circumstances in which there is a somewhat higher chance that biochemical abnormalities may be involved, for example if two or more of the factors listed in Table 20.11 are present.

When biochemical abnormalities are detected, it is often by no means easy to establish a causal relationship to the mental retardation. For example, a girl with hyperprolinaemia type 2 seen at Queen Mary's Hospital for Children, Carshalton, UK, has epilepsy and nearnormal intelligence. However, her parents, who are first cousins, are both graduates, as are her three unaffected sisters. Patients with this disorder are often intellectually normal. An untreated phenylketonuric patient presented in adolescence with mild mental retardation and an anxiety state, whereas another phenylketonuric patient presented at age 6 years with a behaviour disorder and autistic features. A mildly retarded girl with γ-glutamyl transferase deficiency had severe behavioural problems. Clearly, patients with inborn errors may be only mildly retarded and some with late-onset degenerative disorders will also be initially so classified. Minor intellectual deficits may be produced by exposure to harmful substances in the environment.

So far there is no conclusive evidence that mild to moderate malnutrition in otherwise healthy and well-cared-for children will produce lasting intellectual deficits (Susser, 1989). Claims that dietary supplements improve performance in psychometric tests do not carry conviction, except in special cases such as preterm babies (Lucas *et al.*, 1992; Wardle, 1992).

The pattern of laboratory investigations is essentially the same as that for the severely retarded. In epidemiological studies, a high prevalence has been found amongst the mildly retarded of abnormalities of the sex chromosomes (Göstason *et al.*, 1991). Prominent amongst these is the fragile X syndrome with a prevalence of 1 in 2000 to 1 in 3000. Caused by a hereditary unstable DNA sequence, it accounts for nearly 2% of all mental retardation and the majority of cases of sex-linked retardation (Hagerman, 1992). The diverse psychiatric manifestations of this syndrome have been reviewed by Turk (1992) and its molecular biology has been outlined by Sutherland & Richards (1993). Single gene mutations, chromosomal abnormalities and well-defined adverse environmental factors account for the majority of cases of mild mental retardation. The hypothesis that mild retardation is largely due to the action of 'polygenes' is no longer tenable (Åkesson, 1986).

The child with learning disability without physical signs

A problem which recurs in mental deficiency practice is that of what biochemical tests to request for a child who presents with mental retardation but no obvious physical signs. There is no simple answer to this question. Damage to the brain may occur at any stage of its development and affect any of its functions. Cerebral palsy and marked hydrocephalus may coexist in some children with normal intelligence. Conversely, severely retarded patients may exhibit no abnormal neurological signs, and no biochemical abnormalities may be detectable. After death, most of these patients are found to have had abnormal brains, but in a few no lesions can be demonstrated with current neuropathological techniques. In fact, neuropathological changes correlate no more closely with intellectual deficits than do biochemical abnormalities (Crome & Stern, 1972).

Damage sustained by the brain during early development may not become apparent until months or even years later, when demands on the nervous system increase and the ability for abstract reasoning is tested. The agent or agents responsible for the handicap may have been active over a limited period only and be no longer detectable when the child is seen. Again, some metabolic defects may only be unmasked during infections or other environmental stress. When biochemical abnormalities are recorded in the history, caution is necessary before assuming causal relationships, and the more so if abnormal levels in a single blood constituent are followed by slow development. Irreversible brain damage often results from multiple causes, effective in combination, although tolerated singly (Eisenberg, 1977), and should not be attributed too readily to whichever abnormality happens to have been recorded.

Some investigations which may detect mild and atypical cases are listed in Table 20.12. It is unlikely that all will be requested for every patient. Clinical experience and intuition will be the main factors determining the choice of tests.

Selected topics

Phenylketonuria

Neonatal screening and effective dietary treatment have dramatically reduced the number of phenylketonuric children who are mentally retarded. Unresolved or incompletely solved problems include the nature of the pathogenetic process or processes, the factors other than dietary control that may affect the results of treatment, at what age and to what extent, if at all, the diet may be relaxed in older patients, and how to ensure normal development for the fetuses of phenylketonuric mothers.

Progress cannot be solely measured by IQ scales: some patients with neurometabolic disorders do not perform in school and later life as well as expected from their psychometric assessment. Hindley and Owen (1978) found that a quarter of normal children showed changes in IQ exceeding 22 points between the ages of 3 and 17 years. Assessing the results of treatment in individuals is therefore subject to considerable uncertainty and longitudinal studies of large cohorts are necessary for reliable assessment of the results of treatment (Smith *et al.*, 1990a, 1991; Azen *et al.*, 1991). The effects of genotype and of diet may be affected by other variables such as social class or family size. Often, careful dietary control and a stimulating environment go hand in hand, as do dietary failure and unsatisfactory home conditions.

There is now conclusive evidence that raised blood phenylalanine levels are harmful in later childhood as well as in infancy. Currently, the recommended treatment range in the UK is 120–360 μmol/L for pre-schoolchildren, 120–480 μmol/L at school, and preferably below 700 μmol/L thereafter (MRC Working Party, 1993a). In the USA, higher levels are sometimes allowed (Lang *et al.*, 1989).

The pathogenetic mechanisms in phenylketonuria

Table 20.12 Investigations in patients with unexplained, non-progressive learning disability and negative physical findings

Investigation	Comments
Blood and urine amino acids	Mild, non-specific renal amino acidurias common in retarded children
Urine organic acids	In some intermittent disorders, modern GC–MS techniques will identify characteristic metabolites, although patient asymptomatic
Urine purines and pyrimidines	Uric acid : creatinine ratio for mild variants of Lesch–Nyhan syndrome; pyrimidines for late-onset, atypical urea cycle disorders
Blood glucose, lactate, calcium and ammonia	Minor abnormalities in these metabolites are in most cases of no aetiological significance
Liver function tests	Elevation of ALT may be earliest manifestation of Wilson's disease
Thyroid function tests	Clinical signs of hypothyroidism may be absent
Serum CK	Early stages of some disorders of muscle may be asymptomatic; collection of samples must be standardized
Urine screen	Cells and culture; pH, osmolality, protein, glucose and other reducing substances; ketones (if present, assess severity and persistence); odour; nitroprusside test to detect presymptomatic cases of homocystinuria
Urine mucopolysaccharides, TLC of sugars and oligosaccharides	In some mucopolysaccharidoses and mucolipidoses (e.g. mannosidosis) clinical characteristics only develop in late childhood
Blood lead	To allay parental anxiety; blood levels below 1.25 μmol/L are unlikely to be of any aetiological significance
TORCH	Blood TORCH titres in the younger child
Blood film	Vacuoles or metachromatic inclusions in lymphocytes; screen for late-onset lysosomal disorders
Karyotype	Dysmorphic features may be mild or absent, for example in affected female heterozygotes with the fragile X syndrome
Neurophysiological investigations	EEG, electroretinogram, evoked potentials (visual, auditory, sensory), electromyography, nerve conduction velocity studies
Neuroimaging	X-ray of skull and spine, CT scan, ultrasound, MRI scan (particularly for white-matter disorders), as appropriate

GC–MS, gas chromatography–mass spectrometry; ALT, alanine aminotransferase; CK, creatine kinase; TLC, thin layer chromatography; TORCH, toxoplasma, rubella, cytomegalovirus and herpes; EEG, electroencephalogram; CT, computerized tomography; MRI, magnetic resonance imaging.

are still ill defined. There is evidence for an effect of excess phenylalanine on neurotransmitter synthesis, and on myelin metabolism, particularly in the developing brain (Scriver *et al.*, 1989; Fernandes *et al.*, 1990; Hommes 1991). Recently, imaging studies, using nuclear magnetic resonance (NMR) and NMR proton spectroscopy (Lou *et al.*, 1992), and electrophysiological studies have demonstrated white-matter lesions in early-treated adult phenylketonuria patients. No close correlation is evident from most of these studies between the severity of the lesions and neuropsychiatric symptoms, nor between the quality of the dietary control and the extent of the often asymptomatic lesions.

In the UK, classical phenylketonuria is defined by a blood phenylalanine level on a free diet above 1200 μmol/L and atypical phenylketonuria by a level below 900 μmol/L. Levels between 900 and 1200 μmol/L denote possibly atypical phenylketonuria. The terms non-phenylketonuric hyperphenylalaninaemia and mild hyperphenylalaninaemia are sometimes used for levels in the range 250–500 μmol/L. Care must be exercised in using this classification. For example, while breastfed some infants may have blood phenylalanine levels below 500 μmol/L, but on weaning and receiving an increased protein intake the level may rise to 1000 μmol/L or even higher. A molecular basis has recently been established for the phenotypic heterogeneity of phenylketonuria (Okano *et al.*, 1991). A limited number of mutations cover the great majority of the mutant alleles. Screening procedures for detecting further mutations and mutant-specific oligonucleotides for the detection of these mutations are being introduced. This, it is hoped, will make it possible to predict the phenotype from the genotype of the neonate, and to optimize treatment. Even here, caution is necessary as many factors in addition to the genotype determine the expression of a gene (MRC Working Party, 1993b).

Blood phenylalanine levels have been assayed and treatment monitored by thin layer chromatography, the Guthrie bacterial inhibition test, fluorimetry, spectrophotometry, ion exchange chromatography and high pressure liquid chromatography. With proper care, all these methods have been considered adequate in most cases. Since strict control of the blood phenylalanine level is now considered necessary, ion exchange chromatography and high pressure liquid chromatography are the methods of choice, not least because they also offer values for the levels of tyrosine and other amino acids, often essential in achieving optimal therapy.

In phenylketonuria, urine tests are in general not useful. Phenylpyruvate and *o*-hydroxyphenylacetate, its characteristic urinary metabolites, are not excreted by some patients with moderately raised blood phenylalanine levels (<1000 μmol/L) in amounts sufficient for detection by screening tests. Furthermore, phenylpyruvate disappears rapidly from alkaline urine. Phenylalanine itself, effectively reabsorbed by the kidneys, is often present in phenylketonuric urine in amounts easily missed on chromatograms by inexperienced observers.

The conversion of phenylalanine to tyrosine requires the liver enzyme phenylalanine hydroxylase, dihydropteridine reductase and the cofactor tetrahydrobiopterin (BH4). In most phenylketonuric patients, liver phenylalanine hydroxylase activity is reduced but 1–3% have a deficiency of BH4. This cofactor is also required for the hydroxylation of tyrosine to dopa, and of tryptophan to 5-hydroxytryptophan. Deficiency of BH4 may be due to impaired synthesis of dihydrobiopterin (BH2) or to impaired activity of dihydropteridine reductase, leading to a failure of conversion of BH2 to BH4. If untreated, patients usually develop a severe progressive neurological illness, malignant phenylketonuria. Treatment with a low-phenylalanine diet does not arrest the progress of the disease. The severity of the disorder is in large measure due to deficiency of the neurotransmitters dopamine and 5-hydroxytryptamine. Treatment is designed to normalize blood and CSF neurotransmitter levels and to correct the hyperphenylalaninaemia, which is not always pronounced. Usually, treatment consists of cofactor supplements and the neurotransmitter precursors levodopa and 5-hydroxytryptophan protected from decarboxylation by carbidopa. Early detection of these disorders is essential. In the UK a simple diagnostic scheme forms part of the neonatal screening programme for phenylketonuria (Table 20.13). The long-term effectiveness of treatment remains to be established. For recent reviews of this field refer to Matalon *et al.* (1989) and Fernandes *et al.* (1990).

The devastating effects of untreated maternal phenylketonuria are well known. They include intrauterine growth retardation, microcephaly, mental retardation and congenital malformations. The risk to the fetus can be minimized by starting a strict low-phenylalanine diet before conception, and by controlling blood phenylalanine levels during pregnancy at between 60 and 250 μmol/L (Smith *et al.*, 1990b; Thompson *et al.*, 1991; MCR Working Party, 1993a). Monitoring should be both accurate and frequent, ideally by full amino acid profiles.

Table 20.13 Screening phenylketonuric patients for tetrahydrobiopterin deficiency

Defect	Plasma biopterin*
Hydroxylase defect	
Phe ↑	↑
Phe N (on diet)	↓
BH4 deficiency due to impaired biosynthesis of BH2	
Phe ↑	↓
Phe N (on diet)	↓
BH4 deficiency due to DHPR deficiency	
Phe ↑	↑
Phe N (on diet)	↑

* Crithidia fasciculata assay (BH4 + BH2 + sepiapterin); Phe, phenylalanine; DHPR, dihydropteridine reductase; BH4, tetrahydrobiopterin; BH2, dihydrobiopterin; N, normal, ↑, increased, ↓, decreased.

Women with atypical phenylketonuria should be treated if their blood phenylalanine exceeds 300 μmol/L.

There are still in the community a few undiagnosed women of child-bearing age who were born before the nationwide screening programme was introduced. Unlike 95% of untreated phenylketonuric patients, they tend to be only mildly retarded, often with additional psychiatric problems. No case can be made for screening all pregnant women, but a maternal blood phenylalanine assay is indicated for any woman of child-bearing age who: (a) is herself mentally retarded or has neuropsychiatric symptoms; (b) has a family history of phenylketonuria, mental retardation or microcephaly; or (c) has one or more children with the clinical signs of maternal phenylketonuria. Unlike phenylketonuric women who have been diagnosed and treated soon after birth, are of normal intelligence and are used to the diet, late-diagnosed women are often unable or unwilling to cooperate in the management of their pregnancies, and, in our experience, may not be receptive to genetic counselling.

Histidinaemia

The phenotype of histidinaemia has many analogies with phenylketonuria. A block in the first step of the catabolism of histidine results in the accumulation of histidine and other imidazoles in the body fluids. The first few cases detected were patients in psychiatric units, and their neuropsychiatric deficits were attributed to the biochemical abnormality. Histidinaemia is readily detected by neonatal screening, and treatment with a diet low in histidine was started in some centres. Subsequently, however, both prospective and retrospective studies led to the conclusion that in most cases treatment is unnecessary (Scriver *et al.*, 1989). However, there is evidence of enhanced vulnerability of the nervous system in histidinaemia, in particular of a lowered seizure threshold, and treatment should be considered in histidinaemic infants with neurological symptoms. Three out of four histidinaemic patients seen in Queen Mary's Hospital for Children in recent years had epilepsy or a grossly abnormal electroencephalogram, and their IQs ranged from 80 to 120.

Homocystinuria

Excessive excretion of homocystine is found in several inborn errors of metabolism, of which cystathionine synthase is the most important (Scriver *et al.*, 1989). Patients usually have fine fair hair, a malar flush and skeletal abnormalities reminiscent of Marfan's syndrome. The most important clinical sign is dislocation of the lenses. The characteristic appearance develops only gradually, and dislocation of the lenses is usually present by 3 years of age.

The defect results in a block of the trans-sulphuration pathway from methionine to cystine with accumulation of methionine and homocystine in the body fluids, and excessive excretion in the urine of homocystine and other sulphur amino acids, some formed by transamination and disulphide exchange reactions. Detection and screening present problems because biochemical abnormalities are often minimal. The role of the laboratory may be limited to confirmation of the clinician's suspicion aroused by the patient's clinical signs. In suspected homocystinuria, a negative urinary nitroprusside screening test does not exclude the diagnosis. The blood methionine level remains markedly elevated in homocystinuric patients 24 h after a dose of 100 mg L-methionine/kg body weight, and a load test will usually be diagnostic. This test is probably not justifiable in a child who is mentally retarded but has none of the other signs of homocystinuria.

Fewer than half the patients with homocystinuria are severely retarded. Epilepsy and a variety of psychiatric disorders are found in many patients with homocystinuria, including the minority of normal or superior intelligence. Of five patients seen in Queen Mary's Hospital for Children, two children were mildly retarded and had behaviour problems, one adult of normal intelligence had a schizophrenia-like illness, and another suffered from depressions and a personality disorder. Homocystinuria carries a high risk of life-threatening thromboembolic crises, which often occur when the patients undergo minor surgery. Cerebral thromboses probably contribute significantly to the neuropsychiatric symptomatology. Homocysteic acid is present in excess in the body fluids. It is an excitatory amino acid and may be responsible for some of the neurological complications in homocystinuria.

The disorder is genetically heterogeneous. In nearly half the patients, the biochemical abnormalities can be corrected by pyridoxine (100–500 mg/day). A diet low in methionine and supplemented with cystine, betaine and sometimes folate corrects the biochemical abnormalities in all variants of the disorder and is the only effective treatment for pyridoxine-unresponsive patients. Normal or near-normal somatic and intellectual development has been claimed for early-treated patients. The effects of treatment on the long-term sequelae – thromboembolisms, dislocation of the lenses and osteoporosis – are not yet known.

Monitoring of the diet is not easy. Homocystine is unstable and in patients on the diet the upper limit of the acceptable range for homocystine is near the limit of detection by many amino acid analysers. The urinary excretion pattern of the sulphur amino acids may be obscured by artefacts. Specialized advice should be sought whenever homocystinuric patients are to be treated, and particularly when surgery is contemplated. Homocystinuria has been well documented by Mudd *et al.* (1985) and Abbott *et al.* (1987).

Other amino acidurias

Selected amino acidurias related to mental retardation are listed in Table 20.14. Details of clinical and laboratory aspects of the amino acidurias will be found in Bremer *et al.* (1981), Edwards *et al.* (1988), Scriver *et al.* (1989) and Fernandes *et al.* (1990).

Hypoglycaemia

Hypoglycaemia has long been recognized as a cause of brain damage and mental retardation (Table 20.15; see Chapter 5(a)). In the neonate, hypoglycaemia is often associated with additional adverse factors such as anoxia, jaundice, hypocalcaemia, acidosis and septicaemia. The individual contributions of these and other factors to symptoms and any damage to the brain are difficult to disentangle. Neurological dysfunction has been demonstrated by brainstem auditory evoked potentials in children at blood glucose levels between 2.6 and 2.0 mmol/L, in some cases in the absence of symptoms (Aynsley-Green, 1992). Brain damage might ensue if dysfunction is repeated and prolonged (Lucas *et al.*, 1988).

Pre-existing brain damage may exacerbate any tendency to develop hypoglycaemia. In the author's experience, the fasting blood glucose level in older, severely retarded, institutionalized children has often been in the range 2.2–2.6 mmol/L. In these children poor homeostatic control is, of course, not confined to glucose.

In inherited disorders of gluconeogenesis and

Table 20.14 Selected amino acidurias in relation to learning disability

Disorder	Comments
*Associated with mental defect**	
Phenylketonuria	
Homocystinuria	Cystathionine synthase deficiency
Methylene tetrahydrofolate reductase deficiency	Homocystinuria, fits, folate and neurotransmitters involved, low methionine
Methyl tetrahydrofolate homocysteine methyltransferase deficiency	Homocystinuria, megaloblastic anaemia, low methionine, defect in methylcobalamin synthesis, responds to B_{12} therapy
Non-ketotic hyperglycinaemia	Defects in glycine cleavage system
Urea cycle disorders	
Hyperornithinaemia (HHH)	Ornithine transport into mitochondria deficient; hyperammonaemia and homocitrullinuria; mimics urea cycle defects
Hypervalinaemia	Valine aminotransferase deficiency, very rare; distinct from MSUD
Hyperleucinaemia	Defective transamination of leucine and isoleucine, very rare; distinct from MSUD
Tyrosinaemia type II	Richner–Hanhart syndrome; cytoplasmic tyrosine aminotransferase deficiency; oculocutaneous involvement
γ-Glutamyl transpeptidase deficiency	Glutathioninuria, behaviour disorders, very rare
Glutathione synthase deficiency	Pyroglutamic aciduria (oxoprolinuria); hypoglutathionaemia, variable phenotype
β-Alaninaemia	β-Alanine aminotransferase deficiency; somnolence, fits, excess GABA in body fluids, extremely rare
Sulphite oxidase deficiency	Fits, dystonia and dyskinesia; excess sulphite, S-sulphocysteine and taurine in urine; may be due to apoenzyme or cofactor defects; dislocation of lenses
Pipecolic acidaemia	L-Pipecolic acid oxidase deficiency; part of peroxisomal Zellweger syndrome
Lowe's syndrome	Generalized renal amino aciduria, cataracts, buphthalmos, rickets, dwarfism
No proven link to mental defect†	
Histidinaemia	
Sarcosinaemia	Sarcosine oxidase deficiency, benign trait
Hyperprolinaemia type I	Proline oxidase deficiency; benign trait
Hyperprolinaemia type II	δ-1-Pyrroline-5-carboxylic acid dehydrogenase deficiency; some cases mildly retarded with epilepsy
Hydroxyprolinaemia	Hydroxyproline oxidase deficiency; probably a benign trait
Cystathioninuria	Cystathioninase deficiency; benign trait
Tyrosinaemia type I	Fumarylacetoacetase deficiency; hepatorenal dysfunction, peripheral neuropathy; not mentally retarded

continued

Table 20.14 *Continued*

Disorder	Comments
Hyperlysinaemia and saccharopinuria	Defects in bifunctional 2-aminoadipic semialdehyde synthase; variable phenotypes, often asymptomatic
2-Aminoadipic aciduria	2-Ketoadipic acid dehydrogenase deficiency; no consistent clinical pattern
Lysinuric protein intolerance	Transport defect of dibasic amino acids; lethargy, vomiting, growth failure with involvement of muscle, liver and bone unless protein intake is restricted
Glutamate formimino transferase deficiency	Marked excretion of formiminoglutamic acid; variable phenotype
Hartnup disease	Transport defect of monoamino-monocarboxylic amino acids; occasionally, symptoms of pellagra

* Without treatment most patients show variable degrees of learning disability.
† Causal relationship to learning disability not proven; some disorders enhance the vulnerability of the nervous system; interpretation of data complicated by ascertainment bias and phenotypic variability.
MSUD, maple syrup urine disease; GABA, γ-aminobutyric acid.

Table 20.15 Some aetiological factors in hypoglycaemia in infants and children; see also Chapter 5(a)

Environmental
Asphyxia, anoxia, hypothermia, respiratory distress syndrome
Shock, haemorrhage, intracranial injury
Septicaemia, meningitis, Reye's encephalopathy
Hepatocellular failure
Alcohol, salicylate, paracetamol ingestion
Iatrogenic, unstable diabetes

Endocrine
Hyperinsulinism (maternal diabetes, nesidioblastosis)
Deficiency of insulin antagonists (thyroid, adrenal, pituitary)
Beckwith–Wiedemann syndrome
Leprechaunism

Hereditary
Defects in gluconeogenesis
Defects in glycogenolysis, glycogen synthetase deficiency
Galactosaemia, fructosaemia
Defects in oxidation of fatty acids
Defects in ketone body formation
Organic acidurias (propionic, methylmalonic, branched-chain ketoaciduria)

medium-chain acyl dehydrogenase deficiency, careful dietary control of the blood glucose level can prevent the mental retardation which has, in the past, often been associated with these disorders (Cornblath & Schwartz, 1991).

Lactic acidosis

Pyruvate and lactate occupy a key position in metabolism, and even a moderate impairment of their pathways may result in neurological symptoms and mental handicap. Most patients with pyruvate dehydrogenase deficiency succumb in infancy with primary lactic acidosis (Brown, 1992; see also Chapter 7(b)), but children with the subacute form of the disorder, which may include dysmorphic features, may present in the developmental clinic.

In disorders of pyruvate metabolism elevated levels of lactate (>2.0 mmol/L) and pyruvate (>0.1 mmol/L) may be only intermittent, but pyruvate is nearly always raised (>0.115 mmol/L) 1 h after a glucose load. Urinary lactate: creatinine ratios of greater than 0.1 are diagnostically useful, as is an elevated CSF lactate concentration, particularly when the blood lactate level is normal. Blood and urine alanine levels tend to be increased in

Table 20.16 Some causes of lactic acidosis

Cause	Comments
Environmental	
Shock, cardiopulmonary disease, microangiopathy	Poor delivery of oxygen to tissues
Uraemia, liver disease	Impaired clearance, impaired metabolism
Acute infections, septicaemia	CSF lactate has been used to diagnose meningitis
Diabetic ketoacidosis	
Inhibition of enzymes by drugs or poisons	Phenformin, cyanide, acetaldehyde, ethanol, methanol, fructose
Hereditary	
Disorders of gluconeogenesis	Hypoglycaemia and ketosis are present in most cases
Disorders of pyruvate dehydrogenase complex	Increased pyruvate, normal ratio of lactate : pyruvate
Disorders of Krebs citric acid cycle	Fumarase deficiency, 2-ketoglutarate dehydrogenase deficiency
Respiratory chain disorders	Clinical presentation and prognosis variable; nervous system involvement and mental retardation common
Organic acidurias	'Secondary' lactic acidosis seen in branched-chain organic acidurias, fatty acid oxidation defects

CSF, cerebrospinal fluid.

patients in whom gluconeogenesis is impaired. The lactate : pyruvate ratio and 3-hydroxybutyrate : acetoacetate ratio may help to distinguish between disorders of pyruvate metabolism, the respiratory chain and Krebs citric acid cycle (Bonnefont *et al.*, 1992). Disorders of the respiratory chain have been regarded as largely neuromuscular (encephalomyelopathies). They may, however, present multiorgan involvement with features of hepatic, renal, endocrine and haematological dysfunction (Munnich *et al.*, 1992).

The assay of blood lactate and pyruvate is only a screening test. Raised levels indicate that an abnormality exists somewhere in a large area of metabolism. This abnormality is not even specific for a particular pathway or cause (Table 20.16). Enzyme deficiencies may be primary or secondary. For example, pyruvate dehydrogenase activity may be reduced in thiamine deficiency, heavy metal poisoning, or indeed by any toxic substance or metabolite which causes mitochondrial damage.

Organic acidurias

Organic acidurias are characterized by excessive excretion in the urine of aliphatic or aromatic amino acids. They are often hereditary and may or may not be associated with metabolic acidosis and abnormalities of carbohydrate or amino acid metabolism. Mostly, these disorders present soon after birth (see Chapters 3 to 5). However, some patients with milder variants may present at a later age, sometimes with mental retardation. Gas chromatography–mass spectrometry is the method of choice for investigating these disorders. Improved techniques, such as fast atom bombardment tandem mass spectrometry, can reliably detect and assay metabolites present in trace amounts only (Tanaka & Coates, 1990).

Using appropriate techniques, Watts *et al.* (1980) examined urine from 2000 severely retarded patients living in institutions. They found that nearly all the abnormalities observed in urinary organic acids could be attributed to drugs, diet or other environmental factors, or were artefacts produced, for example, by bacteria. Thus, organic acidurias do not contribute significantly to the prevalence of severe mental handicap. The question arises how far organic acidurias should be considered in the differential diagnosis of the mentally retarded child, as opposed to the sick neonate. In the latter, symptoms are usually severe (see Chapter 3). Patients with late-presenting organic acidurias may exhibit a variety of clinical signs. Examples are macrocephaly and extrapyramidal signs in glutaric aciduria type I, multiple malformations in 3-hydroxyisobutyryl-CoA deacylase deficiency, progressive muscular atrophy (without acidosis) in a child with 3-methylcrotonyl-CoA carboxylase deficiency, and a slowly progressive encephalopathy in two siblings with 3-methylglutaconic aciduria. Unexplained metabolic acidosis will, at any age, suggest a screen for organic acidurias. However, affected patients are not always acidotic and any clinical signs tend to be non-specific. Ultimately, the clinician must decide how much weight to give to the possibility that the patient has an organic aciduria. Selected organic acidurias are shown in Table 20.17, and Table 20.8 lists preliminary investigations which should be undertaken before, or while, samples are referred to centres with the specialized facilities essential in this field.

The long-term outlook for patients with classical maple syrup urine disease, propionic aciduria and methylmalonic aciduria is not good; survivors have varying degrees of learning disability. Not unexpectedly, children with intermittent, partial and cofactor-responsive enzyme deficiencies do better. The results of treatment of the fatty acid oxidation disorders have not yet been systematically evaluated, but the prospects of avoiding mental handicap, particularly in medium-chain acyl-CoA dehydrogenase deficiency, the most frequent of these disorders, appear good for the majority of patients (Hale & Bennett, 1992).

Galactosaemia

In most infants with galactosaemia, life-threatening symptoms are seen by the end of the first week of life. The disorder may, however, be milder, patients presenting at a later age with cataracts and mental retardation, or remaining virtually symptom-free and mentally normal. Galactose-1-phosphate uridyl transferase, the enzyme affected in galactosaemia, exhibits polymorphism and this may in part explain the wide range of clinical manifestations.

Galactosaemia may be detected by neonatal screening (see Chapter 8); genotyping usually requires determination of the activity, stability and electrophoretic mobility of the enzyme. Treatment by an essentially galactose-free diet has been monitored by assay of red-cell galactose-1-phosphate or urinary galactitol, but was hampered by inadequate knowledge of the pathogenesis of the disorder (Holton, 1990). Recently, interest has focused on uridine diphosphate-galactose, a galactosyl donor in the synthesis of galactolipids, in the pathogenesis of the disorder. Unfortunately, data interpretation is bedevilled by methodological problems (Kirkman, 1992).

Dietary treatment is life-saving in classical galactosaemia, but long-term results have been disappointing (Waggoner *et al.*, 1990), and there have been deficits in mental development, speech, motor function and ovarian function. Delays in diagnosis or poor dietary compliance do not appear to affect the outcome significantly. In view of the uncertain outlook, genetic counselling is mandatory so that at-risk couples can have the benefit of informed choice in planning their families.

Lead

Mental retardation as the result of an acute lead encephalopathy is rare in the UK. There is, however, considerable concern as to whether or not impaired psychological development and disturbed behaviour may result from low-level lead exposure associated with blood-lead levels below 1.8 μmol/L, which for many years was accepted as the paediatric upper limit of acceptable exposure.

Taylor (1991) concluded from meta-analysis of 19 studies that there is a real effect on intelligence; however, this is small by comparison with that of other factors determining intelligence, such as parental IQ and family environment. Interpretation

Table 20.17 Selected organic acidurias and learning disability (see also Chapter 6)

Disorder	Comments
Involving gluconeogenesis and pyruvate metabolism	
Von Gierke's disease	Deficiency of glucose-6-phosphatase; hypoglycaemia, lactic acidosis, hyperuricaemia, hepatomegaly; three genotypes; mental retardation preventable by dietary control of hypoglycaemia
Fructose-1,6-diphosphatase deficiency	Hypoglycaemia, lactic acidosis and ketosis after fasting; metabolic crises can be prevented by dietary management
Pyruvate carboxylase deficiency	Neonatal variant with lactic acidosis, citrullinaemia and hyperammonaemia, and late-onset variant with lactic acidosis and mental retardation
Pyruvate dehydrogenase complex disorders	Several variants, X-linked subunit E1 deficiency most common variant; heterozygous females usually severely affected; lactic acidosis in some patients, occasionally dysmorphic features; few exceptions to very poor prognosis
Phosphoenolpyruvate carboxykinase deficiency	Hypotonia, hepatomegaly, hypoglycaemia, lactic acidosis, mental retardation, extremely rare
Involving amino acid derivatives	
Maple syrup urine disease (branched-chain amino acids)	Deficiency of branched-chain ketoacid dehydrogenase complex; several variants, mental retardation common in surviving patients with severe variants, even with treatment
Isovaleric acidaemia (leucine)	Isovaleryl-CoA dehydrogenase deficiency; acute neonatal or chronic intermittent presentation, characteristic odour; some patients mentally retarded but outlook much improved by treatment (protein restriction; glycine and carnitine)
3-Methylcrotonyl-CoA carboxylase deficiency (leucine)	Variable, sometimes intermittent neurological presentation, characteristic odour; acidosis, hypoglycaemia inconstant findings; some patients respond to biotin, some are mentally retarded; restriction of protein intake beneficial
3-Methylglutaconic aciduria (leucine)	At least four genotypes, some with severe involvement of the nervous system and mental retardation; primary metabolic disorder unknown in most genotypes
3-Hydroxy-3-methylglutaryl-CoA lyase deficiency (leucine)	Episodes of lethargy, vomiting and hypotonia, with severe acidosis, hypoglycaemia and deficient ketone body formation; normal development on low-protein diet
Mevalonic aciduria (leucine)	Mevalonate kinase deficiency; variable neurological deficits, mental retardation, CNS malformations in some patients
2-Methylacetoacetyl-CoA thiolase deficiency (isoleucine)	Episodes of vomiting, ketosis and metabolic acidosis; treated by protein restriction; most patients not retarded
3-Hydroxyisobutyryl-CoA deacylase deficiency (valine)	Early onset, multiple congenital malformations: methacrylyl-CoA, a metabolite of valine, may be teratogenic; very rare
Propionic acidaemia (valine, isoleucine)	Usually acute neonatal onset; deficiency of either of two carboxylase subunits; ketoacidosis, hyperammonaemia, 'ketotic hyperglycinaemia'; long-term prognosis generally, but not invariably, poor
Methylmalonic acidaemia (valine, isoleucine)	Usually neonatal onset with 'ketotic hyperglycinaemia' but severity and age of onset variable; two mutations involve methylmalonyl-CoA mutase, two synthesis of adenosylcobalamin and three synthesis of adenosyl- and methylcobalamin; nearly half the patients are retarded; cobalamin supplements helpful in some cases
2-Ketoadipic aciduria (lysine)	2-Ketoadipic acid dehydrogenase deficiency; variable clinical presentation, causal relation to mental retardation unproven

continued

Table 20.17 *Continued*

Disorder	Comments
Glutaric aciduria type I (lysine, tryptophan)	Glutaryl-CoA dehydrogenase deficiency; macrocephaly, extrapyramidal symptoms, mental deterioration; treatment by lysine- and tryptophan-restricted diet and with carnitine and riboflavin supplements may help; some homozygotes asymptomatic
4-Hydroxybutyric aciduria (glutamic acid)	Succinic semialdehyde dehydrogenase deficiency; hypotonia, ataxia, mild mental retardation; vigabatrin may be beneficial
2-Oxoprolinuria (glutathione)	(a) Glutathione synthetase deficiency; acidosis, haemolysis, variable neurological involvement (b) 5-Oxoprolinase deficiency; no involvement of the nervous system reported so far
Involving fatty acid disorders	
Acyl-CoA dehydrogenase deficiencies	Long-chain, medium-chain and short-chain variants; variable severity of presentation, may mimic Reye's syndrome; episodes of hypoketotic hypoglycaemia, vomiting and coma may occur; brain damage prevented in many cases by vigorous treatment of crises and careful management of the diet
Long-chain 3-hydroxyacyl-CoA dehydrogenase deficiency	Presentation often mimics Reye's syndrome; liver, heart and brain involved; high mortality, many survivors mentally retarded, but milder cases occur; fat-restricted, carbohydrate-enriched diet plus extra carnitine helpful
Multiple acyl-CoA dehydrogenase deficiency	'Glutaric aciduria type II'; defects in ETF or ETF-QO; presentation severe neonatal with congenital malformations, or milder with later onset; liver, kidney, brain and muscle affected, developmental delay; hypotonia, acidosis, hypoglycaemia, pungent odour; mostly very poor prognosis; exceptionally, a milder case may respond to riboflavin
Miscellaneous	
Multiple carboxylase deficiency (propionyl-CoA, pyruvate and 3-methylcrotonyl-CoA carboxylases)	(a) Holocarboxylase synthetase deficiency; early onset, hypotonia, fits, skin rashes, lactic acidosis and ketoacidosis; some cases improve on biotin supplements (b) Biotinidase deficiency; late onset, skin rashes, hearing loss, optic atrophy; dyskinesia; developmental regression prevented and many symptoms reversed by treatment with biotin; biochemical abnormalities may be absent in remission
L-2-Hydroxyglutaric aciduria	Ataxia, fits, macrocephaly, leucodystrophy, mental retardation; basic defect unknown
Fatty acid transport defects	Brain at risk from hypoketotic hypoglycaemia in primary carnitine deficiency, in the hepatic form of palmitoyl transferase I, in the severe form of palmitoyl transferase II, and in carnitine/acylcarnitine translocase deficiency; the urinary organic acid profile is often inconclusive in these disorders
Fumarase deficiency	Severe neurological impairment, fits, mental retardation; massive organic aciduria
Aspartoacylase deficiency	Spongy degeneration of the white matter, megaloencephaly, mental retardation, abnormal excretion of *N*-acetylaspartic acid
D-Glyceric acidaemia	D-Glycerate kinase deficiency; highly variable presentation includes mental retardation, causal relationship doubtful
Succinyl-CoA:3-ketoacid-CoA transferase deficiency	Episodes of severe ketoacidosis with or without hypoglycaemia; mental retardation preventable by prompt treatment

CNS, central nervous system; ETF, electron transfer protein; ETF-QO, ETF-ubiquinone oxidoreductase.

of this effect is problematic. Taylor (1991) raises the possibility of reverse causality, i.e. the less intelligent are more likely to ingest environmental lead. This mechanism clearly operates in some severely retarded patients housed in old buildings (Bicknell *et al.*, 1968). There are also difficulties in standardizing the psychological and behavioural measurements. Other topics requiring further study are the timing of the exposure, possible synergistic effects of other heavy metals (Lewis *et al.*, 1992), and the extent to which any harmful effects can be reversed or mitigated. It is not clear whether the effects of lead on intelligence result from small changes in the majority of individuals or from larger changes in a minority of vulnerable individuals. The latter appears more probable.

The problems are rather different in disturbed, severely retarded patients, in whom pica is common. Access to lead in old paint or soil is sometimes difficult to prevent and these patients are liable to ingest excessive amounts of the metal. A majority of severely retarded patients with pica studied in Queen Mary's Hospital for Children by Bicknell *et al.* (1968) had a markedly elevated body burden of lead. There was no evidence that lead played any part in the aetiology of the mental defect in most of these patients, resulted in further intellectual deterioration, or resulted in other symptoms attributable to lead. However, an increased body burden of lead is dangerous in these inherently vulnerable patients, and regular monitoring of blood-lead levels of children with pica, and treatment with chelating agents for those with elevated blood-lead levels (significantly above 1.8 μmol/L) is mandatory.

Finally, we must mention the child with mild learning disability or behaviour disorder whose parents believe that their child's problems are due to lead exposure. If the blood-lead level is on several occasions between 1.0 and 1.8 μmol/L, avoidable sources of lead exposure should be looked for to reassure the parents while other, more likely, causes of their child's disability are sought.

Conclusion

In a minority of cases it is possible to attribute a learning disability to a single hereditary or environmental cause. More often it is a case of several adverse factors acting synergistically to overwhelm the adaptive capacity of the brain. The rate at which a metabolic disturbance develops, the integrity of the blood–brain barrier, the developmental stage of the brain and the selective vulnerability of its formations enhance or mitigate the impact of a metabolic insult.

Diverse pathogenetic mechanisms may act via a common pathological pathway; diverse disorders of the brain may give rise to closely similar behavioural phenotypes. In all cases, the quality of the environment in which a retarded child is brought up materially affects the ultimate outcome. Ready access to comprehensive laboratory services is an important part of a favourable environment for these children.

References

Abbott, M.H., Folstein, S.E., Abbey, H. & Pyeritz, R.E. (1987) Psychiatric manifestations due to cystathionine beta-synthetase deficiency: prevalence, natural history and relationship to neurologic impairment and vitB6-responsiveness. *Am J Med Genet* **26**, 959–969.

Adams, R.D. & Lyon, G. (1982) *Neurology of Hereditary Metabolic Disease of Children*. McGraw-Hill, London.

Aicardi, J. (1992) *Diseases of the Nervous System in Childhood*. Blackwell Scientific Publications, Oxford.

Åkesson, H.O. (1986) The biological origin of mild mental retardation: a critical review. *Acta Psychiatr Scand* **74**, 3–7.

Aynsley-Green, A. (1992) Hypoglycaemia. In: *Recent Advances in Paediatrics*, No. 10 (ed. J.T. David). Churchill Livingstone, Edinburgh.

Azen, C.G., Koch, R., Friedman, E.G. *et al.* (1991) Intellectual development in 12-year-old children treated for phenylketonuria. *Am J Dis Child* **145**, 35–39.

Baumann, N., Federico, A. & Suzuki, K. (eds) (1991) Late onset neurometabolic genetic disorders from clinical to molecular aspects of lysosomal and peroxisomal disease. *Dev Neurosci* **13**, 185–376.

Bicknell, J., Clayton, B.E. & Delves, H.T. (1968) Lead in mentally retarded children. *J Ment Defic Res* **12**, 282–293.

Bonnefont, J.-P., Chretien, D., Rustin, P. *et al.* (1992) Alpha-ketoglutarate dehydrogenase deficiency presenting as congenital lactic acidosis. *J Pediatr* **121**, 255–258.

Bremer, H.J., Duran, M., Kamerling, J.P., Przyrembel, H. & Wadman, S.K. (1981) *Disturbances of Amino Acid Metabolism: Clinical Chemistry and Diagnosis*. Urban & Schwarzenberg, Munich.

Brown, G.K. (1992) Pyruvate dehydrogenase E1α deficiency. *J Inher Metab Dis* **15**, 625–633.

Brunner, H.G., Nelen, M., Breakfield, X.O. *et al.* (1993) Abnormal behavior associated with a point mutation in the structural gene for monoamine oxidase A. *Science* **262**, 578–580.

Corbett, J.A. & Campbell, H.S. (1981) Causes of severe self-injurious behaviour. In: *Frontiers of Knowledge in Mental Retardation*, Vol. 2 (ed. P. Mittler), pp. 285–292. University Park Press, Baltimore.

Cornblath, M. & Schwartz, R. (1991) *Disorders of Carbohydrate Metabolism in Infancy and Childhood*, 3rd edn. Blackwell Scientific Publications, Oxford.

Crome, L. & Stern, J. (1972) *Pathology of Mental Retardation*, 2nd edn. Churchill Livingstone, Edinburgh.

Dubowitz, V. (1980) *The Floppy Infant*, 2nd edn. *Clinics in Developmental Medicine*, No. 76. Heinemann Medical, London.

Edwards, M.A., Grant, S. & Green, A. (1988) A practical approach to the investigation of amino acid disorders. *Ann Clin Biochem* **25**, 129–141.

Eisenberg, L. (1977) Development as a unifying concept in psychiatry. *Br J Psychiatr* **131**, 225–237.

Fernandes, J., Saudubray, J.-M. & Tada, K. (eds) (1990) *Inborn Metabolic Diseases: Diagnosis and Treatment*. Springer-Verlag, Berlin.

Fuggle, P.W., Grant, D.B., Smith, I. & Murphy, G. (1991) Intelligence, motor skills and behaviour at 5 years in early treated congenital hypothyroidism. *Eur J Pediatr* **150**, 570–574.

Gieselmann, V. & von Figura, K. (1990) Advances in the molecular genetics of metachromatic leucodystrophy. *J Inher Metab Dis* **13**, 560–571.

Gillberg, C. (1992) Autism and autistic-like conditions: subclasses among disorders of empathy. *J Child Psychol Psychiatr* **33**, 813–842.

Göstason, R., Wahlström, J., Johannison, T. & Holmquist, D. (1991) Chromosomal aberrations in the mildly mentally retarded. *J Ment Defic Res* **35**, 240–246.

Hagberg, B. & Hagberg, G. (1984) Aspects of prevention of pre-, peri-, and postnatal brain pathology in severe and mild mental retardation. In: *Scientific Studies in Mental Retardation* (ed. J. Dobbing), pp. 43–64. The Macmillan Press, London.

Hagerman, R.J. (1992) Fragile X syndrome: advances and controversy. *J Child Psychol Psychiatr* **33**, 1127–1139.

Hale, D.E. & Bennett, M.J. (1992) Fatty acid oxidation disorders: a new class of metabolic diseases. *J Pediatr* **121**, 1–11.

Hindley, C.B. & Owen, C.F. (1978) The extent of individual changes in IQ for ages between 6 months and 17 years in a British longitudinal sample. *J Child Psychol Psychiatr* **19**, 329–350.

Hirst, M.C., Knight, S.M., Nakahori, Y., Roche, A. & Davies, K.E. (1992) Molecular analysis of the fragile X syndrome. *J Inher Metab Dis* **15**, 532–538.

Holton, J.B. (1990) Galactose disorders: an overview. *J Inher Metab Dis* **13**, 476–486.

Hommes, F.A. (1991) On the mechanism of permanent brain dysfunction in hyperphenylalaninaemia. *Biochem Med Metab Biol* **46**, 277–287.

Jaeken, J. (ed.) (1993) Inherited metabolic diseases and the brain. *J Inher Metab Dis* **16**, 613–812.

Jones, K.L. (1988) *Smith's Recognizable Patterns of Human Malformation*, 4th edn. WB Saunders, Philadelphia.

Kirkman, H.N. (1992) Erythrocytic uridine diphosphate galactose in galactosaemia. *J Inher Metab Dis* **15**, 4–16.

Laidlaw, J., Richens, A. & Chadwick, D. (eds) (1993) *A Textbook of Epilepsy*, 4th edn. Churchill Livingstone, Edinburgh.

Lang, J.M., Koch, R., Fishler, K. & Baker, R. (1989) Non-phenylketonuric hyperphenylalaninemia. *Am J Dis Child* **143**, 1464–1466.

Lewis, M., Worobey, J., Ramsay, D.S. & McCormack, M.K. (1992) Prenatal exposure to heavy metal: effect on childhood cognition, skills and health status. *Pediatrics* **89**, 1010–1015.

Lou, H.C., Toft, J., Andresen, J. *et al.* (1992) An occipito-temporal syndrome in adolescents with optimally controlled hyperphenylalaninaemia. *J Inher Metab Dis* **15**, 687–695.

Lucas, A., Morley, R. & Cole, T.J. (1988) Adverse neurodevelopmental outcome of moderate neonatal hypoglycaemia. *Br Med J* **297**, 1304–1308.

Lucas, A., Morley, R. Cole, T.J., Lister, G. & Leeson-Payne, C. (1992) Breastmilk and subsequent intelligence quotient in children born preterm. *Lancet* **339**, 261–264.

Matalon, R., Michals, K., Blau, N. & Rouse, B. (1989) Hyperphenylalaninaemia due to inherited deficiencies of tetrahydrobiopterin. *Adv Pediatr* **36**, 67–90.

Meijer, J.W.A. (1991) Knowledge, attitude and practice in anti-epileptic drug monitoring. *Acta Neurol Scand* **83** (Suppl. 134), 1–128.

Moser, H.W., Moser, A.B., Smith, K.D. *et al.* (1992) Adrenoleucodystrophy: phenotypic variability and implications for therapy. *J Inher Metab Dis* **15**, 645–664.

MRC Working Party (1993a) Recommendations on the dietary management of phenylketonuria. *Arch Dis Child* **68**, 426–427.

MRC Working Party (1993b) Phenylketonuria due to phenylalanine hydroxylase deficiency: an unfolding story. *Br Med J* **306**, 115–119.

Mudd, S.H., Skovby, F., Levy, H.L. *et al.* (1985) The natural history of homocystinuria due to cystathionine beta-synthase deficiency. *Am J Hum Genet* **37**, 1–31.

Munnich, A., Rustin, P., Rötig, A. *et al.* (1992) Clinical aspects of mitochondrial disorders. *J Inher Metab Dis* **15**, 448–455.

Okano, Y., Eisensmith, R.C., Guttler, F. *et al.* (1991) Molecular basis of phenotypic heterogeneity in phenylketonuria. *N Engl J Med* **324**, 1232–1238.

Pedley, T.A. & Meldrum, B.S. (eds) (1992) *Recent Advances in Epilepsy*, No. 5. Churchill Livingstone, Edinburgh.

Rutter, M. & Casaer, P. (eds) (1991) *Biological Risk Factors in Psychosocial Disorders*. Cambridge University Press, Cambridge.

Scriver, C.R., Beaudet, A.L., Sly, W.S. & Valle, D. (eds) (1989) *The Metabolic Basis of Inherited Disease*. McGraw-Hill, New York.

Smith, I., Beasley, M.G. & Ades, A.E. (1990a) Intelligence and quality of dietary treatment in phenylketonuria. *Arch Dis Child* **65**, 472–478.

Smith, I., Beasley, M.G. & Ades, A.E. (1991) Effect on intelligence of relaxing the low phenylalanine diet in phenylketonuria. *Arch Dis Child* **66**, 311–316.

Smith, I., Glossop, J. & Beasley, M. (1990b) Fetal damage due to maternal phenylketonuria: effects of dietary treatment and maternal phenylalanine concentrations around the time of conception. *J Inher Metab Dis* **13**, 651–657.

Susser, M. (1989) The challenge of causality: human nutrition, brain development and mental performance. *Bull NY Acad Med* **65**, 1032–1049.

Sutherland, G.R. & Richards, R.I. (1993) Dynamic mutations on the move. *J Med Genet* **30**, 978–981.

Tanaka, K. & Coates, P.M. (eds) (1990) *Fatty Acid Oxidation: Clinical, Biochemical and Molecular Aspects*. Alan R. Liss, New York.

Tanzi, R.E., Petrukhin, K., Chemou, I. *et al.* (1993) The Wilson disease gene is a copper transporting ATPase with homology to the Menkes gene. *Nature Genet* **5**, 345–350.

Taylor, E. (1991) Developmental neuropsychiatry. *J Child Psychol Psychiatr* **32**, 3–47.

Thompson, G.N., Francis, D.E., Kirby, D.M. & Compton, R. (1991) Pregnancy in phenylketonuria: dietary treatment aimed at normalising maternal plasma phenylalanine concentrations. *Arch Dis Child* **66**, 1346–1349.

Turk, J. (1992) The fragile X syndrome. On the way to a behavioural phenotype. *Br J Psychiatr* **160**, 24–35.

Waggoner, D.D., Buist, N.R.M. & Donnell, G.N. (1990) Long-term prognosis in galactosaemia: results of a survey of 350 cases. *J Inher Metab Dis* **13**, 802–818.

Wallace, S.J. (1990) Anti-epileptic drug monitoring: an overview. *Develop Med Child Neurol* **32**, 923–926.

Wardle, J. (ed.) (1992) Nutrition and IQ. *Psychologist* **5**, 399–413.

Watts, R.W.E., Baraitser, M., Chalmers, R.A. & Purkiss, P. (1980) Organic acidurias and aminoacidurias in the aetiology of long-term mental handicap. *J Ment Defic Res* **24**, 257–270.

WHO (1980) *International Classification of Impairments, Disabilities and Handicaps*. World Health Organization, Geneva.

21: Malignant Disease

A. CRAFT & J. PRITCHARD

Introduction

Children account for only 1% of all patients with malignant disease but taking into account work-years lost, child cancer assumes much greater importance in economic terms. It is third in terms of life-years lost from cancer and second to breast cancer in terms of life-years saved (SEER, 1990). Death rates from cancer in children are falling and they now rank third behind accidents and congenital abnormalities. It is also becoming increasingly important in terms of inpatient and outpatient medical care. About 1400 children in Britain develop malignant disease each year: accurate incidence data are now available, since 1954 and 1968 respectively, for the Manchester (Birch *et al.*, 1980) and Newcastle (Craft *et al.*, 1987) areas. With a combined total population of almost nine million covering the whole of the north of England, there is a remarkably constant incidence rate, in these two areas, of 105 cases per million children per year (Table 21.1). This means that around 1 in 600 children in the UK develop cancer before their 15th birthday.

In contrast to adults with cancer, in whom carcinomas (epithelial tumours) predominate, most childhood malignancies are of mesodermal origin (embryonal tumours, sarcomas or haemopoietic tumours). The frequency of specific tumour types occurring in the Manchester and Newcastle regions is shown in Table 21.1. Leukaemia accounts for almost one-third of registered cases and brain tumours for one-fifth. International comparisons of the incidence of various types of malignant disease in childhood show similarities, but also some interesting differences (Parkin *et al.*, 1988). There is, for instance, an inverse relationship between the incidence of acute leukaemia and of lymphoma. In several 'developing' countries, lymphomas – especially Burkitt-type non-Hodgkin's lymphomas – are more common than leukaemia, whereas in the 'developed world', acute lymphoblastic leukaemia (ALL) is three times more frequent than lymphoma.

Ewing's sarcoma of bone is rare in black children both in Africa and North America, perhaps suggesting an inborn genetic 'resistance' to the development of this tumour, whereas retinoblastoma occurs twice as frequently (6–8% of all tumours) in Africa as in the UK or USA. There are two peak ages of incidence of childhood tumours: the first at 1–4 years and the second during the pubertal growth spurt. Each type of cancer occurs at a more-or-less characteristic age. ALL, Wilms' tumour and neuroblastoma are usually diagnosed between 2 and 4 years, whilst Ewing's tumour and non-Hodgkin's lymphoma (NHL) peak later. Acute myeloid leukaemia, osteosarcoma and Hodgkin's disease occur more commonly in adolescents and into the young adult years. Several types of cancer (lymphomas, liver tumours and osteosarcomas) are more common in boys than girls.

The 'two-hit' theory of Knudsen (1976) is now widely acknowledged to explain the final common pathway of tumour development in children. The 'hits' are mutations in a pair of 'allelic' genes, one on the maternal and the other on the paternal chromosome. In the case of inherited malignancy, for example familial retinoblastoma, the first mutation is constitutional, i.e. inherited in the sperm or ovum, and the second is sporadic, i.e. in the single cell from which the tumour is derived. In 'sporadic' childhood cancer, the 'two-hits' also occur in allelic genes but both are 'sporadic'. In familial

Table 21.1 Annual incidence of malignant disease per million children in the northern and northwest regions in the UK

Category of tumour	Newcastle	Manchester
Leukaemia	34.7	31
Lymphoma	9.7	10
Glioma	23.2	19
Neuroblastoma	6.2	8
Retinoblastoma	4.8	3
Soft-tissue sarcoma	10.1	12
Wilms'	5.8	6
Gonadal	1.3	1
Teratomas	0.4	3
Epithelial	2.9	4
Ewing's	1.6	2
Miscellaneous	4.7	6
Total	105.4	105

retinoblastoma, for example, one mutation is inherited (sometimes as a visible deletion) from the long arm of chromosome 13 ($13q^-$), whilst in sporadic retinoblastoma both hits are 'sporadic'. Knudsen's theory neatly explains the earlier mean age of onset of inherited, compared with 'sporadic', tumours, but is less readily applicable to those embryonal tumours (Wilms' and neuroblastoma) whose familial incidence is very low, probably because in these tumours two or more pairs of alleles are involved (i.e. four-hit, six-hit and upwards). Children with certain other chromosomal abnormalities also have a higher incidence of cancer. Acute leukaemia, usually lymphoblastic, occurs 15 times more often in children with Down's syndrome than in the general population. Patients with congenital immunodeficiency (e.g. Wiskott–Aldrich syndrome) or 'chromosomal instability syndromes' (e.g. ataxia telangiectasia, Bloom's syndrome, Fanconi anaemia) are particularly susceptible to the development of lymphoreticular malignancies or, in the case of xeroderma pigmentosum, skin cancer.

Large-scale epidemiological studies have been, and are being, carried out to try to identify environmental factors important in the aetiology of childhood cancer. Parental occupation, for example working in the radiation industry or paint spraying, has been implicated in some studies but, as yet, there are no proven associations. Bilateral retinoblastoma is transmitted in an autosomal dominant fashion; some patients have a constitutional deletion of deoxyribonucleic acid (DNA) from the long arm of chromosome 13, representing Knudsen's 'first-hit'. Prenatal administration of stilboestrol was identified during the 1970s as the cause of infant adenocarcinoma of the vagina and was discontinued, so this form of childhood cancer no longer occurs. Prenatal exposure to X-rays carries a slight increase in the risk for development of acute leukaemia. The role of exposure to more generalized ionizing and non-ionizing radiation (radon) is uncertain and is under investigation. The association of Epstein–Barr virus with Burkitt lymphoma is well known and it may also have a role in Hodgkin's disease, but whether the virus is oncogenic or merely a 'passenger' is still uncertain. Radiation-induced thyroid cancer (Refetoff *et al.*, 1975) and second malignancies in patients cured of their first tumour by radiation or drugs (Mike *et al.*, 1982) are the only clearcut examples of child cancer that are recognizably induced by environmental factors.

Whereas, 20 years ago, a diagnosis of malignant disease in a child was an almost inevitable death sentence, well over half the patients diagnosed now will be long-term survivors and probably cured of their disease. Table 21.2 shows the encouraging 5-year survival rates for 10 of the more common

Table 21.2 Expected 5-year disease-free survival for children with the 10 most frequently diagnosed childhood cancers

Tumour	Survival (%)
Retinoblastoma	95
Hodgkin's disease	85
Wilms' tumour	85
Non-Hodgkin's lymphoma	75
Rhabdomyosarcoma	60
Acute lymphoblastic leukaemia	65
Ewing's sarcoma	60
Medulloblastoma	55
Neuroblastoma	50
Acute myeloid leukaemia	45

Table 21.3 Factors contributing to improved survival of children with cancer

Introduction of combination chemotherapy
Introduction of 'new agents', e.g. *cis*-platinum, epipodophyllotoxins
Use of 'delayed' rather than immediate surgery
Refinements of radiotherapy techniques, e.g. megavoltage equipment, interstitial radiation, [^{131}I]-MIBG
Better 'imaging' investigations
Refinements in histopathological subclassification
Use of 'tumour markers'
Better supportive care, e.g. biochemical support, blood products' support, antibiotics, antivirals, antifungals
Development of multidisciplinary teams within paediatric oncology centres
Successful organization of national and international collaborative groups, e.g. United Kingdom Children's Cancer Study Group, International Paediatric Oncology Society
Early diagnosis (retinoblastoma only)

MIBG, metaiodobenzylguanidine.

tumours. Factors contributing to this encouraging state of affairs are listed in Table 21.3. The introduction and rational use of chemotherapy, along with the continual introduction of new 'active' agents, has been of central importance. Refinements of surgical and radiotherapy techniques have also made a contribution. Careful screening of families with a known predisposition to retinoblastoma has led to early diagnosis and more effective therapy in this particular group.

The establishment of multicentre cooperative groups, such as the United Kingdom Children's Cancer Study Group (UKCCSG), means not only that information about these rare diseases is carefully collated and shared, but also that clinical trials can reach statistically significant conclusions much more quickly. Improved 'support' facilities have also been crucial, and cooperation with the clinical chemistry laboratory is of special importance. 'Tumour markers' can be used to monitor the progress of patients with certain tumours (see p. 429) and a better understanding of the biochemical complications of the diseases and their treatment has enabled more aggressive therapy to be given with greater safety (p. 445).

With higher cure rates there is now heightened concern about the medium- and long-term effects on the child of both disease and treatment (p. 449 *et seq*.). As cancer patients, children are unique because they are growing and are therefore particularly susceptible to developmental, emotional and physical handicap. 'Quality of survival' of both the child *and* family is now at the forefront of the concerns of the paediatric oncologist: thus, current strategies focus on 'cure at least cost' rather than 'cure at any cost'.

General principles of diagnosis, staging and management

Diagnosis

Optimal management depends upon precise pretreatment diagnosis and staging. A diagnosis of cancer in childhood may be strongly suggested by the history, physical examination and imaging investigations, but confirmatory histopathological tissue diagnosis is almost always needed. Exceptions, where reliance is placed upon clinical and radiological diagnosis, include brainstem tumours. Tissue for microscopic diagnosis of solid tumours can be obtained by either an open surgical procedure (biopsy or resection) or closed 'needle' biopsy. As well as conventional histological studies, special histochemical and immunohistochemical procedures are used. Histopathological subclassification is often crucial because, in some tumours, the subtypes have major prognostic significance. The treatment plan for patients with 'favourable histology' Wilms' tumour, for example, is quite different from that for patients with 'unfavourable histology'. More and more centres perform cytogenetic studies on fresh tumour tissue: the DNA content ('ploidy') can be an important prognostic variable, whilst specific chromosome abnormalities can clarify an otherwise uncertain diagnosis. For example, a translocation between the long arms of chromosomes 11 and 22 (t 11:22) in tumour cells identifies a small round cell malignancy as a Ewing's sarcoma or a PNET (primitive neuroectodermal tumour).

Once diagnosis is secure, staging investigations are carried out. Staging has three main functions: it

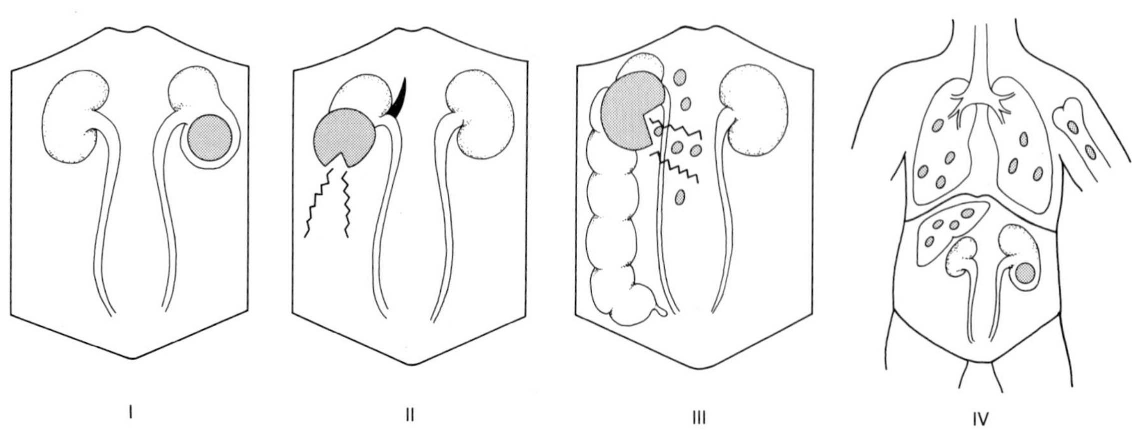

Fig. 21.1 Staging system used in the third US National Wilms' Tumour Study and in the first UK Children's Cancer Study Group Wilms' Trial. Stage II includes tumour rupture localized to the flank and inferior vena cava involvement. Patients with lymph node involvement and those with measurable residual disease are classified as having stage III disease.

indicates the extent of disease at diagnosis, and thereby aids treatment planning; it enables an estimate of prognosis to be made; and it allows for comparison of the results of treatment between different centres.

At present, there is variation between staging systems used in different centres but much effort is being directed towards the adoption of internationally accepted systems (e.g. the International Neuroblastoma Staging System) (Brodeur *et al.*, 1988). For ALL, the height of the white-cell count at diagnosis is considered the most important 'staging' investigation (Palmer & Hann, 1980) and determines the intensity of treatment used for individual patients. As an example of solid-tumour staging, the system used for Wilms' tumour is shown in Fig. 21.1. In general, the 'higher' the stage, the more extensive the disease.

In the individual child, the aims of staging investigations are to indicate the extent of the primary tumour and to seek evidence of metastatic spread. The limits of the primary tumour may be defined surgically or by 'imaging' investigations using diagnostic X-rays, ultrasound, isotope scanning (including metaiodobenzylguanidine (MIBG) scanning for neuroblastoma), computerized tomography (CT) and magnetic resonance imaging (MRI). Most tumours have a characteristic pattern of metastatic spread: investigation can thus be rationalized, saving the child unnecessary trauma and the hospital unnecessary expense. Even where disease appears to be confined to the primary site, the presence of micrometastases – too small to be shown by presently available techniques – is very common. Systemic therapy is therefore usually indicated, even for apparently localized disease.

Treatment

Chemotherapy is now the main treatment for the majority of children with cancer. Even when surgical excision of the 'primary' is indicated, chemotherapy is often used first, to make surgery easier. Complete eradication of some 'sensitive' tumours, especially leukaemias and lymphomas, can often be achieved with chemotherapy alone, but excision of residual masses is indicated, providing the surgery is not mutilating. Thus, it is acceptable to perform a nephrectomy for Wilms' tumour, but total pelvic extenteration for rhabdomyosarcoma of the bladder or vagina is now rarely indicated since this tumour can often be eradicated by chemotherapy alone or by combined chemotherapy and radiotherapy. To reduce the radiation dose to normal tissue, radiotherapists now often use a 'shrinking field' technique, concentrating the highest doses on the tumour residue rather than on the whole of the original tumour-bearing volume. Chemotherapy is usually the most appropriate initial treatment for patients with obvious metastatic disease. Unless the secondaries can be controlled, an attack on the primary is clearly illogical. With precise knowledge of the histological subtype and stage, it is possible to plan a treatment strategy for each patient using surgery, radiotherapy and chemotherapy, in whatever combination provides maximum benefit with a minimum of immediate and late toxicity.

'Routine' biochemical studies

Clinicians, without thinking carefully, often ask for regular biochemical studies as part of their 'routine' management of patients. Such an attitude should be discouraged. Repeated liver function tests may be indicated in children with leukaemia receiving regular methotrexate (MTX) (see p. 452) but are superfluous in children with 'solid' tumours, unless there is a clinical indication. The choice of investigations should be tailored to the needs of the individual patient or to the requirements of thoughtfully designed investigational protocols. Many patients are now entered into national clinical trials and data are sent to a central reference point. This may entail the submission of sera for special investigations under standardized conditions. New ideas are constantly evolving and it is a valuable practice to store pretreatment serum, at −70°C and in aliquots if possible, for future studies.

Clinical trials and organization of services

Sensitivity testing *in vitro* of tumour cells against cytotoxic agents is not yet of proven clinical value, so new drugs have to be tested in patients with tumours already resistant to other drugs. Promising results in phase II trials are extended by selecting 'responsive' tumours and comparing results in larger numbers of patients with these tumours in a phase III trial. In this way an accurate 'response rate' can be measured and compared with the existing standard treatment. The dramatic improvement in the prognosis for children with leukaemia, lymphoma, liver tumour, Wilms' tumour, malignant germ-cell tumours (MGCT) and sarcomas can be directly attributed to such a sequence of studies.

The prognosis for children with cancer is significantly better when diagnosis, staging and treatment planning are carried out in a recognized paediatric oncology centre, comprising, for example, one of the 21 centres of the UKCCSG. Survival is better and treatment-related morbidity less when patients are entered, via these centres, into well-designed clinical trials (Stiller, 1988; Pritchard *et al.*, 1989a). It is therefore no longer appropriate for these children to be treated in small units without the personnel and facilities required for optimal medical and psychosocial care.

So long as clinicians are discriminating in their requests, the clinical biochemistry laboratory in each paediatric oncology centre should be prepared to offer, on an emergency basis when necessary, facilities for the investigations described in the following sections.

Diagnosis and monitoring of individual tumours

General aspects of 'tumour markers'

The search for a 'test for cancer' provides much material for media speculation but no such test exists. Assays of carcino-embryonic antigen (CEA) and the epitopes CD1 and CD125, whilst valuable in some 'adult' carcinomas, are of no value in monitoring childhood cancer. Techniques as diverse as proton nuclear magnetic resonance spectroscopy and polymerase chain reaction (PCR) amplified transcripts coding for a cell-surface glycoprotein (CD44) have been devised recently but have not yet been evaluated in children (Matsumara *et al.*, 1992). With very rare exceptions, such as the abundant production of monotypic immunoglobulins by myeloma or follicular lymphoma cells, tumour-specific 'markers' simply do not exist. For specific tumour types, however, it has been established that amplified expression of secreted antigens, expressed in normal tissues by only rare cells or at low frequency, does occur after malignant transformation. The characteristics of the malignant clone determine the type of protein produced. If transformation has occurred in a 'committed' fetal cell the 'tumour marker' is often a normal fetal product, for example fetal haemoglobin (HbF) in juvenile chronic myeloid leukaemia and α-fetoprotein (AFP) in liver and yolk-sac tumours. Such 'onco-fetal' proteins are easily detected because of their very low normal blood levels after the first few months of life. If, on the other hand, transformation occurs in a 'committed' adult cell, a marker will be detectable only if it is readily separable in the laboratory from other normal components of the plasma (e.g. 'M-spike' in myeloma) or if it has physiological activity (adreno-corticotrophic hormone (ACTH) from adrenal carcinoma). A third cellular mechanism – aberrant gene activation in tumour cells – leads to 'ectopic' production of tumour markers, for example ACTH by occasional Wilms' or liver tumours. Available, clinically useful 'tumour markers' are listed in Table 21.4 and described in the following sections.

Table 21.4 Chemical 'markers' for paediatric tumours

Tumour	Marker
Brain tumours	
Medulloblastoma	None known
Ependymoma	None known
Germ-cell tumour	AFP/βhCG (also in CSF)
Glioma	None known
Pinealoma	Melatonin (?)
Carcinoid	5-HIAA*
Endocrine tumours	
Adrenal adenoma/ carcinoma	Cortisol/androgens*
Parathyroid tumours	Calcium/PTH*
Phaeochromocytoma	Urinary VMA* Plasma catecholamines*
Thyroid (C-cell) tumour	Calcitonin*
Thyroid carcinoma	Thyroglobulin
Granulosa cell tumours	Oestrogen/inhibin
Ewing's sarcoma	LDH
Langerhans cell histiocytosis	None known
Leukaemia/lymphoma	LDH (for B-cell lymphoma) Lysozyme (AMMoL only)
Haemaphagocytic lympho-histiocytosis	Fibrinogen ↑ Triglycerides ↓ Ferritin Pterins
Liver tumours	
Hepatoblastoma	AFP*
Hepatocellular carcinoma	AFP*/transcobalamin I*
Neuroblastoma	Urinary HVA/VMA* Ferritin VIP LDH NSE
Osteogenic sarcoma	Alkaline phosphatase
Rhabdomyosarcoma	None known
Wilms' tumour	Erythropoietin Glycosaminoglycan
Malignant germ-cell tumours including yolk-sac tumours	AFP*/βhCG*

'Tumour markers' are in serum, unless otherwise stated.
* Indicates markers of *established* clinical value.
Abbreviations are defined in the text.

Brain tumours

Primary brain tumours are, as a group, the most common 'solid' tumours of childhood. CT and MRI scanning have made preoperative assessment much easier. Improvements in surgical technique, especially the use of the operating microscope, and in radiotherapy have recently been reflected in an improving prognosis, particularly for patients with medulloblastoma and ependymoma. Chemotherapy is now widely used, especially in children less than 3 years old at the time of diagnosis, because of the damaging effects of high-dose radiotherapy in this age group. Initial results are promising.

Attempts have been made to identify possible 'tumour markers' in the cerebrospinal fluid (CSF) of patients with various types of brain tumour (Seldenfeld & Marton, 1979). Isocitrate dehydrogenase, desmosterol and polyamines are possible candidate compounds, but, as yet, none can be regarded as having clinical value.

The radioimmunoassay of serum melatonin may be of diagnostic help in patients with inoperable pineal masses for whom the clinician would prefer to avoid surgical biopsy. In pinealoma, serum melatonin levels may be significantly raised (Barber *et al.*, 1978). So-called nyctohemeral variation in melatonin production by pineal cells, even in tumours, indicates the need for measurement in serial samples over a 24-h period (Wurtman & Moskowitz, 1977). The value of serial serum melatonin monitoring in patients after treatment for malignant pineal tumours has still to be established. The monitoring of intracranial germ-cell tumours using serial measurements of serum and CSF AFP and/or the β subunit of human chorionic gonadotrophin (βhCG) (Neuwelt *et al.*, 1979) is discussed in the section on MGCT (p. 434).

Carcinoid tumour

In children, a diagnosis of carcinoid tumour is most frequently made incidentally during the histopathological examination of appendicectomy specimens. Tumours sometimes arise in other locations, especially the large and small intestines and the bronchus. The presence of metastases seems to depend less on the histopathological subtype of a carcinoid than on its size at diagnosis. It is rare to find evidence of secondary deposits when the primary tumour is less than 1 cm in diameter but more than 80% of tumours of 2-cm diameter or more have already spread to the liver, lung and sometimes other sites. The 'carcinoid syndrome', which only develops when a tumour has spread, usually to the liver, is the result of the production of one or more physiologically active agents by the tumours. Serotonin secretion causes diarrhoea, hypertension, arthropathy and endomyocardial fibrosis; bradykinin and prostaglandins give rise to flushing; and over-production of histamine can lead to bronchospasm. Elevated levels of some of or all these substances can be documented in carcinoid patients, but measurement of 24-h secretion of the 5-hydroxytryptamine (5-HT, serotonin) breakdown product 5-hydroxyindole acetic acid (HIAA) is the most useful single diagnostic test (Melmon, 1975; Sampson, 1987). A 24-h excretion of >47 μmol/24 h (>9 mg HIAA/24 h) is considered diagnostic of carcinoid tumour. With the introduction of high performance liquid chromatography (HPLC) methodology, dietary restrictions are no longer necessary during urine collection.

When small tumours are completely resected, surgery is almost always curative. In adults, successful control of unresectable or metastatic disease, with or without symptoms of the carcinoid syndrome, has been achieved with chemotherapeutic agents (e.g. streptozotocin, 5-fluorouracil) (Brennan, 1982) and long-acting somatostatin analogues (Kvols *et al.*, 1986). Metastatic carcinoid tumours in children are exceedingly rare.

Endocrine tumours

Adrenal carcinoma and adenoma

Most adrenal primary tumours are neuroblastomas and only around 5% are tumours of the adrenal cortex (Stewart *et al.*, 1974). The clinical features and histopathological appearance of carcinoma and adenoma are similar, and the distinction between them is often made on the basis of tumour size (smaller masses being more commonly benign) and apparent invasiveness. Presentation is most frequently with an endocrine syndrome, Cushing's, virilization or feminization (very rarely), or a 'mixed' endocrine picture, but nearly half these tumours are functionally 'silent' and present as an abdominal mass. Some patients with adrenal carcinoma have features of the so-called Li–Fraumeni cancer family syndrome (Li *et al.*, 1988). Many often will also have a strong family history of cancer, especially sarcomas and breast cancer in first-degree relatives. The value of investigations used in the diagnosis of Cushing's syndrome has been reviewed (Golde, 1979). Primary adrenal pathology is suspected when, in the presence of a fasting plasma cortisol >276 nmol/L (>10 μg/dL), the plasma ACTH is reduced. The distinction between adrenal hyperplasia and tumour is readily made by abdominal ultrasound, intravenous urography (IVU) or CT scanning. The following findings make a diagnosis of a tumour more likely than adrenal hyperplasia: (a) a palpable mass; (b) no suppression of plasma cortisol by high-dose dexamethasone; and (c) increased urinary 17-oxosteroids. Abdominal tumours (Wilms', hepatoblastoma or phaeochromocytoma) secreting ACTH 'ectopically' may mimic an adrenal cortical tumour; in these cases [^{131}I]-19-iodocholesterol scanning, after dexamethasone suppression, or selective venous sampling may aid tumour localization.

Whilst surgery is curative for patients with adenoma, the survival of children with adrenal carcinoma is only 20–30% after surgery alone, despite the fact that few patients have overt metastases at diagnosis. The drug *o,p'*-DDD (mitotane), which acts as a metabolic 'blocker' and a cytocidal agent, is the first choice for treatment of widespread disease (Hogan *et al.*, 1978) but some tumours respond to chemotherapy, especially etoposide and cisplatin. The role of radiation therapy is uncertain.

Parathyroid tumours

Fasting hypercalcaemia in children is rare. Children with metastatic neuroblastoma or leukaemia/ lymphoma, especially those with numerous bony deposits, may develop hypercalcaemia resulting from the action of soluble bone-resorbing factors released locally by tumour cells. Several rare types of renal tumour can be associated with the production of prostaglandins (PGE_2, PGF_2 and prostacyclin) or other 'bone-resorbing' factors circulating in sufficient concentrations to cause 'distant' bone resorption.

Parathyroid carcinoma (Toledano, 1979) and adenoma are both extremely uncommon. Carcinoma is more likely to be visible or palpable (Holmes *et al.*, 1969), and metastatic spread to lymph nodes or distant organs may be present at diagnosis. Hypercalcaemia resulting from high circulating levels of parathyroid hormone (PTH) occurs in most instances, but 'non-secreting' tumours have also been described. When serum PTH levels are borderline, the measurement of PTH in 'selective' venous blood samples, obtained by catheterization of the great vessels, may aid tumour localization (Gluckman *et al.*, 1977). The possibility of a multiple endocrine neoplasia (MEN) syndrome (Brennan, 1982) should be considered when a diagnosis of parathyroid hyperplasia, or tumour, is made. Surgery is the treatment of choice; the role of radiation therapy or chemotherapy is unknown.

Phaeochromocytoma

In children, phaeochromocytomas are usually benign. Sweating and symptoms of hypertension are the most common presenting features. The tumours are rarely palpable. Hypertension can be intermittent or continuous (Stringel *et al.*, 1980). The presence of multiple tumours (20% of cases) or a family history of phaeochromocytoma or other endocrine tumour may suggest the diagnosis of a MEN syndrome (Brennan, 1982); in other families, only phaeochromocytoma occurs. Studies in adults indicate that measurements of catecholamines and their metabolites are more likely to be accurate when plasma rather than urine is tested (Bravo *et al.*, 1979). Most phaeochromocytomas secrete noradrenaline; a raised level of vanillylmandelic acid (VMA, HMMA) and a normal homovanillic acid (HVA) level are common in 24-h urine specimens from these patients. In adults, excretion of both HVA and VMA is said to indicate that the tumour is malignant but the same excretion pattern has been recorded in several children with benign phaeochromocytoma (Gitlow *et al.*, 1972). Statistically, neuroblastoma is a much more likely diagnosis when the urinary excretion of both HVA and VMA is elevated.

Plasma for catecholamine estimation should ideally be taken at rest, but this can be difficult in young children, without unacceptably heavy sedation. Inhibition tests, for example with clonidine (Bravo *et al.*, 1981), may be helpful, but false-negative results have been recorded. Stimulation tests are likely to provoke dangerous hypertension in a child and are contraindicated. There is poor correlation between the degree of hypertension and the amount of circulating total plasma catecholamines and metabolites; in practice, this means that plasma need not be sampled during a hypertensive episode.

Tumour immunolocalization with ^{131}I-labelled MIBG scanning is now regarded as the most useful method to localize phaeochromocytoma. This non-invasive technique is an essential adjunct to angiography and serial venous sampling for accurate tumour localization (Sisson *et al.*, 1981). Neuropeptide Y may also be a useful marker (Adrian *et al.*, 1983).

Successful management of patients with phaeochromocytoma requires excellent collaboration between the paediatrician, surgeon and anaesthetist. Adequate preoperative control of blood pressure is achieved using a combination of α- and β-blocking drugs, usually phenoxybenzamine and propranolol. At operation, the surgeon should scrutinize not only both adrenal glands but also other possible sites, e.g. pararenal areas, sympathetic chains and the organ of Zuckerkandl. Multiple tumours can be familial and part of a MEN syndrome.

Thyroid carcinoma

Thyroid cancer in young people is often iatrogenic, for instance the result of neck radiation in early childhood (Refetoff *et al.*, 1975). There are four histological subtypes; the follicular and papillary varieties are more common than anaplastic and medullary carcinomas. Thyroid cancer usually presents as a visible neck swelling in a euthyroid patient. Thyroid scanning, using [^{99m}Tc]-pertechnetate, shows a cold nodule in the gland. Lymph node and lung secondaries are present in only around 5% of patients. There are no associated biochemical abnormalities. Monitoring of serum thyrocalcitonin is of great value in members of families at risk for the development of medullary carcinoma of the thyroid (MCT), especially those with a MEN 2 syndrome (Goltzman *et al.*, 1974; Brennan, 1982). Whereas patients with sporadic MCT characteristically develop the tumour during middle age, MEN 2 patients develop medullary carcinoma much earlier, at a median of 15–20 years. Calcitonin is a sensitive 'tumour marker' and monitoring of MEN 2 patients should start early in the second decade of life. When resting serum calcitonin levels are borderline, or when there is clinical concern, provocative tests – using intravenous pentagastrin – can be carried out.

The principal treatment for thyroid cancer is surgical, either full or partial thyroidectomy, with or without neck node dissection. Since the growth of some tumours is thyroid-stimulating hormone (TSH) dependent, thyroxine treatment can be effective in controlling recurrent disease. Measurement of thyroglobulin may be useful both at diagnosis and at follow-up to detect early recurrence. External beam radiation therapy and ^{131}I are less often used in children than adults because of the risk of induction of a second tumour, especially in the thyroid when only partial thyroidectomy has been performed.

Very rare endocrine tumours in childhood

There have been occasional reports of malignant behaviour of tumours that are usually considered benign. Thus, occasional examples of malignant aldosteronoma, gastrinoma, glucagonoma, somatostatinoma and vasoactive intestinal polypeptidoma (VIP) have been described, principally in adults. Residual or metastatic disease may respond to chemotherapy (Bloom, 1973).

Ewing's sarcoma

The histogenesis of Ewing's tumour is still uncertain, although recent evidence suggests an origin from neural cells rather than from vascular endothelium, as suggested by Ewing himself. Immunocytochemistry has revealed a new entity, PNET or Askin tumour, which is histopathologically indistinguishable from Ewing's tumour. In practical and prognostic terms the distinction is, as yet, of no significance because both tumours are treated identically. Primary Ewing's tumour most commonly arises in bone (long and flat bones are affected with about equal frequency) but can also arise in soft tissue. Presentation is usually with pain and/or associated soft-tissue swelling. MRI scanning is invaluable for establishing the exact extent of soft-tissue involvement. Nearly all patients have overt or occult metastases at diagnosis, most commonly in the lymph nodes, lungs, bones and bone marrow. Investigation of a newly diagnosed patient should therefore include a chest X-ray, CT scan of the lungs, isotope bone scan and bone-marrow biopsy. The important prognostic factors at diagnosis are the presence of metastases, size of primary tumour, site (distal limb tumours have the best outcome, then proximal limb, then axial) and serum lactate dehydrogenase (LDH) (Bacci *et al.*, 1988). Pulsed intensive chemotherapy, using vincristine, doxorubicin, cyclophosphamide and/or ifosfamide and actinomycin D, often leads to impressive shrinkage of the primary and control of metastases (Rosen *et al.*, 1981). Local control of bulky disease, however, is virtually never achieved without either high-dose radiation (45–60 Gy) or, where the cosmetic result is acceptable, surgical excision of the bone in which the primary has arisen. Apart from serial LDH estimations, biochemical studies are not indicated as a 'routine' during follow-up, although if ifosfamide has been given, studies of renal tubular function may be indicated (see p. 450).

Germ-cell tumours

The terminology of this group of tumours is complex. Teratoma is a generic term used to describe all germ-cell tumours. Many, especially those arising in the sacrococcygeal area in children, are histologically benign, but other frankly malignant tumours may either be undifferentiated or contain elements of embryonal or extra-embryonal (yolk sac/trophoblast) origin (Altman & Schwartz, 1978). Immunoperoxidase studies show production of AFP by yolk-sac (endodermal sinus tumour) elements and βhCG by malignant trophoblasts. In adults, 90% of MGCTs produce one or both 'markers'. In children, elevated serum AFP levels are found in more than 90% of cases and βhCG in 10–15% of cases at diagnosis (Mann *et al.*, 1978). As these figures suggest, childhood malignant teratomas almost always contain yolk-sac components. In the case of infant MGCT of the testis (orchioblastoma), 'pure' yolk-sac histology is the rule. Thus, AFP is more commonly a useful 'tumour marker' than βhCG, although both should be measured in each patient at diagnosis. Apart from the gonads, malignant teratoma can arise in the sacrococcygeal region, genitourinary tract, mediastinum, ovary, liver, extradural space and brain (Altman & Schwartz, 1978). Lungs, lymph nodes, bones and liver may be invaded by metastatic disease.

Regular monitoring of 'markers' is of established value in the management of MGCTs of adults (Javadpour, 1980) and children (Mann *et al.*, 1981). As for liver tumours, positive levels of AFP and/or βhCG should be followed at weekly intervals for 4–6 weeks from the start of treatment in order to establish a 'decay curve'. Graphic comparison of the gradient of this curve with a standard curve calculated from the known half-life of AFP (4–5 days) and βhCG (24 h) (Javadpour, 1980) accurately reflects the response to treatment. Too shallow a decay curve or a new rise above the normal range indicates either a suboptimal response to therapy or recurrence. Early diagnosis of recurrence increases the chance of successful therapy and possible cure using 'second-line' agents. In adults, 20% of recurrences of previously AFP- or βhCG-positive tumours are 'marker negative'. This phenomenon has also been recorded in children and presumably reflects relative chemosensitivity of the yolk sac and/or trophoblastic elements of the tumour, whilst the undifferentiated 'non-secretory' component is resistant. As for liver tumours, slight elevations in AFP should lead to a search for evidence of intercurrent inflammatory liver disease.

Neonates with sacrococcygeal teratoma present special problems for the clinician and the clinical biochemist. Over 90% of these tumours are benign, but tumours recur in the other 10% of patients. Most patients undergo initial surgical excision of the bulk primary disease within the neonatal period. Serum AFP monitoring is hampered by the extremely high physiological levels of AFP found in infant serum. As with levels of HbF, 'adult' levels of AFP are reached only between 6 and 12 months of age (Wu *et al.*, 1981). There is virtually no information concerning normal βhCG levels in infant serum; however, the rarity of trophoblast histology in sacrococcygeal teratoma makes it very unlikely that βhCG can be a useful 'tumour marker' in this group of patients. Each centre must decide for itself whether it is cost-effective to monitor serum AFP levels postoperatively in these infants.

Up to 5% of childhood MGCTs arise in the central nervous system, in either the pineal or pituitary region. If yolk-sac elements are present, raised AFP levels may be detected in CSF and serum (Norgaard-Pedersen *et al.*, 1978; Neuwelt *et al.*, 1979). These tumours are hardly ever resectable and are now managed with chemotherapy with or without radiotherapy, and the cure rate is improving. Estimations of CSF and serum AFP and βhCG are therefore urgently indicated in children with newly diagnosed masses in the pineal and pituitary regions.

Among boys with apparently localized orchioblastoma 70% will be cured by surgery (radical orchidectomy) alone. Chemotherapy is indicated for almost all other patients. The use of combination chemotherapy regimes containing vinblastine or etoposide, bleomycin and cisplatin or carboplatin, together with surgical removal of primary and residual secondary tumours, where technically feasible, has led to dramatically improved survival rates. More than 80% of these patients are now curable (Pinkerton *et al.*, 1986). However, many of the drugs used are toxic and there is a need for close monitoring of, for example, renal toxicity of cisplatin (see p. 449). The clinical chemistry laboratory is of the utmost importance in providing an efficient service for monitoring AFP and/or βhCG levels and this cannot be overstressed. As MGCTs are rare in childhood, supraregional laboratories are often well placed to provide the rapid and reliable service required for the proper management of these patients.

Granulosa cell tumours of the ovary

These tumours usually occur in adults and are most often benign. However, there is an 'aggressive juvenile' form which sometimes occurs in childhood. The peptide hormone inhibin is useful in diagnosis and follow-up of adults (Lappöhn *et al.*, 1989). The same may be true for granulosa cell tumours in childhood, although, as yet, no hard data are available.

Acute leukaemia

The acute leukaemias are the most common single type of childhood malignancy. ALL occurs about four times more often than acute non-lymphoblastic leukaemia (ANLL). Recently, immunological approaches, including the development of heterologous and monoclonal antisera to leukaemia-associated antigens, have considerably improved understanding of the biology of the acute leukaemias and have led to greater diagnostic accuracy in individual cases. Management is with multiagent chemotherapy and central nervous system (CNS)-directed therapy with either cranial irradiation, high-dose systemic chemotherapy, intrathecal chemotherapy or a combination of these treatments (Eden, 1992). As a result, the overall cure rate for ALL has now risen to 60–70%. Recently, there has also been a slight but definite improvement in the outlook for children with ANLL (Hann, 1992). Intensive induction therapy, although very toxic, may give cure rates of 30–50%. Bone-marrow transplantation following ablative chemoradiotherapy may offer an even higher chance of cure for ANLL patients (around 70% long-term survival), but because only 10–20% of all children have a histocompatible sibling marrow donor, there is only a minor impact on overall survival. This type of treatment may also be considered as first-line therapy for 'pre-leukaemia' syndrome (e.g. bone-marrow monosomy 7) and following the first relapse of ALL. Biochemical investigations are crucial in the management of the early phase of induction chemotherapy, particularly the 'tumour lysis syndrome' (see p. 445), but only rarely of use in diagnosis.

The monomyelocytic (AMML) and monocytic (AMoL) types of acute ANLL, as well as the related but extremely rare 'malignant histiocytosis', are sometimes associated with hyperlysozymuria (Wiernik & Serpick, 1979). Lysozyme is synthesized and secreted by monomyeloblasts and is excreted by renal tubules: its action on the renal tubular epithelium can lead to wastage of K^+. Thus, hypokalaemia may be a pointer towards a diagnosis of this particular subtype of AML. It has been suggested that measurement of CSF β_2-microglobulin may help in the early diagnosis of CNS infiltration by leukaemia or lymphoma (Mavligit *et al.*, 1980). However, significant elevations of β_2-microglobulin have also been noted during the post-radiotherapy 'somnolence syndrome' (Demeocq *et al.*, 1981).

Lymphoma

Lymphomas in children, as in adults, are classified as either Hodgkin's disease or non-Hodgkin's lymphoma (NHL).

Hodgkin's disease

Of children with this form of lymphoma 85% present with painless lymphadenopathy, most commonly in the neck and/or mediastinum (Kaplan,

1980). Only a few patients have 'B symptoms', i.e. weight loss, fever or night sweats. After biopsy confirmation of the diagnosis, patients are clinically staged using chest X-rays, abdominal ultrasound and CT scanning of the chest and abdomen. Lymphangiography and staging laparotomy are no longer used in children because chemotherapy, rather than radiotherapy, is now the treatment of first choice for stages II, III and IV disease, and also because of the risks of splenectomy in young children resulting in especially serious infections with encapsulated bacteria, e.g. *Pneumococcus pneumoniae* and *Haemophilus influenzae*. Although serum levels of liver enzymes (alanine aminotransferase (ALT), aspartate aminotransferase (APT) and alkaline phosphatase) were considered helpful in identifying liver involvement (Abt *et al.*, 1974), current opinion is that liver function tests are not of value. Serial serum copper and zinc estimations have been regarded as useful 'tumour markers' in Hodgkin's disease and other lymphomas, but recent evidence suggests that careful clinical examination, with regular chest X-rays, is just as effective for the early detection of recurrent disease (Kaplan, 1980).

Non-Hodgkin's lymphoma

NHL and ALL have many features in common. First, approximately 30% of NHL patients present with a mediastinal mass (immunological studies show the tumour to consist of T-derived lymphoblasts) and, second, the natural history of patients with T-cell NHL is very similar to that of children with T-cell ALL. Another 30% of childhood NHL patients present with nasopharyngeal or abdominal masses; these tumours are usually of B-cell origin, as evidenced by the presence of monotypic cell membrane immunoglobulin. In contrast to the situation in adults with multiple myeloma, neither whole immunoglobulin nor free light and/or heavy Ig chains are secreted by the cells, so electrophoresis of serum or urine reveals no evidence of either Bence–Jones protein or an M-spike. As in Hodgkin's disease and ALL there are no known 'tumour markers'. Serum LDH levels provide an accurate reflection of tumour burden, especially in patients with B-cell lymphoma, but routine serial measurements cannot be recommended. Serial estimations of β_2-microglobulin have been used in adults with NHL to monitor disease activity (Cooper & Plesner, 1980), but there are no available equivalent data for childhood NHL.

Liver tumours

Liver cancers are rare in childhood. The most 'common' histological variety, hepatoblastoma (HB), is diagnosed only 10–15 times every year in the UK, and other varieties of liver tumour – hepatocellular carcinoma (HCC), MGCT and undifferentiated sarcoma – are even rarer. Most children with HB are aged 3 years or less, whilst HCC occurs in later childhood and adolescence (Exelby *et al.*, 1975). Children with HCC, like adults, usually have evidence of hepatitis (especially hepatitis B) or inherited metabolic liver disease (e.g. tyrosinosis or glycogen storage disease (GSD) type I). Patients with GSD often develop multiple hepatic adenomata (Howell *et al.*, 1976) but enlargement of these benign nodules may be decelerated by biochemical control of hypoglycaemia (see Chapter 5), and transformation to HCC is very rare indeed. A combination of abdominal ultrasound and CT scan with or without MRI usually allows the clinician to differentiate between hepatic and non-hepatic origin of upper abdominal masses. Lung metastases are by far the most common site of tumour spread. Cisplatin/doxorubicin (PLADO) chemotherapy is so effective in HB that the 5-year survival rate has risen to 65–75% over the past 5 years and even patients with metastatic disease have a real chance of cure. Chemotherapy is used first, and surgery is delayed until the primary tumour has shrunk sufficiently to reduce operative risks to a minimum. Even some HCCs in childhood respond to PLADO chemotherapy and an initial curative attempt is certainly indicated.

Conventional liver function tests are normal, except in the case of HCC complicating cirrhosis. Tumour-derived serum AFP estimation, however, is invaluable in the diagnosis and follow-up of hepatocellular tumours. At diagnosis AFP levels are grossly elevated in the serum of more than 95% of

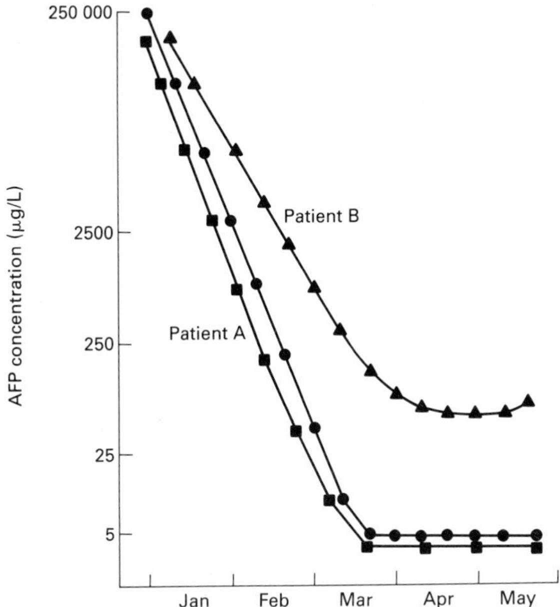

Fig. 21.2 Serum α-fetoprotein (AFP) levels (log scale) after resection of AFP-secreting tumours. The 'ideal' fall of AFP levels, indicating complete tumour control by therapy, is shown (circles). Patient A underwent complete surgical resection of a primary hepatoblastoma and micrometastatic disease was successfully controlled by chemotherapy. Patient B's tumour was incompletely resected; clinical recurrence was correctly predicted by the 'shallow' AFP decline curve.

children with HB. HB is sometimes diagnosed in infancy: the interpretation of AFP levels measured during the first 6 months of life is discussed in the section on germ-cell tumours (p. 434). After diagnosis, serum AFP should be monitored at least weekly for 4 weeks so that a decay curve can be established. Patients whose levels consistently fall with a half-life of 4–6 days whilst they are receiving PLADO chemotherapy almost all have a good prognosis (e.g. patient A in Fig. 21.2). When levels fall more slowly, the outlook is worse. Patient B (Fig. 21.2) with an inoperable primary tumour and lung secondaries showed a transient response to chemotherapy but then succumbed to drug-resistant metastatic disease.

The specificity and sensitivity of AFP immunoassay have been invaluable in diagnosis, monitoring of response and detection of recurrence. Figure 21.3 shows the value of serial monitoring in a child whose AFP level rose to 100–200 µg/L 9 months after a right hepatic lobectomy for an AFP-synthesizing HB. Conventional chest X-rays were negative, but CT lung scans revealed three small lung secondaries. The chemotherapy was changed, the lung secondaries disappeared and the child is a long-term survivor. 'False-positive' elevation of serum AFP levels during liver regeneration after partial hepatectomy or episodes of inflammatory liver disease is rarely a problem. If hepatitis is suspected, liver enzyme studies should reveal the diagnosis. 'False-negative' AFP estimations are rare. We have managed one patient with an AFP-producing tumour whose recurrence was associated with normal AFP levels, but this is a rare phenomenon. However, monitoring of the pre- and post-operative progress of these patients should include chest X-rays and abdominal ultrasound, as well as serial AFP measurements.

In HCC, the incidence of elevated serum AFP levels at diagnosis in children is higher (around 80%) than in adults (65–70%). Rarely, a histologically distinctive form of HCC, the so-called fibrolamellar variant, is diagnosed, usually in females. Serum AFP levels (Paradinas *et al.*, 1982) are normal but serum transcobalamin II levels are grossly elevated and can be useful as an early marker of tumour recurrence. Cystathionine levels are elevated in the urine of some children with liver tumours, but serial monitoring suggests, as in neuroblastoma, that cystathionine is a poor 'tumour marker' (Geiser & Shih, 1980).

Other very rare malignant childhood liver tumours include sarcomas, for which no 'tumour marker' is known, and MGCTs. The presence of yolk-sac (endodermal sinus) or trophoblastic elements in a germ-cell tumour is implied if serum AFP or βhCG levels are elevated at presentation.

Neuroblastoma

It is over 35 years since increased quantities of catecholamine metabolites were demonstrated in the uriine of patients with neuroblastoma (Mason *et al.*, 1957; Labrosse *et al.*, 1976). Figure 21.4 shows

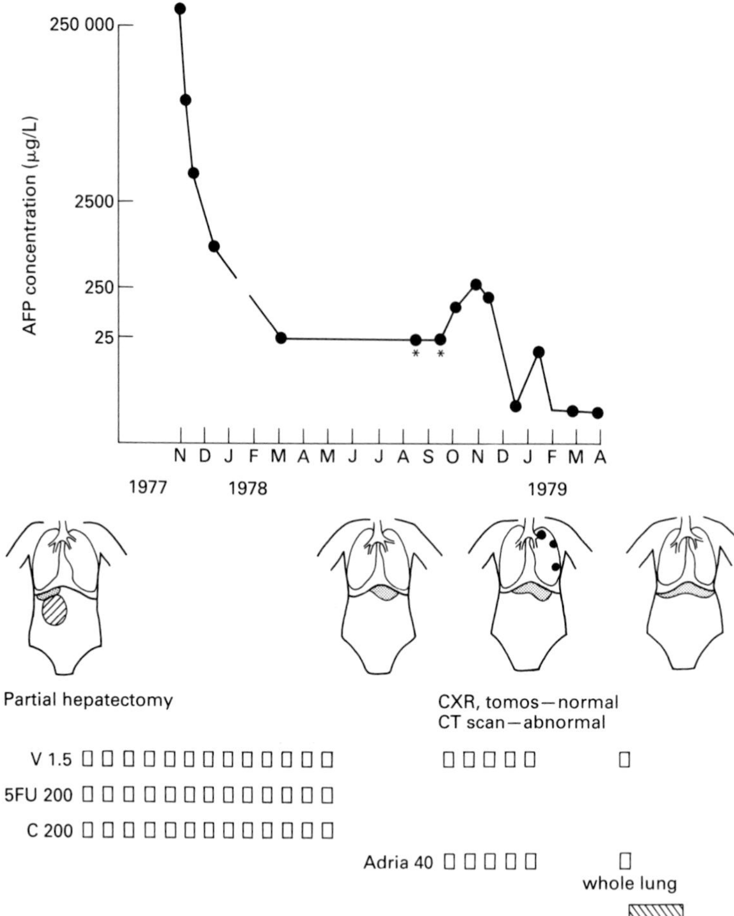

Fig. 21.3 Usefulness of serum α-fetoprotein (AFP) monitoring in the detection of recurrence of hepatoblastoma (see text). Asterisks indicate the lowest limit of detection of AFP by immunoassay at the time (1978). V, Vincristine; 5FU, 5-fluorouracil; C, cyclophosphamide; Adria, adriamycin; figures indicate dose/m^2. CXR, Chest X-ray; CT, computerized tomography; RAD, radiation.

the principal biochemical pathways of catecholamine metabolism. In over 90% of newly diagnosed cases, increased amounts of VMA, HMMA (a metabolite of adrenaline and noradrenaline) and/or HVA (a dopamine metabolite) can be measured in serum or urine. The figure is higher if dopamine itself and/or other metabolites (e.g. 3,4-dihydroxyphenylacetic acid (DOPAC)) are measured. The percentage with elevated values rises with the stage, as shown in Table 21.5 (Pritchard *et al.*, 1989b). When the urinary VMA/HVA is normal but electron microscopy reveals neurosecretory granules in tumour cells, a diagnosis of PNET, rather than neuroblastoma, should be considered. The distinc-

Table 21.5 Patients with neuroblastoma who have elevated urinary catecholamine metabolites (uCATs) at diagnosis. From Pritchard *et al.* (1989a)

Stage (Evans)	Number of children	Number (%) with uCATs > 2 SD above the mean
I	21	14 (66)
II	56	35 (63)
III	122	95 (78)
IV	379	341 (90)
IVs	41	40 (98)
Total	619	525 (85)

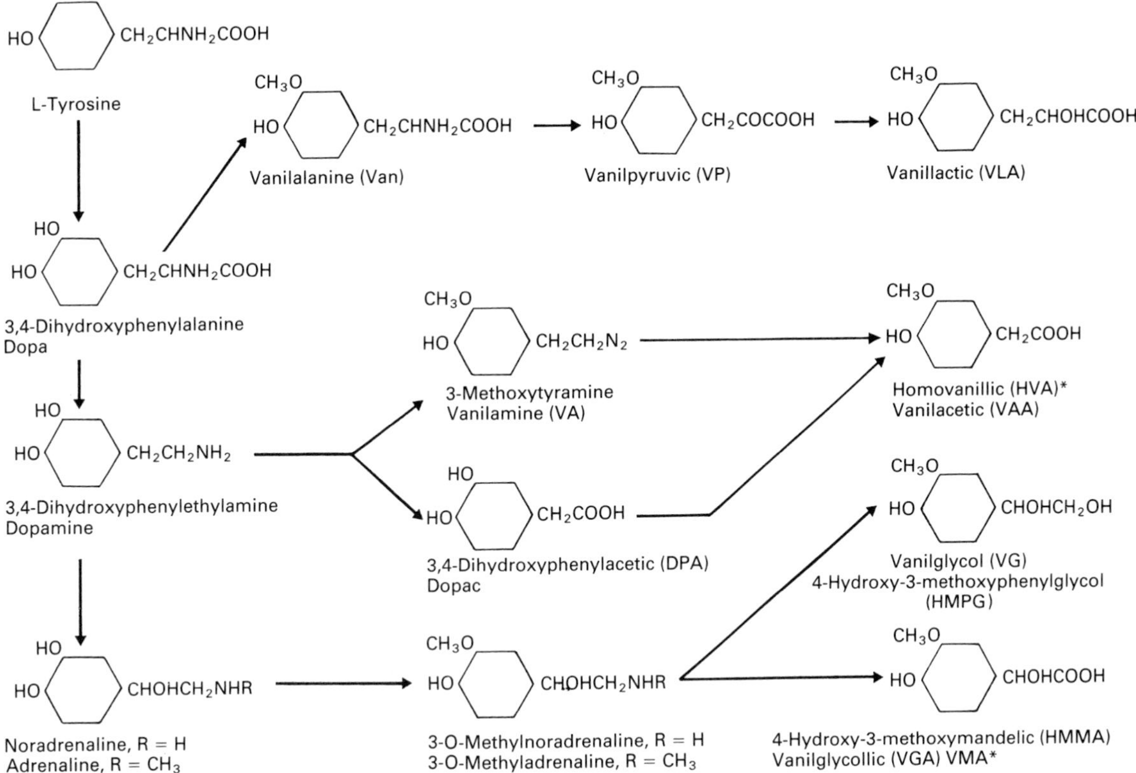

Fig. 21.4 Principal pathways of catecholamine synthesis and degradation. The terms 4-hydroxy-3-methoxymandelic (HMMA), vaniglycollic (VGA) and vanillylmandelic acid (VMA) are interchangeable, as are the terms homovanillic acid (HVA) and vanilacetic acid (VAA) (indicated by asterisks). (Redrawn courtesy of Dr J.W.T. Seakins.)

tion is crucial because the management of PNET is different, with a particular emphasis on 'local control', including radical radiotherapy for residual disease. Rare neuroblastomas produce acetylcholine rather than catecholamines, indicating alternative autonomic commitment.

The Labrosse spot test, which can yield both false-negative and false-positive results (Evans *et al.*, 1971), has been superseded as the standard test for estimation of urinary catecholamine metabolites by chromatographic methods. Using gas chromatography (GC) or HPLC, the metabolites can be separated from dietary contaminants and drugs (Niehaus *et al.*, 1979). The Pisano method is most commonly used for quantitation of VMA and vanilglycol (VG), whilst HVA can be quantitated fluorimetrically (Sato, 1965). The development of population screening programmes for neuroblastoma has heightened interest in the methodology for measurement of catecholamine metabolites. The Japanese have pioneered these screening techniques (Sawada *et al.*, 1984). Initially, they used a crude Labrosse test which detected VMA only but most centres in Japan now use an HPLC method. GC has also been used with either flame ionization detection (GC–FID) (Dale *et al.*, 1988) or mass spectrometry (GC–MS) (Seviour *et al.*, 1988). GC–MS obviates all the problems of interfering peaks due to dietary and other contaminants, and although it is probably not optimal for routine laboratory use, because of its major capital expense, it is ideal for mass screening. Immunoassays currently under development

(Yokomori *et al.*, 1989) may well be the method of choice for mass screening in the future.

Although a 24-h urine collection is theoretically important because of the diurnal variation in catecholamine excretion, in practical terms it is much easier to collect a single urine specimen and relate the concentration of catecholamine metabolites to creatinine. For screening programmes single specimens are the only feasible option. Several studies have shown a very good correlation between 24-h urine collection and single urine specimens (Kellie *et al.*, 1986; Tuchman *et al.*, 1987). HVA and VMA are not independent of creatinine, as shown in Table 21.6, and results are most precisely expressed as micrograms VMA or HVA per milligram creatinine (Cole *et al.*, 1993). HVA and VMA levels are also age dependent, a factor that must be taken into account when determining reference ranges.

Table 21.6 Upper limit of normal values* for homovanillic and vanillylmandelic acids (μmol/mmol creatinine) in 6-month-old infants

Creatinine (μmol/L)	Homovanillic acid: Boys	Homovanillic acid: Girls	Vanillylmandelic acid
100	104.1	45.6	35.5
150	70.5	41.4	30.0
200	55.1	38.4	26.6
250	46.1	36.1	24.0
300	40.4	34.3	22.1
350	36.3	32.7	20.6
400	33.3	31.4	19.3
450	31.1	30.2	18.2
500	29.2	29.2	17.3
550	27.7	28.3	16.5
600	26.5	27.5	15.8
650	25.5	26.7	15.2
700	24.5	26.1	14.6
750	23.8	25.5	14.1
800	23.1	24.9	13.6
850	22.5	24.3	13.2
900	22.0	23.8	12.8
950	21.5	23.3	12.5
1000	21.1	22.9	12.1
1050	20.7	22.5	11.8
1100	20.4	22.1	11.5
1150	20.1	21.7	11.2
1200	19.9	21.4	11.0

* Mean plus 3 SD for normally distributed data.

Can urine screening improve the outcome for children with neuroblastoma? There is no doubt that screening at 6 months of age is effective in identifying some cases of neuroblastoma, but there is now persuasive evidence that many of these tumours would have spontaneously regressed, and therefore otherwise gone undetected (Craft & Parker, 1992). To determine whether or not screening is effective, one very large-scale study of screened and unscreened populations is under way in North America (Woods *et al.*, 1990) and others are planned in Europe (Parker *et al.*, 1991).

The VMA:HVA ratio may have prognostic significance. In some series, patients with a ratio of >1 have a more favourable short-term prognosis than do those with the reverse ratio (Laug *et al.*, 1978). Whether or not VMA:HVA ratio measurement proves to be of importance as a single variable factor awaits more detailed studies and multivariate analysis, including other prognostic factors such as age, stage, nMyc oncogene copy number, tumour-cell DNA content ('ploidy'), serum levels of neurone-specific enolase (NSE) and histological differentiation. There is a considerable body of data, especially from Germany, indicating that plasma catecholamine levels are also of value, but most centres still rely on urinary assays.

Other biochemical markers which show promise as prognostic factors in neuroblastoma are ferritin, LDH and NSE. In each case, high levels are associated with a worse prognosis. The nMyc oncogene copy number in tumour cells is also of importance. Tumours with multiple copies have a worse prognosis than those with a single copy. Cytogenetic studies can also be helpful in prognosis: children with tumours that show aneuploidy and/or deletions from the short arm of chromosome 1 (1p−) fare worse than those whose neuroblastoma is hyperdiploid, with no chromosome 1 deletion (Castleberry, 1992).

Of children with neuroblastoma 10–15% have hypertension at diagnosis. There is no consistent pattern of catecholamine secretion that distinguishes these children from those with normal blood pressure. No child with a solid retroperitoneal mass should undergo anaesthesia without estimation of 24-h urinary VMA and HVA excretion, because the cardiovascular instability that results from the removal of a mass of catecholamine-secreting tumour may have serious, or even fatal, consequences (Creagh-Barry & Sumner, 1992). Since neuroblastomas occasionally originate from within the kidney, and on abdominal ultrasound and CT scanning mimic Wilms' tumour, this rule should always be observed. Slight elevations in urinary catecholamines are occasionally seen in patients with abdominal masses other than neuroblastoma, presumably the result of the 'stress'. Catecholamine-induced hypertension requires control by both α- and β-blocking agents or α-methyl tyrosine before surgery. Results of urinary VMA and HVA excretion should therefore be made available as soon as possible, and certainly within 48 h of the child's admission to hospital.

Compared with serum AFP levels, serial VMA/HVA monitoring during treatment has been disappointing. In our experience, it is unusual for rising urinary VMA and HVA to be the sole manifestation of tumour recurrence; usually, there is also clinical evidence of persistent or recurrent disease. Rarely, florid relapse can recur without elevation of urinary catecholamine metabolite levels, presumably because the resistant tumour population is 'non-secretor'. Occasionally, patients will have a persistent mass with elevated levels of urine catecholamine metabolites yet remain in clinical remission, indicating a maturation of the tumour to benign ganglioneuroma. Most paediatric oncologists follow urinary VMA and HVA excretion in patients on a regular basis and find, as do the patients' parents, that normal levels are reassuring. However, critical cost–benefit analysis of such a policy is needed.

Diarrhoea is often mentioned in textbooks as a symptom of neuroblastoma but is actually rare. Occasionally, a child with a neurogenic tumour presents with profuse watery cholera-like stools, with hypokalaemia and gastric achlorhydria (Jansen-Goemans & Engelhardt, 1977; Tiedemann *et al.*, 1981b). This presentation is similar to that of adults with VIP-secreting pancreatic tumours and other 'VIPomas', and adds to the evidence suggesting that neural crest cells and 'APUD' cells have a common histogenetic origin (Bloom *et al.*, 1973). Histological examination of such tumours usually shows 'differentiation' (to ganglioneuroblastoma or ganglioneuroma) and VIP can be demonstrated immunohistochemically in the cytoplasm of the more mature tumour cells, especially ganglion cells (Mendelsohn *et al.*, 1979). Plasma VIP levels are mildly elevated, without evidence of diarrhoea, in some children with newly diagnosed neuroblastoma (Tiedemann *et al.*, 1981a) and may be helpful in differential diagnosis.

Children with localized neuroblastoma have an excellent prognosis (95% 2-year survival for stage 1 patients, 85% for stage 2), but unfortunately most patients have stage 3 or 4 disease when diagnosed. Because of the potential prognostic importance of biological studies (nMyc gene complement, DNA content and chromosome analysis) on tumour cells, patients should have a biopsy for tissue diagnosis. Well over 50% of patients with advanced neuroblastoma respond, often dramatically, to intensive combination chemotherapy.

The survival rate for stage 4 disease has been very poor until recently but the use of very intensive induction chemotherapy, containing the platinum compounds cisplatin and carboplatin, appears promising. The use of high-dose chemotherapy with autologous bone marrow 'rescue' has now been shown to prolong survival.

Osteogenic sarcoma (osteosarcoma)

This primary bone tumour is almost unheard of in children under 5 years and is most common in the adolescent age group. Presentation is invariably with a painful swelling at the primary site, most commonly in the lower femur, upper tibia or upper humerus, but axial bones can also be affected. A majority of patients have macro- or micrometastases at diagnosis, most commonly in the lungs, but sometimes in other bones. The main prognostic reported variables were age at diagnosis, sex, duration of symptoms, tumour site, tumour size and histological subtype (Taylor *et al.*, 1978), but modern aggressive chemotherapy has abolished many of these factors. The serum alkaline phosphatase level at diagnosis is also an important prognostic indicator:

serial measurements correlate with disease status and high values indicate recurrence. Multivariate analysis indicates that serum alkaline phosphatase levels are of independent prognostic significance (Bacci *et al.*, 1990).

Compared with Ewing's sarcoma, osteogenic sarcoma is not particularly radioresponsive, so surgical removal of the primary is needed. In many patients local excision with limb preservation, rather than amputation, is possible, but this decision depends upon the size and local extent of the tumour and the age of the child. The value of chemotherapy in this disease is now proven. Active agents include cisplatin, doxorubicin, ifosfamide and high-dose MTX. The optimum regimen has yet to be determined, but in the experience of the large European collaborative group cisplatin/doxorubicin seems at least as effective as more complex regimens (Bramwell *et al.*, 1992). Patients should receive high-dose MTX only in centres where facilities for routine monitoring of serum MTX levels and serum creatinine are available. Liver function tests, especially AST, ALT and bilirubin estimations, may also be required in an emergency where acute MTX toxicity is suspected. Each high-dose MTX treatment involves several days in hospital and major disturbances of the family's domestic routine. Although clinicians should schedule therapy so as to avoid public holidays, when clinical chemistry laboratories are staffed only for emergency services, families may have a valid preference for starting or completing treatment at weekends, when it may be easier for them to transport the child to or from hospital. Children with osteosarcoma are often at a crucial stage of their education and weekend treatment also reduces the time lost from school.

Rhabdomyosarcoma

Rhabdomyosarcoma is the most common histopathological type of soft-tissue sarcoma in childhood. Liposarcoma, synovial cell sarcoma and fibrosarcoma are much rarer but are investigated and treated similarly, although generally they behave in a more 'low-grade' fashion. Of all rhabdomyosarcomas, approximately one-third arise in the head and neck, one-third in the trunk and limbs, and one-third in the genitourinary system. Before the 1970s, these tumours were almost always fatal, although some patients were 'salvaged' by very major surgery, e.g. orbital or pelvic extenteration with or without irradiation. Since the advent of chemotherapy, survival rates have greatly improved (60–70% at 5 years from diagnosis). The most active drugs are vincristine, actinomycin D and cyclophosphamide or ifosfamide. In addition to trying to increase the overall cure rate, major emphasis is now focused upon improving the quality of survival. Mutilating surgery is avoided wherever possible, especially in children with head and neck or genitourinary primaries, and efforts are being made to reduce the radiation dosage so that retardation of bone growth is lessened, with the aim of a better final cosmetic result (Glicksman *et al.*, 1979).

No particular biochemical investigations are indicated in the diagnostic phase or follow-up. There are no known 'tumour markers' for any soft-tissue sarcoma.

Wilms' tumour

The high cure rates achieved in children with this, the most common renal tumour of childhood, are possible because the tumour cells are highly sensitive to several chemotherapeutic agents (in particular, vincristine, actinomycin D and doxorubicin) and radiation therapy (D'Angio *et al.*, 1976). Patients most commonly present with abdominal swelling, noted by the parents, health visitor or general practitioner. Because overt metastatic disease is uncommon at diagnosis, most children have no systemic upset and specific symptoms such as haematuria are also uncommon. Between 10 and 20% of patients with Wilms' tumour have raised blood pressure at the time of diagnosis. The most common cause of hypertension in Wilms' tumour is renovascular, and is due to displacement or compression by the tumour of major vessels. Usually, this form of hypertension is mild, and careful examination may indicate dependency on the child's posture, presumably reflecting varying pressure on, or 'kinking' of, the renal vessels when the tumour is in different positions. Some children with Wilms' tumour, especially those with bilateral kidney involvement and other congenital abnormalities such as pseudohermaphroditism, have a histopathologically characteristic nephropathy with associated albuminuria; if the renal lesion is severe, there may be biochemical evidence of renal decompensation with elevated blood urea and creatinine levels (Drash *et al.*, 1970). There are several reports of renin secretion by Wilms' tumour (Mitchell *et al.*, 1970; Ganguly *et al.*, 1973) as well as by the much less common mesoblastic nephroma and juxtaglomerular tumour of childhood. Because of the anaesthetic and surgical risks of severe hypertension, surgery should be delayed whilst investigations, including urinalysis, plasma renin and urinary VMA and HVA, are carried out and blood pressure should be subsequently controlled with appropriate drug therapy.

Patients with Wilms' tumours often excrete increased amounts of glycosaminoglycans (GAG) in the urine (Hopwood & Dorfman, 1978). No systematic studies of GAG have been reported; our own experience has been that urinary GAG estimation and fractionation are unhelpful in the differential diagnosis of the child with a newly diagnosed abdominal mass. Very rarely, massive concentrations of GAGs can be found in the blood and may cause a life-threatening hyperviscosity syndrome.

Multivariate analysis of prognostic features in the US National Wilms' Tumour Studies (NWTS) has shown that (a) histological subtype, (b) involvement of lymph nodes and (c) presence of metastases (stage 4) are the most important prognostic factors at diagnosis (Breslow *et al.*, 1978). As a result, it has been possible to refine therapy for children with localized 'favourable' histological subtypes of Wilms' tumours, especially if the lymph nodes are negative for tumour. Patients with stages 1 and 2 'favourable histology' are therefore now treated in the UK and USA without radiation and with only single- or double-agent chemotherapy, whilst intensive three- or four-drug treatment is recommended for children with 'unfavourable histology' or stages 3 and 4 tumours. Overall treatment results are extremely encouraging. Recent analysis of patients entered into the NWTS and UKCCSG studies shows above 85% survival for all patients with stages 1, 2 and 3 disease. Well over half of patients with stage 4 or 5 (bilateral Wilms') tumours can also be 'cured'.

The histiocytoses

With the exception of 'malignant histiocytosis', which is extremely rare in childhood, the histiocytoses are now regarded as non-malignant disorders. To emphasize the distinction from the childhood cancers, the histiocytoses are discussed separately in this section. For a more detailed discussion, see Pritchard and Malone (1993).

Langerhans cell histiocytosis

The pathogenesis of Langerhans cell histiocytosis (LCH), a multisystem disorder characterized histopathologically by accumulation or proliferation of Langerhans-type histiocytes ('LCH cells'), is still poorly understood but its natural history, with frequent spontaneous remission, is not typical of a malignancy. Patients can present with a seborrhoeic or purpuric skin rash, chronic ear discharge, hepatosplenomegaly, oral involvement, respiratory difficulties, diarrhoea, anaemia, diabetes insipidus, bone pain, scalp swellings, proptosis or any combination of these features. Because of clinical and histopathological overlap between them, subclassification into eosinophilic granuloma, Hand–Schuller–Christian disease and Letterer–Siwe disease is inappropriate. A cell with most of the features of the cutaneous Langerhans cell is pathognomonic: the designation 'Langerhans cell histiocytosis' is therefore preferred to the vague term 'histiocytosis'. The natural history is characteristically fluctuant, with periods of quiescence and, sometimes, total spontaneous regression, but most children with multiple organ involvement eventually require systemic therapy. Corticosteroids, a vinca alkaloid (either vincristine or vinblastine) or an epipodophyllotoxin (etoposide) are most commonly used; administration of more 'aggressive' combination chemotherapy regimens can cause complications without an improved response rate (Broadbent & Pritchard, 1985). Responses to therapy are generally prompt but may not be long lasting. Multisystem disease has a significant mortality, especially among young children (up to 20%), and morbidity (up to 40%). Evidence of 'failure' of a vital organ system, especially bone marrow, liver or gut, is ominous.

Jaundice is uncommon and the most sensitive indices of liver dysfunction in this disease are the level of serum albumin and coagulation studies. A serum albumin concentration of less than 20 g/L and prolonged prothrombin time, partial thromboplastin time and thrombin time are of unfavourable prognostic significance. Serum levels of liver enzymes (APT, ALT) are not usually raised in LCH, even with extensive infiltration; elevated values suggest a complicating hepatitis. Liver failure virtually never occurs in patients with normal-sized livers, so 'routine' liver function tests are unnecessary. Persistent hyperbilirubinaemia suggests cirrhosis – an occasional late manifestation of liver involvement.

Diabetes insipidus (DI), owing to infiltration of the hypothalamus–pituitary axis by LCH, occurs in up to 40% of all patients. In about one-half of cases, DI is present at diagnosis, and in one-half it develops at a later stage, sometimes in patients otherwise in complete remission. All children with newly diagnosed LCH should have an early-morning urine specimen tested for specific gravity (SG). An SG < 1.030 should be followed up by a formal water deprivation test. In borderline cases, or where compulsive water drinking is a possibility, measurement of urinary arginine vasopressin (AVP) may be invaluable (Dunger *et al.*, 1989). Replacement therapy with pitressin is effective. The parents of patients without DI at the time of diagnosis should be warned to look out for an altering pattern of thirst or micturition. Some clinicians retest patients' early-morning urine SG only if there is a clinical indication, whilst others test routinely every 4–6 months.

Haemaphagocytic lympho-histiocytosis

Haemaphagocytic lympho-histiocytosis (HLH) is a new, inclusive term for two conditions with very similar clinical characteristics, previously known as familial erythrophagocytic lympho-histiocytosis (FEL) (Gross-Kieselstein *et al.*, 1981) and viral-induced haemaphagocytic syndrome (VAHS). In 30–40% of cases, HLH is inherited as an autosomal recessive disorder, whilst in the remainder of cases it appears to be 'sporadic', often precipitated by an infectious pathogen (e.g. Epstein–Barr virus, cytomegalovirus, human immunodeficiency virus (HIV), mycobacteria). The inherited form usually (although not invariably) presents in the first year of life, and the sporadic form presents later (but not invariably) in life. The major clinical features are hepatosplenomegaly, anaemia, bleeding and fits or encephalopathy. Pancytopenia, due to phagocytosis of formed blood elements, and coagulopathy, due to severe hypofibrinogenaemia, are characteristic laboratory findings. The cell count may be raised in the CSF and cytophagocytosis may be evident.

Common biochemical abnormalities include hypertriglyceridaemia and an increased concentration of neopterins in the CSF, owing to monocyte 'activation'.

Some patients with 'sporadic' HLH recover spontaneously, but in the others, the disease is progressive and usually fatal. Treatment has improved recently and remissions can often be achieved using systemic corticosteroids, etoposide (VP16) and intrathecal MTX. In remission, plasma triglyceride concentrations are normal, suggesting that raised levels at diagnosis are a secondary phenomenon and presumably due to liver dysfunction. There is some evidence that ablation of the patient's bone marrow, followed by allogeneic transplantation (if a donor is available), may be curative.

Monitoring of drug levels

Side effects of cytotoxic drugs are usually dose related. Pharmacokinetic data from human studies are available for most drugs and there is now an increasing interest in monitoring the pharmacokinetic profile of several drugs (e.g. cisplatin (Reece *et al.*, 1987) and carboplatin (Newell *et al.*, 1993)) to try to achieve maximum therapeutic effect with minimum toxicity. There is considerable variation in the gastrointestinal absorption of oral MTX in children with leukaemia (Craft *et al.*, 1981). However, attempts to relate specific levels to outcome have been contradictory (Pearson *et al.*, 1991), and routine monitoring of levels has not entered clinical practice. The metabolism of 6-mercaptopurine (6MP) is of especial current interest because it appears that there are genetically distinct individuals who handle this drug in different ways. Identification of the phenotype and appropriate modification of the dosage may lead to more effective treatment in ALL (Lennard & Lilleyman, 1989).

Because the current strategy is to treat patients to 'tolerance', monitoring of blood levels is not standard practice. Thus, although methods for the assay of alkylating agents, cytosine arabinoside, 6MP, MTX and other drugs are available, only the MTX assay has direct clinical relevance at present. The cytotoxic action of MTX is the result of inhibition of the intracellular enzyme dihydrofolate reductase. As the reductase is critical to DNA synthesis, MTX acts in the S-phase of the cell cycle. Large doses of MTX may be given in the treatment of lymphoblastic leukaemia/lymphoma and osteogenic sarcoma. At 6 to 24 h after the administration of a sublethal or lethal dose (0.5–20 g/m^2) the patient is biochemically 'rescued' by administration of folinic acid (citrovorum factor). By competitive inhibition, folinic acid restores the blocked enzyme pathway within normal and tumour cells, hopefully only after the latter have been irreversibly damaged. Therapy must be continued until the serum MTX falls to nontoxic levels ($<10^{-8}$ mol/L or 4.5 ng/mL). The excretion of MTX, principally by the kidneys, is enhanced if the urine flow is high and the urine pH > 7.5. Liberal intravenous fluids and sodium bicarbonate are therefore usually given with folinic acid. At an acid urine pH, MTX may crystallize in the renal tubules and daily monitoring of renal function is therefore necessary. Estimation of serum creatinine 48 h after MTX administration is especially worth-while: if this value is >50% higher than the pre-MTX level of creatinine, delayed MTX excretion can be predicted and prolonged folinic acid 'rescue' should be prescribed. Besides renal compromise, other factors delaying excretion of MTX include intercurrent infection, dehydration and the presence of a 'third space', for example pleural or pericardial fluid. High-dose MTX therapy should only be given in those centres where biochemistry laboratories can offer an urgent service for monitoring of renal function and serum MTX levels. Radioimmunoassay is most commonly used to assay MTX (Aherne & Quinton, 1981).

Treatment complications — biochemical aspects

Biochemical abnormalities arising as a consequence of childhood cancer can be acute or of a medium- or long-term nature. Disturbances in newly diagnosed patients, particularly those with lymphoblastic leukaemia/lymphoma or a tumour involving or obstructing the genitourinary tract, may be exacerbated by the introduction of therapy and may become life threatening, but mortality can be reduced if appropriate precautions are taken. Once remission has been achieved, patients receive continuing therapy for periods of 4–24 months, depending on the type of malignancy; during this phase, problems can arise from dysfunction of a specific organ or system as a result of treatment-induced damage. Improving survival rates in childhood cancer have focused attention on the late effects of therapy. Whereas treatment effects in adults are usually obvious immediately, manifestations of tissue damage in children may become evident only during later growth, especially at puberty. Only those long-term treatment consequences of direct relevance to the biochemist are discussed here, but other unwanted late effects, including, for example, impaired skeletal growth within radiation fields, psychological impairment as a result of cranial radiation and induction of second tumours, are also of major concern to the clinician. Although new and unforeseen long-term complications may arise in the future, problems (for example, renal failure resulting from high-dose radiation therapy) occurring in some of today's survivors will be seen less frequently because of recent refinements of treatment.

Acute problems

In Table 21.7, the major factors contributing to biochemical disturbances at the time of diagnosis are listed. The exact causation of renal insufficiency is of particular importance to both clinician and biochemist. In some circumstances, such as renal

Table 21.7 Metabolic abnormalities at diagnosis and their causes

Substance	Increased/present	Decreased/absent
Amylase	'Tumour lysis pancreatitis'	–
Calcium	Extensive bony metastases 'Distant' production of bone-resorbing factors	Secondary to hyperphosphataemia
Creatinine	Urate nephropathy Renal infiltration by leukaemia/lymphoma Obstructive uropathy due to bladder base or pelvic tumour	Malnutrition
Lactic acid	Hypoxic bulky tumour	–
Magnesium	–	Secondary to hypercalcaemia
Metabolic acidosis	Renal failure Secondary to sepsis Lactic acidosis	–
Phosphate	Tumour lysis syndrome Renal failure	–
Potassium	Renal failure Tumour lysis syndrome	Hyperlysozymuria (AMMoL) Diarrhoea
Sodium	Dehydration Other causes (as for creatinine)	Diabetes insipidus Over-hydration
Uric acid	Tumour lysis syndrome	–

AMMoL, acute myelomonocytic leukaemia.

infiltration by leukaemia or uric acid nephropathy, medical treatment may be indicated urgently; in other cases, for example bladder-neck obstruction by prostatic rhabdomyosarcoma, urgent surgical treatment may be required.

Tumour lysis syndrome

Dramatic biochemical changes may follow the treatment of lymphoblastic lymphoma and ALL. Lymphoblasts are abruptly lysed when treatment starts and intracellular components (especially uric acid, potassium and phosphate) may be released into the plasma suddenly and in massive quantities. Figure 21.5 indicates how rapidly such tumours can 'dissolve'; the biochemical consequences are collectively known as the tumour lysis syndrome. The syndrome is hardly ever seen after therapy of non-lymphoblastic malignancies; the explanation presumably lies in the rather slower lysis rate of such tumours.

Uric acid, the final product of nucleic acid degradation, is only weakly protein bound in plasma and is filtered at the glomerulus, reabsorbed in the proximal convoluted tubules and resecreted more distally. At high concentrations uric acid is precipitated in the distal tubules and collecting ducts and may then cause obstructive nephropathy. At pH < 5.4 uric acid is mostly in the un-ionized, less soluble form, but at high pH its solubility is considerably greater and is also inversely related to the urine sodium concentration. Urinary excretion of uric acid is almost always greatly elevated before treatment in patients with lymphoblastic leukaemia/lymphoma, especially those with 'bulky' solid masses or a high circulating blast-cell count (Hande *et al.*, 1981). Introduction of chemotherapy or radiotherapy leads to a sudden increase in the rate of uric

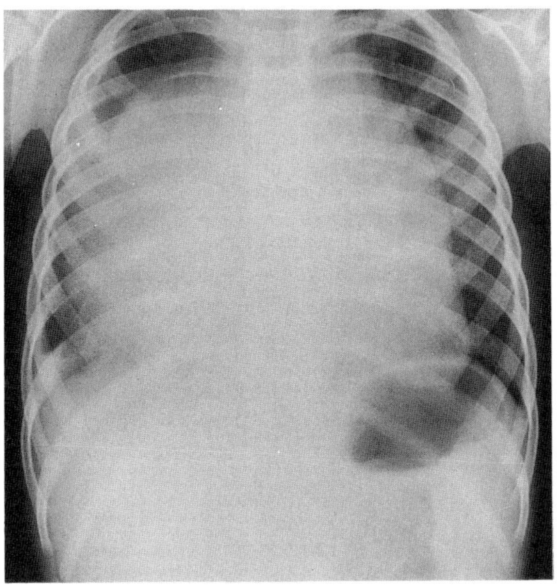
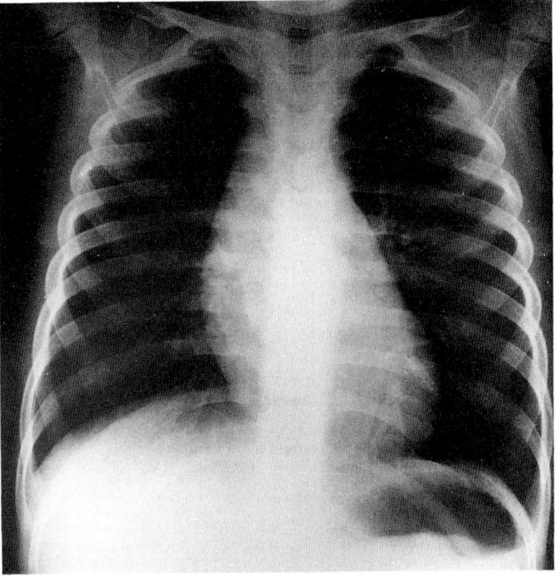

Fig. 21.5 Dramatic response of a T-cell non-Hodgkin mediastinal lymphoma (Sternberg type) to combination chemotherapy. A mild urate nephropathy developed but the patient showed no other features of the tumour lysis syndrome. The radiographs were taken on the first day (left) and fifth day (right) after treatment.

Fig. 21.6 Biochemical inter-relationships of hypoxanthine, xanthine and uric acid; allopurinol inhibits xanthine oxidase activity.

acid excretion (Lynch *et al.*, 1977). The clinical features of hyperuricaemia, including anorexia, nausea, vomiting and lethargy, are hard to appreciate in patients who are already systemically unwell because of their tumour; gout and renal colic are rare. Urate nephropathy can usually be avoided by vigorous preventative measures. All children with lymphoblastic tumours should receive the xanthine oxidase inhibitor allopurinol orally for at least 24 h before anticancer therapy is started. In dire emergencies, such as acute airway obstruction, intravenous allopurinol may lead to adequate xanthine oxidase blockade within 6 h. A fluid intake of 2–3 L/m^2 surface area daily should be combined with sufficient intravenous or oral sodium bicarbonate to keep the urine alkaline. Over-alkalinization can lead to tetany, especially when hyperphosphataemia (see below) coexists. Allopurinol blockade leads to the accumulation of xanthine and its derivatives (Fig. 21.6). Xanthine nephropathy is rare because xanthine and its derivatives are more soluble than uric acid. If, despite maximal 'preventative' therapy, progressive renal decompensation occurs, dialysis is needed. In children with ALL or NHL, allopurinol should be stopped before the commencement of either 6MP or thioguanine (both purines). Allopurinol inhibits the catabolism of these two drugs and concurrent administration of the cytotoxic agents can lead to high serum levels with consequent serious myelotoxicity.

Hyperkalaemia may be the consequence either of renal failure or the liberation of large amounts of potassium from dying cells as part of the tumour lysis syndrome (Lynch *et al.*, 1977; O'Regan *et al.*, 1977). As a consequence, life-threatening cardiac arrhythmias can occur. The period of maximum risk for this complication is 6–48 h after the start of therapy. Lymphoblasts may lyse spontaneously, releasing intracellular cations in anticoagulated blood samples that have been allowed to stand on the bench before the separation of plasma. A fictitious diagnosis of hyperkalaemia ('pseudo-hyperkalaemia') may be the consequence, so high serum potassium levels should be interpreted cautiously in children with ALL, especially in those with high circulating blast counts. In cases of doubt the measurement should be repeated on promptly separated plasma whilst a search is made for clinical evidence of hyperkalaemia, for example by electrocardiography.

The concentration of phosphate ions in leukaemic or lymphoma blast cells (especially B cells) may be greater than in normal cells. Sudden release of intracellular phosphate, another manifestation of the tumour lysis syndrome, may lead to hyperphosphataemia, which may be exacerbated by renal failure; increased renal clearance of phosphate and decreased tubular phosphate reabsorption lead to hyperphosphaturia. A high plasma phosphate level *per se* probably has few clinical consequences but coexistent hypocalcaemia can be responsible for psychological symptoms and, if severe, frank tetany. No specific therapy is required for hyperphosphataemia, but careful intravenous calcium supplementation may be indicated if symptomatic hypocalcaemia occurs. There is a theoretical risk of metastatic calcium phosphate deposition when the serum product (ionized calcium × phosphate) is high, but in practice the complication is exceedingly rare.

Many of these changes can be predicted and most serious problems can be minimized by the use of appropriate intravenous hydration and early renal dialysis. Where the predicted risk of severe tumour lysis is very high, then it may well be sensible to establish dialysis, either peritoneal or haemodialysis, before treatment is instituted. Most chemotherapeutic regimes offer the opportunity to start 'gently' in circumstances where problems might be predicted, although the tumour cells are usually so sensitive that they will lyse with only very minimal treatment.

Liver

Actinomycin D, alone or combined with irradiation, can cause a 'hepatopathy–thrombocytopenia' syndrome, possibly owing to occlusive disease in small hepatic venules (Raine *et al.*, 1991). For some reason, children with Wilms' tumour seem most susceptible to this problem. Clinically, patients present with abdominal distension, due to hepatomegaly and ascites, jaundice and a coagulopathy. The platelet count is low, coagulation times are prolonged and liver enzymes (APT and ALT) are elevated. Treatment is conservative, with blood product replacement. Although most patients recover over a 1- to 2-week period, fatalities have been recorded. Actinomycin can usually be safely reintroduced into the treatment regime within a month or so.

Other acute biochemical changes

Hypokalaemia is rare, but can be present at diagnosis in patients with AML or AMMoL – the consequence of a renal tubular potassium leak induced by high circulating levels of lysozyme (Wiernik & Serpick, 1979). Therapy with antibiotics, especially carbenicillin and gentamicin, which are often used in the management of the febrile neutropenic patient, may cause increased urinary potassium loss.

Hyponatraemia is most commonly due to excessive losses of sodium, because of vomiting and diarrhoea, or to excessive administration of hypotonic intravenous fluids. The syndrome of inappropriate secretion of antidiuretic hormone (ADH) is a rare complication of treatment with vinca alkaloids (Stuart *et al.*, 1975) or, exceptionally, other chemotherapeutic agents. Hypernatraemia is usually iatrogenic. Rarely, it is a consequence of DI owing to infiltration of the posterior pituitary gland or hypothalamic–pituitary pathways by tumour, occasionally leukaemia or brain tumour but most commonly LCH.

Disturbances of calcium and magnesium balance are common. Extensive bony infiltration occurs in up to 20–30% of children with ALL and neuroblastoma but rarely with other malignancies. In these cases hypercalcaemia is due to local bone destruction by bone-resorbing factors, such as prostaglandins released from metastatic tumour cells. Occasionally, bone-resorbing activity released from a primary tumour is so potent that 'distant' calcium release from bone is induced. Emergency management of hypercalcaemia is with prednisone or mithramycin. Apart from its occurrence as a consequence of hyperphosphataemia in the tumour lysis syndrome, hypocalcaemia occurring in children with cancer has been attributed to malnutrition, antibiotics and therapy with L-asparaginase. Hypocalcaemia may coexist with hypomagnesaemia following therapy with cisplatin (Blachley & Hill, 1981).

Various clinical manifestations, including anorexia, muscle weakness and disorientation, have been attributed to hypomagnesaemia but there is still some controversy about the need for regular magnesium supplementation during cisplatin therapy, especially since serum levels may fluctuate quite markedly without alteration of the dietary magnesium content. Hypophosphataemia has been noted in occasional newly diagnosed children with leukaemia, but is of little clinical consequence.

Lactic acidosis is a rare but well-recognized complication of the production of large amounts of lactic acid by poorly oxygenated tumour cells. It occurs especially where tumour masses are large and when complications, especially septicaemia, further increase tumour hypoxia. This possibility should be entertained in patients in whom a metabolic acidosis cannot be explained by renal failure. Urgent institution of tumour treatment is required (Nadiminti *et al.*, 1980). Elevated blood sugar levels, occasionally requiring insulin therapy, are often found in children with leukaemia or lymphoma and receiving treatment with L-asparaginase. Perhaps because this drug is usually given only for a few weeks, evolution into true diabetes is rare. Some children will also develop hyperglycaemia when being treated with large doses of steroids. L-Asparaginase can also cause pancreatitis, with a raised serum amylase level; abdominal ultrasound or CT scanning, revealing an enlarged pancreas, can be confirmatory.

Medium- and long-term effects

Some biochemical side effects of cancer therapy are listed in Table 21.8.

Nephrotoxicity

Several anticancer agents cause renal damage but abnormalities may not be evident for months or years after delivery of the treatment. Kidney damage may be manifest by decreasing glomerular function with consequent azotaemia or, less commonly, by evidence of tubular damage, e.g. proteinuria or glucosuria. Low-dose MTX causes no renal damage but in doses of more than 100 mg/m^2 surface area the drug can be significantly nephrotoxic, even when folinic acid rescue is also given. Toxicity is due to the crystallization of MTX and its less soluble 7-hydroxy metabolite in renal tubules. Since MTX is primarily excreted by the kidney, a 'vicious circle' is set up and toxicity ensues. With appropriate management renal failure usually resolves within 2–3 weeks but fatalities can occur, underlying the importance of carrying out treatment only in centres acquainted with the possible hazards. It is absolutely mandatory to monitor both renal function and drug levels during therapy with moderate and high doses of MTX. The nephrotoxicity of cisplatin (Blachley & Hill, 1981; Womer *et al.*, 1985) is due to deposition of the metal within the kidney, where measurable levels may persist for years. Glomerular function is more adversely affected than tubular function. Plasma creatinine measurements are not sufficiently sensitive for monitoring cisplatin nephrotoxicity

Table 21.8 Side effects of therapy associated with biochemical abnormalities

Agent	Organ/system involved	Biochemical disturbance
Actinomycin D	Liver	↑ ALT, ↑ AST, ↑ bilirubin, ↓ fibrinogen, ↑ FDPs, ↓ platelets
L-Asparaginase	Liver Pancreas	↓ Clotting factor synthesis ↑ Amylase, ↑ glucose
Cisplatin	Kidney	↑ Creatinine, ↑ urea ↓ GFR ↓ Magnesium, ↓ calcium
Cyclophosphamide	Post-pituitary/hypothalamus	Inappropriate ADH ↓ Sodium, ↓ osmolarity
Cytosine arabinoside	Liver	↑ ALT, ↓ AST
6MP and 6-thioguanine	Liver	↑ Alkaline phosphatase, ↑ ALT, ↑ AST
Methotrexate	Liver Kidney	↑ ALT, ↑ AST, ↑ bilirubin ↑ Creatinine, ↑ urea ↓ GFR
Nitrosoureas	Kidney	↑ Creatinine, ↑ urea
Prednisone	Pancreas	↑ Glucose
Radiation therapy (effects enhanced by doxorubicin or actinomycin D)	Kidney Liver Pituitary Gonad/pituitary Thyroid	↑ Creatinine, ↑ urea ↓ GFR ↑ ALT, ↑ AST, ↑ bilirubin ↓ Growth hormone ↑ Gonadotrophins ↑ TSH
Vincristine Vinblastine	Post-pituitary/hypothalamus	↑ ADH ↓ Serum sodium, ↓ osmolality

↑ Increased; ↓ decreased; ALT, alanine aminotransferase; AST, aspartate aminotransferase; FDP, fibrin degradation products; GFR, glomerular filtration rate; ADH, antidiuretic hormone; 6MP, 6-mercaptopurine; TSH, thyroid-stimulating hormone.

and direct measurement of the glomerular filtration rate (GFR) by the [^{51}Cr]-ethylenediaminetetra-acetate (EDTA) method is preferable. Increased renal loss of magnesium, leading to hypomagnesaemia and hypocalcaemia, can occur (see above). Glomerular damage by cisplatin is reversible in at least half the affected children.

Treatment with ifosfamide can lead to serious long-term renal complications. Although there is some glomerular injury, the proximal and distal limbs of the tubules are most seriously affected. Severe tubular nephropathy manifested by massive urinary leak of electrolytes and amino acids occurs and results in a Fanconi syndrome which may be complicated by renal rickets. Children with only one kidney (e.g. after nephrectomy for Wilms' tumour) and infants are especially at risk. The true extent of this problem in children being treated with ifosfamide is currently under investigation. Early detection, by measurement of specific urinary constituents or serum biochemical abnormalities, may allow therapy to be modified early enough to prevent serious consequences (Skinner *et al.*, 1993). Both radiation therapy and nitrosoureas (BCNU, carmustine; CCNU, lomustine) cause renal damage, but the time course is slow. Therefore, frequent measurements of renal function are not indicated during treatment but periodic plasma creatinine levels may be indicated in survivors.

The assessment of renal function in the survivors of childhood cancer is becoming increasingly important. Previous methods of assessing function by measuring either creatinine or even the GFR are inadequate to detect the multitude of important defects which may be present. A comprehensive, minimally invasive, investigation protocol has been developed and is now being applied on both a national and international basis (Skinner *et al.*, 1991). A summary is given in Tables 21.9 and 21.10. Assessment depends largely on the fractional excretion of various substances, calculated by:

$$FEa = \frac{Ua}{Pa} \times \frac{Pcr}{Ucr} \times 100\%$$

where FE is fractional excretion, a is the substance to be studied, P the plasma concentration, U the urinary concentration and cr represents creatinine.

Table 21.9 Nephrotoxicity investigation protocol – functional aspects

Glomerular
  GFR ([^{51}Cr]-EDTA plasma clearance)

Proximal tubular
  Blood: sodium, potassium, chloride, bicarbonate, urea, creatinine, calcium, ionized calcium, magnesium, phosphate, glucose, urate
  Urine: sodium, potassium, chloride, creatinine, calcium, magnesium, phosphate, glucose, urate, with calculation of the fractional excretion of sodium, calcium, magnesium, phosphate, glucose and urate, and of the renal threshold of phosphate
  Urine: low-molecular-weight proteins
  Urine: bicarbonate (in acidotic patients)

Distal tubular
  Early-morning urine osmolality and pH
  DDA VP test (consider in patients with low urine osmolality)
  Assessment of renal control of acid–base balance (consider in patients with high urinary pH or with acidosis)

Bone chemistry
  Blood: alkaline phosphatase activity

General
  Urine: renal tubular enzymes
  Blood and urine: albumin, protein
  Urine analysis and microscopy
  Blood pressure

GFR, glomerular filtration rate; EDTA, ethylenediaminetetraacetate; DDAVP, desmopressin.

Table 21.10 Nephrotoxicity investigation protocol

GFR measurement
  [^{51}Cr]-EDTA plasma clearance

Early-morning urine sample (10 mL)
  The first morning urine passed at home on the day of investigation is collected and brought to the hospital. It should be collected in a sterile container but does not need to be kept on ice. In hospital the sample is divided for the following tests:
    pH
    osmolality
    low-molecular-weight proteins
    renal tubular enzymes

Blood sample (5 mL)
  This can be collected when the [^{51}Cr]-EDTA is injected and then divided as follows:
    sodium, potassium, chloride, bicarbonate, urea, creatinine
    calcium, magnesium, phosphate, albumin, protein, alkaline phosphatase
    ionized calcium
    glucose
    urate

Urine sample (10 mL)
  This should be collected as soon as possible after the blood sample has been obtained and divided as follows:
    sodium, potassium, chloride, creatinine
    calcium, magnesium, phosphate
    glucose
    albumin, protein
    urate
    sample for urinalysis (using urine-testing reagent strips) and microscopy

General
  Blood pressure should be measured

GFR, glomerular filtration rate; EDTA, ethylenediaminetetraacetate.

The renal threshold (Tm) can be calculated, for example, for phosphate (p) by:

$$\mathrm{Tmp/GFR} = \mathrm{Pp} - \frac{\mathrm{Up} \times \mathrm{Pcr}}{\mathrm{Ucr}}$$

(For a further discussion, see Chapter 11.) Tables 21.11 and 21.12 give the appropriate reference ranges for these investigations.

Liver damage

Chronic liver disease is rare in children who have been treated for malignant disease. Liver damage may occur as a late effect of radiation therapy or of drug treatment, especially 6MP and MTX (Perry, 1982). Because of the need for repeated blood transfusions, hepatitis due to cytomegalovirus or one

Table 21.11 Age- and sex-related variation in reference ranges

Substance	Concentration	Age range (years)
Plasma creatinine	≤75 μmol/L	3–6
	≤90 μmol/L	7–14
	≤105 μmol/L	14–18, male
	≤85 μmol/L	14–18, female
Plasma phosphate	1.10–1.85 mmol/L	2–5
	1.00–1.80 mmol/L	6–12
	0.90–1.75 mmol/L	13–16
Plasma urate	0.05–0.40 mmol/L	1–9
	0.11–0.50 mmol/L	10–19, male
	0.11–0.42 mmol/L	10–19, female
Plasma alkaline phosphatase	≤350 U/L	2–9
	≤450 U/L	10–15
	≤300 U/L	16–18, male
	≤130 U/L	16–18, female
Renal threshold for phosphate	1.21–1.71 mmol/L	1–2
	0.92–1.70 mmol/L	3–4
	1.05–1.54 mmol/L	5–6
	1.07–1.72 mmol/L	7–8
	1.00–1.59 mmol/L	9–10
	0.92–1.66 mmol/L	11–12
	0.75–1.64 mmol/L	13–15
Urine retinol-binding protein	<30 μg/mmol creatinine	4–12
	<15 μg/mmol creatinine	>12
Urine renal tubular enzymes		
Alanine aminopeptidase	<2.0 U/mmol creatinine	4–12
	<1.6 U/mmol creatinine	>12
Alkaline phosphatase	<0.45 U/mmol creatinine	4–12
	<0.35 U/mmol creatinine	>12
Lactate dehydrogenase	<3.0 U/mmol creatinine	4–12
	<2.5 U/mmol creatinine	>12

Normal ranges used by Departments of Child Health and Clinical Biochemistry, Royal Victoria Infirmary and University of Newcastle upon Tyne, Newcastle upon Tyne; except for renal threshold for phosphate (see Brodehl *et al.*, 1982). Plasma alkaline phosphatase is measured by a Technicon SMAC Auto-Analyser, utilizing enzymatic hydrolysis of *p*-nitrophenylphosphate with 2-amino-2-methyl-1-propanol buffer to produce *p*-nitrophenol, which is measured colorimetrically.

Table 21.12 Reference ranges

Property	Range	Source of data
Glomerular function		
Glomerular filtration rate	87–174 mL/min per 1.73 m^2	Barratt (1974)
Proximal nephron function		
Plasma ionized calcium*	1.19–1.37 mmol/L	Dept of Clinical Biochemistry†
Fractional excretion of calcium*	<2%	Sutton and Dirks (1981)
Fractional excretion of magnesium*	About 3%	Sutton and Dirks (1981)
Fractional excretion of phosphate*	<20%	Sutton and Dirks (1981)
Fractional excretion of glucose	<1%	See Skinner *et al.* (1991)
Fractional excretion of urate*	7–12%	Grantham and Chenko (1986)
Urine β_2-microglobulin	<0.01 mg/mmol creatinine	Fielding (1984)
Distal nephron function		
See text		
Bone chemistry		
See plasma alkaline phosphatase in Table 21.11		
General		
Renal tubular enzymes		Depts of Child Health and Clinical Biochemistry†
N-Acetylglucosaminidase	<0.3 U/mmol creatinine	
Urine albumin	<10 mg/mmol creatinine	Barratt *et al.* (1970)

* Data derived from adult studies.
† Departments of Child Health and Clinical Biochemistry, Royal Victoria Infirmary and University of Newcastle upon Tyne.

of the hepatitis viruses is more common in these children than in the general population. Some of the chronic liver damage previously attributed to MTX may in fact be due to a combination of factors, in particular infections. Intermittent-dose MTX is probably less likely to give rise to chronic liver disease than is MTX given on a continuous basis; the same probably applies to 6MP.

Gastrointestinal toxicity

Mucosal cells of the gut are amongst the most rapidly proliferating of all body tissues. As a result, the gastrointestinal tract is particularly susceptible to the effects of chemotherapy and radiation, and acute gut toxicity is one of the most important factors limiting treatment dosages. Perhaps because methods of investigation are often invasive, there are few scientific data but there is a growing body of evidence indicating that some degree of biochemically measurable malabsorption does occur in children on prolonged courses of chemotherapy (Brunetto *et al.*, 1988). However, most children on treatment for malignancy continue along their normal height centile, and although some may show evidence of malnutrition, the cause is as likely to be decreased food intake, because of chemotherapy-induced nausea and vomiting, as malabsorption. Some cytotoxic agents are given by mouth and variable absorption of some, such as MTX (Craft *et al.*, 1981), may be of significance in the management of the tumour. In the 1970s, hypoallergenic diets were used with some success in the management of treatment-induced malabsorption, suggesting that altered intestinal permeability, leading to abnormal absorption of macromolecules, may be one manifestation of treatment-induced gut damage. Standard tests for malabsorption, apart from the xylose absorption test and measurement of iron and folate levels, seem to have little place in the routine management of child cancer patients, but newer non-invasive techniques utilizing non-metabolized sugars, e.g. lactulose, mannitol and 3-methylglucose, may be of value (Pledger *et al.*, 1988).

Pulmonary toxicity

Lung damage may be the result of surgery, radiotherapy, infection or some chemotherapeutic agents, especially bleomycin (Ginsberg & Comis, 1982). Blood gas measurements may be required urgently when a child receiving treatment for a malignant tumour presents with an interstitial pneumonitis. Cytomegalovirus, measles and *Pneumocystis carinii* are common 'opportunistic' organisms; arterial hypoxaemia is an early manifestation, whilst hypercapnia develops later. Serial blood gas measurements are required for the monitoring of children with interstitial pneumonitis. The lung toxicity of bleomycin is dose dependent and almost invariable when more than 500 U of the drug have been given. Measurement of the diffusion capacity is the most sensitive index of lung damage from bleomycin and such damage may be exacerbated if high concentrations of inspired oxygen are given during anaesthesia. Cyclophosphamide, melphalan, BCNU, MTX and busulphan can also cause long-term lung damage, but the overall incidence is low.

Cardiac toxicity

The anthracyclines (doxorubicin and daunorubicin) cause dose-dependent cardiotoxicity. However, measurement of intracellular cardiac enzymes (SGOT, SGPT) is of no value in predicting or following cardiomyopathy in children receiving treatment with these drugs. Two-dimensional echocardiography, radioisotope scanning and endomyocardial biopsy are more helpful. The overall incidence of severe late cardiomyopathy requiring heart transplantation is, as yet, unknown. Major determining factors are dose rate and total cumulative dose of the anthracycline.

Brain damage

Cranial radiation and drugs (especially MTX, vincristine and procarbazine) may have adverse effects on the developing or developed CNS. Biochemical monitoring, with the possible exception of β_2-microglobulin measurements (Mavligit *et al.*, 1980), is not helpful in distinguishing between disease-induced and treatment-induced brain damage.

Pituitary abnormalities

Irradiation of the neurohypophysis may affect the production of several hormones, especially growth hormone (GH). Relatively low doses of radiation may alter the GH response to insulin-induced stress, but physiologically important dysfunction occurs only if higher doses or dose rates are used (Shalet *et al.*, 1979). Children cured of brain tumours have usually received high radiation doses (35–60 Gy) (Bloom, 1982) and GH deficiency is even more common than in children cured of ALL. Other factors, such as impairment of vertebral growth after spinal irradiation, may contribute to small stature in these patients.

Secretion of other pituitary hormones may be affected by cranial irradiation, but the identification of a clinically important trophic hormone deficiency is extremely unusual; abnormalities in luteinizing hormone (LH) and follicle-stimulating hormone (FSH) secretion secondary to gonadal damage are discussed in a later section.

Thyroid dysfunction

The thyroid gland is inevitably within the radiation field during effective treatment for nasopharyngeal or cervical malignancy and for supra-diaphragmatic Hodgkin's disease. Biochemical hypothyroidism occurs in up to 40% of patients with Hodgkin's disease and is easily detected at an early stage by raised levels of thyroid-stimulating hormone (TSH). In patients at risk, routine monitoring of TSH is justified. The management of patients with high TSH levels is controversial, especially since spontaneous recovery occurs in some cases. Some clinicians argue that thyroid hormone replacement is not indicated unless clinical symptoms of hypothyroidism appear, whilst others feel that chronic TSH stimulation increases the incidence of later thyroid malignancy and they recommend replacement therapy.

L-Asparaginase, because of its action in decreasing protein synthesis, may be associated with low thyroxine (T_4) levels, whilst the halide-containing 5-fluorouracil may increase levels of both triiodothyronine and T_4 in the serum. However, patients receiving these drugs remain clinically euthyroid, the abnormalities being due to alterations in the characteristics of binding proteins.

Parathyroid function is hardly ever affected by either radiation therapy or chemotherapy.

Gonadal damage

Since curative chemotherapy for patients with ALL and Hodgkin's disease has been available for over 15 years, more data on gonadal function exist for these patients than for those cured of other childhood malignancies. Both endocrine function and fertility are more severely affected in boys than in girls (Chapman, 1982; Whitehead *et al.*, 1982).

Most boys receiving therapy for ALL or Hodgkin's disease during or after puberty have elevated serum FSH levels, indicating Leydig cell dysfunction, sometimes manifested by gynaecomastia (Shalet *et al.*, 1981). By contrast, prepubertal boys less often have raised gonadotrophin levels and almost all enter puberty quite normally. Ovarian function is less predictable. In one study (Siris *et al.*, 1976) 80% of girls cured of ALL showed abnormal pubertal progression, but some of those treated during puberty showed evidence of ovarian dysfunction, manifested by either raised or lowered gonadotrophin levels and accompanied in several cases by menstrual abnormalities. Early menopause is now recognized in some patients following a variety of types of combination chemotherapy.

Impaired fertility is more evident in boys than in girls and results from damage to the germinal epithelium. Those who have received alkylating agents or direct gonadal irradiation are often permanently sterile. On the other hand, recovery of gametogenesis has been described in patients of either sex who have been treated for leukaemia.

Babies born to mothers who have received pelvic irradiation, with the uterus within the radiation field, tend to be premature and small-for-dates. However, there seems to be no increase in congenital abnormalities, or malignancy attributable to their previous treatment, in babies born to men or women who have previously been cured of childhood cancer.

Conspectus

Recent advances in diagnosis, staging and treatment of childhood malignancies have been heartening but the most 'resistant' tumours now have to be tackled and future progress is likely to be slower. The similarity of several of these 'difficult' tumours to adult cancers, e.g. neuroblastoma and oat cell carcinoma of the lung, suggests that increasing collaboration with medical oncologists might be fruitful.

In addition to the introduction of newly discovered single agents and novel drug combinations, new treatment concepts under investigation for 'bad-risk' tumours include more rational 'kinetic' scheduling of drugs and radiation based on a knowledge of the cell-cycle characteristics of both tumour cells and normal proliferating cell populations, and marrow-ablative high-dose chemotherapy with autologous or allogeneic marrow transplantation, which is already of established value in the treatment of acute leukaemia and stage 4 neuroblastoma.

The dream of 'magic bullet' treatment by monoclonal antibodies armed with cytotoxic agents, such as ^{131}I or toxins, has faded with the growing appreciation that tumour-specific antigens probably do not exist. For the foreseeable future, developments in the treatment of paediatric cancer are likely to be intensive and 'high cost'. The value of tertiary referral of children with cancer for specialized management has now been established beyond doubt. Facilities in such centres, and the budgetary implications for diagnosis and monitoring techniques, especially clinical biochemistry, histopathology and imaging, must be planned accordingly.

On account of the toxicities and sheer unpleasantness of the present methods, the refinement of therapy is a major current objective in paediatric oncology. Many current trials have as their principal objective a study of whether the omission of an individual component from combination therapy has deleterious effects on disease-free survival. The occurrence of second tumours (solid tumours or

ANLL) in up to 8% of long-term survivors is of particular concern; radiation therapy, alkylating agents, procarbazine and the epipodophyllotoxins (teniposide and etoposide) are especially culpable and are excluded from treatment programmes when the risk exceeds the likely benefit. 'Continuing maintenance' chemotherapy seems to have no value except in ALL.

References

Abt, A.B., Kirschner, R.H., Belliveau, R.E. *et al.* (1974) Hepatic pathology associated with Hodgkin's disease. *Cancer* **33**, 1564–1571.

Adrian, T.E., Allen, J.M., Terenghi, G. *et al.* (1983) Neuropeptide Y in phaeochromocytomas and ganglioneuroblastomas. *Lancet* **3**, 540–542.

Aherne, G.W. & Quinton, M.T. (1981) Techniques for the measurement of methotrexate in biological samples. *Cancer Treat Rep* **65** (Suppl. 1), 55–60.

Altman, A.J. & Schwartz, A.D. (1978) Tumours of germ-cell origin: Embryonal carcinomas, extra-embryonal tumours and teratomas. In: *Malignant Diseases of Infancy, Childhood and Adolescence* (eds A.J. Schaffer & M. Markowitz), pp. 427–442. WB Saunders, Philadelphia.

Bacci, G., Avella, M., McDonald, D. *et al.* (1988) Serum lactate dehydrogenase (LDH) as a tumour marker in Ewing's sarcoma. *Tumori* **74**, 649–655.

Bacci, G., Picci, P., Ruggieri, P. *et al.* (1990) Primary chemotherapy and delayed surgery (neoadjuvant chemotherapy) for osteosarcoma of the extremities. *Cancer* **65**, 2539–2553.

Barber, S.G., Smith, J. & Hughes, R.C. (1978) Melatonin as a tumour marker in a patient with pineal tumour. *Br Med J* **2**, 328.

Barratt, T.M. (1974) Assessment of renal function in children. In: *Modern Trends in Paediatrics*, Vol. 4 (ed. J. Apley), pp. 181–215. Butterworth, London.

Birch, J.M., Marsden, H.B. & Swindell, R. (1980) Incidence of malignant disease in childhood. A 24-year review of the Manchester Children's Tumour Registry Data. *Br J Cancer* **42**, 215.

Blachley, J.D. & Hill, J.B. (1981) Renal and electrolyte disturbances associated with cisplatin. *Ann Intern Med* **95**, 628–632.

Bloom, H.J.R. (1982) Intracranial tumours: response and resistance to therapeutic endeavours 1970–1980. *Int J Rad Oncol Biol Phys* **8**, 1083–1113.

Bloom, S.R., Polak, J.M. & Pearse, A.G.E. (1973) Vasoactive intestinal peptide and watery-diarrhoea syndrome. *Lancet* **ii**, 14–16.

Bramwell, V.H.C., Burgers, M., Sneath, R. *et al.* (1992) A comparison of two short intensive adjuvant chemotherapy regimens in operable osteosarcoma of limbs in children and young adults: the first study of the European Osteosarcoma Intergroup. *J Clin Oncol* **10**, 1579–1591.

Bravo, E.L., Tarazi, R.C., Fouad, F.M., Vidt, D.G. & Gifford, R.W. (1981) Clonidine-suppression test. A useful aid in the diagnosis of pheochromocytoma. *N Engl J Med* **305**, 623–626.

Bravo, E.L., Tarazi, R.C., Gifford, R.W. & Stewart, A.H. (1979) Circulating and urinary catecholamines in pheochromocytoma. *N Engl J Med* **301**, 682–686.

Brennan, M.F. (1982) Cancer of the endocrine system. In: *Cancer: Principles and Practice of Oncology* (eds V.T. Jr De Vita, S. Hellman & S.E. Rosenberg), pp. 1019–1024. JB Lippincott, Philadelphia.

Breslow, N.E., Palmer, N.F., Hill, L.R., Buring, J. & D'Angio, G.J. (1978) Wilms' tumour: prognostic factors for patients without metastases at diagnosis. Results of the National Wilms' Tumour Study. *Cancer* **41**, 1577–1589.

Broadbent, V.A. & Pritchard, J. (1985) Histiocytosis X: current controversies. *Arch Dis Child* **60**, 605–607.

Brodehl, J., Gellison, K. & Weber, H.P. (1982) Postnatal development of tubular phosphate reabsorption. *Clin Nephrol* **17**, 163–171.

Brodeur, G.M., Seeger, R.C., Barrett, A. *et al.* (1988) International criteria for diagnosis staging and response to treatment in patients with neuroblastoma. *J Clin Oncol* **6**, 1874–1881.

Brunetto, A.L., Pearson, A.D.J., Price, L., Laker, M.F. & Craft, A.W. (1988) Assessment of gut damage by intestinal permeability following intensive consolidation therapy for acute lymphoblastic leukaemia. *Pediatr Res* **23**, 337.

Castleberry, R.P. (1992) Clinical and biologic features in the prognosis and treatment of neuroblastoma. *Curr Op Oncol* **4**, 116–123.

Chapman, R.M. (1982) Effect of cytotoxic therapy on sexuality and gonadal function. *Sem Oncol* **9**, 84–94.

Cole, M., Parker, L., Craft, A.W. *et al.* (1993) Creatinine related reference range for urinary HVA and VMA at 6 months of age. *Arch Dis Child* **68**, 376–378.

Cooper, E.H. & Plesner, T. (1980) Beta-2-microglobulin review: its relevance in clinical oncology. *Med Pediatr Oncol* **8**, 323–334.

Craft, A.W. & Parker, L. (1992) Poor prognosis neuroblastoma: is screening the answer? *Br J Cancer* **66** (Suppl. XVIII), S96–S101.

Craft, A.W., Amineddine, H.A., Scott, J.E.S. & Wagget, J. (1987) The Northern Region Children's Malignant Disease Registry 1968–1982: incidence and survival. *Br J Cancer* **56**, 853–858.

Craft, A.W., Rankin, A. & Aherne, G.W. (1981) Methotrexate absorption in children with acute lymphoblastic leukaemia. *Cancer Treat Rep* **65** (Suppl. 1), 77–81.

Creagh-Barry, P. & Sumner, A. (1992) Neuroblastoma and anaesthesia. *Paed Anaesth* **2**, 147–152.

Dale, G., McGill, A., Seviour, J.A. & Craft, A.W. (1988) Urinary excretion of HMMA and homovanillic acid in infants. *Ann Clin Biochem* **25**, 233–236.

D'Angio, G.J., Evans, A.E., Breslow, N. *et al.* (1976) The treatment of Wilms' tumour: results of the National Wilms' Tumour Study. *Cancer* **38**, 633–646.

Demeocq, F., Malpuech, G., Raynaud, E.J., Plagne, R. & Gaillard, G. (1981) Elevation of beta$_2$-microglobulin in cerebrospinal fluid after irradiation of the central nervous system. *N Engl J Med* **304**, 1366.

Drash, A., Sherman, F., Hartmann, W.H. & Blizzard, R.M. (1970) A syndrome of pseudohermaphroditism, Wilms' tumour, hypertension, and degenerative renal disease. *J Pediatr* **76**, 585–593.

Dunger, D.B., Broadbent, V., Yeoman, L. *et al.* (1989) The incidence and natural history of diabetes insipidus in children with Langerhans cell histiocytosis (histiocytosis X). *N Engl J Med* **321**, 1157–1162.

Eden, O.B. (1992) Malignant disorders of lymphocytes. In: *Paediatric Haematology* (eds I. Hann & J.S. Lilleyman), pp. 329–366. Churchill Livingstone, Edinburgh.

Evans, A.E., Blore, J., Hadley, R. & Tanindi, S. (1971) The Labrosse spot test: a practical aid in the diagnosis and management of children with neuroblastoma. *Pediatrics* **47**, 913–915.

Exelby, P.R., Filler, R.M. & Grosfeld, J.L. (1975) Liver tumours in children in particular reference to hepatoblastoma and hepatocellular carcinoma: American Academy of Pediatrics Surgical Section Survey – 1974. *J Pediatr Surg* **10**, 329–337.

Fielding, B.A. (1984) Albumin and β_2-microglobulin excretion in normal and diabetic children. MPhil thesis, Manchester Polytechnic.

Ganguly, A., Gribble, J., Tune, B., Kempson, R.L. & Leutscher, J.A. (1973) Renin-secreting Wilms' tumour with severe hypertension. Report of a case and brief review of renin-secreting tumours. *Ann Intern Med* **79**, 835–837.

Geiser, C.F. & Shih, V.E. (1980) Cystathioninuria and its origin in children with hepatoblastoma. *J Pediatr* **96**, 72–74.

Ginsberg, S.J. & Comis, R.L. (1982) The pulmonary toxicity of antineoplastic agents. *Sem Oncol* **9**, 52–64.

Gitlow, S.E., Bertani, L.M., Greenwood, S.M., Wong, B.L. & Dziedzic, S.W. (1972) Benign pheochromocytoma associated with elevated excretion of homovanillic acid. *J Pediatr* **81**, 1112–1116.

Glicksman, A.S., Maurer, H.M. & Vietti, T.J. (1979) Overview of conference on sarcomas of soft tissue and bone in childhood. *Med Pediatr Oncol* **7**, 55–67.

Gluckman, P.D., Ferguson, R.S., Osborne, D. & Evans, M. (1977) Primary hyperparathyroidism in a child. Use of jugular venous catheterization in diagnosis. *Arch Dis Child* **52**, 504–505.

Golde, E.M. (1979) The Cushing syndrome: changing views of diagnosis and treatment. *Ann Intern Med* **90**, 829–844.

Goltzman, D., Potts, J.T. Jr, Ridgway, E.C. & Maloof, F. (1974) Calcitonin as a tumour marker. *N Engl J Med* **290**, 1035–1039.

Grantham, J.J. & Chenko, A.M. (1986) Renal handling of organic anions and cations; metabolism and excretion of uric acid. In: *The Kidney* (eds B.M. Brenner & F.C. Jr Rector), 3rd edn, pp. 663–700. WB Saunders, Philadelphia.

Gross-Kieselstein, E., Navon, P., Branski, A., Abrahamov, A. & Dollberg, L. (1981) Familial erythrophagocytic lymphohistiocytosis in infancy. *Eur J Pediatr* **136**, 223–225.

Hande, K.R., Hixson, C.V. & Chabner, B.A. (1981) Post-chemotherapy purine excretion in lymphoma patients receiving allopurinol. *Cancer Res* **41**, 2273–2279.

Hann, I.M. (1992) Acute myeloid leukaemia. In: *Paediatric Haematology* (eds I.M. Hann & J.S. Lilleyman), pp. 83–98. Churchill Livingstone, Edinburgh.

Hogan, T.F., Citrin, D., Johnson, B.M. *et al.* (1978) *o,p'*-DDD (Mitotane) therapy of adrenal cortical carcinoma. *Cancer* **42**, 2177–2181.

Holmes, E.C., Morton, D.L. & Ketcham, A.S. (1969) Parathyroid carcinoma: a collective review. *Ann Surg* **169**, 631–640.

Hopwood, J.J. & Dorfman, A. (1978) Glycosaminoglycan synthesis by Wilms' tumour. *Pediatr Res* **12**, 52–56.

Howell, R.R., Stevenson, R.E., Ben-Menachem, Y., Phyliky, R.L. & Berry, D.H. (1976) Hepatic adenomata with type 1 glycogen storage disease. *J Am Med Assoc* **236**, 1481–1484.

Jansen-Goemans, A. & Engelhardt, J. (1977) Intractable diarrhea in a boy with vasoactive intestinal peptide-producing ganglio-neuroblastoma. *Pediatrics* **59**, 710–716.

Javadpour, N. (1980) The role of biologic tumour markers in testicular cancer. *Cancer* **45**, 1755–1761.

Kaplan, H. (1980) In: *Hodgkin's Disease* (ed. H. Kaplan), 2nd edn, pp. 116–145. Harvard University Press, Cambridge.

Kellie, S.J., Clague, A.E., McGeary, H.M. & Smith, P.J. (1986) The value of catecholamine metabolite determination on untimed urine collections in the diagnosis of neural crest tumours in children. *Aust Pediatr J* **22**, 313.

Knudsen, A.G. Jr (1976) Genetics and the aetiology of childhood cancer. *Pediatr Res* **10**, 513–517.

Kvols, L.K., Moertel, C.G., O'Connell, M.J. *et al.* (1986) Treatment of the malignant carcinoid syndrome: Evaluation of a long acting somatostatin analogue. *N Engl J Med* **315**, 663–666.

Labrosse, E.H., Comoy, E., Bohoun, C., Zucker, J.M. & Schweisguth, O. (1976) Catecholamine metabolism in neuroblastoma. *J Natl Cancer Inst* **57**, 633–638.

Lappöhn, R.E., Burger, H.G., Boumaj, J. *et al.* (1989) Inhibin as a marker for granulosa cell tumours. *N Engl J Med* **321**, 790–793.

Laug, W.E., Siegel, S.E., Shaw, K.N.F. *et al.* (1978) Initial urinary catecholamine metabolite concentrations and prognosis in neuroblastoma. *Pediatrics* **62**, 77–82.

Lennard, L. & Lilleyman, J. (1989) Variable mercaptopurine metabolism and treatment outcome in childhood lymphoblastic leukaemia. *J Clin Oncol* **7**, 1816–1823.

Li, F.P., Fraumeni, J.F. & Mulvihill, J.J. (1988) A cancer family syndrome in 24 kindreds. *Cancer Res* **48**, 5358–5362.

Lynch, R.E., Kjellstrand, C.M. & Coccia, P.F. (1977) Renal and metabolic complications of childhood non-Hodgkin's lymphoma. *Sem Oncol* **4**, 325–334.

Mann, J.R., Lakin, G.E., Leonard, J.C. *et al.* (1978) Clinical applications of serum carcino-embryonic antigen and alpha-fetoprotein levels in children with solid tumours. *Arch Dis Child* **53**, 366–374.

Mann, J.R., Pearson, D., Barrett, A. *et al.* (1981) Results of the United Kingdom Children's Cancer Study Group's malignant germ cell tumour studies. *Cancer* **63**, 1657–1667.

Mason, G.A., Hart-Mercer, J., Millar, E.J., Strang, L.B. & Wynne, N.A. (1957) Adrenaline-secreting neuroblastoma in an infant. *Lancet* **ii**, 322–325.

Matsumara, Y., Tann, D. & Tarin, D. (1992) Significance of CD44 gene products for cancer diagnosis and disease evaluation. *Lancet* **340**, 1053–1058.

Mavligit, G.M., Stuckey, S.E., Cabanillas, F.F. *et al.* (1980) Diagnosis of leukaemia or lymphoma in the CNS by beta$_2$ microglobulin determination. *N Engl J Med* **303**, 718–722.

Melmon, K.L. (1975) The endocrinologic manifestations of the carcinoid tumour. In: *Textbook of Endocrinology* (ed. R.H. Williams), 5th edn, pp. 1084–1104. WB Saunders, Philadelphia.

Mendelsohn, G., Eggleston, J.C., Olson, J.L., Said, S.I. & Baylin, S.B. (1979) Vasoactive intestinal peptide and its relationship to ganglion cell differentiation in neuroblastic tumours. *Lab Invest* **41**, 144–149.

Mike, V., Meadows, A.T. & D'Angio, G.J. (1982) Incidence of second malignant neoplasms in children: results of an international study. *Lancet* **ii**, 1326–1331.

Mitchell, J.D., Baxter, J., Blair-West, J.R. & McCredie, D.A. (1970) Renin levels in nephroblastoma. *Arch Dis Child* **45**, 376–384.

Nadiminti, Y., Wang, J.C., Chou, S.Y., Pineles, E. & Tobin, M.S. (1980) Lactic acidosis associated with Hodgkin's disease. Response to chemotherapy. *N Engl J Med* **303**, 15–17.

Niehaus, C.E., Ersser, R.S. & Atherden, S.M. (1979) Routine laboratory investigation of urinary catecholamine metabolites in sick children. *Ann Clin Biochem* **16**, 38–42.

Neuwelt, E.A., Glasberg, M., Frenkel, E. & Clark, W.K. (1979) Malignant pineal region tumours. A clinicopathological study. *J Neurosurg* **51**, 597–607.

Newell, D.R., Pearson, A.D.J., Balmanno, K. *et al.* Carboplatin pharmacokinetics in children: the development of a paediatric dosing formula. *J Clin Oncol.* (In press.)

Norgaard-Pedersen, B., Lindholm, J., Albrechtsen, R. *et al.* (1978) Alpha-fetoprotein and human chorionic gonadotropin in a patient with a primary intracranial germ cell tumour. *Cancer* **41**, 2315–2320.

O'Regan, S., Carson, S., Chesney, R.W. & Drummond, K.N. (1977) Electrolyte anu acid–base disturbances in the management of leukaemia. *Blood* **49**, 345–353.

Palmer, M.K. & Hann, I.M. (1980) A score at diagnosis for predicting length of remission in childhood lymphoblastic leukaemia. *Br J Cancer* **42**, 841–849.

Paradinas, F.J., Melia, W.M., Wilkinson, M.L. *et al.* (1982) High serum vitamin B12 binding capacity as a marker of the fibrolamellar variant of hepatocellular carcinoma. *Br Med J* **285**, 840–842.

Parker, L., Craft, A.W. & Dale, G. (1991) Screening for neuroblastoma. Background, preliminary experience in the north of England and proposals for an evaluative study. In: *Screening for Cancer* (ed. A.B. Miller), pp. 337–352. Cambridge University Press, Cambridge.

Parkin, D.M., Stiller, C.A., Draper, G.J. *et al.* (1988) *International Incidence of Childhood Cancers.* International Agency for Research into Cancer, Lyon.

Pearson, A.D.J., Amineddine, H.A., Yule, M. *et al.* (1991) The influence of serum methotrexate concentrations and drug dosage on outcome in childhood acute lymphoblastic leukaemia. *Br J Cancer* **64**, 169–173.

Perry, M.C. (1982) Hepatotoxicity of chemotherapeutic agents. *Sem Oncol* **9**, 65–74.

Pinkerton, C.R., Pritchard, J. & Spitz, L. (1986) High complete response rate in children with advanced germ cell tumours using cisplatin-containing combination chemotherapy. *J Clin Oncol* **4**(2), 194–199.

Pledger, J.V., Pearson, A.D.J., Craft, A.W., Laker, M.F. & Eastham, E.J. (1988) Intestinal permeability during chemotherapy for childhood tumours. *Eur J Pediatr* **147**, 123–127.

Pritchard, J. & Malone, M. Histiocyte disorders. In: *Oxford Textbook of Oncology* (eds M. Peckham, B. Pinedo & U. Veronesi). (In press.)

Pritchard, J., Barnes, J., Germond, S. *et al.* (1989b) Stage and urinary catecholamine metabolite excretion in neuroblastoma. *Lancet* **2**, 514.

Pritchard, J., Stiller, C.A. & Lennox, E.L. (1989a) Overtreatment of children with Wilms' tumour outside paediatric oncology centres. *Br Med J* **299**, 835–836.

Raine, J., Bowman, A., Wallendszus, K. & Pritchard, J. (1991) Hepatopathy-thrombocytopenia syndrome – a complication of actinomycin-D therapy for Wilms' tumour. *J Clin Oncol* **9**, 268–273.

Reece, P.A., Stafford, I., Russell, J., Khan, M. & Grantley Gill, P. (1987) Creatinine clearance as a predictor of ultrafilterable platinum disposition in cancer patients treated with cisplatin: relationship between peak ultrafilterable platinum plasma level and nephrotoxicity. *J Clin Oncol* **5**, 304–309.

Refetoff, S., Harrison, J., Karanfilski, B.T. *et al.* (1975) Continuing occurrence of thyroid carcinoma after irradiation to the neck in infancy and childhood. *N Engl J Med* **292**, 171–175.

Rosen, G., Caparros, B., Nirenberg, A. *et al.* (1981) Ewing's sarcoma: ten-year experience with adjuvant chemotherapy. *Cancer* **47**, 2204–2213.

Sampson, D. (1987) Biochemical methods for diagnosis of carcinoid tumour: what should we really be measuring? *Clin Biochem* **8**, 87–94.

Sato, T.L. (1965) The quantitative determination of 3-methoxy-4-hydroxyphenyl-acetic acid (homovanillic acid) in urine. *J Lab Clin Med* **66**, 517–525.

Sawada, T., Hirayama, M., Nakata, T. *et al.* (1984) Mass screening for neuroblastoma in infants in Japan. Interim report of a mass screening study group. *Lancet* **ii**, 271–273.

Surveillance, Epidemiology and End Result Reporting (1990) Trends in survival for children under age 15, selected sites of cancer, 1960–1985 data from the Cancer Statistics Branch. National Cancer Institute as cited in Ca, a journal for Clinicians **40**, 25.

Seldenfeld, J. & Marton, L.J. (1979) Biochemical markers of central nervous system tumours measured in cerebrospinal fluid and their potential use in diagnosis and patient management: a review. *J Natl Cancer Inst* **63**, 919–931.

Seviour, J.A., McGill, A.C., Dale, G. & Craft, A.W. (1988) Method of measurement of urinary homovanillic acid and vanillylmandelic acid by gas chromatography-mass spectrometry suitable for neuroblastoma screening. *J Chromat* **432**, 273–277.

Shalet, S.M., Hann, I.M., Lendon, M., Morris-Jones, P.H. & Beardswell, C.G. (1981) Testicular function after combination chemotherapy in childhood for acute lymphoblastic leukaemia. *Arch Dis Child* **56**, 275–278.

Shalet, S.M., Price, D.A., Beardswell, C.G., Morris-Jones, P.H. & Pearson, D. (1979) Normal growth despite abnormalities of growth hormone secretion in children treated for acute leukaemia. *J Pediatr* **94**, 719–722.

Siris, E.S., Leventhal, B.G. & Vaitukaitis, J.L. (1976) Effects of leukaemia and chemotherapy on puberty and reproductive function in girls. *N Engl J Med* **294**, 1143–1146.

Sisson, J.C., Frager, M.S., Valk, T.W. *et al.* (1981) Scintigraphic localisation of pheochromocytoma. *N Engl J Med* **305**, 12–17.

Skinner, R., Pearson, A.D.J., Coulthard, M.G. *et al.* (1991) Assessment of chemotherapy-associated nephrotoxicity in children with cancer. *Cancer Chem Pharm* **28**, 81–92.

Skinner, R., Sharkey, I.M., Pearson, A.D.J. & Craft, A.W. (1993) Ifosfamide, mesna and nephrotoxicity in children. *J Clin Oncol* **11**(1), 173–190.

Stewart, D.R., Morris-Jones, P.H. & Jolleys, P. (1974) Carcinoma of the adrenal gland in children. *J Pediatr Surg* **9**, 59–67.

Stiller, C.A. (1988) Centralisation of treatment and survival rates for cancer. *Arch Dis Child* **63**, 23–30.

Stringel, G., Ein, S.H., Creighton, R. *et al.* (1980) Pheochromocytoma in children – an update. *J Pediatr Surg* **15**, 496–500.

Stuart, M.J., Cuaso, C., Miller, M. & Oski, F.A. (1975) Syndrome of secretion of antidiuretic hormone following multiple doses of vincristine. *Blood* **45**, 315–320.

Sutton, R.A.L. & Dirks, J.H. (1981) Renal handling of calcium, phosphate, and magnesium. In: *The Kidney* (eds B.M. Brenner & F.C. Rector Jr), 2nd edn, pp. 551–618. WB Saunders, Philadelphia.

Taylor, W.F., Ivins, J.C., Dahlin, D.C., Edmondson, J.H. & Pritchard, D.J. (1978) Trends and variability in survival from osteosarcoma. *Mayo Clin Proc* **53**, 695–700.

Tiedemann, K., Long, R., Pritchard, J. & Bloom, S.R. (1981a) Plasma vaso-active intestinal polypeptide and other regulatory peptides in children with neurogenic tumours. *Eur J Pediatr* **137**, 147–150.

Tiedemann, K., Pritchard, J., Long, R. & Bloom, S.R. (1981b) Intractable diarrhoea in a patient with vaso-active intestinal peptide secreting neuroblastoma. *Eur J Pediatr* **137**, 217–219.

Toledano, S. (1979) Parathyroid carcinoma in an adolescent. *Med Pediatr Oncol* **7**, 95–102.

Tuchman, M., Ramnaraine, M.R., Woods, W.G. & Krivit, W. (1987) Three years of experience with random urinary homovanillic and vanillylmandelic acid levels in the diagnosis of neuroblastoma. *Pediatrics* **79**, 203.

Whitehead, E., Shalet, S.M., Morris-Jones, P.H., Beardwell, C.G. & Deakin, D.P. (1982) Gonadal function after combination chemotherapy for Hodgkin's disease in childhood. *Arch Dis Child* **47**, 287–291.

Wiernik, P.H. & Serpick, A.A. (1979) Clinical significance of serum and urinary muramidase activity in leukaemia and other hematologic malignancies. *Am J Med* **46**, 330–343.

Womer, R., Pritchard, J. & Barratt, T.M. (1985) Renal toxicity of cisplatin in children. *J Pediatr* **106**, 659–663.

Woods, W.G., Tuchman, M., Robison, L. *et al.* (1990) Screening for neuroblastoma in North America. Preliminary results from the Quebec project. *Med Pediatr Oncol* **18**, 368.

Wu, J.T., Book, L. & Sudar, K. (1981) Serum alpha fetoprotein (αFP) levels in normal infants. *Pediatr Res* **15**, 50–52.

Wurtman, R.J. & Moskowitz, M. (1977) The pineal organ. *N Engl J Med* **296**, 1329–1333 and 1383–1386.

Yokomori, K., Hori, T., Tsuchida, Y., Kuruodo, M. & Yoshioka, M. (1989) A new urinary mass screening system for neuroblastoma in infancy by use of monoclonal antibodies against VMA and HVA. *J Pediatr Surg* **24**, 391.

22: Proteins

P.G. RICHES & J. SHELDON

Introduction

Biological fluids contain a variety of proteins, each with a specific function or combination of functions. Proteins function as enzymes, antibodies, complement components, proteinase inhibitors, clotting factors, kinin precursors and transporters of hormones, vitamins, metals, lipids and many other substances. Variations in protein patterns or amounts of specific proteins compared with an accepted reference range can be attributed to genetic polymorphism, deficiency states or disease activity.

Protein assays are routinely used in the examination of serum, urine, cerebrospinal fluid (CSF), saliva and faeces, and provide information of value in the diagnosis and management of disease.

Proteins are routinely analysed by a combination of qualitative and quantitative techniques. The use of specific protein profiling has been suggested but has not gained wide clinical use, and simple listing of numerical results is of little value. The laboratory should provide interpretation of an electrophoretic pattern and specific protein measurement with regard to relevant clinical information.

The most widely used technique for qualitative examination of serum proteins is electrophoresis on cellulose acetate membrane or agarose gel. Normal human serum is separated by electrophoresis at pH 8.6 into five classically described zones: albumin, α_1-globulin, α_2-globulin, β-globulin and γ-globulin (Fig. 22.1). With the exception of albumin the zones are composed of more than one protein, and with modern high-resolution electrophoretic systems multiple bands are usually seen, particularly in the α_2 and β zones.

The stained completed electrophoretogram should be visually examined in comparison with a reference serum. Routine densitometric scanning of the stained strip has little application because a normal value for one zone does not exclude the possibility of abnormal composition. Densitometry has a unique application in the estimation of paraprotein bands (although these are rare in children) as immunochemical estimations of paraproteins are unreliable.

Other techniques, such as polyacrylamide gel electrophoresis and isoelectric focusing, have applications in specialized areas of protein analysis. Isoelectric focusing is routinely used for the identification of α_1-antitrypsin variants and detection of oligoclonal immunoglobulin bands in CSF. Polyacrylamide gel electrophoresis is used for high-resolution analysis of CSF, urine and other body fluids, and the system has the advantage that concentration of the fluid, prior to electrophoresis, is not always necessary. The high resolution shown by systems such as polyacrylamide gel electrophoresis has less application in the routine analysis of serum because the resulting pattern is complex and more difficult to interpret than the relatively simple pattern obtained with cellulose acetate or agarose.

Total protein and albumin concentrations are most frequently estimated by colorimetric assay; albumin can also be measured by specific immunochemical methods, as is usual for the other serum proteins. Immunochemical assays in common use are shown in Table 22.1 together with those proteins most appropriately assayed by each technique, the approximate assay time and the limit of sensitivity.

Technological advances in immunochemical techniques, particularly in nephelometry and turbidi-

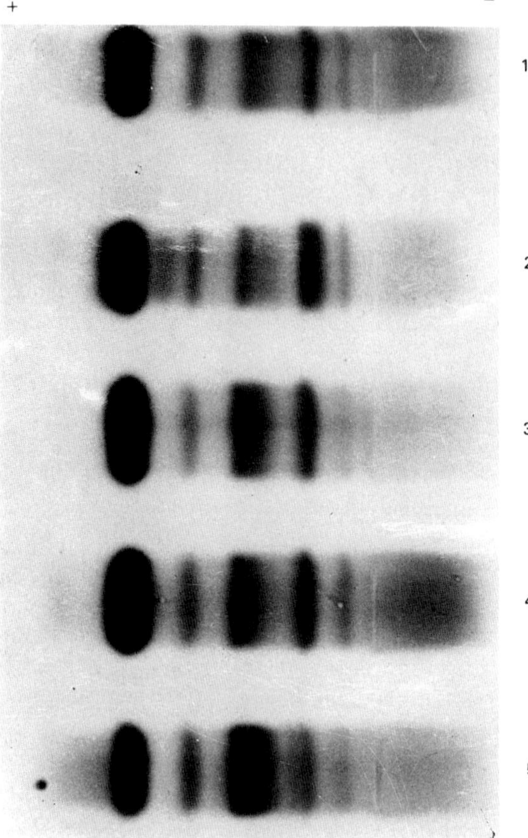

Fig. 22.1 Serum electrophoresis patterns. 1, Normal adult serum; 2, serum of premature infant (note band of α-fetoprotein between albumin and α_1-globulin); 3, 6-week-old infant (note the low γ-globulin); 4, 4-year-old child; 5, typical pattern of serum from 3-year-old patient with nephrotic syndrome (note low albumin and transferrin bands and markedly raised α_2-globulin).

metry, permit very precise and rapid estimation of protein concentrations; this can be helpful when diagnostic problems present as emergencies. It is normally possible to use small sample volumes and this makes these techniques ideally suited to paediatric practice.

Reagents, methods and, most importantly, the standards used in immunochemical assays influence the determined analyte concentration and in recent years internationally accepted standards have become available. Table 22.2 lists the most commonly analysed proteins together with the appropriate international reference preparation and the working standard calibrated against this international reference material. The serum protein standard (SPS) working reagents are available from PRU Procurement, Immunology, Northern General Hospital, Sheffield S5 7YT. Other commercially produced standards are available but it is essential that only those cross-calibrated with the international reference preparation are used. National quality-control schemes exist for most of the routinely analysed proteins and all laboratories offering such assays must participate in an appropriate scheme, details of which can be found in the *Handbook of Clinical Immunochemistry* (SAS Protein Reference Units, 1993), also obtainable from PRU Procurement.

The normal population ranges for many proteins show variations related to ethnic origins, environmental factors and age. The reference ranges quoted throughout this chapter are for Caucasians domiciled in the UK, and unless otherwise stated are derived from the Supraregional Protein Reference Units, and listed in the *Handbook of Clinical Immunochemistry*.

Serum proteins

The concentration of total serum protein in the fetus rises steadily during maturation, reaching concentrations of approximately 15.5 g/L in the second gestational month and 55 g/L at birth.

The establishment of reference ranges which vary with age, is obviously a prerequisite for the interpretation of protein analyses in paediatric samples. Since pregnancy itself significantly changes the levels of many serum protein fractions, relative protein concentrations in the neonate will be different when assessed against maternal, rather than normal adult, concentrations (Gitlin & Gitlin, 1975). As reference ranges in most laboratories relate to a normal adult pool, the details of variation for specific proteins with age, phenotypic expression and pathology will be discussed with respect to normal adult ranges.

Table 22.1 Quantitative methods for specific protein determinations

Method	Limit of sensitivity*	Assay time	Proteins commonly assayed
Nephelometry / Turbidimetry	1 mg/L	End point: 1 h Rate: 1 min	Albumin, α_1-antitrypsin, α_1-acid glycoprotein, caeruloplasmin, transferrin, C3, C4, IgG, IgA, IgM, CRP, haptoglobin, α_1-antichymotrypsin
Gel precipitation	1 mg/L	24–72 h	Albumin, α_1-antitrypsin, α_1-acid glycoprotein, caeruloplasmin, transferrin, C3, C4, IgG, IgA, IgM, haptoglobin, α_1-antichymotrypsin
Radioimmunoassay/ ELISA	1 μg/L	1–5 days	Total IgE, specific IgE, α-fetoprotein, serum ferritin, β_2-microglobulin
ELISA	1 μg/L	1–3 days	IgG subclasses, specific antibodies

CRP, C-reactive protein; ELISA, enzyme-linked immunosorbent assay.
* Using standard methodologies for the proteins listed.

Table 22.2 International and working standards for serum protein estimations

Serum protein	International standard	Working calibrant
Total protein	None	SPS-01 or pure bovine serum albumin
Albumin*	WHO standard for six human serum proteins (1977)	SPS-01
α-Fetoprotein	WHO (1975) 1st reference preparation IARC72/225, NIBSC 72/227	SPS-07
α_1-Antitrypsin*	WHO standard for six human serum proteins (1977)	SPS-01
Caeruloplasmin*	WHO standard for six human serum proteins (1977)	No
Transferrin*	WHO standard for six human serum proteins (1977)	SPS-01
Ferritin	WHO 1st international standard NIBSC-WHO 80/602	No
C-reactive protein*	WHO standard for CRP (1986) 85/506	SPS-03
C3*	WHO standard for six human serum proteins (1977)	SPS-01
C4*	WHO complement standard (1980)	SPS-01
IgG, IgA and IgM*	British working standard 67/99	SPS-01
IgG subclasses	WHO 67/97	SPS-01
IgE	WHO (1980) 2nd reference preparation 75/502	No

* Current international standards will be replaced by a new international reference preparation for protein calibration, CRM470; SPS-01 will be calibrated against this preparation. CRM470 will also be calibrated for α_2-macroglobulin, haptoglobin and prealbumin; IgG subclasses, α_1-antichymotrypsin, κ and λ will probably follow in the near future.

Electrophoretic patterns

The serum protein electrophoretic pattern in infancy is readily distinguishable from that of the adult (Fig. 22.1). In full-term infants, the albumin zone staining is comparable with that of the adult. A faint band of α-fetoprotein (AFP) between albumin and the α_1-globulin can be seen in some newborn premature infants. The low total protein of the neonate is reflected in a low content of α_1- and α_2-globulins but the α_2-macroglobulin, which is present in greater amounts than in the adult, gives a sharp anodal front to the α_2 zone. From birth to 2–6 months, there is a gradual increase in staining in the α_1 and α_2 zones (the α_2 zone maintaining the sharp anodal appearance). The major bands visible in the β zone, transferrin and C3, show comparable staining with the bands from the adult but the diffuse area between the β and γ zone is faint owing to a low concentration of IgA and IgM. Due to active placental transfer of IgG, the γ-zone staining is again comparable with that of the adult. After birth, there is a gradual increase in staining in the β–γ zone as IgA and IgM concentrations mature. There is a decrease in the γ staining, which reaches a nadir between 3 and 6 months, and there is then a progressive increase as the child's own IgG synthesis matures; in the older child the pattern approaches that of the adult.

Significant changes in pattern are associated with nephrosis, liver disease, hypogammaglobulinaemia, abnormal production of immunoglobulin, alloalbuminaemia (bisalbuminaemia) and the acute-phase response.

In nephrosis the pattern shows low staining of albumin, α_1-globulin and transferrin. On cellulose acetate a markedly elevated α_2 band is mainly the result of increased lipoprotein and the relative increase in the concentration of the high-molecular-weight protein α_2-macroglobulin which is not lost into the urine. Depending on the severity of the renal lesion, the γ zone will be either normal or low, predominantly due to changes in IgG. A similar pattern is seen in protein-losing enteropathy, although the protein loss tends to be relatively non-selective with respect to molecular weight.

In chronic liver disease the pattern shows low staining of albumin, raised α_1 and γ zones but a normal α_2 zone, as this fails to show a reactive increase. The γ-globulin increase, attributable to increased synthesis of immunoglobulin, is most marked in hepatic cirrhosis. The increase of IgA seen in many liver diseases gives increased staining in the β–γ region. In infective hepatitis the degree of increased γ-globulin staining correlates with the extent of hepatocellular damage and is an unfavourable finding.

In chronic aggressive hepatitis, the albumin may be relatively unaffected whilst the α_2 zone is raised and the γ zone shows a marked increase, present from the onset of disease; β–γ fusion is usually absent, the predominant increase being attributable to IgG.

In haemolytic jaundice, binding of haemoglobin to haptoglobin results in rapid metabolism of the complex and therefore low circulating concentrations of haptoglobin and a corresponding reduction in the staining of the α_2 zone. Haemoglobin–haptoglobin complexes may result, artefactually, from haemolysis *in vitro*, but since these cannot be metabolized, they remain in the specimen and give rise to a split α_2 zone on electrophoresis.

Hypogammaglobulinaemia with low concentrations of IgG, with or without reduced IgA or IgM, results in a reduced γ-globulin zone on electrophoresis. Hypogammaglobulinaemia involving only IgA or IgM will result in an essentially normal electrophoretic pattern owing to the presence of normal concentrations of IgG.

Paraproteins in childhood are rare and are usually superimposed upon a decreased γ zone associated with primary or secondary immunodeficiency states. They may occur transiently, during infection, in lymphoid neoplasia (particularly after chemotherapy) and occur frequently in children who have had bone-marrow transplantation; in the latter patients, reconstitution usually results in progress to the more usual polyclonal pattern. In very rare circumstances, a paraprotein may be actively transported across the placenta during the normal transfer of maternal IgG to the fetus. The chance finding of a serum paraprotein in the neonatal period is an indication for the electrophoretic investigation of maternal serum.

The appearance of a split albumin zone results from inherited variants of albumin (alloalbuminaemia). Allotypes show different electrophoretic mobility owing to charge differences; both fast and slow variants have been reported. Alloalbuminaemia is rare and of little clinical significance, except where large changes of binding capacity result; the total albumin concentration is usually within the reference range. The anodic charge of albumin is increased by binding of ligands such as bilirubin, antibiotics and sulphosalicylic acid, resulting in a diffuse zone extending forward of the normal albumin zone, rather than the two distinct bands of alloalbuminaemia.

Increased hepatic synthesis of a group of α_1- and α_2-globulins occurs in the acute-phase response which leads to the familiar picture of raised α_1 and

α_2 zones on electrophoresis. Normally, concentrations of C-reactive protein (CRP) are too low to be detected by electrophoresis but during a severe acute-phase response the concentration may increase to a level where it becomes visible as a distinct band in the $\beta-\gamma$ region. The resolution of routine electrophoretic techniques allows the detection of multiple faint bands within the γ region, representing immunoglobulin synthesized as a result of limited clonal expansion during the acute-phase response; these oligoclonal bands are usually transient.

Total protein

In the neonate, variations in total protein are usually reflected in low concentrations of α- and β-globulins. In the healthy newborn, the major fractions, except the immunoglobulins, reach steady state by 6 months of age. Premature infants often show hypoproteinaemia involving all fractions and a slow rate of increase in the first year of life.

Assay of total protein

The biuret reaction is the basis for the total protein assay, standardized against human or other animal serum. It is important that the method used conforms to that most widely accepted (Doumas *et al.*, 1981).

Reference data

Age	g/L
Newborn	45–65
Child	55–80

See also Appendix A1.

Alterations in data

The isolated measurement of total protein in paediatric samples is of limited value; specific protein determinations are more helpful. Estimation as an indicator of hydration status has little application in the laboratory, where other biochemical measurements are available.

Albumin

Albumin is the most important protein in the control of water balance between the intra- and extravascular compartments; it accounts for some 80% of the colloid oncotic pressure. It has a major role in the transport of fatty acids, tryptophan, haem, some hormones, bilirubin and cations including calcium. The physiologically active calcium is the free fraction; as some 50% of the total calcium is bound to albumin, a low total calcium is appropriately seen in hypoalbuminaemia. In all situations, the albumin concentration is essential for the interpretation of total serum calcium. Free concentrations of many substances may increase with low albumin concentrations owing to loss of the albumin-binding capacity, e.g. reduced binding of drugs may result in toxic levels if they are administered in their normal doses.

Assay of albumin

Dye-binding methods are most frequently used for the assay of albumin. The reaction time can cause significant variation and the slower reacting bromocresol green methods are not recommended because non-specific binding leads to significant over-estimation of albumin in the clinically important subnormal range. Specificity can be improved by estimating only the early reaction (less than 20 s). Bromocresol purple methods can be made specific for the estimation of human albumin. Albumin can also be measured by specific immunochemical methods.

Reference data

Albumin is synthesized by the embryo from an early stage and reaches adult concentrations late in gestation. The mean concentration in premature babies of 26 weeks gestation is 19 g/L and at term is 31 g/L (Cartlidge & Rutter, 1986). The concentration may fall in the immediate neonatal period but thereafter should remain relatively stable throughout life (see also Appendix A1).

Alterations in disease

Albumin concentrations above normal are seen only with dehydration, excessive albumin therapy or, artefactually, with prolonged venous stasis. Low albumin concentrations that cannot be accounted for by dilutional factors are of most clinical significance. Oncotic pressure falls as a consequence of acute hypoalbuminaemia, and in these circumstances albumin concentrations below 20 g/L are usually associated with oedema; in chronic conditions low albumin concentrations are not necessarily associated with oedema.

Posture influences albumin concentrations, which may be 5–10 g/L lower in samples collected from patients in a recumbent rather than upright position. Conditions associated with more significant changes in albumin are shown in Table 22.3.

Analbuminaemia results from a rare inherited deficiency in albumin synthesis. The condition is usually asymptomatic, except for transient oedema in early life and persistent hyperlipidaemia. Concentrations of α- and β-globulins are relatively increased such that the total protein concentration, usually above 50 g/L, is higher than might be expected.

The hypoalbuminaemia of malabsorption is due to poor supply of amino acids and subsequent failure of protein synthesis. Prolonged disease results in generalized muscle and tissue wasting, osteoporosis and lowering of all protein concentrations; the low albumin level can result in oedema and the low concentrations of circulating immunoglobulins predispose to infection. Those situations which may give rise to malabsorption are summarized in Table 22.4. In coeliac disease, low albumin concentrations can precede mucosal changes in the gut upon reintroduction of gluten, so that precise monitoring of serum albumin is a sensitive indicator for relapse or poor dietary control.

Inflammatory conditions of the gastrointestinal tract result from a number of complex pathologies. The pathophysiological mechanisms may involve immediate (type I) allergic reaction, mediated by IgE antibodies to ingested food (see section on Immunoglobulins), causing mucosal oedema, spasm of smooth muscle and increased secretions of mucus; type III (immune-complex mediated) and type IV (T-cell mediated) reactions also contribute to the inflammatory process in some patients.

In a number of congenital or acquired conditions, as shown in Table 22.4, enzyme insufficiency results in failure to digest foodstuff proteins, and hence leads to a failure of absorption and lack of amino acid precursors for protein synthesis.

Malnutrition is a general term referring to deficient dietary intake. Kwashiorkor, occurring in developing countries, results primarily from a diet deficient in protein and is characterized by the

Table 22.3 Conditions resulting in hypoalbuminaemia

Mechanism	Disease
Genetic deficiency	Analbuminaemia
Fall in synthetic rate	Malabsorption Malnutrition Liver disease Acute-phase reaction
Increased catabolism	Fever Malignancy Familial hypoproteinaemia
Excessive protein loss	Nephrosis Protein-losing enteropathy Extensive burns Exudative skin disease

Table 22.4 Conditions leading to protein malabsorption

Disease	Mechanism
Coeliac disease (gluten-sensitive enteropathy)	Reduction of absorptive surface
Milk and food allergy Crohn's disease	Inflammation
Fibrocystic disease of the pancreas Chronic pancreatitis Oesophageal atresia Pancreatic atresia Common bile-duct atresia	Failure of digestion

presence of oedema and a very low albumin value. In marasmus, which results from calorie deficiency, there is no oedema and the serum albumin is less reduced, or even normal. A fall in albumin is a relatively late event in protein depletion; prealbumin is a more sensitive indicator owing to its short half-life (2 days compared with 19 days for albumin). A fall in the transferrin concentration, although a relatively sensitive marker, is of less value because the concentrations tend to rise in the iron-deficiency anaemia which may accompany malnutrition. Proteins in malnutrition have been reviewed (Keyser, 1987).

Hypoalbuminaemia resulting from a significantly lowered synthetic capacity is seen in liver disease at a late stage when there is extensive destruction of hepatocytes. In chronic liver disease, fluid retention is the most significant cause of a low serum albumin but, irrespective of the mechanism of hypoalbuminaemia, serum albumin may be one of few abnormal biochemical findings and is an index of the severity of liver damage.

Hypoalbuminaemia following injury and in many diseases results from a redistribution of albumin between the intra- and extravascular compartments as a result of increased vascular permeability and loss of albumin into the tissue spaces (Fleck *et al.*, 1985); in fever and malignancy additional factors, such as increased catabolism and reduced synthesis, may contribute to the hypoalbuminaemia which, in chronic disease, may be severe. In addition to these secondary hypermetabolic states, primary the defect may be, e.g. familial hypercatabolic hypoproteinaemia, or may occur with more complex diseases, such as the Wiskott–Aldrich syndrome.

The nephrotic syndrome is characterized by hypoalbuminaemia, oedema and hyperlipidaemia resulting from protein loss owing to increased glomerular permeability. The clinical syndrome occurs when protein losses exceed 5 g/24 h. In children, the most common cause is a minimal-change glomerulonephritis in which there is extensive inflammatory involvement but the glomerular sieving properties are generally well maintained, such that only lower-molecular-weight proteins (albumin and smaller) are lost, and complete clinical response with minimal residual damage is achieved by appropriate treatment. Selectivity ratios (see section on Urinary proteins) are useful for distinguishing minimal-change lesions from the more progressive, irreversible damage that may be less extensive but results in a loss of selectivity in the glomerular sieving properties, and therefore also affects the higher-molecular-weight serum proteins. Hypoalbuminaemia in the nephrotic syndrome is greater than can be accounted for by urinary loss because increased albumin catabolism in the renal cortex and other organs throughout the body is a further contributing factor.

Hypoalbuminaemia of protein-losing enteropathy is frequently associated with low serum concentrations of the α-, β- and γ-globulins. The protein loss is non-selective, involving proteins of all molecular weights, although its extent is difficult to assess. Measurement of faecal α_1-antitrypsin offers a non-invasive test for intestinal protein loss (see section on Faeces).

Hypoalbuminaemia resulting from protein losses through traumatized skin is frequently accompanied by hypogammaglobulinaemia and reflects the extent of damage.

α-Fetoprotein

AFP is a glycoprotein synthesized by the fetal liver. Maximum circulating concentrations occur at about 20 weeks of gestation; these then decrease and the newborn concentrations fall into the normal adult range by 8 months of age (Wu *et al.*, 1981). The function of AFP is unknown but it may protect the fetus from maternal rejection.

Assay of AFP

Serum AFP is most usually assayed by radioimmunoassay.

Reference data for serum AFP

Age	kU/L
Neonate	50 000–150 000
5 weeks	5–30
>2 months	<10

Alterations in disease

After birth, raised AFP concentrations are found in association with:

regenerating liver (a good prognostic feature following hepatitis);
embryonal tumours;
neonatal hepatitis;
ataxia telangiectasia;
tyrosinosis;
congenital biliary atresia.

Serum concentrations in ataxia telangiectasia are rarely above 800 kU/L. Very high concentrations are seen in tyrosinosis (5×10^3–5×10^6 kU/L). Elevated serum concentrations are seen both in congenital biliary atresia and neonatal hepatitis; concentrations are usually, although not invariably, higher in the latter than in the former disorder.

Serum AFP concentrations and their use in childhood oncology are discussed in Chapter 21.

α_1-Antitrypsin

α_1-Antitrypsin functions as a general proteinase inhibitor and shows a high degree of genetic polymorphism. The inheritance is autosomal codominant. Genetic variants, probably due to a single amino acid substitution, are classified by the Pi (proteinase inhibitor) system. Some 30 or more variants are recognized and allocated an alphabetical connotation, formerly depending on electrophoretic mobility in acid conditions and more recently characterized by isoelectric focusing. PiM is the most frequent allele. Letters earlier in the alphabet (e.g. PiF) refer to variants with more anodic mobility and later letters of the alphabet (e.g. PiZ) to variants with more cathodic mobility.

The variants S, Z, P, W, Mmalton, Mduarte and null are clinically relevant deficiency alleles; the circulating concentrations for some examples of these are shown in Table 22.5 together with the relative incidence within the UK population. Evidence suggests that in deficiency states, with the exception of the null gene (*Pi*–), the protein is synthesized within the hepatocytes but the final stages of maturation, involving glycosylation steps, are defective, leading to failure of secretion. Low circulating α_1-antitrypsin concentrations occur in patients who are homozygous or heterozygous for any two of these deficiency alleles; the finding of such deficiency states indicates a necessity for the investigation of other family members.

Table 22.5 Some of the more common variants of α_1-antitrypsin

Phenotype	Percentage of MM value	Association (%) of population*
MM	100	88.5
MS	80	8.0
MZ	60	1.5
SS	60	0.35
SZ	35	0.22
ZZ	15	0.04
FM	100	0.66
FS	80	0.02
Z–	8	–

* Data from survey of 4565 people (Cook, 1975).

Heterozygosity of deficiency alleles with non-deficiency alleles may be detected by chance, or during investigations of the relatives of deficient patients. Such individuals have intermediate concentrations of α_1-antitrypsin, generally without related clinical problems, but should be counselled concerning genetic inheritance.

In affected families the early diagnosis on cord blood can be used to aid postnatal management. Cirrhosis and end-stage liver disease resulting from severe α_1-antitrypsin deficiency are indications for liver transplantation (Esquivel *et al.*, 1987).

Assay of α_1-antitrypsin

Total α_1-antitrypsin may be assayed by gel immunoprecipitation or turbidimetric or nephelometric techniques.

Pi phenotype identification

The method recommended by the International Nomenclature Committee is isoelectric focusing using polyacrylamide gel at pH 4–5. Specimens with a total α_1-antitrypsin concentration falling below the 50th percentile for the patient's age should have the phenotype identified.

Reference data and 50th percentile of serum α_1-antitrypsin

Age	Range (g/L)	50th percentile (g/L)
Birth	0.9–2.2	1.6
6 months	0.8–1.8	1.2
1 year	1.1–2.0	1.5
5 years	1.1–2.2	1.5
10 years	1.4–2.3	1.7
15 years	1.2–2.0	1.6

See also Appendix A1.

Alterations in disease

Low concentrations of α_1-antitrypsin are observed in the neonatal respiratory distress syndrome, which may result from capillary leakage into the interstitial compartment, and in severe protein-losing conditions. Detection of increased concentrations of α_1-antitrypsin is frequent because it is one of the acute-phase proteins.

Deficiency states are associated with lower lobe emphysema of early onset (20–30 years) and, more importantly in the present context, with transient cholestatic jaundice in the newborn, proceeding in some cases to progressive juvenile cirrhosis. The neonatal hepatitis occurs in about 20% of PiZZ infants and presents within the first 4 weeks of life. The transaminases and alkaline phosphatase are raised and the histology is that of giant-cell hepatitis. About one-third of patients progress to fatal juvenile cirrhosis.

There are two contributing factors to the liver disease associated with α_1-antitrypsin deficiency; the accumulation of α_1-antitrypsin may itself be a destructive process, while irreversible damage occurs as a result of an insult such as viral hepatitis. The situation in the liver is, therefore, analogous to that in the lung, in that there is increased susceptibility to damage rather than inherently progressive changes. Other conditions associated with α_1-antitrypsin deficiency include a number of connective tissue diseases. A convincing association with membranoproliferative glomerulonephritis has been shown in childhood cirrhosis.

α_1-Antitrypsin has an important role in counteracting the proteases released by activated leucocytes; deficiency states predispose to more tissue damage when there is extensive activation of immune complexes; this best explains the more severe effects of hepatitis, nephritis and lung infection.

Several reviews provide more extensive accounts of α_1-antitrypsin (Fagerhol & Cox, 1981).

Caeruloplasmin

Caeruloplasmin is unlikely to function as a transport protein for copper because the binding of copper is so tight. Alternatively, it may function as an oxidoreductase and seems to play an important role in the binding of iron to apotransferrin. In addition, there is evidence that caeruloplasmin may aid in the maintenance of the homeostasis of hepatic copper. Liver transplantation has shown that provision of a site of caeruloplasmin synthesis corrects both the serum level and the manifestations of Wilson's disease, including the clearing of cerebellar dysfunction. It is, of course, much easier to treat Wilson's disease by existing methods.

Assay of caeruloplasmin

Caeruloplasmin may be assayed by immunoprecipitation or turbidimetric or nephelometric techniques.

Reference data

Caeruloplasmin is present in lower concentrations in the neonate than in the normal adult, although concentrations rise rapidly within the first year.

Age	g/L
0–4 months	0.08–0.23
4 months–1 year	0.12–0.35
1–10 years	0.20–0.40
10–13 years	0.15–0.23

Alterations in disease

The serum level may be raised in the acute phase. Low concentrations are found in Wilson's disease, malnutrition, malabsorption and nephrosis.

The principal use of the measurement of caeruloplasmin in serum, together with measurement of serum copper and urine copper, is in the diagnosis of Wilson's disease. Significant findings are:

low concentrations of serum caeruloplasmin (found in 95% of patients);

low concentrations of total serum copper (95% of serum copper is associated with caeruloplasmin);

raised concentrations of urinary copper (more than 1 mol/24 h).

The metabolic defect in Wilson's disease appears to be primarily deranged copper metabolism. Abnormal copper deposition is found in the tissues, particularly in the brain, liver, renal tubules and eye. Deposition of copper at the junction of the cornea and sclera gives rise to the classic Kayser–Fleischer rings which are found in 70% of patients with Wilson's disease. The presence of copper in excess of 250 mg/g dry liver remains the definitive test for the diagnosis. Neurological symptoms of Wilson's disease are common in adults, whereas children usually present with hepatic disturbances.

Transferrin

Transferrin is the specific iron-binding protein of plasma. The iron is carried in the ferric form to stores and to bone marrow. Stored iron is laid down as ferritin or haemosiderin. At least 20 variants of transferrin are documented, but so far only one variant is associated with a quantitative deficiency.

Assay of transferrin

Transferrin may be assayed by immunoprecipitation or turbidimetric or nephelometric techniques. An indirect measurement of transferrin may be obtained by estimation of the total iron-binding capacity.

Reference data

The mean neonatal level of transferrin is about 1.9 g/L. Postnatally, concentrations are maintained in the healthy baby and reach the stable adult range of 2.0–3.2 g/L by 9 months of age.

Alterations in disease

A low transferrin level, resulting from inherited deficiency (atransferrinaemia), is extremely rare and results in marked hypochromic anaemia during early childhood.

Low serum transferrin concentrations are also seen in malnutrition, malabsorption and protein-losing states. Raised concentrations are associated with iron-deficiency anaemia.

Ferritin

Whereas transferrin concentrations do not always correlate well with total body iron stores, there is a direct relationship between serum ferritin concentrations and body iron stores. Iron deficiency is invariably associated with low serum ferritin concentrations. Ferritin is therefore the most reliable protein that can be measured for the assessment of iron status.

Ferritin behaves as an acute-phase reactant and this, along with the varied reference ranges, makes ferritin estimations of little value in monitoring diseases other than iron deficiency.

Assay of ferritin

Commercially produced immunoassay kits are readily available for the assay of serum ferritin.

Reference data

Although there is an international reference preparation for ferritin, there are variations in the standardization of the methods. Individual laboratories should either establish a local reference range or use the paediatric range provided with the reagent kit.

Acute-phase proteins

The homeostatic response of the body to trauma or infection involves local effects at the site of insult; these effects include vasodilatation, release of pre-formed cellular constituents and influx of effector cells. Localized effects are accompanied by a more generalized systemic response characterized by neutrophilia, fever and an increase in hepatic-derived plasma proteins, the acute-phase proteins. Together, these local and systemic effects characterize the inflammatory response, which occurs within hours of the onset of a challenge to the organism and is now known to be mediated by a group of proteins called cytokines. The acute-phase proteins have a variety of functions; as mediators, modulators or inhibitors of the inflammatory and immune responses, they may act as scavengers for products released from damaged tissues or macrophages and are involved in the repair and resolution following tissue damage.

The importance of the acute-phase response is evident from its participation in both innate and acquired responses. The indirect effects of the acute-phase proteins, raised erythrocyte sedimentation rate and raised plasma viscosity, have classically been used to diagnose and monitor inflammation. These provide only non-specific indicators of the acute-phase response and selected proteins will provide a more sensitive, specific and earlier indication of an inflammatory response (International Committee for Standardization in Haematology, 1988). The proteins show great diversity in magnitude of response, speed of response and half-life, and the selection of the most appropriate acute-phase proteins depends upon the diagnostic sensitivity and specificity required in a particular clinical situation.

CRP shows a rapid response (within 6 h) following tissue damage, its plasma concentration rises significantly outside the reference range and it has a short half-life; all these features make CRP an ideal analyte with which to monitor the acute inflammatory response. Chronic inflammatory processes are more appropriately monitored with a combination of CRP and α_1-antichymotrypsin, because the latter protein remains elevated in the plasma longer than CRP. Serum amyloid A is the most responsive of all the acute-phase proteins, increasing with even the most trivial of insults; this, together with a lack of assay reagents and standardization, limits its usefulness as a monitor of the acute-phase response (Thompson *et al.*, 1992).

C-reactive protein

CRP is one of a group of proteins, known as the pentraxins, which binds to a number of ligands in damaged cell membranes or bacterial cell walls, activating the complement pathway and resulting in inflammation, opsonization and phagocytosis of the CRP-containing complexes. The presence of a raised CRP is unequivocal evidence of inflammation and is therefore a useful analyte for the detection and assessment of organic disease; a normal CRP cannot exclude the presence of a mild, chronic inflammatory process. Some synthesis of acute-phase proteins occurs very early in development, even in fetal life, and no genetic deficiencies or inherited variants of CRP have been reported.

Assay of C-reactive protein

It is recommended that assays for CRP should be capable of quantitative measurements down to a serum concentration of 8 mg/L, but if the paediatric population includes neonates it is desirable that the assay be adequately sensitive to 1 mg/L. Qualitative or semi-quantitative latex agglutination methods are insensitive and of little value. For CRP to be used to its full potential, an assay system capable of giving results within 2 h of receipt of the sample is necessary. In hospitalized patients, measurement of acute-phase proteins, and particularly of CRP, is most valuable when used serially; a rise above the baseline gives a clear indication of further complications and a fall from the peak value indicates resolution.

Reference data

The reference range for CRP is 0.068–8.0 mg/L; this wide range covers all age groups, although neonates are unable to induce CRP synthesis to the same extent as older children and adults. The upper limit for a neonatal population is 3.5 mg/L.

Alterations in disease

The acute-phase protein response is an integral part of any tissue-damaging process. The rate of increase in the plasma concentration of any acute-phase protein is dependent upon the sensitivity of the hepatocytes to the cytokine-inducing signals and to the distributional and catabolic properties of the individual protein. The rate of increase in the plasma concentration is also dependent upon the type of inflammation. In some instances, the magnitude of the acute-phase response shows an approximately quantitative relationship to the activity or mass of inflamed tissue.

Elevations (up to 40 mg/L) of CRP occur with a number of mild inflammatory responses. Larger increases (over 100 mg/L) are predominantly seen with bacterial infections but may be present in severe inflammation. In the paediatric population, the most significant use of CRP is in the detection of infection, particularly in patients in whom the more conventional indicators may be unreliable; these groups are considered individually below.

Infection. In the neonate, infection is a major cause of inflammation, and reasons for inflammation other than infection are rare. Raised CRP concentrations indicate bacterial infection; viral, parasitic and mycobacterial infections cause only modest rises of CRP. Serial monitoring of a sick neonate is valuable, as a progressive increase of CRP during the first 48 h of life, even if concentrations are within the reference range, indicates infection or septicaemia.

In suspected meningitis, a serum CRP concentration above 20 mg/L is suggestive of a bacterial aetiology and provides an indication for antibiotic therapy if CSF microbiology is unavailable. The microbiological confirmation of localized infections from blood cultures is insensitive and although the CRP level may not be grossly elevated, if it is raised, this confirms a suspected bacterial infection and therefore has useful application in the diagnosis of low-grade infections.

Chronic inflammatory conditions. In chronic inflammatory diseases, CRP concentrations are variously raised, depending upon the type and extent of the underlying condition; in some instances CRP will be within the reference range, while other acute-phase proteins will remain elevated.

CRP concentrations are raised in rheumatoid arthritis and juvenile chronic polyarthritis (Still's disease). Serum CRP concentrations are generally normal in systemic lupus erythematosus (SLE), and when raised indicate intercurrent infection, except when there is severe serositis complicating the condition; this also causes a raised CRP.

α_1-Acid glycoprotein has been reported to be a marker of disease activity in inflammatory bowel disease. It is less sensitive than a combination of CRP and α_1-antichymotrypsin measurements because a raised CRP will indicate the acute stage and a raised α_1-antichymotrypsin with a normal CRP will indicate a chronic stage.

Immunosuppressed patients. Infection in immunosuppressed patients can be particularly difficult to diagnose because the classical diagnostic features, of fever and neutrophilia, may be suppressed or masked by the therapy. Prompt treatment is essential. CRP provides a sensitive and rapid diagnostic marker of both localized and systemic bacterial infections in any condition where the disease process itself does not result in elevated CRP concentrations. In this group of patients CRP levels should be measured daily. If the patient is receiving antibiotics, increasing concentrations indicate that the therapy is inappropriate. Decreasing concentrations consistent with the half-life indicate appropriate antibiotic therapy. Concentrations should return to normal before antibiotics are discontinued.

Figure 22.2 illustrates the use of CRP in the diagnosis and monitoring of infection in a child with leukaemia.

CRP measurements are particularly useful in the management of children who have had bone-marrow transplantation. Infection and graft-versus-host disease (GVHD) are major contributors to both morbidity and mortality in these patients, and can be difficult to distinguish clinically. The measurement of CRP is useful because it is rare to see significantly raised CRP concentrations even in the most severely acute GVHD, whereas in systemic bacterial infections CRP is invariably raised (Walker *et al.*, 1984). Viral infections, Hickman line infections and localized fungal infections may result in normal or only marginally raised CRP concentrations, although systemic fungal infections result in grossly elevated concentrations (Walker *et al.*, 1985).

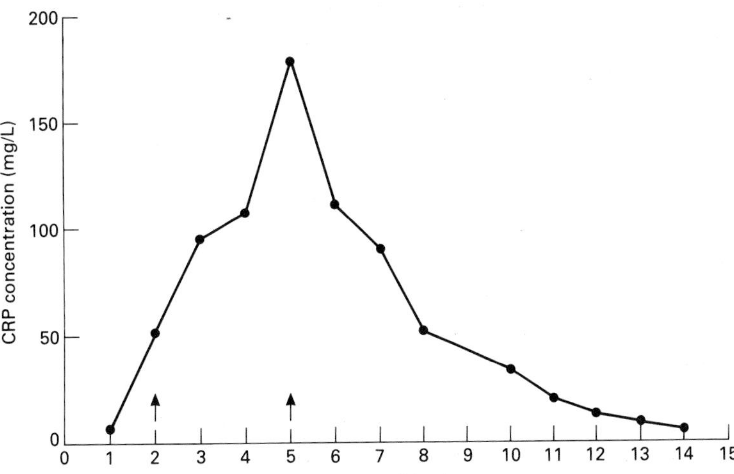

Fig. 22.2 Serum C-reactive protein (CRP) concentrations during an intercurrent infection in a leukaemic child. Antibiotic therapy was started on day 2 of the event, when the child was febrile with a raised CRP; the CRP continued to rise, indicating failure of response. A change of antibiotics on day 5 was followed by a rapid fall in CRP with evident clinical response.

In GVHD, normal CRP concentrations indicate the absence of a proinflammatory component, but abnormal concentrations of other analytes indicate significant immune activation in the disease process. Neopterin is a pteridine intermediate metabolite in the synthetic pathway for biopterin; it is synthesized and secreted by monocytes/macrophages upon stimulation, mainly by γ-interferon produced by activated T cells, and is therefore a valuable marker of immune activation. Simultaneous measurements of CRP and neopterin provide a useful diagnostic tool in differentiating between infection and GVHD when used according to the following guidelines (Sheldon *et al.*, 1991).

Raised neopterin and CRP indicate bacterial infection.

Raised neopterin and normal CRP indicate GVHD.

Normal neopterin and raised CRP indicate non-immunological inflammation.

Normal neopterin and CRP indicate absence of both inflammatory and immune responses.

The level of neopterin cannot be used to distinguish GVHD from viral infections. Renal impairment will elevate plasma concentrations but, together with sound clinical judgement, it remains helpful in distinguishing between the two conditions. Infections, particularly viral, may precipitate GVHD.

α_1-Antichymotrypsin

α_1-Antichymotrypsin is one of the group of acute-phase proteins that shows only a two- to four-fold increase in plasma concentration during tissue injury; it rises rapidly (within 10 h of insult) and is unaffected by genetic variation or differences in catabolism. The plasma concentration of α_1-antichymotrypsin remains elevated for longer than that of CRP; this makes it particularly useful in monitoring more chronic conditions or conditions when the onset of inflammation is unknown.

Assay of α_1-antichymotrypsin

α_1-Antichymotrypsin may be assayed by immunoprecipitation or turbidimetric or nephelometric techniques.

Reference data

No information is available concerning the reference range of α_1-antichymotrypsin in premature babies or neonates. The principal use of this protein is in monitoring chronic inflammatory processes in older age groups, where the adult reference range of 0.3–0.6 g/L applies.

Alterations in disease

α_1-Antichymotrypsin behaves in a similar manner to CRP with a rapid rise at the onset of inflammation, although it has a much smaller magnitude of response and remains elevated after CRP has returned to normal; it is thus a marker of both acute and chronic inflammatory states. α_1-Antichymotrypsin has a longer measured half-life than CRP; however, this cannot entirely explain its persistence in chronic inflammation.

Cytokines

Cytokines are a group of signalling polypeptides produced by a variety of cells. They act as communicating and regulatory factors both within the immune system and between the immune and other systems. They may have autocrine, paracrine or endocrine activity, mediated via specific high-affinity receptors present on many cell types.

Synthesis of the acute-phase proteins, other markers of inflammation and immunity, and the clinical manifestations of the inflammatory response can be directly or indirectly attributed to cytokine activities. Excessive production of cytokines has been implicated in some of the undesirable side effects of acute and chronic inflammatory reactions, e.g. rheumatoid arthritis, septic shock. The measurement of cytokines in biological fluids has the potential to increase both the sensitivity and specificity with which the adverse effect of the acute-phase response can be more clearly defined, monitored and predicted. Unfortunately, the methodologies currently available are slow, expensive and poorly standardized, and to date no real clinical indication for the measurement of cytokines in diagnosis or monitoring of disease has been established (Thompson *et al.*, 1992). Investigations of cytokines do, however, represent an exciting research area for the understanding and definition of normal and pathological inflammatory and immune responses.

Complement

Complement is a multicomponent system of proteins in the blood which is activated by antigen–antibody complexes, bacterial cell walls and endotoxins, and proteolytic enzymes released during inflammation. The individual components are named C1–C9, although additional control proteins are an integral part of the cascade. Sequential activation of complement components leads to three main effects:

1 Release of peptides active in inflammation.

2 Opsonization for phagocytosis.

3 Membrane damage resulting in lysis.

Complement may be activated by either of two pathways. In the classical pathway, activation of C1 by antigen–antibody complexes results in activation of C4 and C2 to form the classical C3 convertase, which is able to activate C3, the central component in both pathways. When alternative pathway (AP) activators are added to serum, small amounts of C3b, which form by spontaneous turnover of C3, are stabilized and three other proteins (factors B, D and P) are then involved in the formation of the AP C3 convertase. A number of inhibitors control both pathways and known deficiency syndromes are associated with three of these, the C1 esterase inhibitor and factors H and I. Details of the complement components, their sequence of activation, inhibitors and activators can be found elsewhere (Whaley, 1985).

Assay of complement

Complement components are investigated by total functional activity, quantification of individual proteins by immunochemical methods, detection of products of activation (conversion products) and functional activity of individual components.

Total functional complement activity via the classical pathway is assessed by haemolytic activity in a system where antibody-coated sheep erythrocytes haemolyse in the presence of the complement from the test serum; results are expressed as an arbitrary CH50 unit, which is defined as the amount of complement (i.e. volume of serum) needed to lyse 50% of the erythrocytes under the defined test conditions. In a similar manner, the complete function of the AP can be assessed by the use of rabbit erythrocytes which, when added to human serum, directly activate this complement pathway. The AP-CH50 unit is defined in the same way as the

CH50 unit. Functional activity of individual components is similarly assessed in haemolytic assays where all necessary components, except the one under test, are present; lysis will occur when the test serum is added, only if the missing component is present in that test serum, and failure of lysis indicates a deficiency state. Suspected complement deficiency should firstly be investigated by total functional activity and only when this is deficient is the investigation of the individual component justified. Detailed methodologies are available (Phimister & Whaley, 1990) but these assays are time consuming, expensive and best performed in specialized laboratories.

Measurements of C3, C4, factor B and the C1 esterase inhibitor have the widest routine clinical applications. Assay techniques, such as radial immunodiffusion, immunoturbidimetry or immunonephelometry, may be used.

Complement components are acute-phase proteins and their concentrations increase during an inflammatory response. Measured component concentrations will reflect the consumption due to activation balanced by synthesis, and in inflammatory conditions the increased synthesis can result in concentrations remaining within the reference range despite activation; in this situation, detection of cleavage components generated during activation will give a better assessment of activation than will simple concentration measurements. Conversion products of C3 and factor B are detected by immunofixation (Whicher *et al.*, 1980). Fresh serum or ethylenediaminetetraacetate (EDTA) plasma should be used to minimize activation *in vitro*, and gels used in complement assays should contain EDTA to avoid activation by the gel.

Reference data

Age	C3 (g/L)	C4 (g/L)
Cord blood	0.5–1.2	0.06–0.2
1–5 months	0.6–1.7	0.07–0.4
6–12 months	0.7–1.7	0.09–0.4
1–10 years	0.8–1.7	0.1–0.6
Adult	0.7–1.7	0.1–0.7

Alterations in disease

Low serum complement concentration is a result either of genetic or acquired deficiencies, or of its activation by antigen–antibody complexes in the classical pathway or by endotoxins or other substances in the alternative pathway.

Tables 22.6 and 22.7 summarize the complement profile in the clinical situations where its measurement is of value.

Immunoglobulins

The term immunoglobulin refers to the type of protein in which specific antibody activity can be

Table 22.6 Genetic errors of complement

Disease	Significant complement profile
Hereditary angio-oedema (HAE)	Decreased C1 esterase inhibitor* Low C4 Normal C3 Low CH50
Increased infections with Gram-positive organisms	Inherited deficiency of C3 Inherited or acquired deficiency of C3b inactivator Inherited deficiency of C1q
Increased incidence of immune complex disease, e.g. SLE	Inherited deficiency of C1, C2 or C4
Recurrent or persistent infections with neisserian organisms	Inherited deficiency of C5, C6, C7 or C8
Partial lipodystrophy	Persistently low C3 C4 often normal Nephritic factors often present

* A non-functional C1 esterase inhibitor has been reported. Normal values are found by immunochemical assay but low values are noted in a functional assay. The incidence in the UK is probably less than 10% of HAE patients. Non-functional C1 esterase inhibitor is falsely diagnosed on old samples and must be proved with absolutely fresh serum.
SLE, systemic lupus erythematosus.

Table 22.7 Complement and acquired diseases

Disease	Significant complement profile
Disseminated cytomegalovirus infection	Very low C4* Normal or raised C3
Gram-negative bacteraemia	Low C3 and factor B with conversion products Normal C4
Gram-positive bacteraemia	Low C3 and C4 with conversion products
Subacute bacterial endocarditis	Low C3 and C4 with conversion products
Systemic lupus erythematosus	Low C3 (or normal C3 with conversion products) Low C4
Post-streptococcal glomerulonephritis	Low C3 returning to normal after several months
Mesangiocapillary glomerulonephritis	Persistently low C3 C4 often absent Nephritic factors often present

* Probably the result of IgM antibodies working directly through the classical pathway. C4 consumption outstrips its replacement, whereas C3 acute-phase response maintains the C3 level.

found; immunoglobulins are synthesized by plasma cells in the primary and secondary lymphoid tissues. The immunoglobulin molecule consists of a basic subunit of two heavy chains joined to two light chains, this being the monomeric unit; IgA can occur in a dimeric form and IgM occurs predominantly as a pentamer. In a given immunoglobulin, only one class of heavy chain and one type of light chain is found. There are five heavy-chain classes: γ, α, μ, δ, ε; the γ chain exists as one of four subclasses (γ 1, 2, 3 or 4) and the α chain as one of two subclasses (α 1 or 2). There are two light-chain types (κ and λ). Allotypic variants are found within immunoglobulin molecules and although these variants may be important in certain immune responses, the typing of these has at present no clinical application.

IgG, IgA and IgM

IgG is the major immunoglobulin in normal serum and accounts for approximately 73% of the total immunoglobulin. In addition, IgG is present in the extravascular, extracellular compartment where one of its major functions is protection of the tissue spaces by neutralization of diffusible toxins.

Approximately 20% of the circulating immunoglobulin pool consists of IgA and, in the serum, this is almost entirely of monomeric immunoglobulin. IgA is the major immunoglobulin in saliva and some 60% of all IgA is synthesized at the secretory surface of the gastrointestinal, respiratory and urinary tracts; secretory IgA is a dimer, the formation of which is facilitated by the presence of a J chain, also synthesized within plasma cells. A secretory piece is added by the epithelial cells as the dimer is exported across the mucosa into the secretions where it plays its major protective role. Secretory IgA reaches adult levels in the saliva by the age of 6 weeks. Serum IgA is mainly derived from bone marrow, slowly populated until adolescence by memory cells, and seems to play an important role in the regulation of immune responses.

IgM is a pentamer and due to its high molecular weight (980 000) is predominantly intravascular. IgM constitutes some 5–10% of the total circulating immunoglobulin pool and plays a major role in protection of the bloodstream by its great efficiency in directly fixing complement to rapidly inactivate organisms.

Assay of immunoglobulins

IgG, IgA and IgM may be assayed by immunoprecipitation or turbidimetric or nephelometric techniques.

Rarely in children, immunoglobulins may form cold insoluble precipitates (cryoglobulins); other proteins, particularly fibrinogen, may also show this phenomenon. Specimens with suspected cryoproteins must be collected and processed at 37°C to avoid precipitation of the protein (Riches, 1990).

Reference data for IgG, IgA and IgM

Of all the serum proteins, the immunoglobulins show the widest variations with age. The full-term neonate has a plasma γ-globulin concentration of about 9–12 g/L, which consists almost entirely of IgG derived from the maternal circulation by placental transfer. The fall of IgG in the first months of life can be attributed to the decay of this maternal IgG prior to the slow rise corresponding to production of γ-globulin by the infant.

Age	IgG (g/L)	IgA (g/L)	IgM (g/L)
Cord serum	5.2–18.0	Below 0.02	0.02–0.2
0–2 weeks	5.0–17.0	0.01–0.08	0.05–0.2
2–6 weeks	3.9–13.0	0.02–0.15	0.08–0.4
6–12 weeks	2.1–7.7	0.05–0.4	0.15–0.7
3–6 months	2.4–8.8	0.1–0.5	0.2–1.0
6–9 months	3.0–9.0	0.15–0.7	0.4–1.6
9–12 months	3.0–10.9	0.2–0.7	0.6–2.1
1–2 years	3.1–13.8	0.3–1.2	0.5–2.2
2–3 years	3.7–15.8	0.3–1.3	0.5–2.2
3–6 years	4.9–16.1	0.4–2.0	0.5–2.0
6–9 years	5.4–16.1	0.5–2.4	0.5–1.8
9–12 years	5.4–16.1	0.7–2.5	0.5–1.8
12–15 years	5.4–16.1	0.8–2.8	0.5–1.9

IgG subclasses

It is only since the early 1980s that reliable antibodies for the measurement of IgG subclasses have become widely and commercially available. IgG1 is the predominant subclass, so its presence may mask deficiencies of one or more of the other three subclasses if only total IgG is measured. There is appreciable IgG subclass restriction of antibody responses against both protein (mainly IgG1 with IgG3 or 4) and polysaccharide antigens (mainly IgG2, although this is slow to mature in children); consequently, the lack of a particular subclass may severely affect a patient's antibody repertoire and the ability to protect against infection from specific organisms.

Assay of IgG subclasses

IgG subclasses are measured using monoclonal antibodies in radial immunodiffusion or enzyme-linked immunosorbent assay (ELISA) methods. The unique specificity of monoclonal antibodies may make some of them unreliable in certain systems (Jefferis *et al.*, 1985); therefore, it is essential that the recommended monoclonal antibodies, appropriate to the technique, are used.

Reference data of IgG1, IgG2, IgG3 and IgG4

IgG1 and IgG3 mature earlier than IgG2 and IgG4, which do not reach adult concentrations until adolescence.

Age	IgG1 (g/L)	IgG2 (g/L)	IgG3 (g/L)	IgG4 (g/L)
Cord blood	3.4–13.7	0.5–5.8	0.2–1.2	0.15–0.95
6 months	1.5–6.8	0.3–1.5	0.1–0.6	0.05–0.45
2 years	4.3–9.8	0.3–3.9	0.1–0.8	0.10–0.65
5 years	5.6–12.7	0.4–4.4	0.3–1.0	0.14–0.84
6 years	6.2–11.3	0.5–4.0	0.3–0.8	0.20–0.94
7 years	5.4–10.5	0.9–3.5	0.3–1.1	0.20–1.10
8 years	5.6–10.5	0.7–4.5	0.2–1.1	0.14–0.80
9 years	3.9–11.4	0.7–4.7	0.4–1.2	0.16–0.98
10 years	4.4–10.8	0.6–4.0	0.3–1.2	0.14–0.89
11 years	6.4–10.9	0.9–4.3	0.3–0.9	0.16–0.97
12 years	6.0–11.5	0.9–4.8	0.4–1.0	0.21–0.93
13 years	6.1–11.5	0.9–7.9	0.2–1.1	0.14–0.83
15 years	5.8–10.7	1.3–7.5	0.2–1.0	0.18–1.11

Alterations in disease

The clinical situations in which serum immunoglobulin measurements are essential for diagnosis are primary immune deficiency, transient hypogammaglobulinaemia of infancy, secondary immune deficiency and monitoring of γ-globulin replacement therapy. Initially IgG, IgA and IgM concentrations should be determined; where these do not indicate an obvious deficiency state and where there is a history of repeated infections, measurement of IgG subclass concentrations is appropriate. IgG subclass measurements are not indicated when monitoring γ-globulin replacement therapy. Low immunoglobulin values may occur transiently in infancy, be associated with primary inherited deficiencies or occur in association with, or secondary to, a number of other pathologies (discussed below).

There are certain instances where raised serum immunoglobulin concentrations, particularly IgM, provide valuable clinical information. The neonate usually has very low levels of IgA and IgM, and elevated levels in cord serum may be associated with a number of intrauterine infections, including congenital syphilis, toxoplasmosis, cytomegalovirus and rubella. At any age, trypanosomiasis and tropical splenomegaly are associated with increased serum IgM concentrations.

In a number of conditions there are characteristic patterns of polyclonal raised immunoglobulins; such conditions include lymphoproliferative diseases, liver cirrhosis, human immunodeficiency virus (HIV) infection and SLE, but in these situations immunoglobulin measurements rarely provide essential diagnostic information.

Cryoproteins may occur due to genetically altered proteins or as a consequence of disease; the clinical symptoms include intolerance to cold with pain in exposed areas and many skin manifestations, e.g. purpura, urticaria and ulcers.

Primary immunodeficiency states. Primary immunodeficiency can be broadly classified into defects predominantly affecting antibody production, combined B- and T-lymphocyte function, complement, phagocytes, or associated with other multisystem disease; within each of these categories there are many variants, as classified by the World Health Organization Scientific Group (WHO, 1986) and outlined in Table 22.8. The clinical presentations of recurrent infections, gastrointestinal disturbances and failure to thrive are common to many of the immunodeficiency states, and a suggested scheme of investigation of immunodeficiency where the underlying defect is not immediately obvious at presentation is given in Fig. 22.3. Common immunoglobulin patterns and clinical manifestations in those defects where immunoglobulin deficiencies are important are summarized in Table 22.9. Once an antibody deficiency has been identified, the patient should be referred to a specialist centre or laboratory for further investigation of lymphocyte populations and functions, and for assay of those enzymes where an absence underlies the deficiency.

IgA deficiency is the most common primary immunodeficiency with a reported frequency of greater than 1 in 500, and is the only deficiency which will be regularly encountered in the non-specialist laboratory. Clinical presentation is very variable and some affected individuals are symptom free. In many patients, IgA-bearing B cells are detected but in those patients where a genetic deletion leads to

Table 22.8 Primary immune deficiency states

Predominantly antibody defects
X-linked agammaglobulinaemia
X-linked agammaglobulinaemia with hormone deficiency
Autosomal recessive agammaglobulinaemia
Immunoglobulin deficiency with hyper-IgM (and IgD)
IgA deficiency
Selective deficiency of other immunoglobulin isotypes
κ-chain deficiency
Transient hypogammaglobulinaemia of infancy
Combined immunodeficiency
Common variable immunodeficiency (CVI)
Predominant antibody deficiency
Predominant cell-mediated deficiency
Severe combined immune deficiency (SCID)
Low T-, low B-cell numbers
Low T-, normal B-cell numbers
SCID with adenosine deaminase (ADA) deficiency
SCID with purine nucleoside phosphorylase (PNP) deficiency
MHC class I deficiency ('bare' lymphocyte syndrome)
MHC class II deficiency
Complement deficiencies
Individual factors
C1 esterase inhibitor, factor I, factor H, properdin
Defects of phagocyte function
X-linked chronic granulomatous disease (CGD)
Autosomal recessive CGD
Phagocyte membrane defect (β 95 deficiency)
Neutrophil glucose-6-phosphate dehydrogenase deficiency
Chediak–Higashi syndrome
Myeloperoxidase deficiency
Secondary granule deficiency
Shwachman's disease
Immune deficiency associated with other major defects
Wiskott–Aldrich syndrome
Ataxia telangiectasia
3rd and 4th pouch/arch syndrome (Di George)
Transcobalamin II deficiency
Immune deficiency with partial albinism
Defective response to Epstein–Barr virus (Duncan syndrome, X-linked immunoproliferative disease)

MHC, major histocompatibility complex.

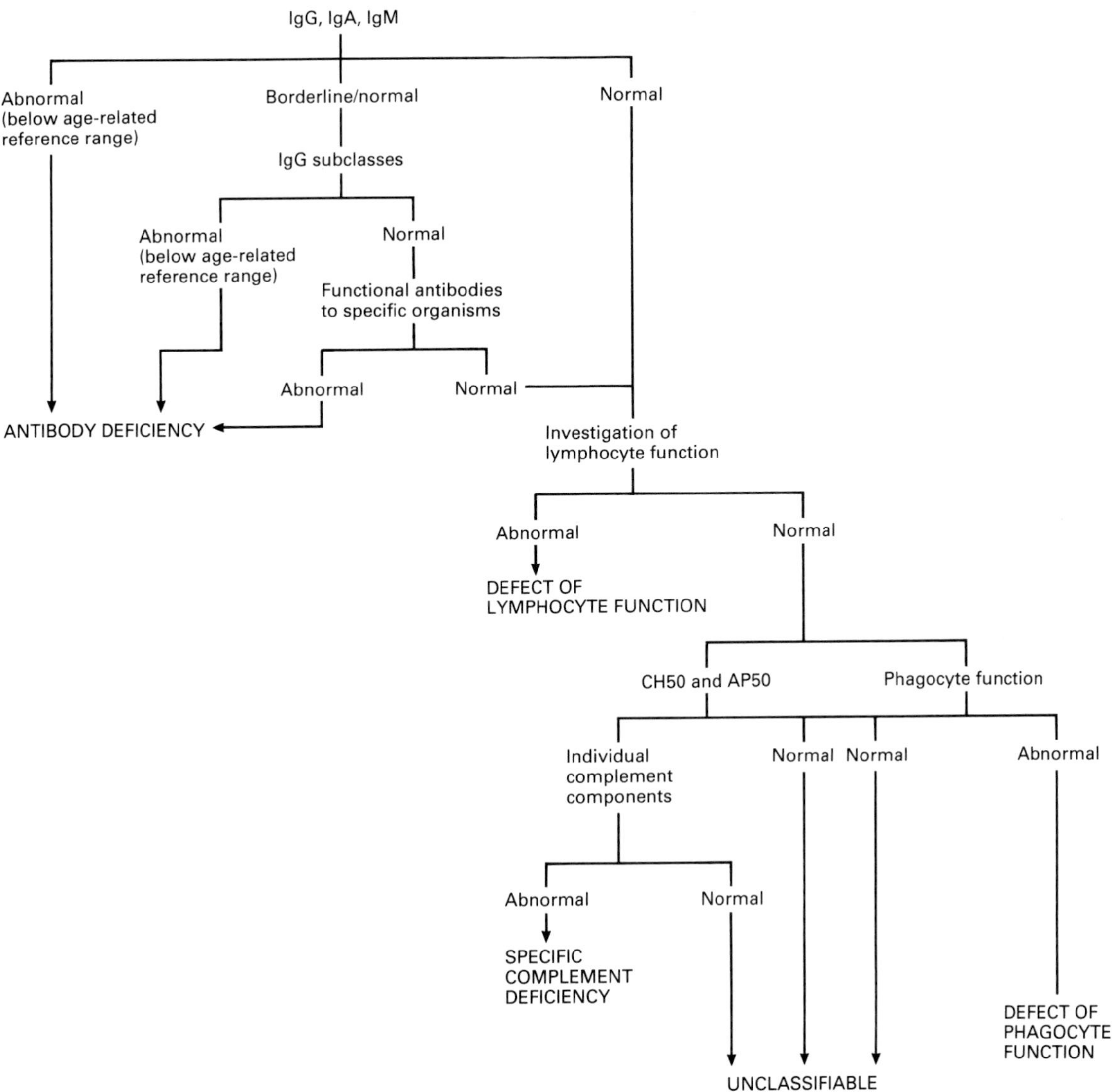

Fig. 22.3 Guidelines for the investigation of immunodeficiency where the underlying defect is not immediately obvious at presentation. After IUIS/WHO 1981, 1988.

no IgA expression there is the potential to develop antibodies to even the very low concentrations of IgA which may be present in blood products. These antibodies need to be identified before any further IgA-containing products are given to the patient. Patterns of infection may be complicated by IgG subclass deficiency associated with IgA deficiency (Oxelius *et al.*, 1981), and bacterial infections dominate the clinical picture; the finding of IgA deficiency in a symptomatic patient is thus an indication for IgG subclass measurements.

IgG subclass deficiency may also occur without IgA deficiency, but only IgG2 deficiency can as yet be regarded as a distinct entity causing an increased susceptibility to infections with encapsulated organisms. Failure to produce an antibody to a specific

Table 22.9 Primary immune deficiencies affecting immunoglobulin levels

Disease	Immunoglobulin pattern	Clinical manifestations
Selective IgA deficiency	Both secretory and serum IgA absent*	Repeated infections at mucosal surfaces, 30% affected individuals are symptom free
Selective IgM deficiency	IgM absent or low†	Septicaemia
Severe combined immune deficiency (SCID)	Maternal IgG, IgM variable, IgA absent	Patients with recurrent infections without treatment do not usually survive childhood
X-linked agammaglobulinaemia (Bruton's)	Maternal IgG, IgM variable, IgA absent; membrane immunoglobulin absent	Recurrent infections with extracellular pyogenic pathogens
Immune deficiency with thrombocytopenia and eczema (Wiskott–Aldrich)	Low IgM, frequently raised IgA; deficient response to carbohydrate antigens; absent isohaemagglutinins	Ear, nose and throat (ENT) infections, thrombocytopenia and eczema
Ataxia telangiectasia	Variable but may be IgA deficient	Cerebellar ataxia appears around 3 years of age
IgG subclass deficiency	Borderline IgG levels; IgG2 subnormal‡	Pyogenic infections
IgA and IgM deficiency	Some IgG	Predominantly gut disease (lymphoid nodular hyperplasia)
IgA and IgG deficiency	Raised IgM	Recurrent infections; tonsils and adenoids recur after removal; hepatosplenomegaly
Other immunoglobulin deficiencies	Low or absent IgG or IgM; functional deficiencies	Recurrent infections (staphylococcal, etc.)
Common variable immunodeficiency	All reduced; membrane immunoglobulins present	Infections with various organisms
Immunodeficiency and thymoma	All markedly reduced	Eosinophilia; pure red cell aplasia in some cases
Transcobalamin II deficiency	All markedly reduced	Frequent bacterial infections

* Exclude use of antiepileptic drugs.
† Exclude lymphoid neoplasia, cytotoxic drugs, etc.
‡ Exclude protein-losing states, etc.

antigen will not always result in subnormal IgG subclass concentrations; in these circumstances functional subclass antibody activity to the suspected organisms should be evaluated.

Transient hypogammaglobulinaemia of infancy. Transient hypogammaglobulinaemia of infancy is not sex-linked, occurs at a rate of about 4% of live births (1% suffer severe hypogammaglobulinaemia) and normal levels of circulating immunoglobulins are achieved, in most instances, after a slow start. Although it is usually classified as a primary deficiency, the heterozygous state of genetic hypogammaglobulinaemia can account for only about 5% of the observed cases. Patients with transient hypogammaglobulinaemia present at 4–6 months of age when the infant's own IgG synthesis should normally be replacing the maternal IgG. In most cases the mechanism of the 'delayed maturity' is unknown; contributing factors may be the transfer of alloantibodies or paraprotein from mother to fetus. Premature babies are also at risk of transient hypogammaglobulinaemia as a result of incomplete transfer of maternal IgG (Fig. 22.4); the use of a

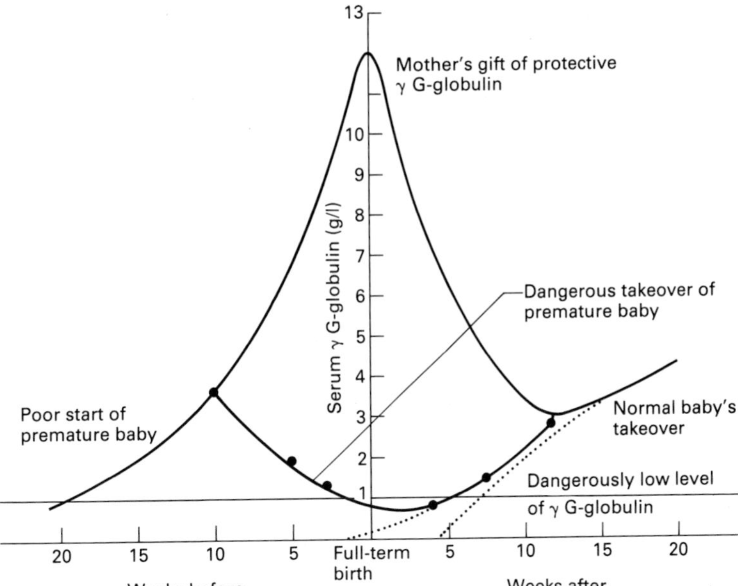

Fig. 22.4 Mechanism of hypogammaglobulinaemia in the premature infant. The maternal IgG may fall to low levels before the infant's own IgG takeover. Courtesy of *J Clin Pathol* (Riches & Hobbs, 1979).

single injection of prophylactic γ-globulin for such infants does not delay subsequent maturation but does reduce both morbidity and mortality.

Secondary hypogammaglobulinaemia. This is 10–100 times more common than the primary forms. The clinical situations giving rise to secondary hypogammaglobulinaemia are shown in Table 22.10. In general, marrow disorders and hypercatabolism (including protein-losing states) result predominantly in low values of IgG and have less effect on IgA and IgM. In toxic suppressions and lymphoid neoplasia, IgM is usually more affected than IgA, which in turn is affected more than IgG. In children, immune suppression associated with lymphoid neoplasia is not as marked as in adults, which may be related to the relative rates of tumour growth.

Table 22.10 Clinical conditions giving rise to secondary hypogammaglobulinaemia

Marrow disorders
Hypoplasia
Bony metastases
Myelosclerosis
Short survival of IgG
Nephrotic syndrome
Protein-losing enteropathy
Myotonic dystrophy
Toxic factors
Prolonged uraemia
Gluten-sensitive enteropathy
Cytotoxic and radiation therapy
Severe infection
Lymphoid neoplasia

IgA subclasses

There is considerable interest in the development of assays for the measurement of IgA subclasses and their specific antibody responses in relation to a number of disease states; however, to date, these remain only of research interest and have no clinical application.

IgD

IgD is predominantly a membrane-bound immunoglobulin and its measurement in paediatric samples has, as yet, had no clinical application, although it seems an interesting marker of tonsillar and adenoidal B cells.

IgE

Reaginic antibodies in man are predominantly of the IgE class. These antibodies bind to cell membranes of mast cells and basophils. Reaction of IgE with specific antigen triggers the release of a number of amines (bradykinin, histamine, serotonin, etc.) which mediate the local anaphylactic reaction. IgE is mainly synthesized at mucosal surfaces and sensitizes mast cells locally, suggesting that it plays a protective role in mucosal defence. Immediate hypersensitivity reactions of this type play an important role in the defence against helminth infections; however, in regions where parasitic infestations are less common, the adverse effects of IgE-mediated responses, i.e. atopy, are clinically of greater importance. In addition to the measurement of total IgE, it is possible to adapt the analytical system to detect IgE directed against a specific antigen.

Assay of IgE and specific IgE

Total IgE may be measured by radioimmunoassay or a solid-phase immunoassay, e.g. paper radioimmunosorbent test (PRIST), where the antibody to IgE is coated onto a paper disc. Specific IgE may be measured by an adaptation of this immunoassay; in this case the specific antigen is coated onto a solid phase, e.g. radio-allergosorbent test (RAST).

Reference data for total IgE

The normal values of serum IgE found in children are shown below. Geographical factors may be of importance, although investigators in many countries have found similar ranges.

Age	Median (kU/L)	95th percentile (kU/L)
Newborn	0.5	5
3 months	3	11
1 year	8	29
5 years	15	52
10 years	18	63

Levels above those of the 95th percentile indicate a high probability of atopic disease.

Mode of reporting specific IgE

The mode of reporting specific IgE and guidelines on the interpretation of the results are shown below.

Grade	Units (kU/L)	Interpretation
0	<0.35	Negative
1	0.35–0.7	Weak positive
2	0.7–3.5	Positive
3	3.5–17.5	Positive
4	17.5–50	Strong positive
5	50–100	Strong positive
6	>100	Strong positive

The potential significance of grades 1–3 will vary with the allergen concerned. Grade 1 positive to mould or food allergens may be regarded as significant, but is of doubtful significance for inhaled allergens.

Alterations in disease

Determination of serum IgE concentrations is only essential in the diagnosis of the rare hyper-IgE syndrome associated with eosinophilia and recurrent infections; total and specific IgE measurements can be useful in other conditions, as shown in Table 22.11. Primary deficiencies of IgE are recorded, although these are of little clinical significance; therefore, routine estimation of serum IgE is used for the

Table 22.11 Indications for determination of serum IgE

Total IgE
Distinguishing IgE-mediated from non-IgE-mediated disorders (rhinitis, bronchitis, asthma, dermatitis, chronic urticaria, food intolerance)
Indication of high risk for atopic disease

Specific IgE
Severe dermatitis that excludes skin testing
Patients receiving symptomatic treatment (i.e. antihistamine)
Allergens that cannot be used for skin testing (toxic, insoluble, etc.)
High sensitivity where testing *in vivo* could be dangerous
Food allergies

Table 22.12 Causes of increased serum IgE

Allergic disease
Extrinsic asthma
Hay fever
Seasonal allergic rhinitis
Atopic eczema
Atopy-related disorders
Minimal-change nephrotic syndrome
Interstitial nephritis due to drugs
Wiskott–Aldrich syndrome
Many T-lymphocyte deficiencies (primary or secondary)
Parasitic disorders
Tropical eosinophilia
Ascariasis
Echinococcosis
Hookworm
Intestinal capillariasis
Bilharziasis
Trichionosis
Schistosomiasis
Toxocara canis infestation

detection of raised values. The conditions associated with raised values are shown in Table 22.12.

It is important to realize that many of the symptoms associated with atopy may also result from non-IgE-mediated reactions. The suspicion of atopy in a child usually begins with the clinical manifestations, and the investigation of such patients is best undertaken with regard to the following major points.

Clinical history. The investigation of atopy should always begin with a full clinical history. Specific clinical symptoms, in addition to the time of year, time of day and place when they are most frequent, must be noted. Other relevant information includes the presence of atopy in any other family members, whether the patient is on any treatment and whether provocation tests or desensitization have been attempted.

Skin-prick testing. In the investigation of atopy, a full clinical history accompanied by skin-prick tests will usually determine the responsible agent; skin-prick tests are easy to perform and are an inexpensive clinical test of hypersensitivity. In some situations skin-prick tests may be difficult to carry out or the results may be difficult to interpret, e.g. in patients with extensive eczema. Furthermore, food antigens can give unreliable results on skin-prick testing and therefore in young children, where most cases of food sensitivity occur, specific IgE determination may be more useful.

Total serum IgE. IgE does not cross the placenta and the finding of a raised serum IgE level in a neonate or child under the age of 2 years is strongly associated with atopic disposition; this is a useful screening test, particularly for children of atopic parents.

Specific IgE testing (RAST). Once a patient has been found to be atopic, the identification of the specific allergen or allergens may help in the clinical management, usually through the modification of allergen exposure. In recent years the detection of allergen-specific IgE *in vitro* has become available and the correlation of non-food-allergen-specific IgE with direct skin testing is good. When skin-prick test results correspond with the clinical history, antigen-specific IgE is of no additional use; however, antigen-specific IgE is of particular use in the situations outlined above, where skin-prick test results are unreliable or when such tests cannot be carried out. Table 22.11 outlines the indications for the assay of allergen-specific IgE.

The range of commercially produced antigens for the detection of specific IgE is both extensive and ever-increasing, and we must resist the temptation to test for specific IgE to all possible allergens because a false-positive finding may indicate low levels of specific IgE that are not clinically important; similarly, specific IgE may be negative when provocation testing clearly indicates clinically significant hypersensitivity. In these cases it is possible that there is only regional synthesis, sufficient to sensitize mast cells but not to give detectable serum levels of specific IgE.

RAST results with regard to food allergens are difficult to evaluate. There is good correlation between testing *in vitro* and sensitivity to inhaled foods, but it is rarely possible to implicate ingested foods unequivocally in the pathogenesis of chronic bronchial asthma. The problems are even more complex when dealing with suspected allergic gastroenteritis; damaged mucosal surfaces may allow the passage of complex antigens, leading to detection of specific IgE to these antigens, which are unrelated to the antigen responsible for the allergic reaction.

The wide variety of clinical manifestations of atopic responses make hard-and-fast rules concerning their investigation inappropriate, and each clinical request must be considered with an adequate clinical history in order that the appropriate investigations are undertaken (Thompson & Bird, 1983).

The finding of a positive allergen-specific IgE indicates only that the patient has the potential to react to that antigen and the result may be of no relevance to the clinical state of the patient.

Antibodies to α-gliadin

The gluten-sensitive enteropathy of coeliac disease is associated with an altered immunological responsiveness to the α-gliadin fraction of gluten, causing raised concentrations of circulating IgA and IgG antibodies to α-gliadin. Criteria for the diagnosis of coeliac disease have been recommended (Working Group of European Society for Paediatric Gastroenterology and Nutrition, 1990); the two diagnostic requirements are the characteristic histological appearance of the duodenal–jejunal mucosa, and definite clinical remission within a few weeks of the patient being placed on a strict gluten-free diet. The investigation of circulating IgA and IgG antibodies to α-gliadin provides a non-invasive test which can be used to support the initial histological diagnosis of coeliac disease, and may also be used to monitor the efficacy of, and compliance with, a gluten-free diet. The finding of IgA antibodies to endomysium and reticulin may add further weight to the diagnosis of coeliac disease, particularly in adults, as these too are highly sensitive markers of the disease. Whenever the diagnosis is uncertain by any of the clinical or laboratory criteria, a gluten challenge is essential and is monitored by further biopsy when there is a noticeable clinical relapse after challenge, or in any event after 3–6 months. Antibody tests are a valuable adjunct and may provide a guide to the timing of biopsies.

Assay of antibodies to α-gliadin

There are available commercially produced kits, based on ELISA techniques, to detect IgA and IgG antibodies to α-gliadin.

Reference data

There is no reference preparation for either IgA or IgG antibodies to α-gliadin and most methods report the results in arbitrary units with respect to the patient's age. Each laboratory is recommended to establish its own range of expected values or use those quoted with the commercial reagents.

Alterations in disease

The presence of circulating IgA antibodies to α-gliadin shows a high specificity, but a relatively low sensitivity, for coeliac disease. Circulating IgG antibodies to α-gliadin show much better sensitivity than do IgA antibodies, but show lower specificity as they are seen in some patients with ulcerative colitis, Crohn's disease and cows' milk intolerance; the simultaneous detection of IgG and IgA antibodies to α-gliadin gives a specificity of approximately 90% and a sensitivity of 95%. In patients with no detectable IgA antibodies to α-gliadin but with clinical manifestations of coeliac disease, IgA deficiency must be excluded.

Urinary proteins

The protein content of urine is determined by the filtration of plasma across the glomerular capillary membrane, tubular reabsorption and tubular secretion. The glomerulus normally excludes proteins with a molecular mass of greater than 45 000 Da and the smaller proteins which are filtered through the glomerulus are predominantly reabsorbed by the proximal tubules. Trace amounts of protein are normally secreted into the urine by the tubules. The net result of these processes is a protein excretion of less than 150 mg/day. Proteinuria, significantly exceeding this value, can be classified into one of three major groups: glomerular, tubular or overflow.

Assay of urinary proteins

For a detailed account see Chapter 11.

Cerebrospinal fluid

CSF is formed primarily from the plasma by filtration across the blood–brain barrier. The amount of any individual protein filtered into the CSF is dependent upon its molecular mass and its relative serum concentration, such that low-molecular-weight proteins pass more readily into the CSF than do high-molecular-weight proteins. The circulation of CSF as it proceeds down the spinal canal is very slow; the CSF proteins tend to equilibrate increasingly with the plasma, leading to as much as a threefold difference in the concentration of total protein between the lumbar sac and the ventricles. The lumbar CSF is characterized by containing proteins of higher molecular mass than those in the ventricular CSF.

In neonates, the blood–brain barrier is more permeable than in later life, and the protein concentration of the CSF is greater.

There is some local production of immunoglobulin within the CSF, although normally this is a very small contribution to the total protein (about 1 mg/day).

Assay of CSF proteins

Assay of CSF total protein may be made by dye-binding methods or turbidimetry. Specific proteins of the CSF can be estimated by any of the usual immunochemical methods, although these often have to be used in a more sensitive range than for the corresponding serum proteins. The protein patterns are investigated by electrophoretic techniques. Using a combination of isoelectric focusing and immunofixation, specific antibodies in the CSF may be detected; these techniques are usually reserved for specialist centres.

Reference data for CSF total protein

Age	g/L
Newborn	0.4–1.2
Up to 1 month	0.2–0.8
1–12 months	0.2–0.7
1–16 years	0.1–0.4

Alterations in disease

In pathological conditions the factors contributing to increased protein levels are:

1 Increased permeability of the blood–brain barrier, leading to more protein entering the CSF.

2 Synthesis of proteins by plasma cells or inflammatory cells within the cerebrospinal canal (intrathecal synthesis).

3 Local destruction of immunoglobulins to produce fragments (including light chains).

CSF total protein may be raised in infections, inflammation or with space-occupying lesions of the central nervous system; a raised CSF total protein indicates an organic lesion but provides no differential diagnosis. A normal CSF total protein does not rule out organic disease as this may occur in some cases of meningitis and demyelinating diseases. The local production of IgG occurs in a number of neurological conditions, e.g. multiple sclerosis, subacute sclerosing panencephalitis, viral meningitis and viral encephalitis, and the detection of this increased IgG synthesis within the CSF can aid in the diagnosis of these conditions.

Intrathecal synthesis of IgG can be detected by isoelectric focusing and subsequent immunofixation of paired serum and CSF samples; a characteristic pattern of oligoclonal banding in the CSF, even with normal IgG quantification, which is not present in the matched serum, is consistent with intrathecal IgG synthesis (Fig. 22.5). In situations where there is a systemic inflammatory response or where increased CSF total protein results from increased transudation, the serum and CSF IgG oligoclonal banding pattern, if present, will be identical. An adaptation of the isoelectric focusing and immunofixation technique is an ideal method for detecting antibodies (particularly IgG and IgM) to specific

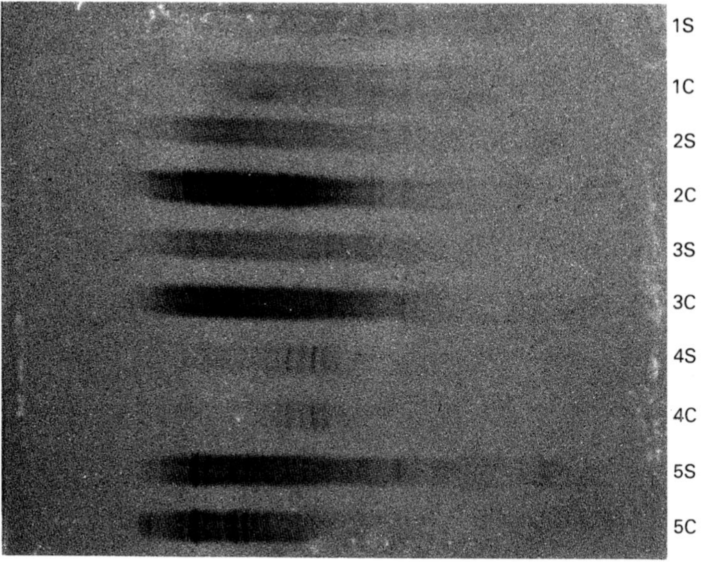

Fig. 22.5 Isoelectric focusing pattern, stained for IgG, of paired serum (S) and cerebrospinal fluid (CSF) (C) from five patients showing (1) normal pattern in both, (2) oligoclonal banding in CSF only, (3) polyclonal increased IgG in CSF, (4 and 5) oligoclonal banding in both serum and CSF.

infectious organisms relevant to various diseases, e.g. measles, syphilis, herpes; these techniques are best performed in a specialist centre.

Saliva

The principal application of protein measurements in saliva is to establish the absence of IgA.

Normal range for salivary IgA

6 weeks to adult: 0.03–0.2 g/L.

Alterations in disease

Levels in neonates should reach adult salivary IgA concentrations by 6 weeks of age, and the absence of salivary IgA at this age can give an early indication of immune deficiency. Mixed saliva shows wide variation in volume and protein content, and other protein measurements are of little value.

Faeces

Most plasma proteins entering the gastrointestinal tract in protein-losing enteropathy are subjected to proteolytic digestion by bacterial proteases. Owing to its natural function as a protease inhibitor, α_1-antitrypsin is more resistant to proteolytic degradation and its measurement is a useful non-invasive test for protein-losing conditions.

Assay of proteins in faeces

A random stool sample should be frozen below −20°C as soon as possible after collection to minimize protein degradation, and should be thawed only immediately before extraction. Aqueous extracts of faeces (1% w/v in 0.9% sodium chloride) are prepared by mixing for 15 min at room temperature, and following centrifugation (as for blood separation) supernatants are assayed by the usual immunochemical techniques.

Reference ranges

0.05–0.48 mg α_1-antitrypsin/g wet weight faeces.

Alterations in disease

Raised concentrations are associated with protein-losing enteropathy; care should be taken to exclude any gastrointestinal bleeding that may give false-positive results. Although fairly specific, the test may be insensitive owing to protein degradation, particularly when samples are not stored correctly.

References

Cartlidge, P.H.T. & Rutter, N. (1986) Serum albumin concentrations and oedema in the newborn. *Arch Dis Child* **61**, 657–660.

Cook, P.J.L. (1975) The genetics of alpha-1-antitrypsin: a family study in England and Scotland. *Ann Hum Genet (London)* **38**, 275–287.

Doumas, B.T., Bayse, D.D., Carter, R.J., Peters, T. & Schaffer, R. (1981) A candidate reference method for determination of total protein in serum. 1. Development and validation. *Clin Chem* **27**, 1642–1650.

Esquivel, C.O., Iwatsuuki, S., Gordon, R.D. *et al.* (1987) Indications for paediatric liver transplantation. *J Paediatr* **111**, 545–548.

Fagerhol, M.K. & Cox, D.W. (1981) The Pi polymorphism: genetic, biochemical and clinical aspects of human α_1-antitrypsin. *Adv Hum Genet* **11**, 1–62.

Fleck, A., Raines, F., Hawker, J. *et al.* (1985) Increased vascular permeability: a major cause of hypoalbuminaemia in disease and injury. *Lancet* **i**, 781–783.

Gitlin, D. & Gitlin, J.D. (1975) Fetal and neonatal development of human plasma proteins. In: *The Plasma Proteins*, Vol. II (ed. F.W. Putnam), pp. 263–319. Academic Press, New York.

International Committee for Standardization in Haematology (Expert Panel on Blood Rheology) (1988) Guidelines on selection of laboratory tests for monitoring the acute phase response. *J Clin Pathol* **41**, 1203–1212.

IUIS/WHO Working Group (1981) Use and abuse of laboratory tests in clinical immunology: critical consideration of eight widely used diagnostic procedures. Report of an expert committee. *Clin Exp Immunol* **46**, 662–674.

IUIS/WHO Report (1988) Laboratory investigations in clinical immunology: methods, pitfalls and clinical indications. *Clin Exp Immunol* **74**, 494–503.

Jefferis, R., Riemer, C.B., Skvaril, F. *et al.* (1985) Evaluation of monoclonal antibodies having specificity for human IgG subclasses: results of an IUIS/WHO collaborative study. *Immunol Lett* **10**, 223–252.

Keyser, J. (1987) *Human Plasma Proteins: Their Investigation in Pathological Condition*, 2nd edn. John Wiley & Sons, Chichester.

Oxelius, V.A., Laurell, A.B., Lindquist, B. *et al.* (1981) IgG subclasses in selective IgA deficiency: importance of IgG2-IgA deficiency. *N Engl J Med* **304**, 1476–1477.

Phimister, G.M. & Whaley, K. (1990) Measurement of complement. In: *Clinical Immunology: A Practical Approach* (eds H.C. Gooi & H. Chapel), pp. 81–109. IRL Press, Oxford.

Riches, P. (1990) Immunochemical investigations of lymphoid malignancies. In: *Clinical Immunology: A Practical Approach* (eds H.C. Gooi & H. Chapel), pp. 111–138. IRL Press, Oxford.

Riches, R.G. & Hobbs, J.R. (1979) Mechanisms in secondary hypogammaglobulinaemia. *J Clin Pathol* **32** (Suppl. 13), 15–22.

SAS Protein Reference Units (1993) *Handbook of Clinical Immunochemistry* (eds A. Milford-Ward, P.G. Riches, R. Fifield & A.M. Smith), 4th edn. PRU Publications, Sheffield.

Sheldon, J., Riches, P.G., Soni, N. *et al.* (1991) Plasma neopterin as an adjunct to C-reactive protein in assessment of infection. *Clin Chem* **37**, 2038–2042.

Thompson, D., Milford-Ward, A. & Whicher, J.T. (1992) The value of acute phase protein measurements in clinical practice. *Ann Clin Biochem* **29**, 123–131.

Thompson, R.A. & Bird, A.G. (1983) How necessary are specific IgE antibody tests in allergy diagnosis. *Lancet* **i**, 169–172.

Walker, S.A., Rogers, T.R., Riches, P.G., White, S. & Hobbs, J.R. (1984) The value of serum C-reactive protein measurement in the management of bone marrow transplant patients. I. The early transplant period. *J Clin Pathol* **37**, 1018–1021.

Walker, S.A., Rogers, T.R., Riches, P.G., White, S. & Hobbs, J.R. (1985) Value of C-reactive protein measurement in the management of bone marrow recipients. *Exp Haematol* **13**, 107.

Whaley, K. (1985) Laboratory investigation of complement disorders. In: *Clinics in Immunology and Allergy* (ed. R.A. Thompson), pp. 407–424. W.B. Saunders Company, London.

Whicher, J., Higginson, J., Riches, P.G. & Radford, S. (1980) Clinical application of immunofixation: detection and quantitation of complement activation. *J Clin Pathol* **33**, 781–785.

WHO (1986) Primary immunodeficiency diseases: report of a WHO scientific group. *Clin Immunol Immunopathol* **40**, 166–196.

Working Group of European Society for Paediatric Gastroenterology and Nutrition (1990) Revised criteria for diagnosis of coeliac disease. *Arch Dis Child* **65**, 909–911.

Wu, J.T., Book, L. & Sudar, K. (1981) Serum alpha fetoprotein (AFP) levels in normal infants. *Pediatr Res* **15**, 50–52.

23: The Essential Trace Elements

P.J. AGGETT

Introduction

Nine elements (iron, zinc, copper, manganese, cobalt, chromium, selenium, molybdenum and iodine) and, possibly, fluoride are thought to be essential. Each has its own specific functions and systemic metabolic control and, therefore, needs to be considered independently. Iron and iodine are considered elsewhere (see Chapters 10 and 22).

Systemic control of trace element distribution, exploitation and homeostasis is effected by their oxidation states and affinities for organic molecules to create selective physicochemical compartments which together achieve specific pathways that deliver the elements to their functional operative sites in appropriate forms and concentrations. The accumulative specificity of these pathways is important because if each step of the metabolic sequence for the cationic trace metals is examined in isolation, numerous interactions amongst the elements can be detected.

In this 'metabolic' context, the essential trace elements can be regarded as forming three groups:

1 Cationic elements such as zinc, manganese and copper. These are transferred and utilized as inorganic ions; they need specific carriers to transfer them across lipid membranes and to maintain their solubility at physiological pH within the intracellular and extracellular environment. Their homeostasis is controlled mainly by the gastrointestinal tract and liver.

2 Anionic elements such as molybdenum, selenium, chromium and possibly fluoride. These have a greater ability to cross lipid membranes spontaneously and are more soluble at physiological pH; these elements have a highly efficient gastrointestinal uptake and transfer, and their systemic utilization and compartmentalization are effected by manipulating their many oxidation states. Their homeostasis is dependent upon renal excretion.

3 Other elements, such as cobalt (in vitamin B_{12}), molybdenum and, possibly, chromium, which have many oxidation states and are metabolized and used in organic complexes.

Table 23.1 summarizes the recommended nutrient intakes (in the UK) for those elements for which enough information exists to enable these values to be set (Department of Health, 1991). A reference nutrient intake (RNI) is calculated to meet the needs of 97% of the relevant population. Thus, failure of a child to achieve an intake matching the RNI cannot be used as indicative of any real, or possible, deficiency. Furthermore, the degree of confidence in the RNI is limited by an inadequate knowledge of the interplay between systemic, luminal and dietary factors which affect the intestinal absorption of trace elements (Table 23.2). This applies particularly to the cationic elements (Aggett, 1991).

Conventional units, where customary, are used in this chapter.

Diagnostic tests

The diagnosis of deficiency or an excess for essential trace elements is difficult. No single test reliably indicates whether an individual is at risk of a deficiency or excess of any particular element. All data need to be interpreted in the context of the homeostatic mechanisms for the elements, an appreciation of the clinical state of the child and of how this would alter the metabolism of the element, and the distribution in tissues such as those selected for analysis.

Table 23.1 UK reference nutrient intakes for trace elements

Age	Zinc (mg/day)	Copper (mg/day)	Selenium (μg/day)
0–3 months	4.0	0.2	10
4–6 months	4.0	0.3	13
7–9 months	5.0	0.3	10
10–12 months	5.0	0.3	10
1–3 years	5.0	0.4	15
4–6 years	6.5	0.6	20
7–10 years	7.0	0.7	30
11–14 years	9.0	0.8	45
15–18 years (males)	9.5	1.0	70
15–18 years (females)	7.0	1.0	60

Table 23.2 Factors influencing the intestinal uptake and transfer of trace elements

Systemic factors
Anabolic demands
- Growth in infancy and childhood
- Pregnancy and lactation
- Post-catabolic states

Endocrine effects
Infection and stress
Specific systemic reserves of metal
Genetic influences, inborn errors of metabolism
Nutritional status for other nutrients

Luminal and dietary
Chemical form and oxidation state of element in the diet
Presence of:
- Antagonistic ligands (phosphate, carbonate, tannates, polyphenols, oxalate)
- Facilitatory ligands (ascorbate (for iron), carboxylic acids, some sugars, amino acids, fatty acids)
- Competing metals

Intestinal redox state
Luminal redox state

The conditions which are associated with an increased risk of deficiency of trace elements are summarized in Table 23.3. In most instances more than one of the listed mechanisms may be active. For example, in malabsorptive states, as well as there being a failure of the intestinal uptake and transfer of the element, there may also be an increased loss of the element arising from catabolism of lean tissue, increased bone turnover and protein-losing enteropathy (Crofton *et al.*, 1990). Clearly, in such patients the intake of elements could also be compromised by an accompanying anorexia. In the discussion of individual elements only specific aspects of deficiencies will be mentioned.

Table 23.3 Mechanisms and conditions which predispose to dietary and systemic deprivations of trace elements

Inadequate intake and absorbability
Malnutrition states
Vegetarianism
Therapeutic and synthetic diets (e.g. enteral and parenteral nutrition, and diets for the management of inborn errors of metabolism)
Intestinal infestation (bacteria, protozoa, helminths)
Nutrient interactions with dietary components and drugs

Maldigestion and malabsorption
Immaturity of absorptive mechanisms
Selective inborn errors of absorption
Surgery: gastric and intestinal resection
Gastric atrophy
Enteropathies
Inflammatory bowel disease
Exocrine pancreatic insufficiency
Biliary obstruction
Hepatic disease

Increased loss and/or degradation
Catabolic states
Protein-losing enteropathies
Renal failure, renal dialysis and diuretic therapy
Chronic blood loss and haemolysis
Chelating agents (specific and non-specific)
Exfoliative dermatoses
Induced hepatic degradation

Increased utilization
Inborn errors of metabolism
Rapid tissue synthesis
Post-catabolic convalescence
Neoplastic disease
Resolving anaemias

Zinc

Zinc has a single oxidation state; this relative stability enables it to stabilize organic polymers, participate in hormone binding to nuclear and cell-membrane

Table 23.4 Some mammalian zinc metalloenzymes

Activity	Role of zinc	Comment
Alcohol dehydrogenase	C,S	Also retinol dehydrogenase
Superoxide dismutase	S	Cytosolic activity
Alkaline phosphatase	C,S	? Intestinal mucosal phytase
Fructose 1,6-bisphosphatase	R,S	Gluconeogenesis
Aminopeptidases	C, (?R)	Hydrolysis of protein
Angiotensin-converting enzyme	C	Specific protease
Endopeptidase	C	Post-translational protein modification; enkephalinase
Collagenase	C	
Carboxypeptidases	C	Probably including folate deconjugase for folate absorption
Carbonic anhydrase	C	Carbon dioxide transport
δ-Aminolaevulinic dehydratase	C	Haem synthesis
Glyceraldehyde-3-phosphate dehydrogenase		Pyridine nucleotide-dependent oxidoreductases
Lactate dehydrogenase	C	Glycolysis
Malate dehydrogenase		

C, catalytic; R, regulatory; S, structural.

receptors, gene transcription factors, and to have a regulatory and catalytic role in enzymes. Numerous zinc metalloproteins have been identified in various species. Some of those relevant to human metabolism are listed in Table 23.4 (Vallee & Galdes, 1984; Coleman, 1992). From this it can be appreciated that zinc participates in gene expression, and in both the control and mechanisms of major metabolic pathways involving the metabolism of protein, carbohydrate, energy, nucleic acids and lipids.

Metabolism

In adults, the body contains about 30 mg zinc/g fat-free mass; this amounts to a total content of 1.4–2.0 g. The tissue distribution of this is shown in Table 23.5. About 90% of the entire zinc content of the body is in muscle, bone, skin and hair: these pools are not effective reserves of zinc. The metal in the latter two tissues probably does not re-enter the body, and the turnover of that in bone and muscle is so slow that it is only released significantly secondary to other causes of muscle catabolism or increased bone turnover and remodelling; the most labile pools are found in the plasma and liver (Jackson, 1989).

Table 23.5 Relative sizes of tissue zinc pools in adults

Tissue	Content (g)	Distribution (%)
Muscle	1.5	60
Bone	0.5–0.8	20–30
Skin and hair	0.21	8.0
Liver	0.10–0.15	4–6
Gastrointestinal tract and pancreas	0.03	2.0
Kidneys	0.02	0.8
Spleen	0.003	0.1
Central nervous system	0.04	1.6
Blood	0.02	0.8
Plasma	0.003	0.1

Zinc is absorbed most efficiently in the proximal small intestine. However, because there is a large enteropancreatic circulation of the metal (two to three times the daily dietary intake) net intestinal absorption of the metal is probably achieved in the distal small intestine (Matseshe *et al.*, 1980). Zinc absorption is increased by dietary ligands, such as protein, amino acids and, possibly, lactose, and it is reduced by complexes with phytate, calcium and magnesium. The interaction between zinc and other intraluminal low-molecular-weight ligands such as lipids and organic acids is unclear (Cousins, 1985). The mechanisms and regulation of the intestinal uptake and transfer of zinc are not known precisely. Both specific and non-specific carrier- and binding-mediated processes are involved.

In the circulation, zinc is bound to albumin, α_2-macroglobulin, low-molecular-weight proteins, and possibly to transferrin and a histidine-rich protein (Morgan, 1981). Normally, more than 80% of plasma zinc is associated with albumin, about 15% with α_2-macroglobulin and less than 1% with amino acids or metalloenzymes. Virtually no zinc would be present in an ionic form.

The uptake of zinc by peripheral cells, such as hepatocytes, fibroblasts and pancreatic acinar cells, shows characteristics of carrier mediation as well as possible passive diffusional non-saturable mechanisms.

Homeostasis at low zinc intakes is achieved by increases in net intestinal absorption and, to a lesser extent, by renal conservation of the element (Taylor *et al.*, 1991). With inappropriately high zinc intakes, the systemic burden is regulated by increased intestinal secretion of endogenous zinc and by reduced mucosal uptake. With particularly high and non-physiological intakes, uptake and transfer of the element is also regulated by the induction, within enterocytes, of a low-molecular-weight cysteine-rich protein, metallothionein, which sequesters zinc within the mucosa and prevents its transfer.

Metallothionein is the generic name for a group of isometallothioneins which are ubiquitous intracellular monomeric polypeptides (relative molecular mass (M_r) 6500) comprising about 60 amino acids, of which 30% are cysteine residues (Coleman, 1992). The function of metallothionein is unknown. Each molecule can bind 6–10 atoms of copper, zinc, cadmium and other metals, but only these three can induce its synthesis. Cadmium is a more effective inducer of metallothionein than is zinc, which in turn is better than copper. Metallothionein is not inducible in zinc deficiency and this suggests that its primary role may be in the metabolism of zinc. Its production is induced also by endotoxaemia, infection, adrenalectomy, oestrogens, glucagon, hypothermia, exercise, starvation and laparotomy. Hepatic synthesis of metallothionein is associated with a hepatocellular accumulation of zinc and with a fall in the plasma concentrations of the metal.

The proposed functions of metallothionein include: (a) the provision of an intracellular zinc depot for the activation of apoenzymes; (b) a regulator of zinc and copper metabolism, which by binding copper more avidly than it does zinc prevents copper from interfering with zinc metabolism; (c) the homeostasis of zinc; (d) a sequestrator protein to protect against potential intracellular damage arising from excessive accumulations of metals such as cadmium and zinc; (e) an intracellular source of cysteine in neonates; and (f) an intracellular source of sulphydryl antioxidant (reducing) groups. Evidently, the protein may have more than one of these functions. Since copper metabolism is not disturbed by the absence of metallothionein in the zinc deficiency, it is possible that this protein does not have a primary role in copper metabolism.

Maturation of zinc metabolism

For a fuller account see Aggett and Barclay (1991).

The concentration of zinc in the term neonate is 20 mg/kg fat-free tissue. The fetus has been calculated to accumulate zinc at a rate of 249 μg/kg daily during the last trimester. The body distribution and metabolism of zinc in the neonate differ from that of the adult. During gestation, although the zinc contents of the heart, kidney and brain remain relatively constant, that in muscle increases from 110 to 160 μg/g dry weight between 20 and 40 weeks of gestation. On the basis of studies in neonatal animal models it is possible that zinc will continue to accumulate in muscles as they become adapted to extrauterine aerobic activity. At birth, the liver contains 25% of the total body zinc and the skeleton contains 40%, compared with 10% and 25% respectively in these tissues in the adult. The hepatic zinc is bound to metallothionein and postmortem studies on human infant liver show that hepatic metallothionein and zinc concentrations decline rapidly to reach constant levels at about 4 months postnatal age (Zlotkin, 1988). This may represent a systemic redistribution of a hepatic zinc reserve. Although many balance studies suggest that preterm, and some term, neonates have net intestinal loss of zinc, studies using stable isotopic markers show that these babies are able, if need be, effectively to increase the uptake of dietary zinc and reduce intestinal losses of endogenous zinc.

Zinc released adventitiously from bone during skeletal remodelling serves as an additional source of the element during infancy. Although systemic redistribution of zinc may meet the requirements of the infant, human breast milk is also an abundant source of the element. Zinc is absorbed more efficiently from human breast milk than from cows' milk or from cows'-milk- or soy-protein-based formulas.

Deficiency

It is noteworthy that tissues with a rapid turnover, such as those involved with cell-mediated immunity, mucosa and skin, are particularly vulnerable in zinc deficiency. Because of the pervasiveness of metabolic roles dependent on zinc, it is not surprising that the features of zinc deficiency (Table 23.6) are so protean. They can resemble the features associated with deficiencies of other essential nutrients, such as amino acids, fatty acids and vitamins.

Table 23.6 Features of zinc deficiency

Infants
Anorexia
Failure to thrive, weight loss
Tremor, jitteriness, hoarseness
Dermatitis (periorificial and extensor), vesiculobullous, pustular, hyperkeratotic, stomatitis, glossitis, paronychia, nail dystrophy
Fine brittle hair, tapered tips, alopecia
Loose frequent stools, malabsorption (disaccharide intolerance)
Increased susceptibility to infection
Additionally in older children
Pica, impaired taste and smell
Height retardation
Depression, mood lability, impaired cerebration
Intention tremor, nystagmus, ataxia dysarthria
Photophobia, night blindness, blepharitis, delayed puberty, hypogonadism

Indeed, features of zinc deficiency can be ameliorated or exacerbated by the supply of these other nutrients (Aggett, 1989). Since zinc is predominantly an intracellular element and is essential for the synthesis of tissue, it may not be a limiting nutrient during catabolic states, but can become so during recovery, as has been seen in patients on total parenteral nutrition or in children recovering from protein energy malnutrition. Malnourished children receiving a marginal zinc supply have impaired synthesis of lean tissue and, as a consequence, an increased energy cost for new tissue deposition representing deposition of adipose tissue (Golden & Golden, 1992).

Zinc deficiency has been described in preterm and term infants. Preterm infants varied between 26 and 34 weeks gestation with birth weights between 710 and 2200 g. Most infants presented at about 3 months of age. With the exception of one report, all had been breast fed. In some cases there had been a preceding period of parenteral nutrition. Severe zinc deficiency in term infants manifested at 3–5 months postnatally. In these cases the zinc content of maternal breast milk was found to be low, but maternal zinc supplementation failed to increase it. The onset of symptoms whilst being breast fed counts against the diagnosis of acrodermatitis enteropathica (AE) (see below) in these babies. Furthermore, they respond to smaller supplements of zinc (5–10 mg elemental zinc daily) than would be expected of a child with AE (Aggett, 1989).

Zinc-responsive defects have been seen in most of the situations outlined in Table 23.3. Responses include improved weight and height gain, improved cellular immune function and healing of skin lesions. Zinc-responsive growth (both height and weight) retardation has been described in infants and preschool children whose only obvious abnormality was short stature (<10th centile) (Anonymous, 1989; Gibson *et al.*, 1989; Walravens *et al.*, 1992).

Inborn errors of zinc metabolism

Acrodermatitis enteropathica

(Aggett, 1989; van Wouwe, 1989)

This is a rare autosomal recessive syndrome of zinc deficiency, arising most probably from a defect in intestinal absorption. Affected infants present during early infancy, but usually after the neonatal period. Babies fed on infant formulas develop syndromes sooner than those fed human breast milk; in the latter group problems do not appear until solids or formulas have been introduced. This difference arises from the more efficient intestinal absorption of zinc from human milk than from other products.

The earliest recognized feature of AE is usually the symmetrical, circumorificial, retroauricular and acral dermatitis which may also involve the cheeks, trunks and limbs. Failure to thrive, anorexia and irritability are the other major presenting features, as may be diarrhoea.

Clinical features may deteriorate during infections and physiological stress, and during the growth spurts of early childhood and puberty. The systemic metabolism of zinc in AE appears to be normal and the relative deficit arising from the intestinal defect is relatively small. As a consequence, the severity of symptoms can vary considerably, particularly after the adolescent growth spurt. Some cases may be undetected during childhood, with the diagnosis being delayed until adolescence or adulthood, when patients present with chronic skin disorders, neuropsychiatric features or a history of abnormal pregnancies.

AE responds rapidly to treatment with oral zinc supplements. Between 35 and 100 mg elemental zinc daily is usually adequate. Zinc sulphate heptahydrate (50 mg or 0.77 mmol of elemental zinc in 220 mg) is used commonly but other salts, such as the acetate, ascorbate or gluconate, or effervescent preparations may be more palatable. Whereas males may not need continued large doses after adolescence, zinc supplements are still advisable, and females should certainly be advised to maintain their supplements, especially if they plan to become pregnant.

Familial hyperzincaemia

In this condition, plasma concentrations may be elevated up to five times normal values. The additional zinc is associated with plasma albumin; affected individuals are asymptomatic and have no other evidence of other disturbances of zinc metabolism (Failla *et al.*, 1982).

Toxicity

Acute toxicity has occurred following the drinking of water and other beverages which have been stored in galvanized containers, or after such water has been used for renal dialysis, and following ingestion of zinc supplements and zinc alloys. Symptoms included anorexia, nausea, vomiting, bleeding gastric erosions, diarrhoea, dizziness, lethargy and fever. In adults such features occur after acute ingestion of 2 g or more of the element. Prolonged intakes of 75–300 mg/day have been associated with impaired copper utilization, manifested by microcytic anaemia and neutropenia. Intravenous overdosage has progressed to acute renal failure and death. Even short-term intakes of 50 mg zinc daily have been shown to interfere with the metabolism of both iron and copper in adults (Yardrick *et al.*, 1989).

Diagnostic tests

There is no single reliable indicator of zinc deficiency. Plasma and serum contain between 10 and 20% of circulating zinc and their zinc content (reference range about 9–22 μmol/L) is often used to screen for deficiency. However, it is noteworthy that these values are depressed by infections and other stresses, such as those which induce the synthesis of metallothionein. Nonetheless, these factors do not depress levels to those seen in severe hypozincaemic zinc deficiency, and values below 8 μmol/L are strongly suggestive of zinc deficiency. It should also be noted that because of the release of zinc from tissue catabolism, plasma and serum zinc levels can be within the reference range even though the patient has clinical and biochemical features of zinc deficiency (Mack *et al.*, 1989).

Plasma and serum zinc concentrations are similar. Previously reported discrepancies between the two have now been attributed to delays in separating the fluids from the blood cells and in the use of citrate, rather than deionized heparin anticoagulant, for plasma collection. Plasma and serum zinc concentrations are increased by collection with venous stasis, and care must always be taken to avoid haemolysis and contamination of the samples by rubber. Samples collected after fasting give higher values than those taken after meals. Assays of leucocytes are difficult to perform reliably and to interpret in attempts to detect deficiency. The various subsets of the leucocyte population have different half-lives and zinc contents. Consequently, assays of mixed leucocytes will be heavily influenced by other factors which affect the composition of the circulating white cell population. Analyses on selected separated subsets may prove to be more useful but are, for the moment, too impractical for routine or emergency use.

Concentrations of zinc in erythrocytes are also difficult to interpret and, in isolation, are not helpful clinically; neither are those in urine and hair, unless zinc toxicity is suspected, but even so there are no adequate reference ranges with which to compare data. If zinc deficiency is suspected, the best policy is to assess the plasma concentrations in the context of the patient's condition and monitor the clinical and biochemical (e.g. alkaline phosphatase activity) response to zinc supplements (2 mg or 930 μmol zinc/kg body weight daily).

Copper

The transition between Cu(I) and Cu(II) enables the element to participate in a variety of electron transfer enzyme activities (Table 23.7). The features of copper deficiency are largely attributable to impairment of these activities (Danks, 1988; O'Dell, 1990). Caeruloplasmin, a glycoprotein with a molecular weight of 135 000, contains six copper atoms which are incorporated during its hepatocytic synthesis. Most of the circulating copper is bound to caeruloplasmin, but the precise function of the metallocomplex is unknown. Caeruloplasmin is an efficient donor of copper to apoenzymes. Cellular mechanisms have been described for the endocytic uptake of caeruloplasmin, the release of some of its copper, and the re-excretion of the protein. Caeruloplasmin has numerous oxidase activities, substrates for which include biogenic amines, adrenaline, serotonin, ascorbate and sulphydryl groups, as well as Fe(II) for its incorporation as Fe(III) into transferrin. It has also been proposed that caeruloplasmin may serve as a free radical scavenger.

Adults contain 80–120 mg copper; much of this is in the liver, brain and muscle (Table 23.8). The daily intake of copper is 1–2 mg. Absorption of copper from free solution occurs predominantly in the small intestine and, as with other elements, a specific saturable carrier-mediated process is involved as well as a non-saturable non-specific process. Luminal ligands, such as glutathione, cysteine, lactose, starch and glucose, are thought to facilitate the intestinal absorption of copper. However, it is not known whether any specific cotransport pathways exist for this. Newly absorbed copper is transported on albumin, in binary complexes with low-molecular-weight ligands, such as amino acids (histidine, threonine and glutamine). Two other

Table 23.7 Some mammalian cuproenzyme activities

Enzyme	Comment
Cytochromic oxidase	Mitochondrial; requires iron; oxidative phosphorylation
Superoxide dismutase	Cytosolic antioxidant: $2O_2^- + 2H^+ \rightarrow H_2O_2 + O_2$
Dopamine-β-monooxygenase	Synthesis of epinephrine and norepinephrine noradrenergic tissues
Tyrosinase	Tyrosine → dopa → dopaquinone in pigment production in choroid and epidermis
Uricase	Renal and hepatic metabolism of uric acid
Lysyl oxidase (and related enzymes)	Oxidative deamination of peptidyl-lysine residues; condensational cross-link formation in elastin and collagen
Amine oxidases	Plasma and connective tissues
Thiol oxidase	Formation of disulphide linkages
Caeruloplasmin	Multiple activities
Ferroxidase II	Fe(II) to Fe(III); ?vascular compartment

Table 23.8 Approximate tissue copper content in adults, infants and Menkes' syndrome

	Copper content (μg/g wet weight)		
Tissue	Adults	Infants	Menkes' syndrome
Placenta	–	4.1–7.5	8.3–14.5
Liver	4.2–16.9	29.5–78.7	2.8–11.8
Brain	3.6–7.5	0.27–1.20	0.17–1.04
Intestine	1.2–3.4	4.1–7.5	6.4–12.4
Muscle	0.6–1.4	0.25–1.02	1.7–2.6
Spleen	0.90–1.68	0.6–1.9	6.4–15.4
Kidney	2.10–3.74	0.5–1.9	5.9–36.8
Lung	1.02–1.98	0.35–1.00	1.8–4.6

proteins, an intermediate-sized protein (M_r 280 000) called transcuprein and a histidine-rich protein (M_r 60 000), have also been implicated in the vascular transport of copper.

Hepatocytic uptake of copper occurs via specific hepatocytic membrane-binding sites which are independent of those involved with the uptake of copper from desialylated caeruloplasmin returning from peripheral tissues (Kressner *et al.*, 1984).

There are at least three major hepatic pools of the metal: (a) a pool involved in the production of caeruloplasmin; (b) a presumed storage depot; and (c) a pool destined for biliary excretion. In addition, there is a small functional pool of copper enzymes and one in which copper is bound to metallothionein. Exchange of copper occurs between all these pools, except the biliary excretory pool, which is the main homeostatic mechanism. Plasma contains about 3 mg copper, of which 60–70% or more is bound to caeruloplasmin, 15–20% to albumin, about 10% to transcuprein and a similar amount, or a little less, with low-molecular-weight ligands.

Maturation of copper metabolism

(Aggett & Barclay, 1991)

The fetus accumulates about 51 μg (0.8 μmol) copper/kg body weight daily, increasing its copper content from 3.5 mg/kg fat-free tissue at 20 weeks gestation to approximately 4.6 mg/kg at term, when the total body copper content is about 20 mg. The proportionate distribution of this copper in the neonate differs from that in adults. At term, the liver contains 10–12 mg copper (i.e. 50–60% total body copper). Much of this copper is bound to metallothionein and is localized in the lysosomal fraction; this is in contrast to the more nuclear and cytoplasmic distribution of zinc metallothionein. Assuming that the neonate requires about 25 μg copper/kg body weight daily, the term neonate has sufficient copper stores for up to 6 months, whereas a preterm neonate has sufficient for only half this period. As the hepatic copper content declines during infancy the adult pattern of systemic distribution is acquired.

The efficiency of copper absorption from infant formulas may be less than that from human milk because of their different protein and mineral compositions.

In the newborn, plasma concentrations of copper and caeruloplasmin are low. They gradually increase to adult values during infancy (Tables 23.9 & 23.10). They are lower in preterm than in term neonates and are related to post-conceptional age (Halliday *et al.*, 1985; Sutton *et al.*, 1985). This fact may represent developmental maturation of the hepatic synthesis of caeruloplasmin, since immunoreactive apocaeruloplasmin concentrations are also low. However, this phenomenon does not appear to interfere with the systemic distribution and metabolism of copper.

Deficiency

Features of copper deficiency are shown in Table 23.11 (Danks, 1988).

In general, the conditions which predispose to copper deficiency are preterm delivery, very low birth weight, total parenteral nutrition with inadequate copper supplements, malnutrition and malabsorption syndromes, and the use in infancy of inappropriate diets such as unmodified cows' milk, or a combination of these factors (Hillman, 1981). In an adult, alkali therapy has caused copper deficiency.

Infants with copper deficiency present between 4 weeks and 8 months of age. The median ages of presentation are about 3 months and 6 months postnatal age for preterm and term infants, respec-

Table 23.9 Values (μmol/L) for plasma or serum copper, by post-conceptional age

Post-conceptional age (weeks)	Barclay *et al.* (plasma)*	Sutton *et al.* (plasma)	Halliday *et al.* (serum)	Hillman (serum)
25–28				4.5 (2.7)
29–30	3.7 (2.8–4.8)	5.5 (1.9–15.8)		4.3 (2.3)
31–32	4.8 (4.2–5.4)	5.6 (2.5–12.4)	5.9 (2.8–12.4)	4.9 (2.3)
33–34	5.1 (4.5–5.8)	5.5 (2.1–14.6)		5.7 (2.3)
35–36	5.8 (5.1–6.6)	6.1 (3.0–12.4)		6.1 (2.2)
			7.8 (3.8–16.0)	
37–38	6.3 (5.7–7.0)	7.3 (4.3–12.4)		7.4 (1.2)
39–40	7.6 (6.4–9.1)			8.2 (1.8)
41–42	8.9 (8.0–9.8)	9.8 (6.9–13.9)	11.1 (6.7–18.4)	9.4 (3.3)
43–44	9.4 (8.4–10.5)	10.2 (6.2–16.7)		11.0 (4.4)
45–46	11.4 (10.4–12.5)	11.5 (7.4–17.9)		10.2 (2.5)
			12.5 (8.1–19.3)	
47–48	11.00 (9.8–12.4)	13.9 (6.9 – 28.1)		12.8 (2.8)
49–50	11.5 (10.0–13.1)			
51–54	12.4 (11.2–13.6)		13.5 (9.9–18.4)	
55–59	12.9 (11.3–14.6)		16.0 (9.1–28.2)	

Mean and 95% confidence level, except for Hillman's data, which are mean and SD.
* Unpublished data.

Table 23.10 Plasma caeruloplasmin and copper concentrations (μmol/L) in healthy infants

Postnatal age (months)	Caeruloplasmin*	Copper
Birth	0.90 (0.07–2.24)	4.57 (2.05–10.9)
2	1.64 (0.52–3.58)	11.2 (4.6–21.7)
4	2.09 (1.04–4.25)	13.1 (6.9–22.0)
6	2.54 (1.19–5.97)	15.28 (8.03–25.2)
12	3.21 (1.64–5.90)	19.7 (10.4–32.9)

Results are the mean and range (in parentheses).
* SEM at each postnatal age was 0.07 μmol/L.

tively. None have presented during the first month of life and there have been no reports of copper deficiency occurring in exclusively breast-fed or appropriately formula-fed term and preterm infants (Anonymous, 1987).

Because of their variable and increasing plasma

Table 23.11 Clinical features of copper deficiency

Failure to thrive
Pallor
Hypothermia, apnoeic attacks
Hypotonia, poor feeding
Skeletal changes (radiographic generalized and symmetrical)
  Osteoporosis, fractures
  Metaphyseal irregularities, flaring and cupping, spurs and chip fractures
  Epiphyseal porosis and separation
  Periosteal reaction and subperiosteal new bone formation
  Wormian bones, retarded bone age
Abnormal elastic and connective tissues, hernias, tortuous vasculature, varices and aneurysms
Sideroblastic anaemia
  Bone marrow: maturation arrest of erythroid and myeloid series
  Vacuolate cells, ringed sideroblasts
  Altered iron metabolism
  Hypochromic anaemia, anisocytosis, microcytosis
  Neutropenia ($<1 \times 10^9$/L)
Fish odour (?trimethylaminaemia)
Hypocupraemia, hypocaeruloplasminaemia
Hypoproteinaemia with oedema

copper and caeruloplasmin concentrations the reference ranges may have a limited value in diagnosing suspected deficiencies in babies. The diagnosis will ultimately depend on monitoring the clinical, haematological and biochemical response to a therapeutic trial of copper (copper acetate or copper sulphate; 2–5 μmol elemental copper/kg body weight daily). A reticulocytosis within 4–7 days of starting treatment is an early indicator for positive response to copper therapy. Radiological resolution of skeletal changes is normally apparent after 3 weeks of treatment and retarded bone age is one of the last features to resolve.

As with zinc, in patients recovering from catabolism copper may not be a limiting nutrient until other major nutrients are supplied and net tissue synthesis has been established (Castillo-Duran *et al.*, 1983).

Inborn errors of copper metabolism

Menkes' syndrome

This is an X-linked recessive disorder characterized by hypocupraemia, hypocaeruloplasminaemia and features of gross copper deficiency (Danks *et al.*, 1972). There are reduced concentrations of copper in the brain and liver, whereas in other tissues the level is increased or normal (Table 23.8). The Menkes' gene is located in the subregion of band Xq-13.3 proximal to the *PGK1* locus (Verga *et al.*, 1991) and has been found to code for a protein which is homologous with membrane-bound cation-transporting adenosine triphosphatase (ATPase) proteins (Davies, 1993). Various deletions of the gene occur and these may contribute to the heterogeneity of this syndrome and, possibly, to other defects associated with deranged metabolism of copper. However, it is not yet clear how a defect in a copper-transporting protein can create a condition in which a cellular excess of copper is associated with a functional copper deficiency.

The condition affects all races and has a calculated incidence in western European countries of 1 per 254 000 live births. Most cases present by 2 months of age. Presenting features include the slowing of development, the loss of milestones, such as smiling and social responsiveness, convulsions, apnoea, infections and failure to thrive. Many have non-specific histories of low Apgar scores, neonatal hypotonia, hypothermia and prolonged jaundice. At birth, the hair may be normal; characteristic changes of depigmentation, irregular calibre (monilethrix), node-like fracture points (trichorrhexis nodosa) and axial twisting (pili torti) are evident at presentation. Affected babies have placid expressionless faces with frontal and orbital bossing. In addition, they have micrognathia, 'cupids bow-lip', gingival hyperplasia and a high arched palate. Secondary scalp hair and eyebrows may be sparse. At birth, plasma copper and caeruloplasmin concentrations may be normal or even elevated compared with those normally seen in term neonates: levels decline or remain within normal neonatal limits until about 14 days of age but they do not subsequently rise to adult concentrations. When the diagnosis of Menkes' syndrome is suspected, it can be confirmed by demonstrating hypocupraemia (e.g. less than 10 μmol/L at 2 months), hypocaeruloplasminaemia and a low hepatic copper content on needle biopsy. The usual range in early infancy is 50–150 mg/g (0.2–0.4 mmol/g) dry weight, and the demonstration of an increased uptake and retention *in vitro* of copper by lymphocyte or fibroblast cultures.

Prenatal diagnosis can be achieved by measuring the elevated copper content and uptake of radiocopper in amniocytes, cultured chorionic villi and fibroblast cultures. These are techniques which are best referred to specialist laboratories (Tonnesen *et al.*, 1987). A genetic diagnostic technique may soon become available.

Prognosis

There is probably no effective treatment for classical Menkes' disease. Parenteral copper salts have been found to correct hypocupraemia and hypocaeruloplasminaemia, and to induce a cupruresis. They also prolong life and modify some features, but they are less successful in preventing neurodegenerative changes. However, because the condition has some variability it has been suggested that parenteral copper-histidine, supplemented by oral D-penicillamine, may be of benefit to early-treated patients with Menkes' disease (Nadal & Baerlocher, 1988).

Some heterozygotes have features of the disorder. Many have abnormal scalp hair and altered lymphocytic metabolism of copper. Occasionally, typical Menkes' syndromes occur in females (Verga *et al.*, 1991). Variant forms of the condition have been reported; some severely affected patients have a long survival, and other males present later and with less severe impairment of growth and intellect and less obvious hair abnormalities, but with neurological defects such as ataxia, choreoathetosis, speech difficulties, hypotonicity of the upper limbs, hyperreflexia and myoclonus: all have less severe, but nonetheless characteristic, abnormalities of copper metabolism (Gerdes *et al.*, 1988; Inagaki *et al.*, 1988; Nadal & Baerlocher, 1988).

Familial benign copper deficiency

This term has been applied to a 21-month-old boy who presented with failure to thrive, fits and hypotonia, frequent infections, a hypochromic anaemia and skeletal changes but with normal psychomotor development and a normal white blood cell count. He was hypocupraemic (7–8 mmol/L), although his immunoreactive serum caeruloplasmin concentration was normal. The child's mother and a paternal uncle also had features consistent with copper deprivation. An autosomal dominant form of inheritance has been proposed (Mehes & Petrovicz, 1982).

Wilson's disease (hepatolenticular degeneration)
(Sternlieb, 1990)

This syndrome represents a systemic copper toxicosis in which copper accumulates initially in the liver but subsequently becomes dispersed throughout the body. The affected allele for this recessive defect in copper metabolism is on chromosome 13. The hepatic incorporation of copper into caeruloplasmin and the biliary excretion of copper are impaired. The basic metabolic defect is unknown, but the characteristic hypocaeruloplasminaemia and hypocupraemia, together with high hepatic copper concentrations, have stimulated suggestions that the disease reflects an abnormal persistence of fetal and neonatal copper metabolism (Bingle *et al.*, 1990). The incidence of Wilson's disease is between 1 in 200 000 and 1 in 300 000. The onset of symptoms is variable and protean; consequently, unless the disease is considered whenever it is conceivable, the diagnosis is often missed. Frequently, the onset is chronic and insidious, but many such patients when diagnosed are found to have had frequent transient icteric episodes. The diagnosis is also overlooked with an acute presentation as hepatic and renal failure accompanied by haemolysis. This latter syndrome rises from an acute release of the free non-caeruloplasmin-bound copper from the liver. This acute fulminant presentation may arise from the additional stress of intercurrent infections. Wilson's disease should be considered in any individual with recurrent jaundice or haemolysis.

The features of Wilson's disease are summarized in Table 23.12. Children and young adults usually present with hepatic phenomena. Neurological presentations are rare, but not unknown, before adolescence, and frequently children with no gross neurological phenomena have histories of deteriorating school performance.

Kayser–Fleischer rings are caused by the deposition of copper in Descemet's membrane. In children, they may be detectable only by slit-lamp microscopic examination. They appear as brown to green discolorations, first at the upper limbus and then at the inferior limbus before they meet bilaterally to encircle the cornea. It is not unusual for these rings to be absent in patients presenting with hepatic involvement, particularly with acute hepatic disease.

The plasma concentration in most patients is <10 μmol/L and is accompanied by hypocaeruloplasminaemia (<200 mg/L). However, in 15% of cases caeruloplasmin levels may be in the lower end of the reference range. Occasionally, subnormal caeruloplasmin concentrations may be elevated into this range by acute-phase reactions. Since in Wilson's disease much of the plasma pool of copper is not bound to caeruloplasmin, it can pass the glomerular membrane, and the urinary excretion of

Table 23.12 Clinical features of Wilson's disease

Hepatic
Transient hepatitis with intercurrent infections; subacute, acute or chronic hepatitis; cryptogenic cirrhosis; hepatosplenomegaly; portal hypertension; oesophageal varices; pancreatitis; fulminant hepatic failure

Neurological
Encephalopathy; clumsiness; deteriorating handwriting and school performance; ataxia; fine tremor in limbs and/or tongue; dysarthria; dysphagia; dystonia; spasticity; athetosis; muscle spasms; coarse flapping tremor at wrists and shoulders; peripheral neuropathy; epilepsy (partial or generalized); pseudo-bulbar palsy (death); 'schizophrenia'; fatuous drooling facies; non-specific electroencephalogram abnormalities

Ophthalmic
Kayser–Fleischer rings; sunflower cataract; strabismus; xerophthalmia; impaired visual acuity; pallor of disc; night blindness

Haematological
Acute haemolysis; haemolytic anaemia; thrombocytopenia; leucopenia; pancytopenia; coagulopathies; bruising; epistaxes; intravascular coagulation

Renal
Renal tubular acidosis; concentration defect; amino aciduria; phosphaturia; hypercalciuria; uricosuria; acute renal failure; renal stones; haematuria

Skeletal
Osteoporosis; patchy osteosclerosis; osteophytes; cysts in long bones; spontaneous fracture; osteomalacia; chondrocalcinosis; osteoarthritis; chondromalacia patellae; osteochondritis dissecans; chondrocalcinosis; arthritis; morning stiffness

Endocrine
Amenorrhoea; increased incidence of miscarriages; gynaecomastia; abdominal pain; colic; bacterial peritonitis; blue lanulae of finger nails

copper is usually elevated to more than 100 μg/24 h compared with usual values of 50 μg/24 h.

Percutaneous needle biopsy is essential because it enables both histological examination of the hepatic parenchyma and direct estimation of the hepatic copper content: usually in Wilson's disease this is greater than 250 mg (3.9 mmol) per gram dry weight compared with normal levels of less than 50 mg (0.8 mmol) per gram dry weight. The diagnosis of Wilson's disease is difficult in patients with hepatic failure and there is a rapid onset of icterus, encephalopathy, coagulopathy, renal failure and, as a result of hepatic necrosis, a serum copper which is within the normal range. There may be several historical clues to the correct diagnosis. In addition, because of a preceding progressive replacement of functional liver tissue by fibrosis, patients frequently have inappropriately small elevations, if any, in serum aminotransferase and alkaline phosphatase activities. Values of less than 2 for the ratio of alkaline phosphatase : total bilirubin and greater than 4 for the aspartate transaminase : alanine transaminase ratio (Berman *et al.*, 1991), and the presence of cirrhosis in biopsy, if one can be undertaken, are suggestive of fulminant hepatic failure secondary to Wilson's disease.

Secondary copper accumulation occurs in conditions, such as cholestasis, primary biliary cirrhosis and chronic active hepatitis, which prevent the hepatobiliary excretion of copper. In some of these disorders microscopically detectable Kayser–Fleischer rings might develop, as may some neurological sequelae.

To detect asymptomatic homozygotes, all family members, including cousins, of known cases should be screened. The serum caeruloplasmin concentrations of the patient's unaffected parents (i.e. presumed heterozygotes) give an indication of the serum caeruloplasmin concentrations to expect in heterozygous relations. About 10% of heterozygotes have hypocaeruloplasminaemia and, ideally, they should be investigated by hepatic biopsy. The youngest reported age of presentation of Wilson's disease is 4 years, but hepatocellular damage has been found at 1 year of age in an asymptomatic sibling of a known case. Increasingly, polymorphic deoxyribonucleic acid (DNA) markers will be available to detect potential heterozygotes and homozygotes with Wilson's disease in families at risk; this approach should also facilitate antenatal diagnosis and diagnosis in presenting cases (Farrer *et al.*, 1991). Heterozygotes cannot be presumed to be free of any risk of copper toxicosis because some have been found to have hypocaeruloplasminaemia, hypocupraemia, non-specific electroencephalogram (EEG) abnormalities and increased body copper burdens (Maracek & Nevsimalova, 1984).

The treatment of Wilson's disease is aimed at reducing the amount of copper in the body and, if necessary, preventing the toxic effects of free copper in the tissues and circulation. The most commonly used agent is the chelator D-penicillamine (dimethyl cysteine). Other treatments include use of the polyamine chelator trientine (triethylene tetra-amine) (Dubois *et al.*, 1990), unithiol and ammonium tetrathiomolybdate (Brewer *et al.*, 1991). The latter compound is thought to block copper absorption as well as form non-toxic complexes with the element systemically; it is well tolerated and may emerge as the favoured primary treatment. Large oral doses of zinc (150 mg elemental zinc daily) are a useful secondary therapy. This has been attributed to the ability to induce the synthesis of metallothionein, which sequestrates copper both systemically and in the gut mucosa, thereby reducing its toxicity (Veen *et al.*, 1991).

The efficacy of chelation therapy can be monitored by measuring urinary copper excretion. As a rule of thumb, this can be maintained in adults at up to 5 mg copper daily. With adequate treatment the progress is good (Schilsky *et al.*, 1991).

Familial hypocaeruloplasminaemia

Familial hypocaeruloplasminaemia, an autosomal recessive hypocaeruloplasminaemia, has been described in asymptomatic kindred in which affected individuals had a low serum copper concentration but normal urinary copper excretion. They had normal hepatic histology and copper content and no features of copper deficiency (Edwards *et al.*, 1979).

Other hypocaeruloplasminaemic neurological syndromes

A variety of other neurological syndromes have been described in which the aetiological role of disturbed copper metabolism is less clear than that in either Wilson's or Menkes' disease. These include syndromes involving dementia, dysarthria, gait disturbances, involuntary movements and low levels of serum copper and caeruloplasmin, but in which copper metabolism does not resemble that seen in Wilson's disease (Willvonseder *et al.*, 1973; Godwin-Austen *et al.*, 1978; Ono & Kurisaki, 1988).

Chronic dietary or idiopathic copper toxicity

Copper toxicity arises from the deliberate ingestion of copper salts or accidentally from contamination of water and drinks. Acutely, the gastrointestinal tract is affected and there is vomiting and diarrhoea. Variable degrees of intravascular haemolysis may occur, and with gross toxicity hepatocellular and renal tubular necrosis develops with hepatorenal failure. Death may ensue.

With chronic exposure, copper accumulates in the liver and the toxicity is insidious. Hepatic necrosis and cirrhosis with liver failure may develop. An example of this is Indian childhood cirrhosis, which presents at 1–3 years of age and possibly results from the ingestion of milk which has been stored in copper or brass utensils (Bhave *et al.*, 1992). A similar syndrome in western infants may arise from the consumption of feeds which have been prepared from acidic well water that contains leached copper from pipes (Spitalny *et al.*, 1984; Muller-Hocker *et al.*, 1988).

Diagnostic tests

Reference ranges for plasma and serum copper concentrations approximate to 11–22 μmol/L for men. Values in women are approximately 10% higher. Furthermore, plasma copper concentrations are increased in late pregnancy and in women taking oestrogens. Concentrations also increase in the acute-phase response. A plasma copper concentration of less than 10 mmol/L should alert one to the possibility of copper deficiency. Hypocupraemia and hypocaeruloplasminaemia occur in malnutrition, malabsorption, protein-losing enteropathies, nephrotic syndrome, advanced hepatic disease, Wilson's disease, Menkes' disease and other inborn errors of copper metabolism as well as in copper deficiency.

Erythrocyte Cu, Zn-superoxide dismutase activity

is a valuable assay in cohort studies, but its value as an indicator of copper depletion for individual patients is uncertain, as is that of hair copper concentrations. The hair copper content is increased in individuals with systemic copper overload. However, in copper-deficient infants, children with protein-energy malnutrition and those with Menkes' syndrome, the hair copper content is normal or even increased relative to reference groups.

Urinary copper excretion is a useful monitor for the treatment of copper overload, and urinary copper levels are reduced in patients on inadequate copper supplements with total parenteral nutrition.

Selenium

Selenium, as selenocysteine, is essential for at least two mammalian oxidase enzymes. One is the cytosolic antioxidant glutathione peroxidase (GSHpx), which uses glutathione to reduce a variety of organic hydroperoxides (e.g. hydrogen peroxide and hydroperoxides of sterols, steroids, prostaglandins, pre-fatty acids, proteins and nucleic acids) to the corresponding alcohol (Levander, 1987). Selenium is also a component of the hepatic microsomal type 1 iodothyronine 5′-deiodinase (Arthur *et al.*, 1990) in rats, which raises the possibility that selenium deprivation may influence the systemic response to marginal iodine intake in humans. Other selenoproteins have been isolated from mammalian tissues; one may be essential for the normal morphology of mammalian sperm (Watanabe & Endo, 1991).

Customary adult daily intakes of selenium vary between 20 and 300 μg, according to the selenium content of the soil from which the foods are derived; for example, in the People's Republic of China, dietary intakes range from 11 to 5000 μg/day, at which extremes deficiency and toxicity syndromes occur (Yang *et al.*, 1988). Such diversity is not customarily encountered elsewhere, although large regional variations in intake occur throughout the world; intakes in New Zealand, Egypt and Finland, for example, are low (Levander, 1987). Inorganic selenium is absorbed efficiently (60–80%) in the small intestine. Seleno-amino acids, i.e. selenocysteine and selenomethionine, are taken up by similar energy-dependent sodium cotransport processors as their sulphur analogues.

Two selenium pools exist in tissues; the biologically active pool of selenium is that in selenocysteine which can be synthesized endogenously from organic selenium and serine and for which a specific transfer ribonucleic acid (tRNA) exists. The other pool is selenomethionine in protein; this pool is subject to factors which influence methionine metabolism, and its constituent selenium is not reliably available for selenium-dependent processes. For example, when methionine intake is inadequate, the seleno-analogue is used instead, even if there is a simultaneous selenium deficiency. However, if the methionine supply is adequate, selenium released from degraded selenomethionine is then available to enter the selenocysteine synthetic pool.

Systemically, seleno-amino acids can be degraded to yield amino acid residues and selenite. Inorganic selenium anions are sequentially reduced from selenate via selenite to selenide by glutathionine reductase systems, possibly including those in erythrocytes. Selenide can be further reduced and methylated to give a variety of methylselonium compounds. These various derivatives are excreted in the urine and this is the principal systemic homeostatic mechanism for selenium. With excessive intakes of selenium a volatile dimethylated compound, $(CH_3)_2Se$, is formed and causes a characteristic garlic odour when exhaled. With low dietary intakes of selenium, homeostatic retention of the element is achieved by increased intestinal absorption and reduced urinary excretion.

In the plasma, most selenium is associated with α_2- and β-globulins and with proteins, one of which, seleno-protein P, may be involved specifically with selenium uptake and transfer of the element to tissues (Motchnik & Tappel, 1990).

Deficiency

The risk of selenium deficiency depends on the relative activity of other potential antioxidants, such as vitamins C, E and A, and oxidant stress imposed by exercise, infection and the intake of oxidizable substrates such as polyunsaturated fatty acids. The most striking selenium-responsive disease in humans is Keshan disease. This is a cardiomyopathy which predominantly affects children, young adolescents and women in the People's Republic of China. Populations with daily intakes below 12 μg are particularly at risk; at intakes above 19 μg/day the risk is minimal. Mild cases respond to selenium supplements, and population prophylaxis with sodium selenite (0–5 years: 0.5 mg; 6–10 years: 1.0 mg weekly) has considerably reduced the incidence of the disease (Yang *et al.*, 1988).

In the West, similar selenium-responsive cardiomyopathies have been described in patients on total parenteral nutrition. In such circumstances less severe deficiencies have also been observed. These involve skeletal myopathy with increased plasma creatinine kinase activities, macrocytosis and lightening of hair pigmentation. The patient's erythrocytes have an increased sensitivity to peroxide haemolysis, reflecting impaired GSHpx activity. Similar changes have been reported with under-nutrition (Mathias & Jackson, 1982). In animal models, selenium deficiency has been associated with defective microsomal oxidation of xenobiotics and disturbed cellular immunity (Dhur *et al.*, 1990); it is not yet known whether these features are reproduced in humans.

Diagnostic tests

The risk of selenium deficiency or excess can be monitored by determining selenium concentrations in the plasma, serum or whole blood, or by determining plasma or erythrocytic glutathionine peroxidase activity. Blood and plasma selenium concentrations reflect an individual's reserve or intake of selenium over a period of relatively constant intake. However, because with increases in intake these values increase rapidly before the selenium is used functionally, they are less helpful than a functional assay based on GSHpx. Plasma GSHpx is thought to be of hepatic origin and a reliable indicator of short-term selenium supply. Erythrocytes and whole blood levels reflect the adequacy of selenium over a longer period.

Because of the geographical variations in selenium intake, there is no universally applicable reference range for plasma, serum or whole blood selenium concentrations; furthermore, values inconsistent with local reference ranges cannot be presumed to represent significant risks of deficiency or toxicity. In addition, there are age-related trends. Values of serum and whole blood selenium concentrations fall shortly after birth to levels that are approximately 30–50% of those seen in adults. This reflects the low selenium content in infant formulas: breast-fed infants have higher values. When weaning is introduced at 3 to 5 months of age the selenium concentrations gradually increase to those of the adult range.

Low selenium concentrations are seen in children with inborn errors of metabolism and who are being treated with synthetic diets. However, these levels are not as low as those seen in patients with symptomatic selenium deficiency, in whom plasma selenium concentrations approximate 0.11 μmol/L. Whole blood threshold levels at which symptomatic selenium deficiency develops have not been well defined. In New Zealand, where the habitual intake of selenium is low, whole blood values of 0.8–0.9 μmol/L are found. Patients with Keshan disease have whole blood selenium values of less than 0.27 μmol/L, whereas unaffected individuals living in the same area have whole blood values of 0.34 μmol/L. Reported values in children on synthetic diets, and who have shown no obvious deficiency features, are 0.13–0.34 μmol/L, with a similar temporal change during infancy as the diet is diversified.

Selenium excess

This is endemic in areas where selenium is geochemically abundant; sporadic cases may arise from supplementation. Clinical features include malodorous breath, an erythematous bullous pruritic dermatitis, dystrophic nails, dry brittle hair, alopecia and neurological abnormalities with neuropathies: discoloured tooth enamel and caries are endemic in affected areas. These features have been described in areas where daily intake is 3.2–6.7 mg selenium. Disturbed metabolism of selenium occurs at intakes above 750 mg daily, and early features of nail dystrophy have been associated with intakes of 900 mg daily (Yang *et al.*, 1989).

Manganese

Manganese is a component of arginase, pyruvate carboxylase and mitochondrial superoxide dismutase. It may also be needed for various hydrolase, kinase, decarboxylase and phosphotransferase activities and for glutamine synthetase. These activities may be more dependent *in vivo* on magnesium, but phosphoenolpyruvate carboxylase, prolidase and glycosyl transferases specifically require manganese (Hurley & Keen, 1987).

Manganese is absorbed throughout the small intestine with an approximate efficiency of 10% but this can increase on low manganese intakes. High levels of dietary calcium, phosphorus and phytate impair the intestinal absorption of manganese, but these are probably of little significance because, as yet, there have been no well-documented cases of human manganese deficiency.

Systemic homeostasis of manganese is maintained by hepatobiliary excretion. It is also lost in intestinal and pancreatic secretions. With low intakes there is an increased efficiency of manganese absorption, and renal excretion of the element is reduced. Studies in animal models suggest that neonates have a delayed hepatobiliary excretion of manganese and an increased intestinal uptake and transfer of the element. This probably also applies to humans and there is some, as yet unsubstantiated, concern that this may predispose infants on soy-based formulas to an excessive intake of the metal.

Evidence of manganese deficiency in humans is poor (Anonymous, 1988a). In animal models, manganese deprivation has produced growth retardation, impaired cartilage formation and defective endochondrial osteogenesis, impaired glucose tolerance and insulin secretion, reduced gluconeogenic response to glucagon and adrenaline, hypocholesterolaemia, and hepatic and renal lipidosis. Hence, interest in possible human manganese deprivation has been stimulated by reports of manganese-responsive carbohydrate intolerance, reduced manganese concentrations in the hair of some mothers whose babies had congenital abnormalities, reduced contents in the blood or hair, or both, of children with skeletal abnormalities, Perthes' disease, osteoporosis and epilepsy. In contrast with this, men fed a low-manganese diet (10 μg daily) developed only a skin rash and hypocholesterolaemia, neither of which clearly responded to manganese supplementation (Freeland-Graves *et al.*, 1988).

Manganese excess

Mine workers exposed to manganese or dust develop 'manganic madness'. This comprises psychosis, hallucinations and extrapyramidal features suggestive of parkinsonism. Manganese toxicity of dietary origin has not been well documented. However, an increased incidence of parkinsonian-like features has been described in adults in areas where the local well water has a high manganese content (Kondakis *et al.*, 1989).

During the first year of life whole blood manganese concentrations of 14–17 μg/L have been reported, with the concentration in erythrocytes being 20–25 times higher than in serum. Erythrocytic manganese is reportedly high at birth (376 ± 62.3 μg/g haemoglobin); this subsequently falls to reach a constant concentration of 151 ± 34 μg/g haemoglobin at 4 months postnatal age. Serum concentrations of manganese in formula-fed infants at 3 months of age (4.7 ± 1.6 μg/L) are similar to those observed in breast-fed infants (4.4 ± 1.8 μg/L). Plasma levels are elevated with toxic exposure and liver disease (Aggett & Barclay, 1991).

Molybdenum

Molybdenum has several oxidation states. The redox couple between Mo(V) and Mo(VI) is exploited for electron exchange with flavin mononucleotides in the activities of sulphite oxidase, xanthine oxidase and aldehyde dehydrogenase, for which molybdate linked with a pterin is a cofactor. The element is therefore needed for the metabolism of purines, pyrimidines, quinolines, sulphite and bisulphite.

Intestinal absorption of molybdenum is highly efficient (80% or more). Systemic homeostasis is controlled by renal excretion.

Molybdenum deficiency has occurred with prolonged parenteral nutrition (Abumrad *et al.*, 1981), and in infants in whom the hepatic synthesis of the molybdenum pterin cofactor is defective. There are two metabolic defects; both are autosomally recessively inherited and they have similar clinical features involving impaired metabolism of sulphur amino acids and nucleotides (Wadman *et al.*, 1983; Johnson *et al.*, 1989). These children present as neonates with dysmorphic features, feeding difficulties, bilateral dislocation of the lens, hypertonicity or, sometimes, hypotonicity, developmental delay, cerebral and cerebellar atrophy with encephalopathy, and epilepsy. In all these instances the biochemical anomalies include hypouricaemia, xanthinuria, sulphituria, thiosulphaturia and reduced urinary excretion of inorganic sulphate. The activity of sulphite oxidase in fibroblasts from these patients is reduced and offers a means of prenatal diagnosis. Acquired molybdenum deficiency has similar clinical and biochemical features.

Chromium

Trivalent chromium (Cr(III)) may have a role in glucose tolerance and lipid metabolism. Possibly, it potentiates the action of insulin by optimizing the number of membrane insulin receptors or their reaction with insulin, or both. Thus, chromium is said to improve both hyper- and hypoglycaemic responses to glucose loads. However, trials of chromium in patients with diabetes mellitus have produced inconsistent results, thereby creating scepticism about the essentiality of chromium. The element may also have a role in the metabolism of lipids and nucleic acids, and it has been suggested that some of its observed effects may arise from a non-specific effect on phosphoglucomutase (Anonymous, 1988b; Stoecker, 1990).

Intestinal absorption of chromium(III) is low. Organic chromium is absorbed more efficiently but it is not certain whether a proposed 'glucose tolerance factor', comprising chromium nicotinic cysteine glycine, is a biological entity.

Chromium-responsive features, including an insulin-resistant hypoglycaemia, elevated serum lipids, impaired nitrogen retention, weight loss, ataxia, peripheral neuropathy and encephalopathy, have been described in adults and in a child receiving prolonged parenteral nutrition. The adult patients responded to intravenous chromium chloride, but the response in the child was less conclusive.

Fluoride

The essentiality of fluoride is debatable but it is accepted as being beneficial to dental health (Schamschula & Barnes, 1981). Both topical and systemic fluoride replaces hydroxyl moieties in enamel to form calcium fluoroapatite which, being less soluble in acid than is calcium hydroxyapatite, is more resistant to demineralization. Fluoride may also inhibit cariogenic oral microflora.

Of systemic fluoride 95% is in the skeleton and teeth. The concentration in bone increases with age and it has been suggested, but not proven conclusively, that fluoride may enhance the mineralization of bone and the maintenance of peak bone mass. Absorption and transfer of fluoride are highly efficient and systemic homeostasis of fluoride is achieved by the kidneys.

Low intakes of fluoride are not associated with any known clinical problems, other than an increased susceptibility to caries. However, fluoride excess (fluorosis) is endemic throughout the world. At its mildest, with intakes approximating 0.1 mg/kg per day (Leverett, 1982), a patchy demineralization (mottling) of the enamel of the permanent dentition develops. In areas where the water contains 1 mg (50 μmol) fluoride/kg, 10% of children are affected. Chronic exposure to high intakes (e.g. 10–25 mg/day, as occurs where the water supplies contain 15 mg/kg) results in a sclerotic calcification of the bones, ligaments, tendons and interosseous membranes, leading to debilitating skeletal deformities and arthropathies (Krishnamachari, 1986). Entrapment neuropathies can ensue. Affected areas include the Middle East, Tanzania, Kenya, South Africa, the Indian subcontinent and China. In these areas, children have presented at 6 years of age with severe fluorosis. Asymptomatic skeletal changes can be detected radiologically. The features vary with calcium intake. Clinical biochemistry shows a urinary fluoride concentration above that in the plasma, evidence of hyperparathyroidism and, because fluoride interacts with iodide, hypothyroidism. Acute toxicity, and perhaps death, has been reported in adults exposed to intakes of 0.5–2.6 g/day (Waldbott, 1981).

References

Abumrad, N.N., Schneider, A.J., Steel, D. & Roberts, L.S. (1981) Amino acid intolerance during prolonged total parenteral nutrition reversed by molybdate therapy. *Am J Clin Nutr* **34**, 2551–2559.

Aggett, P.J. (1989) Severe zinc deficiency. In: *Zinc in Human Biology* (ed. C.F. Mills), pp. 259–280. Springer-Verlag, Berlin.

Aggett, P.J. (1991) The scientific considerations of recommended dietary intakes. *Eur J Clin Nutr* **44** (Suppl. 2), 37–43.

Aggett, P.J. & Barclay, S.M. (1991) Neonatal metabolism of trace elements. In: *Principles of Perinatal–Neonatal Metabolism* (ed. R.M. Cowett), pp. 500–530. Springer-Verlag, Berlin.

Anonymous (1987) Copper and the infant. *Lancet* **1**, 900–901.

Anonymous (1988a) Manganese deficiency in humans: fact or fiction. *Nutr Rev* **46**, 348–352.

Anonymous (1988b) Is chromium essential for humans? *Nutr Rev* **46**, 17–20.

Anonymous (1989) Does zinc supplementation improve growth in children who fail to thrive? *Nutr Rev* **47**, 356–358.

Arthur, J.R., Nicol, F. & Beckett, G.J. (1990) Hepatic iodothyronine deiodinase: the role of selenium. *Biochem J* **272**, 537–540.

Berman, D.H., Leventhal, R.I., Gavaler, J.S., Cadoff, E.M. & Van Thiel, D.H. (1991) Clinical differentiation of fulminant Wilsonian hepatitis from other causes of hepatic failure. *Gastroenterology* **100**, 1129–1134.

Bhave, S.A., Pandit, A.N., Singh, S., Walia, B.N. & Tanner, M.S. (1992) The prevention of Indian childhood cirrhosis. *Ann Trop Paediatr* **12**, 23–30.

Bingle, C.D., Srai, S.K. & Epstein, O. (1990) Developmental changes in hepatic copper proteins in the guinea pig. *J Hepatol* **10**, 138–143.

Brewer, G.J., Dick, R.D., Yuzbasiyan Gurkin, V. *et al.* (1991) Initial therapy of patients with Wilson's disease with tetrathiomolybdate. *Arch Neurol* **48**, 42–47.

Castillo-Duran, C., Fisberg, M., Valzuela, A. *et al.* (1983) Controlled trial of copper supplementation during the recovery from marasmus. *Am J Clin Nutr* **37**, 898–903.

Coleman, J.E. (1992) Zinc proteins: enzymes, storage proteins, transcription factors, and replication proteins. *Ann Rev Biochem* **61**, 897–946.

Cousins, R.J. (1985) Absorption, transport and hepatic metabolism of copper and zinc: special reference to metallothionein and caeruloplasmin. *Physiol Rev* **65**, 238–309.

Crofton, R.W., Aggett, P.J., Gvozdanovic, D. *et al.* (1990) Metabolism of zinc in celiac disease. *Am J Clin Nutr* **52**, 379–382.

Danks, D.M. (1988) Copper deficiency in humans. *Ann Rev Nutr* **8**, 235–257.

Danks, D.M., Campbell, P.E., Stevens, B.J., Mayne, V. & Cartwright, E. (1972) Menkes' kinky hair syndrome: an inherited defect in copper absorption with widespread effects. *Pediatrics* **50**, 188–201.

Davies, K. (1993) Cloning the Menkes' disease gene. *Nature* **361**, 98.

Department of Health (1991) *Dietary Reference Values for Food Energy and Nutrients for the United Kingdom*. HMSO, London.

Dhur, A., Galan, P. & Hercberg, S. (1990) Relationship between selenium, immunity and resistance against infection. *Comp Biochem Physiol* **96**, 271–280.

Dubois, R.S., Rodgerson, D.O. & Hambidge, K.M. (1990) Treatment of Wilson's disease with triethylene tetramine hydrochloride (Trientine). *J Ped Gastr Nutr* **10**, 77–81.

Edwards, C.Q., Williams, D.M. & Cartwright, G.E. (1979) Hereditary hypoceruloplasminemia. *Clin Genet* **15**, 311–316.

Failla, M.L., Van der Verdonk, M., Morgan, W.T. & Smith, J.C. (1982) Characterisation of zinc binding proteins of plasma in familial hyperzincaemia. *J Lab Clin Med* **100**, 943–952.

Farrer, L.A., Bowcock, A.M., Hebert, J.M. *et al.* (1991) Predictive testing for Wilson's disease using tightly linked and flanking DNA markers. *Neurology* **41**, 992–999.

Freeland-Graves, J., Behmardi, F., Bales, C.W. *et al.* (1988) Metabolic balance of manganese in young men consuming diets containing five levels of dietary manganese. *J Nutr* **118**, 764–773.

Gerdes, A.M., Tonnesen, T., Pergament, E. *et al.* (1988) Variability in clinical expression of Menkes' syndrome. *Eur J Pediatr* **148**, 132–135.

Gibson, R.S., Vanderkooy, P.D.S., MacDonald, A.C. *et al.* (1989) A growth limiting, mild zinc-deficiency syndrome in some Southern Ontario boys with low height percentiles. *Am J Clin Nutr* **49**, 1266–1273.

Godwin-Austen, R.B., Robinson, A., Evans, K. & Lascelles, P.T. (1978) An unusual neurological disorder of copper metabolism clinically resembling Wilson's disease but biochemically a distinct entity. *J Neurol Sci* **39**, 85–98.

Golden, B.E. & Golden, M.H.N. (1992) Effects of zinc on

lean tissue synthesis during recovery from malnutrition. *Eur J Clin Nutr* **46**, 697–706.

Halliday, H.L., Lappin, T.R.J., McMaster, D. & Paterson, C.C. (1985) Copper and the preterm infant. *Arch Dis Child* **60**, 1105–1106.

Hillman, L.S. (1981) Serial serum copper concentrations in premature and SGA infants during the first three months of life. *J Pediatr* **98**, 305–308.

Hurley, L.S. & Keen, C.L. (1987) Manganese. In: *Trace Elements in Human and Animal Nutrition*, Vol. 1 (ed. W. Mertz), pp. 185–223. Academic Press, San Diego.

Inagaki, M., Hashimoto, K., Yoshino, K. *et al.* (1988) Atypical form of Menkes' kinky hair disease with mitochondrial NADH-CoQ reductase deficiency. *Neuropediatrics* **19**, 52–55.

Jackson, M.J. (1989) Physiology of zinc: general aspects. In: *Zinc in Human Biology* (ed. C.F. Mills), pp. 1–14. Springer-Verlag, Berlin.

Johnson, J.L., Wuebbens, M.M., Mandell, R. & Shih, V.E. (1989) Molybdenum cofactor biosynthesis in humans: identification of two complementation groups of cofactor-deficient patients and preliminary characterisation of a diffusible molybdopterin precursor. *J Clin Invest* **83**, 897–903.

Kondakis, X., Makris, N., Leotsinidis, M., Prinou, M. & Papapetropoulos, T. (1989) Possible health effects of high manganese concentrations in drinking water. *Arch Environ Health* **44**, 175–178.

Kressner, M.S., Stockert, R.J., Morell, A.G. & Sternlieb, I. (1984) Origins of biliary copper. *Hepatology* **4**, 867–870.

Krishnamachari, K.A.V.R. (1986) Skeletal fluorosis in humans: a review of recent progress in the understanding of the disease. *Prog Food Nutr Sci* **10**, 279–314.

Levander, O.A. (1987) A global view of human selenium nutrition. *Ann Rev Nutr* **7**, 227–250.

Leverett, D.H. (1982) Fluorides and the changing prevalence of dental caries. *Science* **217**, 26–30.

Mack, D., Koletzko, B., Cunnane, S., Cutz, E. & Griffiths, A. (1989) Acrodermatitis enteropathica with normal serum zinc levels: diagnostic value of small bowel biopsy and essential fatty acid determination. *Gut* **30**, 1426–1429.

Maracek, Z. & Nevsimalova, S. (1984) Biochemical and clinical changes in Wilson's disease heterozygotes. *J Inher Metab Dis* **7**, 41–45.

Mathias, P. & Jackson, A.A. (1982) Selenium deficiency in kwashiorkor. *Lancet* **1**, 1312–1313.

Matseshe, J.W., Philips, S.F., Malagelada, J.R. & McCall, J.T. (1980) Recovery of dietary iron and zinc from the proximal intestine of healthy man: studies of different meals and supplements. *Am J Clin Nutr* **33**, 1946–1953.

Mehes, K. & Petrovicz, E. (1982) Familial benign copper deficiency. *Arch Dis Child* **57**, 716–718.

Morgan, W.T. (1981) Interactions of the histidine-rich glycoprotein of serum with metals. *Biochemistry* **20**, 1054–1061.

Motchnik, P.A. & Tappel, A.L. (1990) Multiple selenocysteine content of selenoprotein P in rats. *J Inorg Biochem* **40**, 265–269.

Muller-Hocker, J., Meyer, U., Wiebecke, B. *et al.* (1988) Copper storage disease of the liver and chronic dietary intoxication in two further German infants mimicking Indian childhood cirrhosis. *Pathol Res Pract* **183**, 39–45.

Nadal, D. & Baerlocher, K. (1988) Menkes' disease: long-term treatment with copper and D-penicillamine. *Eur J Pediatr* **147**, 621–625.

O'Dell, B.L. (1990) Copper. In: *Present Knowledge in Nutrition* (ed. M.L. Brown), pp. 261–267. International Life Sciences Institute Nutrition Foundation, Washington.

Ono, S. & Kurisaki, H. (1988) An unusual neurological disorder with abnormal copper metabolism. *J Neurol* **235**, 397–399.

Schamschula, R.G. & Barnes, D.E. (1981) Fluoride and health: dental caries, osteoporosis and cardiovascular disease. *Ann Rev Nutr* **1**, 427–435.

Schilsky, M.L., Scheinberg, I.H. & Sternlieb, I. (1991) Prognosis of Wilsonian chronic active hepatitis. *Gastroenterology* **100**, 762–767.

Spitalny, K.C., Brondum, J., Vogt, R.L., Sargent, H.E. & Kappel, S. (1984) Drinking-water-induced copper intoxication in a Vermont family. *Pediatrics* **74**, 1103–1106.

Sternlieb, I. (1990) Perspectives on Wilson's disease. *Hepatology* **12**, 1234–1239.

Stoecker, B.J. (1990) Chromium. In: *Present Knowledge in Nutrition* (ed. M. Brown) 6th edn, pp. 287–293. International Life Sciences Institute Nutrition Foundation, Washington.

Sutton, A.M., Harvie, A., Cockburn, F. *et al.* (1985) Copper deficiency in the preterm infant of very low birth-weight. *Arch Dis Child* **60**, 644–651.

Taylor, C.M., Bacon, J.R., Aggett, P.J. & Bremner, I. (1991) The homeostatic regulation of zinc absorption and endogenous zinc losses in zinc deprived man. *Am J Clin Nutr* **53**, 755–763.

Tonnesen, T., Horn, N., Sondergaard, F. *et al.* (1987) Experience with first trimester prenatal diagnosis of Menkes' disease. *Prenat Diag* **7**, 497–509.

Vallee, B. & Galdes, A. (1984) The metallobiochemistry of zinc enzymes. In: *Advances in Enzymology* (ed. A. Meister), pp. 283–430. John Wiley & Sons, Chichester.

Veen, C., Van den Hamer, C.J. & de Leeuw, P.W. (1991) Zinc sulphate therapy for Wilson's disease after acute deterioration during treatment with low-dose D-penicillamine. *J Intern Med* **229**, 549–552.

Verga, V., Hall, B.K., Wang, S.R. *et al.* (1991) Localization of the translocation breakpoint in a female with Menkes' syndrome to Xq13.2–q13.3 proximal to PGK-1. *Am J*

Hum Genet **48**, 1133–1138.

Wadman, S.K., Duran, M., Beemer, F.A. *et al.* (1983) Absence of hepatic molybdenum cofactor: an inborn error of metabolism leading to a combined deficiency of sulphite oxidase and xanthine dehydrogenase. *J Inher Metab Dis* **6** (Suppl. 1), 78–83.

Waldbott, G.L. (1981) Mass intoxication from accidental overfluoridation of drinking water. *Clin Toxicol* **18**, 531–541.

Walravens, P.A., Chakar, A., Mokni, R., Denise, J. & Lemonnier, D. (1992) Zinc supplements in breastfed infants. *Lancet* **340**, 683–685.

Watanabe, T. & Endo, A. (1991) Effects of selenium deficiency on sperm morphology and spermatocyte chromosomes in mice. *Mutat Res* **262**, 93–99.

Willvonseder, R., Goldstein, N.P., McCall, J.T., Yoss, R.E. & Tauxe, W.N. (1973) A hereditary disorder with dementia, spastic dysarthria, vertical eye movement paresis, gait disturbance, splenomegaly and abnormal copper metabolism. *Neurology* **23**, 1039–1049.

van Wouwe, J.P. (1989) Clinical and laboratory diagnosis of acrodermatitis enteropathica. *Eur J Pediatr* **149**, 2–8.

Yang, G., Ge, K., Chen, J. & Chen, X. (1988) Selenium-related endemic diseases and the daily selenium requirement of humans. *Wld Rev Nutr Diet* **55**, 98–152.

Yang, G., Yin, S., Zhou, L. *et al.* (1989) Studies of safe maximal dietary Se intake in a seleniferous area in China. *J Trace Elem Electrol Hlth Dis* **3**, 123–130.

Yardrick, M.K., Kenney, M.A. & Winterfeldt, E.A. (1989) Iron, copper and zinc status: response to supplementation with zinc or zinc and iron in adult females. *Am J Clin Nutr* **49**, 145–150.

Zlotkin, S.H. & Cherian, M.G. (1988) Hepatic metallothionein as a source of zinc and cysteine during the first year of life. *Pediatr Res* **24**, 326–329.

Appendix: Reference Data

A1: Reference Data for Newborn Babies

P.H. SCOTT & B.A. WHARTON

Tables A1.1–A1.11 give a variety of reference values found in the neonatal period. Unless otherwise stated they were obtained at the Biochemistry Department, Selly Oak Hospital, Birmingham, from babies in the Sorrento Maternity Hospital, Birmingham. Babies were either of low birth weight, 1.7–2.2 kg and fed on Gold Cap SMA-S26 (a demineralized whey formula; Wyeth Laboratories, Berkshire), or were babies weighing over 2.2 kg at birth and fed either breast milk or a low-solute formula. Some babies exhibited varying degrees of physiological jaundice but were otherwise well.

Note: Lithium heparin is a suitable anticoagulant for most analytes. There are some analytes, such as glucose, lactate, ammonia and toxic elements, for which heparin is not suitable. If in doubt, contact your laboratory for information about the correct anticoagulant and tube or urine collection bottle.

Table A1.1 Reference ranges for plasma electrolytes, glucose, urea, creatinine and osmolality, and for blood ionized calcium and acid–base measurements

Measurement (units)	Description of infants	Age	Mean	Range
Base excess (mmol/L)	Low birth-weight	2 days	−0.1	Base excess greater
	Low birth-weight	Second week	−2.7	than −8.0 is indicative
	Low birth-weight	Third week	+0.2	of significant acidosis
Calcium, total[1,2] (mmol/L)	Low birth-weight formula-fed	First 3 weeks	2.41	1.90–2.85
	Full-term breast-fed	1–2 days	2.33	2.15–2.52
	Full-term breast-fed	2–3 days	2.39	2.10–2.67
	Full-term breast-fed	4–6 days	2.46	2.20–2.72
	Full-term breast-fed	6–12 days	2.47	2.20–2.75
	Full-term breast-fed	At 1 month	2.55	2.15–2.95
Calcium, ionized[3] (uncorrected for pH) (mmol/L)	Preterm on SCBU	0–2 days	1.19	0.78–1.59
	Preterm on SCBU	2–4 days	1.26	0.95–1.58
	Preterm on SCBU	4–6 days	1.41	1.20–1.62
	Preterm on SCBU	6–90 days	1.37	1.14–1.58
	Term	7–14 days	1.31	1.13–1.49
(corrected for pH)	Preterm	6–90 days	1.32	1.12–1.52
Chloride[4] (mmol/L)	Preterm, 30–36 wks gestation	3–5 days	105	90–117
	Preterm, 30–36 wks gestation	8–18 days	108	104–113
	Birth weight >2.2 kg	First 3 weeks	101	92–109

continued on p. 512

Table A1.1 *Continued*

Measurement (units)	Description of infants	Age	Mean	Range
Creatinine[5] (μmol/L)	25–28 wks 0.66–1.3 kg at birth	2 days	116	76–156
	25–28 wks 0.66–1.3 kg at birth	7 days	84	52–116
	25–28 wks 0.66–1.3 kg at birth	14 days	72	40–104
	25–28 wks 0.66–1.3 kg at birth	21 days	60	28–92
	25–28 wks 0.66–1.3 kg at birth	28 days	58	18–98
	29–32 wks 0.75–2.75 kg at birth	2 days	104	66–142
	29–32 wks 0.75–2.75 kg at birth	7 days	83	43–123
	29–32 wks 0.75–2.75 kg at birth	14 days	69	37–101
	29–32 wks 0.75–2.75 kg at birth	21 days	59	27–91
	29–32 wks 0.75–2.75 kg at birth	28 days	52	20–84
	33–36 wks 1.04–2.96 kg at birth	2 days	93	55–131
	33–36 wks 1.04–2.96 kg at birth	7 days	68	24–112
	33–36 wks 1.04–2.96 kg at birth	14 days	55	19–87
	33–36 wks 1.04–2.96 kg at birth	21 days	50	14–86
	33–36 wks 1.04–2.96 kg at birth	28 days	35	11–59
	37–42 wks 1.6–4.46 kg at birth	2 days	75	37–113
	37–42 wks 1.6–4.46 kg at birth	7 days	50	14–86
	37–42 wks 1.6–4.46 kg at birth	14 days	38	18–58
	37–42 wks 1.6–4.46 kg at birth	21 days	35	15–55
	37–42 wks 1.6–4.46 kg at birth	28 days	30	12–48
Glucose[6] (mmol/L)	Term and preterm babies	Day 1		Greater than 2.1
	Term and preterm babies	After day 1		Greater than 2.5
Magnesium[1,7] (mmol/L)	Preterm 1.5–2.0 kg at birth	At birth	0.81	0.62–1.27
	Preterm 1.5–2.0 kg at birth	2 days	0.94	0.77–1.10
	Preterm 1.5–2.0 kg at birth	7 days	0.98	0.71–1.13
	Term	At birth	0.78	0.71–0.89
	Term	2 days	0.94	0.77–1.08
	Term	7 days	0.90	0.76–1.04
	Birth weight >2.2 kg	First week	0.80	0.59–1.05
	Full-term breast-fed	At 1 month	0.83	0.51–1.15
Osmolality[8] (mmol/kg H_2O)	Preterm >2.5 kg at birth	At birth	286	275–300
	Preterm >2.5 kg at birth	7 days	291	276–305
	Preterm >2.5 kg at birth	28 days	290	274–305
$P\text{CO}_2$ (kPa)	Low birth-weight	2 days	4.32	2.00–6.50
	Low birth-weight	2–3 weeks	4.60	2.50–6.00
pH[9]	Term	1–4 days	7.39	7.32–7.45
	Low birth-weight	2 days	7.44	7.32–7.54
	Low birth-weight	2–3 weeks	7.41	7.33–7.47
Phosphate[10] (mmol/L)	Low birth-weight	1 day	1.84	1.00–2.60
	Low birth-weight	2–3 weeks	2.44	1.80–3.20
	Full-term breast-fed	2–3 days	2.40	1.81–3.00
	Full-term breast-fed	3–4 days	2.25	1.74–2.76
	Full-term breast-fed	4–6 days	2.17	1.64–2.70
	Full-term breast-fed	6–12 days	2.21	1.39–3.03
	Full-term breast-fed	21 days	2.20	1.74–2.66
	Full-term formula-fed	21 days	2.36	1.68–3.04

continued

Table A1.1 *Continued*

Measurement (units)	Description of infants	Age	Mean	Range
Potassium (mmol/L)	Low birth-weight	7 days	4.6	3.8–5.1
	Low birth-weight	11 days	4.9	3.9–5.7
	Low birth-weight	21 days	5.0	4.1–5.9
	Birth weight >2.2 kg	1–3 weeks	4.8	3.6–5.8
Sodium (mmol/L)	Low birth-weight	7 days	136	131–141
	Low birth-weight	11 days	138	133–143
	Low birth-weight	21 days	136	130–140
	Birth weight >2.2 kg	1–3 weeks	138	130–145
Urea[1,11,12] (mmol/L)	Low birth-weight	1–3 weeks	3.0	1.0–5.0
	Full-term	At birth	4.9	3.5–6.7
	Full-term	Day 1	5.3	1.2–11.5
	Full-term	Day 2	5.1	2.2–11.3
	Full-term breast-fed	Day 6	3.3	1.6–4.6
	Full-term breast-fed	8–9 days	2.9	1.7–4.2
	Full-term breast-fed	11 days	3.3	1.3–5.3
	Full-term breast-fed	21 days	2.7	1.5–3.9
	Full-term formula-fed	11 days	2.9	1.1–4.7
	Full-term formula-fed	21 days	3.4	1.8–5.0
	Full-term breast-fed	At 1 month	3.0	1.9–5.2

SCBU, special-care baby unit. [1] After Fomon *et al.* (1970); [2] after Thalme (1972); [3] after Tovey and Murphy (personal communication); [4] after Young *et al.* (1941); [5] after Rudd *et al.* (1983); [6] after Srinivasan *et al.* (1986); [7] after Hillman *et al.* (1977); [8] after Davies (1973); [9] after Reardan *et al.* (1960); [10] after Thalme (1972); [11] after McCance and Widdowson (1947); [12] after Acharya and Payne (1965).

Table A1.2 Reference values for serum or plasma enzyme activities (U/L) in neonates and red cell enzyme activities

Measurement	Age	Mean	Range	
Alkaline phosphatase	3 weeks	349	124–574	at 37°C
		(248	88–407	at 30°C*)
γ-Glutamyl-transferase[1]	2 weeks	45	4–261	at 25°C
5′-Nucleotidase[2]	4 weeks	2.8	Up to 10	at 37°C
Creatine kinase[3]	1–3 days	564	63–1137	at 37°C*
		(349	40–747	at 30°C*)
	4–10 days	222	63–1413	at 37°C*
		(142	40–903	at 30°C*)
	11–31 days	107	32–288	at 37°C*
		(69	20–184	at 30°C*)
Lactate dehydrogenase[4]	Birth	874	470–1824	at 37°C*
		(605	326–1263	at 30°C*)
	To 3 weeks	672	326–1507	at 37°C*
		(465	226–1044	at 30°C*)

continued on p. 514

Table A1.2 *Continued*

Measurement	Age	Mean	Range	
α-Hydroxybutyrate dehydrogenase[4]	Birth	377	172–735	at 37°C*
		(342	156–667	at 30°C*)
	To 3 weeks	273	127–669	at 37°C*
		(248	116–608	at 30°C*)
Aspartate[5] aminotransferase	3 weeks	37.8	22.9–72.8	at 37°C*
		(24.9	15.1–47.9	at 30°C*)
Alanine[5] aminotransferase	3 weeks	27.1	9.1–43.7	at 37°C*
		(19.7	6.6–31.7	at 30°C*)
Alkaline ribonuclease	Birth	687	320–918	at 37°C
	2 weeks	726	473–1003	at 37°C
	3 weeks	645	439–918	at 37°C
Galactokinase	Birth to 1 month	2.8	1.6–6.8	
Galactose-1-phosphate uridyltransferase[6] (μmol/h per g Hb)	Birth to 1 month	18.5	12.3–24.6	at 37°C
Glucose-6-phosphate dehydrogenase[7] (μmol/min per g Hb)	2 days	11.4	5.9–21.3	
	Later	7.7	5.5–9.3	

* Value calculated from another temperature. [1] After Shore *et al.* (1975); [2] after Belfield and Goldberg (1971); [3] after Kupke *et al.* (1980); [4] after Veit *et al.* (1975); [5] after Sitzmann *et al.* (1974); [6] after Ellis and Goldberg (1969); [7] after Herz *et al.* (1973).

Table A1.3 Reference ranges for serum protein concentrations in neonates

Measurement (units)	Description of infants	Age	Mean	Range
Albumin[1] (g/L)	Low birth-weight formula-fed	1 day	37	25–49
	Low birth-weight formula-fed	11 days	39	29–49
	Low birth-weight formula-fed	21 days	40	32–48
	Full-term breast-fed	11 days	41	35–47
	Full-term breast-fed	21 days	41	35–47
	Full-term breast-fed	11 days	42	35–48
	Full-term breast-fed	21 days	41	35–47
	Full-term breast-fed	At 1 month	40	33–47
Total protein[1,2] (g/L)	Normal infants	At birth	62	51–75
	Normal infants	1 day	61	42–67
	Normal infants	2 days	57	44–67
	Normal infants	4 days	58	53–63
	Birth weight >2.2 kg	1–2 weeks	56	43–76
	Full-term breast-fed	At 1 month	58	48–69

continued

Table A1.3 *Continued*

Measurement (units)	Description of infants	Age	Mean	Range
IgG[3] (g/L)	Preterm	At birth	9.1	6.3–13.0*
	Term LGA	At birth	12.2	8.4–17.6*
	Full-term	At birth	11.5	7.9–16.9*
	Preterm and term LGA	2 weeks	8.0	4.8–13.0*
	Preterm and term LGA	2 weeks	5.1	3.1–8.5*
	Preterm and term LGA	3 weeks	6.3	3.5–11.0*
IgA[3] (mg/L)	Full-term	At birth	6.6	0.5–87.4*
	Low birth-weight	1–2 weeks	16.9	0–72.2*
	Full-term	2 weeks	151.8	34.6–663.3*
	Low birth-weight	3 weeks	44.0	15.6–124.0*
IgM[3] (mg/L)	Full-term	At birth	104	41–258*
	Low birth-weight	At birth	82	27–200*
	Full-term	2 weeks	502	157–1609*
	Low birth-weight	2 weeks	291	140–603*
	Low birth-weight	3 weeks	451	253–802*
α_1-Antitrypsin[4] (g/L)	Undefined healthy newborns	At birth	2.26	A value less than 1.0 suggests deficiency of the PiZ type*
	Undefined healthy newborns	3 days	2.94	
	Undefined healthy newborns	30 days	1.99	
Transferrin (g/L)	Preterm	At birth	1.60	0.90–2.60*
	Term LGA	At birth	2.09	1.30–2.75*
	Low birth-weight	2–3 weeks	1.71	1.05–2.65*
	Full-term breast-fed	11 days	1.84	1.36–2.32*
	Full-term formula-fed	11 days	1.84	1.38–2.30*
	Full-term breast-fed	21 days	1.98	1.14–2.82*
	Full-term formula-fed	21 days	1.82	1.22–2.42*
Ferritin[5,6] (μg/L)	Term, birth weight >3.0 kg	At birth	160	65–395*
	Preterm, birth weight 1.22 kg	1 week	370	230–770*
	Term, birth weight >3.0 kg	2 weeks	238	90–638*
	Preterm, birth weight 1.22 kg	2 weeks	330	250–950*
	Term, birth weight >3.0 kg	4 weeks	240	144–399*
	Preterm, birth weight 1.22 kg	4 weeks	480	120–820*
Caeruloplasmin[7] (mg/L)	34–43 weeks gestation	At birth	135	20–380*

* See also Chapter 22; LGA, light-for-gestational age. [1] After Fomon *et al.* (1970); [2] after Overman *et al.* (1951); [3] after Cejka *et al.* (1974); [4] after Kueppers and Offord (1979); [5] after Saarinen and Siimes (1978); [6] after Shaw (1982); [7] after Freer *et al.* (1979).

Table A1.4 Concentrations of electrolytes and other components in random urines from low birth-weight babies

Component concentration (mmol/L)	Age	Mean	Range
Sodium	To 1 month	6.0	1–15
Potassium	To 1 month	12.6	2–28
Calcium	To 1 week	0.16	0.05–0.38
	Week 2	0.39	0.23–0.73
	Week 3	0.49	0.25–1.16
	At 1 month	1.26	0.50–1.65
Chloride	To 1 month	13.6	5–30
Phosphate	Weeks 1 and 2	3.6	<0.5–9.9
	Weeks 3 and 4	7.5	0.5–12.0
Urea	To 1 week	34.0	5–108
	Weeks 2 and 3	24.0	4–59
	At 1 month	15.0	9–34
Creatinine	To 1 week	1.61	0.5–3.5
	Weeks 2 and 3	1.06	0.2–2.4
	At 1 month	1.09	0.5–1.8

Table A1.5 Urine nitrogen partition, urinary sulphate and free and peptide hydroxyproline excretion in a group of low birth-weight infants

Measurement	Age (days)	Mean	Range
Total nitrogen (mmol/kg per 24 h)	3	5.9	4.5–8.2
	10	6.1	5.3–6.6
	20	10.1	7.5–14.8
Urea nitrogen (mmol/kg per 24 h)	3	3.5	2.7–4.9
	10	3.7	3.0–4.2
	20	7.5	5.5–11.7
Creatinine nitrogen (mmol/kg per 24 h)	3	0.27	0.23–0.36
	10	0.26	0.22–0.32
	20	0.34	0.20–0.48
Uric acid nitrogen (mmol/kg per 24 h)	3	0.30	0.14–0.58
	10	0.18	0.06–0.41
	20	0.20	0.05–0.48
α-Amino nitrogen (mmol/kg per 24 h)	3	0.40	0.28–0.66
	10	0.62	0.41–1.26
	20	0.60	0.51–0.73
Residual nitrogen (mmol/kg per 24 h)	3	1.2	0.7–2.2
	10	1.3	0.9–1.8
	20	1.5	1.0–2.0
Sulphate (mmol/kg per 24 h)	3	0.14	0.09–0.18
	10	0.16	0.13–0.20
	20	0.34	0.20–0.51
Total hydroxyproline (μmol/24 h)	3	91	51–124
	10	170	113–220
	20	284	217–519
Free hydroxyproline (μmol/24 h)	3	38	13–66
	10	82	58–107
	20	129	96–242
Peptide hydroxyproline (μmol/24 h)	3	54	37–69
	10	88	55–115
	20	155	114–277

Table A1.6 Creatinine clearance and fractional sodium excretion in preterm and full-term infants during the first weeks of life. After Aperia *et al.* (1980)

Measurement (units)	Description of infants	Age	Mean
Creatinine clearance (mL/min per 1.73 m^2)	PreTerm	1–2 days	15.9
	Term	1–2 days	20.8
	PreTerm	4–6 days	24.1
	Term	4–6 days	46.6
	PreTerm	3–5 weeks	37.0
	Term	3–5 weeks	60.1
Fractional sodium excretion ($C_{Na}:C_{Cr}$ ratio as %)	PreTerm	1–2 days	2.11
	Term	1–2 days	0.20
	PreTerm	4–6 days	1.28
	Term	4–6 days	0.30
	PreTerm	3–5 weeks	0.09
	Term	3–5 weeks	0.12

Table A1.7 Reference ranges for hormones and related compounds in serum (unless stated otherwise), given as mean and range

Compound	Description of infants	Full-term Mean	Full-term Range	Low birth-weight Mean	Low birth-weight Range
Growth hormone[1] (mU/L)	At birth	66	9–320		
	0–24 h	52	9–167	59	38–83
	24–48 h	72	28–221	76	17–150
	2–6 days	32	0–69	51	20–133
	7–13 days	20	1–51	30	5–51
	14–20 days	20	5–47	41	16–150
	21–27 days	16	4–26	47	17–125
Insulin (mU/L)	Highly method dependent; consult your laboratory				
Cortisol[2]* (nmol/L)		Full-term		Preterm	
	0–6 h	337	0–695	398	238–558
	7–12 h	113	58–168	692	0–1719
	13–24 h	71	27–115	320	0–706
	1–2 days	81	0–185	516	0–1062
	3–5 days	63	19–107	252	0–644
	6–12 days	134	29–239	171	0–430
	13–30 days	121	0–259	85	0–228
17-Hydroxyprogesterone[3,4]†					
Plasma (nmol/L)	At birth	203	127–279		
	12 h	45	13–77		
	24 h	8	3–14		
	36 h	6	2–10		
	7 days	3.5	1.1–5.9		
Blood spot (nmol/L)	Up to end of first week		Up to 20		Up to 200
Testosterone, Aldosterone, Renin	See Appendix A3				
Sex hormone-binding globulin[5] (nmol/L)	First month		Up to 100		
Thyroxine (free)[6] (pmol/L)	1–3 days	27.4	16.7–48.3	15.9	11.3–24.0
	4–10 days	21.7	13.7–28.0	17.8	10.0–30.0
Tri-iodothyronine (free)[6] (pmol/L)	1–3 days	5.0	2.5–9.3	3.2	1.2–7.3
	4–10 days	4.4	2.8–5.7	2.8	1.2–4.9
Thyroid-stimulating hormone[7,8] (mU/L)	At birth		1–14		2–16
	0–24 h		3–120		1–72
	1–2 days		3–30		1–17
	2–3 days		1–8		2–13
	4–6 days		1–4		1–6
	2–4 weeks		Up to 5		Up to 5

[1] After Cornblath *et al.* (1965); [2] after Rokicki *et al.* (1990); [3] after Hughes *et al.* (1979); [4] after Berry *et al.* (1986); [5] after Bolton *et al.* (1989); [6] after John and Bamforth (1987); [7] after Jacobsen *et al.* (1977); [8] after Jacobsen and Hummer (1979).
* Because of the high degree of oxidation to cortisone, check cross-reaction of cortisone in assays (J. Honour, personal communication). † Preterm infants may have higher values; levels in neonates with 21-hydroxylase deficiency will be in excess of 60 nmol/L by day 3 and after (J. Honour, personal communication).

Table A1.8 Plasma amino acid concentrations from preterm formula-fed, term light-for-gestational age (LGA) formula-fed, term appropriate-weight-for-gestation (AGA) formula-fed and term AGA breast-fed infants. Formula in each case was a demineralized whey formula (Gold Cap SMA, Wyeth Laboratories). After Scott *et al.* (1985, 1990)

Amino acid (μmol/L)	Age (days)	Preterm formula-fed		Term LGA formula-fed		Term AGA formula-fed		Term AGA breast-fed	
		Mean	Range	Mean	Range	Mean	Range	Mean	Range
Alanine	11	254	145–466	241	152–338	254	148–420	229	137–362
	21	241	162–492	263	187–371	253	164–496	199	138–284
Arginine	11	78	43–171	59	38–121	52	30–117	54	11–88
	21	83	31–153	63	50–109	71	31–130	56	23–99
Asparagine	11	34	9–57	30	21–48	44	16–50	38	5–46
	21	33	13–49	29	11–53	44	11–87	38	14–40
Aspartic acid	11	55	23–114	55	38–66	30	29–62	25	16–62
	21	58	23–82	44	26–73	29	29–59	25	24–58
Citrulline	11					43	25–77	40	20–84
	21					43	22–70	36	22–84
Cystine	11	44	31–62	41	27–77	46	39–56	45	33–55
	21	45	30–64	39	20–44	46	42–52	46	40–53
Glutamic acid	11	552*	327–772	558*	438–651	195	100–341	244	76–551
	21	574*	438–794	589*	463–900	225	110–478	187	90–360
Glutamine	11					410	240–723	349	147–623
	21					416	238–917	322	158–711
Glycine	11	289	220–403	225	175–310	117	63–498	192	66–432
	21	277	193–457	228	182–321	184	101–434	140	34–276
Histidine	11	118	71–297	95	58–139	84	62–114	81	25–126
	21	102	71–123	95	77–111	84	53–116	75	50–104
Hydroxyproline	11	65	50–109	52	33–81				
	21	67	42–113	62	45–99				
Isoleucine	11	46	28–74	45	34–80	73	49–99	67	31–124
	21	42	24–69	45	32–65	67	30–65	58	40–100
Leucine	11	97	65–168	92	61–153	126	82–177	119	86–171
	21	83	51–127	91	72–123	123	70–190	105	69–194
Lysine	11	197	122–306	163	135–225	119	70–213	129	65–282
	21	163	119–272	149	90–233	137	52–250	125	44–190
Methionine	11	27	15–39	25	14–41	34	21–55	30	21–52
	21	27	19–46	27	20–44	35	22–52	29	16–41
Ornithine	11					101	49–197	121	39–386
	21					117	36–247	110	52–172
Phenylalanine	11	63	45–90	68	37–93	60	37–94	54	35–112
	21	52	39–78	63	46–94	59	37–98	51	25–103
Proline	11	165	91–282	156	100–194				
	21	136	100–210	137	110–187				

continued

Table A1.8 *Continued*

Amino acid (μmol/L)	Age (days)	Preterm formula-fed		Term LGA formula-fed		Term AGA formula-fed		Term AGA breast-fed	
		Mean	Range	Mean	Range	Mean	Range	Mean	Range
Serine	11	224	150–450	168	146–450	162	107–288	147	79–227
	21	183	113–299	144	111–208	164	102–272	143	86–225
Threonine	11	256	137–391	165	114–250	148	101–232	101	67–143
	21	182	101–271	136	74–206	149	88–215	101	54–137
Tryptophan	11					47	39–175	49	48–122
	21					46	41–131	48	26–110
Tyrosine	11	103	61–159	78	53–125	85	39–175	81	48–122
	21	64	50–108	88	53–131	75	41–131	68	26–110
Valine	11	155	114–265	157	104–252	129	76–206	110	56–154
	21	142	85–230	160	125–202	137	45–223	111	71–145

* Combined glutamine and glutamic acid.

Table A1.9 Other miscellaneous plasma or serum reference values and blood lactate concentrations during the neonatal period

Measurement	Age	Mean	Range
Triglyceride (mmol/L)	Birth (AGA)	0.32	0.06–0.60
	Birth (LGA)	0.47	0.06–1.06
	2–3 weeks	0.86	0.24–1.58
Cholesterol (mmol/L)	Birth	1.81	0.55–3.15
	Weeks 1–3	2.93	1.93–3.93
Uric acid[1] (mmol/L)	Birth	0.42	0.09–0.58
	To 1 week	0.20	0.14–0.34
3-Hydroxybutyrate[2] (mmol/L) fasting	To 1 week	2.45	Value <1.1 suggestive of hyperinsulinaemia
Free fatty acid[2] (mmol/L) fasting	To 1 week	1.53	Value <0.46 suggestive of hyperinsulinaemia
Lactate[3] (mmol/L)	Birth		1.5–4.5
	5 h		0.9–2.0
	1 day		0.8–1.2
	To 1 month		0.5–1.4
Ammonia[4] (μmol/L)	To 2 weeks	89	32–192
Iron[5] (μmol/L)	Birth	26	4–52
	2–3 weeks	18	14–25
Zinc[6,7] (μmol/L)	Birth	12.7	10.7–15.6
	5 days	13.5	11.6–15.6
	1 month	10.1	7.9–12.5

continued on p. 520

Table A1.9 *Continued*

Measurement	Age	Mean	Range
Copper[7,8] (μmol/L)	Preterm postnatal, 30 wks	5.5	1.9–15.8
	Preterm postnatal, 32 wks	5.6	2.5–12.4
	Preterm postnatal, 34 wks	5.5	2.1–14.6
	Preterm postnatal, 36 wks	6.1	3.0–12.4
	Preterm postnatal, 38 wks	7.3	4.3–12.4
	Preterm postnatal, 41 wks	9.8	6.9–13.9
	Preterm postnatal, 43 wks	10.2	6.2–16.7
	Preterm postnatal, 45 wks	11.5	7.4–17.9
	Preterm postnatal, 47 wks	13.9	6.9–28.1
	1 month	9.9	7.8–16.3

AGA, appropriate-weight-for-gestational age; LGA, light-for-gestational age.
[1] After Wharton *et al.* (1971); [2] after Stanley and Baker (1976); [3] after Koch and Wendel (1968); [4] after Beddis *et al.* (1980); [5] after Brozovic *et al.* (1974); [6] after Henkin *et al.* (1971); [7] after Ohtake (1977); [8] after Sutton *et al.* (1985).

Table A1.10

Measurement	Age (days)	Range
Plasma/serum bilirubin*		
Total[1] (μmol/L)	2–6	Up to 217
	6–10	Up to 230
Conjugated[1] (μmol/L)	6–10	Up to 40
Urine calcium[2] (mmol/24 h)	Preterm	0.96 ± 0.25†
	Term, first week	0.51 ± 0.04†
Urine calcium/creatinine[3] (mmol/mol)	Preterm After 2 weeks <34 weeks	>2.0

* During phototherapy results obtained with a bilirubinometer do not correlate well with those obtained by a chemical method in the laboratory and may be misleadingly raised. See also section on Plasma bilirubin, pp. 30–31.
† Corrected to 1.73 m^2.
[1] Method: after Westwood (1982); [1] range: A. Green and K. Richmond (unpublished observations); [2] Karlen *et al.* (1985); [3] timed urines over 6–8 h.

Table A1.11 Galactokinase: [^{14}C]-galactose with chromatography

Measurement	Age	Range
Galactose phosphate (μmol/h per g Hb)	Cord–1 month	1.6–6.0

References

Acharya, P.R. & Payne, W.W. (1965) Blood chemistry of normal full-term infants in the first 48 hours of life. *Arch Dis Child* **40**, 430–435.

Aperia, A., Broberger, O., Elinder, G., Herin, P. & Zetterstrom, R. (1980) Postnatal development of renal function in preterm and full term infants. *Acta Paed Scand* **70**, 183–187.

Beddis, I.R., Hughes, E.A., Rosser, E. & Fenton, J.C.B. (1980) Plasma ammonia levels in newborn infants admitted to an intensive care baby unit. *Arch Dis Child* **55**, 516–520.

Belfield, A. & Goldberg, D.M. (1971) Normal ranges and diagnostic value of serum 5-nucleotidase and alkaline phosphatase activities in infancy. *Arch Dis Child* **46**, 842–846.

Berry, J., Betts, P. & Wood, P.J. (1986) The interpretation of bloodspot 17 hydroxyprogesterone levels in term and pre-term neonates. *Ann Clin Biochem* **23**, 546–551.

Bolton, N.J., Tapanainen, J., Koivisto, M. & Vihko, R. (1989) Circulating sex hormone-binding globulin and testosterone in newborns and infants. *Clin Endocrinol* **31**, 201–207.

Brozovic, B., Burland, W.L., Simpson, K. & Lord, J. (1974) Iron status of preterm low birth weight infants and their response to oral iron. *Arch Dis Child* **49**, 386–389.

Cejka, J., Mood, D.W. & Kim, C.S. (1974) Immunoglobulin concentrations in sera of normal children: quantitation against an international reference preparation. *Clin Chem* **20**, 656–659.

Cornblath, M., Parker, M.L., Reisner, S.H., Forbes, A.E. & Daughaday, W.H. (1965) Secretion and metabolism of growth hormone in premature and full term infants. *J Clin Endocrinol* **25**, 209–218.

Davies, D.P. (1973) Plasma osmolality and protein intake in preterm infants. *Arch Dis Child* **48**, 575–579.

Ellis, G. & Goldberg, D.M. (1969) The enzymological diagnosis of galactosaemia. *Ann Clin Biochem* **6**, 70–73.

Fomon, S.J., Filer, L.J., Thomas, L.N. & Rogers, R.R. (1970) Growth and serum chemical values of normal breast fed infants. *Acta Paed Scand Suppl* **202**.

Freer, D.E., Statland, B.E., Johnson, M. & Felton, H. (1979) Reference values for selected enzyme activities and protein concentrations in serum and plasma derived from cord-blood specimens. *Clin Chem* **25**, 565–569.

Henkin, R.I., Marshall, J.R. & Meret, S. (1971) Maternal–fetal metabolism of copper and zinc at term. *Am J Obstet Gynecol* **110**, 131–134.

Herz, F., Kaplan, E. & Scheye, E.S. (1973) Erythrocyte acetylcholinesterase and glucose-6-phosphate dehydrogenase in newborn infants of low birth weight. *Clin Chim Acta* **46**, 147–152.

Hillman, L.S., Rojanasathit, S., Slatopolsky, E. & Haddad, J.G. (1977) Serial measurements of serum calcium, magnesium, parathyroid hormone, calcitonin and 25-hydroxy vitamin D in premature and term infants during the first week of life. *Pediatr Res* **11**, 739–744.

Hughes, I.A., Riad-Fahmy, D. & Griffiths, K. (1979) Plasma 17OH-progesterone concentrations in newborn infants. *Arch Dis Child* **54**, 347–349.

Jacobsen, B.B. & Hummer, L. (1979) Changes in serum concentrations of thyroid hormones and thyroid hormone-binding proteins during early infancy. *Acta Paed Scand* **68**, 411–418.

Jacobsen, B.B., Andersen, H.J., Peitersen, B., Dige-Petersen, H. & Hummer, L. (1977) Serum levels of thyrotropin, thyroxine and triiodothyronine in full term, small for gestational age and preterm newborn babies. *Acta Paed Scand* **66**, 681–687.

John, R. & Bamforth, F.J. (1987) Serum free thyroxine and free triiodothyronine concentrations in healthy fullterm, preterm and sick preterm neonates. *Ann Clin Biochem* **24**, 461–465.

Karlen, J., Aperia, A. & Zetterstrom, R. (1985) Renal excretion of calcium and phosphate in preterm and term infants. *J Pediatr* **106**, 814–819.

Koch, G. & Wendel, H. (1968) Adjustment of arterial blood gases and acid base balance in the normal newborn infant during the first week of life. *Biol Neonat* **12**, 136–161.

Kueppers, F. & Offord, K.P. (1979) Alpha-1 antitrypsin elevation in healthy neonates. *J Lab Clin Med* **94**, 475–480.

Kupke, I.R., Tritschler, W., Kather, B. & Bablok, W. (1980) Creatinkinase 'NAC-aktiviert': Referenzwerte bei Kindern. *Klin Pediat* **192**, 348–350.

McCance, R.A. & Widdowson, E.M. (1947) Blood urea in the first nine days of life. *Lancet* **1**, 787–788.

Ohtake, M. Tamura (1977) Serum zinc and copper levels in healthy Japanese infants. *Tohoku J Exp Med* **120**, 99–103.

Overman, R.R., Etteldorf, J.N., Bass, A.L. & Horn, G.B. (1951) Plasma and erythrocyte chemistry of the normal infant from birth to two years of age. *Pediatrics* **7**, 565–576.

Reardan, H.S., Baumann, M.L. & Haddad, E.J. (1960) Chemical stimuli of respiration in the early neonatal period. *J Pediatr* **57**, 151.

Rokicki, W., Forest, M.G., Loras, B., Bonnet, H. & Bertrand, J. (1990) Free cortisol in human plasma in the first 3 months of life. *Biol Neonate* **57**, 21–29.

Rudd, P.T., Hughes, E.A. & Placzek, M.M. (1983) Reference ranges for plasma creatinine during the first month of life. *Arch Dis Child* **58**, 212–215.

Saarinen, U.M. & Siimes, M.A. (1978) Serum ferritin in assessment of iron nutrition in healthy infants. *Acta Paed Scand* **67**, 745–751.

Scott, P.H., Berger, H.M. & Wharton, B.A. (1985) Growth velocity and plasma amino acids in the newborn. *Pediatr Res* **19**, 446–450.

Scott, P.H., Sandham S., Balmer, S.E. & Wharton, B.A. (1990) Diet-related reference values for plasma amino acids in newborns measured by reversed-phase HPLC. *Clin Chem* **36**, 1922–1927.

Shaw, J.C.L. (1982) Iron absorption by the premature infant. *Acta Paed Scand Suppl* **299**, 83–89.

Shore, G.M., Hoberman, L., Dowdey, A.B.C. & Combes, B. (1975) Serum gamma-glutamyl transpeptidase activity in normal children. *Am J Clin Pathol* **63**, 245–250.

Sitzmann, N.F.C., Low, C., Kaloud, H. & Prestele, H. (1974) Normalwerte der serumtransaminasen GOT und GPT mit neuen substratoptimierten Standardmethoden. *Klin Pediat* **186**, 346–352.

Srinivasan, G., Pildes, R.S., Cattamanchi, G., Viora, S. & Lillien, L.D. (1986) Plasma glucose values in normal neonates: a new look. *J Pediatr* **109**, 114–117.

Stanley, C.A. & Baker, L. (1976) Hyperinsulinism in infants and children – diagnosis and therapy. *Adv Pediatr* **23**, 315–355.

Sutton, A., Harvie, A., Cockburn, F., Farquharson, J. & Logan, R.W. (1985) Copper deficiency in the preterm infant of very low birthweight. *Arch Dis Child* **60**, 644–651.

Thalme, B. (1972) Calcium, chloride, cholesterol, inorganic phosphorus and total protein in blood plasma during the early neonatal period studied with ultramicrochemical methods. *Acta Paed Scand* **51**, 649–660.

Veit, S., Sitzmann, F. & Prestele, H. (1975) Normalwerte für Lactat- und Glutamatdehydrogenase, sowie Leucinarylamidase, erstellt mit optimierten standardansätzen. *Klin Pediat* **187**, 244–251.

Westwood, A. (1982) Determination of total and direct bilirubin in plasma by means of a biochromatic method on a centrifugal analyser. *Ann Clin Biochem* **19**, 151–156.

Wharton, B.A., Bassi, U., Gough, G. & Williams, A. (1971) Clinical value of plasma creatine kinase and uric acid levels during first week of life. *Arch Dis Child* **46**, 356–362.

Young, W.F., Hallum, J.L. & McCance, R.A. (1941) The secretion of urine by premature infants. *Arch Dis Child* **16**, 243–252.

A2: Reference Data excluding Neonates

A. GREEN & D. ISHERWOOD

Acid phosphatase (serum/plasma)

Method
Fluorimetric/α-naphthylphosphate.

Age (years)	U/L
Up to 12	2.7–10.2
Over 12	0.9–3.0

Notes
1 Activity is age dependent and declines slowly with age to reach normal adult values after the age of 14 years.
2 There is no significant sex-related difference in activity.

See also:

Chen, J., Yam, L., Janckila, A.J., Chin-Yang, L. & Lam, W.K.W. (1979) Significance of 'high' acid phosphatase activity in the serum of normal children. *Clin Chem* **25**, 719–722.

Alanine aminotransferase (ALT) (plasma)

Method
Kodak Ektachem at 37°C.

Age (years)	U/L
1–3	5–45
4–6	10–25
7–9	10–35

See also:

Lockitch, G., Halstead, A.C., Albersheim, S., MacCallum, C. & Quigley, G. (1988) Age- and sex-specific paediatric reference intervals for biochemistry analytes as measured with the Ektachem-700 analyser. *Clin Chem* **34**, 1622–1625.

Albumin

See p. 465.

Albumin : creatinine ratio (urine)

Urine collections over 24 h divided into day-time and night-time specimens.

Method
Immunoturbidimetric.

	A:C ratio (mg/mmol)	Albumin excretion rate (μg/min)
12-h overnight specimens		
Boys and girls	0.26–2.10	0.79–7.8
12-h ambulant day-time specimens		
Boys and girls	0.20–6.9	0.52–35.1

Notes
1 Excretion of protein in urine is affected by posture and activity.
2 Albumin excretion during the day exceeds that at night in both boys and girls.

See also:

Davies, A.G., Postlethwaite, R.J., Price, D.A. *et al.* (1984) Urinary albumin excretion in school children. *Arch Dis Child* **59**, 625–630.

Rowe, D.J.F., Bagga, H. & Betts, P.B. (1985) Normal variations in rate of albumin excretion and albumin to creatinine ratios in overnight and daytime urine collections in non-diabetic children. *Br Med J* **291**, 693–694.

Salardi, S., Cacciari, E., Pascucci, M.G. *et al.* (1990)

Microalbuminuria in diabetic children and adolescents. *Acta Paed Scand* **79**, 437–443.

Alkaline phosphatase (ALP) (plasma/serum)

Method

Nitrophenylphosphate/diethanolamine buffer at 37°C. The reference ranges are *highly* method dependent.

Age	U/L Girls	Boys
6 months–9 years (girls and boys)	250–1000	
10–11	250–950	250–730
12–13	200–730	275–875
14–15	170–460	170–970
16–18	100–250	125–720
Over 18 (girls and boys)	100–250	

Notes

1 ALP activity changes markedly throughout childhood and rises rapidly during the adolescent growth spurt, particularly in boys. Increases in bone ALP of 3.4–4.0 times the adult range can be seen during this time. Values of over 1000 in teenage boys or 800 in teenage girls require investigation.

2 Marked transient increases may be observed in infants/young children (usually under 5 years) in the absence of associated pathology. The activity is often dramatically increased, exceeding 10-fold the upper limit of normal, and returns to normal within 4 months.

See also:

Penttila, I.M., Jokela, A., Viitala, A.J. *et al.* (1975) Activities of aspartate and alanine aminotransferases and alkaline phosphatase in sera of healthy subjects. *Scand J Clin Lab Invest* **35**, 275–284.

Lockitch, G., Halstead, A.C., Albersheim, S., MacCallum, C. & Quigley, G. (1988) Age- and sex-specific paediatric reference intervals for biochemistry analytes as measured with the Ektachem-700 analyser. *Clin Chem* **34**, 1622–1625.

α_1-Antichymotrypsin

See p. 473.

α_1-Antitrypsin

See p. 468.

α-Fetoprotein (AFP)

See p. 467.

Aluminium (serum)

See p. 377.

Amino acids

Heparinized venous plasma (after an overnight fast).

Method

Amino acid analyser.

Amino acid	3 months–10 years (μmol/L)	6–18 years (μmol/L)
Taurine	11–93	47–120
Aspartic acid	3–12	5–13
Threonine	40–139	125–200
Serine	93–176	106–173
Glutamic acid	11–79	12–41
Glutamine	475–746	723–1079
Asparagine	71–149	40–147
Proline	40–332	73–301
Glycine	125–318	144–338
Alanine	148–475	226–646
Citrulline	8–47	14–86
α-Amino-*n*-butyric acid	12–43	
Cystine	23–68	77–111
Valine	85–334	115–325
Methionine	5–34	19–25
Isoleucine	13–81	47–79
Leucine	40–158	90–160
Tyrosine	24–105	39–76
Phenylalanine	34–101	38–80
Histidine	22–108	48–120
Tryptophan	12–69	23–53
Ornithine	27–96	18–100
Lysine	85–218	94–250
Arginine	32–142	63–119

Notes

1 Plasma should be deproteinized as soon as possible and certainly *within 30 min*.

2 Values for 6–18 years from S. Krywawych (UGL Hospitals) (personal communication).

See also:

Applegarth, D.A., Edelsten, A.D., Wong, L.T.K. & Morrison, B.J. (1979) Observed range of assay values for plasma and CSF amino acid levels in infants and children aged 3 months to 10 years. *Clin Biochem* **12**(5), 1173–1178.

Ammonia (plasma)

See p. 104.

Notes

1 Venous/arterial sampling is preferred.

2 Capillary concentrations are generally higher than venous or arterial samples, particularly in the newborn period. If capillary blood is used great care must be taken to avoid contamination from sweat and the environment.

3 Specimens must be transported to the laboratory promptly, on ice.

Amylase (plasma/serum)

Method

Phadebas (Kabi Pharmaceuticals, Davy Avenue, Knowhill, Milton Keynes, MK5 8PH).

Age (years)	U/L
1–15	98–405

Notes

1 Activity is low in the newborn and increases throughout early infancy. Mature levels are observed in some children by 2 months of age and in most cases by 9 months.

2 Activity depends on the method used – consult your local laboratory.

See also:

Aggett, P.J. & Taylor, F. (1980) A normal paediatric amylase range. *Arch Dis Child* **55**, 236–238.

Aspartate aminotransferase (AST) (plasma/serum)

Method

Kodak Ektachem at 37°C.

Age (years)	U/L
1–3	20–60
4–6	15–50
7–9	15–40

Note

1 Activity in newborns and early infancy (up to 6 months) is significantly higher than in childhood (approximately two- to threefold) and shows a gradual decline throughout infancy.

See also:

Lockitch, G., Halstead, A.C., Albersheim, S., MacCallum, C. & Quigley, G. (1988) Age- and sex-specific paediatric reference intervals for biochemistry analytes as measured with the Ektachem-700 analyser. *Clin Chem* **34**, 1622–1625.

Bicarbonate (plasma/serum)

Method

Venous blood, fasting. Ion-selective electrode.

Age (years)	mmol/L
1–4	17–25
4–8	19–27
>8	21–29

Note

1 Bicarbonate concentrations are lower in infants than in children or adults. Adult levels are reached by approximately 7–12 years.

See also:

Burrit, M.F., Slockbower, J.M., Forsman, R.W. *et al.* (1990) Pediatric reference intervals for 19 biologic variables in healthy children. *Mayo Clin Proc* **65**, 329–336.

Bilirubin (total) (plasma/serum)

Method
Kodak Ektachem.

Age (years)	μmol/L
1–19	10–24

Notes
1 Concentration falls if the specimen is exposed to strong sunlight.
2 Bilirubin concentrations normally decline rapidly after the first week. A concentration of 750 μmol/L after 14 days is not normal and should be investigated.

See also:
Lockitch, G., Halstead, A.C., Albersheim, S., MacCallum, C. & Quigley, G. (1988) Age- and sex-specific paediatric reference intervals for biochemistry analytes as measured with the Ektachem-700 analyser. *Clin Chem* **34**, 1622–1625.

Bilirubin (conjugated or 'direct' reacting) (plasma/serum)

Method
Kodak Ektachem.

Age (years)	μmol/L
1–19	<2

Note
1 An increased concentration of conjugated bilirubin must be investigated.
2 Reference data are highly method dependent.

See also:
Lockitch, G., Halstead, A.C., Wadsworth, L. *et al.* (1988) Age- and sex-specific paediatric reference intervals for biochemistry analytes as measured with the Ektachem-700 analyser. *Clin Chem* **34**, 1622–1625.

Bilirubin (unconjugated) (plasma/serum)

Method
Kodak Ektachem.

Age (years)	μmol/L
1–19	3–17

See also:
Lockitch, G., Halstead, A.C., Wadsworth, L. *et al.* (1988) Age- and sex-specific paediatric reference intervals for biochemistry analytes as measured with the Ektachem-700 analyser. *Clin Chem* **34**, 1622–1625.

Caeruloplasmin

See p. 469.

Calcium (total) (plasma)

Venous plasma from fasting children; blood collected between 07.00 and 09.00 h.

Method
Atomic absorption spectrophotometry (up to 10 years).
Cresolphthalein complexone (over 10 years).

Age (years)	mmol/L
Girls	
1–12	2.4–2.65
12–15	2.38–2.60
15–19	2.28–2.58
Boys	
1–15	2.4–2.65
15–17	2.38–2.63
17–19	2.38–2.60

See also Chapter 12.

Calcium (urine)

Method
Atomic absorption spectrophotometry.

Age (years)	mmol/kg per 24 h (upper limit)
1–15	0.1

Note

1 Elevated excretions are observed up to 10 days postoperatively and during prolonged immobilization.

See also:

Ghazali, S. & Barratt, T.M. (1974) Urinary excretion of calcium and magnesium in children. *Arch Dis Child* **49**, 97–101.

Calcium/creatinine (urine)

Method

Atomic absorption spectrometry on *second* urine passed after overnight fast (i.e. *not* the overnight urine).

Age (years)	mol/mmol (mean ± 2SD)
1–15	0.40 ± 0.34

Notes

1 There is considerable diurnal variation in calcium excretion and the use of other random urines may lead to falsely elevated levels.

2 Elevated values may be observed up to 10 days postoperatively and during prolonged immobilization.

See also:

Ghazali, S. & Barratt, T.M. (1974) Urinary excretion of calcium and magnesium in children. *Arch Dis Child* **49**, 97–101.

Carotene (serum)

Method

Spectrophotometry.

μmol/L
0.9–3.7

Notes

1 Normal values show wide variation with method and dietary habits. Slightly higher levels are found in girls than in boys.

2 Hypercarotenaemia can be confused with jaundice. Serum vitamin A and serum lipids may assist with the differential diagnosis.

3 The neonate <1 week has relatively low carotene levels (<0.8 μmol/L).

4 Levels increase during early infancy with the introduction of solid food.

5 The most likely cause of carotenaemia is excessive dietary consumption of carotene-rich foods. More rarely, it can be associated with certain disease states such as diabetes mellitus and hypothyroidism.

6 The inherited defect carotenaemia is associated with high carotene concentrations and low vitamin A levels.

See also:

Leung, A.K.C. (1987) Carotenemia. *Adv Pediatr* **34**, 223–248.

Chloride (plasma/serum)

Method

Ferric thiocyanate/Autoanalyzer (1–7 years).
Chloride meter (7–16 years).

Age (years)	mmol/L
1–16	98–107

Chloride (faeces)

Faecal chloride should be less than the sum of the concentrations of sodium and potassium.

Method

If the faeces are liquid, centrifuge and analyse the supernatant.

If the faeces are solid, homogenize in water, centrifuge and analyse the supernatant.

Note

1 High chloride is suggestive of congenital chloride malabsorption, characterized by a metabolic alkalosis.

See also:

Homberg, C., Perheentupa, J., Launiala, K. & Hallman, N. (1977) Excess loss of chloride compared with sodium and potassium may be familial. *Arch Dis Child* **52**, 255.

Cholesterol

See p. 311.

Cholinesterase (plasma)

Method

Cholinestrase assay is very method dependent and it is essential to consult your laboratory.

Notes

1 The main value of the cholinesterase assay relates to the use of a muscle relaxant suxamethonium (succinyl dicholine, scoline) which can produce prolonged apnoea in certain genetically predisposed individuals.
2 Genetic typing is complex and best done in a specialist laboratory.

Complement

See p. 475.

Copper (plasma/serum)

Method

Atomic absorption spectrophotometry.

Age	µmol/L
1–6 months	9.3–14.6
1–5 years	12.6–23.6
6–9 years	13.2–21.4
10–13 years	12.6–19.0
Over 14 years	11.0–22.0

Notes

1 Neonates have low concentrations, especially during the first week.
2 Concentrations increase, with the highest levels in late infancy, early childhood.

See also:

Henkin, R.I., Schulman, H.D., Schulman, C.B. & Bronzert, D.A. (1973) Changes in total, non-diffusible and diffusible plasma zinc and copper during infancy. *J Pediatr* **82**, 831–837.

Lockitch, G., Halstead, A.C., Wadsworth, L. *et al.* (1988) Age- and sex-specific pediatric reference intervals and correlations for zinc, copper, selenium, iron, vitamins A and E, and related proteins. *Clin Chem* **34**, 1625–1628.

See also Chapter 23.

Copper (urine)

Method

Atomic absorption spectrophotometry.

Age (years)	µmol/mol creatinine	µmol/24 h
5–18	6–119	0.63

Notes

1 In untreated Wilson's disease urinary copper excretion is typically 1.2–15.7 µmol/24 h.
2 After D-penicillamine load:
In patients with Wilson's disease urinary copper exceeds 15.6 µmol/L.
3 In the early presymptomatic phase of the disease, the urinary copper may be normal.

See also:

Sternlieb, I. (1990) Perspectives on Wilson's Disease. *Hepatolology* **12**, 1234–1239.

Cortisol

See p. 540.

Creatine kinase (CK) (serum/plasma)

Method

Kodak Ektachem-700 at 37°C.
Ranges are *highly* method dependent.

Age (years)	U/L
1–3	60–305
4–6	75–230
7–9	60–365
10–11	
(M)	55–215
(F)	80–230
12–13	
(M)	60–330
(F)	50–295
14–15	
(M)	60–335
(F)	50–240

Age (years)	U/L
16–19	
(M)	55–370
(F)	45–230

Notes

1 Activity increases after exercise or muscle trauma.
2 CK activity is high in newborns (up to 10-fold adult levels) with a rapid decline during the first week. Activity continues to fall markedly during the first year. Levels are stable by the age of 1 year and remain so throughout childhood. In males, levels rise at 15–16 years before declining to adult levels. In females, activity continues to decline gradually.
3 Use of CK values for prediction of carrier detection for Duchenne muscular dystrophy requires the use of a sensitive and precise method for which predictive values have been established. The use of this test has been largely superseded by the use of DNA analysis. Consult your local laboratory for advice.
4 African and West Indian individuals have higher CK levels than do Caucasians; whether this applies to children is not known at present.

See also:

Lockitch, G., Halstead, A.C., Albersheim, S., MacCallum, C. & Quigley, G. (1988) Age- and sex-specific paediatric reference intervals for biochemistry analytes as measured with the Ektachem-700 analyser. *Clin Chem* **34**, 1622–1625.

Creatinine (serum)

See p. 255.

C-reactive protein

See p. 471.

Dihydropteridine reductase (DHPR)

Method

Blood/dried blood spots may be used.
Spectrophotometric at 37°C.

	μmol NADH oxidized/min per g Hb	
Age	Mean	SD
<1 month	1.11	0.31
1–12 months	1.17	0.34
1–12 years	0.96	0.25
>12 years	0.87	0.20

Notes

1 Measurement of DHPR is used as part of the differential diagnosis of hyperphenylalaninaemia.
2 Infants >1 year have lower activity than those <12 months.
3 Specimens (as dried blood) can be sent through the post.
4 Heterozygotes have an activity which overlaps with normal but is usually <0.6.
5 Homozygotes with DHPR deficiency usually have undetectable activity.

See also:

Surplice, I.M., Griffiths, P.D., Green, A. & Leeming, R.J. (1990) Dihydropteridine reductase activity in eluates from dried blood spots: automation of an assay for a national screening service. *J Inher Metab Dis* **13**, 169–177.

Ferritin

See p. 470.

Fructosamine (serum)

Method

Nitroblue-tetrazolium reduction (Roche).
The reference range is *highly* method dependent.

Age (years)	mmol/L
0–3	1.56–2.27
3–6	1.73–2.34
6–9	1.82–2.56
9–12	2.04–2.50
12–16	2.02–2.63

See also:

Abe, F., Yano, M., Minami, Y. *et al.* (1989) Alterations in fructosamine and glycated albumin levels during childhood. *Ann Clin Biochem* **26**, 328–331.
Schepper, J. De, Derde, M.-P., Goubert, P. & Gorus, F. (1988) Reference values for fructosamine concentrations

in children's sera: influence of protein concentration, age and sex. *Clin Chem* **34**, 2444–2447.

Galactokinase (erythrocytes)

Method

[^{14}C]-Galactose as substrate with chromatography.

	µmol galactose phosphate/h per g Hb	
Age	Mean	Range
1–6 months	1.9	1.2–3.9
6 months–1 year	1.5	1.1–2.0
1 year	1.0	0.7–1.5

Notes

1 After a blood transfusion at least 6 weeks must elapse before a reliable result can be obtained.

2 Blood specimen taken into lithium heparin.

3 Erythrocytes must be prepared within 30 min of blood being taken.

See also:

Ng, W.G., Donnell, G.N. & Begren, W.R. (1965) Galactokinase activity in human erythrocytes of individuals at different ages. *J Lab Clin Med* **66**, 115–121.

Galactose-1-phosphate (erythrocytes)

Method

Radioenzymatic method with chromatography.

	µmol/L erythrocytes
Children	<10
Galactosaemic children with adequate dietary control	<150

Note

1 Lithium heparin specimens as whole blood must be transported to the laboratory within 24 h of collection.

See also:

Dobbie, J.A. & Holton, J.B. (1986) A modified method for the estimation of galactose-1-phosphate. *Ann Clin Biochem* **23**, 325–328.

Galactose-1-phosphate uridyl transferase (erythrocytes)

Method

[^{14}C]-Galactose-1-phosphate.

	µmol galactose-1-phosphate/h per g Hb
Homozygous/normal	18–28
Normal/galactosaemic heterozygote	8–12
Classical galactosaemic	<0.5

Notes

1 Lithium heparin specimen must be transported as whole blood to laboratory within 24 h of collection.

2 After a blood transfusion at least 6 weeks must elapse before a reliable result can be obtained.

See also:

Monk, A.M., Mitchell, D.W.A., Milligan, D.W.A. & Holton, J.B. (1977) Diagnosis of classical galactosaemia. *Arch Dis Child* **52**, 943.

Ng, W.G., Bergren, W.R. & Donnell G.N. (1967) An improved procedure for the assay of hemolysate galactose-1-phosphate uridyl transferase activity by the use of ^{14}C-labelled galactose-1-phosphate. *Clin Chim Acta* **15**, 489–492.

Glucose (plasma)

Preprandial specimens.

Method

Kodak Ektachem.

Age (years)	mmol/L
1–19	3.9–7.0

Note

1 Plasma glucose concentrations are 11–15% higher than in whole blood.

See also:

Lockitch, G., Halstead, A.C., Albersheim, S., MacCallum, C. & Quigley, G. (1988) Age- and sex-specific paediatric reference intervals for biochemistry analytes as measured with the Ektachem-700 analyser. *Clin Chem* **34**, 1622–1625.

γ-Glutamyl transpeptidase (plasma)

Method
37°C kinetic.

Age	U/L (upper limit, mean + 2SD)
1–2 months	114
2–4 months	81
4–7 years	34
7–12 years	23
1–15 years	24

Note
1 High activity in the neonatal period and early infancy with gradual decline over the first 6 months. After 1 year activity is comparable with adult levels.

See also:
Knight, J.A. & Haymond, R.E. (1981) γ-Glutamyl transferase and alkaline phosphatase activities compared in serum of normal children and children with liver disease. *Clin Chem* **27**, 48–51.

Glucose-6-phosphate dehydrogenase (erythrocytes)

Method
Spectrophotometry at 37°C.

	μmol NADPH reduced/min per g Hb
Boys	7.2–22.8
Girls	7.6–20.5
Hemizygous males	<2.0

Glycosaminoglycans (urine)

Uronic acid:creatinine ratio.

Method
24-h urine. Precipitation of glycosaminoglycans with cetylpyridinium chloride followed by reaction with carbazole in borate/sulphuric acid.

Age (years)	μmol/mmol creatinine (observed range)
≤2	5.8–33
>2	1.2–16.9

Notes
1 Levels decline throughout childhood after 2 years.
2 Screening tests are unreliable, especially if not related to creatinine, and may give false-positive/negative results.
3 Urines with a low creatinine, i.e. <2 mmol/L, may give misleading results and should be repeated.
4 Separation of individual glycosaminoglycans, e.g. cellulose acetate electrophoresis, is of value in the diagnosis of a particular type of mucopolysaccharidosis.

Homovanillic acid and vanillylmandelic acid

See p. 440.

17-Hydroxyprogesterone

See p. 540.

Immunoglobulins (IgGs)

IgG1–4, see p. 477.

Immunoreactive trypsin (IRT) (serum/blood spot)

Method
Fluoroimmunoassay (Delfia).

	μg/L
Blood spot (0–2 weeks)	<70
Serum	10–130

Notes
1 The high blood IRT concentrations found in the neonate with cystic fibrosis decline with age, and beyond very early infancy the blood-spot assay has limited usefulness. Better discrimination is achieved with the serum assay.
2 Up to 2 years of age at least 90% of infants with cystic fibrosis have serum IRT concentrations which exceed the upper limit of normal; beyond this age normal or low levels predominate.

See also:

Heeley, A.F. & Bangert, S.K. (1992) The neonatal detection of cystic fibrosis by measurement of immunoreactive trypsin in blood. *Ann Clin Biochem* **29**, 361–376.

Insulin

Results for paediatric insulin levels are highly method dependent. Correct specimen collection is essential. Contact your Chemical Pathology Department before taking the specimen to find out their reference range. Alternatively, contact your local SAS laboratory.

Iron (serum)

Method
Spectrophotometric (2 months–1 year).
Ferrozine (1–19 years).

Age	Iron (μmol/L)	TIBC (μmol/L)
4–12 months	3–29	40–78
1–10 years (boys + girls)	4–25	48–91
Boys		
10–14 years	5–24	54–91
14–19 years	6–29	52–102
Girls		
10–14 years	8–26	57–103
14–19 years	5–33	52–101

TIBC, total iron binding capacity.

See also:

Lockitch, G., Halstead, A.C., Wadsworth, L. *et al.* (1988) Age- and sex-specific paediatric reference intervals and correlations for zinc, copper, selenium, iron, vitamins A and E and related proteins. *Clin Chem* **34**, 1625–1628.
Saarinen, U.M. & Siimes, N.A. (1977) Developmental changes in serum iron TIBC and transferrin saturation in infancy. *J Pediatr* **91**, 875–877.

Lactate

See p. 90.

Lactate dehydrogenase (plasma/serum)

Method
Assayed at 37°C.
Activity varies with method conditions – consult your local laboratory.

Age	U/L (upper limit)
7 months–1 year	1097
1–3 years	849
4–6 years	615
7–12 years	
(M)	764
(F)	582
13–17 years	
(M)	683
(F)	436
17 years	480

Notes
1 Neonates in first week show a marked fall in activity with a more gradual, but continued, fall throughout infancy and early childhood.
2 Males have higher activity than females during adolescence.

See also:

Fischback, F. & Zawta, B. (1992) Age-dependent reference limits of several enzymes in plasma at different measuring temperatures. *Klin Lab* **38**, 555–561.

Lead (blood)

See p. 379.

Method
Atomic absorption spectrophotometry.

Notes
1 Whole blood should be collected into lithium heparin or ethylenediaminetetraacetate (EDTA).
2 Specimens must not be allowed to clot as most of the lead is in the erythrocytes.
3 *Interference*: lead is widely distributed in dust and specimens must be collected with care to minimize contamination.
4 If excessive exposure is found in a child as indicated by an increased high blood-lead concentra-

tion, the other siblings in the family should have their blood-lead levels measured.
5 In suspected severe/acute lead poisoning an X-ray examination of the abdomen is of value in identifying radio-opaque fragments of ingested source of lead.

Magnesium (plasma/serum)

Method
Capillary sample, Kodak Ektachem (1 month–14 years).
Venous sample, atomic absorption spectrophotometry (7–17 years).

Age	mmol/L
1 month–2 years	0.65–1.05
2–6 years	0.60–1.00
6–14 years	0.65–0.95
7–17 years	0.73–0.99

Notes
1 There is no significant change during puberty.
2 Capillary specimens generally yield higher results than venous specimens.
3 Haemolysis causes increased values.

See also:
Clayton B.E., Jenkins, P. & Round, J.M. (eds) (1980) *Paediatric Chemical Pathology*, pp. 107–108. Blackwell Scientific Publications, Oxford.
Meites, S. (1989) *Paediatric Clinical Chemistry*, 3rd edn, p. 191. American Association for Clinical Chemistry, Washington, DC.

Magnesium (urine)

Method
Atomic absorption spectrophotometry of 24-h urine specimen.

Age (years)	mmol/kg per 24 h (mean ± 2SD)
1–15	0.116 ± 0.065

Note
1 These values are for healthy children receiving normal diets. Changes in intake can produce large variations in excretion.

See also:
Ghazali, S. & Barratt, T.M. (1974) Urinary excretion of calcium and magnesium in children. *Arch Dis Child* **49**, 97–101.

Magnesium/creatinine (urine)

Method
Atomic absorption spectrophotometry.

Age (years)	mmol/mmol (mean ± 2SD)
1–15	0.977 ± 0.93

Note
1 Urine collected was *second* specimen passed after an overnight fast, i.e. *first* specimen passed after the overnight urine had been voided.

See also:
Ghazali, S. & Barratt, T.M. (1974) Urinary excretion of calcium and magnesium in children. *Arch Dis Child* **49**, 97–101.

Osmolality (plasma/serum)

Method
Freezing point depression.

mosmol/kg
275–295

Notes
1 Hyperosmolality, in addition to being caused by hypernatraemia, hyperglycaemia, uraemia and abnormal metabolites, can also result from the presence of exogenous molecules such as prescribed drugs (e.g. mannitol), poisoning and contrast media.

Oxalate (urine)

Method

Enzymatic with immobilized oxalate oxidase; creatinine modified Jaffe reaction with Autoanalyser 2.

Urinary oxalate:creatinine molar ratios:

Age (years)	Mean	Range
<1	0.061	0.015–0.26
1–5	0.036	0.011–0.12
5–12	0.030	0.0059–0.15
>12	0.013	0.0021–0.083

Notes

1 Plasma oxalate:creatinine molar ratios (independent of age): mean 0.033.

2 Plasma oxalate concentration: mean 1.53; range 0.78–3.02 μmol/L.

See also:

Barratt, T.M., Kasidas, G.P., Muirdoch, I. & Rose, G.A. (1991) Urinary oxalate and glyoxylate excretion, and plasma oxalate concentration. *Arch Dis Child* **66**, 500–503.

$P\text{CO}_2$ (blood)

Method

Arterialized capillary blood.

Age (years)	kPa
1–3	3.7–5.3
3–7	4.1–5.3
7–12	4.3–5.5
12–18	4.5–5.7

Notes

1 Capillary gases are unreliable in conditions of hypotension and poor perfusion.

2 After 3 years $P\text{CO}_2$ increases gradually with age to reach adult values after 18 years.

See also:

Dong, Sheng-Huang, Liu, Hui-Min, Song, Guo-Wei, Rong, Zhen-Pen & Wu, Yan-Ping (1985) Arterialized capillary blood gases and acid–base studies in normal individuals from 29 days to 24 years of age. *Am J Dis Child* **139**, 1019–1022.

H^+/pH (blood) (see Fig. A2.1 *opposite*)

Method

Arterialized **free-flowing** capillary blood. (The ideal specimen is arterial blood taken anaerobically into a heparinized syringe.)

Age	H^+ (nmol/L)	pH
1 month	37–46	7.43–7.34
1–18 years	35–44	7.46–7.36

Notes

1 Special tubes should be used for arterialized capillary blood; contact your Chemical Pathology Department for advice.

2 Blood **must** be taken to the laboratory promptly.

See also:

Dong, Sheng-Huang, Liu, Hui-Min, Song, Guo-Wei, Rong Zhen-Pen & Wu, Yan-Ping (1985) Arterialized capillary blood gases and acid–base studies in normal individuals from 29 days to 24 years of age. *Am J Dis Child* **139**, 1019–1022.

pH (faecal)

pH > 7.0 (except when the infant is receiving breast milk, in which case the pH will be acid).

The specimen should be collected directly into a container or on to a polythene liner and sent immediately to the laboratory or deep frozen within a few minutes of being passed. Misleading results can be obtained if the liquid portion of a specimen is allowed to soak into the nappy.

An acid pH (except in breast-fed infants) supports the diagnosis of sugar intolerance but an alkaline pH does not on its own exclude the diagnosis.

pH (urine)

Method

Urine pH should be measured with a glass electrode.

Notes

1 Urine must be fresh; if any infection is present it will invalidate the result.

2 If a narrow-range indicator paper is used in

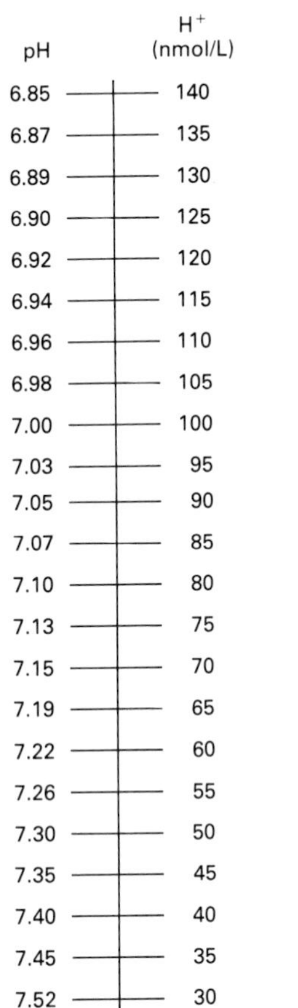

Fig. A2.1 Relationship of pH to hydrogen ion [H^+] concentration. From: Insley, J. (ed.) (1990) *A Paediatric Vade Mecum*, 12th edn. Edward Arnold, Kent.

an emergency, it is essential that the paper should have been stored dry and according to the manufacturer's instructions.

3 *Interpretation*: infants and children usually pass acid urine during the night. If the patient is not acidotic, i.e. H^+ <50 nmol/L and if the pH of the overnight urine specimen is less than 5.5, a defect of hydrogen ion excretion is excluded and no further assessment is necessary.

Phenylalanine (plasma)

	μmol/L
Infants/children	30–100

Notes

1 Plasma phenylalanine concentrations in untreated infants of greater than 200 μmol/L should be followed up.

2 Dietary treatment of phenylketonuria: the aim is to keep phenylalanine concentrations between 120 and 360 μmol/L at least until school age.

See also:

Report of Medical Research Council Working Party on Phenylketonuria (1993) Recommendations on the dietary management of phenylketonuria. *Arch Dis Child* **68**, 426–427.

See also Amino acids, p. 524.

Phosphate (inorganic) (plasma/serum)

Method

Fasting venous blood, Kodak Ektachem (1–7 years). Phosphomolybdate, non-fasting (7–19 years).

Age (years)	mmol/L
Boys and girls	
1–3	1.25–2.10
4–6	1.30–1.75

Changes during puberty

Boys Genital development	mmol/L	Girls Breast maturity	mmol/L
1	1.10–1.46	1	1.05–1.53
2	1.02–1.54	2	1.03–1.55
3	1.19–1.53	3	1.03–1.41
4	1.03–1.49	4	0.92–1.36
5	0.95–1.27	5	0.90–1.30

Chronological age

Years	Boys (mmol/L)	Girls (mmol/L)
8	1.10–1.62	1.06–1.54
9	1.05–1.70	1.06–1.43
10	1.08–1.50	1.09–1.43
11	1.12–1.45	1.01–1.52
12	1.06–1.55	0.98–1.50
13	1.04–1.62	0.91–1.49
14	0.99–1.57	0.85–1.39
15	0.96–1.47	0.88–1.35
16	0.93–1.33	0.87–1.34

Notes

1 There is a gradual decrease in concentration during childhood.

2 Haemolysis causes artefactually increased concentrations.

See also:

Lockitch, G. & Halstead, A.C. (1989) In: *Paediatric Clinical Chemistry* (ed. S. Meites), 3rd edn. American Association for Clinical Chemistry, Washington, DC.

Potassium (plasma/serum)

Method

1 month–7 years capillary blood; 7–16 years venous blood.
Flame photometry.

Age	mmol/L
1 month–7 years	4.1–5.6
7–16 years	3.3–4.7

Notes

1 Capillary concentrations are greater than those in venous blood.

2 Haemolysis and delay in separating plasma from red cells cause a falsely high level.

3 Ion-selective electrodes (ISE): the 'indirect' ISE gives results which are almost identical to those obtained by flame photometry. The 'direct' ISE gives results which are higher than those from flame photometry, especially in cases of gross hyperlipidaemia or hyperproteinaemia.

See also:

Clayton, B.E., Jenkins, P. & Round, J.M. (eds) (1980) *Paediatric Chemical Pathology: Clinical Tests and Reference Values*. Blackwell Scientific Publications, Oxford.

Serotonin (blood)

Method

Blood (ascorbate and EDTA) collected after a 6-h fast. High performance liquid chromatography (HPLC).

Age (years)	μmol/L
5–18	0.64–2.45

Note

1 There is no significant difference between boys and girls and no change with age.

See also:

Goldsmith, B.M., Feinstein, C., Munson, S., Reiss, A. & Borengasser-Caruso, M.A. (1986) Determination of a reference range for whole blood serotonin in a pediatric population using high pressure liquid chromatography with electrochemical detection. *Clin Biochem* **19**, 359–363.

Sodium (plasma/serum)

Method

Flame photometry.

Age (years)	mmol/L
<1	130–145
7–16	135–144

Notes

1 Beware of pseudohyponatraemia from gross lipidaemia or dilutional artefacts.

2 Ion-selective electrodes (ISE): the 'indirect' ISE gives results which are almost identical to those obtained by flame photometry. The 'direct' ISE gives results which are higher than those from flame photometry, especially in cases of gross hyperlipidaemia or hyperproteinaemia.

See also:

Clayton, B.E., Jenkins, P. & Round, J.M. (eds) (1980) *Paediatric Chemical Pathology: Clinical Tests and Reference Values*. Blackwell Scientific Publications, Oxford.

Sodium and potassium (urine)

Changes in dietary intake produce large variations in the excretion of sodium and potassium, but see also p. 261.

Sugar (urine)

Method
Thin layer chromatography on a random urine specimen collected 1–2 h after a feed; store specimen deep frozen.
In healthy children on a mixed diet:

Glucose, galactose, fructose	<1 mmol/L.
Lactose and sucrose	<0.5 mmol/L.

Notes
1 Marked galactosuria suggests galactosaemia, galactokinase deficiency or liver dysfunction.
2 Fructosuria is seen in liver disease, hereditary fructose intolerance and essential fructosuria.
3 Raffinose and stachyose (α-galactosides) may appear in the urine when milk substitutes made from soya bean are used.

See also:
Ersser, R.G. & Andrew, B.C. (1971) Rapid thin layer chromatography of clinically important sugars. *Med Lab Tech* **28**, 355–359.
Menzies, I.S. & Seakins, W.T. (1976) In: *Chromatographic and Electrophoretic Techniques* (eds I.S. Smith & J.W.T. Seakins), 4th edn, pp. 183–217. Heinemann Medical, London.

Sweat electrolytes
See p. 9.

Total protein

In plasma, see p. 465.
In urine, see p. 257.
In CSF, see p. 485.

Transferrin

See p. 470.

Thyroid stimulating hormone

See p. 540.

Urate (plasma/serum)

Method
Kodak Ektachem-700/uricase.

Age (years)	μmol/L
1–3	105–300
4–6	130–280
7–9	120–295

Age (years)	μmol/L
Boys	
10–11	135–320
12–13	160–400
14–15	140–465
16–19	235–510
Girls	
10–11	180–280
12–13	180–345
14–15	180–345
16–19	180–350

Notes
1 Levels are high at 24 h and then fall over the first week; levels remain stable throughout infancy and childhood (prepubertal).
2 Plasma urate increases during puberty in boys but not in girls.

See also:
Lockitch, G., Halstead, A.C., Albersheim S., MacCallum, C. & Quigley, G. (1988) Age- and sex-specific paediatric reference intervals for biochemistry analytes as measured with the Ektachem-700 analyser. *Clin Chem* **34/8**, 1622–1625.

Urate/creatinine excretion (urine)

Method
Random urine.
Colormetric method.

Age (years)	mmol/mmol
1–2	0.42–1.53
2–6	0.57–1.35
6–10	0.39–0.85
10–18	0.15–0.67
18+	0.11–0.53

Notes

1 Analyse as soon as possible, as urate may precipitate after prolonged standing. If crystals are observed, dissolve by gentle warming or treat with alkali.

2 In sulphite oxidase deficiency level is likely to be <0.03 mmol/mmol.

3 Patients with Lesch–Nyhan syndrome are likely to have levels >1.0 mmol/mmol.

See also:

Clayton, B.E., Jenkins, P. & Round, J.M. (eds) (1980) *Paediatric Chemical Pathology: Clinical Tests and Reference Values*. Blackwell Scientific Publications, Oxford.

Kaufman, J.M., Greene, M.L. & Seegmiller, J.E. (1968) Urine uric acid to creatinine ratio – a screening test for inherited disorders of purine metabolism. *J Pediatr* **73**, 583–592.

Urea (plasma/serum)

Method

Kodak Ektachem-700/urease.

Age	mmol/L
1 month–4 years	1.7–6.7
4–13 years	2.5–6.4
14–19 years	2.9–7.5

See also:

Earle, J. (1989) In: *Paediatric Clinical Chemistry* (ed. S. Meites), 3rd edn. American Association for Clinical Chemistry, Washington, DC.

Lockitch, G., Halstead, A.C., Albersheim, S., MacCallum, C. & Quigley, G. (1988) Age- and sex-specific paediatric reference intervals for biochemistry analytes as measured with the Ektachem-700 analyser. *Clin Chem* **34/8**, 1622–1625.

Vitamin A (retinol) (plasma)

Method

HPLC.

Age (years)	μmol/L
1–6	0.7–1.5
7–12	0.9–1.7
13–19	0.9–2.5

Notes

1 The specimen must not be left in strong sunlight and should be separated promptly and stored deep frozen for assay.

2 Vitamin A concentration in plasma is only an approximate assessment of body stores as >90% of total body vitamin A is in the liver.

See also:

Lockitch, G., Halstead, A.C., Wadsworth, L. *et al.* (1988) Age- and sex-specific paediatric reference intervals and correlations for zinc, copper, selenium, iron, vitamins A and E and related proteins. *Clin Chem* **34**, 1625–1628.

Vitamin D (plasma)

See p. 274.

Vitamin E (tocopherol) (plasma)

Method

HPLC.

Age (years)	μmol/L
1–6	7–21
7–12	10–21
13–19	13–24

Notes mmol/L

1 There is no difference in concentration between boys and girls.

2 There is a correlation between tocopherol concentration and the serum concentration of lipids. Assessment of deficiency therefore requires that lipid concentrations be taken into account.

See also:

Laryea, M.D., Biggemann, B., Cieslicki, P. & Wendel, U. (1989) Plasma tocopherol and tocopherol to lipid ratios in a normal population of infants and children. *Int J Vit Nutr Res* **59/3**, 269–272.

Lockitch, G., Halstead, A.C., Wadsworth, L. *et al.* (1988) Age- and sex-specific paediatric reference intervals and correlations for zinc, copper, selenium, iron, vitamins A and E and related proteins. *Clin Chem* **34**, 1625–1628.

Zinc (plasma)

Method
Atomic absorption spectrophotometry.

Age	mmol/L
Childhood	9–22

See also p. 493.

Acknowledgments

The authors would like to thank the members of their respective departments who have helped with the production of these data, in particular Paul Claridge and Sue Keffler at Birmingham Children's Hospital.

General references

Burritt, M.F., Slockbower, J.M., Forsman, R.W. *et al.* (1990) Pediatric reference intervals for 19 biologic variables in healthy children. *Mayo Clin Proc* **65**, 329–336.

Cheng, M.H., Lipsey, A.I., Blanco, V., Wong, H.T. & Spiro, S.H. (1979) Microchemical analysis for 13 constituents of plasma from healthy children. *Clin Chem* **25**, 692–698.

Fleming, P.J., Speidel, B.D., Marlow, N. & Dunn, P.M. (eds) (1986) *A Neonatal Vade-Mecum*, 2nd edn. Edward Arnold, London.

Gomez, P., Coca, C., Vargas, C., Acebillo, J. & Martinez, A. (1984) Normal reference intervals for 20 biochemical variables in healthy infants, children and adolescents. *Clin Chem* **30**, 407–412.

Hughes, I.A. (1986) *Handbook of Endocrine Tests in Children*. John Wright & Sons, Bristol.

Insley, J. (ed.) (1990) *A Paediatric Vade-Mecum*, 12th edn. Edward Arnold, London.

Lockitch, G., Halstead, A.C., Albersheim, S., MacCallum, C. & Quigley, G. (1988) Age- and sex-specific paediatric reference intervals for biochemistry analytes as measured with the Ektachem-700 analyser. *Clin Chem* **34**, 1622–1625.

Lockitch, G., Halstead, A.C., Wadsworth, L. *et al.* (1988) Age- and sex-specific paediatric reference intervals and correlations for zinc, copper, selenium, iron, vitamins A and E, and related proteins. *Clin Chem* **34**, 1625–1628.

Meites, S. (1989) *Pediatric Clinical Chemistry*, 3rd edn. American Association for Clinical Chemistry, Washington, DC.

Stephenson, J.B.P. & King, M.D. (1989) *Handbook of Neurological Investigations in Children*. Wright, London.

A3: Hormone Reference Ranges

J.W. HONOUR

Single measurements are rarely useful in the investigation of a child with a potential endocrine problem because many events are pulsatile in nature, often at night. The relationship between an adrenal, gonadal or thyroid hormone and its respective trophic hormone is more useful. Timing of all samples is critical. All assays depend on method and choice of calibration material.

Hypothalamic-pituitary-thyroid (H-P-T) axis (serum)

Age	Thyroxine (nmol/L)	T_3 (nmol/L)	TSH (iu/L)	rT_3 (nmol/L)	TBG (mmol/L)
Full-term infants (first week)	100–200	0.2–1.5	Surge after birth; peak values at 30 min at 20–150	0.5–5	6–60
First decade	80–160	1.5–3.5	0.5–5.0	0.2–1.2	18–40
Pubertal	50–160			0.5–1.4	18–33

T_3, Triiodothyronine; TSH, thyroid-stimulating hormone; rT_3, reverse T_3; TBG, thyroxine-binding globulin.

Hypothalamic-pituitary-adrenal (H-P-A) axis

Cortisol (serum/plasma)

Basal		
Children	08.00 h	200–700 nmol/L
	Midnight	<150 nmol/L
Neonates	See Appendix A1, p. 517	
ACTH stimulation	(250 μg Synacthen 1–24) Increment of 220 nmol/L or peak values (30 or 60 min after stimulation) above 550 nmol/L Warning – see p. 229.	

ACTH, Adrenocorticotrophic hormone.

Urine.
Children: 150–400 nmol/L.

ACTH (plasma)

Blood must be cooled, plasma separated rapidly and then frozen.
See p. 222.

17-Hydroxyprogesterone (serum/plasma)

NB: Results are difficult to interpret during the first 36 h of life, but see p. 226.

Dehydroepiandrosterone-sulphate (DHAS) (serum/plasma)

See p. 226.

Androstenedione (serum/plasma)

There is a marked diurnal variation in circulating levels; specimens should be collected before 10.00 h.
Prepubertal children: 0.3–1.8 nmol/L.

During puberty there is a rise into the adult range. 08.00 h: 2–8 nmol/L.

Renin-angiotensin-aldosterone

It is important to judge results for aldosterone as appropriate or inappropriate for the prevailing renin activity.

Age	Aldosterone (serum/plasma) (pmol/L)	Renin (pmol/h per mL)
Term infants (first week)	Up to 5000	Up to 20
First year	300–1500	2–7
1–2 years	200–1500	2–6
2–10 years	100–800	1.5–4
10–15 years	60–600	0.8–2

Hypothalamic-pituitary-gonadal (H-P-G) axis

Testosterone (serum/plasma)

Age	Male (nmol/L)	Female (nmol/L)
Birth	4–14	0.5–2
First week	Fall rapidly to 0.5–1.5	Less than 1
15–60 days	4–10	Less than 0.5
4 months to start of puberty	0.1–0.5	Less than 0.5
Puberty		
Tanner 2*	0.5–5†	Less than 1
Tanner 3*	3.5–11†	Less than 1
Tanner 4*	7–20‡	Less than 1.5

* For pubertal stages Tanner 1–5 see: Tanner, J.M. & Whitehouse, R.H. (1976) Clinical longitudinal standards for height, weight, height velocity and weight velocity and stages of puberty. *Arch Dis Child* **51**, 170–179.
† Peak levels are nocturnal (collect specimens before 10.00 h).
‡ Levels sustained throughout the day.

Follicle-stimulating hormone (FSH) (plasma)

(Highly method dependent).

Age	Male (U/L)	Female (U/L)
First week (levels fall during this period)	0–10	1–35
10–60 days	0–28	1–44
1 year	0–5	0–35
1–3 years	0–8	1–12
Prepuberty	2–7	0.3–8
Puberty		
Tanner 2*	0.5–4	1.6–7
Tanner 3*	2.5–4.5	4–7
Tanner 4†	3–5.5	3–8

* Peak levels are nocturnal (collect specimens before 10.00 h).
† Levels sustained throughout the day.

Luteinizing hormone (LH) (plasma)

Age	Male (U/L)	Female (U/L)
Prepuberty	0–1	0–1
Peripuberty		
Nocturnal peaks	1–4	1–4
Puberty		
Tanner 2*	1–4	1–7
Tanner 4†	2–8	2–8

* Peak levels are nocturnal (before 10.00 h).
† Levels sustained throughout the day.

Parathyroid hormone (intact 1-84 PTH)

PTH estimations are highly method dependent, depending on the antibody used.
2–15 years: 1.14–3.64 pmol/L; 11–35 ng/L.
Intact PTH: molecular weight 9600; 1 ng/L = 0.104 pmol/L.

See also:
Shaw, N.J., Wheeldon, J. & Brocklebank, J.T. (1990) Indices of intact serum parathyroid hormone and renal excretion of calcium, phosphate and magnesium. *Arch Dis Child* **65**, 1208–1211.

Index